D0146591

A NEW HISTORICAL GEOGRAPHY
OF ENGLAND

A NEW
HISTORICAL GEOGRAPHY
OF ENGLAND

Edited by

H. C. DARBY

CAMBRIDGE

AT THE UNIVERSITY PRESS

1973

Published by the Syndics of the Cambridge University Press
Bentley House, 200 Euston Road, London NW1 2DB
American Branch: 32 East 57th Street, New York, N.Y. 10022

© Cambridge University Press 1973

Library of Congress Catalogue Card Number: 72–93145

ISBN: 0 521 20116 0

Printed in Great Britain
at the University Printing House, Cambridge
(Brooke Crutchley, University Printer)

CONTENTS

[v]

591595

LIST OF MAPS AND DIAGRAMS

[vii]

LIST OF MAPS AND DIAGRAMS

LIST OF MAPS AND DIAGRAMS

LIST OF MAPS AND DIAGRAMS

PREFACE

In 1936 the Cambridge University Press published *An historical geography of England before A.D. 1800*. It has been reprinted a number of times but clearly the moment has come not for a further reprint nor for a new edition but for an entirely new volume based upon the enormous amount of work that has been done since 1936, and especially since 1945. Much of the new work has appeared in the pages of three journals, *The Economic History Review*, *The Agricultural History Review* and the *Transactions and Papers of the Institute of British Geographers*. Moreover, the English Place-Name Society has continued to produce its scholarly volumes year by year; the Domesday Book has been analysed geographically, so have the Lay Subsidies of 1334 and 1524–5. Our views of agriculture and industry in later times have also been modified by a variety of monographs.

Not only has there been much exploration of sources, but also much discussion about the method of historical geography. In particular, a contrast has often been drawn between the reconstructions of past geographies and the study of geographical changes through time, between the so-called 'horizontal' and 'vertical' approaches. This present volume seeks to combine both, and the combination was suggested by J. O. M. Broek's *The Santa Clara Valley, California: a study in landscape changes* (Utrecht, 1932).

The new work begins with the coming of the Anglo-Saxons in the belief that so far as there ever is a new beginning in history, that event was such a beginning. It continues beyond the eighteenth century, up to 1900 or so, but this has not been taken as a rigid date because 1914, rather than 1900, marks the effective end of the nineteenth century. In contemplating the result, one can only be very conscious of what remains to be done. As far as sources are concerned, although much recent work has been done on the Tithe Returns of the 1840s, a comprehensive treatment has yet to appear. There have also been interesting studies on enclosure, but no large-scale attack on a geographical basis. Or, to take another example, even so obvious a source as the Census Returns, from 1801 onwards, still awaits comprehensive analysis and interpretation. As far as method is concerned, much further enquiry is needed to see to what extent statistical techniques and locational analysis can be applied to historical data of varying quality and coverage.

[xiii]

In another generation or so the materials for an historical geography of England will not be as we know them now. A wider range of sources will have been explored and evaluated. Fresh ideas about method will have prepared the way for a more sophisticated presentation. And by that time we and our landscape will have become yet one more chapter in some other *Historical Geography of England*.

I am greatly indebted to my fellow contributors for their co-operation and patience. All of us must thank the staff of the Cambridge University Press for their skill and care. We must also thank Mr G. R. Versey. He has not only drawn all the maps and diagrams but has compiled many of them and has also given much general assistance at all stages of the work.

H. C. DARBY

KING'S COLLEGE
CAMBRIDGE

Candlemas, 1973

Chapter 1

THE ANGLO-SCANDINAVIAN FOUNDATIONS

H. C. DARBY

The Anglo-Saxon settlement was a new beginning in the history of Britain. It made a decisive contribution to the peopling of the south-east; it determined the language of the area; and it brought institutions that formed the basis of all later development. In addition, it laid the foundations of the later geography of England. With the coming of the Anglo-Saxons, a new chapter in the history of settlement and land utilisation was begun, and, with but little interruption, the story has been continuous up to the present day.

It is true that the Anglo-Saxons did not come into an empty land, and that many contributions from pre-Saxon days have entered into the making of England. The various strains in the present population may be traced far back into prehistoric times, and the early peoples of those times left remains that are still visible today. These remains are particularly numerous on the downlands of the southern counties. Megalithic monuments (long barrows, tumuli and stone circles) are prominent in the chalk country of Wessex; here is Stonehenge, the most famous of all megalithic monuments, and not far away is the stone circle of Avebury and the mysterious mound of Silbury. Earthworks, or 'camps', like Maiden Castle in Dorset and Windmill Hill in Wiltshire, with their ramparts and ditches, are also prominent local features that remain to excite our imagination. Then again, old trackways can still be traced on the present surface of the ground. The Icknield Way below the Chilterns, and the Pilgrims' Way along the North Downs, may have been already old when the Roman legionaries tramped the country, and there are others like them. Finally, there are traces of primitive cultivation that have also survived the changes of later time. Many have been destroyed by later cultivation, but others, now grass-grown, can still be seen in the present landscape. These early monuments and features are especially characteristic of the chalk downlands, but they are also to be found elsewhere. Thus there are hut circles and primitive corn plots on Dartmoor, and megalithic remains

1

on the Yorkshire Wolds and along many stretches of the western seaboard.

The Romans who came in A.D. 43 bequeathed, in turn, a substantial legacy to the geography of succeeding ages. The lines of many Roman roads are still in use as arterial ways. The strategical locations of many cities were recognised in Roman as in later times. In the north, the remains of the Roman Walls still cross the landscape in a distinctive fashion; and in the south many of the traces of early cultivation date from Romano-British times. The legacy of Rome to the geography of England is no mean one. But, even so, as far as there ever is a new beginning in history, the coming of the Angles, Saxons and Jutes was such a beginning.

The circumstances of this new beginning, however, are far from clear. The period that separates the end of Roman rule in Britain from the emergence of the earliest English states is obscure and baffling. The prelude to the settlement of the newcomers from across the North Sea consisted of plundering raids that began as early as A.D. 300, and even before. In the west and north, too, the raids of the Picts and the Scots left trails of devastation. But it was not until about A.D. 410 that the Roman legions were withdrawn and the cities of Britain told to look to their own defence. Whether this injunction marked the end of Roman rule in Britain is a disputed matter. At any rate, within a generation or so, it is clear that the Romano-British were incapable of withstanding the raiders from overseas. At what date this raiding passed into settlement is uncertain, but it would seem that the so-called 'Adventus Saxonum' took place in the years around the middle of the fifth century.

THE ANGLO-SAXON SETTLEMENT

The arrival of the English

The evidence that relates to the coming of the English is of three kinds – literary, archaeological and place-name.[1] Apart from a few brief and scattered references, the main body of the literary evidence is derived

[1] General sources for the account that follows include: (1) R. G. Collingwood and J. N. L. Myres, *Roman Britain and the English settlements* (Oxford, 2nd ed. 1937); (2) F. M. Stenton, *Anglo-Saxon England* (Oxford, 2nd ed. 1947); (3) P. H. Blair, *An introduction to Anglo-Saxon England* (Cambridge, 1956); (4) R. H. Hodgkin, *A history of the Anglo-Saxons*, 2 vols. (Oxford, 3rd ed. 1952); (5) K. Jackson, *Language and history in early Britain* (Edinburgh, 1953); (6) H. R. Loyn, *Anglo-Saxon England and the Norman Conquest* (London, 1962).

from the writings of Gildas (*circa* 550), Bede (*circa* 730), Nennius (*circa* 800) and from the earlier parts of the Anglo-Saxon Chronicle (*circa* 900). Each of these four sources is unsatisfactory, but taken together, they indicate a sequence of events that may well bear some approximation to the truth. In about A.D. 449 a certain British chieftain named Vortigern, so runs the tradition, enlisted the help, from across the sea, of the Saxons, under two brothers named Hengest and Horsa, to repel the raids of the Picts and the Scots. The mercenaries were then joined by others of their kind, and, after some dispute about their provisions, the newcomers turned against their employers and raided through the land from sea to sea. In the years that followed, others continued to come from across the North Sea. Thus did Nemesis fall upon the Britons, and, in the words of Gildas, the fire of the Saxons burned across the island until 'it licked the western ocean with its red and savage tongue'. The result of the calamity was a devastated countryside, ruined cities, and the complete collapse of British resistance. The newcomers were divided by Bede into Angles, Saxons and Jutes, but recent work has shown that Bede's generalisation was too simple and that the newcomers may also have included some Frisians.[1]

The first collapse was followed by a rally of the British forces and by a period of indecisive warfare, victory alternating with defeat, until a great British victory was won at a place called Mons Badonicus. Its date seems to have been about 500. Its site has been variously identified: one possible location is Badbury Rings to the north-west of Wimborne Minster in Dorset.[2] There is no doubt about its decisive nature for it was followed by a respite that continued for some fifty years, that is up to the time that Gildas wrote. Nennius tells us that among those who had fought against the Saxons was a man named Arthur; 'he fought against them with the kings of Britain, but he himself was the military commander'. Then follow the names of his twelve battles, and the twelfth was none other than that of Mons Badonicus. Later ages were to add legend after legend to the name of Arthur. It is interesting to think that at the core of all the romance of the Arthurian cycle, which has so fascinated generation after generation, lies something of the story of the defence of Britain against the English.

[1] F. M. Stenton, 'The historical bearing of place-name studies: the English occupation of South Britain', *Trans. Roy. Hist. Soc.*, 4th ser, XXII (1940), 5–6; H. R. Loyn, 27.

[2] K. Jackson, 'The site of Mount Badon', *Jour. Celtic Studies*, II (Temple Univ., Baltimore, Md, 1958), 152–5.

Its hero has been called the last of the Romans, the last to use Roman ideas of warfare for the benefit of the British people, and, wrote Professor Collingwood, 'the story of Roman Britain ends with him'.[1]

The last of the four sources, the Anglo-Saxon Chronicle, adds some further details, and describes how the Britons of Kent 'fled from the English like fire'; but, in common with the other sources, it tells us little about the shifting frontier between the advancing English and the Britons. The little it does tell us raises many difficulties. Under the year 571 it records an English victory over the Britons, and the capture of the four settlements of Limbury, Aylesbury, Benson and Eynsham, to the west of the Chilterns. 'No annal in the early sections of the Chronicle', wrote Sir Frank Stenton, 'is more important than this, and there is none of which the interpretation is more difficult.'[2] It is hard to see how the Britons could have been here at this date. Was their territory an enclave surrounded by the English? Or might it have been territory at first over-run by the English, then lost after Mons Badonicus, and then finally re-covered once more in 571? Whatever the interpretation, the entry does suggest that the settlement of the English was no easy appropriation of south-east Britain. The presence of Britons here, well over a hundred years after the traditional date of Hengest's landing, may indicate that the triumph of the Anglo-Saxons was by no means a foregone conclusion. The English occupation of south Britain may well have comprised two phases separated by the fifty years or so that followed their defeat at Mons Badonicus, i.e. by the earlier half of the sixth century.

Revealing as the literary evidence is, it consists only of the fragments of a story that is far from complete. It makes specific mention of the coming of the invaders only to Kent, Sussex and Wessex. It tells us nothing, for example, of the arrival of the East Anglians and the East Saxons, nor any-thing about that of the Mercians, the Middle Anglians and the North Anglians. For an idea of the early spread of the Anglo-Saxons as a whole we must turn to the evidence of archaeology and of place-names. The bulk of the archaeological evidence consists of cemeteries and burial sites belonging to the age before the Anglo-Saxons became Christian, and it dates therefore roughly from between 450 and 650. To the north of the Thames, as Fig. 1 shows, was a widespread occupation both of the Midlands and of southern England by about 570. The finds are strikingly concentrated around the entrance of the Wash, and it is clear that many

[1] R. G. Collingwood and J. N. L. Myres, 34.
[2] F. M. Stenton (1947), 27.

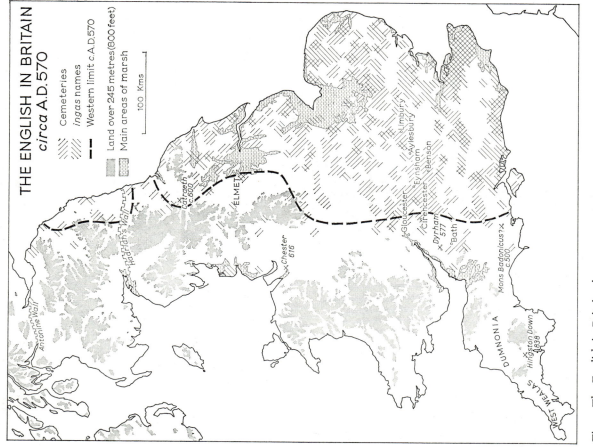

Fig. 1 The English in Britain *circa* A.D. 570

Based on: (1) E. Ekwall, *English place-names in -ing* (Lund, 1935); (2) *Map of Britain in the Dark Ages* (Ordnance Survey, 2nd ed., 1966).

invaders came up the Fenland rivers into the Midlands and East Anglia. To the north of the Wash, the estuary of the Humber provided another entry; some of these invaders turned south, along the Trent and the Soar, to mingle with those from the Wash entry; others followed the Derwent into south Yorkshire, and there is a cluster of finds on the Yorkshire Wolds. Farther north still, the finds are few. The traditional date of the founding of the northern kingdom is 547, but the archaeological evidence does not carry us back very clearly to those days. The English struggle against the Britons was still proceeding here late in the sixth century, and, in view of this, it is not surprising that the archaeological finds of the heathen period are few in number.

This archaeological evidence is, in a general way, confirmed by that of place-names. It has been shown that place-names ending in *ing*, represent-ing the Old English *ingas*, are older than most other place-names, and that they were formed during the earliest phases of the settlement.[1] They were folk-names, and referred not to localities but to communities; thus *Woking* in Surrey is derived from 'Wocc' and 'ingas', and means 'Wocc's people'; Reading likewise 'Reada's people'. The groups designated by such names ranged from what seem to have been small communities to tribes such as the *Sunningas* (Sonning), whose territory covered much of eastern Berkshire and the *Hrothingas* of the Roding valley in Essex. All these names ceased to be formed at an early date when topographical rather than folk-names became current. But not all *ing* names belong to the age of migration, and *ing* names in general must be treated with caution.[2] Another early group of names is that indicating heathen beliefs and prac-tices. Some of these include the names of the old Germanic gods; Woden and Thunor, for example, appear in Woodnesborough in Kent and Thursley in Surrey.[3] Over fifty of these heathen names have been identi-fied, and they, too, appear to belong to pre-Christian days. The distribu-tion of both groups of early names confirms the general impression of a wholesale occupation of south-east Britain before about 570. Theoreti-cally, this distribution should agree with that of heathen burial places, and in a general sense the two patterns do agree (Fig. 1). But in detail some

[1] E. Ekwall, *English place-names in -ing* (2nd ed. Lund, 1962).

[2] See A. H. Smith (1) *English place-name elements* (Cambridge, 1956), 282–293; (2) 'Place-names and the Anglo-Saxon settlement', *Proc. British Academy*, XLII (1956), 73–80.

[3] F. M. Stenton, 'The historical bearing of place-name studies: Anglo-Saxon heathenism', *Trans. Roy. Hist. Soc.*, 4th ser, XXIII (1941), 1–24.

differences are at once apparent. There are, for example, few pagan cemeteries in Essex but many *ing* names. On the other hand, there are few *ing* names in the Upper Thames area where there are early cemeteries.

It is impossible to assess the significance of these discrepancies, but in view of the element of chance involved both in the survival of cemeteries and of place-names, it would be remarkable if they showed a detailed correspondence in every locality. Moreover, some anomalies may be accounted for by our inability to distinguish other types of archaic place-names 'which may have had a greater frequency in some districts'.[1] Many *ing* names, moreover, may belong to a period 'later than, but soon after, the immigration-settlement that is recorded in the early pagan-burials'.[2] Whatever be the answer to these and other riddles, we may be sure that only by a synthesis of the literary evidence with that of archaeology and place-names can some approximation to a probable sequence of events be made.

The Anglo-Saxon kingdoms

Whatever the obscurity of the two centuries following 450, we know that in the course of the sixth century the Anglo-Saxons became organised into states. The process by which the historic kingdoms emerged is lost for ever from our sight. By the seventh century the names of as many as eleven kingdoms are to be found, and there may have been more. Some of these were joined to others until the result was the seven kingdoms of the Heptarchy, the boundaries of which often reflected local features of wood and marsh and heath (Fig. 2). To this Anglo-Saxon age belong some of the numerous linear bank-and-ditch earthworks that are found in many parts of the country, and that commonly have such names as Devil's Dyke or Grim's Dyke. Some were constructed primarily as boundaries of, say, cattle ranches or large sheep walks; such is Grim's Ditch in the north-east of Cranborne Chase. Others were for defensive purposes, and to these belong the striking Cambridgeshire system of dykes that cross the open chalk country between fen and wood and that were probably built by the East Angles against the Mercians (Fig. 3).

The earliest kingdoms to emerge seem to have been those of the south-east – East Anglia, Essex, Kent and Sussex; but they were soon out of

[1] A. H. Smith, 'Place-names and the Anglo-Saxon settlement', 84.

[2] J. McN. Dodgson, 'The significance of the distribution of the English place-name in *-ingas, inga* in south-east England', *Med. Archaeol.*, x (1966), 19. See also J. N. L. Myres, 'Britain in the Dark Ages', *Antiquity*, IX (1935), 455–64.

H. C. DARBY

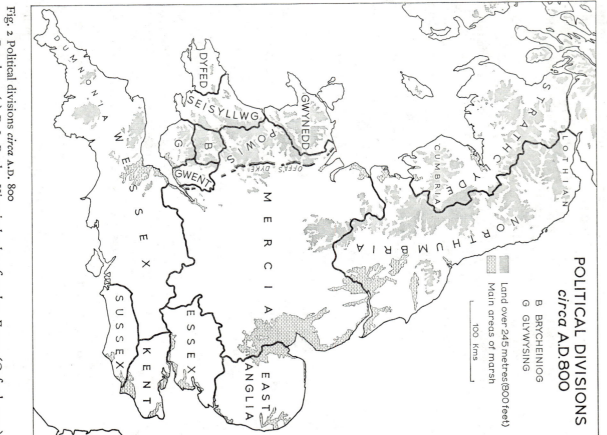

POLITICAL DIVISIONS
circa A.D. 800

B BRYCHEINIOG
G GLYWYSING

Land over 245 metres (800 feet)
Main areas of marsh

100 Kms

Fig. 2 Political divisions *circa* A.D. 800
Based on: (1) R. L. Poole, *Historical atlas of modern Europe* (Oxford, 1902),
plate 16; (2) W. Rees, *An historical atlas of Wales* (2nd ed., Cardiff, 1959),
plate 22.

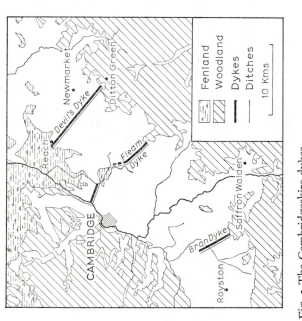

Fig. 3 The Cambridgeshire dykes
Based on C. Fox, 'Dykes', *Antiquity*, III (1929), 137.

touch with the advancing frontier against the Britons, and the great powers of the later Anglo-Saxon period came to be those of Wessex, Mercia and Northumbria. Their rivalry features prominently in the records of English history during the seventh and eighth centuries, but the complexities of their mutual relations did not prevent continued advance to the west.

The evidence for the origin of the kingdom of Wessex has been much debated. On the one hand, the Anglo-Saxon Chronicle, in entries for the years between 495 and 519, speaks of West Saxon chieftains landing on the Hampshire coast and fighting and killing many Britons. On the other hand, the archaeological evidence seems to show that the West Saxons arrived from the east, maybe from the Wash along the Icknield Way to the Upper Thames area and so southwards. The two versions have been reconciled on the assumption that the archaeological evidence bears witness to a mass movement of people, while the literary evidence records the exploits of individuals who were to become the ancestors of the royal house of Wessex. Whatever be the truth, the frontier against the Britons was being vigorously pressed long before the end of the sixth century. Under the year 577 the Chronicle records a victory over the Britons at Dyrham, and it tells how the Saxons took Gloucester, Cirencester and

Bath, Dyrham is six miles north of Bath, and the victory opened up the lower Severn valley to Saxon colonists, thus separating the Britons of the south-west peninsula from those of Wales.

The details of the further advance of the Saxons into the south-west are far from clear. The place-names of Dorset and Somerset suggest Saxon occupation in the seventh century.[1] To the west lay the Welsh kingdom of Dumnonia, the memory of which survives in the name of Devon; and it would seem that the eastern part of this passed into Saxon hands before the end of the seventh century, and that the northern and western parts were occupied after 710 when Ine of Wessex defeated Geraint, king of Dumnonia.[2] The Saxon occupation of Devon was thorough. Its place-names are overwhelmingly English, both in the west as well as in the east of the county; and Welsh names are, surprisingly, fewer than in Somerset or Dorset.[3] The reason may possibly lie in the fact that the Welsh popula-tion of Devon was sparse when the Saxons arrived, partly because, during the fifth and sixth centuries and earlier, there had been a large migration across the sea to transform Armorica into what later became Brittany.[4] Even so, it has been suggested that the Welsh constituted 'a far from negligible element in the population of Devon' in the Dark Ages.[5]

Farther west still, were the Welsh of Cornwall into which Dumnonia had contracted, and there are echoes of warfare between Welsh and Saxon. They were fighting in 710, in 722 and again in 753, and in 815 the Saxon king 'laid waste West Wales from east to west'. Ten years later the Welsh raided into Devon, and in 838 they joined a Danish army against Wessex only to be defeated at Hingston Down, to the west of the Tamar in Cornwall. Henceforward, Cornwall came under Saxon overlordship, but a native dynasty of Welsh kings seems to have survived probably into the early years of the tenth century. By this time, the force of Saxon coloni-sation had spent itself, and Cornwall, except in the extreme east, did not lose its Celtic character with its independence.

The details of the expansion westward across the Midland Plain are lost from our sight. But in the early seventh century various groups in the area emerged as the kingdom of Mercia. A limit to the westward expansion

[1] F. M. Stenton (1947), 63.
[2] W. G. Hoskins, *The westward expansion of Wessex* (Leicester, 1960).
[3] J. E. B. Gover et al., *The place-names of Devon*, pt 1 (Cambridge, 1931), xix.
[4] H. R. Loyn, 47; Norah K. Chadwick, 'The colonization of Brittany from Celtic Britain', *Proc. British Academy*, LI (1965), 235–99.
[5] H. P. R. Finberg, *The early charters of Devon and Cornwall* (Leicester, 1953), 31.

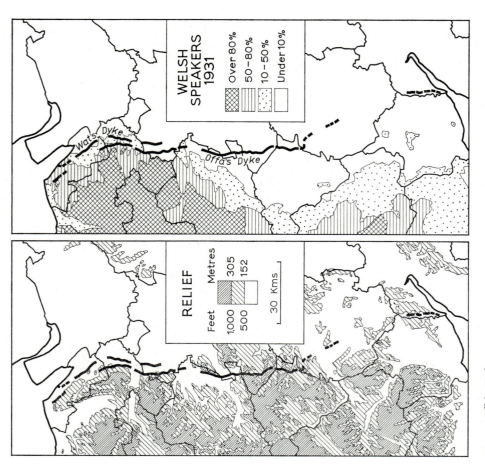

Fig. 4 Offa's Dyke
Based on: (1) C. Fox, *Offa's Dyke* (London, 1955), 166; (2) D. T. Williams,
'A linguistic map of Wales according to the 1931 Census, with some observations
on its historical and geographical setting', *Geog. Jour.*, LXXXIX (1937), 148.
Modern county boundaries are shown.

of the Mercian kings was set by the Welsh foothills. Little is known of the
process by which the boundary between the English and the Welsh
evolved, and only a few echoes of the English advance have come to us
from this period. It was about 790, during the reign of Offa, that the
famous dyke that now bears his name was constructed.[1] It seems to have

[1] C. Fox, *Offa's Dyke* (British Academy, London, 1955).

been the product of negotiation for, in places, it meant the yielding up of English territory to the Britons. It was not the first attempt to provide a frontier. Wat's Dyke which runs parallel to the east of Offa's Dyke in the north may have been an earlier attempt, made in the seventh or eighth century. But Offa's Dyke far surpassed all earlier attempts, and after twelve centuries much of it still remains (Fig. 4). In the north it is a prominent feature of the countryside for some seventy miles, cutting straight across hill and valley alike, and running for nearly twenty miles above the 1,000 feet contour line. Southwards in Herefordshire it becomes intermittent and seems to have been constructed only in clearings made in the wooded area. The last stretch reaches the Wye near Bridge Sollers, six miles or so to the west of Hereford. From here, the Wye replaced the Dyke as a boundary, but to the east of the lower Wye it reappears again until it reaches the Bristol Channel.

To the north, beyond the Humber, the two Anglian states of Deira and Bernicia had been united to form the kingdom of Northumbria by Aethelfrith (593–617). He was, wrote Bede, notable for acquiring territory from the Britons, either making them tributary or driving them out and establishing English settlers on their lands. To these years belongs one of the earliest Welsh poems, but preserved only in a thirteenth-century manuscript. This is Aneirin's elegy called the 'Gododdin' after a British tribe (Ptolemy's Otadenoi) who lived in the area to the south of the Firth of Forth. From here, about the year 600, a band of British warriors raided southward into Northumbria, only to be defeated by greatly superior numbers at the battle of Catraeth, usually identified with Catterick in the North Riding of Yorkshire. Almost all the band were killed, and the poem tells of their distinction and bravery and how 'they shall be honoured until the end of the world'.[1] In 616, Aethelfrith defeated the Britons at Chester. It is not clear whether this was an isolated raid or an important event that brought the English to the western sea as had the battle of Dyrham in the south. In any case, it is clear that by the middle of the seventh century the Britons of the west had been separated into three groups – those of Cornwall and Devon, those of Wales itself and those of the north-west. One obstacle to the westward advance of the Northumbrians had been the presence of the British kingdom of Elmet, between the Aire and the Wharfe roughly north of what is now Leeds. This was

[1] I. Williams, 'The poems of Llywarch Hên', *Proc. British Academy*, XVIII (1932), 270–1; K. Jackson, 'The "Gododdin" of Aneirin', *Antiquity*, XIII (1939), 25–34; C. A. Gresham, 'The book of Aneirin', *Antiquity*, XVI (1942), 237–57.

conquered by Edwin, Aethelfrith's successor, sometime between 616 and 632. Before the end of the century, Anglian settlement and control extended in the north-east over the lowlands of Lothian to the Firth of Forth, and in the north-west into the plain of Carlisle and the lands around the Solway Firth. Affairs in the latter area were complicated by the survival of the Welsh kingdom of Strathclyde with its own royal line. At its greatest extent this seems to have stretched northward from the Ribble well into south-west Scotland. Its political relations and its extent at various times are very confused, but amid much that is obscure it is clear that Britons survived in the Lakeland mountains, leaving the name Cumberland to remind us that this was once the land of the Cymry, as the Welsh still call themselves today.

The survival of the Romano-Britons

To what extent was there continuity between the new Anglo-Saxon villages and those of the Romano-Britons? For some areas, at any rate, the answer to this question is clear, and has been given with the aid of air photography. On the chalk country of Wessex and Sussex the Anglo-Saxon settlement seems to have involved a complete break with the past, and the Romano-British villages ceased to be inhabited. 'Their sites were abandoned,' wrote O. G. S. Crawford, and are now 'a maze of grass-covered mounds. We do not know what happened during that century and a half of darkness and confusion; but when the dawn broke through once more, we find a totally new system, Teutonic in character, and different in every way from its predecessor. New villages with new Saxon names have sprung up along the valleys; the once populous uplands are deserted.'[1]

It does not follow from this that the Romano-Britons had not lived in the valleys. Evidence of early settlement has less chance to survive in the valleys because it is removed by the ploughs of later occupiers or buried beneath their houses, but the gravel terraces of the Thames and of other rivers show that in places the Britons were there before the Saxons. On several sites the association of Roman and Saxon remains seems to indicate 'a virtual continuity of occupation from one age into the other'.[2] And for

---

[1] O. G. S. Crawford, 'Air survey and archaeology', *Geog. Jour.*, LXI (1923), 353. See also C. C. Taylor, 'The study of settlement patterns in pre-Saxon Britain' in P. J. Ucko et al., *Man, settlement and urbanism* (London, 1972), 109–13.

[2] F. M. Stenton (1947), 25. See also S. Applebaum's chapter entitled 'Continuity?' (pp. 250–65) in H. P. R. Finberg (ed.), *The agrarian history of England and Wales*, vol. I, pt 2, A.D. 43–1042 (Cambridge, 1972).

Withington, in the Cotswolds, H. P. R. Finberg has also argued the case for continuity.[1]

One of the questions that inevitably comes to mind is to what extent did the Britons survive the changes of the time. There are various references, in the Chronicle and elsewhere, to the slaughter of Britons, but these can hardly refer to the whole period of the Anglo-Saxon advance. Even so, whatever the degree of survival the fact remains that the English language is almost entirely free of British influence. Celtic words relating to agriculture and to domestic life are virtually absent from English speech.[2] What is more, the place-names of villages and hamlets are almost entirely non-Celtic, except in Cornwall. As might be expected, the counties with the greatest number of Celtic names are those of the western border. In Herefordshire, names like Dinedor and Moccas are anglicised Celtic names; so are Cumwhiton, Blencarn and the like in Cumberland; and Eccles and Haydock in Lancashire. Scattered Celtic place-names are also found in many other counties. Among those of West Riding, there is the name of the village of Wales (i.e. the *Wealas* or Welshmen) to the south-east of Sheffield. There are a few Celtic place-names in the Midlands and some even in the south and east. Wiltshire is a county with a relatively large number. They support the view that the Saxon occupation of Wiltshire did not take place until after the first destructive phase of invasion was over, and they point to a period of peaceful relations between Britons and newcomers.[3]

But the main indication of British survival comes not from names of places but from those of natural features, from the names of hills and woods and, above all, rivers. Celtic names of hills and woods include Cannock, Morfe and Kinver in Staffordshire, Chute and Savernake in Wiltshire, Chiltern in Hertfordshire and Brill and Chetwode in Buckinghamshire. More numerous still are the Celtic names of streams, both large and small. It is true that they are rare in Sussex and in the eastern counties generally, but they become frequent towards the west. The Aire, the Trent, the Derwent and the numerous rivers called Avon or Ouse, are only a few out of a large number. Eilert Ekwall summed up this evidence by saying that 'the old theory of a wholesale extermination or displacement of the British population is no doubt erroneous or exaggerated, but it may come near the truth in the districts first conquered'.[4]

[1] H. P. R. Finberg, *Roman and Saxon Withington* (Leicester, 1955).
[2] Max Forster, *Keltisches Wortgut im Englischen* (Halle, 1921).
[3] J. E. B. Gover et al., *The place-names of Wiltshire* (Cambridge, 1939), xv–xvi.
[4] E. Ekwall, *English river-names* (Oxford, 1928), lxxxix.

There is also other evidence for the survival of the British. The laws of Ine (circa 690) show that within a Wessex that did not yet include Devon and Cornwall, there were not only Welsh slaves but Welsh freemen of some standing. There was even a Welsh strain in the West Saxon royal house.[1] But, in general, the term 'Welsh', derived from the English word 'walh' or 'wealh', was commonly taken to imply a foreigner or a slave; the Welsh called, and still call, themselves 'Cymry'.

At the other end of the country, the persistence of a strong Celtic element in the institutions of Northumbria must surely point to a substantial survival of the British among the Anglian settlers.[2] Furthermore, Glanville Jones has postulated the survival of British arrangements in the system of central manors with satellite hamlets that is a feature of northern England in the early Middle Ages and that meets our eye in the sokelands and berewicks of Domesday Book. He also thinks that the complex manors with dependencies to be found in southern England likewise result from the survival of Romano-British institutions.[3] Here is an hypothesis that needs further investigation.

The conclusion must be that there is not one but many answers to that debated question 'Did the native population survive?'. In some areas the British may have been exterminated. In many other areas they must have survived. In yet other areas they may have been absorbed into the economy and social structure of the newcomers at an early date. No single generalisation will cover the truth.

English place-names

Whatever the degree of British survival, and whatever the precise relationship between the two groups of peoples, the new villages established by the invaders were the villages of later times. Where, for example, Waels and his men settled in Norfolk, there is Walsingham today. Where Babba made a 'stoc' or settlement in Wiltshire, there Baverstock still stands. It is true that some Anglo-Saxon sites have been deserted, and that in Nor-

[1] F. M. Stenton, 'The historical bearing of place-name studies: England in the sixth century', Trans. Roy. Hist. Soc., 4th ser., XXI (1939), 13–14; L. F. R. Williams, 'The status of the Welsh in the laws of Ine', Eng. Hist. Rev., XXX (1915), 271–7.

[2] J. E. A. Jolliffe, 'Northumbrian institutions', Eng. Hist. Rev., XLI (1926), 1–42.

[3] G. R. J. Jones, (1) 'Basic patterns of settlement distribution in northern England', The Advancement of Science, XVIII (1961), 192–200; (2) 'Early territorial organisation in England and Wales', Geofriska Annaler, XLIII (1961), 174–81; (3) 'Settlement patterns in Anglo-Saxon England', Antiquity, XXXV (1961), 221–32. See also F. M. Stenton (1947), 311.

man and later times some new villages came into being. But in general, the village geography of the sixth to the ninth century has formed the basic element in the pattern of much of the English countryside up to the present day. The rise and fall of the kingdoms of the Heptarchy occupy a prominent place in the annals of Anglo-Saxon England, but behind the clash and thunder of Anglo-Saxon warfare, there went on an almost silent, and for the most part unrecorded, process by which the English occupied and cultivated the land that was now theirs.

Two very frequent Anglo-Saxon elements in village-names are 'ham' and 'ton'. As the English advanced westwards, the word 'ham' seems to have become obsolete and, at the same time, the meaning of 'tun' was changing from 'enclosure' to 'farmstead' and then to 'village'. The result is that 'ham' names are most frequent in the south-east and in East Anglia, and that 'ton' or 'tun' names are most frequent in the Midlands and the west. As the settlement was consolidated, there came into being the great mass of terminations that give interest and variety to our place-names — 'cote', 'ford', 'stede', 'worth' and the like.[1] But while the Anglo-Saxons were thus laying the foundations of one village geography they were interrupted by a new group of invaders, who in turn were to continue and intensify the spread of settlement in the north and east. These were the Scandinavians.

THE SCANDINAVIAN SETTLEMENT

The Scandinavian invasions

Under the year 787 the Anglo-Saxon Chronicle records the first coming of Scandinavian raiders to the shores of England: 'then the reeve rode to the place, and would have driven them to the king's town, because he knew not who they were, and they there slew him'. From other sources we know that the place was Portland in Dorset, and that the reeve was the reeve of Dorchester. The Chronicle records the plundering of Lindisfarne in 793 and of Jarrow in 794. Other raids followed upon the coasts of western Scotland and Ireland, and to the litany of the Church in the ninth century was added the prayer: 'From the fury of the Northmen good Lord deliver us.' In 835, raiders appeared at the mouth of the Thames, and plundered the Isle of Sheppey. Others followed, and we hear of their activities in East Anglia, Lindsey, Devon and elsewhere. The Chronicle

[1] The authoritative survey is A. H. Smith, *English place-name elements*, 2 vols. (Cambridge, 1956).

entry for 851 says that the raiders 'for the first time remained over winter in Thanet'; and in 855 they likewise wintered for the first time in Sheppey.

It was not until 865 that invasion really started with the arrival of 'a great heathen army' in East Anglia. Here, in the following year, the Chronicle tells us, they obtained horses; and, thus equipped, they extended their activities during the next three years into southern Northumbria and eastern Mercia. Under the year 876 we read that they shared out the land of Northumbria and that 'they began to plough and make a livelihood'. In the following year they 'shared out' part of Mercia, and gave the rest to a puppet English king. We are not told whether this 'sharing out' was a division into small units for the purpose of cultivation, or one into large districts such as those connected with the fortified centres of what later came to be known as the 'Five Boroughs' of Lincoln, Nottingham, Derby, Leicester and Stamford. In 879, the land of East Anglia was also 'shared out'.

In the meantime, the Danes had advanced westward, to attack Wessex, and the defence of the English devolved upon King Alfred from his base on the Isle of Athelney in the Somerset marshes. There followed a period of confused warfare that culminated in a treaty traditionally ascribed to 886, the year in which Alfred captured London. The treaty defined the boundary between the political power of the English and that of the Danes. The line ran along the Thames and then along the Lea to its source, whence it continued in a straight line to Bedford and so along the Ouse to Watling Street; how far it continued north-westward we cannot say. To the east lay the territory that became generally known as the Danelaw, with its three divisions of East Anglia, Mercia and southern Northumbria.

The treaty of 886 marked not the end of conflict but a very temporary balance of forces, and before his death in 899 Alfred had begun a systematic fortification of southern England by building a series of 'burhs' or forts, a list of which survives in the so-called Burghal Hidage.[1] The work of recovery was carried on by Alfred's son Edward the Elder of Wessex and by his daughter Aethelflaed, 'the Lady of the Mercians'.[2] Acting in concert during the years 907–16, they continued the practice of building burhs until a zone of fortified centres stretched from the Mersey to Essex

[1] See pp. 36–7, below.

[2] F. T. Wainwright, 'Æthelflaed, Lady of the Mercians', in P. Clemoes (ed.), *The Anglo-Saxons* (London, 1959), 53–69.

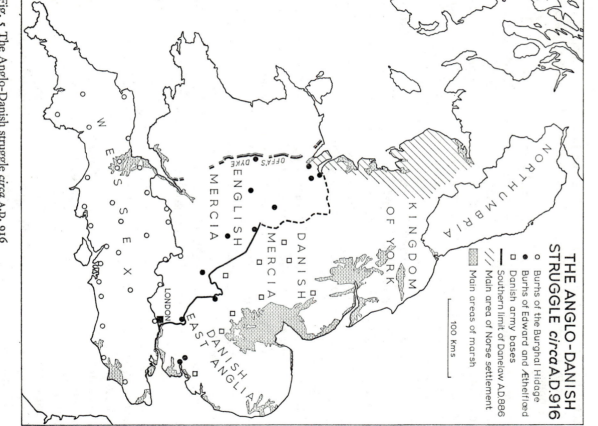

THE ANGLO-DANISH STRUGGLE *circa* A.D. 916

- ○ Burhs of the Burghal Hidage
- ● Burhs of Edward and Æthelflaed
- □ Danish army bases
- —— Southern limit of Danelaw A.D. 886
- /// Main area of Norse settlement
- ▨ Main areas of marsh

100 Kms

Fig. 5 The Anglo-Danish struggle *circa* A.D. 916 Based on: (1) F. T. Wainwright, 'Æthelflaed, Lady of the Mercians', in P. Clemoes (ed.), *The Anglo-Saxons* (London, 1959), 59; (2) J. F. Benton, *Town origins: the evidence from medieval England* (Boston, Mass., 1968), 49. The possible continuation of the English–Danish boundary to the north-west is marked.

(Fig. 5). The campaign of the years 917 and 918 resulted in the recovery of East Anglia and Danish Mercia by the English; and, with the death of Aethelflaed in the latter year, English Mercia was joined to Wessex. When Edward the Elder died in 924, the northern boundary of his kingdom stretched from the Humber to the Mersey.

Moreover, to the north the Scandinavian kingdom of York (southern Northumbria), the English relic of Northumbria beyond the Tees, and the Celtic kingdom of Strathclyde had acknowledged the king of Wessex as their overlord. But this success was followed by rebellion and much confusion, increased, as we shall see, by the arrival of the Norse from Ireland. It was not until 954 that an independent Scandinavian kingdom of York ceased to exist. Any clear chronology of events is made difficult by the extreme obscurity of the sources.

While the Danes had thus been advancing from the east coast, another, and rather different, Scandinavian immigration was taking place from the west. Norsemen from south-western Norway had established settlements in the Shetlands and the Orkneys, and from there one stream of migration went northward to the Faroes, to Iceland and beyond, while another stream came to the northern mainland of Scotland, to the Western Isles and southward into the Irish Sea. The monastery of Iona was plundered in 795, and during the next thirty years or so the Irish annals made many references to Viking raiders. In the tenth century a line of Norse kings reigned in the Isle of Man, and powerful Norse kingdoms arose around the coastal towns of Ireland, around Wexford, Waterford, Cork, Limerick and, above all, around Dublin.

It was from these older colonies in Ireland, the Isle of Man and the Western Isles that the Norse came to Cumberland, Westmorland, Lancashire, Cheshire and to the coasts of North and South Wales. The only literary record (although much later in date) of this movement concerns the expedition of a certain Ingimund from Ireland to establish a Norse colony in the Wirral peninsula of Cheshire about the year 902.[1] The settlement seems to have been very largely a peaceful one, and this may also have been true generally of the Norse settlement to the north, in Lancashire and elsewhere. It would seem that in the tenth century there was still much uninhabited land in north-west England, and that the Norse were content to occupy less attractive districts neglected by the English or

[1] F. T. Wainright, 'Ingimund's invasion', *Eng. Hist. Rev.*, LXIII (1948), 145–69. See also F. T. Wainright, 'North-west Mercia', *Trans. Hist. Soc. Lancs. and Cheshire*, XCIV (1942), 3–55.

Welsh.[1] They spread eastward across the Pennines, so that the areas of Norse and Danish settlement overlapped. Norsemen from Ireland also played a part in the political complications of northern England. They even invaded Northumbria and, in 919, founded a new kingdom of York which continued with intervals until 954. During this period the Norse kings of Dublin at times also ruled in York. To the Danish settlement in Yorkshire was thus added a Norwegian–Irish element.

Before the end of the tenth century there began a second episode of Danish power in England. From the year 980 onwards there was a fresh series of Danish raids, especially along the south and south-east coasts; and recurrent Danegelds from 991 onwards failed to satisfy the raiders. Under the year 1004 the Anglo-Saxon Chronicle tells of the plundering of Norwich and Thetford. Under the year 1006 we hear of 'every shire in Wessex sadly marked by burning and plundering'. In 1009 a large army appeared off the coast of Kent. Repulsed by the city of London, a Danish force made its way through the Chilterns to Oxford, which was burned. After returning to Kent to repair their ships, the Danes appeared in 1010 off East Anglia which they 'ravaged and burned'. Here, 'they were horsed' and they raided through the countryside, destroying men and cattle; they burned Thetford and Cambridge, and continued south to the Thames and westward into Oxfordshire and Buckinghamshire and Bedfordshire, before returning ', to their ships with their booty'. By 1011, says the Chronicle, they had ravaged the whole or part of eighteen counties. Although it was sporadic, local and relatively ephemeral, this repeated devastation was an element of no little importance in the changing economic geography of eleventh-century England.

The invasions culminated in 1013 with the submission of London and the conquest of the realm by Swein who was followed by his son Canute of Denmark in 1016. With Canute's death in 1035 England ceased to be part of the Scandinavian empire, but a Danish dynasty did not cease to rule in England until 1042 when Edward the Confessor of the royal house of Wessex was chosen as king. In the meantime, the Anglo-Scottish frontier was emerging. Strathclyde was united with Scotland about 945, but Carlisle and the district to the south became part of England in 1092. To the east, Lothian was ceded to Scotland in 975 or shortly before. Thus did the boundary between the two kingdoms become the Tweed–Cheviot–Solway line, at any rate in name.

[1] F. T. Wainright, 'The Scandinavians in Lancashire', Trans. Lancs. and Cheshire Antiq. Soc., LVIII (1946), 71–116.

Scandinavian place-names

Although Scandinavian political power vanished from England, the Scandinavians themselves remained. They left little archaeological evidence as compared with the Anglo-Saxons, and this has sometimes been attributed to the fact that the Scandinavians soon adopted the Christian faith and so discarded heathen burial practices. Whatever be the truth, the main evidence for their settlement is derived from place-names. One characteristic Scandinavian place-name element is 'by' implying a village, and there are about 250 Yorkshire village names that end in 'by', and nearly that number in Lincolnshire. Well over a half of these are compounded with personal names and may belong to a period 'within at most a generation or two' of the original settlement; some may even be the names of Danes who had taken part in the conquest.[1] A detailed examination of these 'by' names in relation to geology and soil led Kenneth Cameron to the view that the Danes settled in the less attractive localities left empty by the English and that they came 'predominantly as colonisers, occupying new sites'.[2]

Next in importance is the element 'thorpe' meaning 'hamlet' or outlying settlement from an older village. Many of these are compounded with personal names that suggest they were made by individuals from parent villages; generally they were later than 'by' names.[3] There is also a wide variety of other Scandinavian elements – *toft* (homestead), *garth* (enclosure), *lathe* (barn) and many others. The greater number of these words were common to both Danes and Norse, but there were a few words peculiar to each; thus *thorpe*, *both* (booth) and *hulm* (holme) were Danish, while *gill* (valley), *skali* (hut) and *erg* (shieling or hill pasture) were Norse or Norse borrowings from Irish. One further point must be made. While many Scandinavian names go back to the early days of the settlement, 'there is no safe way of distinguishing those that arose in the ninth century from those that may have arisen at a later date'.[4] The Scandinavian language continued to be spoken until at least the eleventh

[1] F. M. Stenton (1947), 516.

[2] K. Cameron, *Scandinavian settlement in the territory of the five boroughs: The place-name evidence* (University of Nottingham, 1965), 13.

[3] A. H. Smith, *English place-name elements*, pt 2, 209.

[4] F. T. Wainright, 'Early Scandinavian settlement in Derbyshire', *Jour. Derbyshire Archaeol. and Nat. Hist. Soc.*, n.s., xx (1947), 100.

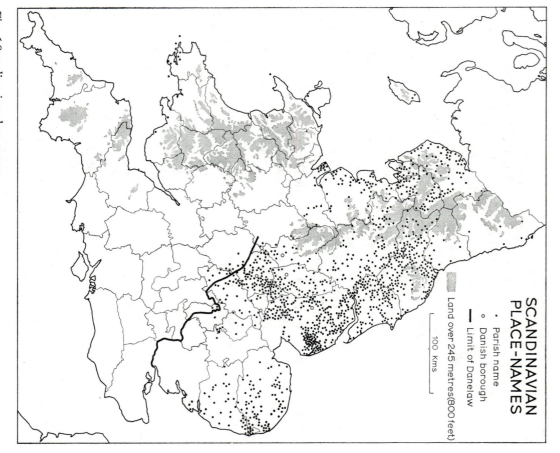

SCANDINAVIAN
PLACE-NAMES

· Parish name
○ Danish borough
━ Limit of Danelaw

▨ Land over 245 metres(800 feet)

⊢━━━┥
100 Kms

Fig. 6 Scandinavian place-names
Based on A. H. Smith, *English place-name elements*, pt I (Cambridge, 1956).
The five boroughs are these of Derby, Leicester, Lincoln, Nottingham and
Stamford.

century,[1] and, moreover, the English language itself became highly Scandinavianised. Many 'Scandinavian' place-names date from the Middle English period, and cannot be regarded 'as affording in themselves evidence of a Scandinavian settlement'.[2]

The distribution of Scandinavian parish names as plotted on Fig. 6 shows that the area of settlement was not as extensive as the Danelaw itself; political control had been wider than actual settlement.[3] But a map such as Fig. 6 can give only a very inadequate idea of the Scandinavian element in place nomenclature. In the first place, not all Scandinavian names are of equal value in indicating the spread of settlement. Some names are Scandinavian in the strictest sense, that is words that were formed by people who spoke a Scandinavian language, names like Laceby and Litherland. Other names are only Scandinavianised versions of English ones, such as Keswick and Louth, which in their pure Anglian forms would have been Cheswick and Loud. Then there are hybrid names such as Grimston which is a combination of the Scandinavian personal name 'Grim' with the English 'tun'. In the second place, a parish with an English name often contains many minor names that are Scandinavian, names of topographical features and of hills. The hilly districts of the western parts of the North Riding and the West Riding, for example, are rich in Scandinavian names of minor features; so is north Lancashire; and the Lake District, especially in its remote parts, is even more rich. Names ending in *erg*, *saetr* and *scale* denote the former presence of Norse shielings. Thousands of such minor names supplement the evidence of village names, but it must be said at once that many, or perhaps most of them, date from a period long after the first settlements of the ninth century. In the present state of our knowledge, it would be impossible to construct a map that

[1] E. Ekwall, 'How long did the Scandinavian language survive in England?' *A grammatical miscellany offered to Otto Jespersen on his seventieth birthday* (Copenhagen and London, 1930), 17–30.

[2] H. Lindkvist, *Middle English place-names of Scandinavian origin*, pt I (Uppsala, 1912), liii.

[3] Convenient summaries of the Scandinavian place-name evidence may be found in the following: (1) E. Ekwall, 'The Scandinavian element', in A. Mawer and F. M. Stenton (eds.) *Introduction to the survey of English place-names* (Cambridge, 1929), 55–92; (2) E. Ekwall, 'The Scandinavian settlement' in H. C. Darby (ed.), *An historical geography of England before A.D. 1800* (Cambridge, 1936), 133–64; F. M. Stenton, 'The historical bearing of place-name studies: the Danish settlement of Eastern England', *Trans. Roy. Hist. Soc.*, 4th ser, XXIV (1942), 1–24. Individual elements are discussed in A. H. Smith (1956). See also the relevant county volumes of the English Place-Name Society.

took both these groups of considerations into account. Fig. 6 must stand as giving a general impression.

The density of the Scandinavian settlement

The general impression given by Fig. 6 leaves at least one major question unsettled. How numerous were the newcomers in relation to the existing English population? We can never answer that question with any certainty, but some scholars have seen clues in two types of evidence. One is the indication given by the personal names that occur in charters of the twelfth and thirteenth centuries. Personal names are always subject to fashion, and some English people may well have adopted Scandinavian names and vice versa. Yet, bearing this in mind, the percentage of Scandinavian personal names in many districts is very high. Sir Frank Stenton declared that in the northern Danelaw more than one half the native personal names that survived the Norman Conquest were Scandinavian, and that they were borne by men and women of widely different social condition.[1] They suggest, he wrote, that 'something like a genuine migration may have taken place in the ninth century, that, in particular, the armies may have sent for their womenkind when they turned from war to agriculture'.[2] The twelfth-century Scandinavian personal names on a series of Norfolk manors amount to about 10%,[3] and on a series of Suffolk manors to about $8\frac{1}{2}$%.[4]

Another indication of the density of the Scandinavian settlement has been seen in the proportion of sokemen to total population as recorded in the Domesday Book. This is on the assumption that they represented 'the rank and file of the Scandinavian armies' of the ninth century. The percentage of sokemen in Lincolnshire wapentakes ranged from 20 to 73%, and it averaged 50%. In Leicestershire it ranged from 27 to 50% and it averaged about 30%. In Nottinghamshire the range was from 11 to 53% and the average was again about 30%.[5] The position in East Anglia was complicated by the presence of freemen as well as sokemen, but taken together they averaged over 40% in Norfolk and in Suffolk, and they, or

[1] F. M. Stenton, *Documents illustrative of the social and economic history of the Danelaw* (British Academy, London, 1920), cxiv–cxvi.

[2] F. M. Stenton, 'The Danes in England', *Proc. British Academy*, XIII (1927), 32.

[3] J. R. West, *St Benet of Holme, 1020–1210* (Norfolk Record Society, 1932), 258–9.

[4] D. C. Douglas, *Feudal documents from the abbey of Bury St Edmunds* (British Academy, London, 1932), cxx.

[5] F. M. Stenton, 'The free peasantry of the northern Danelaw', *Bulletin de la Société Royale des Lettres de Lund, 1925–19 26*, XXVI (1926), 73–185.

many of them, have been thought to be 'descendants of the Danish settlers of the ninth century'.[1]

This belief in an intensive Scandinavian colonisation of much of the north and east of England has not gone without criticism. H. W. C. Davis has pointed to the fact that the distribution of the free peasantry of East Anglia was far from coincident with that of Scandinavian place-names, and that there already was an English free peasantry here before 890. The 'by' names compounded with Scandinavian personal names might well refer 'not to peasants, but to the lords or owners of the villages'. There is no reason why the villagers themselves should not have been English. And as for personal names, he believes it is 'a complete delusion to think that every man with a Scandinavian name was of Scandinavian origin'. Personal names change with fashion as may be seen from the very large number of Norman names in East Anglia at the end of the twelfth century in spite of the fact that the Normans contributed little or nothing to the settlement of the area.[2]

Similar views have been expressed by P. H. Sawyer. He believes that the Danish armies were to be numbered in hundreds, not in thousands, and that 'the Danish settlements were the work of far fewer men than has often been supposed'. The Danish soldiers who settled here 'were probably joined by others about whom we hear as little as we do of Norse settlers in the north-west'. These later arrivals did not overwhelm the English; 'they settled where they could, most often on land left that the English had not yet occupied'. Even these do not account for all the Scandinavian and Scandinavianised place-names, many of which were not formed until after the Norman Conquest and so are 'clearly irrelevant for the study of the original Scandinavian settlements'.[3]

While there may be much in these arguments, particularly in so far as they concern East Anglia, the very large number of Scandinavian place-names and the evidence of Scandinavian influence upon law, language and administrative divisions, have made it difficult for some scholars to believe that an intensive colonisation did not take place.[4] The armies themselves

[1] B. Dodwell, 'The free peasantry of East Anglia in Domesday', *Trans. Norfolk and Norwich Archaeol. Soc.* XXVII (1939), 153.

[2] H. W. C. Davis, 'East Anglia and the Danelaw', *Trans. Roy. Hist. Soc.*, 5th ser., v (1955), 23–29.

[3] P. H. Sawyer, (1) 'The density of the Danish settlement in England', *Univ. o, Birmingham Hist. Jour*, VI (1958), 1–17; (2) *The age of the Vikings* (London, 1962), 145–67.

[4] H. R. Loyn, 55. See also K. Cameron, 10–11.

may well have been small in size, but behind them, and protected by them,
considerable immigration may well have taken place. One can only conclude that the Scandinavian settlement may have been a much more
complicated process than was at one time thought.

THE CHANGING COUNTRYSIDE

The location of Anglo-Saxon and Scandinavian villages indicates a rational
selection of sites – maybe along the juxtaposition of two geographical
formations where springs gave water, or upon gravel terraces above the
flood-levels of streams, or on some other 'dry-point' sites. The territories
attached to these villages show how the needs of largely self-suffising
communities were met by including a variety of soils within the jurisdiction of each community. In changing the geography of Britain, the newcomers were themselves influenced by the varieties of soil and vegetation
that confronted them. The sizes and shapes of the parishes of later times
frequently provide an indication of the conditions under which early
settlement took place. Arable, meadow, wood and pasture were the main
ingredients of the usual village territory.

Exactly how that arable was cultivated we do not know, and there has
been much discussion about the beginnings of the open-field system in
England. To what extent did it exist in Britain before the coming of the
Anglo-Saxons? Did the Anglo-Saxons introduce it from the Continent?
And, in any case, what was the relation of this early system to the fully
developed common-field system of later times? Or again, what was the
influence, if any, of the Scandinavian settlement upon the agricultural
arrangements they found? Any certain answers to these questions must
await further investigation. A well-known passage in the laws of Ine
has often been taken to imply the existence of some kind of open-field
agriculture in Wessex about A.D. 690. It is not, however, until the tenth
century that charters provide fairly numerous glimpses of unmistakable
intermixed strips; but, even so, doubt has again been expressed as to
whether, at this early date, they were grazed in common by the stock of
the villagers after harvest and in fallow seasons. Not until post-Domesday
times does a relatively clear picture emerge.[1]

The existence of open fields must not be taken to imply any uniformity

[1] H. P. R. Finberg (1972), 261–3, 416–20, 487–93; R. G. Collingwood and J. N. L.
Myres, 210–13, 442–3; H. L. Gray, *English field systems* (Cambridge, Mass., 1915),
50–61.

in their organisation. If we may judge from later evidence, the open fields of the south-east and of East Anglia were organised in a different way from those of midland England. The Danes in East Anglia,[1] and either the Romans[2] or the Jutes[3] in Kent, and the Frisians in both areas[4] have been held responsible for the differences but these are very disputable matters. In some areas, natural features reflected themselves in individual agrarian arrangements, in, for example, the Fenland, the Weald and the wooded Chilterns. Then again, not only open fields but other forms of agrarian arrangements were to be found in the south-west and in the upland areas of the north and the north-west of England, where relief, soil and climate made pastoral activities important and where isolated farms and hamlets may have been more important than villages. The Northumbrian charters point to 'the use of summer and winter grazings remote from the agricultural settlements'.[5] In the north-west, Norse and Irish – Norse terminations in *erg*, *booth* and *saetr* also indicate shielings to which cattle were sent in summer,[6] but one must hasten to add that the earliest place-names incorporating these elements belong to a much later period. Although we have to rely upon the uncertainty of retrospective deduction for hints about agrarian arrangements in Anglo-Scandinavian times, there can be no doubt about the variety of those arrangements and about the fact that there were many different types of settlements.

The establishment of these Anglo-Saxon and Scandinavian settlements had profound consequences for the English countryside. In the earlier charters of the Anglo-Saxon period, grants of land are only vaguely defined as lying near this stream or as bounded by that wood. They reflect a time when settlement had not become intensive and when village rights were not precisely defined in the waste and wood around. But the charters of the tenth and eleventh centuries show a remarkable precision of topographical detail. Boundaries of estates are defined with great exactness and the charters abound in the local names of minor natural and other features. This contrast between the earlier and later charters 'indicates the nature of the unrecorded changes which had come over English country

[1] D. C. Douglas, *The social structure of medieval East Anglia* (Oxford, 1927), 50.
[2] H. L. Gray, 50 and 415–16.
[3] J. E. A. Jolliffe, *Pre-feudal England: the Jutes* (Oxford, 1933), 73 et seq.
[4] G. C. Homans, 'The rural sociology of medieval England', *Past and Present*, no. 4 (1953), 32–43.
[5] J. E. A. Jolliffe, 'Northumbrian institutions', *Eng. Hist. Rev.*, XLI (1926), 12.
[6] E. Ekwall in A. Mawer and F. M. Stenton (1929), 89.

life between the eighth and the tenth centuries'.[1] Some later villages bear names which had appeared merely as those of boundary marks in earlier charters. A charter of 1002 defines the boundaries of Little Haseley in Oxfordshire as beginning 'at Roppanford' and as passing 'over against Stangedelf'. The first of these boundary marks gave its name to the later village of Rofford, and the second to the later hamlets of Upper and Little Standhill. Such examples suggest the filling out of an earlier pattern of settlement, and it is clear that new settlements were being formed throughout a long period.[2]

It was F. W. Maitland who drew our attention to 'the numerous hints that our map gives us of village colonisation'.[3] A settlement with a wide territory often came to have another settlement within its limits. Such names as Great and Little Shelford or Guilden Morden and Steeple Morden in Cambridgeshire point to a time when a single territory 'by some process of colonisation or subdivision' became two territories. Not only two, but sometimes three or four or more settlements bear the same basic name, e.g. the three Rissingtons in Gloucestershire, the four Ilketshalls in Suffolk, the five Deverills in Wiltshire, the six South Elmhams in Suffolk, the seven Burnhams in Norfolk and the eight Rodings in Essex. Sometimes it is clear that the division had formally taken place before 1086. The Domesday folios for Hertfordshire refer to 'Hadham' and 'Parva Hadham', i.e. Much and Little Hadham; and those for Northamptonshire refer to 'Heiforde' and 'altera Haiford', i.e. Upper and Nether Heyford; those for Lincolnshire already speak of 'Nortstoches' and 'Sudstoches', i.e. North and South Stoke. Many such divided territories were to acquire the names of landholders of importance in the twelfth and thirteenth centuries. Thus the Domesday settlements of 'Emingeforde' and 'Alia Emingeforde', in Huntingdonshire, became known as Hemingford Abbots and Hemingford Grey after the Abbot of Ramsey and the Grey family. Usually, however, the Domesday text does not reveal whether the division had already taken place. The three Buckinghamshire villages of Bow Brickhill, Great Brickhill and Little Brickhill, separately distinguished by 1200 or so, are represented in the Domesday text only by the single name of 'Brichella'.

The process of colonisation went steadily on behind the clash of warfare

[1] F. M. Stenton (1947), 283.
[2] F. M. Stenton, in A. Mawer and F. M. Stenton (1929), 40 and 48.
[3] F. W. Maitland, *Domesday Book and beyond* (Cambridge, 1897), 365.

and the complications of the Anglo-Danish conflict. The dwellings of both groups of people were so many pioneer settlements battling to reduce the wood and waste around. The frontiers of expansion came to lie not to the west and north against the Welsh, but everywhere against the woods of the heavy clay soils. An echo of this activity comes to us in a remarkable passage in a treatise by King Alfred (871–901):

We wonder not that men should work in timber-felling and in carrying and building, for a man hopes that if he has built a cottage on the *laenland* of his lord, with his lord's help, he may be allowed to lie there awhile, and hunt and fish and fowl, and occupy the *laenland* as he likes, until through his lord's grace he may perhaps obtain some day boc-land and permanent inheritance.[1]

In time the log-hut might become one of a cluster, and this, in turn, might become a group of homesteads. The lumberman with his axe and his pick became the ploughman with his oxen. The cleared land was divided and tilled by the common plough, and thus there grew up a new hamlet or village in the outlying wood of some 'ham' or 'ton'. It is not surprising that the Anglo-Saxon poet described the ploughman as the 'enemy of the wood'.[2]

In spite of four centuries of Roman civilisation, Britain was very largely a wooded land when the Anglo-Saxons arrived. Perhaps the greatest single physical characteristic of most of the Anglo-Saxon countryside was its wooded aspect; the heavier clays in particular still carried great woodlands, which provided pasturage for swine feeding on acorns and beechmast. Of the group of descriptive place-names that reflect the primitive landscape of England, those denoting the presence of wood, and the clearing of wood, are the most frequent and the most important. In recalling these names we are, in a sense, looking at the countryside through the eyes of those who saw it in an unreclaimed condition.

Among the later Anglo-Saxon place-names belong those with such terminations as *ley, hurst, holt* and *hey*. There are also those numerous place-names that make specific references to species of tree, e.g. Oak-hanger, Ashton, Elmstead, Buckholt (beechwood) and Berkhamsted (homestead among the birches). Sometimes a place-name tells its story

[1] 'Blossom Gatherings out of St Augustine', Brit. Museum, Vit A, xv, f. 1; quoted in F. Seebohm, *The English village community* (Cambridge, 1883), 169.
[2] W. S. Mackie (ed.), *The Exeter Book*, pt 2 (Early English Text Soc, London, 1934), 111; R. K. Gordon, *Anglo-Saxon poetry* (Everyman's Library, London, 1954), 295.

by embodying a personal name. The village names of Knowsley and Winstanley in Lancashire are derived from men named Cynewulf and Wynstan respectively, while Chorley, found in Lancashire and elsewhere, was originally the clearing of the ceorls or peasants.[1] These examples happen to be the names of parishes, but the names of hamlets, farms and topographical features within a parish are no less significant. The parish of Chiddingford in Surrey contains the names of Frillinghurst, Killinghurst and Sydenhurst which indicate the woods of Frith, Cylla and Sutta respectively, and there are also other wood names in the parish; the earliest known forms of these names are post-Domesday, but, whether they be older or not, they indicate the former existence of wood.[2] Or again, the parish of Hampton in Warwickshire contains the name of Kinwalsey which, when traced back to its earliest form (twelfth century) is seen to be nothing other than Cyneweald's 'haeg' or clearing.[3]

In the Scandinavian parts of England strange-sounding terminations, such as *lundr*, *skogr* and *viohr*, are incorporated in the place-names; all three mean 'wood'. Along the Pennine valley of the Ure lies Wensleydale. The name Wensley itself is English and means 'Waendel's clearing'. Higher up the valley there is sequence of parish and other names that also imply wood; some are English and some are Scandinavian. West Witton is 'the farm in the wood' (*wudu ton*); Ellerlands is 'alder wood' (*elri lundr*); Aysgarth is 'the open space by the oaks' (*eik scarth*); Lunds is 'wood' (*lundr*); Brindley is 'the clearing caused by fire' (*brende leah*); Litherskew is 'the slope with wood' (*lith skogr*).[4] Similar names are encountered elsewhere in the Pennine valleys and around the North York Moors.

In the Celtic parts of England there is another group of un-English names that indicate the former presence of wood. The place-names Clesketts and Culgaith in Cumberland[5] and Culcheth and Penketh in Lancashire[6] incorporate the Celtic element 'ceto' which is cognate with the modern Welsh 'coed' meaning wood. The exposed plateau surfaces

[1] E. Ekwall, *The place-names of Lancashire* (Chetham Society, Manchester, 1922), 113–14, 104 and 131.

[2] J. E. B. Gover *et al.*, *The place-names of Surrey* (Cambridge, 1934), 186–94.

[3] J. E. B. Gover *et al.*, *The place-names of Warwickshire* (Cambridge, 1936), 61.

[4] A. H. Smith, *The place-names of the North Riding of Yorkshire* (Cambridge, 1928), 257–69.

[5] A. H. Armstrong *et al.*, *The place-names of Cumberland*, pt 1 (Cambridge, 1950), 84 and 184.

[6] E. Ekwall (1922), 97 and 106.

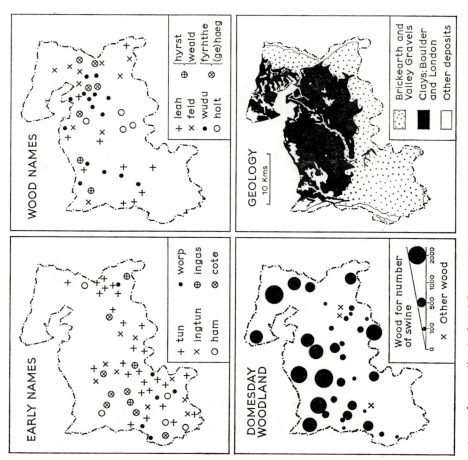

Fig. 7 The woodland of Middlesex
Based on: (1) J. E. B. Gover et al., *The place-names of Middlesex* (Cambridge, 1942); (2) H. C. Darby and E. M. J. Campbell, *The Domesday geography of south-east England* (Cambridge, 1971), 100 and 124.

of Cornwall are not very favourable for the growth of trees, but there was once a fair number in its valleys. The Old Cornish version of 'coed' was 'coit', and it appears in such forms as *cut, quite, coose* and *coys*. Thus the name Trequite in the parishes of St Germans, St Mabyn and St Kew is the same as Tregoose in the more western parishes of Probus, St Erth, Sithney and the like. Penquite and Pencoose, meaning the 'end of the wood', are other common Cornish names.[1]

[1] C. Henderson, *Essays in Cornish history* (Oxford, 1935), 135–51.

The full bearing of such information upon the history of settlement has yet to be investigated. A clear example of its relevance comes from Middlesex where the distribution of different types of names provides a revealing supplement to any deductions from surface geology (Fig. 7).[1] Upon the light gravel and loamy soils in the south of the county, names which do not indicate wood are common, e.g. those ending in 'ton' and 'cote'. They belong to the earlier phase of the Saxon settlement upon open and fertile land. The wood names, on the other hand, lie on the intractable London Clay to the north. In the extreme north-east there are scarcely any names, for here was a great expanse of unsettled wood, the memory of which is preserved by the name Enfield Chase. The wood names register a stage in the process of clearing this northern part of the county. There was still much wood left here by 1086; and even in the twelfth century, it could still be described as 'a great forest with wooded glades and lairs of wild beasts, deer both red and fallow, wild boars and bulls'.[2]

Warwickshire is another county with a marked contrast between north and south.[3] Names ending in *leah* and *ge(haeg)*, for example, indicate the former presence of wood, and they lie almost entirely to the north of the River Avon (Fig. 8). Here again, in spite of some five centuries of clearing there was still much wood left in Domesday times. Later documents show how the arable continued to expand at the expense of the wood, but the wood had far from disappeared even by the seventeenth century. Writers of that time were able to draw a distinction between the southern part of the county, called Feldon or open county, and the northern woodland where lay the Forest of Arden. The earliest one-inch Ordnance map published in the 1830s shows a contrast between the south, characterised by compact villages surrounded by open fields, and the north, characterised by dispersed houses, the dispersion resulting from scattered settlement proceeding piecemeal as the wood was cleared. These features of later times carry us back to a much earlier age.

One of the most well-known stretches of wood was that of the Weald which means 'wood'. The Anglo-Saxon Chronicle described it, in an entry for 893, as 'the great wood which we call Andred', or Andredesweald. It was a region with a distinctive economy. When something can be learned of its condition, from the later Anglo-Saxon charters, it

[1] J. E. B. Gover et al., The place-names of Middlesex (Cambridge, 1942), xv.
[2] F. M. Stenton et al., Norman London (Hist. Assoc. London, 1934), 27.
[3] J. E. B. Gover et al., The place-names of Warwickshire (Cambridge, 1936), xiii–xv, 315, 316.

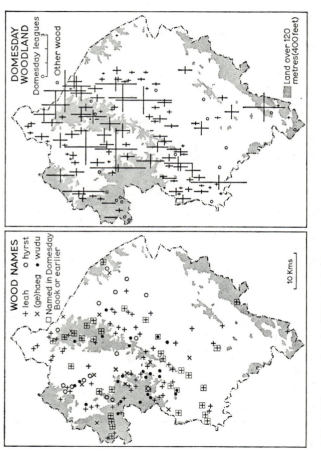

DOMESDAY
WOODLAND

Domesday leagues

o Other wood

Land over 120
metres (400 feet)

WOOD NAMES

+ leah   o hyrst
× (ge)haeg   • wudu
□ Named in Domesday
  Book or earlier

10 Kms

Fig. 8 The woodland of Warwickshire
Based on: (1) J. E. B. Gover *et al.*, *The place-names of Warwickshire* (Cambridge, 1936); (2) H. C. Darby and I. B. Terrett (eds.), *The Domesday geography of Midland England* (2nd ed., Cambridge, 1971), 296.

appears as a land of swine pastures or 'denns' appurtenant to the villages around. The distribution of these denn names poses its own problems.[1] Fig. 9 shows very many denn names in the eastern part of the Weald of Kent; the area of frequent occurrence stops abruptly at the Kent–Sussex border, although some denn names are widely scattered throughout Sussex and there are a few in Surrey. Why should this be so? Does the restricted distribution mean that swine pastures were frequent in the Kentish Weald but not in the Wealds of Sussex and Surrey? This can hardly have been so. Wealden swine pastures were not restricted mainly to Kent, and pre-Conquest charters show that the names of those in Sussex and Surrey frequently did not end in 'den'. The distribution of the element 'denn' thus reflects not only geographical circumstances but also local peculiarities of dialect, and it may imply the settlement of the south-east peninsula by different groups of people. Another termination that almost

[1] H. C. Darby, 'Place-names and the geography of the past' in A. Brown and P. Foote (eds.), *Early English and Norse studies* (London, 1963), 14–18.

2

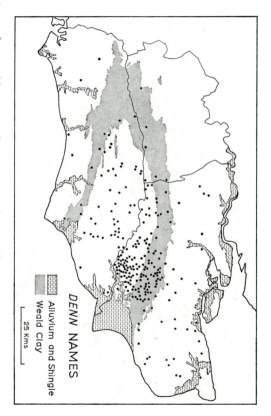

Fig. 9 The Wealden area: place-names ending in 'denn'
Based on: (1) J. E. B. Gover et al, The place-names of Surrey (Cambridge, 1934); (2) A. Mawer et al, The place-names of Sussex, 2 pts. (Cambridge, 1929–30); (3) J. K. Wallenberg, Kentish place-names (Uppsala, 1931); (4) J. K. Wallenberg, The place-names of Kent (Uppsala, 1934).

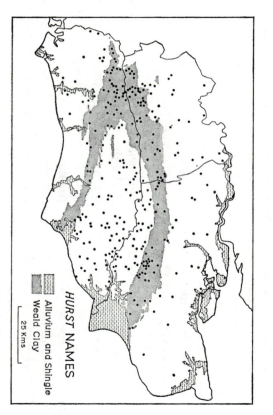

Fig. 10 The Wealden area: place-names ending in 'hurst'
Sources as for Fig. 9.

always implies the presence of wood is 'hurst'. Place-names which embody it are much more widely distributed, but even they tend to show local concentrations (Fig. 10). Then there are other elements (such as *leah* and *feld*) which fill out the picture of the wooded Weald. We can, however, only guess at the variety within this woodland, a variety ranging from the dense oakwoods of the heavy clay to the more open woods of the lighter soils.

Taken together, this mass of place-name evidence shows how the attack on the wooded countryside was prosecuted with vigour between the fifth and the eleventh century. New farms and new hamlets sprang up and grew, until by 1086, a large proportion of the villages we know today had come into being. There was, it is true, still much wood left, but a great part of the clayland had become arable and was tilled by the plough-teams that feature so prominently in the entries of Domesday Book.

Another important element in the Anglo-Saxon landscape was marsh, but references to its extent are very few. The biographer of St Guthlac of Crowland, writing shortly after 700, told of a 'fen of immense size' that stretched from Cambridge northwards to the sea. It was overhung by fog and was characterised by wooded islands and tortuous streams.[1] But it was not without inhabitants or profit; Bede explained how the island of Ely received its name from the great quantity of eels to be found around, and pre-Domesday Fenland charters refer to fishermen and nets and boats.[2] Another expanse of marsh was that in Somerset, and King Alfred's biographer, writing in the ninth century, described the island of Athelney as surrounded by impassable marshes; it could be approached only by boat until, after much effort, a causeway was built.[3] Apart from these two accounts and a few scattered references in charters (e.g. to Romney Marsh) we can only infer the existence of many other marshy areas from the distribution of peat and alluvium on the geological map – of such marshes as those of the Humber lowlands, of the Ancholme valley, of the mosslands of Lancashire and the peaty areas of north Shropshire, to say nothing of the marshy tracts along many a river valley and many a stretch of coast. What is clear is that the conquest of the marsh, unlike that of the wood, did not begin until many centuries after the Anglo-Scandinavian period.

[1] B. Colgrave (ed.), *Felix's Life of St Guthlac* (Cambridge, 1956), 87.

[2] H. C. Darby, *The medieval Fenland* (Cambridge, 1940), 24, 29.

[3] W. H. Stevenson (ed.), *Asser's Life of King Alfred* (Oxford, 1904), 79–80.

URBAN LIFE

Allied with the question of the survival of the Romano-Britons is that of the survival of Roman towns and cities. Gildas in the sixth century spoke of the destruction of the twenty-eight cities of Britain, and the evidence, so far as it goes, points to decline and disappearance. A few centres may have survived to carry on what little trade there was, but the trade must have been very small in amount and the towns very rudimentary. The fate of London, for example, has often been discussed. Some have believed that its organised life as a city continued throughout all the stress of the time, and it is possible that the site of the Roman city was far from derelict in the sixth and seventh centuries, but its life must have been at a very low ebb.

Before the end of the eighth century, the glimmerings of some revival can be discerned. There are indications of trans-Channel traffic in such things as glass, pottery, swords, metalware, woollen garments, wine, slaves, and from this time come references to trade between Offa and Charlemagne. In the eighth century, too, Bede described London as 'the market place of many peoples coming by land and sea', and we hear also of Frisian traders in York and of a market in Southampton. These are but stray fragments of information.[1] We know practically nothing of the early Anglo-Saxon towns themselves.

With the coming of the Scandinavians in the ninth century, urban life received a new stimulus. The Danish supremacy in the eastern Midlands rested upon a confederate group of five fortified centres or *burhs* (Fig. 6) — Derby, Leicester, Lincoln, Nottingham and Stamford – and these were not the only Danish boroughs. Here, today, survives an 'exceedingly neat and artificial scheme of political geography'. Each shire has taken its name from a fortified *burh*, which became its administrative centre.[2]

The example of the Danes was not lost upon the English who, in turn, proceeded to establish a series of strongholds. Our information about these comes mainly from two sources which are complementary in that they refer to different areas.[3] One is the curious document known as the 'Burghal Hidage' which seems to date from the years 911–919, and which

[1] G. C. Dunning, 'Trade relations between England and the Continent in the late Anglo-Saxon period', in D. B. Harden (eds.), *Dark Age Britain* (London, 1956), 218–33.

[2] F. W. Maitland, 187.

[3] J. F. Benton, *Town origins: the evidence from medieval England* (Boston, Mass., 1968), 48–53.

names some thirty or so burhs which Alfred had begun to build before his death in 899; all of them, except Buckingham, were to the south of the Thames or along it, and they had formed a defensive system for Wessex against Danish attack. The other source is the Anglo-Saxon Chronicle which mentions over a score of burhs established in the reign of Edward the Elder (899–924) of Wessex, some built by himself and his sister Aethelflaed of Mercia during the years 907–916 to secure English territory in the Midlands.[1] Many burhs were on sites where Roman towns had once flourished, e.g. Bath and Exeter; the place-name element ceaster or caster (from the Latin *castra*) is often embodied in the names of such Roman centres, e.g. Chichester and Towcester, but this must not be taken as necessarily implying continuity of occupation.

The fact that the term 'burh' or 'borough' later came to mean town does not justify the conclusion that all burh-building meant the creation of urban settlements. The Anglo-Danish burhs were forts, but very often the reasons that made them suitable defensive and administrative centres also made them commercial centres. There has been much debate about whether defence or trade provided the basic impetus to early urban development in England, but clearly both were important. The security of fortified centres encouraged trading activity, while, in turn, trading centres needed to be fortified for protection. The potentialities of the burhs varied. Not all survived to become the boroughs of a later age; the names of some cannot even be identified. Conversely, later boroughs developed on sites which had known no burhs. Some sites were especially favourable for trade, and here burhs, or maybe villages, developed into flourishing towns – at bridgeheads on rivers, at the confluences of tributaries, at the crossing-places of routeways, or near convenient gaps in hill country. Long before the Norman Conquest a force had begun to operate which was ultimately to give to the English borough its most lasting characteristic – i.e. that of a trading centre. We hear, for example, of merchants frequenting York about the year 1000, and they were said to have been mainly Danes. An account of the customs of the port of London about the year 1000 shows trade across the Channel and beyond to Lorraine; cargoes included fish, pepper and wine. Anglo-Saxon merchants also were to be found crossing the Alpine passes on their way to Rome.[2] There were Irish traders in Cambridge about 975.[3]

[1] See pp. 17–19 above.
[2] For a summary see J. Tait, *The medieval English borough* (Manchester, 1936), 118–19.
[3] *Ibid*, 10 n.

The association of the early boroughs with markets and mints, difficult as it may be to interpret, indicates a commercial element. The evidence of the coins themselves is striking. The number of moneyers in a borough may be taken to indicate, in a very general way, its importance as a commercial centre. Between 1042 and 1066 London had as many as twenty moneyers working at the same time. York had more than ten; Lincoln and Winchester had at least nine each; Chester had at least eight; Canterbury and Oxford had at least seven each; Thetford, Gloucester and Worcester had at least six each.[1] This kind of evidence depends upon the chance discovery of coins, and may well do less than justice to many cities; but at any rate it prepares us for the indications of urban life to be found in Domesday England.

[1] F. M. Stenton (1947), 529. See also (1) G. C. Brooke, *Catalogue of English coins in the British Museum: the Norman Kings* I (London, 1916), clx–clxxxviii; (2) R. H. M. Dolley (ed.), *Anglo-Saxon Coins* (London, 1960).

Chapter 2

DOMESDAY ENGLAND

H. C. DARBY

The Norman Conquest in 1066, unlike the Anglo-Saxon and Scandinavian invasions, was not a mass movement of people but the work of a small power group. Twenty years after their coming, the Normans instituted the enquiry that resulted in the Domesday Book. With hindsight we can say that it came at a fortunate moment for us because it enables us to inspect the economic and social foundations of the geography of England after the Anglo-Saxons and Scandinavians had firmly established themselves in the land to which they had come.

We speak of the Domesday Book, but it is really in two volumes. The smaller volume describes Essex, Norfolk and Suffolk; the larger volume, in somewhat less detail, describes the rest of England with the exception of the four northern counties. There are also a number of subsidiary documents that must have been composed from the original returns and that sometimes help in the elucidation of the main survey. Among these is the so-called Exeter Domesday Book dealing with the counties of Cornwall, Somerset, most of Devonshire, parts of Dorset and a solitary Wiltshire manor. It was from this Exeter Domesday Book that the relevant portions of the main Domesday Book were made. Obviously an account that is nearer to the original returns than the main Domesday Book must be of especial interest. The word 'Domesday' does not occur in the Book itself, and there are varying opinions about how it got this name. The Treasurer of England, writing in 1179, said: 'This book is called by the natives Domesday – that is, metaphorically speaking, the day of judgement.' So thorough was the enquiry that its result may well have seemed comparable to the Book by which one day all will be judged (*Revelations*, 20:12).

Domesday Book is far from being a straightforward document. The exact method of its compilation is the subject of controversy, and many of its entries bristle with difficulties. Even so, it is probably the most remarkable statistical document in the history of Europe. Its information is arranged under the heading of each county. Within each county, it is set out under the names of the principal landowners, beginning with the king himself and continuing with the great ecclesiastical lords, then with the lay lords. The holding of each landowner is described village by village. If, say, three lords held land in a village, the three sets of information must be brought together to obtain a picture of the village as a whole.[1] The account of Buckden in Huntingdonshire is fairly representative. It was held entirely by the bishop of Lincoln, and so is described in a single entry on folio 203b:

In Buckden the bishop of Lincoln had 20 hides that paid geld. Land for 20 plough-teams. There, now on the demesne 5 plough-teams, and 37 villeins and 25 bordars having 14 plough-teams. There, a church and a priest and one mill yielding 30s [a year] and 84 acres of meadow. Wood for pannage one league long and one broad. In the time of King Edward [i.e. in 1066] it was worth £20 [a year], and now £16. 10s.

The variety of detail in such entries as this falls into two categories. In the first place there are those items that recur in almost every entry: hides (or carucates in the Danish districts) which were units of taxation, plough-lands, plough-teams, various categories of population, and annual values usually for 1066 and 1086, but sometimes also for an intermediate date. The second group comprises such items as the mill, meadow and wood entered for Buckden, and also, where relevant, pasture, salt-pans, fisheries, waste and vineyards. It is from these two groups of information that the geography of England in 1086, in all its regional diversity, can be reconstructed.

COUNTIES, HUNDREDS, WAPENTAKES AND VILLS

The land that William the Conqueror took over already possessed a highly developed territorial organisation. England south of the Tees was completely divided into shires, and these were divided into hundreds or wapentakes, and these, in turn, comprised vills. The Inquest itself, and the Book that resulted from it, was primarily organised on the basis of shires or, to use the Norman word, of counties. Lancashire, it is true, did not appear under that name until towards the end of the twelfth century; the area south

[1] For a general account of the making and contents of the Domesday Book, see R. Welldon Finn, *An introduction to Domesday Book* (London, 1963).

of the Ribble was described in a kind of appendix to the Domesday account of Cheshire, and the northern part of the county was included with York-shire. Rutland likewise did not appear as a county until the thirteenth cen-tury. Its eastern part was in Domesday Northamptonshire, and its western part formed an anomalous unit called 'Roteland' which was described at the end of the Nottinghamshire section of the Book. There have also been various other less important changes in the inter-county boundaries. Be-yond Domesday England, the four northern counties were in the nature of border provinces, and 'it is probable that responsibility for their internal order, as for their defence, rested with the great lords of the county'.[1]

The Old English word 'scir' meant division, and it came to be applied to administrative divisions. Today, the word has survived not only in the names of many counties but also in those of some smaller districts that were also once administrative units, e.g. Hallamshire in the West Riding and Richmondshire in the North Riding. The shire system first emerged in Wessex, apparently by the early years of the ninth century.[2] Some shires corresponded to the areas occupied by groups of people organised in dependence upon a local capital – Dorset upon Dorchester, Somerset upon Somerton, Hampshire upon Southampton, Wiltshire upon Wilton. Former independent kingdoms also appeared as shires. In the west were Devon and Cornwall, the descendants of the old Celtic kingdom of Dumnonia or West Wales. In the east, Sussex and Kent also corresponded to old kingdoms. Other shires to emerge were Surrey and Berkshire. Thus it was that the supremacy of Wessex in the ninth century had made possible 'the establishment of a uniform scheme of local administration throughout southern England'.[3]

The shires to the north of the Thames did not appear until a later date. The expansion of Wessex at the expense of the Danes in the tenth century resulted in the emergence as shires of the three ancient units of the Danish kingdom of East Anglia – Norfolk, Sussex and Essex. The Danish kingdom of York likewise became a shire. The remaining shires of England are largely artificial in the sense that they bear no relation to the earlier territorial divisions of the Anglo-Saxon period.[4]

---

[1] F. M. Stenton, *Anglo-Saxon England* (Oxford, 1947), 496.

[2] *Ibid*, 289–90 and 332–4.

[3] *Ibid*, 290.

[4] C. S. Taylor, 'The origin of the Mercian shires', *Trans. Bristol and Gloucs. Archaeol. Soc.*, XXI (Bristol, 1898), 32–57. Reprinted in a modified form in H. P. R. Finberg, *Gloucestershire studies* (Leicester, 1957), 17–51.

Although their names as shires did not appear until the early years of the eleventh century, their origins were older. Those of the west Midlands seem to have resulted from a reorganisation of existing arrangements by the Wessex kings, Edward the Elder (899–924) and Athelstan (924–39), who had destroyed Mercian independence. Thus Shropshire, centred on Shrewsbury, was an artificial union of lands belonging to two groups of people – the Wreocensætan in the north and the Magesætan in the south. Other districts such as Warwick, Oxford, Gloucester were likewise created and named after central fortresses. The shires of the east Midlands were equally artificial and represented districts occupied by the various Danish armies, each district being centred on a town – Derby, Nottingham, Cambridge and the like. What Maitland described as an 'exceedingly neat and artificial scheme of political geography' was then incorporated into the Wessex scheme of things.[1]

Shires were subdivided into smaller units called 'hundreds'. These appeared as units of local government in the tenth century, and they provided a basis for the administration of justice and of public finance. In theory they were districts assessed for the purposes of taxation at 100 hides but this correspondence was rare. The assessments of the West Saxon hundreds ranged from about 20 to 150 or so hides, and the complexities that lie behind these variations are lost to us. In the Midlands, hundred assessments more often approximated to 100 hides, which may suggest that the division was imposed as part of the organisation into shires in the first half of the tenth century. It is impossible to say to what extent it replaced a more primitive arrangement.[2] The place of the hundred was taken by the wapentake in the highly Scandinavianised counties of Derby, Leicester, Lincoln, Nottingham and in the North and West Ridings (but not in the East Riding). The word is of Scandinavian origin and was first used to denote a territorial division in 962. This use seems to have been an innovation, 'for no divisions so named are known from Scandinavia'.[3] Functionally, the wapentake was the same as the hundred, and the terms are occasionally used interchangeably, e.g. twice in the Domesday folios for Northamptonshire and once in those for the East

[1] F. W. Maitland, *Domesday Book and beyond* (Cambridge, 1897), 187.
[2] O. S. Anderson, 'On the origin of the hundredal division', being pp. 209–17 of *The English hundred names: the south-eastern counties* (Lund, 1939). See also F. M. Stenton, 296–7.
[3] O. S. Anderson, *The English hundred-names* (Lund, 1934), xxi.

Riding.[1] The four northern counties were without hundreds or wapentakes, but in their place the unit of the 'ward' appeared in the thirteenth century.

Other territorial units are also named in the Domesday Book. Some were units intermediate between those of the shire and the hundred or wapentake. Thus Yorkshire was divided into three ridings (the Scandinavian word 'riding' implies a third part). There were also ridings in Lindsey, itself a third part of Lincolnshire; the other parts were those of Kesteven and Holland. Kent had 'lathes' and Sussex had 'rapes', which may have represented older divisions of these ancient kingdoms.

Hundreds were composed of villages or vills, but alongside the physical reality of the village there was the institutional reality of the manor which features so prominently in the Domesday folios. Sometimes, a vill coincided with a manor. Sometimes it contained two or three manors belonging to different lords. Sometimes vills themselves were components of a large manorial complex which their lord treated as one unit. Clearly, manors varied greatly in size. There were enormous manors, such as that of Rothley in Leicestershire with holdings or members (*membra*) in twenty-two separate villages. Other manors were minute, such as Fernhill in Devonshire with only one recorded man and a solitary ox. Manors also varied in character; much has been written about the place of the manor in the feudal system and about its varying nature in different parts of the country. The normal Domesday entry distinguishes between the plough-teams of the manorial demesne, on which the local peasantry were obliged to work, and the plough-teams of the peasantry themselves. But there are many variations, and any attempt to assess the varying degree of manorialisation over the country as a whole is fraught with difficulties.[2] Important as the manor was in the social and legal organisation of the realm, the village itself was the unit prominent in the agrarian landscape of Domesday England.

About 13,000 separate vills are named in the Domesday Book, but this

[1] Throughout the chapter no references are given to Domesday folios. These may be found in the relevant volumes of the Domesday geography of England: H. C. Darby, *The Domesday geography of eastern England* (3rd ed. 1971); H. C. Darby and I. B. Terrett (eds.), *The Domesday geography of midland England* (2nd ed. 1971); H. C. Darby and E. M. J. Campbell (eds.), *The Domesday geography of south-east England* (1962); H. C. Darby and I. S. Maxwell (eds.), *The Domesday geography of northern England* (1962); H. C. Darby and R. Welldon Finn (eds.), *The Domesday geography of south-west England* (1967). These volumes also include bibliographies.
[2] R. Lennard, *Rural England, 1086–1135* (Oxford, 1959), 213–36.

cannot be the total number in the area it surveys. Some Domesday names covered more than one settlement. A number of these came to be represented in later times by groups of two or more adjoining places with distinguishing appellations such as Great or Little and East or West; and some of these subdivisions were already in existence by 1086. Or again, the constituent members of some large manors were not named; they were described as *berewichae*, *appendicii* or *membra*, or they were not even indicated at all. Thus the large manor of Sonning in Berkshire seems to have included a number of places unnamed in Domesday Book;[1] so did the large manor of Farnham in Surrey.[2]

Where we can be reasonably sure that one name stood for one settlement, it is clear that there was much variation in size even in the same locality. The recorded populations of the dozen vills in the Cambridgeshire hundred of Wetherley ranged from 12 to 82. Whatever factor we use (whether 5 or some other) to obtain the actual populations, the fact of variety remains. Some places were very small indeed. Radworthy in Devon had only four recorded people with $1\frac{1}{2}$ plough-teams, and Lank Combe had only one without a team. Other places were completely uninhabited in 1086 and some were said to be waste. It would appear that village churches constituted a familiar feature of the countryside and that most were built of wood. They were, however, only irregularly entered in the Domesday Book. They appear for 352 out of 634 recorded places in Suffolk, but for only 17 out of 440 places in Essex, and for only two out of 334 places in Staffordshire.

Vills varied greatly not only in size but in form. Maitland, following Meitzen, drew attention to the fact that 'at least two types of vill must be in our eyes when we are reading Domesday Book' – nucleated and dispersed.[3] The number of place-names mentioned for Devonshire amounts to 983, but in this land of dispersed settlement, the number of hamlets and isolated farmsteads in 1086 has been placed at many times this number.[4] We can only argue retrospectively from the evidence of later ages about the varying types of settlement in Domesday England. The same is true of the size and shape of the parishes or tracts of territory attached to these names, which may preserve in a fossilised form vestiges of older arrangements.

1 A. Mawer and F. M. Stenton, *Introduction to the study of English place-names* (Cambridge, 1929), 39; *V.C.H. Berkshire*, 1 (1906), 301.
2 F. W. Maitland, 13–14.
3 F. W. Maitland, 16.
4 W. G. Hoskins, *Provincial England* (London, 1963), 21.

In spite of these limitations, what the details recorded for the Domesday vills can do is provide information about their respective resources.

POPULATION

The details of population in successive Domesday entries cover a whole series of categories, ranging from freemen to slaves. Between these extremes come villeins, bordars, cottars and many other groups. The total number of people so entered amounts to about 275,000. This recorded population is usually taken to refer to heads of households, and, in order to obtain the actual population, we must multiply it by some factor according to our ideas of the size of a medieval family. Various multipliers have been suggested, but the most likely seems to be 5.[1] The total that results, however, is subject to many doubts. The information about towns, for example, is often fragmentary. Then there are occasional entries which seem to be incomplete; we hear, for example, of plough-teams and other resources but not of people. There are also places about which we are told practically nothing beyond their names. There are, moreover, a number of entries which refer to unspecified numbers of men, and these cannot be included in a total; and there are also a few blanks where figures were never inserted. Such omissions and imperfections seem to indicate the presence of unrecorded households. Taking these doubts into considera- tion, it may be fair to say that the total population of England, excluding the four northern counties, amounted to about $1\frac{1}{4}$ million in 1086; the northern counties must have been very sparsely peopled. Merely to be able to make such an estimate for such a remote date is in itself an indica- tion of the value of the Domesday Book as a source of information.

On Fig. 11 East Anglia stands out as the most densely populated area, much of it with over 15 recorded people per square mile, and sometimes with over 20. The concentrations along the fertile coastlands of Sussex and Kent are also notable. On the other hand, there are a number of agriculturally unrewarding districts with very low densities in southern and eastern England – the Weald, the Bagshot Sands area of Surrey and north-eastern Hampshire, the Burnham gravel area of the southern

[1] Maitland (p. 437) suggested 5 'for the sake of argument'. J. C. Russell more recently suggested 3.5 – *British medieval population* (University of New Mexico, Albuquerque, U.S.A., 1948), 38 and 52. But for evidence in support of the traditional multiplier of 5 or something near it, see J. Krause, 'The medieval household: large or small', *Econ. Hist. Rev.*, 2nd ser., IV (1957), 420–32.

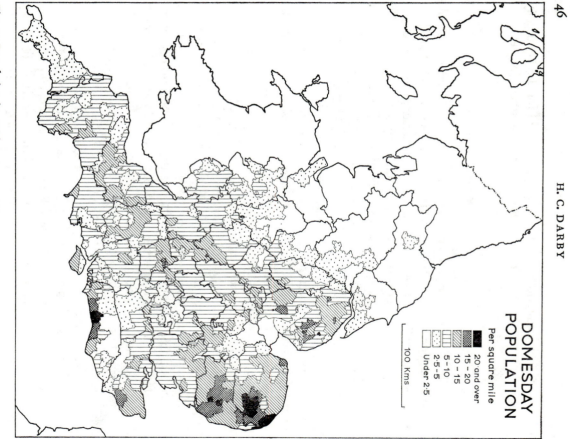

Fig. 11 Population in 1086
Based on H. C. Darby (ed.), *The Domesday geography of England*, 5 vols.
(1952–71). See p. 43 n above.

Chilterns, the Fenland, the Breckland, the New Forest, the Dorset heath-lands, Dartmoor and Exmoor. Apart from these sparsely occupied areas, the density of population, in general, decreases towards the west and the north. Nowhere in the Midlands is a density of over 15 reached, and northward the figure falls to 5 and in places even to below 2.5. The variations in density of population reflected to a large extent the fertility of the land. Upland areas were inherently infertile, so were the light soils of the lowlands before the days of improved farming. But natural conditions do not explain all of the low density areas. Some of them, towards the northern parts of the realm, owed their poverty to the fact that they had been deliberately devastated, particularly by the armies of the king.[1]

THE COUNTRYSIDE

Arable land

From numerous hints in late Saxon documents, we may suppose that the arable, or much of it at any rate, was arranged in open-field strips. There is, however, only a solitary Domesday entry (that for Garsington in Oxfordshire) that seems to refer to scattered strips, and we have to rely upon later evidence for details about variations in field arrangements from district to district. Although Domesday Book itself gives us no information about these matters, it does enable us to indicate the relative distribution of the arable over the face of the country.

Entry after entry states: (a) the amount of land for which there were teams (*terra n carucis*); (b) the number of teams (*carucae*) actually at work, both on the demesne and on the land of the peasants. The first statement sounds very straightforward, especially as the Exeter version expands it to indicate the number of teams which could plough the land of an estate (*hanc terram possunt arare n carucae*). But, in fact, the implications of the Domesday teamland are far from clear and have provoked considerable discussion. The number of teams sometimes exceeds that of teamlands; the formula varies in detail; it is absent for some counties; and for other counties (e.g. Lincoln, Northampton and Nottingham) the figures are often related to those for the assessments in such a way as to suggest that they are artificial. Some scholars have thought that teamlands refer to the teams of 1066; others that they indicate the arable (actual and potential) of 1086; and yet others that, for some counties certainly, they refer to an earlier and obsolete assessment. In view of these difficulties and the present state of our knowledge, it would be unwise to regard teamlands as reflecting geographical realities generally over the face of the country in 1086. It is reasonable to suppose that we are being given an idea of the land

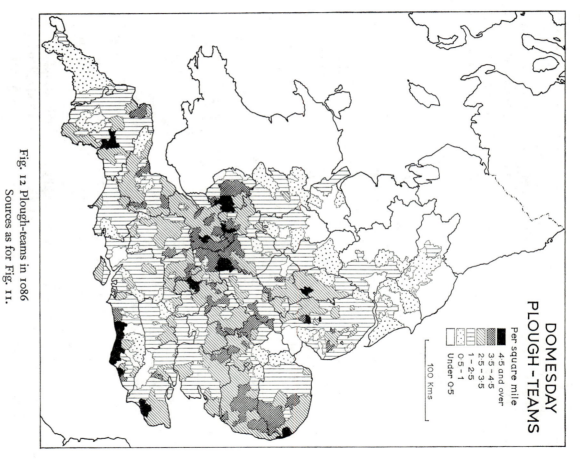

DOMESDAY
PLOUGH-TEAMS

Per square mile
■ 4·5 and over
▨ 3·5 – 4·5
▧ 2·5 – 3·5
▤ 1 – 2·5
⠿ 0·5 – 1
☐ Under 0·5

100 Kms

Fig. 12 Plough-teams in 1086
Sources as for Fig. 11.

actually cultivated. Although the teams at work in the field may have
varied in size on different soils, a comparison of parallel Domesday
entries show that, for the purposes of computation, eight oxen made
a Domesday team just as, wrote Maitland, 'twelve pence make a shilling'.[1]

[1] F. W. Maitland, 417.

It is true that this has been challenged, particularly for the south-western counties for which a variable Domesday team has been postulated, but the balance of evidence seems to be in favour of a standard eight-ox team.[1]

On Fig. 12 variations in the distribution of plough-teams reflect, to a great extent, the fertility of the land. The areas with the most arable, those with, say, over $3\frac{1}{2}$ teams per square mile included the coastal plain of Sussex, the eastern part of East Anglia, and districts in the Midlands and other places where, in the context of the time, soils were favourable – districts such as the 'red lands' of north Oxfordshire, the Vale of Evesham, south-east Herefordshire, south Warwickshire and the lower Exe Basin. Conversely, districts with less than a team per square mile included the Weald, the New Forest, the Dorset and Surrey heathland, the Fenland and the Breckland. To these must be added much of the northern counties of Domesday England – Yorkshire, Cheshire and the northern parts of Staffordshire and Derbyshire. Natural conditions do not explain all these low density areas; those of the north, in particular, owed their lack of teams very largely to the fact that they had been deliberately devastated.

This statistical picture can be supplemented by very little other evidence about the agriculture of eleventh-century England. Dating from a generation or so before the Conquest, there is the remarkable *Rectitudines* which deals with the management of an estate and with the duties of the peasantry. Supplementing this is the *Gerefa* which sets out the duties of a reeve throughout the year. Both documents note that conditions of husbandry varied from place to place, but without giving any details.[2] Crops and people were very much at the mercy of bad weather. The year of the Inquest, according to the Anglo-Saxon Chronicle, was 'very disastrous and sorrowful'; grain and fruits ripened late and there was a murrain amongst cattle. The following year was also a bad one in which 'many hundreds died of hunger'. Then again, 1089 was a backward year in which crops were sometimes not reaped until Martinmas (11 November).

[1] R. Lennard, 'Domesday plough-teams: the south-western evidence', *Eng. Hist. Rev.*, LX (1945), 217–33; H. P. R. Finberg, 'The Domesday plough-team', *Eng. Hist. Rev.*, LXVI (1951), 67–71; R. Lennard, 'The composition of the Domesday caruca', *Eng. Hist. Rev.*, LXXXI (1966), 770–5.
[2] H. R. Loyn, *Anglo-Saxon England and the Norman Conquest* (London, 1962), 189–94.

Grassland

The distinction between meadow and pasture was clear. Meadow denoted land bordering a stream, liable to flood, and producing hay; pasture denoted land available all the year round for feeding cattle and sheep. The two varieties of grassland merged into one another, but they were usually entered separately. The amount of meadow recorded for the average Domesday vill was not great, but it was important; thirteenth-century evidence shows that an acre of meadow was frequently two, three or more times as valuable as an acre of arable. Meadow, like wood, was measured in a variety of ways. For the majority of counties it was measured in terms of acres, but we do not know what area was implied by such an 'acre'; for the counties of Bedford, Buckingham, Cambridge, Hertford and Middlesex, it was entered in terms of the teams of oxen which its hay could support; occasionally, it was expressed in terms of linear dimensions or of money payments or in some other way. As in the case of wood, it is impossible to reduce these various types of measurement to a common denominator. Very little meadow was entered for Cornwall which may reflect geographical conditions, and none at all for Shropshire which may indicate some local idiosyncrasy in the record.

Fig. 13 shows the varying incidence of meadow over a great part of England. Much lay along the alluvial valleys of the larger river courses —in the south, along the Thames itself and along the Lea and the Kennet; in the north, along the Nene and the Trent together with its tributaries the Soar and the Dove. It was also widely distributed along the many streams of the clay belt that extends from north Berkshire through Oxfordshire, Buckinghamshire, Bedfordshire and Huntingdonshire. In northern Berkshire, in the western part of the Vale of White Horse, renders of cheese were entered for three places, at two of which dairies are also mentioned; here may be the hint of a local dairying economy. Elsewhere, dairies and renders of cheese were entered for only a few scattered places.

Pasture was much more irregularly entered. It never, or only rarely, appears in the folios for the counties of the north and the Midlands. For the three counties of Cambridge, Hertford and Middlesex, its presence is indicated by the formula 'pasture for the cattle of the village'. For Oxfordshire and the south-western counties, it was measured in terms either of acres or of linear dimensions. Other variants, including money renders, occasionally appear for the south-eastern counties. This irregular record makes it impossible to reconstruct the distribution of pasture over the

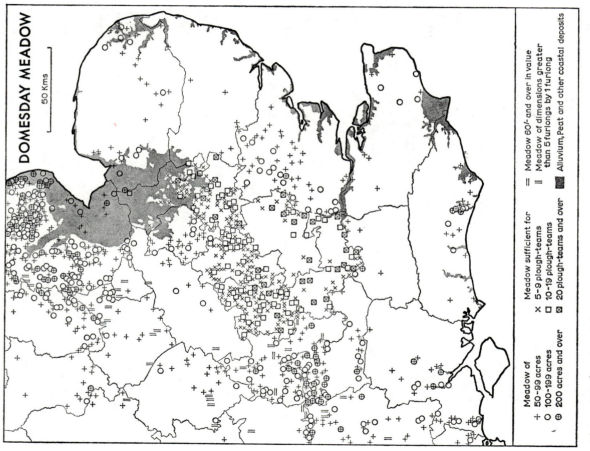

DOMESDAY MEADOW

50 Kms

Meadow of
+ 50–99 acres
○ 100–199 acres
⊕ 200 acres and over

Meadow sufficient for
× 5–9 plough-teams
□ 10–19 plough-teams
⊠ 20 plough-teams and over

= Meadow 60/- and over in value
‖ Meadow of dimensions greater
    than 5 furlongs by 1 furlong
▓ Alluvium, Peat and other coastal deposits

Fig. 13 Meadow in 1086

Sources as for Fig. 11.

There was some meadow in almost every village. Only the larger quantities
are plotted in order to emphasise the areas with considerable amounts. It is
impossible to equate the four methods of enumeration.

country as a whole. The Essex entries are unusual in that they mostly record not pasture as such but pasture for so many sheep (*pastura oves*). The villages with such pasture lay in a belt along the coast, and it is clear that this Domesday pasture corresponded with the Essex marshes, famous in later times for the making of cheese from ewes' milk.

There are occasional glimpses of the arrangements under which the pasture was used. Scattered entries, especially for Devonshire, refer to 'common pasture'. Two Somerset entries reveal what seems to have been an arrangement for intercommoning between the neighbouring villages of Hardington and Hemington. The burgesses of Oxford held pasture in common 'outside the wall'; we also hear of common pasture at Cambridge, and common pasture may be implied at Colchester. Then, again, there is the famous reference to the pasture of the Suffolk hundred of Colneis which was 'common to all the men of the hundred'. On two occasions we hear of pasture converted to arable; at Swyre, in Dorset, land which had formerly been pasture was 'now sown', and at Bourne in Kent there were six acres of pasture 'which men from elsewhere had ploughed up'.

We are told very little of the animals that grazed on these pastures. Apart from the ploughing oxen and the woodland swine and a few scattered references to sheep, the stock of a manor was passed over in silence by the Great Domesday Book. For seven counties only, described in the Little Domesday Book and the Exeter Domesday Book, is there information about animals, and then only about the demesne livestock; they included, amongst others, sheep, goats, 'animals' and horses. From such incomplete information, it is impossible to obtain any idea of their distribution over the face of the country.

Woodland

One of the outstanding facts about the landscape of eleventh-century England was its wooded aspect. The Anglo-Saxons and Scandinavians, it is true, had pierced the woodland and broken it everywhere with their 'dens' and 'leahs' and 'skogrs', but, even so, almost every page of Domesday Book shows that a great deal of wood still remained. One of the questions put by the Domesday commissioners was 'How much wood?'. Broadly speaking, the answers fell into one of five categories. Sometimes, they said that there was enough wood to support a given number of swine, for the swine fed upon acorns and beechmast. A variant of this was a statement not of total swine but of annual renders of swine in return for pannage. A third type of answer gave the length and breadth of wood in terms of leagues, furlongs and perches, but whether the information referred to mean diameters or to extreme diameters or conveyed some other notion, we cannot say. A fourth type stated the size of the wood in terms of acres, but we do not know what area was implied by such an eleventh-century 'acre'. The fifth category of answer was a miscellaneous one that included a number of variants and idiosyncrasies occasionally encountered in the text, e.g. wood for fuel, for the repair of houses or for the making of fences. Normally, each county was characterised by one main type of entry but also included a few other entries of a different character.

The difficulty presented by this array of information can be simply stated. It is impossible satisfactorily to equate swine, acres and linear dimensions, and so reduce them to a common denominator. Any map of Domesday woodland covering a number of counties must therefore suffer from this restriction, and we cannot be sure that the visual impression as between one set of symbols and another is correct. There are also other difficulties such as the fact that some woods may not have been recorded. Despite these limitations, much can be learned from Fig. 14. With all its problems, it leaves us in no doubt about the wooded aspect of large tracts of England in 1086. One surprising feature is the absence of wood from the Weald. This arises from the fact that much of the wood of the Weald was entered under the names of surrounding villages. It must therefore be 'spread out', so to speak, by eye. But we may well suspect the existence of unrecorded wood here, and also in other districts such as northern Berkshire.

Domesday Book tells us little or nothing about the process of clearing. The fuller information of the Little Domesday Book for Essex, Norfolk and Suffolk enables us to know that wood had been cut down between 1066–86 in at least 109 out of the 1,807 villages in those three counties. The circumstantial evidence about Wealden 'denes' or swine pastures in the folios for Kent also indicate clearing; so does some circumstantial evidence for Sussex; so also do the references to cartloads of wood that fed the Droitwich salt industry in Worcestershire. Ironworks must have consumed more wood, but we are told nothing of this. It is certain that clearings for cultivation were already known as 'assarts', a word derived from the French *essarter*, meaning to grub up or clear land of bushes and trees. Hereford is the only county for which the Domesday entries mention them, but there is every reason to believe that what was happening in Herefordshire was also happening in other counties.

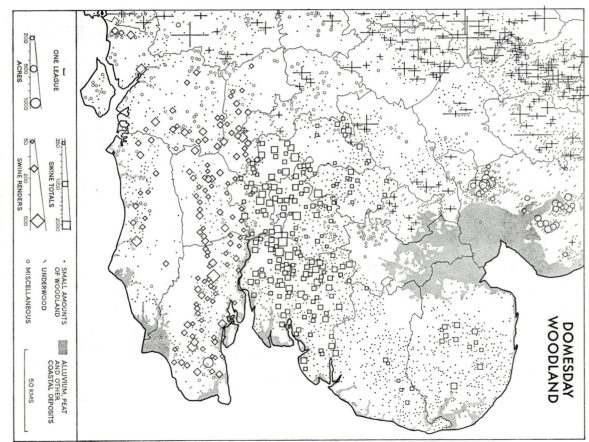

DOMESDAY
WOODLAND

ONE LEAGUE

SWINE TOTALS

350   1000   2000

SWINE RENDERS

50   200   500

ACRES

200   500   1000

· SMALL AMOUNTS
  OF WOODLAND

＼ UNDERWOOD

○ MISCELLANEOUS

ALLUVIUM, PEAT
AND OTHER
COASTAL DEPOSITS

50 KMS

Fig. 14 Woodland in 1086
Sources as for Fig. 11.

Forests

Pre-Domesday kings had hunted and possessed game preserves, but with the Norman Conquest these royal activities greatly increased. The forest law and forest courts of Normandy were introduced into England and there was a rapid and violent extension of forest land.[1] The Norman kings had a passionate love of the chase and, as the Anglo-Saxon chronicler wrote under the year 1087, King William 'made large forests for deer'. The word 'forest' is neither a botanical nor a geographical term, but a legal one. It implied land outside (*foris*) the common law and subject to a special law that safeguarded the king's hunting. Forest and woodland were thus not synonymous terms, for the forested areas included land that was neither wood nor waste, and they sometimes covered whole counties. Even so, a forested area usually contained some wood and often large tracts of wood. Within the forests, no animals could be taken without express permission and the right to cut wood and to make assarts was severely restricted. In this way, much forest land was kept free from the plough and maintained in its primitive condition.

As forests were not liable for renders or geld, they were rarely mentioned and specifically described in Domesday Book. But we frequently hear that part of a manor had been placed 'in the king's forest', or 'in the forest', or 'in the king's wood' or 'in the king's enclosure' (*in defensione regis*). Sometimes it was the wood itself that had been taken out of a manor and afforested (*Silva hujus maneriae foris est missa ad silvam regis*). Or again at *Haswic* in Staffordshire there was land for eight teams, but the entry adds, 'it is now waste on account of the king's forest'; and at Ellington in Huntingdon one out of ten hides was waste on account of the king's wood (*per silvam regis*). Such references occur in the entries for many counties. There is also mention of 'hays' or enclosures (*haiae*) for regulating the chase, particularly in entries relating to Cheshire, Herefordshire, Shropshire and Worcestershire. Finally, parks were recorded for 33 places; and sometimes these were specifically for wild beasts (*parcus bestiarum silvaticarum*). Some belonged to the king; others to various lay or ecclesiastical lords.

There is one forest, and only one, about which the Domesday Book gives a wealth of detail. It is the New Forest, and the account of it occupies a special section of the Domesday description of Hampshire. The amount

[1] C. Petit-Dutaillis, *Studies and notes supplementary to Stubbs' Constitutional History*, 2 vols. (Manchester, 1915), II, 166–70.

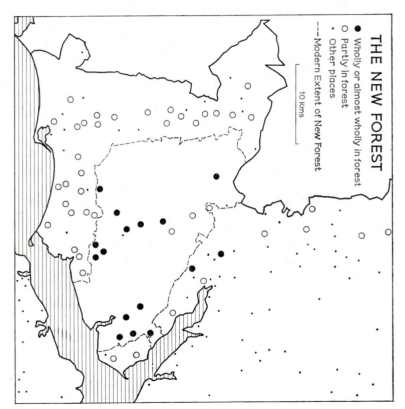

Fig. 15 The making of the New Forest
Based on H. C. Darby and E. M. J. Campbell (eds.), *The Domesday geography of south-east England* (Cambridge, 1971), 325.

THE NEW FOREST

● Wholly or almost wholly in forest
○ Partly in forest
· Other places
--- Modern Extent of New Forest

10 Kms

of destruction involved in the making of the Forest was once a matter of controversy. The chroniclers of the twelfth century declared that William reduced a flourishing district to a waste by the wholesale destruction of villages and churches. But this is not entirely borne out either by the poor soils of the district or by the evidence of the Domesday Book itself. It would appear that the area was but thinly occupied in 1066; there were some villages of moderate size but many were very small. About 30 to 40 of these villages were placed wholly or almost wholly under forest law, leaving no record of their former population. Only about one-half of these appear on Fig. 15 because the rest are unidentified. They lay almost entirely in the centre of the forest or to the south-east. Varying portions of about 40 other villages on the borders of the main part of the forest were also included. The limits of this eleventh-century forest have shrunk but it still covers some 92,000 acres and remains a unique memorial of Norman England.

Miscellaneous resources

Fish, salted or otherwise, must have played an important part in the life of an eleventh-century community, and the presence of fisheries is noted in many Domesday entries. They are sometimes recorded in association with mills, being derived from mill-ponds, and sometimes separately. It is very unlikely, however, that the Domesday record gives a complete picture of the fishing activity of the time. Maybe some fisheries were unrecorded because they returned no profit to the lord, or perhaps because of the absence of any contrivance such as a weir (*gurgites*) for catching fish.[1] The species normally mentioned is that of eels, and these often appear as eel renders from mills. Salmon were recorded for places along the Severn, the Dee and the Dart.

The fisheries were located along the main rivers and their tributaries – the Thames, the Severn, the Nene, the Trent, the Great Ouse, the Dee, the Medway, the Avon and others. But the chief assemblage was in and around the Fenland. Here, some of the north Cambridgeshire vills returned great numbers of eels to their lords – as many as 33,260 from Wisbech; 27,150 from Doddington; 24,000 from Stuntney; and 17,000 from Littleport. Clearly, here was a district with distinctive regional economy. Few of the entries tell us anything about the operations involved. There is, it is true, a very interesting description of Whittlesey Mere, on the western edge of the Fenland, with its fishermen and its fishing boats owned by the abbots of Ramsey, Thorney and Peterborough. We hear also of nets, fish-traps and weirs in the Fenland and elsewhere. Of the marsh itself, there is only occasional indication in entries relating to the Fenland, the Isle of Axholme, the Somerset Levels and Romney Marsh; but the evidence of these entries provides scarcely more than a hint of the economies of these regions.

Sea-fisheries were hardly ever mentioned. Herring renders were recorded for a number of villages along or near the Suffolk coast; there is also reference to a sea weir (*heiemaris*) at Southwold and to 24 fishermen at Yarmouth. Herring renders were also occasionally entered for London and Southwark and for places along the coasts of Kent and Sussex, and there is mention even of porpoises for Southease in Sussex. We can only

[1] R. Lennard (1959), 248–9.

conclude that these are but stray indications of what must have been a considerable activity around the coasts of the country.

Among the other miscellaneous resources of a manor were vineyards. They are recorded for 55 places, the most northerly being Ely. J. H. Round argued that the culture of the vine was reintroduced (since Roman times) by the Normans, and he based his view upon the fact that Domesday vineyards were normally on holdings in the direct hands of Norman tenants-in-chief; that they were usually measured by the foreign unit of the arpent; and that some had but lately (noviter or nuperrime) been planted.[1] But this view cannot be maintained, because we hear of vineyards in England in the eighth, ninth and tenth centuries.[2] It looks as if the Normans had not so much reintroduced the vine as extended, possibly greatly extended, its culture. We may see in the Domesday references to vineyards, as in the references to forests, indications of the new order that William brought to the land he conquered.

Devastated land

Devastated land had formed an important element in the geography of England during the period of Anglo-Danish warfare; and it did not cease to be important after 1042 when a member of the royal house of Wessex, Edward the Confessor, once more ruled over all England. During his reign, the complicated politics of the great earldoms of the realm resulted in much conflict. Earl Godwin and his sons, banished overseas, returned in 1052 to raid along the south coast. There were also raids from Ireland, from Wales and from Norway that resulted in the devastation of various parts of the realm. The sources are obscure, and we can only repeat what the Anglo-Saxon Chronicle says under the year 1058: 'it is tedious to relate how it all happened'. An echo of some of the complications comes in the Domesday Book which tells us that Edward the Confessor had reduced the geld liability of Fareham in Hampshire from 30 hides to 20 'on account of the Vikings, because it is by the sea'. From the north, Earl Morcar of Northumbria raided south, in 1065, to Northamptonshire, and killed and plundered so that, says the Anglo-Saxon Chronicle, that shire and the other shires around 'were for many years the worse'. One indication of this is seen in the low 1066 values for many Northamptonshire estates, and in the Northamptonshire Geld Roll

[1] J. H. Round in H. R. Doubleday and W. Page (eds.), *V.C.H. Essex*, 1 (1903), 282–3.

[2] G. Ordish, *Wine growing in England* (London, 1953), 20–1.

(not later than 1075) which records about a third of the county as waste;[1] by 1086, however, recovery was general. Then again, the Domesday folios for Cheshire, Shropshire and Herefordshire show that many villages lay waste in 1066, the result of Welsh raiding. Along the northern border the raids of the Scots also left much devastation.

The Norman Conquest brought further devastation. William landed at Pevensey and, after the battle of Hastings, marched to Dover, then via Canterbury towards London. He did not cross the Thames into the city itself but, after some of his forces had burnt Southwark, he began an en-circling movement westward and then northward across the Thames at Wallingford. From here he continued in a general north-east direction, and then turned to approach London from the north. The Anglo-Saxon Chronicle summarises these events by saying that 'he harried all that part which he over-ran until he came to *Beorh-hamstede*'. This is generally identified with Berkhamsted but a case has also been made for Little Berkhamsted to the east. An army such as this lived largely on the country-side it passed through, and inevitably left a trail of destruction. The differences in the values of many manors just before and then after the Conquest have been held to indicate the 'footprints' of the Conqueror's march.[2] The evidence does not lend itself to any rigid interpretation for there was not one army but several forces, and also there may have been foraging bands. A number of villages around Hastings and Dover still lay entirely waste in 1070 or so, and there were many more around London generally that had been reduced in value. But by 1086 most had recovered their former values, wholly or in part.

The south-eastern counties might have borne the first burden of the Conquest, but they did not bear the heaviest brunt. To a large extent, wasting by armies and raiders was the reason for the low densities of plough-teams and population in the west and particularly in the north. The term waste (*wasta*) found in many Domesday entries implies not the natural waste of mountain, heath and marsh, but land that had gone out of cultivation mainly as a result of deliberate devastation, but also perhaps because of some local vicissitude that is lost to us. Some of the wasted

[1] J. H. Round, *Feudal England* (1895), 147–56. See also D. C. Douglas and G. W. Greenway, *English historical documents, 1042–1189* (London, 1953), 483–6.

[2] F. H. Baring, 'The Conqueror's footprints in Domesday', *Eng. Hist. Rev.*, XIII (1898), 17–25. Reprinted with 'some additions and alterations' in F. H. Baring, *Domesday Tables* (London, 1909), 207–16. For a detailed account of the changes in values, see R. Welldon Finn, *The Norman Conquest and its effects on the economy: 1066–86* (London, 1971).

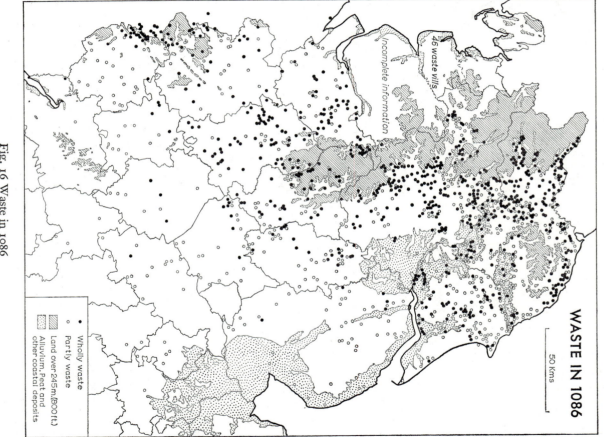

Fig. 16 Waste in 1086
Sources as for Fig. 11.

villages of Herefordshire, Shropshire and Cheshire were, we may suppose, the result of Welsh raiders, but many were the result of the crushing of rebellion by the king's armies during the years 1068–70. Much of the waste of those years was still evident in 1086. The account of Cheshire is unusual in that the Domesday folios give details of waste not only for 1066 (i.e. before the Conquest) and for 1086 (the year of the enquiry) but also for the date when the existing owner received an estate, i.e. about 1070. Out of a total of 264 there were 52 villages wholly or partly waste in 1066, presumably as a result of Welsh raiding. By 1070, the figure had increased to 162, the result of William's campaign of 1069–70. By 1086, there were still 58 villages wholly or partly waste. What we hear of in 1086 is a countryside not yet recovered but on its way to recovery.

It was in the north that William took the most terrible revenge, and he left the countryside in a condition in which it could give him no trouble again. The entry under the year 1069 in the Anglo-Saxon Chronicle is brief but eloquent; it merely says of Yorkshire that the king 'laid waste all the shire'. Yorkshire suffered most, but we are told in one early account that the harrying also extended over Derbyshire, Cheshire, Shropshire and Staffordshire, nor did Nottinghamshire escape. The general statements of the chronicles are borne out in a vivid way when the Domesday entries are plotted on a map (Fig. 16). Seventeen years or so had not been enough to obliterate the effects of the harrying. Entry after entry for these northern villages reads: *Wasta est.* As well as villages specifically described as waste, there were others without population, and most of these, too, we must suppose, were the result of William's campaigns.

There was also other post-Conquest devastation. Thus, in an entry for 1067, the Anglo-Saxon Chronicle tells how one of the sons of King Harold, from whom William had won the kingdom, came with a fleet from Ireland to the mouth of the Avon, and plundered the countryside around. That there were also other raiders across the western seas we know from a marginal note in the Exeter Domesday Book which refers to nine manors in the extreme south of Devon 'devastated by Irishmen'; seven of the manors had far from recovered by 1086, and three of these were each worth only a quarter of their 1066 values. Whatever its cause. Welsh raiding or Norman strategy, the sum total of the evidence set out in Fig. 16 leaves us in no doubt about the importance of devastation as an element in the economic geography of eleventh-century England. Were it possible to reconstruct a similar map for, say, 1070, it is certain that the wasted villages would be more numerous and more widespread. The picture we see on Fig. 16 is that of a countryside already on its way to recovery.

INDUSTRY

Industry was not important in the eleventh century, but, even so, we might expect to hear more of it than we do in the Domesday Book. Although iron was worked in Roman times, there is remarkably little evidence of Anglo-Saxon working, and only a few scattered Domesday references. An ironworks (*ferraria*) is entered for a holding in East Grinstead hundred in Sussex, and there had been others at Corby and Gretton in Northamptonshire. Renders of iron in Gloucestershire may have reflected the presence of iron-working in the Forest of Dean; the renders at Gloucester itself included 'iron rods for making nails for the king's ships'. The *ferrariae* and *fabricae ferri* at Stow and at Castle and Little Bytham in Lincolnshire were not at places on iron-bearing outcrops. Likewise there were *ferrariae* at Chertsey (Surrey), Fifield Bavant (Wilts), Stratfield (Hants), Wilnecote (Warwick), and iron-workers (*ferrarii*) at Hessle (W. Riding) and North Molton (Devon); these may imply the presence of forges. Smiths (*fabri*) are also mentioned for a few places, but the number is not large and, presumably, many smiths and iron-workers were entered as villeins or bordars. At Rhuddlan, described in the Cheshire folios, we hear specifically of mines (*mineriae ferri*).

Lead works (*plumbariae*) are entered for six places in the Peak District of Derbyshire; all lie on or near the outcrop of Carboniferous Limestone which contains veins of metalliferous ores from which the lead must have been derived, but we are told nothing of the methods by which the finished lead was produced. There is also mention of a *fabrica plumbi* in the entry for Northwick and Tibberton, possibly a leadworks for making the vats used in the manufacture of salt at Droitwich nearby. That there was trade in lead throughout the country we may infer from a casual but interesting reference in the *Inquisitio Eliensis* which is near-contemporary with the Domesday Book; this compares the weight of a 'fodder' of Peak lead (*carreta plumbi del pec*) with that of a London 'fodder', and it suggests that Ely Cathedral may have had its supply of lead for roofing from the Peak District.[1] Of other lead-mining areas such as Shropshire, the Pennine Dales and the Mendips, known to have been worked in Roman

[1] N.E.S.A. Hamilton, *Inquisitio Comitatus Cantabrigiensis...subjicitur Inquisitio Eliensis* (London, 1876), 191.

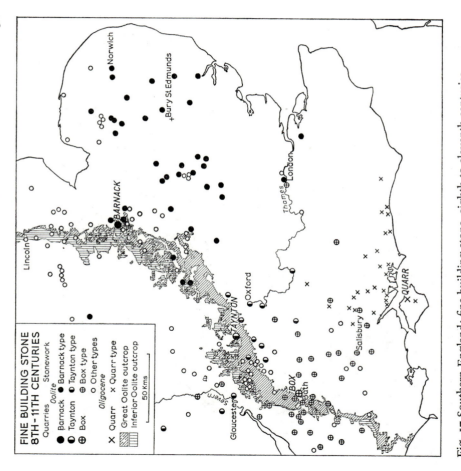

FINE BUILDING STONE
8TH–11TH CENTURIES

Quarries	Stonework
Oolite	
● Barnack	◑ Barnack type
● Taynton	◑ Taynton type
⊕ Box	⊕ Box type
	○ Other types
Oligocene	
✕ Quarr	✕ Quarr type

▨ Great Oolite outcrop
▥ Inferior Oolite outcrop

50 Kms

Fig. 17 Southern England: fine building stone, eighth to eleventh centuries
Based on E. M. Jope, 'The Saxon building-stone industry in southern and
Midland England', *Medieval Archaeology*, VIII (1964), 92; and on additional
information provided by Professor Jope. Navigable portions of the Thames
and Severn are shown.

times, the Domesday Book says nothing. Nor does it mention Cornish
tin.

Quarries are occasionally mentioned. That at Taynton in Oxfordshire
was well known for its Great Oolite freestone which is to be found in the
surviving pre-Domesday masonry in villages up to thirty miles and more
from the outcrop (Fig. 17). The quarries at Whatton in Nottinghamshire
and Bignor in Sussex produced millstones; but these, and a few other
quarries mentioned for other places, can only have been a fraction of the

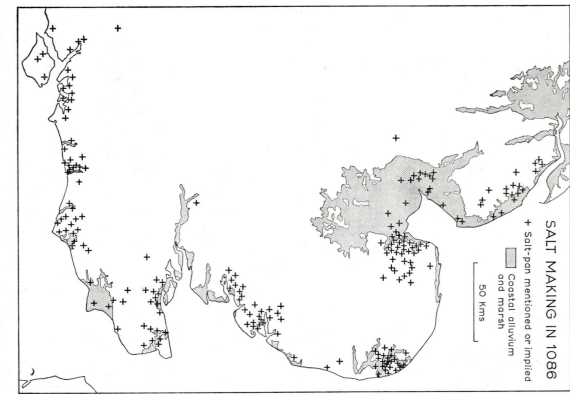

SALT MAKING IN 1086

+ Salt-pan mentioned or implied

⬚ Coastal alluvium and marsh

50 Kms

Fig. 18 Southern and eastern England: coastal salt-making in 1086
Sources as for Fig. 11.

total worked throughout the country. So too, the potteries and potters entered for Bladon (Oxon), Haresfield (Gloucs) and Westbury (Wilts) must stand for a large number in the country as a whole.

There is one industry, however, about which information is relatively abundant. Salt was an indispensable item in medieval economy, especially for the preservation of meat and fish. Rock salt was not worked in Britain until 1670, so that the commodity was obtained by evaporation either from sea water or from inland brine springs. The chief areas for the production of maritime salt were along the marshes and estuaries of the east and south coasts. The Domesday record usually states the number of salt-pans on a holding, and sometimes also includes their render either in money or in fish or in loads of salt. Caister in eastern Norfolk, for example, had as many as 45 salt-pans. Some villages for which salt-pans appear are situated inland, from which we must conclude that their lords held the pans along the coast some distance away (Fig. 18).

The inland centres of production were in Worcestershire and Cheshire, and the brine springs were derived from the Keuper Marl beds of the Triassic system. The centre of the Worcestershire industry was at Droitwich for which brine-pits (*putei*) and salt-pans (*salinae*) are recorded. There is no information about the processes of manufacture, but we do hear of leaden pans or vats (*plumbi*) and of furnaces (*furni*); we also hear occasionally of fuel for the latter in the form of 'wood for the salt-pans'. Some of the Droitwich salt-pans belonged not only to nearby Worcestershire villages, but to villages of other counties (Fig. 19). We even hear of a salt-worker of Droitwich rendering loads of salt as far away as Princes Risborough in Buckinghamshire. All this suggests some interesting reflections about the movement of commodities in Domesday England. The salt was presumably transported by pack-horse; names such as Saltway, Salter's Corner and Salford are preserved on the One-Inch Ordnance maps of today, and it is possible to reconstruct a system of ways leading from Droitwich.[1] The Cheshire industry was centred at Northwich, Middlewich and Nantwich which had only partly recovered from the disturbances of 1070. We hear of salt-pans and of 'boilings of salt' (*bulliones salis*). We are also told of a few nearby manors with salt-pans at one of the Wiches, but that the trade in salt extended into other counties is apparent from the details of the tolls levied on those who transported the salt away. The tolls increased with the distance from which a purchaser

---

[1] A. Mawer et al., *The place-names of Worcestershire* (Cambridge, 1927), 4–9; A. H. Smith, *The place-names of Gloucestershire*, pt I (Cambridge, 1964), 19–20.

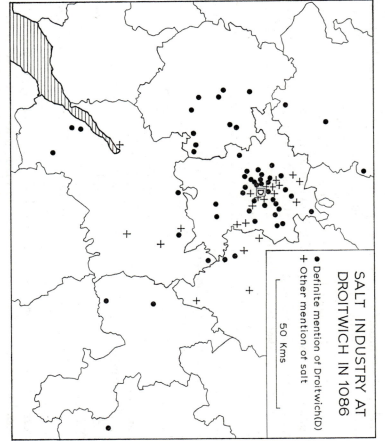

SALT INDUSTRY AT
DROITWICH IN 1086

● Definite mention of Droitwich(D)
+ Other mention of salt

50 Kms

Fig. 19 The Droitwich salt industry in 1086
Based on H. C. Darby and I. B. Terrett, *The Domesday geography of Midland
England* (2nd ed., Cambridge, 1971), 258.

and his wagons (*carri*) came, whether from the same hundred as that of the
Wich, or from another hundred, or from another shire. There is also
reference to the toll paid by those who carried salt about the county to
sell. These glimpses, like those for Droitwich, imply an organised trade
in salt that extended far beyond the vicinity of the Wiches themselves.

TOWNS AND COMMERCE

Whatever the difficulties of interpretation, the Domesday information for
rural England is systematically presented and is remarkable for the information it provides. When we turn from the countryside to the towns, all is
different. The information is so incomplete and so unsystematic that it is
often impossible to form any clear idea of the size of a town or of the

economic and other activities that sustained it. Altogether 112 places (including Rhuddlan in North Wales) seem to have been boroughs in 1086 (Fig. 20). The Anglo-Saxon word 'burh' signified a fortified centre, and the test of burghal status was neither size nor general prosperity. There has been debate about whether defence or trade provided the impetus to urban development, but surely both were important, and it is clear that by Domesday times a commercial element had become an important feature of many boroughs. It is also clear that burghal status was not constant. Not all the boroughs of the tenth century survived to appear as such in the Domesday Book, and not all Domesday boroughs retained their standing as such in the twelfth century. Conversely, in each century, new boroughs emerged.

By far the largest borough must have been London. It is therefore particularly unfortunate that Domesday Book contains no account of it. The 126th folio, where this should have come, is blank, and there are only a few incidental references to it elsewhere in the Book. It had been a Roman city, and, after the confusion of the Anglo-Saxon invasions was over, its advantages of site and location re-asserted themselves. By the eighth century it had become, in the words of Bede, 'the market place of many peoples coming by land and sea'; and, for the eleventh century, there is evidence of its wide trading connections with the Continent. It is true that the idea of a capital city had not yet become current in western Europe, and that the centre of government moved about with the court of the king. But London already had a distinct place among the boroughs of England, a place emphasised by its role as a centre of resistance against Danish invasion in the early years of the eleventh century.[1] A guess might place the number of its inhabitants at over 10,000, but any attempt to estimate how much over becomes even more hazardous. Fire was always a danger in London, as in other cities. The Anglo-Saxon Chronicle, under the year 1077, records an extensive fire in London, greater than any 'since the town was built'; and ten years later there was another fire which destroyed 'the greatest and fairest part of the whole city'.

There are likewise only incidental Domesday references to Winchester, the city to which the results of the Inquest were brought in and in which the Domesday Book was at first kept. There are, however, two early surveys of the city, dating from 1103–5 and 1148, which tell us a little about it. We hear, among other things, of a merchants' guildhall, of a market, of shops and stalls, of mints, of forges and of a prison. Its maximum

[1] F. M. Stenton, §31.

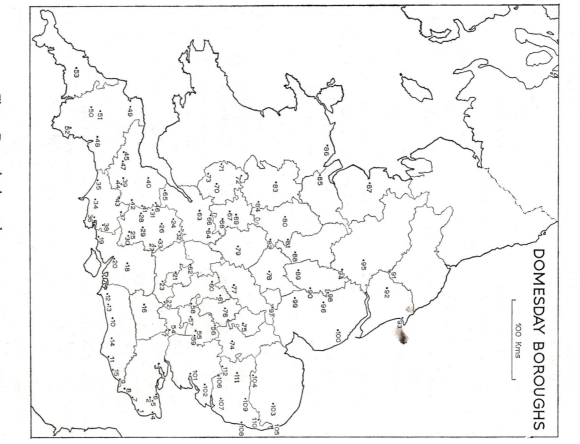

DOMESDAY BOROUGHS

100 Kms

Fig. 20 Domesday boroughs
Sources as for Fig. 11.
Names of boroughs are given on pp. 69-70.

Key to Fig. 20

The counties and the boroughs are set out in the order in which they appear in the Domesday Book, and then in the Little Domesday Book.

Kent
1 Dover
2 Canterbury
3 Rochester
4 Sandwich
5 Fordwich
6 Seasalter
7 Hythe
8 Romney

Sussex
9 Rye
10 Steyning
11 Pevensey
12 Chichester
13 Arundel
14 Lewes
15 Hastings

Surrey
16 Guildford
17 Southwark

Hampshire
18 Winchester
19 Southampton
20 Twynham

Berkshire
21 Wallingford
22 Windsor
23 Reading

Wiltshire
24 Malmesbury
25 Wilton
26 Cricklade
27 Bedwyn
28 Warminster
29 Tilshead
30 Salisbury
31 Bradford on Avon
32 Calne
33 Marlborough

Dorset
34 Dorchester
35 Bridport
36 Wareham
37 Shaftesbury
38 Wimborne Minster

Somerset
39 Langport
40 Axbridge
41 Frome
42 Bruton
43 Milborne Port
44 Ilchester
45 Milverton
46 Bath
47 Taunton

Devonshire
48 Exeter
49 Barnstaple
50 Lydford
51 Okehampton
52 Totnes

Cornwall
53 Bodmin

Middlesex
54 London

Hertfordshire
55 Hertford
56 Ashwell
57 St Albans
58 Berkhamsted
59 Stanstead Abbots

Buckinghamshire
60 Buckingham
61 Newport Pagnell

Oxfordshire
62 Oxford

[continued overleaf

Gloucestershire
63 Gloucester
64 Winchcomb
65 Bristol
66 Tewkesbury

Worcestershire
67 Worcester
68 Pershore
69 Droitwich

Herefordshire
70 Hereford
71 Clifford
72 Wigmore
73 Ewias Harold

Cambridgeshire
74 Cambridge

Huntingdonshire
75 Huntingdon

Bedfordshire
76 Bedford

Northamptonshire
77 Northampton

Leicestershire
78 Leicester

Warwickshire
79 Warwick

Staffordshire
80 Stafford
81 Tutbury
82 Tamworth

Shropshire
83 Shrewsbury
84 Quatford

Cheshire
85 Chester
86 Rhuddlan

(Lancashire)
87 Penwortham

Derbyshire
88 Derby

Nottinghamshire
89 Nottingham
90 Newark

(Rutland)
Nil

Yorkshire
91 York
92 Pocklington
93 Bridlington
94 Dadsley
95 Tanshelf

Lincolnshire
96 Lincoln
97 Stamford
98 Torksey
99 Grantham
100 Louth

Essex
101 Maldon
102 Colchester

Norfolk
103 Norwich
104 Thetford
105 Yarmouth

Suffolk
106 Sudbury
107 Ipswich
108 Dunwich
109 Eye
110 Beccles
111 Bury St Edmunds
112 Clare

population at this time has been estimated at between 6,000 and 8,000, but this seems large.[1] Of the other boroughs, we can conjecture from the unsatisfactory Domesday evidence that the following had at least 4,000 and maybe over 5,000 inhabitants in 1086: York, Lincoln, Norwich and possibly Thetford. Below them came a group with at least 2,000 and maybe over 3,000 each. This group included:

Bury St Edmunds	Huntingdon
Canterbury	Lewes
Colchester	Oxford
Dunwich	Stamford
Exeter	Wallingford

How many others should be counted in the group, we cannot say. Of towns such as Bedford, Bristol, Gloucester and Salisbury, the Domesday evidence is too fragmentary to allow us to make a guess. In the case of Gloucester, the evidence of a subsidiary document (the Evesham Abbey Survey) shows that the city had ten churches and a population of certainly over 3,000. Below the 2,000 mark came a variety of boroughs; some were very small and poor, such as Bruton and Langport in Somerset, both of which ceased to be boroughs in the twelfth century. We must also remember that most, if not all, boroughs, small or large, had an agricultural flavour. Arable, meadow and pasture are sometimes entered for them, and also such categories of population as villeins and bordars as well as burgesses. Thus Cambridge, with a population of at least 1,600, was a substantial settlement in the context of the time, yet its burgesses had been accustomed to lend their plough-teams to the sheriff at least three times a year, and in 1086 he was demanding that this be increased to nine times.

The immediate effect of the Norman Conquest on some boroughs was the destruction of houses for the building of castles. At York, one of its seven wards had been so wasted; at Lincoln, 166 houses had been destroyed; at Norwich, 98 houses; at Shrewsbury, 51 houses; and there had been similar destruction for castles on a smaller scale at Cambridge, Canterbury, Gloucester, Huntingdon, Stamford, Wallingford and Warwick. We hear also of houses destroyed by fire at Exeter, Norwich and Lincoln. At these and other boroughs there were also wasted or unoccupied houses for which no reason was given. Waste, due to one cause or another, was entered for as many as 33 of the 112 boroughs. Oxford, for example, had suffered badly. Out of a total of about 1,000 properties in 1066, about one half were so wasted or destroyed that they rendered nothing; it does not follow that all their inhabitants had disappeared, but, whatever view we take, the Oxford of 1086 was clearly in reduced circumstances. The cause is not stated; some of the waste may have been the result of clearance to provide a site for the castle which had been built in 1070 but about which the Domesday Book says nothing.

Alongside these setbacks there were long-term tendencies of a different kind of work. Some people have believed that the Anglo-Saxon borough had little or no really urban character and that the origin of commercial towns (as distinct from fortified boroughs) was essentially a development of the years after 1066 and owed much to the Normans.[1] Others have challenged this view which, they have said, underestimates the amount of trade and urban growth before 1066; they point to the development of the Anglo-Saxon coinage and to various hints of internal and foreign trade, and they have concluded that there was no new urban concentration in the years immediately after the Conquest.[2] While this may be true in a general sense, the fact remains that there are many indications of commercial activity within the Domesday boroughs, some of it certainly new. The urban trend of the later Anglo-Saxon period continued to develop and perhaps even to accelerate.[3]

One indication is the establishment of new boroughs alongside the old at Norwich, Northampton and Nottingham; that at Norwich included 125 French burgesses, and at Nottingham we hear of 48 merchants' houses. A new borough (*novus burgus*) had also been established at Rye in the manor of *Rameslie* in Sussex; and, incidentally, another at Rhuddlan in North Wales. Groups of Frenchmen or French burgesses had also settled at Dunwich, Hereford, Shrewsbury, Southampton, Stanstead Abbots, Wallingford and York. In the borough around the castle (*In burgo circa castellum*) at Tutbury, in Staffordshire, there were 42 men who devoted themselves wholly to trade and who rendered £4. 10s. At Eye in Suffolk there were 25 burgesses who lived in or around the market place (*in*

[1] C. Stephenson, *Borough and town: a study of urban origins in England* (Cambridge, Mass., 1933), 70–1.

[2] J. Tait, *The medieval English borough* (Manchester, 1936), 132–8.

[3] For a convenient summary of the problem, see J. F. Benton, *Town origins: the evidence from medieval England* (Boston, Mass., 1968). See also R. P. Beckinsale, 'Urbanization in England to A.D. 1420', being ch. 1 (pp. 1–46) of R. P. Beckinsale and J. M. Houston (eds.), *Urbanization and its problems* (Oxford, 1968).

mercato manent xxv burgenses). Then again, there were 25 houses (render-ing 100s. a year) in or around the market place of Worcester (*in foro Wirecestre*). At York, we hear of stalls in the provision market (*banci in macello*). Furthermore, there are hints of external trade at other boroughs. The account of Chester refers to the coming and going of ships, and to a trade in marten pelts which we know, from other sources, came from Ireland.[1] In the south, we are told of dues from ships at Southwark, of foreign merchants (*extranei mercatorum*) at Canterbury, and of a guildhall (*gihalla burgensium*) at Dover and of its harbour and ships. At Frostenden, in Suffolk, there was a seaport (*i portus maris*), and hithes or harbours are mentioned at four places in Kent along the shores of the Thames estuary – at Dartford, Gravesend, Milton and Swanscombe.

Of the 58 markets specifically mentioned in the Domesday Book, only fourteen were entered for boroughs, an example of the incompleteness of the Domesday urban record. Thus of the four boroughs in Gloucester-shire (Gloucester, Bristol, Tewkesbury and Winchcomb) a market was recorded only for Tewkesbury, and it was said to have been established recently by Queen Matilda; but there were also three other places with markets – Thornbury, Cirencester with 'a new market', and Berkeley where there was a *forum* in which dwelt 17 men who paid a rent, but we are not told how much. We hear, moreover, not of a market but of 10 traders at Abingdon dwelling in front of the door of the church, i.e. Abingdon Abbey; there were also 10 traders at Cheshunt. There are only two Domesday references to fairs – one at Aspall in Suffolk, and an 'annual fair' at Methleigh in Cornwall. There must have been many others. There is, for example, non-Domesday evidence of a fair and of a Thursday market at or near St Michael's Mount in Cornwall.[2]

The outstanding Domesday example of new commercial growth comes from what was recorded merely as the *villa* 'where rests enshrined Saint Edmund, King and Martyr of glorious memory'. Accretion around the monastery at Bury St Edmunds had been considerable between 1066 and 1086:

Now the town [*villa*] is contained in a greater circle, including [*de*] land which then used to be ploughed and sown; whereon are 30 priests, deacons, and clerks together [*inter*]; 28 nuns and poor persons who daily utter prayers for the king

[1] J. H. Round (1895), pp. 465–7.
[2] H. P. R. Finberg, 'The castle of Cornwall', *Devon and Cornwall Notes and Queries* XXIII (1949), 123.

and for all Christian people; 75 bakers, ale-brewers, tailors, washer-women, shoemakers, robe-makers [*parmentarii*], cooks, porters, stewards together. And all these daily wait upon the Saint, the abbot, and the brethren. Besides whom there are 13 reeves over the land, who have their houses in the said town, and under them 5 bordars. Now, 34 knights, French and English together, and under them 22 bordars. Now altogether [there are] 342 houses in the demesne on the land of St Edmund which was under the plough in the time of King Edward.

This is about a place which was not technically a borough in 1086, but which has been included in the total of 112. Obviously the town as an economic reality was capable of growing up in places other than legal boroughs. Soon the time was to come when the creation or confirmation of markets was an essential element in the granting of borough charters.

This catalogue of the miscellaneous hints of the commercial life that appears in the Domesday Book points to the new age which the Norman Conquest had inaugurated. While the Conquest can hardly have meant a revolution in commerce, it must have given a great impulse to that trade and urban development that had already begun to grow in Anglo-Saxon times.

Chapter 3

CHANGES IN THE EARLY MIDDLE AGES

R. A. DONKIN

If the compilers of Domesday Book had been able to retrace their steps in the early 1300s they would have been greatly impressed by the changes that had taken place in the English countryside and in English life generally in the course of two centuries or so. Everywhere there was evidence of growth – in population and in the area of improved land, in industry and trade, and in the number and size of towns. The power of the central government also had increased and, while churchmen still played an important part in affairs of state, lay officials were becoming ever more numerous and influential.[1] This partly reflected changes in the system of education: many grammar schools[2] and two universities had come into being, taking the place of the monastic schools. Finally, wider external contacts – through increased commerce, the spread of the new monastic Orders, and participation in the 'twelfth-century renaissance'[3] – strengthened those ties with continental Europe that had been established by the Norman Conquest.

POPULATION

The population of England more than doubled between 1086 and the beginning of the fourteenth century, from approximately 1½ million to 4 or even 4½ million. There were, however, wide variations in rates of increase between one area and another. This is apparent in comparing the

[1] J. R. Strayer, 'Laicization of French and English society in the thirteenth century', *Speculum*, xv (1940), 76–86; H. G. Richardson and G. O. Sayles, *The governance of medieval England* (Edinburgh, 1963), 167, 283, 319.

[2] L. Thorndike, 'Elementary and secondary education in the Middle Ages', *Speculum*, xv (1940), 400–8; A. B. Emden, 'Learning and education' in A. L. Poole (ed.), *Medieval England*, ii (Oxford, 1958), 515–40.

[3] R. W. Southern, 'The place of England in the twelfth-century renaissance', *History*, n.s., xLv (1960), 201–16; C. H. Haskins, 'England and Sicily in the twelfth century', *Eng. Hist. Rev*, xxvi (1911), 433–47; A. C. Crombie, *Robert Grosseteste and the origins of experimental science, 1100–1700* (Oxford, 1953), especially 16–43.

evidence of Domesday Book with that of surveys of estates dating from the twelfth and thirteenth centuries.[1] Many long-established places probably grew at far less than the national rate; the villages of south Warwickshire, for example, appear to have been much the same size in 1279 as in 1086.[2] On the other hand, the Forest of Arden, to the north of the river Avon, was attracting population at this time.[3] In the Fenland, where reclamation and population growth were clearly interdependent, some phenomenal increases occurred. Between 1086 and the late thirteenth or early fourteenth century, the number of tenants in Spalding township multiplied six and a half times, in Pinchbeck township over eleven times, and in Fleet sixty-one times.[4] A particularly rapid upward trend in developing areas was very likely the result, not merely of immigration, but of a lowering of the average age of marriage and thus an increase in fertility.

Some of the highest rural population densities were found in association with the custom of partible inheritance.[5] The latter was a notable feature of Norfolk and Kent, and was also of some importance in parts of Cambridgeshire, Leicestershire, Nottinghamshire, Lincolnshire, Suffolk, Essex and Middlesex. Partition could not go on indefinitely, but it went furthest where there were extensive common rights and/or new land that could be reclaimed. Where a man inherited, or was certain to inherit, a viable holding, even though the nucleus of it was very small, he tended to marry and

[1] J. C. Russell, *British medieval population* (Albuquerque, 1948), 70–6, 246–9. See also R. R. West (ed.), *The eleventh and twelfth-century sections of Cott. MS. Galba E. ii: The register of the abbey of St Benet of Holme*, Norfolk Record Soc., 2 (1932), 244, 246; J. A. Raftis, *The estates of Ramsey Abbey* (Toronto, 1957), 66; F. Baring, 'Domesday Book and the Burton Cartulary', *Eng. Hist. Rev.*, XI (1896), 98–100.

[2] J. B. Harley, 'Population trends and agricultural developments from the Warwickshire Hundred Rolls of 1279', *Econ. Hist. Rev.*, 2nd ser., XI (1958), 13.

[3] J. B. Harley, 'The settlement geography of early medieval Warwickshire', *Trans. and Papers, Inst. Brit. Geog.*, XXXIV (1964), 115–30; B. K. Roberts, 'A study of medieval colonization in the Forest of Arden, Warwickshire', *Agric. Hist. Rev.*, XVI (1968), 101–13.

[4] H. E. Hallam, 'Some thirteenth-century censuses', *Econ. Hist. Rev.*, 2nd ser., X (1958), 340; 'Population density in the medieval Fenland', *Econ. Hist. Rev.*, 2nd ser., XIV (1961), 79; and *Settlement and society: a study of the early agrarian history of south Lincolnshire* (Cambridge, 1965), 198–200.

[5] G. C. Homans, 'Partible inheritance of villagers' holdings', *Econ. Hist. Rev.*, VIII (1937–8), 48–56; D. S. Pitkin, 'Partible inheritance and the open fields', *Agric. Hist.*, XXV (1961), 65–9; R. J. Faith, 'Peasant families and inheritance customs in medieval England', *Agric. Hist. Rev.*, XIV (1966), 77–95; B. Dodwell, 'Holdings and inheritance in medieval East Anglia', *Econ. Hist. Rev.*, 2nd ser., XX (1967), 53–66.

to settle down. The custom of primogeniture and of Borough English (succession through the youngest son) stabilised the population earlier and then encouraged migration. At the same time, an undivided holding might support more than the recognised tenant and his immediate family.[1] By the addition of a cottage or simply a few extra rooms (the usual solution in towns), aged parents, unmarried brothers and sisters, and even undersettlers and co-parceners were sometimes accommodated.[2]

English towns probably grew at about the same rate as the population as a whole in the 250 years after the Conquest. But, as the urban population may not even have been reproducing itself, any substantial increase in numbers depended upon immigration. 'The villages were the primary seedbeds of population'[3] which generally moved first towards local or regional centres and thence from smaller to larger places. Population movement in the twelfth and thirteenth centuries is difficult to quantify, but the place elements in personal names and the many references to *chevagium*, a tax on those living away from their home manor, suggest that it was considerable around towns and within areas of agricultural opportunity.[4] On the estates of Ramsey abbey, the emigration of serfs 'was a regular feature of manorial life from the time of the earliest extant court rolls'.[5]

Also of considerable social significance was the rapid increase in the clerical population.[6] There were possibly twenty times as many regular clergy in 1300 as in 1066. The rate of increase in the number of parish clergy, although less than this, was still comparatively high. The Church drew heavily upon the sons of the lesser nobility and, more important, it was one of the very few avenues of advancement open to the peasant. At one time in the village of Weston in Lincolnshire, 12 out of 68 known adult sons became clergymen.[7]

[1] J. Krause, 'The medieval household: large or small?', *Econ. Hist. Rev.*, 2nd ser., IX (1957), 420–32; H. E. Hallam (1961), 71.

[2] J. Amphlett et al. (eds.), *Court rolls of the manor of Hales*, Worcester Hist. Soc., 28 (1910), xliv, 167; M. Morgan, *The English lands of the abbey of Bec* (Oxford, 1946), 92.

[3] J. C. Russell, 'Late medieval population patterns', *Speculum*, XX (1945), 164.

[4] See pp. 127, 135 below.

[5] J. A. Raftis, *Tenure and mobility: studies in the social history of the medieval English village* (Toronto, 1964), 139.

[6] J. C. Russell, 'The clerical population of medieval England', *Traditio*, II (1944), 177–212.

[7] H. E. Hallam (1958), 356.

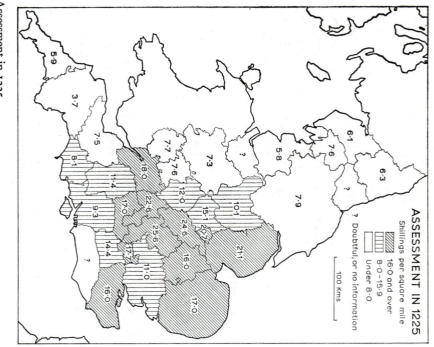

ASSESSMENT IN 1225

Shillings per square mile

16·0 and over

8·0–15·9

Under 8·0

Doubtful, or no information

100 Kms

Fig. 21 Assessment in 1225
Based on F. A. Cazel, 'The fifteenth of 1225', *Bull. Inst. Hist. Research*,
XXXIV (1961), 66–81.

The famines of 1315–17 and 1321[1] are generally regarded as marking
the end of the 'economic thirteenth century' and as a significant turning
point in the population history of England. The incidence of famine is
also relevant to the question of short-term fluctuations in growth and to
the problem of 'overpopulation'. The numerous 'pestilences' before the
Black Death were mainly the result of famine,[2] and affected, first and

[1] H. S. Lucas, 'The great European famine of 1315, 1316 and 1317', *Speculum*, v
(1930), 341–77; J. C. Russell, 'Effects of pestilence and plague, 1315–85', *Comparative
Studies in History and Society*, VIII (1966), 464–73; S. L. Thrupp, 'Plague effects in
medieval Europe', *ibid.*, 474–83.
[2] C. Creighton, *A history of epidemics in Britain*, I (Cambridge, 1891), 8.

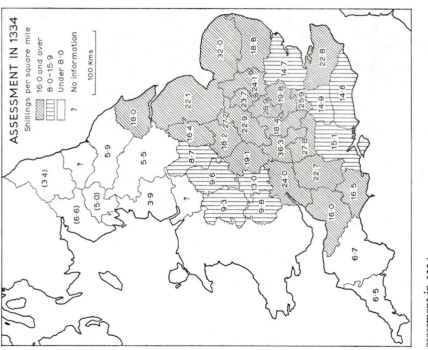

ASSESSMENT IN 1334

Shillings per square mile

16·0 and over
8·0–15·9
Under 8·0
? No information

100 Kms

Fig. 22 Assessment in 1334

Based on J. F. Willard, 'The taxes upon moveables in the reign of Edward III', *Eng. Hist. Rev.*, XXX (1915), 73.

For the purpose of comparison with Fig. 21 the assessment at a tenth has been converted to, and added to, that at a fifteenth. The figures for Cumberland, Westmorland and Northumberland are for 1336; see p. 137 below. London has been excluded in order to make the map comparable with Fig. 21; the inclusion of London would bring the figure for Middlesex up to 77s. For a more detailed map of the assessed wealth itself, see Fig. 35.

foremost, the poor. Famine, in turn, has been partly correlated with adverse weather conditions,[1] especially heavy rainfall in late autumn and early winter, leading to low yields or even to no yields at all. An inadequate return meant skimping on seed corn the following year, so it

[1] J. Z. Titow, 'Evidence of weather in the account rolls of the bishopric of Winchester, 1209–1350', *Econ. Hist. Rev.*, 2nd ser., XII (1960), 360–407.

might take several years to recover from a disastrous harvest. Severe shortages of food must have reduced the physical resistance of the working population and made it more liable to infection. In addition, exceptionally dry summers were sometimes followed by epidemics, probably of typhus-type fevers and dysentery, which hit both rich and poor alike, in town and country; such, apparently, were the 'pestilences' of 1288–9 and 1328–9. The death rate on the manors of the bishop of Winchester was very high, and apparently increasing, between about 1250 and 1350 – an average of 40 per thousand adults over the entire period and 52 per thousand over the last 55 years, 1292–1347.[1] Population responded very quickly to harvest failure; and this, together with the evidence of many small peasant holdings and of rising entry fines,[2] suggests that substantial parts of the country were indeed overpopulated in the half-century or so before the Black Death.

From the middle of the twelfth century, certain taxes took the form of a proportion of the assessed value of a man's 'movable goods' – chiefly grain and the larger domestic animals, and, in towns, household effects and articles of personal use. Fig. 21 shows the county assessments (i.e. excluding the contributions of the religious) for the fifteenth of 1225, expressed in terms of shillings per square mile.[3] The palatinates were exempt and London made no contribution. The assessment for Sussex is obviously much too low and has therefore been ignored; Kent, too, was undervalued by reason of the exemption of the Cinque Ports. With these qualifications, however, the map probably gives a fair overall impression of the geographical distribution of personal wealth and, indirectly, of the density of population in the early thirteenth century. It does not, of course, show the average wealth of individual taxpayers. The most notable feature is a band of medium and high values stretching from Gloucestershire, through the south Midlands to Lincolnshire and East Anglia, with very high values in a central block of country comprising Northamptonshire,

[1] M. M. Postan and J. Titow, 'Heriots and prices on Winchester manors', *Econ. Hist. Rev.*, 2nd ser., XI (1959), 399. See also S. L. Thrupp, 'The problem of replacement rates in late medieval English population', *Econ. Hist. Rev.*, 2nd ser., XVIII (1965), 107.

[2] M. M. Postan, 'Medieval agrarian society in its prime: England', in M. M. Postan (ed.), *Cambridge Economic History of Europe*, I (2nd ed., Cambridge, 1966), 553; J. Z. Titow, *English rural society, 1200–1350* (London, 1969), 73–8. See also B. F. Harvey, 'The population trend in England between 1300 and 1348', *Trans. Roy. Hist. Soc.*, 5th ser., XVI (1966), 23–42.

[3] F. A. Cazel, 'The fifteenth of 1225', *Bull. Inst. Hist. Research*, XXXIV (1961), 66–81; S. K. Mitchell, *Studies in taxation under John and Henry III* (New Haven, 1914), 159–69; and *Taxation in medieval England* (New Haven, 1951), 20ff.

Buckinghamshire and Bedfordshire (the last two were, however, grouped together, like some other counties).

In 1334, rural communities were taxed at the rate of one-fifteenth and boroughs and ancient demesnes at the rate of one-tenth.[1] For the purpose of Fig. 22, the tenth has been converted to a fifteenth and added to the rural assessment. Cheshire and Durham were exempt; Northumberland, Cumberland and Westmorland were excused as they had recently been devastated by the Scots; Kent and Sussex again were undervalued; and London has been excluded. In 1334, the axis of greater 'wealth' still lay diagonally across the country, from the Severn to the Wash. More significantly, the zone of higher values had expanded since 1225, notably towards the north-west and the south-west; only the counties of the north of England and of the extreme south-west stood apart, chiefly, perhaps, on account of the comparatively low density of population in these areas.

THE COUNTRYSIDE

Field arrangements

At the beginning of the fourteenth century, field arrangements fell into two broad categories: (i) those in which the arable was grouped into two or three, or occasionally more, large 'fields', and (ii) those in which such 'fields' were either absent or were organised on very special lines. The area covered by the latter included Kent[2] and East Anglia,[3] both settled very early and both characterised by large numbers of freemen and a comparatively fluid social structure. Here a man's holding, although probably divided, usually lay in one part of the total arable area. This arrangement perhaps developed from family tenements that were originally compact, subsequently breaking up under the influence of partible inheritance, and later being regrouped within the same general area. In Kent, where hamlets were characteristic, most holdings appear to have comprised small, enclosed pieces of land that were worked in severalty. Furthermore, in both

[1] J. F. Willard, 'The taxes upon moveables in the reign of Edward III', *Eng. Hist. Rev.*, xxx (1915), 69–74, and *Parliamentary taxes on personal property 1290 to 1334* (Cambridge, Mass., 1934); R. S. Schofield, 'The geographical distribution of wealth in England', *Econ. Hist. Rev.*, 2nd ser, XVIII (1965), 483–510.

[2] A. R. H. Baker, 'The field system of an East Kent parish', *Archaeologia Cantiana*, LXXVIII (1963), 96–117.

[3] D. C. Douglas, *The social structure of medieval East Anglia* (Oxford, 1927), 17–67; M. R. Postgate, 'The field systems of the Breckland', *Agric. Hist. Rev.*, x (1962), 80–101.

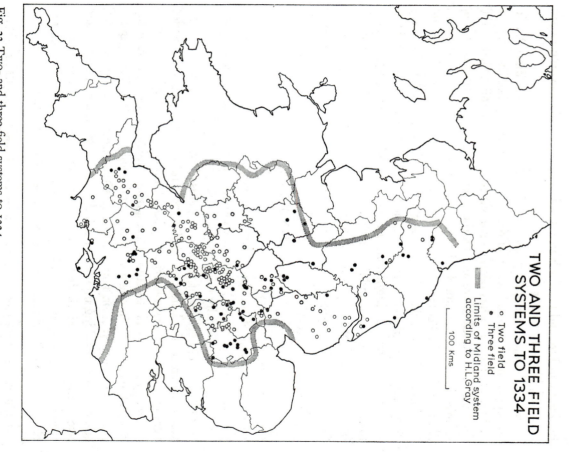

TWO AND THREE FIELD
SYSTEMS TO 1334

○ Two field
● Three field

Limits of Midland system
according to H.L.Gray

100 Kms

Fig. 23 Two- and three-field systems to 1334
Based on H. L. Gray, *English field systems* (Cambridge, Mass., 1915), 450–509.

Kent and East Anglia, as well as in some northern and western districts, varieties of infield-outfield cultivation existed. Under this system, an intensively cultivated central area was surrounded by periodically cultivated 'intakes' or 'breaks'.

The main area of the two- and three-field system was the midland belt of England (Fig. 23). The primary unit of plough-land was the 'land' or selion (*selio*), of varying width and length, but usually less than half an acre in area;[1] alone, or grouped with others, it formed a 'strip' – a tenurial term – where property was intermixed. Selions were often ridged, which had the effect of throwing off water, and Walter of Henley referred, about 1250, to the ridging of wet land.[2]

The next division up the scale was the furlong or *cultura*, a bundle of 'lands', uniformly cultivated. And, finally, over a large part of medieval England, furlongs were grouped into two or three open and common fields. How and when this came about are questions that are still unresolved.[3] The level of co-operation implied by the existence of two or three great 'fields' may be no older than the eleventh or twelfth century. Only then or later, it can be argued, did various *ad hoc* agreements between tenants with intermixed holdings give way, under the influence of population growth, to community-regulated cropping and the joint grazing of fallow. But in any event, where most of the village arable was arranged in 'fields', a man's strips were fairly equally distributed between them. At first, strips may also have lain in the same relation to one another within furlongs, but if so, this was all but obscured by partition and exchange by the late thirteenth century.[4]

In the twelfth century there were usually only two 'fields', half the cultivated area being fallow at any one time, but towards 1200 there is evidence of village territories organized into three 'fields'.[5] H. L. Gray was probably right in thinking that the one system normally developed from the other, and that the change first assumed importance during the thirteenth

[1] H. M. Clark, 'Selion size and soil type', *Agric. Hist. Rev*, VIII (1960), 91–8.

[2] E. Lamond (ed.), *Walter of Henley's Husbandry* (London, 1890), 17.

[3] J. Thirsk, 'The common fields', *Past and Present*, XXIX (1964), 3–25; J. Z. Titow, 'Medieval England and the open field system', *Past and Present*, XXXII (1965), 86–102; J. Thirsk, 'The origin of the common fields', *Past and Present*, XXXIII (1966), 142–7; G. C. Homans, 'The explanation of English regional differences', *Past and Present*, XLII (1969), 32–4.

[4] S. Göransson, 'Regular open-field pattern in England and Scandinavian *Solskifte*' in *Morphogenesis of the agrarian cultural landscape, Geografiska Annaler*, XLIII (1961), 80–104; G. C. Homans, *English villagers in the thirteenth century* (Cambridge, Mass., 1942), 97.

[5] K. Major (ed.), *The registrum antiquissimum of the cathedral church of Lincoln: VII*, Lincoln Record Soc., 46 (1953), 58; D. M. Williamson, 'Kesteven villages in the Middle Ages', *Lincolnshire Historian*, II (1955), 11; D. Roden, 'Demesne farming in the Chiltern Hills', *Agric. Hist. Rev*, XVII (1969), 17.

century.[1] Certainly by 1334 the three-field system had been widely adopted, as can be seen from the sample of evidence plotted on Fig. 23.

Three 'fields', implying a three-course husbandry, had several advantages over two.[2] Assuming that the fallow field was ploughed twice in both cases, the three-field system increased the area a peasant could cultivate by one-eighth and his productivity by one-third, or even by a half if he could apply his 'surplus' ploughing capacity to newly cleared land; it distributed field work more evenly over the year; and by leading to a more nearly balanced pattern of sowing in autumn and spring, it lessened the risk of famine. On the other hand, the three-field system, by reducing the period of fallow, might lead in time to impoverished soils and falling yields.

It is necessary to distinguish clearly between a three-course system of husbandry on the one hand, and arable arranged in three 'fields' on the other. Obviously the latter lent itself to the former, but a three-course rotation was also found in East Anglia[3] and Kent,[4] and indeed in places where there were more than three 'fields'. From the late thirteenth century, the term 'field' was often used very loosely. References to from five to forty 'fields' are fairly common, especially, but not exclusively, in areas of late colonisation. These 'fields' do not necessarily imply a more elaborate field course. A four-course rotation including leguminous crops, which were becoming more important,[5] was sometimes practised, but a three-course system was normal – winter wheat/rye, spring barley/oats/ legumes, and fallow; legumes were occasionally planted in place of fallow, a procedure known as *inhoking*. This three-course rotation could be imposed on any number of 'fields', from two upwards, provided that they were not regarded as fixed cropping or rotational units. During the thirteenth century, the furlong emerges as functionally the most important division of the arable. Furlongs might be grouped differently in different years to vary the proportions of spring and winter corn, or

[1] H. L. Gray, *English field systems* (Cambridge, Mass., 1915), 80.

[2] E. Lamond (ed.) (1890), 8; L. White, *Medieval technology and social change* (Oxford, 1962), 72.

[3] F. G. Davenport, *The economic development of a Norfolk manor, 1086–1565* (Cambridge, 1906), 27.

[4] T. A. M. Bishop, 'The rotation of crops at Westerham, 1297–1350', *Econ. Hist. Rev.*, IX (1938), 41.

[5] F. B. Stitt (ed.), *Lenton priory estate accounts, 1296–98*, Thoroton Soc., Record Series, 19 (1959), xl; W. O. Ault, 'Open field husbandry and the village community', *Trans. American Philos. Soc.*, LV (1965), 19.

a poor furlong might be put down to permanent pasture. Indeed where spring-sown crops predominated, as was commonly the case, a fixed pattern of 'fields' could not possibly have been maintained.

If, by exchange or purchase, a peasant managed to acquire an entire furlong, he might cultivate the land independently. To enclose it was, however, another matter, for this would interfere with grazing rights, arranged either between neighbours or involving the community as a whole. There was undoubtedly a great deal of regrouping of strips for convenience sake, but for a peasant to be able to enclose part of an open field was probably very rare. On the other hand, demesne that had been sufficiently consolidated might be enclosed and withdrawn from common field arrangements by the lord of the manor.[1]

Open and common fields also existed (and were possibly typical in the early fourteenth century) on the margins of the Celtic west – towards the south-west and along the Welsh border – and in parts of Lancashire and Northumberland. Furthermore, where lordship was strong in areas settled or resettled, even in post-Conquest times (for example, parts of the Vale of York and east Cheshire), the demesne was carved out and cast with peasant holdings into two or three 'fields'. The multiplication of 'fields' beyond this was significant only in that it implied greater flexibility of management; and such flexibility, clearly perceptible in the late thirteenth century, tended to reduce the contrast between the south-east and the rest of the country.

Not only arable but pasture was essential to the economy of a village. Animals were turned on to the harvest stubble; and beyond the open fields rights of rough grazing and common over the 'waste' were usually linked with arable holdings. Where, however, there was an abundance of pasture, as in the Fenland and parts of the south-west and the north, such arrangements were unnecessary and intercommoning between villages was also practised.[2]

In the absence of rotation grasses and root crops, the hay from meadows, bordering streams and enriched by floods, was a valuable commodity. Meadow was often worth four or five times as much as arable. Thus on the manor of Laughton (Sussex) in 1325, arable was valued at between 3d. and

[1] See, for example, S. R. Scargill-Bird (ed.), *Custumals of Battle abbey in the reigns of Edward I and Edward II*, Camden Soc, n.s. 41 (1887); A. M. Woodcock (ed.), *Cartulary of the priory of St Gregory, Canterbury*, Camden Soc, 3rd ser, 88 (1956), xvi, xvii, 83, 106.

[2] N. Neilson (ed.), *A terrier of Fleet, Lincolnshire* (British Academy, London, 1920); and 'Early English woodland and waste', *Jour. Econ. Hist.*, II (1942), 57–8.

$6d.$ an acre, pasture at between $6d.$ and $18d.$, and meadow at between $24d.$ and $30d.$[1] Meadow was usually redistributed annually and held in severalty until the hay harvest when it reverted to common use.

Demesne and peasant holdings

An important distinction for the majority of people in the early Middle Ages was that between the strips held by the lord, the demesne, and the strips held directly by the villeins who also provided the demesne with the greater part of its labour. Freeholders paid rent for their holdings and rendered little or no services on the demesne. E. A. Kosminsky estimated from the Hundred Rolls of 1279 (covering parts of seven counties, mainly in the south Midlands) that demesne occupied 32% of the total arable, villein land 40%, and freeholdings 28%.[2] The proportions, however, varied greatly between manors. Moreover some manors included several vills and others amounted to only a fraction of a vill. To add to the complexity, there were sub-manors, formed by persons other than the *capitalis dominus* dividing a substantial holding into 'demesne' and allotments for sub-tenants. Finally, in East Anglia, the Danelaw and Northumbria, where there were large numbers of freemen and sokemen, the village rather than the manor was 'the essential form of rural organization'.[3]

By 1300 there were also wide differences in the size of holdings of both freemen and villeins. This was the combined result of: (i) sub-letting and the buying and selling of freehold land, even by villeins;[4] (ii) inheritance

[1] A. E. Wilson (ed.), *Custumals of the manors of Laughton, Willingdon and Goring*, Sussex Record Soc., 60 (1961), 79.

[2] E. A. Kosminsky, *Studies in the agrarian history of England in the thirteenth century* (Oxford, 1956), 89. See also E. A. Kosminsky, 'The Hundred Rolls of 1279-80 as a source for English agrarian history', *Econ. Hist. Rev.*, III (1931-2), 16-44; and 'Services and money rents in the thirteenth century', *Econ. Hist. Rev.*, V (1935), 24-45; M. M. Postan, 'The manor in the Hundred Rolls: essays in bibliography and criticism, XV', *Econ. Hist. Rev.*, 2nd ser., III (1950-1), 119-25.

[3] F. M. Stenton (ed.), *Documents illustrative of the social and economic history of the Danelaw* (British Academy, London, 1920), lxi.

[4] C. N. L. Brooke and M. M. Postan, *Carte nativorum: a Peterborough abbey cartulary of the fourteenth century*, Northamptonshire Record Soc., 20 (1946), xxixff; D. C. Douglas (ed.), *Feudal documents from the abbey of Bury St Edmunds* (British Academy, London, 1932), cliv; E. Toms (ed.), *Chertsey abbey court rolls abstract*, Surrey Record Soc., 21 (1937), xvi; D. G. Watts, 'A model for the early fourteenth century', *Econ. Hist. Rev.*, 2nd ser., XX (1967), 543-7; E. Miller, *The abbey and bishopric of Ely* (Cambridge, 1951), 144; P. R. Hyams, 'The origins of the peasant land market in England', *Econ. Hist. Rev.*, 2nd ser., XXIII (1970), 18-31.

customs, marriage settlements and piecemeal gifts of land; (iii) assarting and reclamation, which, however, did not generally keep pace with the growth of population. While a few peasants managed to acquire substantial holdings of up to 100 acres, the majority of holdings were getting smaller or at least were supporting an increasing number of folk.[1] Many were very small indeed, especially in south-eastern England. Thus at Wykes in Bardwell, Suffolk, in the latter half of the thirteenth century, there were 76 freeholders of whom 7 held 116 acres between them; 6 had messuages without land, and the remaining 63 possessed little more than 2½ acres apiece.[2] Again at Chippenham, Cambridgeshire, in 1279, 59 of the 143 tenants (more than four times the number recorded in Domesday Book) had less than 2 acres apiece and must have depended on wage labour.[3] Forty-six per cent of the 13,500 peasant holdings examined by Kosminsky in the Hundred Rolls amounted to a quarter-virgate (a *ferling*, about 8 acres) or less.[4] In almost any part of the country, 8 acres of arable was probably near or below the minimum required for subsistence.[5]

The whole or part of the demesne of a manor was either worked directly, under a bailiff, by the immediate feudal lord, or it was leased or 'farmed' (*ad firmam*) for an agreed period at a fixed rent in money or kind or both. The former alternative was generally adopted when agricultural prices were rising, the latter when they were declining and a stable income was desirable. Demesnes were mostly being leased and/or were contracting in area in the late eleventh and the twelfth centuries.[6] Over this period there

---

[1] J. Z. Titow, 'Some evidence of thirteenth century population increase', *Econ. Hist. Rev.*, 2nd ser, XIV (1961), 222–3; and 'Some differences between manors and their effects on the condition of the peasants in the thirteenth century', *Agric. Hist. Rev.*, X (1962), 4.

[2] W. Hudson, 'Three manorial extents of the thirteenth century', *Norfolk Archaeology*, XIV (1901), 17.

[3] M. Spufford, *A Cambridgeshire community: Chippenham from settlement to enclosure* (Leicester, 1965), 29–30.

[4] E. A. Kosminsky (1956), 228; M. M. Postan (1966), 619, 625.

[5] J. Z. Titow (1969), 79.

[6] M. M. Postan, 'The chronology of labour services', *Trans. Roy. Hist. Soc.*, 4th ser, XX (1937), 169–93; 'Glastonbury estates in the twelfth century', *Econ. Hist. Rev.*, 2nd ser, V (1953), 359; and 'Glastonbury estates in the twelfth century: a reply', *Econ. Hist. Rev.*, 2nd ser, IX (1956–7), 106–18; R. Lennard, 'The demesnes of Glastonbury abbey in the eleventh and twelfth centuries', *Econ. Hist. Rev.*, 2nd ser, VIII (1955–6), 355–63; W. H. Hale (ed.), *The Domesday of St Paul's of 1222*, Camden Soc, 69 (1858), xxii; B. A. Lees (ed.), *Records of the Templars in England in the twelfth century* (British Academy, London, 1935), cix, cxx; B. Lyon, 'Encore le problème de la chronologie des corvées', *Le Moyen Age*, LXIX (1963), 615–30; J. A. Raftis (1957), 58, 86, 89.

Table 3.1. *The Great Pipe Roll of 2 Henry II*

Shires	Danegeld due			In waste			Proportion of waste to total
	£	s.	d.	£	s.	d.	
Warwick	128	12	6	80	11	0	nearly 2/3
Notts and Derby	112	1	11	58	11	6	over 1/2
Leicester	99	19	11	51	8	2	over 1/2
Oxford	249	6	5	96	2	10	about 2/5
Bucks and Beds	316	6	8	107	14	3	over 1/3
Berks	205	11	4	77	16	7	over 1/3
Cambridge	114	14	9	34	3	0	about 1/3
Gloucester	184	1	6	59	3	6	nearly 1/3
Northampton	119	10	9	38	12	1	nearly 1/3
Hereford	93	15	6	19	3	6	over 1/4
Worcester	102	5	9	27	14	3	over 1/4
Wilts	389	13	0	99	16	9	about 1/4
Essex	236	8	0	61	4	0	about 1/4
Herts	110	1	3	29	17	4	nearly 3/11
Huntingdon	70	5	0	14	0	6	about 1/5
Stafford	44	1	0	8	8	0	nearly 1/5
Somerset	277	10	4	54	5	0	nearly 1/5
Surrey	184	16	0	30	12	9	nearly 1/6
Middlesex	85	0	6	10	0	0	nearly 1/8
Sussex	216	10	6	9	2	0	nearly 1/23
Kent	105	16	10	0	8	0	nearly 1/270

was much that was 'specially unfavourable to the direct management of the demesne'.[1] The years of civil war in Stephen's reign (1137–54) wrought havoc in the countryside. Areas were devastated at different times and the rate at which they recovered also varied, but the figures in table 3.1, taken from the Pipe Roll of 1156, leave no doubt about the wasted character of much of the country around the middle of the century.[2]

Towards the end of the twelfth century, the prices of grain and live-stock began to rise[3] (Fig. 24). Bad weather and low yields, as in 1201 and

[1] M. M. Postan (1966), 585.
[2] H. W. C. Davis, 'The anarchy of Stephen's reign', *Eng. Hist. Rev.*, XVIII (1903), 630.
[3] D. L. Farmer, 'Some grain price movements in thirteenth-century England', *Econ. Hist. Rev.*, 2nd ser., X (1957), 214; 'Some livestock price movements in the thirteenth century', *Econ. Hist. Rev.*, 2nd ser., XXII (1969), 1–16; and 'Some price

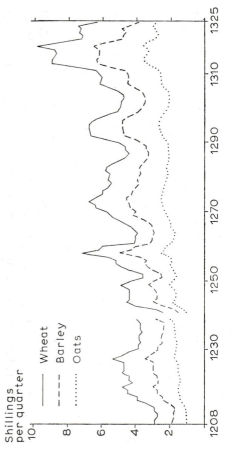

Shillings
per quarter

——— Wheat
– – – Barley
·········· Oats

Fig. 24 Movement of grain prices, 1208–1325
Based on D. L. Farmer, 'Some grain price movements in thirteenth-century
England', *Econ. Hist. Rev.*, 2nd ser., x (1957), 214.
Seven-year moving averages for sales of major grains.

1205,[1] led to sudden rises; then after a good year or two, prices dropped, but not usually to their previous level. The broad inflationary tendency has been attributed sometimes to the increasing amount of money in circulation, sometimes to a greater increase in population than in production, and sometimes to both.[2] But whatever the cause, as prices rose and real wages fell, demesne cultivation became more profitable. Lords began to recover lost portions of their demesnes and to expand them at the expense of any available villein holdings. The thirteenth century generally was an age of prosperous demesne farming when labour services were re-enforced and when the value of demesne arable as much as doubled. 'It was, above all, an age when landlords managed their estates as speculative enterprises geared to expanding markets, when "buoyant"

fluctuations in Angevin England', *Econ. Hist. Rev.*, 2nd ser., IX (1956), 34–9; A. L. Poole, 'Livestock prices in the twelfth century', *Eng. Hist. Rev.*, LV (1940), 285.

[1] D. L. Farmer (1956), 38; C. Easton, *Les hivers dans l'Europe occidentale* (Leyden, 1928), 57.

[2] M. M. Postan, 'The rise of a money economy', *Econ. Hist. Rev.*, XIV (1944), 123–34; J. Schreiner, 'Wages and prices in England in the later Middle Ages', *Scandinavian Econ. Hist. Rev.*, II (1954), 61–73; W. C. Robinson, 'Money, population and economic change in late medieval Europe', *Econ. Hist. Rev.*, 2nd ser., XII (1959), 63, 75; C. M. Cipola, 'Currency depreciation in medieval Europe', *Econ. Hist. Rev.*, 2nd ser., XV (1963), 413–22.

demesnes sustained a high level of agricultural production, and when all the resources of villein and wage labour in growing villages were exploited to the full.'[1]

But even before 1300 there were signs of a recession; demesnes in different parts of the country were being leased as a whole or in part at economic rents, and labour services were being commuted for money payments.[2] Landlords, it has been argued, were beginning to pay the price of earlier overcropping, and productivity levels, and thereby profits, were falling.[3] Peasants, faced with the same situation, had little choice but to take up land that was released from the demesne. In any event, there is no doubt that an unprecedented number of harvest failures and famines – in 1272, 1277, 1283, 1292, 1311 and especially in 1315–18 and 1321 – accentuated the general difficulties of the time.

Arable farming

Perhaps 20% of English peasants in the last quarter of the thirteenth century had holdings of approximately one virgate or 30 acres, i.e. at best 20 acres in cultivation at any one time. Such a holding probably produced at least 70 bushels of grain (over and above seed required for the following year), of which approximately 30 bushels would have been available for sale and to meet any fixed payments in kind.[4] During the thirteenth and early fourteenth centuries, demesne land, as we have seen, was producing for the market,[5] and sales of well over 50% of a harvest were common.

[1] E. Miller, 'England in the twelfth and thirteenth centuries: An economic contrast?', *Econ. Hist. Rev.*, 2nd ser., XXIV (1971), 2. See also N. Denholme Young, 'The Yorkshire estates of Isabella de Fortibus', *Yorks. Archaeol. Jour*., XXXI (1934), 398; F. M. Page (ed.), *Wellingborough manorial accounts, 1258–1323*, Northants. Record Soc., 8 (1936), xxx; R. H. Brinell, 'Production for the market on a small fourteenth-century estate', *Econ. Hist. Rev.*, 2nd ser., XIX (1966), 380–7; J. A. Raftis (1957), 97, 103, 110–12.

[2] See, for example, T. A. M. Bishop, 'The distribution of manorial demesne in the vale of Yorkshire', *Eng. Hist. Rev.*, XLIX (1934), 395; R. H. Hilton, *The economic development of some Leicestershire estates in the fourteenth and fifteenth centuries* (Oxford, 1947), 40, 88; F. B. Stitt (ed.), *Lenton priory estate accounts, 1296–98*, Thoroton Soc., Record Series, 19 (1959), xviii.

[3] M. M. Postan (1966), 556–9, 588; J. Z. Titow (1969), 52–4.

[4] A. Ballard, 'Woodstock manor in the thirteenth century', *Vierteljahrschrift für Social- und Wirtschaftsgeschichte*, VI (1908) 439; A. T. Gaydon (ed.), *The taxation of 1297 for Barford, Biggleswade and Flitt hundreds*, Beds. Hist. Record Soc, 39 (1959), xx.

[5] R. H. Hilton, *A medieval society: the west Midlands at the end of the thirteenth century* (London, 1966), 78–9; J. A. Raftis (1957), 114.

Yet even lords of manors closely connected with the market, such as the bishop of Winchester, were drawing more from rents than from the sale of produce. 'This shows that the market was supplied in the first place by the peasant holdings.'[1] Money rents could be met only by selling surplus products, of which grain was by far the most likely.

Wheat and barley were the preferred bread grains, but, as they were the principal cash crops too, it is difficult to say how far they – and particularly wheat – were consumed in peasant households. Oats also was widely grown, but its use as human food probably varied more than that of the other two. It was the leading crop in Lancashire, and perhaps in other parts of the north and west, in the early fourteenth century. Rye was generally confined to marginal land or was mixed with other grain. Peas, beans, vetch, and occasionally flax,[2] made up the common range of field crops. In house gardens, cabbages, onions, leeks, lettuce, spinach, parsley and a few herbs were grown.[3]

Open-field husbandry and relatively inflexible rotations must often have made it difficult to take full advantage of local variations in soil and terrain.[4] Nevertheless, the value of the arable of a particular township might vary from a few pence per acre to several shillings, depending on whether the land was held in severalty or in common, and, presumably, on its accessibility and upon the nature of the soil. Work that has been done on the question of yields (based on manorial accounts of demesne) suggests (i) that average yields were very low by present standards, and (ii) that there was probably a downward trend from about the middle of the thirteenth century.[5] Lord Beveridge, using the account rolls of nine manors (in seven counties) covering the period 1200 to 1450, found that the average yield in bushels per acre was 9.36 for wheat, 14.32 for barley,

[1] E. A. Kosminsky, 'The evolution of feudal rent in England from the eleventh to the fifteenth centuries', *Past and Present*, VII (1955), 20; and E. A. Kosminsky (1956), 324.

[2] C. W. Foster and K. Major (eds.), *The registrum antiquissimum of the cathedral church of Lincoln: IV*, Lincoln Record Soc., 32 (1937), 36; B. Dodwell (ed.), *Feet of fines, Suffolk, 1199–1214*, Pipe Roll Soc., n.s., 32 (London, 1958), 259; L. C. Loyd and D. M. Stenton (eds.), *Sir Christopher Hatton's Book of Seals* (Oxford, 1950), 98.

[3] J. Clapham, *A concise economic history of Britain* (Cambridge, 1949), 85; J. J. Hunt, 'Two medieval gardens', *Proc. Somerset Archaeol. and Nat. Hist. Soc.*, CIV (1960), 91–101.

[4] See, however, A. T. Gaydon (ed.), xxxi; and A. Smith, 'Regional differences in crop production in medieval Kent', *Archaeologia Cantiana*, LXXVIII (1963), 147–60.

[5] M. M. Postan (1966), 557; J. Z. Titow (1969), 52–3; J. Z. Titow, *Winchester yields: a study in medieval agricultural productivity* (Cambridge, 1972), 12–33.

and 10.56 for oats, the ratios of seed to yield being 1:3.9, 1:3.8, and 1:2.4 respectively.[1] Allowing for a margin of error of 10 to 15%,[2] the general ratio may be put at about 1:3.5, and there is no reason to suppose that it differed much over the country as a whole. The average yield, then, was only one-quarter to one-third of what might reasonably be expected today, and the inevitable fluctuations from year to year were enough to make all the difference between profit or loss on the demesne and between bare subsistence or hunger for the small peasant. It is against a background of low and fluctuating yields from overcropped land that much of the work of medieval reclamation must be viewed.

Here and there, efforts were made to maintain and even to increase yields on demesne arable. Forms of centralised accounting were adopted,[3] and officials might refer to several treatises on estate management.[4] Targets were sometimes set; these varied from manor to manor, and occasionally from year to year for the same manor, and officials were 'fined' when they were not met.[5] The amount of seed sown per acre was related to the importance of the crop and to the quality of the land,[6] and records, even of yield ratios, were kept. Marling was frequently mentioned (circa 1095 is the earliest known date[7]) and chalk, sea sand and seaweed were applied when appropriate and available. We read of animal dung being collected and sold[8] and also being specially reserved when pasture was granted or

1 W. Beveridge, 'The yield and price of corn in the Middle Ages' in E. M. Carus-Wilson (ed.), *Essays in economic history*, I (London, 1954), 16.

2 R. V. Lennard, 'Statistics of corn yields in medieval England', *Econ. Hist.*, III (1934–7), 173–92, 325–49; M. K. Bennett, 'British wheat yield per acre for seven centuries', *Econ. Hist.*, III (1934–7), 19.

3 E. Stone, 'Profit-and-loss accountancy at Norwich cathedral', *Trans. Roy. Hist. Soc.*, 5th ser., XII (1962), 25–48; R. A. L. Smith, 'The central financial system of Christ Church, Canterbury, 1186–1512', *Eng. Hist. Rev.*, LV (1940), 353–69; and 'The *Regimen Scaccarii* in English monasteries', *Trans. Roy. Hist. Soc.*, 4th ser., XXIV (1942), 73–94.

4 D. Oschinsky, 'Medieval treatises on estate management', *Econ. Hist. Rev.*, 2nd ser., VIII (1955–6), 296–309.

5 J. S. Drew, 'Manorial accounts of St Swithun's priory, Winchester', *Eng. Hist. Rev.*, LXII (1947), 29ff.

6 A. E. Wilson (ed.) (1961), xxviii.

7 L. F. Salzman (ed.), *The chartulary of the priory of St Pancras of Lewes*, Sussex Record Soc., 38 (1932), 21.

8 C. G. O. Bridgeman, 'The Burton abbey twelfth-century surveys', *William Salt Collections for Staffordshire* (1916), 273; H. E. Butler (ed.), *The chronicle of Jocelin of Brakelond* (London, 1949), 103; F. W. Weaver (ed.), *A cartulary of Buckland priory*, Somerset Record Soc., 25 (1909), 149; R. H. C. Davis (ed.), *The Kalendar of Abbot*

leased.[1] Grain straw was either ploughed in or burned in the fields. Composting was practised; and on assarts it was apparently a sign of permanent ownership. Arable might even be irrigated in particularly dry years.[2] The significance of all this must not, however, be exaggerated. For most of the time and in most places, not enough mineral and organic matter was being returned to the soil to produce any rise in output. Even the efforts of the most progressive landlords of the period of 'high farming' in the thirteenth century, efforts that have attracted much attention,[3] do not appear to have involved any really considerable capital investment in stock, buildings and equipment.[4]

Lay lords may be said to have eaten their way round their estates. A great lord would not, however, regularly visit all his manors in demesne, and produce was sometimes sent up to chosen centres. This was very characteristic of monastic estates. A religious house would always find it necessary to purchase some commodities, but the bulk came from its estates. A manor or group of manors would be responsible for a week or two's supplies.[5] The canons of St Paul's, about 1300, received each year 5,760 bushels of wheat, the same quantity of oats, and 1,080 bushels of barley.[6] Apart from such renders (*firmae*), grain was also moved between manors to make up deficiencies of one sort or other – the consequence, perhaps, of specialisation within the framework of a great estate. In some cases, it was more convenient or profitable to sell the entire surplus of one

Samson of Bury St Edmunds and related documents, Camden Soc., 3rd ser., 84 (1954), 76, 137–8; E. L. Sabine, 'City cleaning in medieval London', *Speculum*, XII (1937), 24–5.

[1] W. T. Lancaster (ed.), *Abstract of charters of Bridlington priory* (London, 1912), 94.

[2] K. Ugawa, 'The economic development of some Devon manors in the thirteenth century', *Trans. Devon Assoc.*, XCIV (1962), 644.

[3] D. Knowles, *The religious orders in England*, I (Cambridge, 1956), 45–54, 314; R. A. L. Smith, 'The Benedictine contribution to medieval English agriculture' in *Collected Papers* (London, 1947), 103–16, and *Canterbury cathedral priory* (Cambridge, 1943), 133–8.

[4] R. H. Hilton, 'Rent and capital formation in feudal society', *Deuxième Conférence Internationale d'Histoire Economique, Aix-en-Provence, 1962* (Paris, 1965), 33–68; M. M. Postan, 'Investment in medieval agriculture', *Jour. Econ. Hist.*, XXVII (1967), 576–87.

[5] R. H. C. Davis (ed.), *The Kalendar of Abbot Samson of Bury St Edmunds and related documents*, Camden Soc., 3rd ser., 84 (1954), I.

[6] H. H. Hale (ed.), *The Domesday of St Paul's of 1222*, Camden Soc., 69 (London, 1858), xlviii.

manor and to bring in seed grain from outside which, in any case, accorded with what the agricultural writers advised.

Grain milling using water power was made more efficient by the introduction of the overshot wheel, probably in the late twelfth century.[1] The first mention of a windmill occurs in 1191 on the estates of the abbot of Bury St Edmunds.[2] It became common in the thirteenth century, particularly in eastern England. Hand mills were used by the peasants although they were usually expected to take their grain to the lord's mill.[3]

Pasture farming

Peasant livestock was of domestic rather than commercial importance; and peasant holdings, like much demesne, were usually understocked. M. M. Postan, using detailed tax assessments for parts of southern Wiltshire (1225) and the double Hundred of Blackbourne in Suffolk (1283), came to the following conclusions: (i) A bias towards sheep farming among some vills in both areas (one traditionally pastoral, the other arable) was more apparent than any broad contrast between the two. (ii) A disproportionate number of sheep were in the hands of a few men. (iii) In Wiltshire, a much higher proportion of downland was being used for grain, and a smaller proportion for sheep, than in the eighteenth century and perhaps before the thirteenth century. (iv) In both areas, cows were very evenly distributed and numbers were not necessarily smaller where there were comparatively large flocks of sheep. (v) The average number of all animals per taxpayer was low in relation to the needs of thirteenth-century husbandry – 15.6 sheep and 2.8 cows or calves in Wiltshire; 10.5 and 3.2 respectively in Suffolk.[4] Understocking appears to have been even more marked in parts of Bedfordshire,[5] and almost all enquiries show that very few peasants could put out a ploughing team of four oxen or horses, and

[1] M. Bloch, 'Avènement et conquêtes du moulin à eau', *Annales d'Histoire Economique et Sociale*, VII (1935), 539–40, 557–8; B. Gille, 'Les développements technologiques en Europe, de 1150 à 1400', *Cahiers d'Histoire Mondiale*, III (1956), 63–108.

[2] H. E. Butler (ed.), *The chronicle of Jocelin of Brakelond* (London, 1949), 59; J. Salmon, 'The windmill in English medieval art', *Jour. British Archaeol. Assoc.*, 3rd ser., VI (1941), 88–9; L. Delisle, 'On the origin of windmills in Normandy and England', *Jour. British Archaeol. Assoc.*, VI (1851), 403–6.

[3] T. Stapleton (ed.), *Chronicon Petroburgense*, Camden Soc., 47 (1849), 66–7.

[4] M. M. Postan, 'Village livestock in the thirteenth century', *Econ. Hist. Rev.*, 2nd ser., XV (1962), 219–49.

[5] A. T. Gaydon (ed.), *The taxation of 1297 for Barford, Biggleswade and Flitt hundreds*, Beds. Hist. Record Soc., 39 (1959), 107–8.

that they were forced to combine with others or make do with fewer beasts.[1] The chief problem, over much of eastern, central and southern England, was shortage of pasture, and this was felt increasingly as village 'waste' and 'pasture' retreated before the plough.[2]

Demesne livestock comprised beasts of burden and traction, and animals kept primarily for their milk, hides, wool and flesh. The ox was the traditional field animal and it remained ubiquitous.[3] The nailed horseshoe had appeared in western Europe about the beginning of the tenth century, and the solid horse collar and modern tandem harness just a little later. Together they made the horse an economic as well as a military asset,[4] and in England the horse appears to have graduated from the harrow to the plough about the close of the twelfth century.[5] Thereafter, we hear of horses used alone, but more commonly in mixed teams, as Walter of Henley recommended.[6]

As demesnes contracted during the twelfth century, the numbers of livestock they carried also decreased. But this is not to say that the total number of animals, within the country or even within the limits of a particular estate, declined. In fact, the grand total was probably rising, while being differently distributed between lords, tenants and 'farmers', and, to some extent, between different kinds of lords. The houses of the new religious Orders, for example, were building up their flocks and herds at this time. During the thirteenth century, the numbers of demesne livestock increased with the trend towards 'high farming'.[7] Reproduction rates were low by modern standards and disease took a periodic toll, but there is little evidence of heavy autumn killings.[8] With what hay was available, supplemented occasionally by legumes and evergreens,

---

[1] H. G. Richardson, 'The medieval plough team', *History*, n.s., XXVI (1941–2), 287–96.

[2] See, for example, D. M. Stenton (ed.), *The earliest Lincolnshire assize rolls*, Lincoln Record Soc., 22 (1926), 10; W. Farrer (ed.), *Cartulary of Cockersand abbey*, Chetham Soc., n.s., 38 (1898), 85.

[3] J. H. Moore, 'The ox in the Middle Ages', *Agric. Hist.*, XXXV (1961), 90–3.

[4] G. Duby, 'La révolution agricole médiévale', *Revue de Géographie de Lyon*, XXIX (1954), 362; L. des Noettes, *L'Attelage et le cheval de selle à travers les âges* (Paris, 1931), 122, 237.

[5] B. A. Lees (ed.), *Records of the Templars in England in the twelfth century* (British Academy, London, 1935), lxxxii.

[6] E. Lamond (ed.) (1890), 11.

[7] See, for example, J. A. Raftis (1957), 117.

[8] F. M. Page, 'Bidentes hoylandie', *Econ. Hist.*, I (1929), 609.

R. A. DONKIN

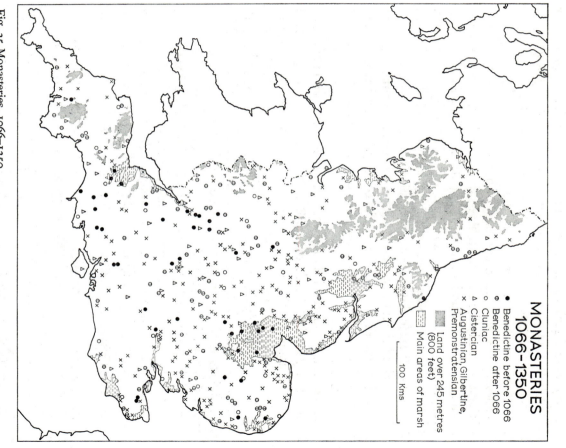

MONASTERIES
1066-1350

- Benedictine before 1066
- ⊖ Benedictine after 1066
- ○ Cluniac
- △ Cistercian
- × Augustinian, Gilbertine,
  Premonstratensian
- ▨ Land over 245 metres
  (800 feet)
- ▦ Main areas of marsh

100 Kms

Fig. 25 Monasteries, 1066–1350
Based on D. Knowles and R. N. Hadcock, *Medieval religious houses: England
and Wales* (London, 1953).

the great majority of animals were brought safely through the winter.[1]

Sheep were kept chiefly for their wool, and some of the larger land-owners grazed well over 10,000 in the late thirteenth and early fourteenth centuries. The bishop of Winchester had 15,000 or more as early as 1208–9, and nearly double this number by 1259.[2] Only Fountains among the Cistercian houses ever possessed anything like 15,000 sheep. The Cistercians (Fig. 25) were probably more deeply committed to wool growing, and certainly to supplying the overseas trade, than any of the other monastic Orders, but they by no means overshadowed all contemporary producers. At best, they supplied only 3 to 4% of all the wool exported at the close of the thirteenth century. Much of the overseas demand, not to speak of the continuing home market, must have been met by the small and middle-order producers, ranging from the prosperous peasant with a hundred or so sheep to the lord of several manors with two or three thousand. Descriptions of grants of pasture strongly suggest that a high proportion of these animals were fed on arable stubble, village commons, and enclosed blocks of lowland demesne pasture. The Cistercians grazed the high Pennines, but their arable granges also played an important part in sheep rearing.

The attention accorded sheep in connection with the overseas wool trade has tended to mask the importance of cattle as milk and hide producers. Cattle thrived on the rich sward of freshwater marsh; and in areas like Romney Marsh, the Isle of Thanet and the Fens, the *vaccaria* or cattle farm was an important feature of demesne farming. Cattle were also prominent in certain areas of woodland, scrub and rough pasture. In private chases and the royal forests they were considered less harmful than either sheep or goats. An unusually interesting development was the siting of vaccaries within and around the Pennines.[3] So far as we know, this commenced late in the twelfth century when there were 15 cattle stations in Wyresdale, and several in Nidderdale (belonging to Fountains) and Wensleydale (belonging to Jervaulx). About the same time, another

[1] W. Farrer (ed.), *Early Yorkshire charters*, III (Edinburgh, 1916), 205; J. Radley, 'Holly as winter feed', *Agric. Hist. Rev.*, IX (1961), 89–92; J. Tait (ed.), *The chartulary or register of the abbey of St Werburgh, Chester*, Chetham Soc., n.s., 82 (1923), 422; W. Farrer (ed.), *Lancashire inquests, extents and feudal aids, 1205–1307*, Record Soc. Lancs. and Cheshire, 48 (1903), 285.

[2] E. Power, *The wool trade in English medieval history* (London, 1941), 34.

[3] R. A. Donkin, 'Cattle on the estates of medieval Cistercian monasteries in England and Wales', *Econ. Hist. Rev.*, 2nd ser., XV (1962), 31–53.

Cistercian house, Stanlaw, began to use the south-eastern corner of the forest of Rossendale for cattle rearing, and by 1295–6 the valleys of Blackburnshire, which included Rossendale, were dotted with 28 vaccaries handling nearly 2,500 head of cattle belonging to the earl of Lancaster alone.[1] In the early fourteenth century, the Pennine chases had a characteristically 'pioneer' economy, consisting of lead and iron working, stone quarrying, and cattle and horse rearing.

Among the smaller animals, the hare was native to England, and there is a Domesday reference to 'warren for hares' at Gelston in Lincolnshire. The rabbit, on the other hand, was a newcomer. It seems to have been introduced from France or possibly Spain in the twelfth century.[2] The consequences of its introduction for stretches of the English countryside were important. Rabbits were valued for their meat and skins, and in the thirteenth century we begin to hear of grants of 'free warren' from the king. That they were also destructive is seen, for example, in the complaint that 100 acres of arable at Ovingdean, in Sussex, were 'lying annihilated' as a result of their activities, and again that in West Wittering wheat had been devoured 'year after year by the rabbits of the bishop of Chichester'.[3]

The expansion of improved land

Clearing the wood and waste. Around the arable nucleus of the medieval township there normally lay clumps of high woodland and blocks of unimproved pasture and scrub (known collectively as 'waste'). These provided such natural products as firewood and turbary and common grazing for cattle, sheep and pigs. But throughout much of the twelfth and thirteenth centuries, and in virtually every part of the country, woodland and 'waste' were also being cut or reclaimed to provide, in particular, more arable. It has been estimated, for example, that about 1,000 acres were cleared in the manor of Witney (Oxfordshire) in the first half of the thirteenth century, and a further 660 acres and 680 acres respectively between 1256 and 1306, when for all practical purposes reclamation may be said to have ended on both manors'.[4] In Laughton (Sussex), 975 acres were added to

[1] P. A. Lyons (ed.), *Two compoti of the Lancashire and Cheshire manors of Henry de Lacy, Earl of Lincoln, 24 and 33 Edward I*, Chetham Soc, 112 (1884), 129–4; R. B. Smith, *Blackburnshire* (Leicester, 1961), 8.

[2] E. M. Veale, 'The rabbit in England', *Agric. Hist. Rev.*, v (1957), 85–90.

[3] W. H. Blaauw, 'Remarks on the *Nonae* of 1340, as relating to Sussex', *Sussex Archaeol. Coll.*, I (1848), 62.

[4] J. Z. Titow (1962), 8.

the cultivated area between 1216 and 1325.[1] As the margins of possible cultivation were reached, in these and other places, township boundaries and common rights became more closely defined.

Not all inter-village waste was freely available to the peasantry. Lords of manors had rights there too, and perhaps the clearest witness to the pressure on unimproved land, and to the land shortage generally, was the Statute of Merton (1236) which permitted lords to enclose waste on condition that adequate amounts were left to their freeholders. Monasteries, especially those of the new Orders, were also given extensive grazing rights, even in places where they possessed no arable. And, finally, covering large parts of the country, there were the royal forests and private parks within which grazing and asserting were controlled in the interests of the chase. The royal forests were probably most extensive at the end of Henry II's reign (1189). Only Norfolk, Suffolk and Kent – areas of generally high population density and considerable numbers of freemen – were then without afforested land. During the thirteenth century, the larger tracts tended to disintegrate through piecemeal disafforestations and the granting of special franchises, and the overall area of forest was reduced by approximately one-third between 1250 and 1325.[2]

Even apart from the village lands that lay within their bounds, forests and chases (which included open moor and marsh as well as woodland) were by no means left unexploited.[3] The monasteries greatly benefited by grazing privileges, and by grants of land, timber, underwood and stone, or simply 'necessities'.[4] Eight of the ten royal foundations of the Cistercian Order were sited within or very close to forest bounds.[5] Permission

[1] J. S. Moore, *Laughton: a study in the evolution of the Wealden Landscape* (Leicester, 1965), 41–2. See also J. A. Raftis (1957), 71–4.

[2] The following contain useful maps: M. L. Bazeley, 'The extent of the English forest in the thirteenth century', *Trans. Roy. Hist. Soc.*, 4th ser, IV (1921), 140–72; N. Neilson, 'The forests' in J. F. Willard and W. A. Morris (eds.), *The English government at work, 1327–36*, I (Cambridge, Mass., 1940), 394–448; J. C. Holt, *The Northerners* (Oxford, 1961).

[3] J. R. Birrell, 'The forest economy of the honour of Tutbury in the fourteenth and fifteenth centuries', *Univ. Birmingham Hist. Jour.*, viii (1962), 114–34, and 'Peasant craftsmen in the medieval forest', *Agric. Hist. Rev.*, XVII (1969), 91–107. The most elaborate study of a particular forest is C. E. Hart, *Royal forest: a history of Dean's woods as producers of timber* (Oxford, 1966).

[4] H. A. Cronne, 'The royal forest in the reign of Henry I' in H. A. Cronne (ed.), *Essays in British and Irish history in honour of J. E. Todd* (London, 1949), 10.

[5] R. A. Donkin, 'The Cistercian settlement and the English royal forests', *Citeaux: Commentarii Cistercienses*, XI (1960), 39–55, 117–32.

to 'assart and cultivate' was given to these and other houses, and also to many individuals; but, judging by the fines imposed for assarting, the amount of land actually cleared far outstripped what was permitted in advance. Where forest rolls of the late twelfth and early thirteenth centuries survive, 'they present an impressive picture of agrarian expansion at the expense of the king's rights'.[1] Yet it is probably true to say that the Crown, in its usual impecunious state, was not unwilling that assarting should continue, provided that privileges be obtained, provided that they were defined and supervised and, of course, duly paid for.

Open fields were often enlarged during the twelfth and thirteenth centuries. Shares in assarts and assarts held 'in common' are occasionally mentioned. More frequent are references to assarts 'in the fields' or to furlongs bearing 'clearing' names. The addition of new furlongs, in which not necessarily the whole village had an interest, appears to have been the usual way in which fields were extended at this time.[2] This meant taking in some 10 to 30 acres at a time to form extra cropping units. The spread of three-field arrangements during the thirteenth century, permitting more land to be cultivated for the same amount of labour as before, and the more flexible grouping of furlongs at this and later times, may have encouraged joint assarting.

Areas that were once heavily wooded (the Weald, Wychwood, Feckenham, Arden and the like) or where there was much exploitable 'waste' (for example, parts of Cannock and of the Pennine and Cheviot foothills) often contain an unusually large number of hamlets and villages with name elements suggesting secondary settlement or 'clearing' (-stoc, -rydding, -leah, -feld, amongst others), and a fair proportion of these probably date from the twelfth or early thirteenth century.[3] In some parts of the country, for example in Suffolk, Essex and north Warwickshire, the presence of 'moated settlements' also appears to be associated with the colonisation of woodland.[4]

There is not much evidence of lords deliberately introducing settlers,

[1] J. C. Holt, 160; G. Wrottesley, *The pleas of the forest, Staffordshire*, William Salt Collections for Staffordshire, 5 (1884), 137ff.

[2] T. A. M. Bishop, 'Assarting and the growth of open fields', *Econ. Hist. Rev*, VI (1935), 17.

[3] B. C. Redwood and A. E. Wildon (eds.), *Custumals of the Sussex manors of the archbishop of Canterbury*, Sussex Record Soc., 57 (1958), xxxi–xxxii; B. F. Brandon, 'Medieval clearances in the east Sussex Weald', *Trans. and Papers, Inst. Brit. Geog*, XLVIII (1969), 136; J. B. Harley (1964), 122–5; B. K. Roberts, 101–13.

[4] F. V. Emery, 'Moated settlements in England', *Geography*, XLVII (1962), 385.

apart from monastic communities, to develop their estates, but some hamlets that appear to have been entirely informal growths may well have been encouraged, just as asserting by individuals was sometimes encouraged, by low initial rents.[1] About 1212–13, Robert Arsic settled a group of freemen on his demesne at Cogges in Oxfordshire, and each man was required to erect a house on his new holding.[2] The monks of La Charité (Caen) had *hospites* on their lands in Northamptonshire in 1107.[3] Temple Bruer, Lincolnshire, seems to have been promoted by the Templars about 1165.[4] In a late twelfth-century confirmation of land near Kniveton, Derbyshire, to one Sewall de Mungei we read, 'if Sewall wishes to settle (*erburgare*) this land, the men who dwell there shall have free common in wood and fields'.[5] Early in the following century (*circa* 1215), the earl of Chester permitted his barons to 'settle strangers' on their lands.[6] Some Cistercian houses also endeavoured to foster existing settlement by introducing new men.[7]

Lay and ecclesiastical lords cleared fresh land partly in order to expand or to consolidate their demesnes, and most of the large blocks of land that appear to have been cleared as a whole and kept in severalty turn out to be additions to the demesne. Lords organised asserting in several ways: (i) directly, by using permanent estate workers, hired labour or, in the case of the monasteries, lay brothers; (ii) by leasing cultivable land to peasants for a single lifetime;[8] (iii) by permitting peasants, villeins as well as freemen, to assart on condition of receiving back a proportion, usually one-third, of the improved land.[9] Land granted for the purpose of clearing

[1] P. A. Lyons (ed.), *Two compoti of the Lancashire and Cheshire manors of Henry de Lacy, Earl of Lincoln, 24 and 33 Edward I*, Chetham Soc., 112 (1884), xxi, 150. See also N. Neilson (1942), 61.   [2] L. C. Loyd and D. M. Stenton (eds.) (1950), 76–8.
[3] C. Johnson and H. A. Cronne (eds.), *Regesta Henrici Primi* (Oxford, 1956), 70.
[4] B. A. Lees (ed.), *Records of the Templars in England in the twelfth century* (British Academy, London, 1935), cxxxviii, clxxxii.
[5] F. M. Stenton, 'The free peasantry of the northern Danelaw', *Bulletin de la Société Royale des Lettres de Lund*, XXVI (1926), 164.
[6] J. Tait (ed.), *The chartulary or register of the abbey of St Werburgh, Chester*, Chetham Soc., n.s., 79 (1920), 103. See also J. E. A. Joliffe, 'Northumbrian institutions', *Eng. Hist. Rev.*, XLI (1926), 13.
[7] R. A. Donkin, 'Settlement and depopulation on Cistercian estates during the twelfth and thirteenth centuries', *Bull. Inst. Hist. Research*, XXXIII (1960), 141–65.
[8] R. Holmes (ed.), *The chartulary of St John of Pontefract*, Yorks. Archaeol. Soc., Record Series, 30 (1902), 409.
[9] N. Neilson (ed.) (1920), lxxxvi; R. P. Littledale (ed.), *Pudsay deeds*, Yorks. Archaeol. Soc., Record Series, 56 (1916), 168.

often had to be cultivated within a specified time and/or be made available for common grazing after cropping.[1]

An important example of clearing by direct methods is provided by the work of several Cistercian houses in the Vale of York about 1140. Here, devastated village lands, in some cases 1,000 acres or more in extent, were reclaimed and converted into large arable granges.[2] Much piecemeal assarting was also undertaken to consolidate holdings;[3] and scores of charters, in referring to 'old' and 'new' assarts, to assarts 'next to each other' or 'next to the moor', testify to the progress of reclamation. Furthermore, a good deal of the land granted to the Cistercian monks had only recently been cleared by peasant families.

Assarting and improvement chiefly benefited the older Benedictine foundations through increases in rent, whether or not the initiative had been theirs or the peasants. But a new house had to assemble an adequate demesne. The Benedictine community at Battle in Sussex, founded in 1066, vigorously developed its *banlieu*, to the extent of about 550 acres of fresh arable within a generation, of which just under half may have been recovered from peasants who had cleared and first cultivated the land.[4] The Church generally encouraged the extension of cultivation for, apart from what it gained directly, the amount of tithe was thereby increased. In the closing years of the twelfth century, Abbot Samson of the Benedictine house of Bury St Edmunds, 'cleared many lands and brought them into cultivation', so the abbey chronicle tells us.[5]

Peasant assarts tended to be small, a few acres at most. They could however be accumulated, and this helps to explain the occasional references to large peasant holdings.[6] In areas of late colonisation, assarts, hedged and dyked and worked in severalty, were frequently more im-

---

1 A. Saltman (ed.), *The chartulary of Tutbury priory*, Hist. MSS. Commission (London, 1962), 133, 194; E. Toms (ed.), *Chertsey abbey court rolls abstract*, Surrey Record Soc., 21 (1954), 72.

2 R. A. Donkin, 'The Cistercian grange in England in the twelfth and thirteenth centuries, with special reference to Yorkshire', *Studia Monastica*, VI (1964), 95–144; and 'The Cistercian Order and the settlement of the north of England', *Geog. Rev.*, LIX (1969), 403–16.

3 R. A. Donkin, 'The English Cistercians and assarting, c. 1128–1350', *Analecta Sacri Ordinis Cisterciensis*, XX (1964), 49–75.

4 E. Searle, 'Hides, virgates and tenant settlement at Battle abbey', *Econ. Hist. Rev.*, 2nd ser., XVI (1963), 290–300.

5 H. E. Butler (ed.), *The chronicle of Jocelin of Brakelond* (London, 1949), 27.

6 See, for example, J. F. Nichols, 'The extent of Lawling, A.D. 1310', *Trans. Essex Archaeol. Soc.*, XX (1931), 184, 187; B. K. Roberts, 112.

portant than intermixed strips, and elsewhere they formed a normal appurtenance to open-field holdings. Around the margins of many older settlements, and where peasant communities were still spreading into areas of woodland and waste, the landscape can hardly have appeared open. There is perhaps a general tendency to underestimate the amount of enclosed land in the England of 1300.

Most references to assarts do not mention the kind of land cleared. When they do, it is sometimes 'heath' or 'waste' or 'pasture' or what had earlier been abandoned – in other words, land that was probably 'marginal not only in location but also in quality'.[1] Nevertheless, the most common description is 'woodland'. A good stand of timber would generally indicate superior soils, although not all assarts were cultivated, and there was also the wood itself to be used or burned. But the most important reason why woodland was more often mentioned is to be found in the fact that regulations about its clearing, and above all about the clearing of oak,[2] were comparatively strict. Before 1300, in areas thinly wooded or closely settled, some concern was being expressed for supplies of constructional timber. There are one or two references to planned cutting, as at Charlbury in Oxfordshire where the woods were divided into seven parts, one to be cleared each year.[3] In places, even hedgerows, a ready source of firewood had to be protected.[4] Fortunately, in this respect, the tide of medieval colonisation was about to turn.[5] 'A progressive fall in the return for capital and labour expended' on clearing and reclamation may have been widely experienced.[6] Lord and peasant alike had sometimes been over-ambitious (or too hard pressed), and soon after 1300, parcels of poorer land, of the demesne at least, appear to have dropped out of cultivation altogether.[7]

[1] M. M. Postan (1966), 551.   [2] L. C. Loyd and D. M. Stenton (eds.) (1950), 322.
[3] H. E. Salter (ed.), Eynsham Cartulary, Oxford Hist. Soc., 51 (1908), xxxvii. See also P. L. Hall (ed.), The cartulary of St Michael's Mount, Cornwall, Devon and Cornwall Record Soc., 5 (1962), xxii; E. W. Crawley-Boevey (ed.), Chartulary of Flaxley abbey (Exeter, 1887), 32; C. W. Foster (ed.), Final concords of the county of Lincoln, II, Lincoln Record Soc., 17 (1920), 46; Calendar Close Rolls, 1234–37 (London, 1908), 416.
[4] H. H. Hale (ed.), cxxvi.
[5] A. R. H. Lewis, 'The closing of the medieval frontier, 1250–1350', Speculum, xxxiii (1958), 475–83.
[6] E. Miller, 'The English economy in the thirteenth century: implications of recent research', Past and Present, xxviii (1964), 31.
[7] A. R. H. Baker, 'Evidence in the Nonarum Inquisitiones of contracting arable lands during the early fourteenth century', Econ. Hist. Rev., 2nd ser., xix (1966), 518–32; E. Miller (1951), 100; M. M. Postan (1966), 558–9; B. F. Harvey, 23–42.

Draining the marsh. About 1135, the monks of Tutbury priory were allowed to make drainage ditches in the moor of Uttoxeter, 'and to take branches off the willows and osiers overhanging the water for the improving of marshy fields'.[1] Work of this kind was going on all over the country in the twelfth and thirteenth centuries, sometimes more than doubling the value of small pieces of land. But the main scenes of reclamation were the great stretches of marsh in eastern and southern England: Holderness and the Humber Levels, the Fenland, the Somerset Levels, and certain coastal marshes, in particular Romney Marsh and Pevensey Levels. Altogether much land was improved and protected at great expense, and the chief driving force was the buoyant market for agricultural produce during the thirteenth century. Reclaimed marsh was either rented out on a contractual basis or kept in demesne. Common-field arrangements were very rarely introduced, but a special form of co-operation between tenants and landowners, combined with specific responsibility for the upkeep of drains and walls, was everywhere essential.

The improvement of Pevensey Levels proceeded piecemeal from the late twelfth century. But after 150 years the grazing of sheep on salt marsh and the production of salt in coastal pans were still typical features of the area.[2] New Romney Marsh, to the south-west of the Rhee Wall, and adjacent parts of Walland Marsh, were reclaimed in a series of 'innings' in the late twelfth and thirteenth centuries.[3] To protect the *terra conquesta*, walls of packed clay, stiffened with timber and hurdles and faced with stone and turf, were constructed. Already in the twelfth century, a body of marsh custom, the *consuetudo marisci*, concerned with responsibilities for water control and general maintenance, was in being.[4] In the late thirteenth century special officers administered this code, which was confirmed by the Crown in 1252 and used as a model for other areas, such as Holderness.[5] Before royal interest manifested itself in the middle of the thirteenth century, the work of organising reclamation fell largely on a group of ecclesiastics headed by the archbishop of Canterbury, and

---

[1] A. Saltman (ed.) (1962), 71.
[2] L. F. Salzmann, 'The inning of Pevensey Levels', *Sussex Archaeol. Coll.*, LIII (1910), 32–60.
[3] N. Harvey, 'The inning and the winning of the Romney Marshes', *Agriculture*, LXII (1955), 334–8; R. A. L. Smith, 'Marsh embankment and sea defence in medieval Kent', *Econ. Hist. Rev.*, X (1940), 29–37.
[4] M. T. Derville, *The Level and Liberty of Romney Marsh* (London, 1936), 6ff.
[5] S. G. E. Lythe, 'The organization of drainage and embankment in medieval Holderness', *Yorkshire Archaeol. Jour.*, XXXIV (1939), 282.

including the abbot of St Augustine's, Canterbury, the abbots of Roberts-bridge and Lesnes, and the prior of Bilsington. In all about 23,000 acres of New Romney and Walland Marsh were recovered by the end of the thirteenth century.[1]

Churchmen were also prominent improvers elsewhere. Meaux abbey was responsible for a whole network of drains in Holderness, some to make cultivation possible, but mostly to improve pasture and meadow.[2] A large part of the Somerset Levels was controlled by the cathedral church of Wells and the abbeys of Glastonbury, Muchelney and Athelney, and here again effort was mainly directed towards improving pasture. Agreements giving exclusive rights within blocks of land to particular villages were often a preliminary to systematic improvement by burning and by cutting drains. At the same time, the natural resources of the fen might be increased by the planting of osiers, alder and willow.[3]

In the Fenland itself, where perhaps as much was achieved as in all the other areas put together, the work of winning new land proceeded from the closely settled siltlands and from points within and around the margins of the peat fen. Seaward reclamation in the Lincolnshire Fenland commenced before the Norman Conquest, but the earliest recorded intake belongs to the period 1090–1110.[4] Work continued throughout the twelfth and thirteenth centuries – notwithstanding frequent extensive flooding – new land being divided into strips and held in severalty. Sea banks were co-operative ventures, though not necessarily the work of entire villages. The successive banks built out into the fen were, however, sometimes the work of several villages. H. E. Hallam in his study of the wapentake of Elloe has stressed the role of the peasantry. The local monasteries – Spalding, Crowland, Swineshead – were 'not alone in taking the initiative. They were regarded, at most, as very senior partners in the adventure.'[5] New land went largely into rent-paying tenancies

[1] N. Neilson (ed.), *The cartulary and terrier of the priory of Bilsington, Kent* (British Academy, London, 1928), 41.

[2] R. A. Donkin, 'The marshland holdings of the English Cistercians before c. 1350', *Citeaux in de Nederlanden*, IX (1958), 262–75.

[3] P. J. Helm, 'The Somerset Levels in the Middle Ages', *Jour. British Archaeol. Assoc*, 3rd ser., XII (1959), 39, 49; M. Williams, *The draining of the Somerset Levels* (Cambridge, 1970), 25–81.

[4] H. E. Hallam, *The new lands of Elloe* (Leicester, 1954), 18. See also H. E. Hallam (1965).

[5] H. E. Hallam (1954), 34. See also B. Lyon, 'Medieval real estate development and freedom', *American Hist. Rev*., LXIII (1957–8), 47–61.

rather than demesne. Where the opportunities for reclamation were greatest, there partible inheritance worked best, and population was encouraged to remain. In the fens of Elloe wapentake, approximately 50 square miles were won between 1170 and 1240, and more than double this amount in the Lincolnshire Fenland as a whole. Nevertheless, population increased at such a rate that in Elloe about 1300 there were only 1 to 1¼ acres of enclosed arable, pasture and meadow per head, after making due allowance for tithe payments and land in demesne and fallow.[1] Clearly many peasants must have had to rely heavily on open grazing and the natural products of the fen – on fish and dairy products, turbary and reeds and salt.[2] The technical resources of the time were limited, and great stretches of peat lay unreclaimed in the southern Fenland, in the Somerset Levels and elsewhere.

INDUSTRY

England was clearly a more industrial country in the fourteenth century than it had been in the eleventh. Most villages and all towns included some craftsmen. The manufacture of pottery and the tanning of leather were widespread and growing activities; so too was metal working and contemporary surnames suggest considerable specialisation.[3] The production of pewter was an essentially English craft, for which detailed regulations were drawn up in 1348. Bell-founding developed locally but served distant markets, and it led to the casting of cannon in the fourteenth century.[4] Shipbuilding also was making progress. In about 1300 a sea-going vessel of 200 tons was exceptional, but by the second half of the fourteenth century the average size was 250 to 300 tons.[5] Other developing industries included glass-making, and we hear of substantial quantities being produced at Chiddingford and neighbouring villages in Surrey and Sussex in the thirteenth and fourteenth centuries.[6]

[1] H. E. Hallam (1961), 78.

[2] D. J. Siddle, 'The rural economy of medieval Holderness', *Agric. Hist. Rev.*, XV (1967), 40–5.

[3] See G. Fransson, *Middle English surnames of occupation, 1100–1350* (Lund, 1935).

[4] L. F. Salzman, *English industries of the Middle Ages* (Oxford, 1923), 151, 156.

[5] K. M. E. Murray, 'Shipping' in A. L. Poole (ed.), *Medieval England*, I (Oxford, 1958), 185.

[6] L. F. Salzman (1923), 183. See also *Calendar Patent Rolls, 1307–13* (London, 1894), 129.

Salt, a vital commodity in every household, was obtained either from sea water or from brine springs. The chief producing areas of maritime salt seem to have been in Lincolnshire[1] and Norfolk, but salterns (*salinae*) were widely distributed along the south and east coasts.[2] Men from the Cinque Ports supplied the London market.[3] The inland centres were in Cheshire[4] and Worcestershire;[5] they were frequently mentioned in the twelfth century and almost certainly increased in relative importance later.

Among the variety of industrial activities, three deserve to be singled out for special attention – mining, building, and the making of cloth.

Mining

Iron mining and smelting were located mainly in the Forest of Dean until the second half of the thirteenth century (Fig. 26). From the forges of St Briavels, picks, shovels, axes and nails were dispatched to Ireland on the occasion of Henry II's expedition in 1172, and horseshoes and other ironware went with Richard I's crusade in 1189–91.[6] While Forest of Dean iron remained important long after this, other centres of production, notably the central and southern Pennines, Cleveland[7] and the Weald of Sussex,[8] gradually came to the fore. Nevertheless, the total production

[1] E. H. Rudkin and D. M. Owen, 'The medieval salt industry in the Lindsey marshland', *Rep. and Papers, Lincs. Archit. and Archaeol. Soc.*, n.s., VIII (1959–60), 76–84; H. E. Hallam, 'Salt making in the Lincolnshire Fenland during the Middle Ages', *ibid*, 85–112.

[2] See, for example, L. Fleming (ed.), *The chartulary of Boxgrove Priory*, Sussex Record Soc., 59 (1960), 57, 161; W. Farrer (ed.), *Early Yorkshire charters*, II (Edinburgh, 1915), 69, 96, 98.

[3] H. T. Riley (ed.), *Munimenta Gildhallae Londoniensis, I (Liber Albus)* (Rolls Series, London, 1859), lxxxv.

[4] J. Tait (ed.), *The chartulary or register of the abbey of St Werburgh, Chester*, Chetham Soc., n.s., 79 (1920), 215–16; H. J. Hewitt, *Medieval Cheshire* (Manchester, 1929), 108–21.

[5] R. R. Darlington (ed.), *The cartulary of Worcester Cathedral priory*, Pipe Roll Soc., n.s., 38 (1968), I, 42, 47, 83, 90, 225, 308; E. K. Berry, 'The borough of Droitwich and its salt industry, 1215–1700', *Univ. Birmingham Hist. Jour.*, VI (1957–8), 39–61.

[6] A. L. Poole, *From Domesday Book to Magna Carta* (Oxford, 1951), 82.

[7] B. Waites, 'Medieval iron working in northeast Yorkshire', *Geography*, XLIX (1964), 33–43.

[8] G. S. Sweeting, 'Wealden iron ore and the history of its industry', *Proc. Geol. Assoc.*, LV (1944), 2.

R. A. DONKIN

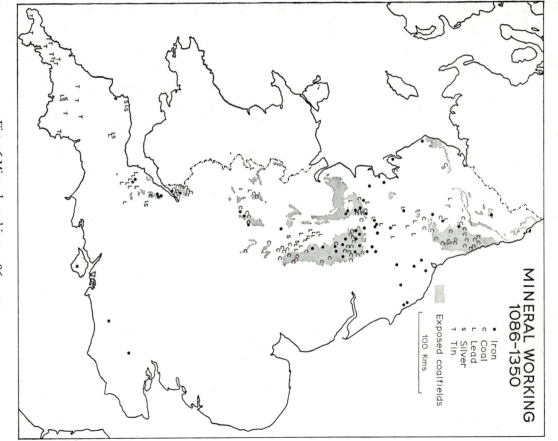

MINERAL WORKING
1086-1350

- Iron
- c Coal
- L Lead
- s Silver
- T Tin

 Exposed coalfields

100 Kms

Fig. 26 Mineral working, 1086–1350
Based on miscellaneous sources.

fell well below the country's needs, and in the late thirteenth century England was still importing iron from Spain[1] and Sweden.

Coal had been worked in Roman times, but it was not until the end of the twelfth century that it seems to have been re-discovered.[2] It is mentioned among the exports to Bruges in 1200.[3] The thirteenth century saw a great expansion of coalmining, and coal was used increasingly in forging, in burning lime, in evaporating brine and in brewing and baking. Almost all the English coalfields had some mining activity by the end of the century. The coal of southern Northumberland was most often referred to on account of its export from the Tyne, but Fig. 26 shows that the Pennine fields also were important.

Lead was much used for roofing as well as for piping. There were four main producing areas: (i) Alston Moor on the borders of Cumberland and Northumberland; (ii) Derbyshire, from which lead was carried to Boston and Lynn for shipment; (iii) the Mendips; and (iv) the district around Bere Ferrers in South Devon where, between 1290 and 1340, there was a silver-lead mining boom to which miners came from other lead centres, notably Derbyshire.[4] Lead was also mined to some extent at other places – at Combe Martin in North Devonshire, in Shropshire and in Flintshire in North Wales.

Tin was produced in the south-west peninsula. It was worked in west Cornwall in pre-historic and Roman times. Documentary evidence begins in 1156 when tin was worked in Devon on the south-western flank of Dartmoor. 'From then until the end of the century the rich alluvial deposits of South-West Devon produced nearly all the tin of Europe.'[5] By the early thirteenth century, however, the main centres of production were in Cornwall, where they remained. Tin working was organised into 'stannaries', each with its own warden or supervisor, five in Cornwall and four in Devon.[6]

Quarrying and building

The Norman Conquest was followed by an enormous amount of new building, from humble and unrecorded extensions to most ambitious work

[1] H. F. Salzmann in W. Page (ed.), *V.C.H., Sussex*, II (London, 1907), 242.
[2] W. Greenwell (ed.), *Boldon Buke*, Surtees Soc., 25 (1852), 5, 24.
[3] A. L. Poole, 81.
[4] W. G. Hoskins, *Devon* (London, 1954), 136.
[5] *Ibid*, 131.
[6] C. R. Lewis, *The Stannaries* (Cambridge, Mass., 1924).

on castles,[1] churches[2] and monasteries. All this represented an incalculably large investment of labour and wealth. The use of stone revolutionised military architecture, producing, first, the rectangular keep, and later, in the thirteenth century, the great curtain wall with half-round mural towers. English architecture generally was now fully exposed to powerful influences emanating from Normandy, the Paris Basin, and Burgundy. Much fine Anglo-Norman work still survives at Winchester, Norwich, Durham and Tewkesbury, for example, but even more was replaced during the thirteenth and fourteenth centuries by succeeding, and distinctively English, forms of Gothic. The records of the period abound in references to the right to take and to transport stone, often by water, and toll exemptions on building materials were freely given to religious communities. These included the Cluniacs, the Cistercians and the Canons Regular, who together were largely responsible for a more than tenfold increase in religious houses of men, excluding friaries, in the two and a half centuries after Domesday (Fig. 25). The building accounts of one Cistercian monastery for the period 1278–81 refer, in all, to more than 40,000 cartloads of stone.[3] Even in domestic building of a superior kind, stone was increasingly used for more than foundation work.[4]

The great majority of English parish churches were in existence by about 1200, but outlying chapels continued to be built long after this. Thus late in the thirteenth century, a certain Adam of Arden was allowed ', a private chapel in his manor house at Gayton [?-le-Marsh, Lincolnshire] since the road to his parish church was long and difficult to traverse, especially in winter'.[5] Churches and chapels, originally mainly of wood,

[1] R. A. Brown, 'Royal castle-building in England, 1154–1216', Eng. Hist. Rev. LXX (1955), 353–98; and 'A list of castles, 1154–1216', Ibid., LXXIV (1959), 249–80; A. J. Taylor, 'Military architecture' in A. L. Poole (ed.), Medieval England, I (Oxford, 1958), 98–127; H. M. Colvin (ed.), The history of the king's works, 2 vols. (London, 1963).

[2] C. R. Cheney, 'Church building in the Middle Ages', Bull. John Rylands Library, XXXIV (1957); H. T. Johnson, 'Cathedral building and the medieval economy', Explorations in Entrepreneurial History, IV (1967), 191–210.

[3] J. Brownbill (ed.), The ledger book of Vale Royal abbey, Record Soc. Lancs. and Cheshire, 68 (1914), 198–203.

[4] M. E. Wood, Thirteenth-century domestic architecture in England, Supplement to Archaeol. Jour., CV (1950); P. A. Faulkner, 'Domestic planning from the twelfth to the fourteenth century', Archaeol. Jour., CXV (1958), 150–83.

[5] R. M. T. Hill (ed.), The rolls and register of bishop Oliver Sutton, 1280–99, Lincoln Record Soc., 48 (1954), 47. See also M. Gibbs (ed.), Early charters of the cathedral church of St Paul, London, Camden Soc., 3rd ser., 58 (1939), 148; F. Bradshaw,

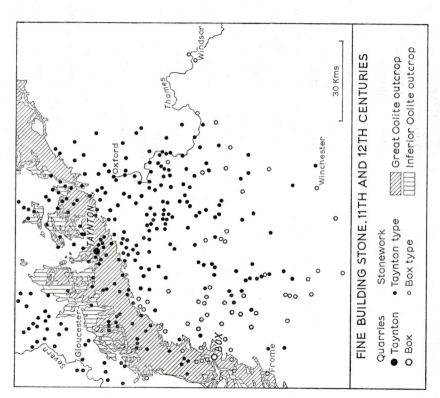

FINE BUILDING STONE, 11TH AND 12TH CENTURIES

Quarries   Stonework
● Taynton   • Taynton type
○ Box   ○ Box type

▨ Great Oolite outcrop
▥ Inferior Oolite outcrop

30 Kms

Fig. 27 Fine building stone from the Taynton–Box area in the eleventh and twelfth centuries

Based on the map by E. M. Jope in A. F. Martin and R. W. Steel (eds.), *The Oxford region* (Oxford, 1954), 114, and on additional information provided by Professor Jope. Navigable portions of the Thames and Severn are shown.

were generally rebuilt in stone between the twelfth and the fourteenth centuries, and in this work the influence of local schools of craftsmen can sometimes be detected.[1]

The quarrying of building stone to meet distant demand can be seen

'The lay subsidy roll of 1296: Northumberland at the end of the thirteenth century', *Archaeologia Aeliana*, 3rd ser., XIII (1916), 256.

[1] S. A. Jeavons, 'The pattern of ecclesiastical building in Staffordshire during the Norman period', *Trans. Lichfield and South Staffs. Archaeol. and Hist. Soc.*, IV (1962–3), 9–11.

in the distribution of surviving eleventh- and twelfth-century masonry containing Great Oolite stone (Fig. 27). Quarries in the Inferior Oolite at Barnack near Peterborough were worked by the abbeys of Peterborough, Ramsey, Sawtry and Crowland the Fen District, and by Bury St Edmunds in East Anglia.[1] Purbeck marble became fashionable towards the end of the twelfth century and remained so for the next 200 years.[2] It was used, for example, in the cathedrals of Chichester and Lincoln, and in St Paul's and Westminster abbey in London.[3] The Boldon Book, a survey of the see of Durham in 1183, refers to 'Lambert the marble cutter' (*marmorarius*) in the village of Stanhope.[4]

Cloth manufacture

In the twelfth century, the manufacture of cloth for sale, although undoubtedly widespread, centred upon a dozen or so towns mainly in eastern and southern England[5] (Fig. 28). Weavers' gilds in London, Oxford, Lincoln and Huntingdon were mentioned in the Pipe Roll for 1130,[6] and before the close of Henry II's reign in 1189 there were others at Nottingham, Winchester and York. We also hear of weavers and fullers at Beverley, Gloucester, Marlborough and Stamford, amongst other places. Much of their output was of high quality and part of it was exported, but England was still very largely an importer of cloth and probably remained so until the striking growth of the industry in the 1330s and 1340s.[7]

The statistical sources do not permit an adequate description of the changes that occurred during the thirteenth century. The demand for cloth was certainly rising, but so too were costs under the restrictive influence of the gilds; furthermore, the great Flemish centres of manu-

[1] L. F. Salzman (1923), 85; and *Building in England down to 1540* (Oxford, 1967), 119–39; C. Johnson and H. A. Cronne (eds.), *Regesta Regum Anglo-Normannorum*, II (Oxford, 1956), 189.

[2] D. Knoop and G. P. Jones, *The medieval mason* (Manchester, 1949), 11.

[3] L. F. Salzman (1923), 91–4.

[4] W. Greenwell (ed.), *Boldon Buke*, Surtees Soc., XXV (1852), 30.

[5] E. M. Carus-Wilson, 'The English cloth industry in the late twelfth and early thirteenth centuries', *Econ. Hist. Rev.*, XIV (1944–5), 32–50; E. Miller, 'The fortunes of the English textile industry in the thirteenth century', *Econ. Hist. Rev.*, 2nd ser., XVIII (1965), 64–82. For an early (1202) list of cloth towns, see D. M. Stenton (ed.), *Pipe Roll 4 John*, Pipe Roll Soc., n.s., XV (London, 1937), xx.

[6] J. Hunter (ed.), *Pipe Roll 31 Henry I* (Record Commission, London, 1929), 2, 48, 109, 194.

[7] E. M. Carus-Wilson, 'Trends in the export of English woollens in the fourteenth century', *Econ. Hist. Rev.*, 2nd ser., III (1950–1), 162–79.

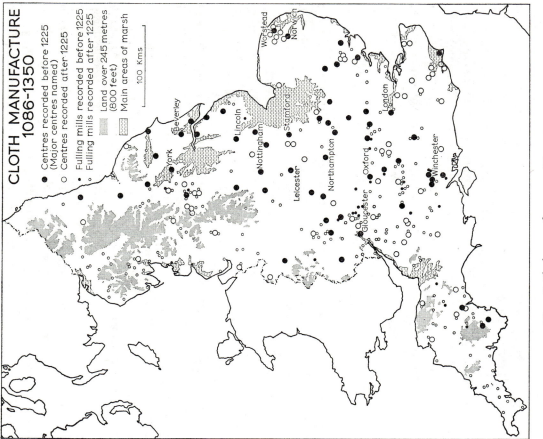

Fig. 28 Cloth manufacture, 1086–1350
Based on miscellaneous sources.

facture were almost certainly increasing their share of the English market. Home production probably failed to expand very much and may even have declined around the middle of the century. The most significant development was the gradual shift in the distribution of cloth-making away from the old-established towns and towards a much larger number of smaller places, many in fact mere villages. The gilds of textile workers in the older centres naturally tried to monopolise manufacture, but in the end they failed.[1] Cloth merchants began actively to encourage the rural industry with its lower costs and freer organisation, and the arrival in England of the water-driven fulling mill (Fig. 28) also contributed to the dispersal of the industry.[2] It was first recorded in 1185 on lands belonging to the Templars;[3] indeed most of the mills that date from the late twelfth and early thirteenth centuries were on royal or ecclesiastical estates, and it seems likely that the 'industrial revolution of the thirteenth century' was closely connected, in this respect, with the large, fully developed manor. Fulling mills were not unknown in the older cloth towns but, on grounds of siting, they could not so easily arrange for a sufficient head of water to employ the more efficient overshot wheel. Cloth produced in London was being fulled in the countryside in the middle of the thirteenth century.[4] The introduction of the fulling mill both strengthened the position of the existing rural industry and stimulated entirely new growth by attracting people prepared to engage in other processes of manufacture. The chief developing areas were the eastern Pennines, the Cotswolds, the middle Severn valley, the upper Kennet, and the south-west peninsula.

TRADE AND TRANSPORT

Merchants and trade associations

Not all town dwellers were burgesses. There were also people of very limited means, the stall holders, the *minuti homines*, who had sometimes quite recently moved in from the countryside; there were the Jews; and there were those who lived within the privileged sokes. The ranks of the burgesses were gradually augmented from below, and, in the larger towns particularly, a distinction developed between the lesser and the greater merchants. The important capitalist emerges as early as the twelfth

[1] See E. E. Hirshler, 'Medieval economic competition', *Jour. Econ. Hist.*, XIV (1954), 52–8.

[2] E. M. Carus-Wilson, 'An industrial revolution of the thirteenth century', *Econ. Hist. Rev.*, XI (1941), 39–60; R. A. Pelham, 'The distribution of early fulling mills in England and Wales', *Geography*, XXIX (1944), 52–6.

[3] B. A. Lees (ed.), *Records of the Templars in England in the twelfth century* (British Academy, London, 1935), 51, 127.

[4] R. R. Sharpe (ed.), *Calendar of letter books of the City of London*, C (London, 1901), 52–3; H. T. Riley (ed.) (1859), 127–9.

century,[1] and wealth derived from commerce was often invested in rural property, thus reinforcing the association between town and country.

A great part of the country's internal trade in the twelfth and thirteenth centuries was in the hands of Englishmen. On the other hand, control of the principal items of international trade – the export of wool and the import of wine – rested largely with the Italians, the Flemings and the Gascons. During the second half of the thirteenth century, the Italians also acted as financiers to the Crown and to monasteries and lay lords. The Jews performed a similar service, and many monasteries were at some time or other in their debt for large sums, the security offered being usually wool or land. Jewries were located almost exclusively in the larger towns. Names mentioned in twelfth-century records indicate important colonies at London, Lincoln, Norwich, Gloucester, Northampton and Winchester. To these should be added Canterbury, Worcester, Oxford and York, which figure prominently in the tallage lists of the thirteenth century.[2] The Jews facilitated the circulation of wealth and the transfer of land; but as usurers they were unpopular, and steps were taken as early as about 1230 to prevent their acquiring further urban property. They were expelled from England in 1290.[3]

The earliest specific reference to the Gild Merchant (*Gilda Mercatoria*) is dated about 1100.[4] Although the association continued to be sanctioned throughout the twelfth and thirteenth centuries, it was not found in the majority of boroughs. The principal advantages of membership were freedom of trade at all times and exemption from the borough tolls; and toll exemptions enjoyed by a borough elsewhere in the country might be confined to gildsmen. The Jews were virtually excluded; they had their counterpart, but it is mentioned only once in England, at Canterbury in 1266.[5] The term 'merchant' appears at first also to have included most, if not all, craftsmen, but weavers and fullers, unlike dyers, were not members in the thirteenth century; on the other hand they had their own craft gilds at this time, whereas the dyers apparently did not. The craft

[1] H. A. Jenkinson, 'Money lenders' bonds of the twelfth century', in H. W. C. Davis (ed.), *Essays in history presented to R. L. Poole* (Oxford, 1927), 190–210.

[2] H. G. Richardson, *The English Jewry under the Angevin kings* (London, 1960), 1–22.

[3] P. Elman, 'The economic causes of the expulsion of the Jews in 1290', *Econ. Hist. Rev.*, VII (1936–7), 145–54.

[4] C. Gross, *The Gild Merchant*, I (Oxford, 1890), 5.

[5] L. Rabinowitz, 'The medieval Jewish counterpart to the Gild Merchant', *Econ. Hist. Rev.*, VIII (1937–8), 180–5.

gilds,[1] first mentioned in the reign of Henry I (1100–35), came into prominence in the late thirteenth century when the Gild Merchant was already beginning to decline as an economic, if not as a social, force. Its influence and power passed, however, not so much to the gilds as to the borough courts and councils which were largely controlled by the same merchant families.

Internal trade

The chief commodities of internal trade were grain, fish and salt, wool and cloth, and metals. Large quantities of grain were moved across country to satisfy manorial needs and obligations.[2] That the trade in grain was also considerable can be inferred from the recurring references to its sale in manorial accounts.[3] Towards the close of the thirteenth century, grain sales commonly accounted for 20 to 30% of manorial profits. The trade was organised primarily on a local and regional basis, with cornmongers, first mentioned in the early thirteenth century, operating between producers and rural markets and the main centres of demand – the towns.[4] There was also some inter-regional movement, particularly coastwise.

Much of the internal trade of the country was carried on at markets and fairs. The right to establish either was a royal prerogative, and during the twelfth and thirteenth centuries approximately 2,500 market charters were granted or confirmed.[5] Every borough eventually had a market, but the larger number of charters were to rural communities. A market lasted a single day and usually took place once a week. The thirteenth-century

1 E. Power, 'English craft gilds of the Middle Ages', *History*, n.s., IV (1919–20), 211–14.

2 R. Lennard, 'Manorial traffic and agricultural trade in medieval England', *Proc. and Jour. Agric. Economics Soc.*, V (1938), 259–77.

3 See, for example, E. M. Halcrow, 'The decline of demesne farming on the estates of Durham cathedral priory', *Econ. Hist. Rev.*, 2nd ser., VII (1954–5), 356.

4 N. S. B. Gras, *The evolution of the English corn market* (Cambridge, Mass., 1926); E. Kneisel, 'The evolution of the English corn market', *Jour. Econ. Hist.*, XIV (1954), 46–52.

5 *Report on markets and fairs in England and Wales*, pt I, Min. of Agric., Econ. Ser., 13 (London, 1927), 7; O. S. Watkins, 'The medieval market and fair in England and Wales', *Y Cymmrodor*, XXV (1915), 21–74; J. L. Cate, 'The Church and market reform in England during the reign of Henry III' in J. L. Cate and E. N. Anderson (eds.), *Medieval historiographical essays in honor of J. W. Thompson* (Chicago, 1938), 27–65; R. A. Donkin, 'The markets and fairs of medieval Cistercian monasteries in England and Wales', *Cistercienser-Chronik*, LXIX (1962), 1–14.

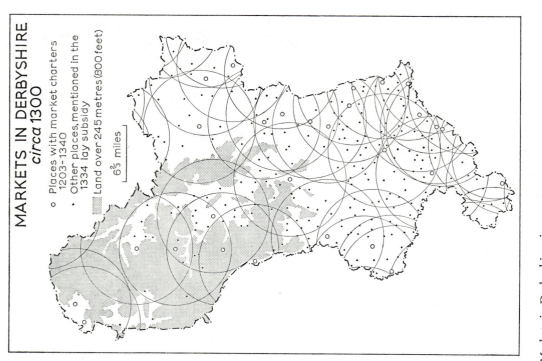

Fig. 29 Markets in Derbyshire *circa* 1300
Based on B. E. Coates, 'The origin and distribution of markets and fairs in medieval Derbyshire', *Derbyshire Archaeol. Jour.*, LXXXV (1965), 92–111.

lawyer Bracton held that markets should not be less than '6 miles and a half and a third part of a half' (i.e. $6\frac{2}{3}$) miles apart, but his argument is difficult to follow.[1] In fact, markets were often closer than six miles, as

[1] L. F. Salzman, 'The legal status of markets', *Cambridge Hist. Jour.*, II (1928), 210.

can be seen from the example of Derbyshire[1] (Fig. 29). Neighbouring markets on different days might be mutually beneficial by attracting merchants into an area; this was the argument put forward in 1252 to justify a Tuesday market at Wingham in Kent, close to Wednesday markets at Canterbury and Sandwich.[2] Nevertheless, there was frequently keen rivalry between neighbouring market centres. Edward III's charter to London in 1327 stated 'that no market from henceforth should be granted...to any within seven miles in circuit of the said city';[3] and a market at Lyme in Dorset was condemned because it was 'more than 5 miles but less than 6' from that of Bridport.[4]

Fairs served much wider areas.[5] The greatest fairs in the early thirteenth century were at Bristol, St Ives (Huntingdon), Winchester, Boston, Stamford and Stourbridge (Cambridge). As English trade expanded, fairs became increasingly important; they commonly lasted for a week, but some extended for a month or more. Booths and stalls were set up and each trade had its allotted place. Here were to be seen not only tin from Cornwall, salt from Worcestershire, lead from Derbyshire, iron from the Forest of Dean, and wool from many parts of the country, but silks, cloth and spices from abroad. Between 1270 and 1315, merchants from Cologne, Douai, Ypres, Ghent, Rouen, Bruges, St Omer, Caen, Dinant, Louvain and Malines frequented the fair at St Ives.[6] But with the fourteenth century came changes in fortune. In 1335, it was said that 'foreigners do not come to St Botolph's fair (at Boston) as they used to do'; and in 1363, we hear that the fair of St Ives had not been held 'for twenty years and more' because of the absence of foreign traders.[7] It is clear that fairs were declining in importance. As towns grew, they themselves became permanent marts. In the fourteenth century, the provincial

[1] B. E. Coates, 'The origin and distribution of markets and fairs in medieval Derbyshire', *Derbyshire Archaeol. Jour.*, LXXXV (1965), 92–111. See also R. H. Hilton, *A medieval society: the west Midlands at the end of the thirteenth century* (London, 1966), 172.

[2] L. F. Salzman, *English trade in the Middle Ages* (Oxford, 1931), 129; H. S. A. Fox, 'Going to town in thirteenth-century England', *Geog. Mag.*, XLII (1970), 666.

[3] W. de G. Birch (ed.), *The historical charters and constitutional documents of the City of London* (London, 1887), 58.

[4] L. F. Salzman (1928), 211.

[5] See, for example, A. H. Thomas (ed.), *Calendar of early mayors' court rolls, 1298–1307* (Cambridge, 1924), 154.

[6] E. Lipson, *The economic history of England*, 1 (London, 1929), 222.

[7] *Ibid.*, 234.

trade of London merchants 'was by no means confined to the great fairs; they were constantly trading all over the country'.[1]

Goods were moved by pack-horse and by cart.[2] The four-wheeled wagon (longa caretta) became common in the thirteenth century. Livestock were sometimes driven several hundred miles between estates or to meet the king's needs.[3] Although not much is heard of the making of roads[4] or even of their maintenance,[5] 'there is little evidence that men complained of bad roads and slow travel in the thirteenth century'.[6] This was partly because rivers were used to convey such bulky commodities as grain, wool, reeds,[7] timber and stone. Roads were also characteristically broad and diversions could easily be arranged. Suitable fording or bridging points, on the other hand, were few and far between, and it is not surprising that there was a considerable amount of bridge-building or re-building in stone from the late thirteenth century onwards. Some idea of the main arterial roads and how they focused on London may be obtained from the fourteenth-century Gough map (Fig. 39). As early as the twelfth century, there are references to 'the London Road' in various towns, for example the *Via Londiniensis* in Missenden (Buckinghamshire) and *Londenestret* in Gamlingay (Cambridgeshire).[8] It is also significant that the Gough map, compiled with practical ends in mind, showed a large number of rivers. The most important were those which converged on the Humber estuary and those of the Fenland together with the Thames and the Severn.

[1] S. Thrupp, 'The grocers of London, a study of distributive trade' in E. Power and M. M. Postan (eds.), *Studies in English trade in the fifteenth century* (London, 1933), 273.

[2] J. F. Willard, 'The use of carts in the fourteenth century', *History*, n.s., XVII (1932–3), 246–50.

[3] M. H. Mills (trans.) and R. S. Brown (ed.), *Cheshire in the Pipe Rolls, 1158–1301*, Record Soc. Lancs. and Cheshire, 92 (1938), 174.

[4] F. M. Stenton, 'The road system of medieval England', *Econ. Hist. Rev.*, VII (1936), 6.

[5] D. S. Bland, 'The maintenance of roads in medieval England', *Planning Outlook*, IV (1957), 5–15.

[6] D. M. Stenton, 'Communications' in A. L. Poole (ed.), I, 201.

[7] N. Neilson, *Economic conditions on the manors of Ramsey abbey* (Philadelphia, 1899), 80. See also M. W. Barley, 'Lincolnshire rivers in the Middle Ages', *Rep. and Papers, Lincs. Archaeol. and Architect. Soc*, n.s., I (1938), 1–22.

[8] F. M. Stenton *et al.*, *Norman London* (Hist. Assoc, London, 1934), 21.

Overseas trade

Imports included woad from Picardy,[1] cloth from Flanders, alum from the Mediterranean lands, iron from Spain and Sweden, and a wide range of luxury goods. In early Norman times, wine appears to have been imported mainly from the Seine Basin, Burgundy and the Rhineland, but after the acquisition of Gascony in 1152, this area provided an ever-increasing quantity.[2] By the early thirteenth century, the greater part of England's supplies came from Gascony. Viticulture in England declined as the trade with Gascony expanded; the latter involved around 20,000 tuns annually in the 1330s, by which time English merchants had captured the lion's share.[3]

Among exports, lead and tin,[4] hides and dairy products[5] were shipped in significant quantities. Newcastle was the leading port for hides; they were probably the basis of its prosperity at the close of the thirteenth century,[6] which suggests that the scattered references to cattle rearing in the Cheviots and along the slopes of the northern Pennines do not reflect its real importance. Grain also was exported. Even as early as 1198, heavy duties on grain sent to Flanders indicate 'an export trade of real magnitude', especially through East Anglian ports.[7] Lynn lay at the north-eastern end of what was probably the main area of surplus grain, and it was the most important shipping point in the early fourteenth century when port statistics first become available.

The principal item in England's foreign trade in the thirteenth century was wool. In 1273, export licences for nearly 33,000 sacks – perhaps

1 E. M. Carus-Wilson, 'La guède français en Angleterre: un grand commerce du Moyen Age', *Revue du Nord*, XXXV (1953), 89–105.
2 E. M. Carus-Wilson, 'The effects of the acquisition and of the loss of Gascony on the English wine trade', *Bull. Inst. Hist. Research*, XXI (1946–8), 145–54; M. K. James, 'The fluctuations of the Anglo-Gascon wine trade during the fourteenth century', *Econ. Hist. Rev.*, 2nd ser., IV (1951), 170–96.
3 M. K. James, 'The medieval wine dealer', *Explorations in Entrepreneurial History*, X (1957), 45–53; N. S. B. Gras, *The early English customs system* (Cambridge, Mass., 1918), 210. *Studies in the medieval wine trade* (ed. E. M. Veale), Oxford, 1971, brings together M. K. James's work on this subject.
4 D. M. Stenton (ed.), *Pipe Roll, 10 Richard I, 1198*, Pipe Roll Soc., n.s. (London, 1932), xxxii, 181–2.
5 H. T. Riley (ed.), *Memorials of London, 1276–1419* (London, 1868), xli.
6 F. Bradshaw, 273.
7 D. M. Stenton (ed.), *Pipe Roll, 10 Richard I, 1198*, Pipe Roll Soc., n.s., IX (London, 1932), xiv.

three-fifths of the country's total production – were issued.[1] Between Michaelmas 1304 and Michaelmas 1305 actual shipments reached a record total of over 46,000 sacks. Our knowledge of the organisation of the trade comes largely from the records of transactions between monasteries and other landowners on the one hand, and shippers, or their agents, on the other. The foreign merchants were mainly Flemish or northern French before about 1275, and thereafter Italian.[2] Buyers travelled about inspecting stocks, and producers were encouraged to sort their wool into three grades.[3] Forms of credit – deferred payments for goods sold or advances on future delivery – were used from the second half of the thirteenth century.[4] The Italians also pioneered the use of 'bills of exchange', and 'assignments' or transferred obligations. Although such refinements were much more fully developed in the later fourteenth and in the fifteenth centuries, they were already facilitating overseas financial transactions by 1334.

The bulk of the wool of the larger producers appears to have been dispatched directly to agreed shipping points (if it had already been bought) or to a central depot, rather than disposed of through local markets. By the close of the thirteenth century many Englishmen were involved in the overseas trade, and their position was strengthened in 1303 by the imposition of an additional duty, the 'New Custom', on foreigners importing or exporting goods, including wool. Individual English merchants seem generally to have been associated with particular areas of the country, and they were probably in closer touch with the small and medium producers than were the Italians and the Flemings.

The distribution of the home towns of English merchants licensed to export wool between 1273 and 1278, and of the religious houses included in a late thirteenth-century list as having a surplus of wool, suggest that

[1] A. Schaube, 'Die Wollausfuhr Englands vom Jahr 1273', *Vierteljahrschrift für Sozial- und Wirtschaftsgeschichte*, VI (1908), 68. Export figures are tabulated in E. M. Carus-Wilson and O. Coleman, *England's Export Trade, 1275–1547* (Oxford, 1963).
[2] E. von Roon-Bassermann, 'Die erste Florentiner Handelgesellschaften in England', *Vierteljahrschrift für Sozial- und Wirtschaftsgeschichte*, XXXIX, 1952, 97–128; R. A. Donkin, 'The disposal of Cistercian wool in England and Wales during the twelfth and thirteenth centuries', *Cîteaux in de Nederlanden*, VIII (1957), 109–31, 181–202.
[3] R. A. Donkin, 'Cistercian sheep farming and wool sales in the thirteenth century', *Agric. Hist. Rev.*, VI (1958), 2–8.
[4] M. M. Postan, 'Credit in medieval trade', *Econ. Hist. Rev.*, I (1928–9), 234–61; and 'Private financial instruments in medieval England', *Vierteljahrschrift für Sozial- und Wirtschaftsgeschichte*, XXIII (1930), 26–75.

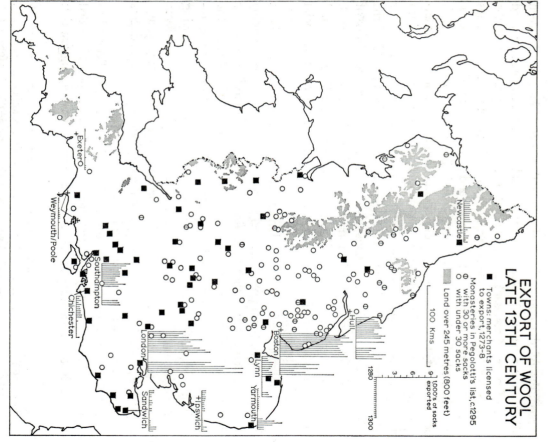

EXPORT OF WOOL
LATE 13TH CENTURY

■ Towns: merchants licensed
   to export, 1273-8
⊖ Monasteries in Pegolotti's list, c.1295
   with 30 or more sacks
○ with under 30 sacks
▨ Land over 245 metres (800 feet)

100 Kms

1280
3
6
9
1300
1000's of sacks
exported

Newcastle
Exeter
Weymouth/Poole
Southampton
Chichester
London
Sandwich
Ipswich
Yarmouth
Lynn
Boston
Hull

Fig. 30 Export of wool in the late thirteenth century
Based on E. M. Carus-Wilson and O. Coleman, *England's export trade, 1275–
1547* (Oxford, 1963). The sources are: (1) *Calendar Patent Rolls, 1272–81*
(London, 1901), 13–27; (2) *Calendar of various Chancery Rolls, 1277–
1326* (London, 1912), 2–11; (3) A. Evans (ed.), F. B. Pegolotti: *La
pratica della mercatura* (Cambridge, Mass., 1936).

interest in the export trade lay mainly in the south and east of the country, with an extension into the southern Welsh border (Fig. 30). This cannot generally be explained in terms of the qualities of wool available;[1] it owes more to the siting of the traditional centres of cloth working, and most to the fact that, apart from the Welsh border, which produced the very best wool, no part of the area was inaccessible from the regular shipping points. These stretched from Newcastle round to Exeter, but lay mainly between Hull and Southampton. Boston, serving a large part of eastern and central England, was most important in the 1270s and 1280s,[2] but towards the end of the century it yielded place to London where English shippers were most prominent. Soon after 1300, cloth supplanted raw wool as the country's chief export. In this, as in other ways, the opening decades of the fourteenth century mark the first major watershed in the history of the English landscape and economy since the Norman Conquest.

TOWNS AND CITIES

In England, as in other parts of Europe, the twelfth and thirteenth centuries were characterised by a considerable urban development. Existing centres of trade increased in size; villages grew to urban rank; and entirely new towns were founded. Such growth was promoted by increased commerce, both internal and international, and by a steady expansion of industry. This expansion did not, however, involve any major advance in organisation. Urban industry at the beginning of the fourteenth century was still based on the workshop, a form of enterprise unlikely to attract and to benefit from large amounts of capital. The wealthier citizens were merchants rather than industrialists and they reinvested in trade and property rather than in new production.

Boroughs and towns

The status of borough (*liber burgus*) was either prescriptive (enjoyed 'time out of mind') or conferred by charter granted either by the king for places on the royal demesne or by a feudal lord for other places. The grant

---

[1] For an early list of wool prices, see *Calendar Patent Rolls, 1334–38* (London, 1895), 480–2.

[2] See E. M. Carus-Wilson, 'The medieval trade of the ports of the Wash', *Med. Archaeol.*, VI–VII (1962–3), 182–201; P. Thompson, 'The early commerce of Boston', *Rep. and Papers, Associated Architect. Socs.*, II (1852–3), 362–81.

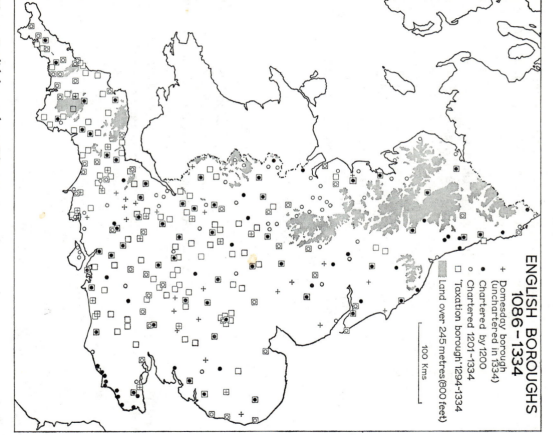

ENGLISH BOROUGHS
1086–1334

+   Domesday borough
    (unchartered in 1334)
●   Chartered by1200
○   Chartered 1201–1334
□   'Taxation borough'1294–1334

 Land over 245 metres(800 feet)

100 Kms

Fig. 31 English boroughs, 1086–1334
Based on: (1) A. Ballard, *British borough charters, 1042–1216* (Cambridge, 1913; (2) J. Tait and A. Ballard, *British borough charters, 1216–1307* (Cambridge, 1923); (3) M. Weinbaum, *British borough charters, 1307–1660* (Cambridge, 1943); (4) J. F. Willard, 'Taxation boroughs and parliamentary boroughs, 1294–1336' in J. G. Edwards et al. (eds.), *Historical essays in honour of James Tait* (Manchester, 1933), 417–35.

of a charter had two aspects. On the one hand, it brought privileges of an administrative, tenurial and legal character. The most important of these was the right of burgage or freehold tenure,[1] by which tenants held their lands on payment of fixed money rents and exercised the right of free alienation.

The other aspect was that the burgesses paid for their privileges, and boroughs were therefore sources of profit from rents and tolls; moreover, the burgesses were often ready to pay further for the confirmation and extension of their privileges. The policy of the Crown varied from reign to reign, but the financial exigencies of Richard I (1189–99) and still more of John (1199–1216) led to a readiness to make concessions. Feudal lords, bishops, abbots and laymen alike, also were eager to share in the profits that flowed from the granting of a borough charter. At the death of John, nearly 120 places had secured charters, either to confirm existing rights or to establish new ones;[2] and of these over 60 had not been described as boroughs in the Domesday Book. Between 1216 and 1334, another 90 or so townships received charters[3] (Fig. 31).

Not only did some villages acquire borough status, but some boroughs were created or 'planted' where no previous settlement had existed; their charters, too, came from kings, bishops,[4] abbots and laymen. M. W. Beresford has identified about 160 plantations in England between 1066 and 1334, and another two followed in 1345 and 1368; there were also about 80 plantations in Wales by 1334 (Fig. 32). The rate of founding fell during the decades 1130 to 1150, the years of civil war between Stephen and Matilda, and rose between 1189 and 1230. After 1368, there is no certain record of a new foundation during the remainder of the Middle Ages.

With few exceptions, such as New Salisbury alongside Old Sarum, the new towns did not supplant the old. 'They were planned additions to a stock of towns at a time when the economic situation gave opportunities both to new and to old. Just as agricultural expansion displayed itself

[1] M. de W. Hemmeon, *Burgage tenure in medieval England* (Cambridge, Mass., 1914); E. W. W. Veale (ed.), *The Great Red Book of Bristol*, Bristol Record Soc., 2 (1931), Introduction.

[2] A. Ballard, *British borough charters, 1042–1216* (Cambridge, 1913).

[3] A. Ballard and J. Tait, *British borough charters, 1216–1307* (Cambridge, 1923); M. Weinbaum, *British borough charters, 1307–1660* (Cambridge, 1943).

[4] M. Beresford, 'The six new towns of the bishops of Winchester, 1200–1255', *Med. Archaeol.*, III (1959), 187–215; and *History on the ground: six studies in maps and landscape* (London, 1957), 127–49.

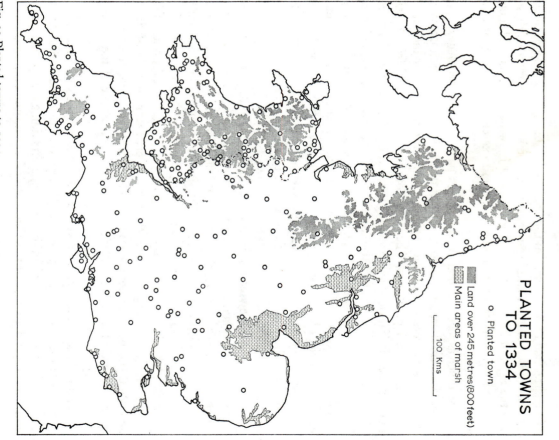

PLANTED TOWNS
TO 1334

o   Planted town

Land over 245 metres (800 feet)

Main areas of marsh

100 Kms

Fig. 32 Planted towns to 1334
Based on M. W. Beresford, *New towns of the Middle Ages* (London, 1967), 275,
282, 328, 342.

through an increased number of plough-teams and farm animals, so
urban expansion bore witness to great developments in the warehouse,
in the workshop and on the wharves.'[1] About a third of the plantations

[1] M. Beresford, *New towns of the Middle Ages* (London, 1967), 231.

were by the sea or on estuaries, reflecting the increasing coastwise traffic and the growing export trade in wool, grain and minerals. Some of these places were destined to become prominent seaports – Boston, King's Lynn and Newcastle founded about 1100, Harwich, Liverpool and Portsmouth founded about 1200, and Kingston upon Hull founded about 1300.

On the other hand, many boroughs came to little or nothing, and it would seem that local lords were not always good judges of the economic possibilities of the places to which they granted charters. Thus of the 23 boroughs created in Lancashire, only four – Lancaster, Liverpool, Preston and Wigan – were in any real sense urban at the end of the Middle Ages.[1] Some of the plantations hardly developed beyond mere villages; such was Mitchell in Cornwall. Others like Newtown in Burghclere (Hampshire) started to grow only to decay into nothing. Some declined as a result of physical changes; New Romney and New Winchelsea[2] were deserted by the sea, and Ravenserodd was washed away by the sea. A total of 23 (13%) plantations in England never succeeded, and 18 (21%) of those in Wales. The failures were widely scattered, and they included a high proportion of later foundations.

The boroughs that grew did so largely through immigration from the surrounding countryside. The Domesday village of Stratford upon Avon was granted a charter by the bishop of Worcester in 1196, and by 1252 there were 250 burgages as well as many shops and stalls.[3] About one-third of the burgesses bore the names of towns and villages whence, presumably, they or their ancestors had come (Fig. 33). Interestingly enough, in the late thirteenth or early fourteenth century there was a similar proportion of place-name surnames among the inhabitants of such long-established towns as Bristol, Gloucester and Worcester.[4]

In 1294 differential taxation on moveable property was introduced;[5] rural areas contributed a tenth and the urban centres a sixth. This differential rating, in varying proportions, was continued, and from 1334

[1] A. Ballard and J. Tait, lxxxviii.

[2] W. M. Homan, 'The founding of New Winchelsea', Sussex Archaeol. Coll., LXXXVIII (1949), 22–41.

[3] E. M. Carus-Wilson, 'The first half-century of the borough of Stratford upon Avon', Econ. Hist. Rev, 2nd ser, XVIII (1965), 51.

[4] R. H. Hilton (1966), 183–4.

[5] J. F. Willard, 'Taxation boroughs and parliamentary boroughs, 1294–1336', in J. G. Edwards et al. (eds.), Historical essays in honour of James Tait (Manchester, 1933), 417–35.

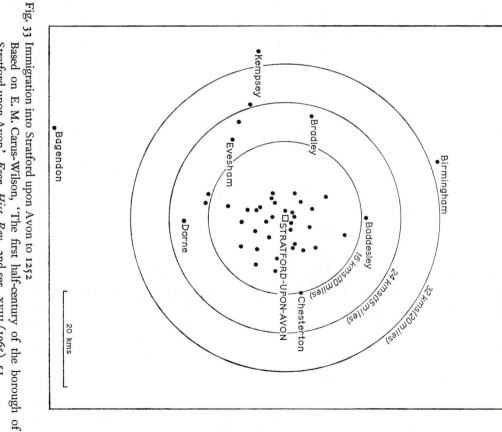

Fig. 33 Immigration into Stratford upon Avon to 1252
Based on E. M. Carus-Wilson, 'The first half-century of the borough of
Stratford upon Avon', *Econ. Hist. Rev*, 2nd ser., XVIII (1965), 51.

onwards it became a fifteenth and a tenth. But the list of 226 places[1]
described as boroughs on the enrolled accounts at various times, and so
taxed at the higher rate, differs from that of the chartered boroughs. On
the one hand, over 40 seignorial boroughs never appeared as taxation
boroughs; thus Bury St Edmunds in Suffolk and Abbots Bromley, New-
borough and Uttoxeter in Staffordshire were always taxed at the rural

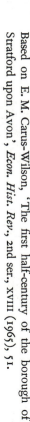

[1] M. W. Beresford (1967), 258.

rate; others appeared only intermittently, for example, Morpeth (Northumberland) in 1307 and Burton on Trent in 1307 and 1313. On the other hand, many places without charters were often, and sometimes usually, taxed at the higher rate. They included places on the royal demesne; places designated as boroughs for taxation purposes (there were eleven in Somerset alone); and places described as *villae mercatoriae* (i.e. vills with merchants), such as Old Windsor, Cheltenham and Andover. The taxation boroughs and merchant vills were selected by the 'chief taxers' in each county. No general criterion based on size can be discerned. Thus Yeovil in Devon, a chartered borough with a tax-paying population of 45 in 1327 never appeared as a taxation borough, while Chard in Somerset, also chartered but with only 11 tax payers in 1327, did appear. Probably 'trading activity', measured by local standards, 'had much to do with the selection or omission of seignorial boroughs'.[1]

There is a further complication – that of the parliamentary boroughs. On a number of occasions before 1265, representatives of certain towns had been called before king and council for specific purposes; but in that year, for the first time, such representatives were summoned for a general political purpose.[2] In 1268, men from 27 towns were summoned. In 1275 and again in 1283, *villae mercatoriae* were included, but after the latter date the distinction between boroughs and 'merchant vills' disappeared. From about 1310, the presence of burgesses in parliament was 'coming to be regarded as a matter of course'.[3] The cities and towns represented varied from parliament to parliament, but they usually numbered between 80 and 90 and were chosen for each county by the sheriffs. The places so selected included (i) most of the royal boroughs, (ii) many seignorial boroughs, (iii) a number of unchartered taxation boroughs, and (iv) some towns that were neither chartered nor taxation boroughs. Of the total of about 180 parliamentary boroughs selected at various times between 1295 and 1336, as many as 44 were not taxation boroughs.[4] The history of borough and town in the early Middle Ages presents many puzzling features.

Urban form and planning

A considerable number of towns grew up in the shadow of a castle or a monastery. Further protection was achieved by walling, but this was

[1] J. F. Willard (1933), 424.
[2] M. McKisack, *The parliamentary representation of the English boroughs during the Middle Ages* (Oxford, 1932), 1.
[3] *Ibid*, 24.
[4] J. F. Willard (1933), 426–7.

an expensive business and the majority of smaller places had either no defence works at all or only ditches and palisades.[1]

There were at least 30 monastic towns – places where an abbot or prior was lord.[2] About three-quarters of these stood immediately adjacent to the abbey, which was usually Benedictine; among the new Orders, only the Austin canons were in control of towns. On the whole, they were small places – Abingdon, Malmesbury, Peterborough and Burton, for example; Coventry was perhaps the largest and the most thriving. In fact, the bulk of the Church's urban property lay elsewhere, in the larger towns, the ports, and above all in London.[3] At the beginning of the fourteenth century, about one-third of the rental of London was in the hands of the Church.[4] Many religious communities had property there or in Southwark – a *hospitium* for use when visiting the capital, warehouses, quays, and land and buildings that simply brought in rent.[5] They were also strongly entrenched in most of the shire towns,[6] and, long before the passing of the Statute of Mortmain in 1279, action had been taken in several places to prevent more land and buildings falling into the hands of the monasteries.[7]

The central areas of most English towns took form during the early Middle Ages, and the street lines and property boundaries then adopted have guided building patterns ever since. The unit of medieval urban building and planning was the burgage plot (*placea*). The buildings on

---

[1] T. D. Hardy (ed.), *Rotuli litterarum clausarum*, 1 (Record Comm, London, 1833), 193a, 421b; H. M. Colvin, 'Domestic architecture and town planning', in A. L. Poole (ed.), *Medieval England*, 1 (Oxford, 1958), 65–7. See also H. L. Turner, *Town defences in England and Wales* (London, 1970).

[2] N. M. Trenholme, *The English monastic borough*, University of Missouri Studies, 2 (1927); D. Knowles, *The Monastic Order in England* (Cambridge, 1963), 444–7.

[3] E. Bradley, 'London under the monastic Orders', *Jour. British Archaeol. Assoc.*, 2nd ser, IV (1898), 9–16; M. B. Honeybourne, 'The extent and value of the property in London and Southwark occupied by the religious houses', *Bull. Inst. Hist. Res.*, IX (1931–2), 52–7; H. M. Cam, 'The religious houses of London and the Eyre of 1321' in J. A. Watt (ed.), *Medieval studies presented to Aubrey Gwynn SJ* (Dublin, 1961), 320–9.

[4] H. T. Riley (ed.), *Memorials of London, 1276–1419* (London, 1868), 98.

[5] See, for example, C. D. Ross (ed.), *Cartulary of Cirencester abbey* (London, 1964), 1, 173–5, II, 399; A. Watkin (ed.), *The great cartulary of Glastonbury*, Somerset Record Soc., 59 (1947), 128.

[6] R. A. Donkin, 'The urban property of the Cistercians in medieval England', *Analecta Sacri Ordinis Cisterciensis*, XV (1959), 104–31.

[7] C. Gross, 'Mortmain in medieval boroughs', *American Hist. Rev.*, XII (1907), 739.

each plot, house and workshop,[1] clustered, gable-on, towards the front; the area behind was usually open and might be worked as a garden or orchard or be used for livestock. In the planted boroughs, plots were sometimes forty to sixty feet wide,[2] instead of the usual ten to twenty feet around the market places of older centres.[3] Although particular trades tended to be concentrated in certain streets or districts,[4] there is not much evidence of any purely social segregation; large and small properties stood side by side.

By the close of the thirteenth century, buildings had often encroached on streets and on the market place. For example, at Pontefract a line of stalls down the centre of the 'new market' gradually (1236–78) became permanent;[5] and at Gorleston in 1275, the market place had been 'improperly' narrowed by 30 feet all round.[6] A shop carried forward into the street by means of a stall was generally the first step in such encroachment. There is also evidence of the building-up of the backs of burgages, of their lengthwise subdivision, and of much sub-letting.[7] Meanwhile, the town was sometimes spreading beyond its walls or official limits (*in suburbio, extra burgum*),[8] partly because of congestion near the centre, and partly as a consequence of attempts to escape industrial regulations and borough fees. In the second half of the thirteenth century, the market in land and property was as active in the towns as in the country.

[1] See W. O. Hassall, *The cartulary of St Mary Clerkenwell*, Camden Soc., 3rd ser., 71 (1949), 140–1.

[2] E. M. Carus-Wilson (1965), 57.

[3] See W. A. Pantin, 'Medieval English town house plans', *Med. Archaeol.*, VI–VII (1962–3), 202–39.

[4] M. Curtis, 'The London Lay Subsidy of 1332' in G. Unwin (ed.), *Finance and trade under Edward III* (Manchester, 1918), 35–92; W. T. Lancaster (ed.), *Abstract of charters of Bridlington priory* (London, 1912), 322.

[5] R. Holmes (ed.), *The chartulary of St John of Pontefract*, I, Yorks. Archaeol. Soc., Record Series, 25 (1899), 125–6, 182. See also, T. B. Dilks (ed.), *Bridgwater borough archives*, Somerset Record Soc., 48 (1933), liii; H. E. Butler (ed.), *The Chronicle of Jocelin of Brakelond* (London, 1949), 77; E. M. Carus-Wilson (1965), 60.

[6] W. Illingworth and J. Carey (eds.), *Rotuli Hundredorum*, II (Record Comm., London, 1818), 169.

[7] R. H. Hilton, 'Some problems of urban real property in the Middle Ages' in C. H. Feinstein (ed.), *Socialism, capitalism and economic growth: essays presented to Maurice Dobb* (Cambridge, 1967), 329, 334.

[8] S. R. Wigram (ed.), *The cartulary of the monastery of St Frideswide at Oxford*, Oxford Hist. Soc., 28 (1894), 215; M. Bateson, *Records of the borough of Leicester*, I (London, 1899), 389; A. Cronne, *The borough of Warwick in the Middle Ages*, Dugdale Society, Occasional Papers, 10 (1951), 18.

The many planted towns of the twelfth and thirteenth centuries were, almost by definition, planned. Twenty-six or more were organised about a grid,[1] for example Salisbury, Hull, Stratford upon Avon, New Winchelsea and King's Lynn; but in most cases the axis of development was a single, comparatively broad street which also served as a market. New house plots spread over waste, pasture, and even arable holdings. Alnmouth in Northumberland absorbed 296 acres of the common pasture and open fields of neighbouring Lesbury, and at Knutsford in Cheshire the burgages were actually measured in *seliones*.[2] At Eynsham, the abbot planted a new borough in 1215 by enclosing a *cultura* of 20 to 22 acres and dividing it into burgage plots of an acre or less.[3] The whole scheme took the form of a rectangular projection into the open fields.

The changing importance of towns

Any attempt to measure the changes in the relative importance of towns in the early Middle Ages is fraught with difficulty. Table 3.2 attempts to set out the ranking order of the thirty most wealthy provincial towns in the middle of the twelfth century and again in 1334. The earlier list is based on the average of the 'aids' levied during the reign of Henry II (1154–89),[4] but the returns are uneven and some boroughs appear only intermittently in the lists of places taxed. Bristol, being in baronial hands, does not appear at all; nor do Chester and Leicester. The Cinque Ports (Dover, Hastings, Hythe, Romney and Sandwich) were exempted from taxation. But even with its limitations, the list does provide a basis for comparison with the ranking order of towns in the 1334 Subsidy.

By 1334, Exeter, Winchester, Gloucester, Canterbury and Cambridge, all Domesday boroughs, had lost their places among the first twelve towns. Grimsby, Scarborough and Colchester also had declined relatively, and Dunwich had been entirely destroyed by the sea, but other east coast ports had improved their positions – Newcastle, Boston, Lynn, Yarmouth and Ipswich. Southampton was the leading south-coast port in the twelfth century and in 1334, and by the latter date Plymouth also was prominent.

---

[1] M. W. Beresford (1967), 151.

[2] A. Ballard and J. Tait, lxxviii.

[3] H. E. Salter (ed.), *Eynsham cartulary*, Oxford Record Soc., 49 (1906–7), 60; 51 (1908), xli, xliv, 50. See also J. S. Brewer (ed.), *Chronicon monasterii de Bello*, Anglia Christiana Soc. (1946), 12ff.

[4] C. Stephenson, *Borough and town*, Medieval Academy of America (Cambridge, Mass., 1933), 225 (with corrections); and 'The aids of the English boroughs', *Eng. Hist. Rev.*, XXXIV (1919), 457–75.

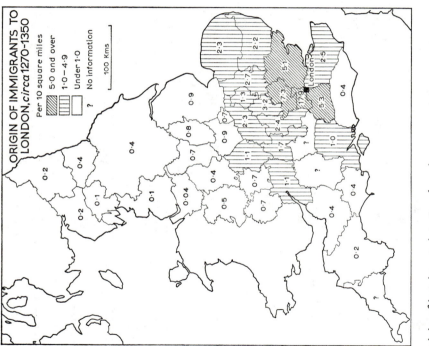

ORIGIN OF IMMIGRANTS TO
LONDON, *circa* 1270–1350

Per 10 square miles
5·0 and over
1·0 – 4·9
Under 1·0
No information

100 Kms

Fig. 34 Origin of immigrants into London *circa* 1270–1350
Based on E. Ekwall, *Studies in the population of medieval London* (Stockholm, 1956).

York, Norwich and Lincoln remained among the chief inland towns of the realm.

Towering above all was London. From 1183 comes the famous description of the city by William Fitz Stephen.[1] There were then 13 great conventual churches and 126 parish churches. To the port of London 'merchants from every nation that is under heaven' brought 'their trade in ships'. Beyond the ancient walls a 'populous suburb' stretched as far

[1] F. M. Stenton *et al*, *Norman London* (Hist. Assoc., London, 1934). See also U. T. Holmes, *Daily living in the twelfth century, based on observations of Alexander Neckam in London and Paris* (Madison, 1962), 18–43.

Table 3.2   *The ranking of provincial towns, 1154–1334*

Based on the average of 'aids' under Henry II (1154–89) and the Lay Subsidy of 1334. (D) indicates a Domesday Borough.

Town	Henry II (1154–89)	1334	Town	Henry II (1154–89)	1334
York (D)	1	2	Southampton (D)	24	18
Norwich (D)	2	7	Caister (Norfolk)	25	
Lincoln (D)	3	6	Marlborough (D)	26	
Northampton (D)	4		Colchester (D)	27	
Dunwich (D)	5		Godmanchester	28	
Exeter (D)	6	26	Hereford (D)	30	14
Winchester (D)	7	17	Huntingdon (D)	29	
Gloucester (D)	8	16	Bristol (D)		1
Oxford (D)	9	8	Boston		4
Canterbury (D)	10	15	Great Yarmouth (D)		5
Cambridge (D)	11	20	Lynn (King's and South)		10
Grimsby	12		Salisbury (D)		11
Newcastle upon Tyne	13	3	Coventry		12
Doncaster	14		Beverley		19
Berkhamsted (D)	15		Newbury		21
Nottingham (D)	16	25	Plymouth		22
Bedford (D)	17		Newark (D)		23
Worcester (D)	18		Peterborough *cum membris*		24
Scarborough	19		Bury St Edmunds (D)		27
Carlisle	20		Stamford (D)		28
Ipswich (D)	21	13	Ely *cum membris*		29
Corbridge	22		Luton		30
Shrewsbury (D)	23	9			

as the palace and church of Westminster; and all around lay 'the gardens of the citizens that dwell in the suburbs, planted with trees, spacious and fair, adjoining one another'. There was a wooden bridge on the site of the present London Bridge; a stone bridge, begun in 1176, was not completed until about 1209. By this time, there were also some stone houses; in 1215, stone from the houses of Jews was taken to repair the walls of London.[1] A series of disastrous fires in the middle of the twelfth century was followed by the Building Assize of 1189 (the work of London's first mayor) which stated that party walls should be of stone and that tiles

[1] C. L. Kingsford (ed.), *A survey of London by John Stow*, 1 (Oxford, 1908), 38.

rather than thatch should be used in roofing.[1] Needless to say, the Assize was not fully observed until several centuries later.

If we may trust the taxation figures, London, at the time that Fitz Stephen wrote, was about three times as wealthy as York; and below York came Norwich, Lincoln and Northampton. By 1334, the capital city was about five times as wealthy as Bristol; and next to Bristol came York and Newcastle. London was not only growing absolutely, its relative importance in the urban hierarchy of the country was also increasing. But of its actual size, we are very poorly informed. An intelligent guess might put the population at about 30,000 for the early part of the thirteenth century and at something under 50,000 in 1334.

Whatever its size, immigrants made a considerable contribution to its growth (Fig. 34). E. Ekwall found about 7,000 place-name surnames among Londoners during the period *circa* 1270 to 1350.[2] We cannot be sure that all people with such surnames were themselves immigrants. Some may have been the sons or even the grandsons of immigrants; others may have been apprentices who took the surnames of their masters. On the other hand, we know of some immigrants without local surnames. Allowing for the uncertainty of the evidence, it seems clear that a substantial number of people came from the neighbouring counties and from East Anglia and the east Midlands. As might be expected, the numbers who came from the north, the west and the north-west were very much smaller. Fig. 34 tells its own tale, but were the information available, it might have been better to present the figures in terms of the total population of each county. The names were those of clerks, of merchants, of lawyers, of priests, of physicians and of tradespeople of various occupations – mercers, woolmongers, hatters, goldsmiths, vintners, skinners and the like. Not only was the attraction of London far-reaching, but those who came were making a significant contribution to the growing trade and industry of England's greatest city.

[1] H. T. Riley (ed.) (1859), xxix–xxxiii, 319–36.

[2] E. Ekwall, *Studies on the population of medieval London* (Stockholm, 1956); and *Early London personal names* (Lund, 1947), 119–22. See also J. C. Russell, 'Medieval midland and northern migration to London, 1100–1365', *Speculum*, XXXIV (1959), 641–5; G. A. Williams, *Medieval London* (London, 1963), 18, 138–43; S. L. Thrupp (1948), 206–10, 389–92.

Chapter 4

ENGLAND circa 1334

R. E. GLASSCOCK

It is now generally agreed that an age of increasing population and prosperity had come to an end by the 1330s. The Black Death, once held responsible for the onset of decline, has been relegated to the less dramatic role of accelerating forces which had already begun to show in the economy of some parts of the country well before 1348.[1] M. M. Postan, having erected in 1950 a general framework for the interpretation of the economy of the thirteenth and fourteenth centuries, looked forward to the time when his 'mere outlines' would be filled out by local studies. That is now happening; twenty years later a number of studies are making us more aware of the regional differences in early fourteenth-century England. They show that economic conditions differed widely; evidence of recession in one place is matched by evidence of expansion elsewhere. What was true of the Winchester estates did not apply to Cornwall; the Weald differed greatly from the Midlands; and the north from East Anglia. Generally speaking it seems that there were fewer signs of recession in regions where there was a small but vital sector of non-agricultural activity within the local economy.[2]

[1] M. M. Postan, 'Some economic evidence of declining population in the later Middle Ages', *Econ. Hist. Rev.*, 2nd ser., II (1950), 221–46; A. R. Bridbury, *Economic growth: England in the later Middle Ages* (London, 1962); B. H. Slicher Van Bath, *The agrarian history of western Europe A.D. 500–1850* (London, 1963), 137–44; E. Miller, 'The English economy in the thirteenth century: implications of recent research', *Past and Present*, XXVIII (1964), 21–40; M. M. Postan in M. M. Postan (ed.), *The Cambridge economic history of Europe*, I (2nd ed., Cambridge, 1966), 548–91; B. F. Harvey, 'The population trend in England between 1300 and 1348', *Trans. Roy. Hist. Soc.*, 5th ser., XVI (1966), 23–42; J. C. Russell, 'The pre-plague population of England', *Jour. Brit. Stud.*, V (1966), 1–21; E. J. Nell, 'Economic relationships in the decline of feudalism: an examination of economic interdependence and social change', *History and Theory*, VI (1967), 313–50; J. Z. Titow, *English rural society 1200–1350* (London, 1969), particularly 64–102.

[2] For studies of specific regions see: (1) R. H. Hilton, *A medieval society: the west Midlands at the end of the thirteenth century* (London, 1967); (2) F. R. H. Du Boulay,

The early mid-fourteenth century is therefore an opportune point in time to pause and to look at the landscape of England, to consider the achievements of the twelfth and thirteenth centuries, and to try to set the back-cloth for events which were to follow. Unfortunately we have no Domesday Book for 1300, possibly the high-water mark of the era of colonisation, nor is there anything so comprehensive that would enable us to take stock of the state of the country in any one year. Only one source, the Lay Subsidy of 1334, although meagre in its detail, gives a fairly complete coverage of the country. This is a valuable date at which to have any kind of survey, as it comes fourteen years before the Black Death, and before the full impact of social and economic disorder. The year 1334 is therefore a convenient date for a description of England; but the lack of sources makes it impossible to reconstruct the geography of England for that single year. It is therefore necessary to examine the state of the country around 1334, that is between say 1320 and 1348.

THE 1334 LAY SUBSIDY

The 1334 Lay Subsidy is a unique source for the early fourteenth century on account of its coverage. The amounts of taxation agreed upon in 1334 were standardised in 1336, and did not finally disappear from use until 1623. This means that the gaps in the 1334 rolls themselves can be filled from later documents which contain the same information.[1] In this way all the counties of England can be covered for 1334, except for the palatine counties of Cheshire and Durham which were not liable for the tax, and for Cumberland, Westmorland and Northumberland which were excused tax in 1334 because of recent damage done by invading Scots; these three latter counties, however, were taxed in 1336 so that their earlier omission can be remedied.

The fifteenth and tenth of 1334 are so called because a fifteenth was asked from rural areas and a tenth from boroughs and ancient demesnes. The tax was one of a series of taxes on movable goods, principally on crops and stock, that had become a standard method of taxation by the

The lordship of Canterbury: an essay on medieval society (London, 1966); (3) J. Hatcher, *Rural economy and society in the duchy of Cornwall, 1300–1500* (Cambridge, 1970).

[1] The rolls of the taxes upon movable goods are all contained within Class E 179, Exchequer Lay Subsidies, at the Public Record Office, London. For studies dealing with the rolls, see p. 138, n. 1.

end of the thirteenth century.[1] The 1334 tax differed in an important way from its immediate predecessors in 1327 and 1332, for whereas those earlier taxations had been based upon the direct assessment of the goods of individuals, the 1334 tax took the form of an agreed sum or quota which each community was expected to pay. This quota was to be not less than the amount paid in 1332, and in most villages the agreed sum turned out to be slightly more.

Before basing any conclusions on the evidence of the 1334 Subsidy we must be under no illusions about some of its weaknesses as a source. Evasion and conventional valuation must have been common. Moreover, fourteenth-century officials of all kinds inherited the knack of lining their pockets by petty extortion. The only consolation is that there is no evidence to suggest that men were more dishonest in some areas than in others, and, therefore, the relative distribution of taxable wealth should not be unduly distorted. Apart from evasion there were several lawful exemptions from the tax.[2] Among the privileged groups so excused were the moneyers of London and Canterbury, the men of the Cinque Ports, and the stannary men of the south-west. Allowance should, therefore, be made for these omissions in Kent, Devon and Cornwall. Then again, a considerable number of people had escaped tax in 1332 because their property was below the taxable minimum, and the 1334 quotas took no account of the small amounts of movables in the hands of a large number of people in every community. A more serious omission was that most of the movable wealth of the Church, separately taxed in Clerical Subsidies, was excluded from the Lay Subsidies.[3] This means that a county with a great amount of clerical property appears less wealthy relative to others than it would have done had a complete assessment been carried out. Nevertheless, even allowing for clerical wealth, it is probable that the pattern of relative prosperity as shown in Fig. 35 would be refined rather than radically altered.[4] It is not likely that counties with low lay assessments would be transformed into counties of great wealth by the inclusion of clerical property.

---

[1] J. F. Willard, *Parliamentary taxes on personal property, 1290 to 1334* (Cambridge, Mass., 1934); M. W. Beresford, 'The Lay Subsidies', *The Amateur Historian*, III (1958), 325–8, and IV (1959), 101–9. A detailed account of the 1334 Subsidy is contained in R. E. Glasscock, *The Lay Subsidy of 1334* (British Academy, London, 1974).

[2] J. F. Willard (1934), 110–82.

[3] *Ibid.*, 96–109.

[4] As demonstrated in R. S. Schofield, 'The geographical distribution of wealth in England', *Econ. Hist. Rev.*, 2nd ser., XVIII (1965), 483–510.

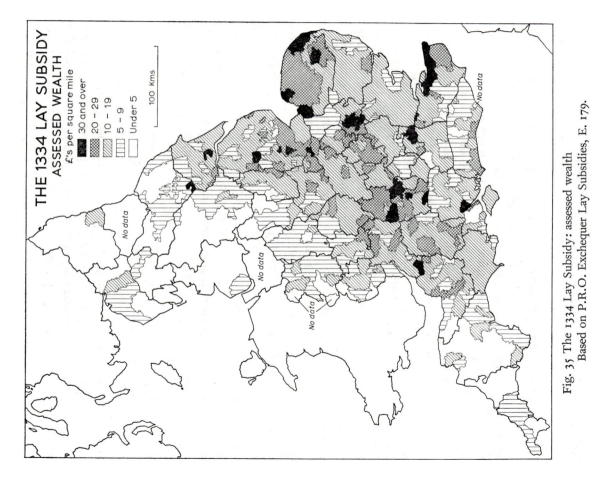

THE 1334 LAY SUBSIDY
ASSESSED WEALTH

£'s per square mile

30 and over
20 – 29
10 – 19
5 – 9
Under 5

100 Kms

Fig. 35 The 1334 Lay Subsidy: assessed wealth
Based on P.R.O. Exchequer Lay Subsidies, E. 179.

The distribution of lay wealth

After a careful examination of the local rolls before 1334, J. F. Willard suggested that the goods taxed as movables represented the surplus over and above the essentials that a family needed to live and work,[1] and this view has been supported by others.[2] Accepting this suggestion it is possible to rank counties in terms of their saleable produce or what has been described as their important 'income-generating resources',[3] For each county, the assessments have been multiplied by either fifteen or ten, as the case may be, to give the amounts at which the movable goods were valued. The sum of these two amounts indicates the total assessed wealth which has been calculated per square mile to produce the ranking shown in table 4.1. This ranking is of course identical with that of the counties on Fig. 22 for which the fifteenth and the tenth have been equated. The ranking shows a slightly different order from one in which the two different rates of tax have not been equated.[4] It must be remembered that there are no data for Cheshire and Durham and that the figures for Cumberland, Westmorland and Northumberland are for 1336.

Interesting though such a table may be, it conceals the geographical differences within each county. For example, the position of Cambridge-shire on the list represents an average of the extremes of its various parts; the assessed wealth of the upland would be near the top of the list whereas that of the Fenland would be near the bottom. This defect is remedied by Fig. 35 which shows the distribution of assessed wealth in terms of small groups of parishes. It must be stressed that the map does not show total wealth but that it is probably a fair indicator of relative wealth.

A line drawn from York to Exeter divided England in 1334 into the poor country of the north and west, with assessed wealth usually below £10 per square mile, and the richer land of the south and east with assessed wealth that was mostly above £10.

Assessed wealth of above £20 per square mile was frequently to be

---

[1] J. F. Willard (1934), 84–5.

[2] A. T. Gaydon, 'The taxation of 1297', *Beds. Hist. Record Soc.*, XXXIX (1959); L. F. Salzman, 'Early taxation in Sussex', *Sussex Archaeol. Coll.*, XCVIII (1960), 29–43, and XCIX (1961), 1–19.

[3] E. J. Buckatzsch, 'The geographical distribution of wealth in England, 1086–1843', *Econ. Hist. Rev.*, 2nd ser., III (1950), 180–202.

[4] As in E. J. Buckatzsch, 187. A table of gross yields which also does not make allowance for the two rates of tax is given in W. G. Hoskins and H. P. R. Finberg, *Devonshire Studies* (London, 1952), 216–17.

Table 4.1   *1334 Lay Subsidy*

Assessed wealth per square mile (by counties)

(The figures for Cumberland, Westmorland and Northumberland are for 1336)

	£		£
Middlesex (including London)	57.7	Somerset	12.0
Oxfordshire	27.2	Hampshire	11.3
Norfolk	24.0	Surrey	11.2
Bedfordshire	21.7	Sussex	11.1
Berkshire	20.8	Essex	11.0
Rutland	20.4	Worcestershire	9.8
Cambridgeshire	18.1	Herefordshire	7.4
Gloucestershire	18.0	Staffordshire	7.2
Huntingdonshire	17.8	Shropshire	7.0
Northamptonshire	17.2	Derbyshire	6.6
Kent	17.1	Devonshire	5.1
Wiltshire	17.1	Cumberland (1336)	4.9
Lincolnshire	16.6	Cornwall	4.9
Hertfordshire	14.8	North Riding	4.4
Warwickshire	14.4	West Riding	4.1
Suffolk	14.1	Westmorland (1336)	3.8
Buckinghamshire	13.8	Lancashire	2.9
Leicestershire	13.7	Northumberland (1336)	2.5
East Riding	13.5	Cheshire	No data
Dorset	12.4	Durham	No data
Nottinghamshire	12.3		

found in a great zone across lowland England, from the Parrett valley of Somerset to the fens and coastlands of Norfolk and Lincolnshire. As the assessments were above average, this must have been an area of considerable agricultural prosperity. That there was a distinct correlation between these areas and the corn growing lands of the country will be shown below. Outside this zone the only areas of comparable wealth were the Thames valley, east Kent, and the coastal plain of Hampshire and Sussex. All these areas carried twice, and in places at least three times as much movable wealth as the west Midlands, the north and the west.

There was also a random scatter of wealthy localities. Values over £30 per square mile were to be found in various parts of Norfolk – around Hunstanton in the north-west, in the silt marshland in the north-east, and to the south-east of Norwich. Upland Cambridgeshire was also very wealthy. So also were mid-Oxfordshire, parts of west Berkshire, parts of the north Kent lowlands, and parts of the Kesteven claylands of Lincolnshire. No common factor can be found to explain such a haphazard distribution. These very high assessments probably resulted from exceptional local agricultural prosperity induced by favourable local economic and social conditions. It is for the local historian to try to unravel the story of each area, and to explain, for example, the curiously high assessment of the north-western corner of Norfolk, as prominent on a map of 1334 as on one of 1086.[1]

By contrast almost all the west and north had assessed wealth below £10 per square mile, but within this area there were variations. The moorlands stand out as blank areas on the map – in the south-west, Exmoor, Dartmoor and Bodmin Moor; and in the north, the Pennines, the North York Moors, the Lake District and the Cheviots. Almost all the settled agricultural lowlands of the west and north were only between £5 and £10 per square mile, and the lowlands of Lancashire, the North Riding and Northumberland were particularly poor. It should be remembered, however, that not only were these areas remote from the centre from which the taxation was administered, but that they had recently been overrun in the war with Scotland.

Almost all the poorer areas of the south-east were also in the £5 to £10 category. These included the infertile sands and the heavy clays of Hampshire, Surrey, and the Weald of Sussex and Kent, and also parts of Essex and Suffolk. Furthermore, parts of the chalkland had similarly-low assessments, for example, in Hampshire, and in the Chilterns. In some areas the assessed wealth fell below £5 per square mile, as, for example, in the New Forest, north-west Surrey and parts of the Weald.

Fig. 35 represents, so to speak, a framework into which might be fitted the mass of evidence for population and agriculture for the different regions of medieval England. One thing, however, is abundantly clear: different amounts of movable wealth were to be found in areas of similar physical conditions, and, on the other hand, similar amounts were found in contrasting areas. For example, values of between £20 and £30 per square mile were to be found on the silts of the Lincolnshire fens, on the clays of western Cambridgeshire, and on the medium soils of much of Norfolk. While climate, relief and soils exerted broad controls over differences in farming, and underlay the strong contrast between the

[1] H. C. Darby, *The Domesday geography of eastern England* (Cambridge, 3rd ed., 1971), 117.

north-and-west and the south-and-east, the wealth of any particular locality can only be explained in terms of its local economic and social conditions. To this end a number of studies of the 1334 assessment for smaller areas and particular counties have already been produced.[1]

POPULATION

While the 1334 Subsidy allows us to grasp some essentials about the distribution of lay wealth over the country it does not enable us to estimate the total population. Estimates of the population of England just before the Black Death vary between $2\frac{1}{2}$ million and just over 6 million.[2] If demographers cannot agree on the population of the country in 1377, when the Poll Tax at least gives some quantitative basis for calculations, it is unlikely that they will agree on the population forty years earlier. In 1948 J. C. Russell estimated the total in 1377 as 2,332,373[3] and, although this figure has been criticised as too low by Stengers[4] and Krause,[5] his recently revised estimate is even lower at 2,199,916.[6] Assuming that the Black Death carried off around 40% of the population, Russell arrived at a pre-plague total of about 3.7 million. While many regard this as far too low, there is little hope of reaching a definite figure from the fragmentary sources available; and as the accepted method is to work backwards from the supposed totals of 1377, estimates for the pre-plague population will

---

[1] F. W. Morgan, 'The Domesday geography of Devon', *Trans. Devon Assoc.*, LXXII (1940), 321; B. Reynolds, 'Late medieval Dorset; three essays in historical geography', unpublished M.A. thesis, University of London, 1958; C. T. Smith, in W. G. Hoskins and R. A. McKinley (eds.), *V.C.H. Leicestershire*, III (1955), 134; H. C. Darby, *The medieval Fenland* (Cambridge, 1940), 134–5; W. G. Hoskins and E. M. Jope, in A. F. Martin and R. W. Steel (eds.) *The Oxford region* (Oxford, 1954), 109; R. E. Glasscock, 'The distribution of wealth in East Anglia in the early fourteenth century', *Trans. and Papers, Inst. Brit. Geog.*, XXXII (1963), 113–23; and 'The distribution of lay wealth in Kent, Surrey and Sussex, in the early fourteenth century', *Archaeol. Cantiana*, LXXX (1965), 61–8; W. G. Hoskins and H. P. R. Finberg, *Devonshire studies* (London, 1952), 212–49.

[2] M. M. Postan (1966), 561–2. A summary of the controversy over the pre-Black Death population of England is given by J. Z. Titow (1969), 66–73.

[3] J. C. Russell, *British medieval population* (Alberquerque, New Mexico, 1948), 146.

[4] J. Stengers, in a review in *Revue Belge de philologie et d'histoire*, XXVIII (1950), 600–6.

[5] J. Krause, 'The medieval household: large or small?', *Econ. Hist. Rev.*, 2nd ser., IX (1957), 420–32.

[6] J. C. Russell (1966), 14.

continue to vary according to the weight given to the mortality of the Black Death. For various reasons very large numbers of people were not recorded on the more detailed Lay Subsidy rolls of 1327 and 1332, and any calculation of total population based on these rolls is bound to be 'purely conjectural and probably erroneous'.[1] Recent work on the local rolls of Sussex puts this beyond any doubt.[2] At best the rolls may be used to infer the relative size of places, such has been done for Warwickshire.[3]

A figure midway between the various estimates gives a population of between say 4 and 4½ million in England before the Black Death. It may therefore have been about two and a half times what it had been in 1086, which indicates the striking growth during the intervening 250 years. Unfortunately, except for some classes of people, for example the clergy,[4] we have no idea of the breakdown of this number, nor is there direct evidence of how this population was distributed. It is reasonable, however, to suppose that the distribution of population bore a fairly close relationship to that of lay wealth (Fig. 35).

By modern standards England in the early fourteenth century might appear to have been sparsely peopled, but it should be remembered that most people lived directly off the land. The early fourteenth century constituted a crucial phase in the relationship between a rapidly growing population and its ability to produce food. Much land had been taken into cultivation since the Norman Conquest, but grain yields remained low and there were extra mouths to feed. As Miller has pointed out, parts of the country became overpopulated relative to food supply, with the result that bad harvests and famines, when they came, were progressively harder to bear.[5] The great European famine of 1315–17, in England as elsewhere, was a critical point in the population cycle. Undoubtedly its severity varied from place to place. Figures for some Winchester manors suggest that the death rate was twice as usual during these years.[6] Despite Russell's view that the population loss of the decade 1310–19 was only a few per

[1] J. F. Willard (1934), 181.
[2] L. F. Salzman (1960), 40–3.
[3] J. B. Harley, 'The settlement geography of early medieval Warwickshire', *Trans. and Papers, Inst. Brit. Geog.*, XXXIV (1964), 115–30.
[4] J. C. Russell, 'The clerical population of medieval England', *Traditio*, II (1944), 177–212.
[5] E. Miller (1964), 37–9.
[6] M. M. Postan and J. Titow, 'Heriots and prices on Winchester manors', *Econ. Hist. Rev.*, 2nd ser., XI (1959), 392–411.

cent and that the population went on increasing until 1347,[1] there is, at present, much more evidence to support the belief that, in the period under consideration, namely between 1320 and the Black Death, the total population was at best static, at worst declining. It is no coincidence that the founding of new towns, representative of an expanding population and flourishing commercial activity, fell rapidly away in the early fourteenth century.[2] As Miller has put it, 'the first half of the fourteenth century was a time of waiting for the blow to fall'.[3]

THE COUNTRYSIDE

Villages and houses

By the fourteenth century almost all the villages known to us today were in existence, together with many more which have long since disappeared.[4] Fortunately we know for almost all of England which villages were in existence from the *Nomina Villarum* of 1316.[5] Added to this is the almost complete coverage of the Lay Subsidy of 1334 with its long lists of vills. Few places of any size can have escaped both lists: 13,089 places are named on the records of the 1334 Lay Subsidy,[6] and in addition the rolls include many unspecified hamlets and other places.

Yet we know relatively little about the size and shape of the villages or about the dwellings of the villagers. Later layouts have replaced the medieval ones, and newer houses have obliterated the traces of their predecessors. Even the shape of a village in the fifteenth or sixteenth century may be no guide to its plan two or three centuries earlier. This has been shown at Wharram Percy in the East Riding, where excavations suggest three distinct stages of growth between the eleventh and fourteenth centuries, and at Wawne in the same county where the village was laid out on a completely new alignment sometime in the fourteenth century.[7] These examples sound a useful cautionary note against the habit of discussing settlement types and origins from the viewpoint of the

---

[1] J. C. Russell (1966), 21.
[2] M. W. Beresford, *New towns of the Middle Ages* (1967), 319–38.
[3] E. Miller (1964), 39.
[4] M. W. Beresford, *The lost villages of England* (London, 1954).
[5] Published in *Feudal aids*, 6 vols. (H.M.S.O., 1899–1920).
[6] This figure excludes places in Durham and Cheshire, palatine counties which were exempt from tax, but includes places in Cumberland, Westmorland and Northumberland taxed in 1336.
[7] M. W. Beresford and J. G. Hurst (eds.), *Deserted medieval villages* (London, 1971), 125–6.

present form of villages or at best from their form in first map evidence. In much the same way as many place-names have changed beyond recognition, so too have villages. No doubt most of the 'green' villages, whose distribution presents such a fascinating problem of interpretation,[1] existed in the early fourteenth century, but we know very little about their organisation. Planned villages were certainly a feature of the early fourteenth-century landscape, as there is increasing evidence, especially from excavation, that many villages and their surrounding fields were laid out anew in the thirteenth century.[2]

The best hope of discovering the form of English villages just before the Black Death lies in the excavation of those villages that were depopulated either in, or soon after, the early fourteenth century, and whose only traces lie 'fossilised' under the soil. Such desertions are not very numerous but there are several examples of local decline and abandonment in the early fourteenth century, for reasons unknown, for example at Beere in Devon,[3] at Pickwick in Somerset,[4] and at Heythrop, Shelswell and Langley in Oxfordshire.[5] Elsewhere the Black Death hastened the decline of many settlements that were already in difficulties; excavated examples are Hangleton in Sussex, and Seacourt in Berkshire.[6] Only a handful of villages seem to have disappeared as a direct result of the plague[7] and as yet none of these has been excavated. Examples of this type are Tilgarsley and Tusmore in Oxfordshire,[8] Middle Carlton in Lincolnshire, and Ambion in Leicestershire.[9]

The only village buildings of the fourteenth century that survive today are those which were built in stone, namely churches, manor houses and

[1] L. Dudley Stamp, 'The common lands and village greens of England and Wales', *Geog. Jour.*, CXXX (1964), 465; and L. D. Stamp and W. G. Hoskins, *The common lands of England and Wales* (London, 1963), 28–34.

[2] See, for example, P. Allerston, 'English village development: findings from the Pickering district of north Yorkshire', *Trans. and Papers, Inst. Brit. Geog.*, LI (1970), 95–109.

[3] E. M. Jope and R. I. Threlfall, 'Excavation of a medieval settlement at Beere, North Tawton, Devon', *Med. Archaeol.*, II (1958), 115.

[4] *Med. Archaeol.*, IV (1960), 156.

[5] K. J. Allison *et al.*, *The deserted villages of Oxfordshire* (Leicester, 1965), 5–6.

[6] E. W. Holden, 'Excavations at the deserted medieval village of Hangleton, Part I', *Sussex Archaeol. Coll.*, CI (1963), 72; M. Biddle, 'The deserted village of Seacourt, Berkshire', *Oxoniensia*, XXVI–XXVII (1961–2), 70–201.

[7] M. W. Beresford (1954), 158–62.

[8] K. J. Allison *et al.*, 44–5.

[9] W. G. Hoskins, *The making of the English landscape* (London, 1955), 93.

an occasional barn. While the churches survive in great numbers there are comparatively few surviving manor houses. There are, however, enough to show that the manor houses of the early fourteenth century were either of the simple, rectangular, first-floor hall type,[1] dating from the twelfth and thirteenth centuries, or of the newer end-hall types.[2] In addition, the moated manorial homestead was in fashion and was especially characteristic of late-colonised woodland areas such as Essex, Suffolk and the Arden of Warwickshire where land had been sub-infeudated to new freeholders.[3] In all manor houses the emphasis was on the hall, the home of the lord and the place of the manorial court. As the nearest equivalent to a public building it usually warranted the use of stone. By 1334 most castles had also become centres of local administration where courts were held, accounts audited and records kept.[4] Only a few, such as those on the borders with Wales and Scotland, retained their importance as military strongholds.

In contrast, while many of the peasant houses had stonewall-footings they were frail structures, and none has withstood the ravages of time. Our knowledge of them depends on archaeological rather than architectural investigation, and, while this has increased in the last twenty years with the new interest in medieval archaeology, there are still relatively few excavated peasant houses of the period. Excavation has shown that the peasant houses of the fourteenth century were of three types, the one-roomed cot, the long-house, and the farm complex.[5]

At Wharram Percy on the chalk wolds of the East Riding, deserted just after 1500, excavations have shown that the houses in use in the fourteenth century were of the long-house type, sometimes up to 90 feet in length, and with the living end separated from the lower end by a cross passage with opposite doorways.[6] Whether, in the early fourteenth century,

[1] As illustrated in M. Wood, *The English medieval house* (London, 1965), plate v and frontispiece.

[2] P. A. Faulkner, 'Domestic planning from the twelfth to the fourteenth centuries', *Archaeol. Jour.*, CXV (1958), 164.

[3] F. V. Emery, 'Moated settlements in England', *Geography*, XLVII (1962), 378–88; and B. K. Roberts, 'A study of medieval colonization in the Forest of Arden, Warwickshire', *Agric. Hist. Rev.*, XVI (1968), 101–13.

[4] N. Denholm-Young, *The country gentry in the fourteenth century* (Oxford, 1969), 32–3.

[5] M. W. Beresford and J. G. Hurst (1971), 104–12.

[6] For example house B1 in the plan of Wharram Percy, Area 10 in *Med. Archaeol.*, VIII (1964), Fig. 95; dated to the fourteenth century in *Med. Archaeol.*, IV (1960), 161.

the byre ends were used for animals, or whether they had been taken over for human use is not always clear. The presence of animals can only be certain where there is evidence of actual mangers, pens, or drains. Undoubtedly land had become precious by this period for at Wharram Percy chalk pits had been filled and the ground levelled off, and even the undercroft of the twelfth-century manor house was filled with rubbish in the first half of the fourteenth century to provide a level surface upon which to build houses.[1]

An idea of the character of a medieval village may be seen from the reconstruction drawing of Wharram Percy.[2] It shows the village as it might have looked in the early fifteenth century, but excavation has revealed that the shape of the houses was much the same a century earlier. The half-timbered dwellings of the fourteenth century rested on low stone foundations. On one house-site a complete reconstruction of the house seems to have taken place every thirty or forty years, and on occasions the alignment of the house was turned through 90 degrees from being long-side to the street to being gable-end to the street; a fourteenth-century house was overlain by three fifteenth-century houses on different alignments.[3] Similar changes took place on other house sites.[4]

One most important fact to emerge through the air survey and excavation of deserted medieval villages[5] is that the long-house was widespread over lowland England and not restricted to the highland zone as was formerly thought.[6] Excavations in many parts of the country, for instance at West Whelpington (Northumberland), Hangleton (Sussex), Holworth (Dorset), Houndtor (Devon), Gomeldon (Wiltshire), Garrow and Treworld (Cornwall),[7] have produced long-houses similar to the classic

[1] Med. Archaeol., II (1958), 206.

[2] Illustrated London News, No. 6485, 16 November 1963, 816–17.

[3] Med. Archaeol., I (1957), Fig. 34; IV (1960), 161; and VIII (1964), Fig. 95.

[4] Med. Archaeol., VIII (1964), 291. For a recent summary of changes in house alignments see M. W. Beresford and J. G. Hurst (1971), 122–3.

[5] Listed in Villages désertés et histoire économique (Ecole Pratique des Hautes Etudes – Vle. Section, S.E.V.P.E.N. Paris, 1965), 573–6.

[6] J. G. Hurst, 'The medieval peasant house', in A. Small (ed.), The Fourth Viking Congress (Edinburgh, 1965), 191.

[7] M. G. Jarrett, 'The deserted village of West Whelpington, Northumberland', Arch. Aeliana, 4th ser., XL (1962), 189–225; E. W. Holden, 73, Fig. 5; and J. G. Hurst and D. G. Hurst, 'Excavations at the deserted medieval village of Hangleton, Part II', Sussex Arch. Coll., CII (1964), 112, Fig. 6; P. A. Rahtz, 'Holworth medieval village excavations', Proc. Dorset Nat. Hist. and Archaeol. Soc., LXXX (1959), 127–47. For Houndtor, see Med. Archaeol., VI–VII (1962–3), Fig. 102 and plate XXXIIB, and

example excavated at Beere (Devon) in 1938–9,[1] mainly of thirteenth-century date, but used into the fourteenth century. The foundations of similar houses show up from the air on deserted village sites in areas of stone building, for example, in Lincolnshire and the Cotswolds. The presence of timber-built long-houses in the Midlands is highly probable but difficult to prove as they have left little trace. While the long-house was obviously not the only kind of medieval peasant house it must have been a common element in the rural landscape of the early fourteenth century. But our knowledge of this and other house types, and of the most interesting problems of the use of wood and stone in peasant house building, await the excavation of more fourteenth-century sites.

The agrarian landscape

Most peasant houses stood within small enclosures. At Wharram Percy all the houses stood in small enclosures which were used for pasture, vegetables, pigs or poultry. Behind these, and stretching back to the edge of the open fields, were the crofts, larger areas held in severalty and used either for arable or pasture, as they were in some Leicestershire villages.[2] On Midland deserted village sites, where timber houses have left no trace, it is often the small enclosures and crofts which show up most clearly from the air.[3]

To present any picture of the common fields in the early fourteenth century is much more difficult. It is hazardous to generalise, if only because we now know that 'field' arrangements varied enormously from place to place according to social organisation, physical conditions and types of farming.[4] While there is evidence for the existence of some kind of open-field cultivation in every English county at some time,[5] the

VIII (1964), Figs. 90 and 91; for Gomeldon, see *Med. Archaeol.*, VIII (1964), Fig. 94, D. Dudley and E. M. Minter, 'The medieval village at Garrow Tor, Bodmin Moor, Cornwall', *Med. Archaeol.*, VI–VII (1962–3), 272–94, esp. Fig. 88; for Treworld see *Med. Archaeol.*, VIII (1964), 282.

[1] E. M. Jope and R. I. Threlfall, Fig. 26.

[2] R. H. Hilton in W. G. Hoskins and R. A. McKinley (eds.), *V.C.H. Leicestershire*, II (1954), 166.

[3] E.g. Cestersover, Warwickshire, in M. W. Beresford (1954), plate 15, and Burston, Buckinghamshire, M. W. Beresford and J. K. S. St Joseph, *Medieval England: an aerial survey* (Cambridge, 1958), Fig. 44.

[4] A. R. H. Baker, 'Howard Levi Gray and *English field systems*: an evaluation', *Agric. Hist.*, XXXIX (1965), 86–91.

[5] J. Thirsk, *Tudor enclosures* (London, 1959), 4, footnote.

overall situation as it was at any one time, at say about 1330, cannot be pictured. We do not yet know, for example, the extent of enclosed land at this time.

Clearly in the early fourteenth century most of the villages in H. L. Gray's 'Midland Zone' were surrounded by their open fields (Fig. 23). The details of field arrangement varied with local conditions, and in some districts two- and three-field systems existed side by side, as for example in Wiltshire,[1] and Leicestershire. Yet in some Leicestershire villages it was often the furlong and not the open field which was the main cropping unit.[2] This implies that even for the Midlands we can no longer envisage, without qualification, villages surrounded by large open fields each given over to one crop or lying fallow. Invariably some arable land was held in the separate closes of holdings created as a result of thirteenth-century assarting.[3] We must therefore allow for considerable flexibility in the system by 1330, and picture some villages with different furlongs under various crops.[4] Relatively advanced crop rotations must have developed earlier than we have hitherto thought. By the early fourteenth century, three-course rotations were practised on both demesne and tenant holdings, as for example in the Chilterns,[5] and there is increasing evidence that a more refined four-course rotation, which included legumes, was becoming increasingly common, for example on the estates of Ramsey Abbey,[6] and at Westerham.[7]

Even within the furlongs, rigidity was further broken down as the holdings of landlord and tenant, perhaps once arranged in a particular order, became scattered irregularly as a result of exchange and amalgamation in the previous two centuries. In addition, by the early fourteenth century much demesne land had been enclosed and was valued more highly than demesne land cultivated in the common fields. For example,

---

[1] W. G. Hoskins in E. Crittall (ed.), *V.C.H. Wiltshire*, IV (1959), 14.

[2] R. H. Hilton in W. G. Hoskins and R. A. McKinley (eds.), *V.C.H. Leicestershire*, II (1954), 159–61.

[3] Well illustrated in C. C. Taylor, 'Whiteparish: a study of the development of a forest-edge parish', *Wilts. Archaeol. and Nat. Hist. Mag.*, LXII (1967), 90–1.

[4] V. H. T. Skipp, and R. P. Hastings, *Discovering Bickenhill* (Birmingham, 1963), 15–18.

[5] D. Roden, 'Demesne farming in the Chiltern Hills', *Agric. Hist. Rev*, XVII (1969), 9–23.

[6] J. A. Raftis, *The estates of Ramsey Abbey* (Toronto, 1957), 220.

[7] T. A. M. Bishop, 'The rotation of crops at Westerham, 1297–1350', *Econ. Hist. Rev*, IX (1934), 38–44.

in Wiltshire, at Wootton Bassett 126 acres of demesne lay in severalty in 1334, and at Draycote Cerne 66 of the 240 acres of demesne had also been enclosed by 1344.[1]

H. L. Gray's recognition that the field systems of the Midlands differed from those of the west and east still holds good, although his ethnic reasons for the origin of these differences are slowly being revised. It has recently been argued that the differences between the areas of regular two- or three-field systems and areas of irregular field arrangements stem from a contrast between arable and pasture farming;[2] and that many pastoral areas of west and east which have hitherto been thought of as 'early enclosed' may be so called not because enclosure proceeded early in these areas but because they had never known a fully developed common-field system. In addition it is now clear that systems of infield and outfield were much more frequent throughout England than was formerly supposed, especially on light and less fertile land.

While there are differences of opinion on field systems[3] (and on their terminology), most authorities would agree that the agricultural landscapes of much of East Anglia, Kent, the Welsh borderlands, Somerset, and the south-west, must have looked very different from those of the Midlands in the early fourteenth century. In both west and east much land was parcelled up into numerous small enclosures; hedgerows were more important in the landscape, and field and tenurial arrangements were more flexible.

Arable farming

The available evidence suggests that the early fourteenth century, a time of rapid inflation, was a period of instability in agriculture. The first phase, before about 1315 was, as it were, a left-over from the thirteenth century, a period of fifteen years in which the overall picture was still one of prosperity. After about 1320, conditions became more confused. We must presume that the effects of the famine of 1315–17 varied from place to place, and it is therefore not surprising that evidence from different estates suggests that agricultural fortunes varied considerably in the period 1320–45. In some places there appears to have been only temporary

[1] W. G. Hoskins in E. Crittall (ed.), *V.C.H. Wiltshire*, IV (1959), 13.

[2] J. Thirsk, 'The common fields', *Past and Present*, XXIX (1964), 3–25.

[3] See J. Z. Titow, 'Medieval England and the open field system', *Past and Present*, XXXII (1965), and the reply by J. Thirsk, *Ibid*, XXXIII (1966); G. C. Homans, 'The explanation of English regional differences', *Past and Present*, XLII (1969), 18–34.

dislocation and a fairly quick return to reasonable agricultural productivity, sometimes accompanied by changes of land use. In others the effect of the famine seems to have been more prolonged and there was no return to early fourteenth-century levels of production. To suggest this contrast before and after the 1320s is to make a hazardous generalisation, and what might be true for one area was not true for another. Moreover because peasant agriculture has left little written record, the evidence must be based on the accounts of a few large estates, mainly in the south-east of the country, where records of profit and loss were meticulously kept. For it was on these estates, where capital and efficiency in administration overcame the difficulties of scattered demesne production, that the main commercial agricultural enterprise of the early fourteenth century was to be found. The monastic houses, in particular, were best suited to efficient production not only because of their better administration,[1] but because in some cases their estates were more compact, for example, those of Battle abbey in Sussex and of St Augustine's priory in Canterbury.

Whereas demesne land gave scope for specialised production we know very little of the fortunes of the peasants tied down to customary practice on small and scattered holdings, and tilling the land for their own needs. Not that we should underestimate the role of the peasant in the food supply of medieval England. On the bishop of Winchester's estates in the thirteenth century, for example, a considerable amount of the town demand was met by local peasant produce.[2] This is likely to have been even more true in the second quarter of the fourteenth century when more demesne was being let out by landlords, and when peasants had need of a higher income to meet increasing rents. Consequently there was an increase in the number of land transactions among peasants, as for example on the estates of Titchfield abbey in Hampshire.[3] The network of local markets in early fourteenth-century England, inherited from the previous century, was partly a response to peasant production.

The first two decades of the century were years of prosperous demesne farming in the best thirteenth-century tradition.[4] On estates as far apart as Durham and Canterbury, Peterborough and Glastonbury, more land

[1] D. Knowles, *The Religious Orders in England,* 3 vols. (Cambridge, 1948–59), I, 32.
[2] E. Kosminsky, 'The evolution of feudal rent in England from the xIth to the xVth Centuries', *Past and Present,* VII (1955), 20.
[3] D. G. Watts, 'A model for the early fourteenth century', *Econ. Hist. Rev.,* 2nd ser., xx (1967), 543–4.
[4] E. Miller (1964), 32.

had come under cultivation, and administration and husbandry were more efficient.[1] The Benedictine houses in particular, by a careful division of the types of farming on their estates, were able to specialise in cereal farming in contrast to the specialisation in wool that was characteristic of the Cistercians.[2] A good example of a commercialised monastic estate was that of Canterbury cathedral priory, midway between the markets of London and the Continent, where grain production reached a peak in the period 1306–24.[3] The agricultural achievement of this priory helped to make north-east Kent one of the most prosperous parts of England in 1334.[4] On the priory estates increasingly efficient commercial agriculture was partly achieved by a drastic change in the system of the food rents, whereby distant manors which had formerly supplied food to the priory took advantage of rising grain prices to sell their produce at favourable prices in local markets.

The cash crop of midland and southern England was wheat. Barley, although more widely grown, fetched lower prices; so did oats and also rye, the crop of the light soils. The market for grain, especially for wheat, in the early fourteenth century was an incentive to improve yields, both of seed and of land. Not that any dramatic overall results were achieved. Yields of grain remained low by modern standards throughout the thirteenth and fourteenth centuries.[5] The statistics are full of difficulties,[6] but a probable average yield of 8–9 bushels of wheat per acre represents little over a quarter of an expected yield today.[7] Similarly, there has been roughly a threefold increase in the yield of seed since the medieval period.[8] Despite the efforts on the estates of Canterbury priory to sow seed more intensively and to act on the advice of Walter of Henley to get their seed from outside, the yields showed no great improvement and remained

[1] D. Knowles, 44–7.

[2] R. A. L. Smith, 'The Benedictine contribution to medieval agriculture', in *Collected Papers* (London, 1947), 103–16.

[3] R. A. L. Smith, *Canterbury cathedral priory* (Cambridge, 1943), 128–45.

[4] R. E. Glasscock (1965), 62–8.

[5] Lord Beveridge, 'The yield and price of corn in the Middle Ages', *Economic History*, I (1927), 155–67, reprinted in E. M. Carus-Wilson (ed.), *Essays in economic history* (London, 1954).

[6] R. Lennard, 'Statistics of corn yields in medieval England', *Economic History*, III (1935), 173–92 and 325–49.

[7] M. K. Bennett, 'British wheat yield per acre for seven centuries', *Economic History*, III (1935), 21.

[8] Lord Beveridge, 158–9.

below expectations.[1] As on many other estates, attempts were also made to improve the soil by draining, dunging and marling. In coastal areas where marl was not available, seaweed and sand were used to sweeten the land.[2]

In the drive to increase cereal production and take advantage of rising prices not only was agricultural efficiency stepped up, but the acreage under crops was increased. Tillage was increased by the conversion of pasture land, and on the Canterbury priory manors by the reclamation of marshland which was particularly suited to oats. The result of all these efforts at Canterbury may be seen in Prior Henry of Eastry's remarkable survey of 1322,[3] which shows the great area under wheat and barley in east Kent, under oats on the marshland manors, and the importance of legumes in the various rotations (Fig. 36).

The Canterbury priory estates enjoyed of course special advantages, firstly in possessing the redoubtable Prior Henry, one of the most enlightened landowners of his time,[4] and secondly in their nearness to the markets of London and the Continent. But at the other end of the country on the estates of Durham cathedral priory, similar improvements were taking place, and there was a corresponding peak in grain production at the beginning of the fourteenth century. Increased grain production was not merely the result of the cultivation of more land, but also of improved techniques.[5] Elsewhere also the monastic houses gave a lead in agricultural improvement at this period, for experiment was only possible where capital was available and where demesne land was held in severalty. On some estates, enclosure and improvement went hand in hand, for example on the lands of the abbey of Evesham.[6]

While improvement may be detected it is not surprising that there was no great breakthrough in medieval agriculture because the level of investment was extremely low. Even on the best run monastic and lay estates gross investment was usually under 5%.[7] The likelihood is that it was lower on smaller estates, although it must be admitted that the role of

[1] R. A. L. Smith (1943), 133–5.
[2] H. P. R. Finberg, *Tavistock Abbey* (Cambridge, 1951), 89.
[3] R. A. L. Smith (1943), 140–1.
[4] D. Knowles, 49–54.
[5] E. M. Halcrow, 'The decline of demesne farming on the estates of Durham cathedral priory', *Econ. Hist. Rev.*, 2nd ser, VII (1955), 345–56.
[6] R. A. L. Smith (1947), 110.
[7] M. M. Postan, 'Investment in medieval agriculture', *Jour. Econ. Hist.*, XXVII (1967), 579.

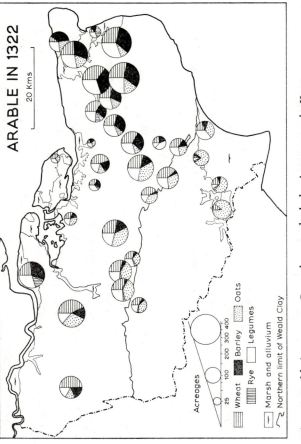

ARABLE IN 1322

20 Kms

Acreages

25 100 200 300 400

Wheat
Rye
Marsh and alluvium
Northern limit of Weald Clay
Barley
Oats
Legumes

Fig. 36 Arable in 1322 on Canterbury cathedral priory estates in Kent
Based on R. A. L. Smith, *Canterbury cathedral priory* (Cambridge, 1943),
140–1.

medium-size estates within medieval agriculture is, as yet, largely un-explored. Capital invariably went into the purchase of more land or the improvement of buildings, neither of which did anything to improve overall economic development or to increase crop yields. Profits went everywhere except into improving agricultural production. And if a low level of investment was characteristic of the lords then it was even more true of the peasants, three-quarters of whom must have lived at near subsistence level with little chance to save or to invest anything in their agriculture.

By about 1330 there were unmistakable signs that agriculture was less prosperous, for not only were the prices of cereals falling and rents and wages rising, but in a few places land was lying uncultivated, all symptoms of a decreasing demand.[1] In addition, after 1330, landlord interest in acquiring new property slackened off (partly because the squeeze between wages and prices reduced profits), and sometimes ceased entirely, a sure sign that land was no longer thought of as a good

[1] M. M. Postan (1950), 225–40.

investment.[1] Certainly the profitability of commercial grain growing declined, and with it demesne farming.[2]

This sudden change in fortunes has not yet been explained, but it seems clear that a number of factors combined to produce a change in European agriculture at this time, although the exact role of each one is still undecided. These included bad harvests with consequent food shortages, famines, losses of population, disruptions and bankruptcies in international finance caused by the two series of Anglo-French wars, the decline of European silver production with its possible effects on wages and prices,[3] and increases in taxation on both people and commodities.

A most important factor was undoubtedly the great European famine of 1315–17.[4] In 1315 and 1316, in England as elsewhere, heavy late-summer rains prevented the ripening and harvesting of corn,[5] and the subsequent scarcity led to great increases in the prices of grain and other foodstuffs.[6] Scarcity and pestilence followed famine in a sequence that must have been all too familiar to the village small-holders and labourers, those most liable to hunger and hardship. But while the crop failures of 1315–16 could have ushered in a short period of agricultural depression, they can hardly be held responsible for a depression that was to last a century and a half. It is not yet known whether the bad weather, and in particular, the heavy summer and autumn rainfall of these years was a local intensification of a wider climatic deterioration at this period. There is some evidence to suggest that it might have been. The severity of the climate at this period is well known,[7] and climatic change may well have

[1] G. A. Holmes, *The estates of the higher nobility in fourteenth-century England* (Cambridge, 1957), 113–14.

[2] For a case study illustrating this see I. Keil, 'Farming on the Dorset estates of Glastonbury abbey in the fourteenth century', *Proc. Dorset Nat. Hist. and Archaeol. Soc.*, LXXXVII (1965), 234–50.

[3] J. Schreiner, 'Wages and prices in England in the later Middle Ages', *Scand. Ec.H.R.*, II (1954), 61–73.

[4] H. S. Lucas, 'The great European famine of 1315, 1316 and 1317', *Speculum*, V (1930), 343–77.

[5] J. Titow, 'Evidence of weather in the account rolls of the bishopric of Winchester, 1209–1350', *Econ. Hist. Rev.*, 2nd ser., XII (1960), 385–6.

[6] J. E. T. Rogers, *A history of agriculture and prices in England*, 7 vols. (Oxford, 1866–1902), I, 230; N. S. B. Gras, *The evolution of the English corn market* (Cambridge, Mass, 1915), 47.

[7] E. Huntington and S. S. Visher, 'The climatic stress of the fourteenth century', being ch. 6 of *Climatic changes* (New Haven, 1922), 98–109; H. H. Lamb, 'Britain's changing climate', *Geog. Jour.*, CXXXIII (1967), 454–60.

been important in the economic life of Europe in the fourteenth and fifteenth centuries.[1] Certainly we cannot dismiss the possibility that it was a climatic deterioration, and especially heavier summer rainfall, that triggered off the chain reaction of poor harvests, lack of seed corn for the following year, famine and pestilence, economic dislocation, a declining population, the reduction of the cultivated area, and political unrest. All these appeared in England well before 1348 when the Black Death caught the country at such a low ebb.

Prices of agricultural produce, especially of wheat and barley, began to decline steadily after about 1320, although there was a temporary recovery in some years, for example in 1330 and 1331.[2] The problem of declining prices and what it meant to an estate dependent upon selling agricultural produce is well illustrated on the manors of the bishop of Ely at Great Shelford and at Wisbech in Cambridgeshire.[3] At Shelford, wheat which had sold at between 10s. 6d. and 14s. a quarter in the early twenties, sold at an average of 5s. 6d. in the period 1325–33, and only just over 4s. in the years around 1340. The revenue on both manors dropped drastically between 1325 and 1348, a drop that was due for the most part, though not entirely, to the great fall in prices. A consequence of this fall was that less grain was marketed locally and more was sent instead direct to the bishop's household. Moreover, less capital was available for the maintenance and improvement of the manors. Higher wages and declining income from sales had to be offset by increasing income from rents, both by letting out more demesne and by commuting labour services. The bishop of Ely benefited from the new high rents which thirteenth-century inflation had made possible, and the letting out of demesne continued during the period before the Black Death. On his Cambridgeshire estates, at Downham, for example, 264 acres were let out during 1299–1337, at Stretham 135 acres during 1316–46, and at Linden End 362 acres during 1316–45. At Wisbech, the income from demesne leaseholds rose from 53s. 4d. in 1320 to £48 10s. in 1345.[4]

The same was happening on the estates of Ramsey abbey,[5] and towards

[1] G. Utterstrom, 'Climate fluctuations and population problems in early modern history', *Scand. Ec.H.R.*, III (1955), 3–47.

[2] J. E. T. Rogers, I, 230–2, and N. S. B. Gras, 60. D. L. Farmer, 'Some grain price movement in thirteenth century England', *Econ. Hist. Rev.*, 2nd ser., X (1957), 207–20.

[3] E. Miller, *The abbey and bishopric of Ely* (Cambridge, 1951), 105–12.

[4] E. Miller (1951), 100.

[5] J. A. Raftis, 241.

the end of the thirteenth century Cistercian holdings began to be regularly leased.[1] Sometimes, as on the abbot of Westminster's estates, leasing involved just a few acres piecemeal, sometimes whole demesnes.[2] Commutation of labour services increased; thus on 15 of 81 manors in twenty counties in 1350, labour services had either entirely disappeared or were insignificant.[3] Both leasing and commutation increased the freedom of the tenantry and in the long term contributed to the break-up of the manorial economy.

While these were general characteristics of the period, the pace of change varied from place to place. Thus on many of the Durham manors the sales of grain increased after 1340, and the decline of demesne farming did not take place until the second half of the fourteenth century,[4] nor did it on the Berkeley estates.[5] On the estates of the archbishopric of Canterbury there was slow contraction of demesne land under cultivation throughout the fourteenth century.[6] In Leicestershire, demesne farming flourished well into the fifteenth century.[7] Generally speaking, however, the 1330s and 1340s were times when landlords, lay and ecclesiastic, found money was much easier to come by through rents than through sales.

References to *terra frisca* (uncultivated land) in the *Nonarum Inquisitiones* of 1342 show that much land was lying uncultivated in various parts of the country in 1341. The record of uncultivated land is very uneven, and there is difficulty in knowing what exactly the term means from one place to another.[8] In some cases it may have implied merely a change in land use or the temporary abandonment of outfield. Nevertheless it seems clear that much land which was formerly cultivated was no longer so, and the same impression is gained from *Inquisitiones Post Mortem* of the period. In counties that have been studied in some detail the contraction

---

1 R. A. Donkin, 'Cattle on the estates of medieval Cistercian monasteries in England and Wales', *Econ. Hist. Rev.*, 2nd ser., XV (1962), 44.

2 B. Harvey, 'The leasing of the abbot of Westminster's demesnes in the later Middle Ages', *Econ. Hist. Rev.*, 2nd ser., XXII (1969), 17-27.

3 E. Lipson, *The economic history of England* (12th ed. London, 1959), I, 95-7.

4 E. M. Halcrow, 347.

5 Lord Ernle, *English farming past and present* (6th ed. London, 1961), 46.

6 F. R. H. Du Boulay, 218-19.

7 R. H. Hilton, *The economic development of some Leicestershire estates in the fourteenth and fifteenth centuries* (London, 1947).

8 A. R. H. Baker, 'Evidence in the *Nonarum Inquisitiones* of contracting arable lands in England during the early fourteenth century', *Econ. Hist. Rev.*, 2nd ser., XIX (1966), 518-32.

of arable land was, as we might expect, more marked on the uplands than the lowlands. In some places waste land was attributed to poverty; for other places the entry was more specific, as for example at Stockland and Compton abbey in Dorset where the tenants had departed and their land lay uncultivated.[1] On the other hand, uncultivated land in seven Wealden parishes amounted to over 2,000 acres yet only for two parishes was it specifically attributed to poverty, and some of the parishes concerned were more prosperous in 1341 than they had been fifty years before.[2] Cambridgeshire had almost 5,000 acres out of cultivation, yet this was exactly the period when the nearby Huntingdonshire estates of Ramsey abbey were enjoying a temporary revival of the prosperity of earlier years.[3] It would seem that the references to uncultivated land in the *Nonarum Inquisitiones* cannot be regarded as indicators of widespread retreat but they may be seen as symptomatic of places which were falling on hard times.

Pasture farming

Livestock formed an integral part of medieval agriculture not only in the pastoral west of the country but in the drier east where the emphasis was on grain production. Oxen for ploughing, plough and cart horses, and cattle for milk and hides were essential to every village. Pasture farming was a useful stand-by in years of bad harvests, and the prices of animals fluctuated less than those for grain which varied sharply from year to year.[4] Sheep were valued throughout the country for their wool, mutton, and skins, as is clear from the frequency with which animals are listed on the local rolls of taxation of the early fourteenth century.[5]

There may have been about eight million sheep in England in the mid-fourteenth century.[6] Peasant flocks often exceeded those of demesne but we still know very little about peasant wool production.[7] In addition to

[1] B. Reynolds, 'Late medieval Dorset; three essays in historical geography', unpublished M.A. thesis, University of London, 1958, 166–7.
[2] J. L. M. Gulley, 'The Wealden landscape in the early seventeenth century and its antecedents', unpublished Ph.D. thesis, University of London, 1960, 345–8.
[3] J. A. Raftis, 241.
[4] D. L. Farmer, 'Some livestock price movements in thirteenth-century England', *Econ. Hist. Rev.*, 2nd ser, XXII (1969), 1–16.
[5] J. F. Willard (1934), 73.
[6] R. A. Pelham in H. C. Darby (ed.), *An historical geography of England before A.D. 1800* (Cambridge, 1936), 240.
[7] E. Power, *The wool trade in English medieval history* (Oxford, 1941), 29–31.

the many thousands of animals in the hands of the peasants there were the large flocks of the lay and ecclesiastical landlords. Sheep-farming as a source of wealth was the oldest of all forms of commercial farming, and retained its importance to such great monastic houses as those of Ely, Peterborough, Glastonbury, Crowland, and Canterbury cathedral priory.[1] The Cistercians, having access to great areas of moorland sheepwalk, specialised in wool production, and on their largest estates, such as Fountains in the West Riding, flocks numbered anything up to 15,000. The limestone uplands of the Pennine flanks, the Peak District, the Cotswolds, Lincolnshire and the chalklands of southern England were the great sheep runs of fourteenth-century England. In the south, the flocks of the bishop of Winchester, and of the priory of St Swithun in Winchester compared in number with those of the great Cistercian houses in the north. Battle abbey ran flocks of up to 3,000 sheep on the Sussex downs but this was only a small proportion of the total in the county. It would seem that in 1340–1 there were about 110,000 sheep in Sussex;[2] while they were most numerous on the chalk downs there were very large numbers on the grain-growing land of south-west Sussex owned by tenant farmers who grazed their animals on open-field fallow and waste. Even on the lands of Canterbury cathedral priory, the greatest of the grain-producing estates, wool production and dairy farming were just as important as cereals, especially on manors with extensive marshland pastures. Here, the peak years of wool production were 1319–21,[3] so that the survey of 1322 comes just at the right time to show the importance of sheep especially on Thanet (Fig. 37).

Numbers of sheep varied, of course, from year to year according to the incidence of disease, but on many Cistercian estates the numbers of sheep began to drop steadily well before the middle of the fourteenth century.[4] At Canterbury also the flocks never recovered from losses by floods and disease in 1324–6. While it is difficult to isolate trends in pasture farming in the early fourteenth century it is clear that the decline in arable farming was not due to expansion in pasture and livestock. With sheep so widely distributed it is not surprising that wool varied greatly in quality from the

[1] D. Knowles, I, 41–2.
[2] R. A. Pelham, 'The distribution of sheep in Sussex in the early fourteenth century', *Sussex Archaeol. Coll.*, CXXV (1934), 128–35.
[3] R. A. L. Smith (1943), 156.
[4] R. A. Donkin, 'The Cistercian Order in medieval England: some conclusions', *Trans. and Papers, Inst. Brit. Geog.*, XXXIII (1963), 191.

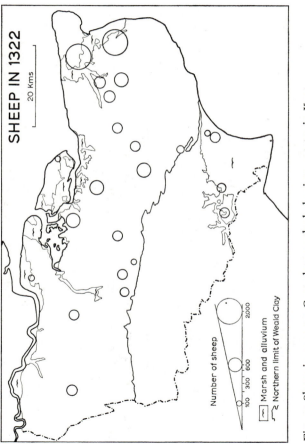

Fig. 37 Sheep in 1322 on Canterbury cathedral priory estates in Kent
Based on R. A. L. Smith, *Canterbury cathedral priory* (Cambridge, 1943), 156.

poor, coarse wools of the south-west to the fine wools of Shropshire and Lincolnshire. To judge from the prices in 1343 the best wools were the short wools of the Ryeland sheep of the Welsh border country and the west Midlands, and the long wools of the Leicester and the Lincoln sheep. Pennine wool, although not highly priced, was nevertheless far more in demand than the medium-quality wools of the chalklands, East Anglia, and the marshlands of south-east England.[1]

While oxen were kept for ploughing and carting, cattle met the need for milk and hides. The scarcity of winter feed no doubt placed restrictions on the number of cattle, for whereas pigs could root around on any waste ground and sheep exist on the poorest grazing, cattle could survive the winter only with sufficient hay, together with legumes when available. Generally, the animals supplied only the needs of the manor and village, but where there was extensive marshland grazing such as on Romney Marsh, the Fens, and the Somerset Levels, some specialised dairying and cheesemaking took place. The Canterbury manors of Romney and east Kent provide examples of efficient stock rearing and dairying,[2] On some

[1] R. A. Pelham (1936), 245.   [2] R. A. L. Smith (1943), 157–8.

6

DHG

of the manors, where dairying was only a sideline, the leasing out of cattle for milking for a cash payment was a custom that grew up in the early fourteenth century. In keeping with arable and pasture farming, the profits from dairy produce at Canterbury waned after 1327. Perhaps the death of Prior Henry of Eastry in 1331 was one of the contributory reasons why these activities never recovered. Cattle were also important on the Cistercian estates.[1] Cattle grazing was an important means of extending the limits of occupation, and the upper parts of Nidderdale and other Yorkshire dales may have been colonised for pasture in the late twelfth century. In the early fourteenth century cattle were more important than sheep on some Cistercian estates in central Wales, and from the mid-fourteenth century cattle farming increased in importance on many other Cistercian lands.[2]

Throughout the early fourteenth century inundations had serious effects on the pasture farming of the coastal areas through the loss of animals and land. On the marsh manors of Canterbury, protection of newly reclaimed land required constant embanking, draining, and the building of sea defences.[3] Around Canvey Island the fourteenth century was a time of continual fight against the sea,[4] and in the Somerset Levels[5] and the fenlands of Norfolk and Lincolnshire,[6] floods and devastation were frequent.

Lay wealth and agriculture

If goods taxed in 1334 represent the surplus that was for sale, then we might expect to find some relationship between areas with high assessments and those of the greatest grain and wool production. The year 1334 was a good year for the sale of both wool and grain. The export of wool was rising following the abolition of the home staples and the restoration of free trade in 1328. Although grain exports were lower in the reign of Edward III (1327–77) than earlier, they were still considerable, and the year 1332, on which the tax quotas of 1334 were based, was a good year and grain prices were low.[7] In 1334 we hear of seven merchants being licensed to export 52,000 quarters of wheat to Gascony.[8]

[1] R. A. Donkin (1962), 31–53.   [2] R. A. Donkin (1963), 191–2.
[3] R. A. L. Smith (1943), 186–9.
[4] B. E. Cracknell, *Canvey Island: the history of a marshland community* (Leicester, 1959), 12.
[5] P. J. Helm, 'The Somerset Levels in the Middle Ages (1086–1539)', *Jour. Brit. Archaeol. Assoc.*, XII (1949), 37–52.
[6] H. C. Darby, *The medieval Fenland* (Cambridge, 1940), 58–9.
[7] J. Titow (1960), 363; and N. S. B. Gras, 60.
[8] M. McKisack, *The fourteenth century, 1307–1399* (Oxford, 1959), 349.

Wheat and barley were the preferred cereals over most of south-east England, although oats were very widely grown and rye was important on the lighter soils. On the basis of average wheat prices between 1259 and 1500 N. S. B. Gras divided the country into somewhat arbitrary price regions, in six of which the price of wheat did not exceed 6s. per quarter.[1] These areas of low wheat prices, all in lowland England, were almost certainly areas of high wheat production. Three of his six regions, the Upper Thames basin, the Cambridge region (apart from the peat fens), and the Norwich region, coincided with the areas of the highest assessments in 1334. The fourth region, the Lower Severn, was an area of medium-to-high assessments, while the remaining two, the Bristol area and east Suffolk, were, on the whole, areas of low assessments. South-east Lincolnshire, north Kent, the Thames valley and the south coast, however, were areas of high assessments in 1334 but not among the lowest-priced wheat areas. There is little doubt that wheat was the main crop in these regions, and the price may have been kept above 6s. a quarter by the demand, especially in Kent and in Sussex from the Continent,[2] and in the Thames valley from London.[2] On the whole there is a definite coincidence between areas of high assessment in 1334 and those of considerable wheat production.

No such relationship existed between the 1334 tax and the important wool-producing areas. A valuation of 1343 (nine years later) shows that the highest priced wools were those of the Welsh border, the west Midlands, Leicestershire, and Lincolnshire. All of these, with the exception of eastern Lincolnshire, were relatively poor areas in 1334, with assessed wealth mostly below £10 per square mile. Even the chalklands, where a greater number of sheep might be expected, had low assessments. Nor was there any coincidence between low wool price areas and high assessments, not that this would be expected because wool was priced on quality not quantity. Some low-price counties, for example Norfolk and Berkshire, had a high 1334 tax; others such as Surrey, Hampshire and Suffolk had a low tax.

The pattern was very irregular and no conclusions can be drawn at present in the face of so many unknown quantities. Certain surviving local taxation rolls of before 1334 show that sheep and lambs were fairly numerous in some districts, yet they were 'far from being as plentiful as

[1] N. S. B. Gras, 41 and 47. See also E. Kneisel, 'The evolution of the English corn market', *Jour. Econ. Hist.*, XIV (1954), 46–52.
[2] R. A. Pelham (1936), 238.

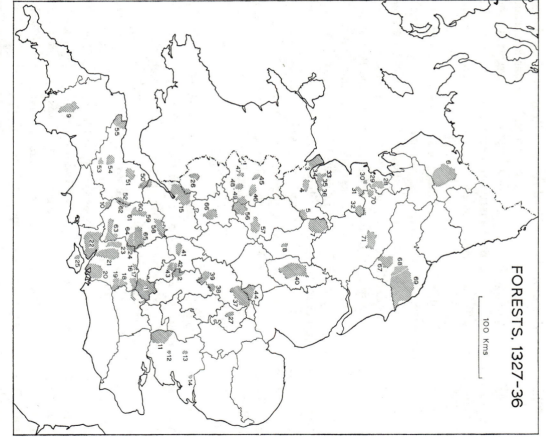

FORESTS, 1327–36

100 Kms

Fig. 38 Forests, 1327–36
Based on N. Neilson, 'The forests', being ch. 9 of J. F. Willard and W. A.
Morris (eds.), *The English government at work, 1327–1336*, I (Cambridge,
Mass., 1940), map v.
Names of forests are given opposite.

Key to Fig. 38

Berkshire
1 Windsor
Buckinghamshire
2 Bernwood
Cheshire
3 Wirral
4 Delamere
5 Macclesfield
Cumberland
6 Inglewood
Derbyshire
7 High Peak
8 Duffield
Devonshire
9 Dartmoor
Dorset
10 Gillingham
Essex
11 Essex
12 Writtle
13 Hatfield
14 Kingswood
Gloucestershire
15 Dean
Hampshire
16 Pamber
17 Bagshot
18 Alice Holt
19 Woolmer
20 Bere by Porchester
21 Bere by Winchester
22 New Forest
23 Buckholt
24 Chute
25 Isle of Wight
Herefordshire
26 Hereford Hay
Huntingdonshire
27 Wauberghe
Lancashire
28 Quernmore
29 Bleasdale
30 Myerscough
31 Fulwood
32 Blackburn
33 Simonswood
34 West Derby

35 Toxteth
36 Croxteth
Northamptonshire
37 Rockingham
38 Salcey
39 Whittlewood
Nottinghamshire
40 Sherwood
Oxfordshire
41 Wychwood
42 Shotover
43 Stowood
Rutland
44 Rutland (or Leighfield)
Shropshire
45 Lithewood
46 Wellington
47 Stretton
48 Shirlet
49 Morfe
Somerset
50 Kingswood
51 Mendip
52 Selwood
53 Neroche
54 North Petherton
55 Exmoor
Staffordshire
56 Kinver
57 Cannock
Wiltshire
58 Braden
59 Chippenham
60 Pewsham
61 Melksham
62 Selwood
63 Clarendon
64 Chute
65 Savernake
Worcestershire
66 Feckenham
Yorkshire
67 Galtres
68 Spaunton
69 Pickering
70 Bowland
71 Knaresborough

the amount of wool raised in medieval England, and the descriptions of large herds would render necessary'.[1] The low numbers would be understandable if they represented a surplus over and above the essentials needed by a family. Nor do we know whether the valuation of sheep varied from place to place. Were the Welsh border sheep, for example, valued at the same sum as marshland sheep? Were they valued on their size or fleece? In addition to these unanswered questions, another serious drawback is the fact that sheep on monastic demesne land were excluded from the subsidy.

In conclusion, therefore, the assessments suggest that the wealthiest part of England in 1334 was the grain country of the south and east Midlands, long settled and intensively farmed. The wealth of coastal Sussex rested on a sheep-and-corn husbandry in exceptionally favourable physical conditions.[2] With the exception of Norfolk and north Kent, these rich lands were situated within H. L. Gray's zone of the Midland field system. In contrast, there was little movable wealth in those parts of the country famous for their wool. It must be remembered, however, that a low assessment did not necessarily mean that an area was poor. While its resources of movable goods (grain and stock) may have been low in comparison with those of other places, it may have had non-movable resources, for example, in timber. Almost all the main forests were areas of very low assessments, e.g. Windsor, Epping, the New Forest and the Forest of Dean. The same was true of the woodlands of Essex, the Weald, the Chilterns and the west Midlands.

Forests, woodland and parks

A considerable area of England was still subject to forest law in the early fourteenth century despite the fact that the extent of royal forest had diminished by about a third since about 1250. There seem to have been at least 71 forests in the period 1327–36.[3] The largest were the New Forest, and the forests of Dean, Sherwood, Essex, Windsor, Inglewood and Pickering. Their distribution was very uneven; whereas one half of Hampshire and about a third of Wiltshire, for example, were subject to forest law, East Anglia was entirely free from it (Fig. 38).

[1] J. F. Willard (1934), 73.
[2] R. E. Glasscock (1965), 61–8; and P. F. Brandon, 'Demesne arable farming in coastal Sussex during the later Middle Ages', *Agric. Hist. Rev*, xix (1971), 113–34.
[3] N. Neilson, 'The forests', being ch. 9 of J. F. Willard and W. A. Morris (eds.), *The English government at work, 1327–1336*, I (Cambridge, Mass., 1940), 394–467.

The forests, which varied in character from open ground to dense woodland, were not only timber reserves but often important grazing areas for deer, cattle, pigs and horses, as at Duffield Frith in Derbyshire and Needwood in Staffordshire.[1] Forest timber served to augment the patches of woodland that still remained after centuries of clearing and colonisation. The value of timber and its general use meant that charcoal-burners, carpenters, smiths and other craftsmen lived and worked in and around almost all the remaining stands of woodland.[2] Local shortage of timber was an additional incentive for landlords to enclose land for small parks in the early fourteenth century; thus in Wiltshire, we know that there were two enclosed parks in Wootton Bassett by 1334, a 200-acre park at Colerne in 1311, a 95-acre park in Oaksey in 1347, and another park in Castle Combe by 1328.[3] Emparking of this kind was also taking place on the Evesham estates, and from such enclosures coppice wood and loppings were sold for firewood. In the Weald the number of parks in-creased from 46 about 1300 to 68 in 1350,[4] and they varied in size from 60 to 4,000 acres. While they were mainly recreational their assets were not wasted, and, in addition to their use as timber reserves, they also pro-vided grazing. In some there was assarting and intermittent cropping, as at Westerham in Kent.[5]

Agricultural regions?

Before concluding this section on the countryside we might ask whether there were signs in the early fourteenth century of the distinctive local economies that were evident by the sixteenth. Can their origins be seen before the Black Death or did they develop only after 1350, when the emergence of regional farming was strengthened by the break up of highly organised demesne farming?[6] Technology and exchange had not pro-gressed far enough by the early fourteenth century to allow much speciali-sation. While it is true that certain parts of the country came under the strong influence of the monastic houses this does not mean that agriculture

[1] J. R. Birrell, 'The forest economy of the Honour of Tutbury in the fourteenth and fifteenth centuries', *Univ. of Birmingham Hist. Jour.*, VIII (1962), 117–19.
[2] J. Birrell, 'Peasant craftsmen in the medieval forest', *Agric. Hist. Rev.*, XVII (1969), 91–107.
[3] W. G. Hoskins in E. Crittall (ed.), *V.C.H. Wiltshire*, IV (1959), 18.
[4] J. L. M. Gulley, 312.
[5] R. A. L. Smith (1943), 186–9.
[6] W. G. Hoskins, 'Regional farming in England', *Agric. Hist. Rev.*, II (1954), 7.

was one-sided.[1] Livestock were essential for the fertility of arable land. On the great grain-growing estates, such as those of Canterbury cathedral priory, cattle formed a vital element in the economy, and conversely even the Cistercian granges, with which we usually associate wool production, could be regarded primarily as arable holdings.[2]

Nevertheless the pattern of agriculture in early fourteenth-century England reflected the broad physical controls of soils, climate and topography. Climatically the south-west and the northern uplands were more suited to pasture, while the drier lowlands of the Midlands and the east were well suited to grain. The chalk and limestone hills were sheep country; the forest and woodland provided grazing for cattle, swine and deer, and the marshland pastures for cattle and sheep. Within this framework the specialisations of the monastic houses were a natural outcome of the areas which they settled, the Benedictines in the south and east, and the Cistercians in the north and west. By 1300 their enterprise in seeing the potential of the land they settled had already underlined the fundamental distinction between the pastoral west and the arable east. Beyond this broad division it was only in areas of very distinctive physical environment, where soils and climate made the land especially suitable to a particular use, that some specialisation in agriculture first emerged, for example on the Essex marshlands (as early as 1086?), on the Fenland pastures, on those of Sedgemoor, and on the limestone uplands. Or again, the sensitivity of farm practice to soil conditions may be seen from details for Kent about 1300. Here, oats predominated on the marshland manors while elsewhere wheat and barley were the predominant crops (Fig. 36).[3] Barley was the dominant crop of the high wolds of Yorkshire where an agriculture based upon barley and sheep had already emerged by the mid-fourteenth century.[4] But until more studies are available, we cannot be clear about the exact degree of specialisation in the fourteenth century over England as a whole.

[1] The diversity of production is well illustrated in B. Waites, *Moorland and vale-land farming in north-east Yorkshire: the monastic contribution in the thirteenth and fourteenth centuries* (York, 1967).

[2] R. A. Donkin (1963), 187.

[3] R. A. Pelham (1936), 241. Ann Smith, 'Regional differences in crop production in medieval Kent', *Archaeol. Cantiana*, LXXVIII (1963), 147–60.

[4] B. Waites, 'Aspects of thirteenth and fourteenth century arable farming on the Yorkshire Wolds', *Yorks. Archaeol. Jour.*, XLII (1967), 136–42.

INDUSTRY

England in 1334 was overwhelmingly rural. The majority of people were engaged in tilling the land, in looking after livestock, and in meeting their everyday needs from whatever materials were near at hand. This is not to say that villages were entirely self-sufficient. Peasants had to sell produce in order to get cash for rents and taxes, and to buy some goods which were not available locally such as cloth, leather, salt, pottery and metal goods. Of the various industries that catered for more than local needs, the following were of especial importance – cloth manufacture, mining and quarrying and the making of salt.

The cloth industry

The 1330s were midway in time between the 'urban' cloth industry of the thirteenth century and the 'country' cloth industry of the fifteenth. Throughout the thirteenth century, England had been exporting wool to the Low Countries and importing finished cloth. The export of raw wool reached a peak of 46,000 sacks in 1304–5[1] and although we do not have an exact figure it is likely that cloth imports reached a peak at about the same time.[2] After this date the output of home-produced cloth undoubtedly increased, paradoxically at a time which most scholars see as a period of contraction. Yet both propositions are tenable; at times when all the outward signs point to general economic difficulty there can often be found expanding sectors in an economy. The difficulty lies in knowing where the increase in production took place. The respective roles of town and country in the making and marketing of cloth in the early mid-fourteenth century is a matter of debate. The new growth seems to have been principally in the rural areas, with the towns, which had been the backbone of the thirteenth-century industry, fluctuating in their fortunes. In the 1330s the difficulties of the weavers in towns had intensified. By 1334 the weavers of Northampton, formerly about 300, had apparently disappeared completely, and their numbers were also dwindling in London, Winchester, Oxford, and Lincoln.[3] Yet we know from the 1334 Lay Subsidy that these same towns and other centres of cloth-making were still among the

---

[1] E. M. Carus-Wilson and O. Coleman, 1963, *England's export trade, 1275–1547* (Oxford, 1963), 41–122.

[2] E. M. Carus-Wilson, *Medieval merchant venturers* (2nd ed. London, 1967), 242–5.

[3] E. Miller, 'The fortunes of the English textile industry during the thirteenth century', *Econ. Hist. Rev.*, XVIII (1965), 70.

wealthiest towns in the country. In 1334, London, York, Newcastle, Bristol, Lincoln, Norwich and Oxford were among the wealthiest ten, and Salisbury, Coventry and Beverley were also wealthy towns. Clearly the wealth that was built on wool and cloth was still evident despite a recession in some places. On the other hand, cloth-making was spreading rapidly in the rural areas, made possible on the streams of the upland valleys by the spread of the fulling-mill,[1] and encouraged elsewhere by urban entrepreneurs who, free from the restrictions of the town gilds, could produce cloth more cheaply in rural areas.[2] By so doing, English cloth-makers could compete with the Flemish who had dominated the trade in 1300. An additional stimulus to home production was provided by the war-time conditions and policies under which the export of raw wool to Flanders was temporarily prohibited and the import of cloth virtually ceased. The Flemish industry never fully recovered from these measures, and, even after their withdrawal, increasingly heavy export duties on raw wool pushed up the cost for the Flemish cloth-makers. As the Flemish industry declined so the English cloth industry, in the late 1330s and 1340s, developed under economic protection and captured the home market. English industry also benefited from Edward III's policy of encouraging foreign cloth-workers to come to England. Many Flemings left the unrest and disturbances of the Low Countries. Alien weavers, dyers and fullers seem to have settled mainly in such centres as London, York, Winchester, Norwich, Bristol and Abingdon, but we also hear of them in the West Riding and in the West Country.[3] But we must not exaggerate their importance; their coming was a symptom rather than 'a cause of the progress of English enterprise'.[4] By the eve of the Black Death England was exporting a considerable amount of cloth each year.[5]

Sustained demand led to the expansion of rural cloth manufacture in the West Riding, in the Lake District, in Wiltshire and the West Country, in the Mendips, the Cotswolds, the Kennet valley and East Anglia. As the market for home produced cloth grew, so the demand for 'Ludlows' and 'Cotswolds', 'Stroudwaters', 'Westerns' and 'Worsteds' began to replace that for 'Lincoln Scarlets', 'Beverley Blues' and 'Stamfords'. Although

[1] E. M. Carus-Wilson (1967), 183–210.
[2] E. Miller (1965), 73–4.
[3] E. Lipson, *The economic history of England: the Middle Ages* (5th ed., London, 1929), 399–400.
[4] E. M. Carus-Wilson in M. Postan and E. E. Rich (eds.), *The Cambridge economic history* II (Cambridge, 1952), 415.
[5] E. M. Carus-Wilson (1967), 245.

the industry was growing it is impossible to estimate the number employed in woollen manufacture in the 1330s; we might guess at something of the order of 20–25,000, having regard to the estimates which have been made for the industry later in the century.[1]

Mining and quarrying

Mining played a small but important part in the economy of the early fourteenth century. Coal, iron, tin, lead and silver were all worked along the fringes of the uplands of the north and west, but except for tin mining in the south-west the numbers employed must have been very small (Fig. 26).

Coal, which was still insignificant as a fuel in comparison with charcoal and wood, was mined in surface workings along the Pennine flanks, in South Wales, and in the north-east, where it achieved its greatest local importance.[2] By the early fourteenth century coal was being exported from Newcastle, and the traffic in 'sea-coal' from the Tyne to London was sufficiently well established for a tax to be levied on it.[3] Coal, however, was not very popular with Londoners who in 1307 complained about the smell,[4] having earlier said that 'the air is infected and corrupted to the peril of those frequenting and dwelling in those parts'.[5] As charcoal was the first fuel for smelting, and as domestic coal burning did not appear until the late fourteenth century, the coal was mainly used in kilns for lime-burning, brewing, baking, and metal work.

The Forest of Dean, where there were still plentiful supplies of wood, was the principal centre of the iron industry, although the developing Wealden industry was already supplying the London market. Iron was also mined in the Cleveland Hills, along the Pennines, and in Furness, where a large quantity of iron formed part of the booty of a Scots raid in 1316.[6] But in many parts of the country fuel supplies for iron-working were becoming very short and for the first time the depletion of woodland resources began to cause anxiety. The iron industry, in the region of Skipton in Craven, closed down; and iron-working in the Forest of Knaresborough, in other parts of the West Riding and in Duffield Frith in

[1] See note in E. M. Carus-Wilson (1967), 261, n. 3.
[2] J. U. Nef, *The rise of the British coal industry*, 2 vols. (London, 1933), 1, 9.
[3] R. A. Mott, 'The London and Newcastle chaldrons for measuring coal', *Archaeol. Aeliana*, XL (1962), 228; R. Smith, *Sea-coal for London* (1961), 2.
[4] *Calendar of the Close Rolls, 1302–7* (H.M.S.O., 1908), 537.
[5] *Calendar of the Patent Rolls, 1281–92* (H.M.S.O., 1893), 207 and 296.
[6] L. F. Salzman, *English industries of the Middle Ages* (Oxford, 1923), 26.

Derbyshire virtually ceased.[1] The early fourteenth century was a period of slump in the industry which resulted in its disappearance from some districts and its increasing concentration in well-wooded areas. Even so, England could not meet the demand for iron for the making of arms and agricultural implements, and ore was being imported from Spain in the early fourteenth century.

Tin production was localised in Devon and Cornwall, where it was obtained by streaming. The early fourteenth century was a period of exceptional activity in tinning, and the charters of Edward I in 1305 had the desired effect of encouraging the industry and confirming the ancient privileges of the stannary men, which included exemption from ordinary taxation.[2] The main producing areas lay in Cornwall where production rose to high levels in the 1330s and 1340s.[3] Production was severely curtailed by the Black Death, but it recovered again by the late fourteenth century.[4] In Devon also the mines were prospering, although administration was chaotic, and there were complaints that the tinners were destroying good farm land at the rate of over 300 acres a year.[5] There is hardly any evidence of the working of the copper deposits of Cornwall and Devon at this time. Most of the copper used in fourteenth-century England must have been imported.

Silver and lead were precious commodities, the one needed for coin and ornament, the other for roofing and piping. The two minerals were mined together in the Mendips,[6] around Bere Alston in Devon,[7] at Alston in Cumberland,[8] in the Pennine dales of Yorkshire[9] and Durham, and in Derbyshire.[10]

Mention must also be made of stone quarrying and the building

[1] H. R. Schubert, *History of the British iron and steel industry* (London, 1957), 111–15.

[2] G. R. Lewis, *The stannaries* (London, 1908), 39.

[3] J. Hatcher, 'A diversified economy: later medieval Cornwall', *Econ. Hist. Rev.*, 2nd ser, XXII (1969), 208–27.

[4] A. R. Bridbury (1962), 25–6.

[5] H. P. R. Finberg, 175–81.

[6] J. W. Gough, *The mines of Mendip* (Oxford, 1930).

[7] L. F. Salzman, 'Mines and stannaries', in J. F. Willard *et al.* (eds.), *The English government at work, 1327–1336*, III (Cambridge, Mass., 1950), 67–104.

[8] J. Walton, 'The medieval mines of Alston', *Cumberland and Westmorland, Antiq. Soc.*, XLV (1946), 22–33.

[9] A. Raistrick and B. Jennings, *A history of lead mining in the Pennines* (London, 1965).

[10] L. F. Salzman (1923), 41–3.

industry. Although building activity was much less than in the preceding century, the early fourteenth century was nevertheless a period of great architectural achievement in the Decorated Gothic style. The west front of York Minster, the choir at Gloucester Cathedral and the nave at Exeter all belong to this period; so does the central tower at Wells, supported on crossed arches that are one of the most spectacular achievements of Gothic building. Quarrying stone was an important local industry, e.g. at Barnack in Northamptonshire; and boatloads of stone moved many miles by river, and around the coast. Thus, in 1317 Kentish rag was used for work on the Tower of London; Ramsgate stone was taken to Westminster in 1324 and 1333, Yorkshire stone to Westminster in 1343, Purbeck marble to Exeter in 1309, and Portland stone to Exeter in 1303, and to London and Westminster in 1349. Caen stone was still being brought across the Channel, for example to Exeter and Norwich.[1]

Salt making

Salt was an essential commodity in fourteenth-century England as in earlier times. By the 1330s, the main coastal producing area was in Lincolnshire, but substantial amounts of salt were also produced elsewhere along the south and east coasts, notably in Norfolk, Kent and Sussex.[2] Inland, salt was produced from the brine springs of Worcestershire and Cheshire whence it was exported through Chester to Ireland. In 1334, export to the Continent was less than it had been early in the century, and, as salt was greatly in demand, England imported some, mainly from Bourgneuf Bay south of the estuary of the Loire, and, to a lesser extent, from Spain and Portugal. In the 1320s and 1330s Scarborough, Lynn and Hull handled the bulk of the imports, and only later did London, Bristol and Yarmouth take over as the main points of distribution. Much of the home-produced and imported salt was used at the coast for salting fish; the rest went inland where it had a multitude of uses in every home including the salting of butter and cheese, and the preserving of meat.

[1] All examples from L. F. Salzman, *Building in England down to 1540* (Oxford, 1952), 119–39.
[2] A. R. Bridbury, *England and the salt trade in the later Middle Ages* (Oxford, 1955), 16–39.

TRADE AND TRANSPORT

Most places were within three or four hours' journey of one or more market centres. The markets ranged from small weekly gatherings in villages to the specialised commodity markets of the larger towns. Many places also had annual fairs, but the great international fairs of the thirteenth century had declined in importance; their functions had been assumed by the permanent and increasingly complex trading arrangements of the towns.

Any consideration of roads in the early fourteenth century must be based upon the Gough map, formerly thought to date from about 1300 but now assigned to *circa* 1360.[1] The original is on a scale of approximately 16 miles to one inch, and the more it is studied the more remarkable appears the knowledge of towns, roads, and routes, of the unknown map-maker (Fig. 39). The essence of the modern road pattern existed in the early fourteenth century, except that very few roads crossed the country from south-west to north-east. London was at the hub as it had long been, and Coventry was the great crossing point in the Midlands. On the roads we must picture travellers on foot and on horseback, merchants with pack-horses, carts, and occasionally with great four-wheeled wagons.[2] Carts were used to transport goods such as fish, grain, flour, wine, salt, cloth, hay, faggots and brushwood, peat and stone, and less frequent loads of iron, tin, and military weapons.[3] The carriage of Exchequer goods from Westminster to York in the early fourteenth century took between 10 to 14 days;[4] the journey from Malmesbury to Carlisle in 1318 took 12 days;[5] London to Gloucester was an eight-day return journey.[6] Travel and methods of transport by road formed one of the few aspects of the

[1] E. J. S. Parsons, *The map of Great Britain circa A.D. 1360, known as the Gough Map* (1958), with facsimile. A photograph of the Gough map is included in Lady Stenton's chapter on Communications in A. L. Poole (ed.), *Medieval England*, 2 vols. (Oxford, 1958), i, 208.

[2] J. F. Willard, 'Inland transportation in England during the fourteenth century', *Speculum*, i (1926), 361–74.

[3] J. F. Willard, 'The use of carts in the fourteenth century', *History*, xvii (1932), 246–50; R. A. Pelham, 'Studies in the historical geography of medieval Sussex', *Sussex Archaeol. Coll.*, lxxii (1931), 167–75.

[4] D. M. Broome, 'Exchequer migrations to York in the thirteenth and fourteenth centuries', in A. G. Little and F. M. Powicke (eds.), *Essays presented to T. F. Tout* (Manchester, 1925), 298.

[5] *Calendar of the Close Rolls, 1313–18* (H.M.S.O., 1893), 548.

[6] J. F. Willard (1926), 367.

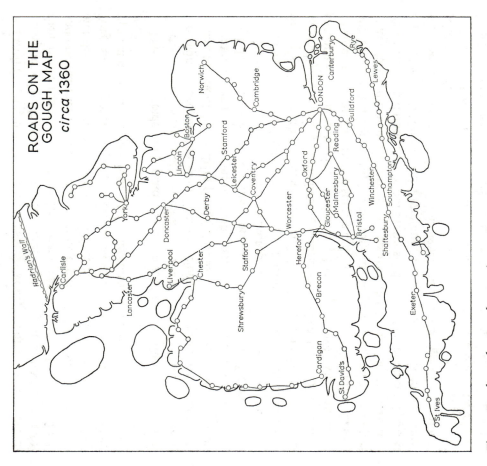

ROADS ON THE
GOUGH MAP
circa 1360

Fig. 39 Roads on the Gough map *circa* 1360
Based on E. J. S. Parsons, *The map of Great Britain circa A.D. 1360, known as
the Gough map* (Oxford, 1958).
It is impossible to provide an accurate scale.

geography of England in 1334 that the Black Death and the next century would scarcely change.

As well as transport by road there was movement along the navigable
rivers and also coast-wise traffic. Thus, boats plied to Southampton with
herrings and stockfish from the east coast, with coal from the north-east,
with wheat, malt, and iron from nearby Kent and Sussex, with ropes, sails,
and cordage from Bridport, with Purbeck stone and marble from Poole,

and with fish, tin, and slates from the south-west. At a time when overland transport of bulky goods was slow and expensive, small boats, most of them under 100 tons, played an important part in the distribution of food and raw materials. They could penetrate the rivers deep into the country, and the Thames, Trent, Yorkshire Ouse, Humber, Severn, and the East Anglian Stour were all important arteries of trade.[1] Goods were taken upstream as far as possible, and only then were they taken on by road.

The export trade of England with the rest of the Continent in the early fourteenth century was almost entirely in raw materials.[2] Wool was still the main export, and the sale of finished cloth abroad had hardly begun.[3] Grain, tin, cloth, hides, dairy produce, coal, and salt were among the main commodities exported from the many ports, large and small, around the English coast. In exchange, came wine, cloth, timber, dyestuffs, and luxury goods from a variety of places in Europe between Scandinavia and the Mediterranean. England's trade links at this time were in four directions, each associated with particular commodities. The ports of the east coast, especially Newcastle, Hull, Boston and Lynn looked to the Baltic, and exchanged coal, salt, cloth and grain, for timber, fish, wax and furs. Boston and Lynn also traded southwards with the Low Countries, Calais and France.[4] Farther south, Yarmouth, Ipswich, London and the Cinque Ports faced the manufacturing industry of the Low Countries across narrow seas, and traded raw wool for finished cloth. The south coast ports, especially Southampton and Plymouth, looked southwards to the wine trade with Gascony, as did Bristol, which also looked west to the trade with Ireland.

The location of individual ports in relation to the European mainland largely determined the nature of their trade and specialisation; only London handled almost every commodity. By the early fourteenth century many ports were associated with a particular trade: Lynn, for example, with wool and grain, Boston with wool and salt, Yarmouth with herring and salt, Newcastle with coal and hides, and Hull with salt. The fortunes of the ports were closely linked with the prosperity of the cities which

[1] R. A. Pelham (1936), 264.

[2] M. M. Postan, 'The trade of medieval Europe: the north', being ch. 4 of M. M. Postan and E. E. Rich (eds.), *The Cambridge economic history of Europe*, II (Cambridge, 1952).

[3] E. M. Carus-Wilson and O. Coleman, 41–122.

[4] E. M. Carus-Wilson, 'The medieval trade of the ports of the Wash', *Med. Archaeol.*, VI–VII (1962–3), 182–201.

they served, Boston with Lincoln, Yarmouth with Norwich, Hull with York, Southampton with Winchester and Salisbury. Wool, the 'sovereine marchandise', was exported in the early fourteenth century at an average rate of 30,000 sacks a year, equivalent to just over 4,000 tons.[1] London handled almost half of this, followed in order of importance by Boston, Hull, Southampton, Ipswich, Lynn, Newcastle and Yarmouth.[2] Cloth was exported in small quantities; Bristol exported from the growing Cotswold woollen industry, and London and Yarmouth from the worsted industry of East Anglia.

Among imports, wine featured prominently, and the early fourteenth century was the peak period of the Anglo-Gascon wine trade. Between 1305 and 1336, when the Gascon trade was relatively stable, the average annual export of wine from the Bordeaux ports was about 83,000 tuns, with up to 100,000 tuns in peak years such as 1308–9. From the evidence of wine imports contained in the English customs accounts, the reliability of which has been debated, it seems that England probably took between a fifth and a quarter of this amount.[3] On the outbreak of the Hundred Years' War in 1337 wine exports dropped, and prices rose on account of limited supplies and the increased cost of escorting the wine vessels safely across the Channel. After a partial recovery the trade was curtailed in 1345 on the resumption of hostilities, and again in 1348 when the Black Death swept through Europe.

TOWNS AND CITIES

Under the double rating plan used in 1334 boroughs were taxed at the higher rate of the tenth, but the selection of boroughs by the taxers does not seem to have conformed to any consistent rules. Consequently some places which were selected as taxation boroughs could hardly have been thought of as towns, either by their size or their economic activity. Conversely some very large towns, for example Boston and Coventry, were not considered taxation boroughs. No doubt some towns were anxious not to be regarded as boroughs so that they might escape the higher rate

[1] One sack = 364 lb. E. M. Carus-Wilson and O. Coleman, 13.

[2] E. M. Carus-Wilson and O. Coleman, 122–37. A detailed study of the business of a leading wool merchant in 1337 is E. B. Fryde, *The wool accounts of William de la Pole* (York, 1964).

[3] M. K. James, 'The fluctuations of the Anglo-Gascon wine trade during the fourteenth century', *Econ. Hist. Rev.*, 2nd ser., IV (1951), 170–96. See also E. M. Carus-Wilson and O. Coleman, 201–7.

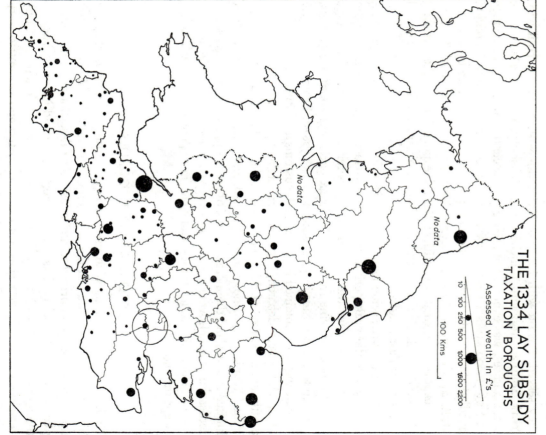

THE 1334 LAY SUBSIDY
TAXATION BOROUGHS

Assessed wealth in £'s

10 100 250 500 1000 1600 2200

100 Kms

No data

No data

No data

Fig. 40 The 1334 Lay Subsidy: taxation boroughs
Sources as for Fig. 35.
London taxed at a fifteenth is shown by an outline circle for comparative
purposes in view of its great size.

of tax. Coventry's protest in the early fourteenth century that it was not a borough[1] was upheld, and in consequence it paid tax at a fifteenth in 1334.

Fig. 40 shows the taxation boroughs in 1334 and their respective wealth. The Leicestershire and Shropshire boroughs are those taxed in 1332 and 1336, and it is presumed that they are the unspecified boroughs mentioned in the 1334 Accounts for these counties.[2] Carlisle, Corbridge, Bamburgh and Appleby were taxation boroughs in 1336 having escaped tax in 1334. The city of London, not strictly a borough as it was taxed at a fifteenth, is included on the map for comparative purposes.

Taxation boroughs

There were surprisingly few taxation boroughs in the main zone of wealth. Most of the Midland counties had only one each. Rutland, Bedfordshire and Buckinghamshire had none in 1334. Of the wealthy counties, Oxfordshire had three boroughs and Norfolk four. Taxation boroughs were far more numerous in the west and south. Shropshire, Staffordshire and Herefordshire had nine between them, Hampshire, Surrey and Sussex had 18; Somerset, Devon and Cornwall had no less than 57 – a number which probably reflects the multiplication of small boroughs as a result of late colonisation. The north had hardly any. The 1334 taxers for the most part followed the choices of previous taxers, and most taxation boroughs had been selected before 1306.[3] A comparison of taxation and parliamentary boroughs for the period 1240–1336 shows that taxers and sheriffs used much the same criteria for selection; but there were differences. Some parliamentary boroughs were omitted by the 1334 taxers, and conversely some taxation boroughs were not parliamentary boroughs. This may have been due to the fact that the meaning of 'market town' (a term used in the instructions to the taxers between 1290 and 1297) varied from place to place. Many small settlements, some hardly more than villages, qualified as market towns in the eyes of the taxers of Devon and Cornwall, but would no doubt have failed in the eyes of Norfolk men.

The choice of some places and not others is frequently most puzzling, and is exemplified in Hampshire where, of the 20 chartered boroughs in

[1] P.R.O., E. 368, L.T.R. Memoranda Roll 77, m. 8.

[2] P.R.O., E. 179/133/3 and E. 179/166/4.

[3] J. F. Willard, 'Taxation boroughs and parliamentary boroughs, 1294–1336', in J. G. Edwards *et al.* (eds.), *Historical essays in honour of J. Tait* (Manchester, 1933), 417–35.

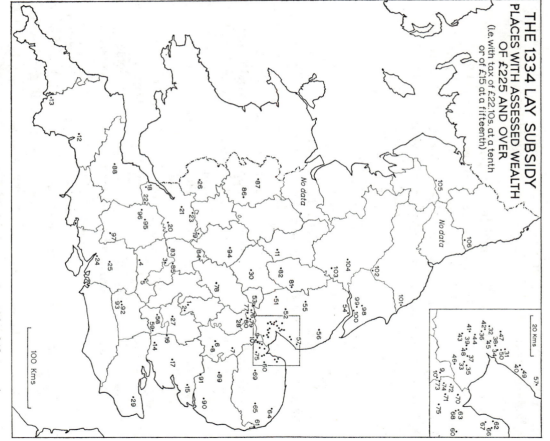

THE 1334 LAY SUBSIDY
PLACES WITH ASSESSED WEALTH
OF £225 AND OVER
(i.e. with tax of £22.10s. at a tenth
or of £15 at a fifteenth)

100 Kms

20 Kms

No data

Fig. 41 The 1334 Lay Subsidy: places with assessed wealth of £225 and over
Sources as for Fig. 35.
Names of places are given on pp. 181–2.

Table 4.2 *Places with assessed wealth of £225 and over*

(i.e. with tax of £22 10s. and over at a tenth or of £15 and over at a fifteenth)

NOTE: The numbers in brackets indicate the places on Fig. 41.

	£		£
Bedfordshire		**Kent**	
Leighton Buzzard (1)	249	Canterbury (29)	599
Luton (2)	349	**Leicestershire.**	
Berkshire		Leicester (30)	267
Abingdon (3)	269	**Lincolnshire (Holland)**	
Newbury (4)	412	Boston (31)	1,100
Reading (5)	293	Donington (32)	250
Cambridgeshire		Fleet (33)	270
Cambridge (6)	466	Frampton (34)	255
Ely (7)	358	Gedney (35)	233
Fulbourn (8)	293	Gosberton (36)	450
Leverington (9)	360	Holbeach (37)	495
Wisbech (10)	410	Kirton (38)	413
Derbyshire		Moulton (39)	465
Derby (11)	300	Old Leake (40)	315
Devonshire		Pinchbeck (41)	675
Exeter (12)	366	Quadring (42)	225
Plymouth (13)	400	Spalding (43)	630
Essex		Surfleet (44)	315
Barking (14)	341	Sutterton (45)	320
Colchester (15)	261	Sutton St James (46)	375
Waltham Holy Cross (16)	262	Swineshead (47)	285
Writtle (17)	267	Whaplode (48)	480
Gloucestershire		Wrangle (49)	235
Bristol (18)	2,200	Wyberton (50)	240
Campden (19)	255	**Lincolnshire (Kesteven)**	
Cirencester (20)	250	Grantham (51)	293
Gloucester (21)	541	New Sleaford (52)	241
Marshfield (22)	270	Stamford (53)	359
Tewkesbury (23)	243	**Lincolnshire (Lindsey)**	
Hampshire		Barton upon Humber (54)	246
Southampton (24)	511	Lincoln (55)	1,000
Winchester (25)	515	Louth (56)	227
Herefordshire		Wainfleet (57)	233
Hereford (26)	605	**Middlesex**	
Hertfordshire		Harrow (58)	257
St Albans (27)	265	London (59)	11,000
Huntingdonshire		**Norfolk**	
Yaxley (28)	227	Gayton (60)	225

Table 4.2 (cont.)

	£		£
Norfolk (cont.)		Shrewsbury (87)	800
Great Yarmouth (61)	1,000	**Somerset**	
Heacham (62)	248	Bridgwater (88)	260
King's Lynn (63)	500	**Suffolk**	
North Walsham (64)	225	Bury St Edmunds (89)	360
Norwich (65)	946	Ipswich (90)	645
Sedgeford (66)	233	Sudbury (91)	281
Snettisham (67)	285	**Surrey**	
South Lynn (68)	270	Bramley (92)	298
Swaffham (69)	300	Godalming (93)	247
Terrington (70)	607	**Warwickshire**	
Tilney (71)	450	Coventry (94)	750
Walpole (72)	533	**Wiltshire**	
Walsoken (73)	396	Bremhill (95)	233
West Walton (74)	345	Corsham (96)	225
Wiggenhall (75)	555	Salisbury (97)	750
Northamptonshire		**Yorkshire (East Riding)**	
Barnack (76)	269	Beverley (98)	500
Castor (77)	276	Cottingham (99)	330
Northampton (78)	270	Hull (100)	333
Paston (79)	251	**Yorkshire (North Riding)**	
Peterborough (80)	383	Scarborough (101)	333
Nottinghamshire		**Yorkshire (West Riding)**	
Newark on Trent (81)	390	York (102)	1,620
Nottingham (82)	371	Doncaster (103)	255
Oxfordshire		Pontefract (104)	270
Bampton (83)	969	**Cumberland**	
Banbury (84)	267	Penrith (105)	398
Oxford (85)	914	**Northumberland**	
Shropshire		Newcastle upon Tyne (106)	1,333
Bridgnorth (86)	244		

the county, only five were considered taxation boroughs in 1334;[1] to these must be added the soke of Winchester which was listed separately as a taxation borough in addition to the city itself. Of the remaining 14, four were treated as ancient demesnes and taxed at a tenth, ten were taxed

[1] M. W. Beresford, 'The six new towns of the bishops of Winchester 1200–55', *Med. Archaeol.*, III (1959), 213.

at a fifteenth, and one was not mentioned. The 1334 taxers must have been as puzzled as we are why Porchester was a taxation borough and not Overton, and why New Alresford and not New Lymington. Whatever the criteria of selection, perhaps a combination of commercial activity, population, or most probably, local reputation, they were not uniformly applied. Many of the taxation boroughs of the south-west were taxed at only a fraction of the amounts paid by some large villages in eastern England.

Fig. 41 shows those places which had assessed wealth of £225 and over, i.e. with quotas of either £15 and over at a fifteenth or with £22 10s. and over at a tenth. Comparison with Fig. 40 shows that only about one-fifth of the taxation boroughs were in this category. Only 3 out of 57 in the west country qualified, only 2 of the 18 in Hampshire, Surrey and Sussex, and only 3 of the 11 in the west Midlands. Even if they were locally important, almost all the boroughs of the west and south had much less wealth than many places in midland and eastern England that were not considered boroughs. Thus almost every fenland township in the Holland division of Lincolnshire and in Norfolk marshland had assessed wealth of over £225. Clearly most of these were not towns. They were wealthy because they included extensive and rich areas of agricultural land.[1] With these exceptions the wealthier places were the large trading centres. Some were taxation boroughs as noted above, others were towns that had somehow escaped the higher rate of tax, for example Coventry, Boston, Bury St Edmunds, and Newbury. A few were probably only large market villages such as Fulbourn (Cambs), Marshfield (Gloucs), Yaxley (Hunts). Most of the larger trading centres were situated in the east Midlands and East Anglia. They were conspicuously lacking in the west Midlands and the border counties, in the south-west, in the north, and on the south coast. The Cinque Ports were excluded from the Lay Subsidy, but even so, the impression given by the map may be correct, for, with the exception of Southampton and Plymouth, the south coast trade was shared between many small ports rather than concentrated in any large one.

The chief towns

Table 4.3 shows all towns with assessed wealth of £300 and over, ranked according to their values. Allowance has been made for the different rates of tax; the yields for taxation boroughs have been multiplied by ten; and

[1] R. E. Glasscock (1963), 120.

Table 4.3 *Ranking list of chief towns in 1334*

It is difficult to construct a satisfactory ranking list of towns in 1334. Many other settlements would qualify for inclusion in the list below, on the basis of their wealth alone. Thus there were nineteen fenland townships with valuations of £300 or more; the largest of these were Pinchbeck (£675) and Spalding (£630). Another addition would be Bampton *cum membris* (in Oxfordshire) with a valuation of £969. Penrith, too, with £398 in 1336 would be included, but it was taxed at a tenth as Ancient Demesne. Furthermore, Chester and Durham were not taxed, and they would certainly come fairly high on the list. Even so, the table may serve to indicate some, at any rate, of the main facts about the relative sizes of English towns in 1334. (F) denotes the lower rate of the fifteenth and so indicates the towns that were not considered to be taxation boroughs in 1334, although some of them were at other dates.

	Assessed wealth £		Assessed wealth £
London (F)	11,000	Beverley	500
Bristol	2,200	Cambridge	466
York	1,620	Newbury (F)	412
Newcastle upon Tyne	1,333	Plymouth	400
Boston (F)	1,100	Newark on Trent (F)	390
Great Yarmouth	1,000	Peterborough *cum membris* (F)	383
Lincoln	1,000	Nottingham	371
Norwich	946	Exeter	366
Oxford	914	Bury St Edmunds (F)	360
Shrewsbury	800	Stamford	359
Lynn (King's and South)	770	Ely *cum membris* (F)	358
Salisbury	750	Luton (F)	349
Coventry (F)	750	Barking (F)	341
Ipswich	645	Hull	333
Hereford	605	Scarborough (F)	333
Canterbury	599	Cottingham, Yorks. E.R. (F)	330
Gloucester	541	Derby	300
Winchester	515	Swaffham (F)	300
Southampton	511		

those for places such as Boston, Coventry, and Spalding have been multiplied by fifteen.[1] Some totals such as those for Ely and Peterborough include suburbs. The figure for Plymouth covers Sutton Prior and Sutton Vautort. Chester and Durham, not taxed, are omitted. So is Bampton, in Oxfordshire, with a huge quota that included many nearby places. Moreover, £300 is an arbitrary minimum and some towns of local importance were not far below this figure. They included Reading, Northampton, Leicester, Colchester, Bridgwater, Cirencester and Bridgnorth.

In spite of these difficulties, the list provides a general indication of the most important towns in England in 1334 and of their relative wealth. Some generalisations emerge from the list. Firstly, most of the towns listed lay south of the Trent and the Severn. In the north, only Newcastle, York, Beverley, Nottingham, Hull and Scarborough were of comparable importance to the towns of the south; and in the west and south-west only Bristol, Shrewsbury, Hereford, Gloucester, Exeter and Plymouth. Secondly, all the leading towns were either ports or centres of cloth manufacture. Thirdly, there was the overwhelming predominance of London, a city of perhaps 50,000 people, which had more wealth than the three leading provincial cities, Bristol, Newcastle and York, combined. It was 'a metropolis, bearing far more resemblance to the great cities of northern Continental Europe than to any other English town'.[2]

---

[1] The ranking shows a different order from that in which the two different rates have not been equated as in W. G. Hoskins, *Local history in England* (London, 1959), 176.

[2] S. C. Thrupp, *The merchant class of medieval London* (Chicago, 1948), 1.

The geographical changes in England between 1334 and 1600 formed part of a far wider transformation. From Norman times English kings had been pre-occupied with France, but after the so-called Hundred Years' War (1338–1453) nothing was left of English territory on the Continent except Calais, and that was lost in 1558. Factional struggles of the Wars of the Roses (1455–85) had brought a new dynasty to the throne; and the crowning of Henry Tudor in 1485, we can see in retrospect, was the beginning of a new age. Savage and ruthless though the Wars of the Roses had been, they had not inflicted such damage on the countryside as, say, the disorders of the twelfth century. Domestic events within the British Isles were overshadowed by changes of a world-wide character. When Columbus crossed the Atlantic in 1492, and when Portuguese mariners rounded the Cape of Good Hope to India in 1497–9, a new epoch had been inaugurated. The trade of the marginal seas of Europe was now to be extended to the great oceans beyond. England at first played but little part in the maritime explorations and expansion of the sixteenth century. Its searches for a north-west and a north-east passage to the East came to relatively little. But other routes were open, even if they involved conflict with the Spanish and the Portuguese. Drake's voyage around the world in 1577–80, Raleigh's settlement in Virginia in 1558–95, the foundation of the East India Company in 1600 – these were among the symptoms of the change in England's position from that of an off-shore island of a continent to the centre of world trade routes. Clearly, the England of 1600 was very different from that of the early fourteenth century.

Not only was the geography of England changing, but men in general were becoming more self-conscious, or at any rate more vocal, about it. The invention of printing in the fifteenth century and its introduction to England by William Caxton in 1476 resulted in an explosion of information. The Tudors, in the words of Charles Whibley, 'recognised that the most brilliant discovery of a brilliant age was the discovery of their country'.[1] At any rate, a rich tradition of topographical writing was launched upon its course. John Leland's 'Itinerary' of the 1530s and 1540s was not printed until many years after his death, but William Harrison's *Historical description of the island of Briain* appeared in 1577, and William Camden's *Britannia* in 1586. Then, too, came the printed maps of Christopher Saxton in the 1570s and 1580s and those of John Norden in the 1590s. There were also other more detailed descriptions – William Lambarde's *Perambulation of Kent* (1576) and John Stow's *Survey of London* (1598) were the precursors of a long line of local studies.

POPULATION

Evidence about population during the later Middle Ages is usually either direct but incomplete, or indirect and controversial, and trends and changing distributions are more easily established than absolute numbers. By about 1330 growth had ceased, and population was possibly declining; the rapid rate of increase in the thirteenth century had certainly been retarded.[2] Population growth had outstripped the means of subsistence, producing widespread malnutrition and increasing susceptibility to famine and disease. Harvest failures and years of summer epidemics were accompanied by exceptionally high death rates on some Winchester manors,[3] and generally rising wage rates and falling food prices suggest a declining population.[4] If the first symptoms of decline appeared before 1348, it was nevertheless the Black Death of 1348–50 which decimated an already vulnerable and unstable population.

The invasion of the British Isles by bubonic plague in 1348 was only an incident in a great epidemic outburst of the disease from its Indian home. Between about 1340 and 1352 this outburst involved most of Asia Minor, much of North Africa, the whole of Europe and some of the islands lying

[1] C. Whibley, 'Chronicles and antiquarians', being ch. 15 of *The Cambridge history of English literature* (Cambridge, 1908), III, 313.

[2] J. E. T. Rogers, *Six centuries of work and wages* (London, 1884), I, 217; J. C. Russell, *British medieval population* (Albuquerque, New Mexico, 1948), 246–60; B. F. Harvey, 'The population trend in England, 1300–1348', *Trans. Roy. Hist. Soc.*, 5th ser, XVI (1966), 23–42.

[3] M. M. Postan and J. Titow, 'Heriots and prices on Winchester manors', *Econ. Hist. Rev*, 2nd ser, II (1958–9), 392–411.

[4] M. M. Postan, 'Some economic evidence of the declining population of the later Middle Ages', *Econ. Hist. Rev*, 2nd ser, II (1949–50), 221–46.

off that continent such as the Channel Islands, the British Isles, and Greenland. The arrival of the plague added to the pattern of mortality; diseases such as smallpox, measles, typhus fever and dysentery were repeatedly epidemic in Britain during the later Middle Ages; pneumonia undoubtedly occurred in epidemic form in the winter months, and whooping cough, the enteric fevers and influenza in all probability were also epidemic at times.[1] After the plague subsided in England in 1350, the country seems to have been free from a major eruption of epidemic disease for ten or eleven years; the epidemics of 1361–2, of 1369 and of 1374 may have been of bubonic plague or another deadly disease.

It has been variously estimated that the Black Death of 1348–50 carried off between one-third and one-half of the population. Studies of particular manors and of particular districts, however, reveal widely differing death rates, and the incidence of the plague was extremely irregular. Plague, both the bubonic and the more deadly pneumonic type, appeared in England in August 1348, entering through the south-western ports; it spread to Bristol and thence by way of Oxford to London, which in early 1349 it spread northwards, reaching Yorkshire in March, by which time it had almost ceased in London, although it raged in York until the end of the summer.[2] Its toll was heaviest in crowded towns, especially in ports. At Bristol, between 35% and 40% of the population were victims.[3] Mortality was heavier among clergy than laity; by the very nature of their profession priests were exposed to contagion, and, as a group, their average age was higher than that of the community as a whole.[4] Mortality rates among parish clergy were highest in the dioceses of Exeter, Winchester and Norwich; the deanery of Kenn, to the south of Exeter, was the worst hit deanery in all England, and lost 86 incumbents from its 17

[1] J. F. D. Shrewsbury, *A history of bubonic plague in the British Isles* (Cambridge, 1970), 37–263; P. Ziegler, *The Black Death* (London, 1969).

[2] C. Creighton, *A history of epidemics in Britain from A.D. 664 to the extinction of plague* (Cambridge, 1891), 116–18; J. F. D. Shrewsbury, 37–53; E. Miller, 'Medieval York' in P. M. Tillott (ed.), *V.C.H. Yorkshire: The city of York* (1961), 85; J. M. W. Bean, 'Plague, population and economic decline in England in the later Middle Ages', *Econ. Hist. Rev*, 2nd ser, xv (1962–3), 422–37.

[3] C. E. Boucher, 'The Black Death in Bristol', *Trans. Bristol and Glos. Archaeol. Soc*, 60 (1938), 31–46.

[4] G. G. Coulton, *Medieval panorama. The English scene from Conquest to Reformation* (Cambridge, 1938), 495–503; J. C. Russell, 218 and 230; Y. Renouard, 'Conséquences et intérêt démographiques de la Peste Noire de 1348', *Population*, III (1948), 459–66.

churches between 1349 and 1351.[1] On manors of the see of Winchester, in central southern England, about a third of the population died, but mortality varied widely, not only from manor to manor but from tithing to tithing. At Witney in Oxfordshire and at Downton in Wiltshire, A. Ballard estimated a mortality rate of about 66% in 1349; at Brightwell in Berkshire, on the other hand, the mortality was less than 30%.[2] In the hundred of Farnham, in Hampshire, 344 heads of households died within three years (185 in 1348–9, 101 in 1349–50 and 58 in 1350–1), representing between one-third and one-half of the total.[3]

In Essex, 70 tenants appear to have died from plague on the manor of Fingreth in Chelmsford hundred during the first six months of 1349, but in the adjacent hundred of Ongar the effects were slight and only two places out of 25 received any tax relief in 1352.[4] Tax reliefs given in 1352, 1353 and 1354 to villages in Norfolk hard-hit by plague show an interesting distribution of stricken communities; the plague appears to have entered by the ports of Yarmouth and Lowestoft and by the smaller Norfolk ports, and it was severe around these places.[5] The first victims on manors of Crowland abbey in Cambridgeshire were reported in October 1348; between mid-May and July 1349 plague was rife; and by January 1350 it was passing away. At Dry Drayton 20 out of 42 tenants (47%) died, at Cottenham 33 out of 58 (57%), and at Oakington 35 out of 50 (70%).[6]

To the west and north, in the diocese of Lincoln (which then stretched to Bedfordshire and Oxfordshire) 40% of all benefices became vacant by death between Lady Day (25 March) 1349 and Lady Day 1350. Mortality was highest in Lincolnshire itself (48% in the archdeaconry of Lincoln and 57% in the archdeaconry of Stow) and lowest in Oxfordshire (34% in the archdeaconry of Oxford).[7] In the large diocese of York (which

[1] G. G. Coulton, 496; W. G. Hoskins, *Devon* (London, 1954), 169–70.

[2] A. Ballard, 'The manors of Witney, Brightwell and Downton' in A. E. Levett (ed.), *The Black Death on the estates of the see of Winchester* (Oxford, 1916), 181–216.

[3] E. Robo, 'The Black Death in the hundred of Farnham', *Eng. Hist. Rev.*, XLIV (1929), 560–72.

[4] J. L. Fisher, 'The Black Death in Essex', *Essex Review*, LII (1943), 13–20; M. W. Beresford, 'Analysis of some medieval tax assessments: Ongar Hundred' in W. R. Powell (ed.), *V.C.H. Essex*, IV (1956), 296–302.

[5] K. J. Allison, 'The lost villages of Norfolk', *Norfolk Archaeol.*, XXXI (1957), 116–62; G. G. Coulton, 496.

[6] F. M. Page, *The estates of Crowland abbey* (Cambridge, 1934), 120–5.

[7] A. H. Thompson, 'The registers of John Gynewell, bishop of Lincoln, for the years 1349–1350', *Archaeol. Jour.*, LXVIII (1911), 301–60.

included all or parts of Nottinghamshire, Lancashire, Westmorland and Cumberland as well as Yorkshire) mortality rates also varied considerably. In the city of York it was 32%, and in the diocese as a whole it was 39%. A. H. Thompson concluded that 'mountainous country on the one hand and marshland on the other were comparatively immune from pestilence, while normal agricultural country and the lower highlands suffered most heavily'; thus in the moorland deanery of Cleveland, mortality was only 21% compared with 61% in that of Dickering, on the Wolds. Mortality was highest where population was most thickly settled.[1]

Although mortality was high there are few unequivocal instances of the total depopulation of villages; the last recorded reference to Ambion in Leicestershire was in 1346, and it may have been completely depopulated by the plague; so, too, Tilgarsley and Tusmore in Oxfordshire and Middle Carton in Lincolnshire.[2] Such phenomena were rare; settlements shrank rather than disappeared.[3] One further point must be made. Although population was generally declining during the third quarter of the fourteenth century, some localities and towns witnessed an increase; the population of York, for example, has been estimated to have been 50% higher in 1377 than it had been just before 1348.[4]

Some idea of the distribution of population during the fourteenth century can be derived from the Poll Tax returns of 1377. This tax was imposed at the flat rate of a groat (4d.) a head on the lay population; only those under 14 years old and those who regularly begged for a living were exempted. Various attempts have been made to calculate the total population of the country from these figures, notably by J. C. Russell, but all involve varying degrees of conjecture and controversy.[5] The Poll Tax is a better guide to relative densities of population than to absolute numbers. But even then, assumptions are involved: that there were no significant differences from one county to another in the proportion of the population

---

[1] A. H. Thompson, 'The pestilences of the fourteenth century in the diocese of York', *Archaeol. Jour.*, LXXI (1914), 97–154.

[2] K. J. Allison, M. W. Beresford and J. G. Hurst, 'The deserted villages of Oxfordshire' (Leicester, 1965), 44–5; W. G. Hoskins, *Essays in Leicestershire history* (Liverpool, 1950), 104, and *The making of the English landscape* (London, 1955), 93.

[3] M. W. Beresford, *The lost villages of England* (London, 1954), 159, 269, 286 and 289.

[4] E. Miller, 84.

[5] J. C. Russell, 132–46; J. Krause, 'The medieval household: large or small?', *Econ. Hist. Rev.*, 2nd ser., IX (1956–7), 420–32; J. Stengers, review note, *Revue Belge de Philologie et d'Histoire*, XXVIII (1950), 600–6.

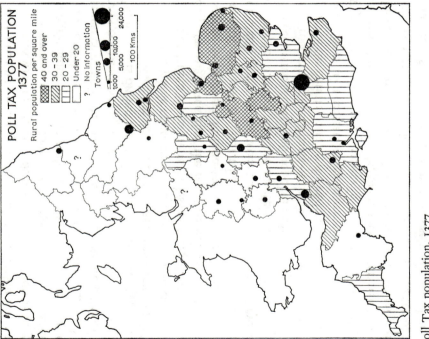

Fig. 42 Poll Tax population, 1377
Based on J. C. Russell, *British medieval population* (Alberquerque, 1948),
132–3, 142–3 (P.R.O. Exchequer Lay Subsidies E. 179).
The Poll Tax figures refer to lay population over 14 years of age.

who were under 14 years old, or in the proportion who were mendicants, or in the proportion who evaded the tax.[1] These assumptions must be borne in mind when looking at any map based on these returns (Fig. 42). There was a sharp contrast between highland and lowland zones; no county to the north of a line joining the Severn and Humber estuaries had thirty or more taxpayers to the square mile. Averages, however, conceal wide variations within individual counties; the average density for Sussex was 25 taxpayers per square mile, but along the coastal plain it was almost

[1] M. W. Beresford, 'The Poll Taxes of 1377, 1379 and 1381', *Amateur Historian*, III (1956–8), 271–8.

certainly 40.[1] Both within individual counties and within the country as a whole, the areas with the greatest densities of population were generally the principal grain-growing regions; the largest towns were those associated with agricultural markets and the cloth industry or with overseas trade. London, with its 23,000 or so taxpayers, stood far above York, the second largest city, with just over 7,000. Next in size were Bristol (6,000), Coventry (5,000) and Norwich (4,000).

Population trends during the half-century or so after 1377 are difficult to discern. J. C. Russell has suggested that between 1377 and about 1400 the population as a whole continued to decline by another 5%, and that between 1400 and 1430 it was more or less stable.[2] M. M. Postan, on the other hand, has suggested that the last two decades of the fourteenth century saw some recovery in industry and agriculture, as did the early years of the fifteenth century, but whether they were also years of rising population seems uncertain. Between about 1410 and 1430 there was some general economic recovery, but the succeeding three decades were years of declining economic activity and population.[3] There were certainly signs of growing prosperity in many parts of the country round about 1400, and in some parts of the country during the early fifteenth century, but growing prosperity does not necessarily mean a growing population.[4] Similarly, declining prosperity does not necessarily mean a declining population; hence in terms of population trends, the numerous tax reliefs of the first half of the fifteenth century are difficult to interpret. But they give clear signs of an economic contraction during the second quarter of the century. In Leicestershire, one of the more densely populated counties in 1377, the tax reduction of 16% in 1445 was distributed unevenly throughout the county; some of the larger settlements of the county appear to have suffered a decline in wealth, and possibly numbers, that was greater than average. The 1334 assessment at Melton Mowbray was cut in 1445

[1] R. A. Pelham, 'Fourteenth-century England' in H. C. Darby (ed.), *An historical geography of England before A.D. 1800* (Cambridge, 1936), 230–65. See also C. T. Smith, 'Population' in W. G. Hoskins and R. A. McKinley (eds.), *V.C.H. Leicestershire*, III (1955), 133–6.

[2] J. C. Russell, 269.

[3] M. M. Postan (1949–50), 245, and 'The fifteenth century', *Econ. Hist. Rev.*, IX (1938–9), 160–7.

[4] P. F. Brandon, 'Arable farming in a Sussex scarp-foot parish during the late Middle Ages', *Sussex Archaeol. Coll.*, C (1962), 60–72; J. A. Raftis, *The estates of Ramsey abbey* (Toronto, 1957), 264–5; R. A. L. Smith, *Canterbury cathedral priory: a study in monastic administration* (Cambridge, 1943), ix.

by 38%, at Wigston Magna by 40% and at Barrow upon Soar by 47%, and the largest reduction of all was one of 60% at Humberstone.[1] Six villages which received tax cuts of between 30% and 40%, and two with cuts of more than 40%, were totally depopulated and deserted later in the century. In these villages a notable falling-off in population may have been produced by successive pestilences.

During the economic contraction between 1350 and 1450, some villages were deserted. There was in particular a retreat from marginal soils as the pressure of population upon land was relaxed. On the Lincolnshire Wolds, for example, large reliefs from tax were granted in 1352–4, and the amalgamation of parishes which had already begun in 1428 suggests a retreat of settlement there.[2] Most deserted villages in Norfolk were situated in the west on the light marginal soils, and even in south and east Norfolk the deserted villages tended to lie on areas of lighter soils, such as the plateau gravels north of Norwich and the sands and gravels of the Wensum valley. Many of these desertions represented a true retreat of settlement from marginal soils.[3] Something similar may have taken place on the Wolds of the East Riding of Yorkshire.[4] But many more villages were deserted later for quite other reasons.[5]

Towards the middle of the fifteenth century, population apparently began to increase again; at the beginning of the next century it was increasing rapidly, and it probably continued to do so throughout the sixteenth century. But growth was selective rather than general. In southeast Lancashire there was a marked expansion in the number of chapelries dependent upon parish churches between 1470 and 1548, more so than elsewhere in the county because of the growing textile industries.[6] Expansion of the cloth industry in England saw the rise of new centres of population, while some of the older centres stagnated or declined. Growth was associated more with rural industry and less with agriculture. The

---

[1] C. T. Smith, 137; W. G. Hoskins, 'The population of an English village, 1086–1801: a study of Wigston Magna', *Trans. Leics. Archaeol. and Hist. Soc*, XXXIII (1957), 15–35.

[2] M. W. Beresford (1954), 164 and 170–2.

[3] J. Saltmarsh, 'Plague and economic decline in England in the later Middle Ages', *Cambridge Hist. Jour*, VII (1941–3), 23–41; K. J. Allison, 138–40.

[4] M. W. Beresford, 'The lost villages of Yorkshire. Part II', *Yorks. Archaeol. Jour*, XXXVIII (1952–5), 44–70; M. W. Beresford (1954), 150, 170 and 241.

[5] M. W. Beresford (1954), 164 and 170–2.

[6] G. H. Tupling, 'The pre-Reformation parishes and chapelries of Lancashire', *Trans. Lancs. and Cheshire Antiq. Soc*, LXVII (1957), 1–16.

Weald of Kent, for example, sparsely peopled in the fourteenth century had, by the middle of the sixteenth century and in consequence of the growth of the cloth and iron industries, joined the arable and sheep lands of north-east Kent as the most densely peopled parts of the county.[1]

Migration probably contributed as much to changing population distributions as differential birth and death rates. During the sixteenth century there was increasing seasonal migration, principally of rural land-owners and their families, to and from London, as well as more permanent migrations. The population of London and its immediate suburbs grew more rapidly than the population of the country as a whole.[2] Population in the provinces was also becoming increasingly mobile, especially in the Midlands and south-eastern England.[3] A comparison of the names of 1544 subsidy payers in the North Clay division of Bassetlaw wapentake, Nottinghamshire, with those of 1557 reveals that 24% of the names in the later list were new to the district.[4] In the hundreds of Godalming, Farnham and Godley in Surrey, more than 50% of the men who answered the muster of 1575 did not answer that of 1583. Some of this change was due to old men passing beyond military age and to young men growing into it, but much of it was a consequence of emigration. Moreover, people were not only leaving the district but also arriving and almost a third of those registered in 1583 bore family names not included in the earlier list.[5]

But in some places natural increase was of paramount importance. For example, the population of Wigston Magna (Leicestershire) increased dramatically between 1563 and 1603 by about 50%, as a result partly of immigration of new families but largely of natural increase consequent upon a rising birth rate (itself due in part to earlier marriages) and a dimini-shing death rate (especially a fall in infant mortality).[6] In Leicestershire as

[1] H. A. Hanley and C. W. Chalkin, 'The Kent Lay Subsidy of 1334/5', *Kent Records*, XVIII (1964), 58–172. See also: E. M. Yates, 'A contribution to the historical geography of north-west Staffordshire', *Geog. Studies*, II (1955), 39–52; J. M. W. Bean (1962–3), 435.

[2] F. J. Fisher, 'The development of London as a centre of conspicuous consumption in the sixteenth and seventeenth centuries', *Trans. Roy. Hist. Soc.*, 4th ser., XXX (1940), 37–50.

[3] E. J. Buckatzsch, 'The constancy of local populations and migration in England before 1800', *Population Studies*, V (1951–2), 62–9.

[4] S. A. Peyton, 'The village population in the Tudor Lay Subsidy rolls', *Eng. Hist. Rev.*, XXX (1915), 234–50.

[5] E. E. Rich, 'The population of Elizabethan England', *Econ. Hist. Rev.*, 2nd ser., II (1949–50), 247–65.

[6] W. G. Hoskins (1957), 18–19 and 32.

a whole the population in 1563 was still far below its level in 1334, and was substantially smaller than it had been in 1377 (as it also was in many other counties). But the county's population was increasing; coalmining was beginning to add to the population of the north-west, and market towns now had relatively larger populations compared with the purely agricultural settlements. Parish registers from many counties indicate a rapid population increase during the second half of the sixteenth century, but with sharp, localised setbacks in years of pestilence.[1]

In some localities, plague and other diseases reversed the general population trend; at Crediton, in Devon, 551 people died during 1571. The average number of burials for preceding normal years was 40 to 45, so that nearly 500 people must have died of plague in one year in this small town – possibly a third of its population. Between the autumn of 1590 and that of 1592 another 535 people also died here. Thus Crediton lost over 1,000 people by pestilence in the space of 21 years.[2] During the later Middle Ages, plague occurred intermittently but with a generally decreasing vehemence nationally, and it became increasingly a regional, particularly an urban (especially a London), phenomenon.[3]

The distribution of wealth in England in the early sixteenth century as reflected in the Lay Subsidies of 1524–5 (Fig. 43) may be taken as an approximate summation of the economic changes, including the population changes, of the later Middle Ages; but it must be borne in mind that wealth was often concentrated in the towns and in the hands of a few individuals.[4] Estimates of the total population vary greatly both for 1334 and for 1600. Whatever the uncertainty about these figures, it may be reasonably assumed that resurgence and growth during the late fifteenth and sixteenth centuries had brought the total population of England back to its pre-Black Death figure of about 4½ million or so.

[1] C. T. Smith, 137–41; J. W. F. Hill, *Tudor and Stuart Lincoln* (Cambridge, 1956), 88; J. Cornwall, 'An Elizabethan census', *Records of Bucks*., XVI (1953–60), 258–73.
[2] W. G. Hoskins (1954), 171.
[3] J. Saltmarsh, 32–40, and J. M. W. Bean (1962–3), 428–32.
[4] J. Sheail, 'The distribution of taxable population and wealth in England during the early sixteenth century', *Trans. and Papers, Inst. Brit. Geog*, LV (1972), 111–26.

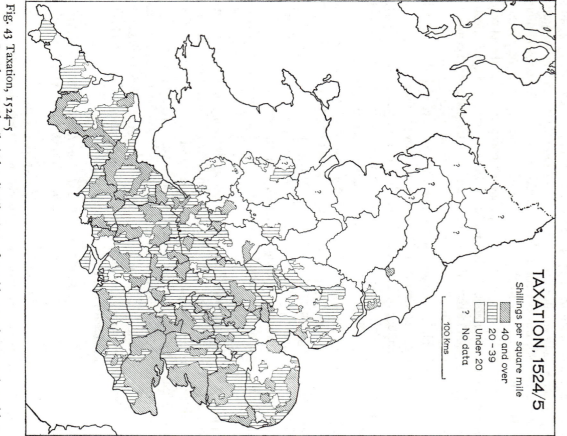

Fig. 43 Taxation, 1524–5
Based on J. Sheail, 'The distribution of taxable population and wealth in England during the early sixteenth century', *Trans. and Papers, Inst. Brit. Geog.*, LV (1972), 120, and on additional information provided by Dr Sheail. (P.R.O. Exchequer Lay Subsidies E. 179.)

TAXATION, 1524/5

Shillings per square mile

40 and over
20 – 39
Under 20
? No data

100 Kms

THE COUNTRYSIDE

Demographic changes were linked with price changes. As the supply of labour and the demand for produce changed, so did prices of products and the profitability of different economic enterprises. Price changes encouraged fundamental changes in both agriculture and industry during the later Middle Ages. A brief descriptive account of these changes therefore precedes discussion of agricultural and industrial developments.

The construction of accurate, meaningful price indices for this period is extremely difficult. Isolated references to prices are found in a host of documentary sources, but their isolation detracts from their utility. Most useful are the records of institutions with regular purchases or regular sales of goods, and so with continuous, or near continuous, series of prices. Moreover, comparability of data from different sources is essential; the use of local measures often bedevils direct comparison, and prices are affected by the time, place and conditions of sale.[1] The index of wool prices compiled by J. E. T. Rogers, who himself admitted its defects, has been described by P. J. Bowden as worthless, for two reasons. In the first place, the number of items utilised was very small and sometimes only two or three price quotations represented an entire decade; secondly, no differentiation was made between different qualities of wool.[2] Despite difficulties of this nature, trends of some prices during the later Middle Ages can be discerned (Fig. 44). Those of wages, consumables and land will be discussed briefly.

By about 1340, *wages* for agricultural workers were rising and continued to rise steeply in the 1340s and 1350s, indicative of a growing scarcity of labour, consequent upon a declining population. The rise both preceded and succeeded the Black Death and cannot be attributed solely to it. Nevertheless, the decades after 1350 saw a sharp rise in wages, a rise that became permanent and that levelled out after about 1370. W. Beveridge's study of wages on eleven manors of the see of Winchester, spread over seven counties in southern England, noted the rise and fall of threshing and winnowing costs between 1362 and 1368, followed by the establishment of a new high level about 1374. These fluctuations were not simultaneous on all manors, and Beveridge regarded them as a delayed

[1] E. V. Morgan, *The study of prices and the value of money*, Helps for History Students, 53 (London, 1950).

[2] P. J. Bowden, 'Movements in wool prices, 1490–1610', *Yorks. Bull. Econ. and Soc. Research*, IV (1952), 109–24.

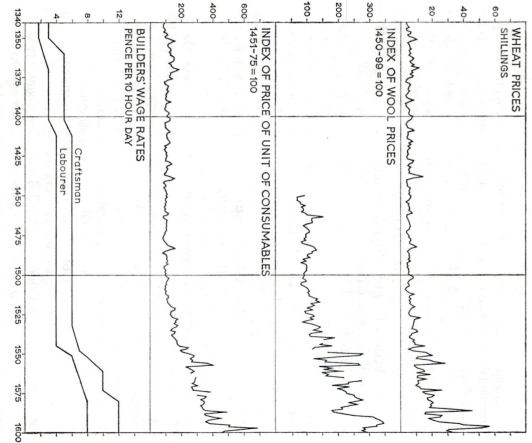

Fig. 44 Movement of some prices, 1340–1600
Based on: (1) J. E. T. Rogers, *A history of agriculture and prices in England*,
7 vols. (Oxford, 1866–1902), I, 228–34; IV, 282–90; VI, 268; (2) P. J. Bowden,
The wool trade in Tudor and Stuart England (London, 1962), 219–20; (3) E. H. P.
Brown and S. V. Hopkins, 'Seven centuries of the prices of consumables
compared with builders' wage-rates', *Economica*, n.s., XXIII (1956), 311–12.
Breaks in the graphs are the result of no data.

reaction to the change of labour conditions begun by the Black Death, although it was the pestilence of 1361 which finally broke down resistance to a change of wage-rates.[1] According to M. M. Postan, wages rose throughout the greater part of the fourteenth and fifteenth centuries.[2] Builders' wages during the later Middle Ages certainly rose sharply during 1350–65, 1400–12 and 1545–80, and showed remarkably level trends during the late fourteenth century and throughout most of the fifteenth century.[3] In both agricultural and industrial occupations, the rise in the wages of unskilled labourers was relatively greater than in those of skilled craftsmen. And, until the sixteenth century, the rise was a considerable one in real wages, for the prices of consumables rose relatively less than wage-rates. After about 1510 there was a considerable fall of wage-rates in relation to prices, indicative of a growing population.[4]

A price index of a composite unit of *consumables*, constructed by E. H. P. Brown and S. V. Hopkins, shows a remarkably level but slightly falling trend from 1350 or so to about 1500. Prices began to creep upwards after about 1510, and climbed steeply during the 1540s.[5] Wheat prices showed a very similar trend: they were generally level in the fourteenth and fifteenth centuries, and increased greatly during the sixteenth century, especially in the second half of the century. The trend of wool prices was somewhat different, the upward movement beginning earlier. The trend of wheat prices was eventually steeper than that of wool prices and the two crossed in the middle of the sixteenth century. Between 1490 and 1552 there were only two transitory periods in which the price of wheat was high relative to the price of wool. It was not until well after 1552 that prices moved relatively in favour of wheat for any length of time. The period 1552–81 was one of transition, in which, on balance, relative prices appear to have slightly favoured wool. From 1581 to 1609 the price of wheat

[1] W. Beveridge, 'Wages in the Winchester manors', *Econ. Hist. Rev*, VII (1936–7), 22–43, and 'Westminster wages in the manorial era', *Econ. Hist. Rev.*, 2nd ser., VIII (1955–6), 18–35.

[2] M. M. Postan (1949–50), 226–33.

[3] E. H. P. Brown and S. V. Hopkins, 'Seven centuries of building wages', *Economica*, n.s., XXII (1955), 195–206.

[4] W. Beveridge (1936–7), 31–5, and (1955–6), 26; E. H. P. Brown and S. V. Hopkins, 'Wage-rates and prices: evidence for population pressure in the sixteenth century', *Economica*, n.s., XXIV (1957), 289–306; M. M. Postan (1938–9), 166, and (1949–50), 226–7.

[5] E. H. P. Brown and S. V. Hopkins, 'Seven centuries of the prices of consumables compared with builders' wage-rates', *Economica*, n.s., XXIII (1956), 296–314.

moved steeply upwards and except for two short periods was markedly high in relation to the price of wool.[1]

For a century after the Black Death, *land values* were at a low and often falling level. There was in general terms a land surplus, as vacant holdings and lapsed rents testified. According to M. M. Postan, the agricultural slump, which began in the early fourteenth century (in some places before the Black Death), continued, with only a slight halt around 1410, until the late 1470s and 1480s.[2] Towards 1500, however, the picture changed and rents rose as demand for land increased, and the sixteenth century witnessed a considerable increase in land values.[3]

An acceptable explanation of these price changes remains to be firmly established. M. M. Postan has related the low and falling prices of consumables, the high and rising wages, and the low land values to reduced population pressure during the fourteenth and most of the fifteenth century. This view has, for the most part, been accepted by English although not by some foreign scholars.[4] The inflationary price rises of the sixteenth century are now similarly being attributed to increased population pressure, and to 'the fact that population outdistanced supplies'.[5] The view that the price changes of the sixteenth century were caused by a large influx of precious metals from the New World has been challenged, principally on two grounds: (1) that a fall in the value of money began to be marked, in some instances, several decades before the influx from the New World became appreciable, and (2) that not all commodities were similarly affected as might be expected with a change in the value of money brought about purely by an expansion in the circulating medium. Similarly, Henry VIII's devaluation and debasement of the coinage after 1543 certainly aggravated, but did not initiate, the inflationary processes. When the currency was restored in 1561, inflation continued, although at a slower rate.[6]

[1] P. J. Bowden (1952), 116–23; W. G. Hoskins, 'Harvest fluctuations and English economic history, 1480–1619', *Agric. Hist. Rev*, XII (1964), 28–46.

[2] M. M. Postan (1938–9), *passim*, and (1949–50), 236–9.

[3] A. Simpson, *The wealth of the gentry, 1540–1660* (Cambridge, 1961).

[4] M. M. Postan (1938–9) and (1949–50), *passim*. For dissenting views, see for example: J. Schreiner, 'Wages and prices in England in the later Middle Ages', *Scand. Econ. Hist. Rev*, II (1954), 61–73; E. Kosminsky, 'The evolution of feudal rent in England from the XIth to the XVth centuries', *Past and Present*, VII (1955), 12–36; W. C. Robinson, 'Money, population and economic change in late medieval Europe', *Econ. Hist. Rev*, 2nd ser., XII (1959–60), 63–76.

[5] J. Hurstfield, *The Elizabethan nation* (London, 1964), 15.

[6] J. Hurstfield, 13–15. The older view stemmed especially from E. J. Hamilton,

A further factor inducing short-term fluctuations in agricultural prices and in population levels were climatic changes. Patterns of alternating good and bad runs of harvests have been identified in England between 1480 and 1619, and on the estates of the abbey of Battle in particular between 1340 and 1444. The years of harvest failure were associated with periods of excessive rainfall. But there is also increasing evidence for a possible long-term deterioration in the climate during the later Middle Ages at certain periods, notably between about 1420 and 1445 and between 1555 and 1600.[1]

Whatever the exact chronology and causes of changes in prices and population levels, their consequences for agriculture were enormous. The later Middle Ages witnessed three fundamental agricultural changes; first, the decline of demesne farming; secondly, the growth of peasant farming; and thirdly, changes in the use of land, notably enclosure for livestock farming.

The decline of demesne farming

By 1330 or so the age of medieval high farming had ended – in some places long ended – and during the century, as costs rose with wage-rates and as receipts fell with prices, profits of large-scale production declined. At dates differing according to the locality and the policy of a particular lord, grain production was curtailed and demesnes were increasingly leased, at first in portions and then in their entirety, on ever longer leases to decreasing numbers of aspiring tenants. The usual reaction of lords to declining profits was an attempt to reduce costs and to stabilise revenues. The establishment of rentier economies was a major economic change of the later Middle Ages.[2]

Even when demesnes were being cultivated directly for profit by lords

'American treasure and the rise of capitalism', *Economica*, XXVII (1929), 338–57. Current objections to this view are summarised in J. D. Gould, 'The price revolution reconsidered', *Econ. Hist. Rev*, 2nd ser, XVII (1964–5), 249–66.

[1] W. G. Hoskins (1964), 28–46; H. H. Lamb, 'Britain's changing climate', *Geog. Jour.*, CXXXIII (1967), 444–66; G. Manley, 'Climate in Britain over 1,000 years', *Geog. Mag.*, XLIII (1970–1), 100–7; P. F. Brandon, 'Late-medieval weather in Sussex and its agricultural significance', *Trans. and Papers, Inst. Brit. Geog.*, LIV (1971), 1–17; G. Utterström, 'Climatic fluctuations and population problems in early modern history', *Scandinavian Econ. Hist. Rev.*, III (1955), 3–47.

[2] J. A. Raftis, 217–50; E. M. Halcrow, 'The decline of demesne farming on the estates of Durham cathedral priory', *Econ. Hist. Rev*, 2nd ser, VII (1954–5), 345–56; A. R. Bridbury, *Economic growth. England in the later Middle Ages* (London, 1962), 18.

or their officials, rents of various kinds provided a substantial proportion of the total manorial income. A major part of the archbishop of Canterbury's income from land, throughout the medieval period, was from rents, as it was on other estates, both lay and ecclesiastical.[1] It was not the introduction but the growth of a rentier economy (i.e. the renting of demesne lands), and the almost total disappearance of direct demesne farming, that was so momentous. The cultivated acreages of demesnes diminished, partly because some arable land was abandoned and more especially because lands were increasingly leased to tenants.

Contraction of demesne farming was apparent some decades before the Black Death, which for many landlords came as the culmination of a period of shaken prosperity. This retrenchment was to continue, with occasional and local revivals, until about 1450 or so. Marginal lands brought into cultivation at the height of the medieval arable expansion now tumbled to grass.[2] Decreasing money profits were a disincentive to arable farming. Thus in the years immediately after the Black Death total grain production was cut by about one half on the Huntingdonshire manors of Ramsey abbey, and the acreage of crops sown on the demesne at Hutton (Essex) in 1389 was not one half that in 1342, having especially diminished since 1369.[3] On many manors, however, reduction in demesne cultivation immediately after the Black Death was only temporary, most of the loss being made up again in the following decade or so.[4] On other manors, land lay waste because the lord considered it unprofitable to cultivate, and because he could find no tenant willing to lease it.[5] Some lords turned more towards pasture farming to reduce their costs.[6]

[1] F. R. H. Du Boulay, 'A rentier economy in the later Middle Ages: the Archbishopric of Canterbury', *Econ. Hist. Rev.*, 2nd ser., XVI (1963–4), 427–38; G. A. Holmes, *The estates of the higher nobility in fourteenth-century England* (Cambridge, 1957), 88–9 and 109–13; J. S. Donnelly, 'Changes in the grange economy of English and Welsh Cistercian abbeys', *Traditio*, X (1954), 399–458.

[2] P. F. Brandon (1962), 71; M. M. Postan (1938–9), 161; F. R. H. Du Boulay, 'Late-continued demesne farming at Otford', *Archaeol. Cantiana*, LXXIII (1959), 116–24; M. Morgan, *The English lands of the abbey of Bec* (London, 1946), 98–104.

[3] J. A. Raftis, 253; K. G. Feiling, 'An Essex manor in the fourteenth century', *Eng. Hist. Rev.*, XXVI (1911), 333–8.

[4] For example, on the East Anglian manors of the bailiwick of Clare: G. A. Holmes, 91.

[5] F. R. H. Du Boulay (1959), 121.

[6] K. G. Feiling, 336.

Diminishing grain production and arable cultivation on demesnes was accompanied by a movement to lease them, at first piecemeal for short periods to a few tenants, sometimes to a single tenant. Leasing was an attempt to maximise and stabilise manorial incomes. It was not an innovation of the later Middle Ages, but the ubiquity and the finality of the process during that period was new. On some manors leasing of parcels of demesne for short periods in the two decades before and after the Black Death suggests that the process was intended to be temporary, an expedient method of raising ready cash.[1] But with continued economic stagnation, more landlords adopted a policy of leasing out whole demesnes.[2] From the 1360s and 1370s onwards, the process quickened, until by the mid-fifteenth century it was generally completed. The forty demesnes of the see of Canterbury, widely distributed over south-east England, began to be leased out permanently between the 1380s and 1420s, and most had been leased by 1422; the manors of Durham cathedral priory were all leased by 1451; the leasehold system was established on almost all the estates of Canterbury cathedral priory by 1396; most Cistercian granges had been broken up into tenant holdings and leased either in part or as units by 1410 or so.[3] The process was equally frequent on lay estates; a rentier economy was established on the estates of the Percy family in Yorkshire, Northumberland, Cumberland and Sussex by 1416; and all the demesne lands of the bailiwick of Clare, a group of manors in East Anglia, passed out of the lord's hands between 1360 and 1400.[4] Short-term leases, taken up by tenants anxious for quick profits and unconcerned with the state of buildings or with maintaining soil fertility, were gradually replaced by long-term leases incorporating conditions for the maintenance of buildings and soils; thus it was hoped to stabilise rent incomes and to preserve property. With this type of leasing, the early

[1] K. G. Feiling, 335; F. M. Page, 114.

[2] R. H. Hilton, *The economic development of some Leicestershire estates in the 14th and 15th centuries* (London, 1947), 105; J. A. Raftis, 281–301.

[3] F. R. H. Du Boulay (1959), 116, and (1963–4), 426; E. M. Halcrow, 355–66; R. A. L. Smith, 192; J. S. Donnelly, 451. See also: J. S. Drew, 'Manorial accounts of St Swithun's priory, Winchester', *Eng. Hist. Rev*., LXII (1947), 20–41; R. H. Hilton (1947), 85–91; F. W. Maitland, 'The history of a Cambridgeshire manor', *Eng. Hist. Rev*., IX (1894), 417–39; M. Morgan, 113–24; J. A. Raftis, 251–80; N. S. B. and E. Gras, *The economic and social history of an English village (Crawley, Hampshire), A.D. 909–1928* (Cambridge, Mass., 1930), 80–3; F. G. Davenport, *The economic development of a Norfolk manor, 1086–1565* (Cambridge, 1906), 49–55.

[4] J. M. W. Bean, *The estates of the Percy family 1416–1537* (Oxford, 1958), 13; G. A. Holmes, 92.

medieval relationship of lord and villein changed to that of landowner and tenant farmer.[1]

Leasing increased the proportion of a landlord's income derived from rent without necessarily increasing his total income. In fact, for most lords the century after 1334 was a period of declining incomes from land. Incomes from the customary lands of tenants as well as from leased demesnes showed an overall decline; rents per acre fell, the demand for tenements which lapsed into the lord's hands was slack, and arrears of rent accumulated. The rural landscape assumed in many parts a neglected aspect as buildings decayed, as tenements were abandoned and as land reverted to rough pasture. Landlords and landscape were the twin sufferers from the agricultural slump which characterised the period 1350–1450. It was not until after about 1460 that rents again began to rise, and on some estates the rise did not come until much later.[2]

This general trend (contracting demesne cultivation, spread of leasing, declining incomes from land) was typical, but it was not universal, and the effects of the retrenchment were not felt with equal severity throughout the country. The experiences of lay and ecclesiastical estates were essentially similar, but it seems that small estates were more able than large ones to weather the economic storm.[3] On large estates, conversion to wholesale demesne leasing took place earlier on distant manors than on 'home-farms' retained in cultivation to produce not for sale but for consumption.[4] Nor was the general trend continuous; the years around 1400 witnessed a measure of economic recovery throughout England, more especially in the south-east. The incomes of Canterbury cathedral priory and of the see of Canterbury, both of whose estates were widely spread over the south-east, were apparently more stable during the later Middle Ages than those of most other large estates, and this owed something to an expansion of the cultivated areas. But the peculiarity of these estates was one of degree not

[1] E. M. Halcrow, 352–4; M. M. Postan, 'The rise of a money economy', *Econ. Hist. Rev.*, XIV (1944–5), 123–34; J. M. W. Bean (1958), 55–6; R. H. Hilton (1947), 126–8; R. A. L. Smith, 200; R. Scott, 'Medieval agriculture' in E. Crittall (ed.), *V.C.H. Wiltshire*, IV (1959), 7–42.

[2] J. M. W. Bean (1958), 17–48; R. H. Hilton (1947), 85–8; G. A. Holmes, 114–20; W. G. Hoskins, *The Midland peasant: the economic and social history of a Leicestershire village* (London, 1957), 84–5; F. M. Page, 147–9; M. M. Postan (1938–9), 161–2; F. G. Davenport, 56–9; J. A. Raftis (1957), 285–94.

[3] R. H. Hilton (1947), 117–21.

[4] R. A. L. Smith, 200–1; R. H. Hilton (1947), 88, 91 and 132; P. F. Brandon (1962),

of kind, for both witnessed a decline of demesne farming. Both prior and archbishop became landlords pure and simple.[1]

The growth of peasant farming

The emergence of rentier landlords during the later Middle Ages was paralleled by the appearance in increased numbers of both rich peasant farmers and poor landless labourers. Throughout this period land became available on terms freer than the customary terms governing traditional holdings. Demesnes were increasingly leased for money rents; land taken in from the waste was almost always leased out for a money rent only; and peasant tenements were being divided up even before the Black Death, and taken piecemeal by other peasant lessees at money rents.[2]

Peasants who leased demesnes often found themselves in possession of potentially prosperous farms because demesnes were often situated on the best soils in a locality and had often been meticulously manured and cultivated.[3] When worked by lessee peasants utilising little hired labour, incurring no managerial expenses, and working for their own profit, former demesne lands again became viable economic units, operated now by a prosperous section of the peasantry. Occasionally, the lessees were newcomers to a district, but more often they were enterprising local men, sometimes former manorial officials, who had built up substantial holdings by purchasing and leasing usually small plots, and who now augmented their holdings further by leasing part or all of a demesne.[4] The process of demesne leasing gave added momentum to an already fluid land market and accentuated the growing inequality in the sizes of peasant holdings. Holdings were far from being equal in size at the beginning of the fourteenth century, and they became decreasingly so during the later Middle Ages. From an economic point of view, the commutation of villein services was one of many ways of lightening the terms on which customary lands were let, and, as the demands of demesnes declined so lords could dispense with some of the services previously required. The leasing of demesnes was essentially complete by about 1450 and the exaction of labour services had all but ceased by 1500.[5]

---

[1] F. R. H. Du Boulay (1963–4), 438.

[2] R. H. Hilton, 'Medieval agrarian history', in W. G. Hoskins (ed.), V.C.H. Leicestershire, II (1954), 145–98.

[3] E. M. Halcrow, 349; A. R. Bridbury (1962), 91.

[4] R. H. Hilton (1954), 94 and 157–62; M. Morgan, 111–12 and 115–18.

[5] M. M. Postan (1949–50), 238, and 'The chronology of labour services', Trans. Roy. Hist. Soc, 4th ser, xx (1937), 169–93; L. C. Latham, 'The decay of the manorial

Social status came to be determined more by the amount of land occupied than by the nature of its tenures.[1] One direct consequence of the decline in population during the fourteenth century was an increase in the average size of peasant holdings. Demesne land and vacant tenant land provided ample opportunity for the enterprising and industrious peasant to build up an estate. Lands which became vacant during and after the Black Death were frequently taken up by land-hungry tenants.[2] The land market became increasingly active, and holdings were frequently sub-divided and let to new tenants. In Leicestershire, the disintegration of tenements and regrouping of lands produced a sort of polarisation process in which on the one hand richer peasants built up farms above the size of the 'normal' virgate holding of 20–30 acres and, on the other hand, the poorer lost what land they had and so became labourers. This stratification of the peasantry was one of the most important developments in the English countryside during the fourteenth and fifteenth centuries and was accompanied by the rise of yeomen and husbandmen from bondage.[3]

By the first quarter of the sixteenth century, as evidenced in Leicester-shire assessments of 1524 and 1525, the rough economic equality of the early medieval community had been shattered beyond recognition. Approximately one man in five was assessed on his wages and could be regarded as being dependent on them as a source of income rather than on income from land or possessions, so that he belonged undoubtedly to the labouring classes.[4] Economic inequality was well-marked in Leicestershire in 1524–5, for (omitting the squirearchy, who were less wealthy than many a yeoman, in personal estate at least) 4% of the rural population owned a quarter of the personal estate and 15½% owned half of it. And even in villages where personal estate was more evenly distributed, as at Wigston Magna, 20% of the taxpayers owned half of the personal estate.[5]

[1] K. C. Newton, *Thaxted in the fourteenth century* (Chelmsford, 1960), 20–32; R. H. Hilton (1954), 96.

[2] G. A. Holmes, 90–2; F. M. Page, 123–4 and 152; J. A. Raftis, 252–3.

[3] R. H. Hilton (1947), 79–105, and (1954) 185–8; W. G. Hoskins (1957), 39–52; N. S. B. and E. Gras, 95–8; F. G. Davenport, 70–97.

[4] R. H. Hilton (1954), 94–5; A. R. Bridbury (1962), 103.

[5] W. G. Hoskins (1950), 127–30.

system during the first half of the fifteenth century, with special reference to manorial jurisdiction and to the decay of villeinage, as exemplified in the records of twenty-six manors, in the counties of Berkshire, Hampshire and Wiltshire', *Bull. Inst. Hist. Research*, VII (1929–30), 113–16.

The remainder of the sixteenth century saw an accentuation of this trend towards inequality and with it, the final obliteration of the medieval framework of landholding. Changes which had been taking place during the previous century and a half were hastened by inflation during the sixteenth century. The redistribution of land continued, and the dissolution of the monasteries added to the pool of land in the market from 1536 onwards, reducing further the direct influence of the Church upon the English landscape, and setting in motion the emergence of such ecclesiastical relict features as the ruins of Fountains abbey (Yorkshire).[1] One interesting cartographic by-product of the selling and exchanging of land and of the boundary disputes that often accompanied such transactions was the production of estate maps by improved techniques of land surveying.[2]

Changing land use and rural settlement

The *Nonarum Inquisitiones* of 1341 show that the abandonment of arable lands was under way before the Black Death.[3] Some lands, notably in the north (especially in Lancashire), were abandoned because of political disturbances; others were abandoned because of flooding by the sea (especially along the south coast); others because climatic hazards had reduced their supply of seed corn (especially in the Chilterns); others because soils were thought to be exhausted (Fig. 45). But probably the most common cause of this agricultural retrenchment was a shrinkage of village populations. Arable reverted to pasture and dwellings tumbled into ruins; many arable acres *solebant seminari et modo jacent ad pasturam* and many dwellings *sunt derelicta sine habitoribus*.[4] Such abandonment was not a linear retreat of settlement; it revealed itself 'not in a long thin high-water mark, the whole length of a shore, but in many scattered

[1] H. J. Habakkuk, 'The market for monastic property, 1539–1603', *Econ. Hist. Rev.*, 2nd ser., x (1957–8), 362–80; R. A. Donkin, 'The Cistercian Order in medieval England: some conclusions', *Trans. and Papers, Inst. Brit. Geog.*, XXXIII (1963), 181–98.

[2] H. C. Darby, 'The agrarian contribution to surveying in England', *Geog. Jour.*, LXXXII (1933), 529–35.

[3] G. Vanderzee (ed.), *Nonarum Inquisitiones in Curia Scaccarii* (London, 1807).

[4] A. R. H. Baker, 'Evidence in the *Nonarum Inquisitiones* of contracting arable lands in England during the early fourteenth century', *Econ. Hist. Rev*, 2nd ser, XIX (1966), 518–32; A. R. H. Baker, 'Some evidence of a reduction in the acreage of cultivated lands in Sussex during the early fourteenth century', *Sussex Archaeol. Coll.*, CIV (1966), 1–5.

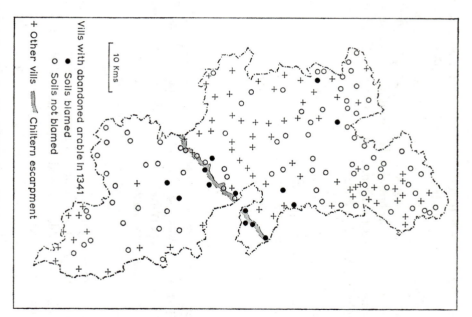

Vills with abandoned arable in 1341

● Soils blamed

○ Soils not blamed

+ Other vills

⟳ Chiltern escarpment

10 Kms

Fig. 45 Abandoned arable land in Buckinghamshire in 1341. Based on A. R. H. Baker, 'Evidence in the *Nonarum Inquisitiones* of contracting arable lands in England during the early fourteenth century', *Econ. Hist. Rev.*, 2nd ser., XIX (1966), 527–8.

rock-pools'.[1] Land reverted to pasture because it was not wanted as arable. This process was particularly significant in the East Riding of Yorkshire, in the Lincolnshire Wolds and in the light soil areas of Norfolk, but to some extent it was characteristic over a much wider area and led to the establishment of a new balance between grain lands and grass lands which may have been reached sometime between 1420 and 1440.' It was clearly

[1] M. W. Beresford (1954), 204.

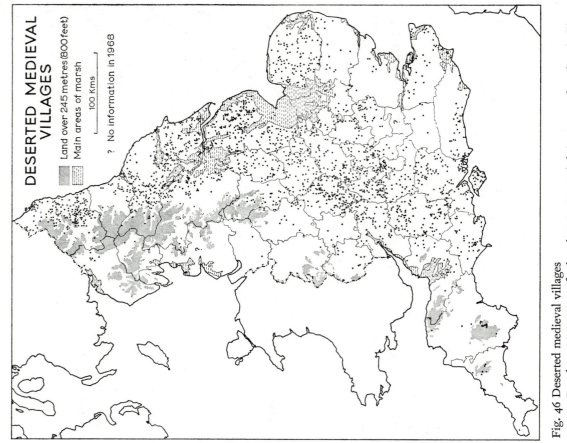

DESERTED MEDIEVAL
VILLAGES

Land over 245 metres (800 feet)

Main areas of marsh

100 Kms

? No information in 1968

Fig. 46 Deserted medieval villages
Based on M. W. Beresford and J. G. Hurst (eds.), *Deserted medieval villages*
(London, 1971), 182–212.

a different balance from the balance of the late thirteenth century' and it was soon to be altered again.[1]

From about 1450 onwards grass, instead of being the residual land use on abandoned arable land, became the desired land use because of the growing profitability of sheep farming. The demand for wool grew and its price rose as cloth exports expanded and as the requirements of a growing home population increased. The relationship between wool and corn prices was responsible for the conversion of much arable land to pasture, for the enclosure of many open fields, for the eviction of numerous tenants and for the desertion of many villages in the period 1450–1520.[2] An earlier view that enclosure and conversion of land during this period was a consequence of declining crop yields and diminishing soil fertility has been found wanting.[3] Most of the two thousand or so deserted medieval villages now identified in England were not associated with poor quality land, or with a retreat of settlement, but with land that was equally good for arable and pasture, and with a changing emphasis in agricultural production (Fig. 46). In terms of the landscape, this enclosure movement was responsible for the fossilisation of many strips and furlongs of the common fields in the form of ridge-and-furrow, for the creation of large enclosed fields (larger than those associated with later and less revolutionary enclosure) and for the transformation of many active settlements into lost villages marked in the modern landscape, if at all, only by isolated or ruined churches, sunken roadways and grass-covered earthworks.

After about 1520, the death rate of villages slackened, partly because the movement of grain prices made depopulating enclosure for pasture less attractive, and partly because of state action, in response to outcries against enclosure, which made it less possible. The villages most liable to be depopulated were those with small populations, those with small proportions of freeholders and those with landlords having connections in the wool trade or an eager acquisitive appetite.[4] Most enclosure for

[1] M. W. Beresford (1954), 20; See also 150, 170–2, 198–204 and 241; J. G. Hurst, 'Deserted medieval villages and the excavations at Wharram Percy, Yorkshire', being ch. 10 of R. L. S. Bruce-Mitford, *Recent archaeological excavations in Britain* (London, 1956), 251–73.

[2] M. W. Beresford (1954), 27–77, 177–89 and 207–15.

[3] H. Bradley, *The enclosures in England: an economic reconstruction* (New York, 1918); R. Lennard, 'The alleged exhaustion of the soil in medieval England', *Econ. Jour.*, XXXII (1922), 12–27; M. K. Bennett, 'British wheat yields per acre for seven centuries', *Econ. Hist.*, III (1934–7), 12–29.

[4] M. W. Beresford (1954), 102–33, 212–14 and 228; M. W. Beresford and J. G. Hurst (eds.), *Deserted medieval villages* (London, 1971), 3–75.

livestock farming took place on the densely settled claylands of the Midlands, where mixed farming was practised and where crop rotations which included leys enabled farmers to keep more stock when market conditions favoured them. Enclosure here was often for cattle as well as for sheep, for beef and leather as well as for wool and mutton. In western Leicestershire, for example, in the neighbourhood of Charnwood Forest, enclosures were principally for dairying and cattle-rearing but in the eastern uplands enclosures were principally for sheep.[1] Other areas which underwent much enclosure in the Tudor period were the less fertile chalk and limestone uplands, the traditional sheep-rearing lands. A large farmer wanting to enlarge his sheep flock found it easier to enclose great tracts of common or to force enclosure of the open fields here, because his tenants were often few and their opposition weak. The economic incentive to enclose land in the uplands was the same as that which encouraged enclosure in the lowlands – the high prices of animal products, particularly wool and mutton. And these uplands produced a fine short wool which became more scarce and sought after as the sixteenth century wore on.[2]

Enclosure and its accompanying depopulation produced much controversy, many pamphlets, a number of government enquiries (in 1517–18, 1548–9, 1566 and 1607), a number of ineffective Acts of Parliament (e.g. in 1489, 1515, 1536 and 1563), and a revolt in the Midlands in 1607. The evidence is incomplete and difficult to interpret, but it would seem that between 1455 and 1607 about 8 to 9% of the area of some Midland counties were enclosed, and smaller percentages in a number of other counties (Fig. 47). The total area enclosed over 24 counties may have amounted to about half a million acres or less than 3% of the total area of England, a small amount when compared with the enclosures of the eighteenth century.[3] M. W. Beresford has suggested that the Midlands were particularly susceptible to depopulating enclosures for livestock farming because lands there were equally suitable for grain and grass; no such land use change was to be expected where the balance was set firmly towards the one or the other.[4]

Many parts of England did not see much depopulating enclosure.

[1] J. Thirsk, *Tudor enclosures* (Hist. Assoc., London, 1959), 20. [2] *Ibid.*, 18–19. [3] E. Gay, 'The inquisitions of depopulation in 1517 and the Domesday of inclosures', *Trans. Roy. Hist. Soc.*, n.s., XIV (1900), 231–67, and 'Inclosures in England in the sixteenth century', *Quarterly Jour. Econ.*, XVII (1902–3), 576–97; A. H. Johnson, *The disappearance of the small landowner* (London, 1909), 47–59. [4] M. W. Beresford (1954), 242.

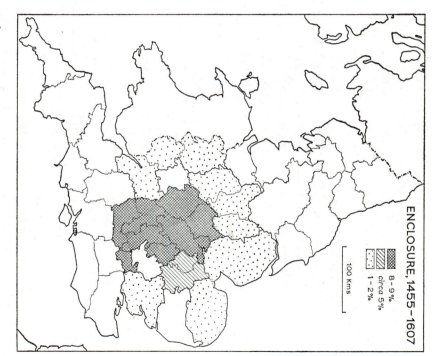

ENCLOSURE, 1455–1607

8 – 9 %
circa 5 %
1 – 2 %

100 Kms

Fig. 47 Enclosure, 1455–1607
Recorded in the proceedings of the Courts of Exchequer, the Court of Chancery, the Star Chamber and the Court of Requests. Based on: (1) A. H. Johnson, *The disappearance of the small landowner* (Oxford, 1909), 48–9, and map at end; (2) E. F. Gay, 'Inclosures in England in the sixteenth century', *Quart. Jour. Economics*, XVII (1903), 585–6.

Little took place in the highland zone where pastoral pursuits had long dominated the economy – none of the 113 villages in three upland hundreds of the West Riding of Yorkshire was abandoned. In such places there was little corn to be displaced, and plenty of room to expand grazing grounds without destroying villages. Elsewhere, there were wooded areas whose agriculture was biased away from arable husbandry, as in the Weald, the Forest of Dean, Sherwood Forest and much of Essex and the Chilterns. Within the Midlands, individual counties showed important contrasts because of their far from uniform terrains; settlements in the Forest of Arden, to the north of the Avon, were relatively immune from depopulation, unlike villages to the south, in the classical common field country. Marshland and fenland settlements were also usually immune – not a single vill in the Holland division of Lincolnshire was deserted, although in Romney Marsh ruined churches testify to the swing from arable to pasture which occurred there during the later Middle Ages.[1]

There was, in addition to the dramatic enclosure for pasture during 1450–1520, a slow and piecemeal reclamation and enclosure for both pasture and arable throughout the later Middle Ages. Some new enclosures were made direct from the waste – the story, even in the first half of this period, was not entirely one of contraction. Some additional woodland was cleared, for example in the Weald and in the Forest of Arden, and some more marshland reclaimed, for example in Romney Marsh and in the fenlands of Lincolnshire.[2] In open-field districts, such as the west Midlands, the disintegration and regrouping of tenements in a fluid land market made possible a consolidation of scattered parcels so that eventually large enough blocks of land could be made into private enclosures. Certain districts which in 1334 lay within the region of the Midland open-field system had by 1600 passed silently into enclosed field systems. Such in particular were the counties of the west – Herefordshire, Shropshire, parts of Staffordshire, Worcestershire, Warwickshire, Gloucestershire and Somerset.[3] Enclosure by private agreements, often confirmed by the Court of Chancery or the Court of Exchequer, gathered pace during the sixteenth century, but such agreements required the unanimous consent of those involved.[4] Simultaneously, the pattern of old enclosed land was changing;

---

[1] M. W. Beresford (1954), 217–61.

[2] J. L. M. Gulley, 'The Wealden landscape in the early seventeenth century and its antecedents', unpublished Ph.D. thesis, University of London (1960), 259–67; P. F. Brandon, 'Medieval clearances in the East Sussex Weald', *Trans. and Papers, Inst. Brit. Geog.*, XLVIII (1969), 135–53; R. H. Hilton, 'Old enclosures in the west Midlands: a hypothesis about their late medieval development', *Geographie et histoire agraires* (Nancy, 1959), 272–83; R. A. L. Smith, 203; J. Thirsk, *Fenland farming in the sixteenth century* (Leicester, 1953).

[3] H. L. Gray, *English field systems* (Cambridge, Mass., 1915), 108.

[4] K. J. Allison, 'The sheep-corn husbandry of Norfolk in the sixteenth and seventeenth centuries', *Agric. Hist. Rev.*, v (1957), 12–30; M. W. Beresford, 'Glebe terriers and open-field Buckinghamshire. Part II', *Records of Bucks.*, XVI (1953–60), 5–28; J. Cornwall, 'Agricultural improvement, 1560–1640', *Sussex Archaeol. Coll.*, XCVIII (1960), 118–32; R. H. Hilton (1959), 274–80, and (1954), 189–94; E. Kerridge, 'Agriculture c. 1500– c. 1793', in E. Crittall (ed.), *V.C.H. Wiltshire*, IV (1959), 43–64;

in the Chilterns and in parts of Kent, for example, the great demesne fields of the early fourteenth century had been divided into smaller enclosures by the end of the sixteenth century, and many small closes of the peasantry and yeomen had been amalgamated into larger fields as holdings became concentrated into fewer hands.[1]

Not all enclosure involved a change from arable to pasture. Landlords could increase their flocks without recourse to converting arable land; they could extend their 'fold-courses' or set up new ones; they could enclose their own arable land to prevent livestock of tenants feeding there after harvest; they could overstock commons; and they could enclose commons for their own benefit. Enclosure in Norfolk in the sixteenth century was almost entirely of this kind, and did not involve conversion to pasture.[2] Enclosure here was a means of increasing productivity, for it enabled land to be more efficiently cultivated and stock to be more carefully bred. Arable closes were better manured and sometimes carried crops in the fallow year. There is some evidence from many different parts of England that towards the end of the sixteenth century, under the stimulus of high grain prices, there was both enclosure of existing open fields which were kept as arable and also renewed reclamation to extend the cultivated area. But the progress of enclosure of arable fields was limited in common field areas by the system of common grazing. In lowland England, especially in the densely settled Midlands, rights of common both in the fields and on the wastes were of vital importance, for the supply of grazing was very limited. What common grazings there were had to be carefully stinted; and ultimately, when land shortage and food supplies became critical, the wastes were enclosed and divided among individuals. In highland England, where population pressure was generally less acute

[1] D. Roden and A. R. H. Baker, 'Field systems of the Chiltern Hills and of parts of Kent from the late thirteenth to the early seventeenth century', Trans. and Papers, Inst. Brit. Geog., XXXVIII (1966), 73–88.

[2] K. J. Allison (1957), 12–30.

M. R. Postgate, 'The field systems of Breckland', Agric. Hist. Rev., X (1962), 80–101; G. Youd, 'The common fields of Lancashire', Trans. Hist. Soc. Lancs. and Cheshire, CXIII (1961), 1–41; R. H. Hilton, 'Social structure of rural Warwickshire in the Middle Ages', Dugdale Soc. Occasional Papers, IX (1950); M. W. Beresford, 'Habitation versus Improvement: the debate on enclosure by agreement', in F. J. Fisher (ed.), Essays in the economic and social history of Tudor and Stuart England (Cambridge, 1961), 40–69; S. A. Johnson, 'Some aspects of enclosure and changing agricultural landscapes in Lindsey from the sixteenth to the nineteenth century', Rep. and Papers, Lincs. Archit. and Archaeol. Soc., IX (1962), 134–50.

than in the lowlands, where farming had a pastoral bias, and where there was abundant waste awaiting improvement, enclosures raised no great opposition. Here, enclosure of open fields proceeded piecemeal and by agreement, and much enclosure took the form of the division of the commons between intercommoning parishes, the introduction of stints on formerly unregulated waste and the taking in of waste in order to increase the arable acreage.[1]

Enclosure was one but not the only important technical change in agriculture during the later Middle Ages. Cereal production – in terms of yields per acre – certainly increased.[2] But probably the greatest single problem facing farmers was that of animal feed, and there were a number of attempts to increase fodder supplies. In open-field areas, the number of fields cultivated communally was often increased, thus reducing the fallow acreage; and rotations themselves, often based on furlongs rather than on fields, became more complicated than the number of open fields in any township might suggest. The fallow was further reduced by being sown, wholly or in part, often with legumes.[3] In some localities, field systems were rationalised during the later Middle Ages; thus the multiple open fields which existed at Bickenhill (Warwickshire) in 1350 had been simplified into three common arable fields by 1612.[4] In other localities, such as the dales of Cumberland, growing population pressure in the sixteenth century resulted in communal reclamation of waste lands, and in a multiplication of open fields.[5] Another important change was the more widespread adoption of convertible husbandry, single parcels and sometimes whole furlongs being laid down as temporary grass. This practice increased fodder supplies but not productivity per acre; ley

[1] J. Thirsk (1959), 4–8; R. H. Tawney, *The agrarian problem in the sixteenth century* (London, 1912), 237–53; W. G. Hoskins and L. D. Stamp, *The common lands of England and Wales* (London, 1963), 44–52.

[2] M. K. Bennett, 26–8.

[3] P. F. Brandon (1962), 65–66; H. L. Gray, 9, 73–81 and 109–56; F. G. Gurney, 'An agricultural agreement of the year 1345 at Mursley and Dunton – with a note upon Walter of "Henley"', *Records of Bucks.*, XIV (1941–6), 246–8; R. H. Hilton (1954), 159–61; (1947), 65 and 152–6; E. Kerridge (1959), 52; F. M. Page, 119; R. Scott, 15.

[4] V. H. T. Skipp and R. P. Hastings, *Discovering Bickenhill* (Birmingham, 1963), 15–29. See also J. Thirsk, 'The common fields', *Past and Present*, XXIX (1964), 3–25.

[5] G. G. Elliott, 'The system of cultivation and evidence of enclosure in the Cumberland open fields in the sixteenth century', *Geographie et histoires agraires* (Nancy, 1959), 118–36. See also A. J. Roderick, 'Open-field agriculture in Herefordshire in the later Middle Ages', *Trans. Woolhope Naturalists' Field Club*, XXXIII (1949–51), 55–67.

husbandry improved soil texture but not soil fertility, because on balance grazing beasts took out what they put in.[1]

Agriculture during the later Middle Ages was marked more by local diversity than by scientific experiment. Manuring of fields became more intensive and, as far as animal manure was concerned, more controlled and therefore more effective. Marling and liming were common practices by the end of the sixteenth century, and resulted in many pits and depressions over the surface of the ground.[2] But fertilising was still far from being scientific. The development of printing saw the appearance of agricultural writings such as John Fitzherbert's *Boke of Husbondrye* (1523) and Thomas Tusser's *Hundreth good pointes of Husbandrie* (1557), but new ideas spread only very slowly.[3] Turnips had long been grown as a garden vegetable, but Barnaby Googe, in his *Foure Bookes of Husbandrie* (1577), advised cultivation of turnips as a fodder crop, although it was not until the seventeenth century that field cultivation of turnips for livestock fodder began in England. After the middle of the sixteenth century experiments began in the floating of water-meadows to improve and increase fodder supplies, in the vales of Taunton and Hereford and elsewhere.[4] Experiments in fen drainage also became frequent, especially after 1560.[5] Agriculture in general was marked by increasing local specialisation, as the emergence of hop and fruit cultivation in Kent for the London market exemplifies. The commercialisation of agriculture made most advances in the neighbourhood of London and in the developing industrial areas, where there were large and growing populations employed partially at least in non-agricultural pursuits.[6]

[1] R. H. Hilton (1954), 197–8; W. G. Hoskins (1950), 140–4, and (1957), 67, 95, 152 and 164; E. Kerridge (1959), 52; E. L. Jones, 'Agriculture and economic growth in England, 1660–1750: agricultural change', *Jour. Econ. Hist.*, XXV (1965), 1–18.

[2] J. Cornwall, 121–4; H. C. Prince, 'The origin of pits and depressions in Norfolk', *Geography*, XLIX (1964), 15–32.

[3] G. E. Fussell, *The old English farming books from Fitzherbert to Tull, 1523 to 1730* (London, 1947), 1–20.

[4] E. Kerridge, *The agricultural revolution* (1967), 252–4   [5] *Ibid*, 222–39.

[6] M. Campbell, *The English yeoman under Elizabeth and the early Tudors* (New Haven, 1942), 156–220; F. J. Fisher, 'The development of the London food market, 1540–1640', *Econ. Hist. Rev.*, V (1934–5), 46–64; D. C. D. Pocock, 'Some former hop-growing areas', *Agric. Hist. Rev.*, XIII (1965), 17–22; N. S. B. Gras, *The evolution of the English corn market* (Cambridge, Mass., 1915), 95–129 R. H. Hilton (1959), 276 and 281–3; E. Kneisel, 'The evolution of the English corn market', *Jour. Econ. Hist.*, XIV (1954), 46–52; C. W. Chalklin, *Seventeenth-century Kent: a social and economic history* (London, 1965), 88–95.

During the fourteenth and fifteenth centuries derelict buildings – products often of neglect and sometimes of destruction – were prominent relict features in the English landscape, not only in the countryside but also in the towns. Towards the end of the sixteenth century, a contrary process known as the Great Rebuilding gathered momentum.[1] Many of the edifices erected or altered in towns during this Great Rebuilding have been removed by subsequent developments; but in rural districts, particularly in areas already characterised in the sixteenth century by isolated farmsteads and small hamlets (such as the Weald), many structures were rebuilt in the sixteenth century and remain today.[2] Reconstruction usually took the form of inserting a ceiling in a medieval hall, formerly open to the rafters, so producing a living room and parlour on the ground floor and bedrooms above, reached by a staircase. Many farmsteads of the Cotswolds, built or rebuilt in local stone, date from 1570–1640, as do the characteristic 'black-and-white' timber-framed houses of the west Midlands and the Wealden houses of south-east England.[3] Indeed, J. L. M. Gulley has claimed that in the Weald many more houses were built between 1570 and 1640 than during any period of comparable length before or since, and that most of the older surviving buildings incorporate substantial structural alterations carried out between 1570 and 1640.[4] In addition, as population pressure increased during the late sixteenth century, numerous cottages were erected on newly reclaimed waste, for example on the moors and heaths of Devon, in the Lancashire forest of Rossendale, and on the commons of the Weald.[5] Most cottage building was by small farmers and labourers whereas the building and rebuilding of more substantial structures was the work of husbandmen, yeomen and lesser gentry with money accumulated from the gap between relatively fixed expenses (rents and wages largely) and rapidly rising prices of farm products.[6]

[1] W. G. Hoskins, 'The rebuilding of rural England, 1570–1640', Past and Present, IV (1953), 44–59; M. W. Barley, The English farmhouse and cottage (London, 1961), 55–125.

[2] J. L. M. Gulley, 'The Great Rebuilding in the Weald', Gwerin, III (1961), 1–16.

[3] W. G. Hoskins (1953), 46–7.

[4] J. L. M. Gulley (1961), 2.

[5] W. G. Hoskins (1952), 328–9; G. H. Tupling, The economic history of Rossendale (1927), 47, 62, 64 and 66–7; J. L. M. Gulley (1960), 105–14.

[6] W. G. Hoskins (1953), 50–5.

Forests, woodland and parks

The acreage of land subject to forest law declined during the later Middle Ages as the result both of wholesale disafforestations – the removal of their territories from forest jurisdiction and from control by royal officials – and of piecemeal enclosure. The Tudors increased their revenues at times by disafforesting or by selling areas of forest. Destruction of timber was heavy during the sixteenth century, and fears of a shortage were frequently voiced. In 1543 came the first important timber preservation act, one provision of which was that, wherever woods were cut down, at least twelve young trees must be left on every acre. Elizabeth's first parliament passed an act to preserve ship timber within fourteen miles of navigable water. The quantity of acts and the diversity of their provisions made it clear that during the sixteenth century real alarm over the timber situation was developing. Destruction of timber was going on apace, and local and regional shortages were attracting national attention.[1]

On the other hand, the area of parkland had greatly increased. Money from farming, industry and commerce was not only invested in improved housing but also in deer parks, particularly in the south-east, a reflection of the concentration of wealth in the London region (Fig. 58). By the end of the sixteenth century at least 69 parks had been established in the Chiltern Hills and not less than 10% of the surface area of the Weald was covered by parks. They were widely scattered, often remote from villages or public highways, and most seem to have been located on land of little value for agriculture. But deer parks were esteemed not only as social and recreational retreats; they were also valued for the income to be derived from their woods and trees at a time of growing timber shortage.[2]

INDUSTRY

The later Middle Ages were years of fundamental change in the industrial and commercial geography of England. They witnessed a radical transformation of the export trade, from the export of raw wool to the export of manufactured textiles. In addition, these years saw both a wider spread and a growing regionalisation of industry, and also increasing complexity in manufacturing processes and a broadening of domestic and overseas markets.

---

[1] R. G. Albion, *Forests and sea power* (Cambridge, Mass., 1926), 95–127.

[2] H. C. Prince, 'Parkland in the Chilterns', *Geog. Rev.*, XLIX (1959), 18–31; J. L. M. Gulley (1960), 61–72.

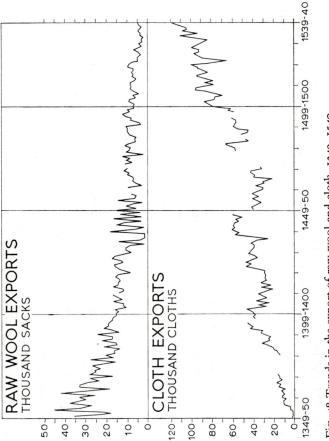

RAW WOOL EXPORTS
THOUSAND SACKS

CLOTH EXPORTS
THOUSAND CLOTHS

Fig. 48 Trends in the export of raw wool and cloth, 1349–1540
Based on E. M. Carus-Wilson and O. Coleman, *England's export trade,*
1275–1547 (Oxford, 1963), 122–3, 138–9.
Breaks in the graphs are the result of no data.

The textile industries

A remarkably precise picture of England's exports is provided by the
Enrolled Customs Accounts between 1347 and 1547, when there was
a sharp break in the continuity of both the customs and their records. The
cloth custom established in 1347 applied to cloth exported by natives as
well as by aliens, unlike the custom of 1303 which applied only to the
latter. The accounts show clearly the replacement of exports of raw wool
by exports of manufactured textiles (Fig. 48).[1] A turning point was reached
in the years around 1450, but the decline in wool and the increase in cloth
exports had begun a century earlier. *Exports of raw wool* suffered a serious
but temporary setback in the years immediately after the Black Death,
but in the 1350s and early 1360s they recovered to almost their previous
level of about 30,000 sacks annually. Renewed depression began in the

[1] E. M. Carus-Wilson and O. Coleman, *England's export trade, 1275–1547*
(Oxford, 1963), 1–33.

late 1360s, and exports of raw wool took a markedly downward turn; by 1440 the annual export averaged only some 18,000 sacks. *Cloth exports* showed a temporary setback in the years immediately after the Black Death; but recovered and showed an annual growth rate of 18% between 1353 and 1368. By 1392–5 exports averaged some 43,000 cloths, nearly tenfold the 1347–8 figure of 4,422. The increase in cloth export more than compensated for the decrease in that of wool. English cloth, moreover, also captured the home market and led to a decrease in imported foreign cloth.[1]

Total wool exports – raw and manufactured – were virtually unaffected by the considerable loss of population during the fourteenth century, the loss of population being counterbalanced by a higher level of productivity *per capita*. It was not until the first half of the fifteenth century that the total wool export declined substantially as a consequence of drastically reduced raw wool exports. The decade 1431–40 was probably the first in which the export of cloths exceeded in sack equivalents the export of raw wool.[2] Thereafter, with the exception of 1461–70, the amount of cloth exported exceeded the equivalent quantity of raw wool exported. The gap between them increased, as raw wool exports declined and cloth exports multiplied, until the mid-sixteenth century when the cloth export market contracted sharply, for reasons to be discussed shortly.

This radical transformation of the export trade had a number of related causes which sprang from a growing economic incentive to export cloth rather than wool. In the first place, of some negative importance were sporadic and temporary embargoes placed on the export of wool in war-time, but these could be evaded by special licence. Secondly, but more important, was the positive encouragement given to the domestic cloth industry when wars interrupted the import of cloth. In the period between the outbreak of the Hundred Years' War and the Black Death (i.e. be-tween 1338 and 1348) war-time conditions provided a direct stimulus to the domestic industry; the government placed large orders for clothing the armed forces (2,000 pieces of cloth were bought for the navy alone

[1] H. L. Gray, 'The production and exportation of English woollens in the fourteenth century', *Eng. Hist. Rev.*, XXXIX (1924), 13–25; E. M. Carus-Wilson, 'The aulnage accounts: a criticism', *Econ. Hist. Rev.*, II (1929–30), 114–23; 'Trends in the export of English woollens in the fourteenth century', *Econ. Hist. Rev.*, 2nd ser., III (1950–1), 162–79, and 'The woollen industry', being ch. 6 of M. M. Postan and E. E. Rich (eds.), *The Cambridge economic history*, II (Cambridge, 1952), 355–428.

[2] Reckoning 4½ cloths to one sack of wool: A. R. Bridbury (1962), 23–38.

in 1337). Moreover, the temporary closing of the English market to Flemish cloth, and the prohibition of the export of English wool to Flanders, gave protection to the English industry. When these extreme measures were withdrawn, heavily increased export duties on raw wool (amounting to some 33%) burdened foreign manufacturers with greatly increased costs and gave more permanent protection to the home industry; cloth exports, on the other hand, paid a duty of less than 2%. Cloth manufacturers in Flanders and Florence, to a great extent dependent upon supplies of English wool, thus had increased costs added to internal social problems, and they became decreasingly able to face the competition of English manufacturers.[1] A third factor, acting as a further disincentive to the export of English raw wool, was the establishment after 1337, and more especially after 1350, of a quasi-monopoly of the wool trade by the English Company of the Staple, replacing the relatively free export of wool by both aliens and natives and restricting the channels of export; cloth exports were not similarly restricted.[2] A fourth factor was the rising price of domestic wool during the fifteenth century which was an added brake on raw wool exports, as also in this and in the following century was a deterioration in the quality of many English wools which made them less sought after by foreign buyers.[3]

The boom period of English raw wool exports before 1334 had been characterised by large-scale production of wool based on demesne flocks: that period was one of wholesale contracts between large producers and exporters, of free trade, of low taxation and of the predominance of foreign merchants in the English trade. The period of declining raw wool exports after 1334, more especially after 1400, was characterised by smaller scale production based on peasant flocks; this period was one of the middleman dealer, of monopoly, of high taxation and of the English Company of the Staple.[4] The sixteenth century, however, saw a progressive decline in the relative importance of the small producer, as sheep ownership became concentrated in fewer hands and the economies of large-scale production came to be realised. By then, wool production was primarily to supply the domestic textile industry.[5]

[1] E. M. Carus-Wilson (1952), 413–15.

[2] E. Power, *The wool trade in English medieval history* (London, 1941), 36.

[3] P. J. Bowden, *The wool trade in Tudor and Stuart England* (London, 1962), 5, 26–7 and 109–10.

[4] E. Power, 37–40.

[5] P. J. Bowden (1962), 2.

During the later Middle Ages, the English cloth industry witnessed not only a prodigious increase in production but also important changes in location and in the organisation of manufacture and marketing. On a national scale, the industry became more widely distributed than at any time before or since. The urban exodus, initiated during the early Middle Ages, gained momentum with the more widespread adoption of the fulling-mill.[1] Some of the long-established urban centres of the industry, such as Salisbury, remained important, but the period was characterised by the growth of a rural industry, particularly during the fifteenth century.[2] Although cloth manufacture during the later Middle Ages was widespread, from Devon to Yorkshire and from Westmorland to Kent, it came to be concentrated in three main areas – the West Country, East Anglia and the West Riding (Fig. 49).[3] Within these areas it prospered principally along river valleys at sites providing water for cleansing the cloth and working the fulling-mills. Of secondary importance were local supplies of fine-quality wool and fuller's earth.

The West Country (Devon, Somerset, Gloucestershire and Wiltshire) emerged as England's prime cloth-producing region; here numerous new settlements developed in narrow valleys, and agricultural villages and hamlets acquired industrial proletariats. The Stroud valley provides a striking example of this industrial development, where expansion took place not in existing upland villages but in valley hamlets. Proximity to abundant supplies of high-quality Cotswold wool, to beds of fuller's earth, to Bristol as an outlet for exports, and, most important of all, to water of a quality suited to the finest dyes and of a quantity sufficient for driving large numbers of mills – all combined with a favourable manorial structure to encourage the growth of a prosperous cloth industry in the Stroud valley and its adjacent valleys. Growth was particularly noticeable in the mid-fifteenth century at such places as Stroudwater and Castle Combe. At Castle Combe, 50 new houses, most built by clothier tenants, were erected between 1409 and 1454.[4] Similar new centres arose in the kersey-

[1] E. M. Carus-Wilson, 'An industrial revolution of the thirteenth century', *Econ. Hist. Rev.*, XI (1941), 39–60.   [2] A. R. Bridbury (1962), 39–51.

[3] E. M. Carus-Wilson (1952), 412 and 417–22; E. Lipson, *A short history of wool and its manufacture (mainly in England)* (London, 1953); J. L. M. Gulley (1960), 204–10, 279–82 and 377–8; P. J. Bowden (1962), 46; G. D. Ramsay, 'The distribution of the cloth industry in 1561–2', *Eng. Hist. Rev.*, LVII (1942), 361–9; R. A. Pelham, 250.

[4] R. P. Beckinsale, 'Factors in the development of the Cotswold woollen industry', *Geog. Jour.*, XC (1937), 349–62; E. M. Carus-Wilson, 'Evidences of industrial growth on some fifteenth-century manors', *Econ. Hist. Rev.*, 2nd ser., XII (1959–60), 190–205;

producing region of Dorset, Devon and Cornwall and in the broadcloth- and kersey-producing areas of Berkshire, Hampshire and Kent.[1] In East Anglia, growth occurred along the Stour and its tributaries, in Essex and Suffolk, at such places as Sudbury, Long Melford, Clare and Lavenham.[2] In Norfolk, on the other hand, the manufacture of worsted – a light cloth of high quality made of long, not short wool, combed not carded, and requiring little milling – was, by the mid-fifteenth century, in serious decline. The manufacture of worsteds in Norfolk had probably reached its zenith in the late fourteenth century and thereafter declined as overseas markets in France, Spain and Portugal were slowly lost to the growing continental light cloth industry, especially to Dutch manufacturers.[3] But with the introduction of new types of cloth by Dutch and Walloon refugees after 1550, the East Anglian textile industry once more flourished. In the West Riding, industrial development was checked in the fourteenth century as much by Scottish devastations as by plague, but in the following century there was considerable growth, notably on the upper reaches of the Aire and Calder and this region emerged as an area of kersey pro- duction.[4] To the west of the Pennines, the coarse woollen cloths produced in Lancashire had gained outlets on the Continent while linens were, for the most part, consumed by the home market. Here, in the years around 1600, new branches of manufacture were introduced, cotton among them, which were in due course to become paramount.[5]

R. H. Kinvig, 'The historical geography of the West Country woollen industry', *Geog. Teacher*, VIII (1916), 243–54 and 290–306; R. Perry, 'The Gloucestershire woollen industry, 1100–1690', *Trans. Bristol and Gloucs. Archaeol. Soc.*, LXVI (1945), 49–137; K. G. Ponting, *A history of the west of England cloth trade* (London, 1957), 1–59; G. D. Ramsay, *The Wiltshire woollen industry in the sixteenth and seventeenth centuries* (London, 1943), 1–84; E. M. Carus-Wilson, 'The woollen industry before 1550' in E. Crittall (ed.), *V.C.H. Wiltshire*, IV (1959), 115–47.

[1] R. P. Beckinsale, 357–61; P. J. Bowden (1962), 50–1.

[2] B. McClenaghan, *The Springs of Lavenham and the Suffolk cloth trade in the XV and XVI centuries* (Ipswich, 1924), 1–28; J. E. Pilgrim, 'The rise of the "new draperies" in Essex', *Univ. Birmingham Hist. Jour.*, VII (1959–60), 36–59; P. J. Bowden (1962), 52–3; G. A. Thornton, *A history of Clare, Suffolk* (Cambridge, 1928), 141–211.

[3] K. J. Allison, 'The Norfolk worsted industry in the sixteenth and seventeenth centuries: 1. The traditional industry', *Yorks. Bull. Econ. and Soc. Research*, XII (1960), 73–83.

[4] H. Heaton, *The Yorkshire woollen and worsted industries* (Oxford, 1920), 1–88.

[5] A. P. Wadsworth and J. de L. Mann, *The cotton trade and industrial Lancashire* (Manchester, 1931), 3–23.

A L A N R. H. B A K E R

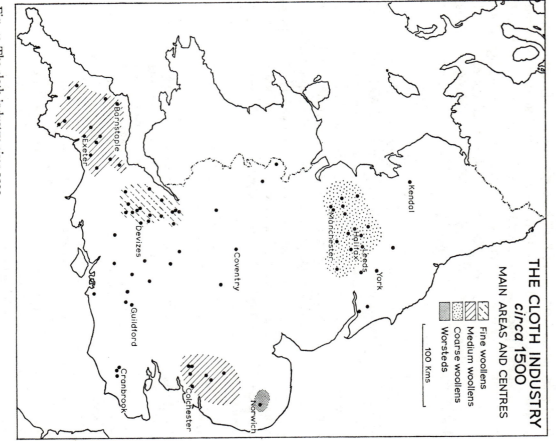

THE CLOTH INDUSTRY
circa 1500
MAIN AREAS AND CENTRES

Fine woollens
Medium woollens
Coarse woollens
Worsteds

100 Kms

Barnstaple
Exeter
Devizes
Guildford
Cranbrook
Colchester
Norwich
Coventry
Kendal
Manchester
Halifax
Leeds
York

Fig. 49 The cloth industry *circa* 1500
Based on P. J. Bowden, *The wool trade in Tudor and Stuart England* (London,
1962), 46.

The location of the cloth industry was not related directly to wool supplies – the cloth-producing counties were not identical with the wool-producing ones, although the nature of local wool supplies strongly in-fluenced the type of cloth produced. The highest quality wools in 1454 came from Shropshire and Herefordshire, which were not among the principal cloth-producing counties; the West Country broadcloth area relied more on these wools than on wools from the Cotswolds, while the serge industry of Devon and south-west Somerset was based on local long-woolled sheep. The industry does, on the other hand, seem to have been related directly to a supply of running water.[1] But not all areas with favourable sites for mills developed a cloth industry. Social factors may also have been important in influencing the emergence of Wiltshire, Suffolk and Yorkshire as the main cloth areas; J. Thirsk has suggested that rural industries may have developed where density of population was relatively high and farming essentially pastoral, so that there was both the need and the opportunity to supplement a meagre farming income by part-time industry.[2]

Industrial growth was associated with changes not only in location but also in the processes of manufacture and marketing. Fulling-mills were increasingly employed, as were other labour-saving devices such as the spinning wheel and the gig-mill for raising the nap on cloth; the first recorded instance of a gig-mill in England comes from Castle Combe in 1435.[3] Both fulling-mills and gig-mills aroused opposition from estab-lished cloth-workers, afraid of losing their livelihoods. But such protests, as well as legislation intended to limit the use of mills, were ineffectual. Indeed, an important feature of the cloth industry was its increasingly free operation, uncontrolled either by gild or state regulation. The only really effective regulation of the industry was that controlling the length and breadth of cloth put on the market, and even to this there were many permitted exceptions. Attempts to control the quality of cloth – for example, to prevent undue stretching of cloth and the use of waste or inferior wools – were never very effective, because they were not en-forced by a regular system of inspection.[4]

[1] E. Lipson, 14; R. H. Kinvig, 249–54 and 296–300; R. A. Pelham (1936), 255.
[2] J. Thirsk, 'Industries in the countryside', in F. J. Fisher (ed.), 70–88.
[3] R. A. Pelham, 'The distribution of early fulling mills in England and Wales', *Geography*, xxix (1944), 52–6; E. M. Carus-Wilson (1959–60), 201–2.
[4] K. J. Allison (1960), 80–1; H. Heaton, 124–40; G. D. Ramsay, 50–64; R. H. Tawney and E. Power (eds.), *Tudor economic documents*, iii (London, 1924), 210–25 (Leake's treatise on the cloth industry, 1577).

8

The cloth manufacturer, the 'clothier', came to exercise great control over the rural industry. Clothiers were especially important in the West Country, where the industry was organised by capitalists who bought and owned the wool but who had all the work (except fulling) done in the workers' homes, usually with their own implements although some clothiers provided these as well. The clothiers then marketed the finished product.[1] These capitalist entrepreneurs, producing primarily undyed broadcloths for export, differed considerably from the 'meaner clothiers' of the West Riding, who made the cloth in their own homes, employing a few labourers, sometimes only their own family.[2] In the West Riding, cloth-workers were often also agricultural workers, thus differing from the West Country where both capitalist organisers and industrial pro- letariat came to be divorced from agriculture.[3] Some East Anglian clothiers, such as the Springs of Lavenham, became very prosperous, using part of their wealth to build or restore beautiful churches. A few clothiers, such as William Stumpe of Malmesbury (Wiltshire), operated on a grand scale, employing many workers in a single industrial establish- ment. But these were not typical.[4]

A widening of markets and of sources of wool supply was typical of the period. By the late fifteenth century 'Stroudwaters' and 'Castle Combes' were as well known on the Continent as in England.[5] In the sixteenth century, the internal market for wool began to develop on a wider-than- regional basis as the character of both English wool and cloth manu- facture changed.[6] This widening of markets was more a consequence of the rise of the professional middleman (the economic catalyst of the later Middle Ages) and carrier than of improvements in the means of communi- cation.[7]

The early Tudor period has been called 'the golden age of traditional

[1] E. M. Carus-Wilson (1959–60), 195 and 200–2; R. Perry, 110–19.

[2] H. Heaton, 89–123.

[3] E. M. Carus-Wilson (1959–60), 202; H. Heaton, 24–5 and 93.

[4] B. McClenaghan, 30–41 and 60–88; G. Unwin, 'Woollen cloth – the old draperies' in W. Page (ed.), V.C.H. Suffolk, II (1907), 254–66; R. Perry, 49–50; G. D. Ramsay, 31–49; P. H. Ditchfield, 'Cloth making' in W. Page (ed.), V.C.H. Berkshire, I (1906), 387–95; J. G. Oliver, 'Churches and wool: a study of the wool trade in 15th century England', History Today, I (1951).

[5] E. M. Carus-Wilson (1959–60), 190.

[6] P. J. Bowden (1962), 56–63 and 72–6.

[7] P. J. Bowden (1962), 77–106; T. C. Mendenhall, The Shrewsbury drapers and the Welsh wool trade in the XVI and XVII centuries (London, 1953), 1–119.

broadcloth manufacture'.[1] Cloth exports trebled in the first half of the sixteenth century, then fell during the second half. Earlier boom conditions never returned, although broadcloth manufacture did enjoy something of an 'Indian summer' at the end of the century.[2] During the sixteenth century wool from central England (which had produced much of the country's fine, short staple wool) became coarser and longer. P. J. Bowden has suggested that this was a direct consequence of the increased feed resulting from enclosure; as sheep pastures improved so fleeces deteriorated and fibres lengthened. M. L. Ryder, on the other hand, has argued that the change in the type of wool came about as a result of a change in the type of sheep; better nutrition allowed selective breeding of the primitive long-wool for increased length. It is clearly the case that the changing character of wool and of the textile industry in England were related: long and coarse wool was better suited to production of worsted than of woollen fabrics. The decline of the woollen and the revival of the worsted branches of the textile industry during the sixteenth century would seem to have been a cause rather than a consequence of the changing nature of the wool supply.[3] The Midlands became a source of supply for long-fibre wool for the 'new draperies' of East Anglia, introduced by immigrants from the Netherlands fleeing from Spanish persecution. In 1554 a few weavers were persuaded to settle in Norwich. In 1565 another 300 Dutchmen and Walloons came to the city. The harsh policy of the Duke of Alva in the Netherlands after 1567 increased the flow. By 1569 the number of refugees in Norwich had increased to 2,826; by 1571 to 3,900; and by 1582 to 4,678. The immigrants introduced two main types of fabric: (1) Walloon 'caugeantry' which differed from traditional worsteds only in size, colour, or in number of threads used in the warp or weft, and which were akin to worsteds in being fine, light-weight cloths; and (2) Dutch bays, which were heavier, more akin to woollens than to worsteds, and like woollens they were fulled. Only slowly did English workers in Norwich, or in Colchester (the second major concentration of Flemish immigrants), turn to producing these 'new draperies'. But by the

---

[1] P. J. Bowden (1962), 43; G. D. Ramsay, 65.

[2] P. J. Bowden (1962), 43–4; G. D. Ramsay, 65–7; F. J. Fisher, 'Commercial trends and policy in sixteenth-century England', *Econ. Hist. Rev.*, x (1940), 95–117.

[3] P. J. Bowden (1962), 25–7 and 44–5, and 'Wool supply and the woollen industry', *Econ. Hist. Rev.*, 2nd ser., IX (1956–7), 44–58; M. L. Ryder, 'The history of sheep breeds in Britain', *Agric. Hist. Rev.*, XII (1965), 65–82. See also correspondence between P. J. Bowden and M. L. Ryder in *Agric. Hist. Rev.*, XIII (1965), 125–6.

end of the century these new fabrics, in their many varieties, were being made both in the old worsted district of Norwich and in the traditional broadcloth district of Colchester.[1] Whereas the arrival of Flemish immigrants in England in the early and mid-fourteenth century was a symptom rather than a cause of the progress of English woollen manufacture, the arrival of immigrants in the mid-sixteenth century resulted in the introduction of new fabrics and in the increased importance of East Anglia in the English textile industry.

The iron industry

Unlike the cloth industry, which during the Middle Ages added a rural distribution to an existing urban distribution, the charcoal iron industry throughout this period maintained its rural location. But it became increasingly a regional industry; widely distributed itinerant bloomeries were superseded by more narrowly located furnaces and forges. The scale of iron-working continued to be limited throughout the fourteenth and fifteenth centuries. There is some evidence of a smaller production in the fourteenth than in the preceding century, but the fortunes of the industry between 1334 and 1500 are difficult to trace, because of its peripatetic nature.[2] *Forgiae errantes* had short lives and an irregular production of blooms. Accounts of the Tudeley works near Tonbridge, in the Weald, began in 1330 and show that they were rebuilt in 1343 but lay unused in 1346. By 1350 they were operating again, although production costs had risen because of plague, and another onset of plague closed the works finally in 1363. Output had varied greatly from year to year, being about 200 blooms per annum in 1330–4, 600 blooms in 1335, and 252 blooms in 1350–1.[3]

Many bloomeries have left no documentation, only cinder heaps; from sites of others, even their cinders have been removed for road-building.[4] In all of the main areas of production – the Forest of Dean, the Weald and the Cleveland Hills – it seems that working was limited more by the

[1] K. J. Allison, 'The Norfolk worsted industry in the sixteenth and seventeenth centuries: 2. The new draperies', *Yorks. Bull. Econ. and Soc. Research*, XIII (1961), 61–77; J. E. Pilgrim, *passim*; P. J. Bowden (1962), 52–4; D. C. Coleman, 'An innovation and its diffusion: the "new draperies"', *Econ. Hist. Rev.*, 2nd ser., XXII (1969), 417–29.

[2] H. R. Schubert, *History of the British iron and steel industry from c. 450 B.C. to A.D. 1775* (London, 1957), 112–13.

[3] J. L. M. Gulley (1960), 375.

[4] J. L. M. Gulley (1960), 278–9.

local exhaustion of supplies of wood fuel for charcoal-making than of supplies of iron ore.[1] There is increasing evidence of coppicing during the fifteenth century and of a growing use of young and small trees in preference to dead wood and the branches and roots of trees felled in woodland clearance. Coppices were planted even in remote places where iron production was on a small scale and there was no lack of wood; such were the coppices near Barnard Castle (Co. Durham) in 1437. Timber shortages were to become of more consequence during the sixteenth century when production expanded significantly.[2]

Increased iron production was made possible by technological advances, and was stimulated by the demands of war. The application of water power to bellows, providing an artificial draught for smelting, was more widely adopted after about 1350, and may, in some instances, have been encouraged by labour shortages after the Black Death.[3] The use of water power to work bellows was not new in the later Middle Ages, but its application to hammers was an innovation. The first unequivocal reference to water-powered hammers comes from the Weald in the 1490s, where the works at Newbridge, in the parish of Hartfield in Ashdown Forest, included a 'great water hammer', but it was probably in use some decades before this.[4]

The most important change in the iron industry during the later Middle Ages, however, was the introduction of the blast furnace. Instead of obtaining malleable iron with a very low carbon content directly, as did bloomeries, the blast furnaces produced a highly carbonised or pig-iron too brittle for the smith's hammer; this pig-iron had then to be freed from the surplus of carbon and other impurities by smelting in a hearth or 'finery' before the iron could be drawn out and shaped by the forge hammer. Refining introduced an intermediate stage between the two stages of smelting the ore into a bloom and hammering the reheated bloom – hence the term 'indirect process' of iron-working. This new process needed larger and more complex plant, and greater capital investment, than did the old direct process; and the use of water power made necessary a local separation of furnace and forge, because separate water

[1] F. T. Baber, 'The historical geography of the iron industry in the Forest of Dean', *Geography*, XXVII (1942), 54–62; B. Waites, 'Medieval iron working in northeast Yorkshire', *Geography*, XLIX (1964), 33–43; J. L. M. Gulley (1960), 189–200, 275–9 and 375; E. Melling, 93–104; E. Straker, *Wealden iron* (London, 1931), 101–40.
[2] H. R. Schubert, 122–3 and 145. [3] H. R. Schubert, 133–4.
[4] H. R. Schubert, 134–40 and 147; J. L. M. Gulley (1960), 277–8.

wheels were needed for blast production, one for the furnace, and one each for the 'finery' and power hammer. The sites of iron production consequently became more permanent than hitherto, and furnaces were often distinct from forges.[1]

The first definite reference to a blast furnace comes from Newbridge, in the Weald, in 1496, an ironworks commissioned by the Crown to manufacture iron for armaments to be used in the Scottish war. An account for 1496–7 tells of axle-trees, wheel-rims, cast-iron bullets and shot to be carried to the Tower of London. Here at Newbridge, too, in 1509, guns of cast-iron were successfully manufactured for the first time in England. By 1542 two groups of blast furnaces had appeared in the Weald, one in south-east Sussex and the other around Ashdown Forest.[2]

At the time of the introduction of the blast furnace, bloomeries using the direct process existed in sufficient numbers to satisfy local demands for iron in rural districts of England, and native production was supplemented by imports of better quality iron for use in tool-making crafts and armaments. Furthermore, blast furnaces and forges were comparatively expensive structures requiring large ponds, dams and equipment. The indirect process spread only slowly, and new bloomeries using the direct process were still being erected well into the sixteenth century. The fact that a blast furnace could produce seven times as much iron per day as a bloomery was one factor encouraging the diffusion of the innovation. Another was that throughout the early sixteenth century the chief market for the new industry was the demand for ordnance, and for cars and other military equipment. Adoption of the indirect process was, however, slow, because of difficulties encountered in producing a pig-iron completely suitable for casting and conversion into malleable iron, and because, whereas Henry VII had stimulated a domestic iron industry with the skill of French founders, Henry VIII placed large orders for arms abroad, especially in the Low Countries. There were in 1542 only nine English blast furnaces, all in Sussex. Proximity to London encouraged the development of the iron industry in the Weald.[3]

During the 1540s the productive capacity of the industry grew enormously, and eleven new furnaces were built in the Weald in 1543–8. Threat of war with France greatly enlarged the demand for ordnance. By 1548 there were some 20 furnaces and 28 forges in the Sussex Weald, and in the

[1] H. R. Schubert, 157–8.
[2] H. R. Schubert, 162–6; J. L. M. Gulley (1960), 277–8; E. Straker, 38–52.
[3] H. R. Schubert, 158–61 and 166–70; J. L. M. Gulley (1960), 278.

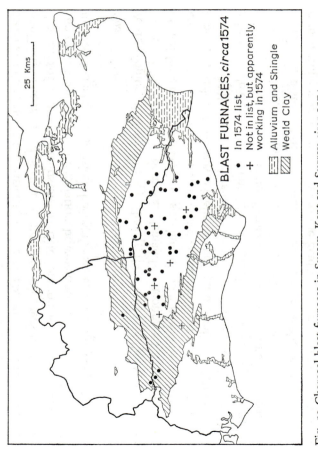

BLAST FURNACES, *circa* 1574

• In 1574 list
+ Not in list, but apparently working in 1574

▨ Alluvium and Shingle
▩ Weald Clay

Fig. 50 Charcoal blast furnaces in Sussex, Kent and Surrey *circa* 1574
Based on: (1) E. Straker, 'Wealden ironworks in 1574', *Sussex Notes and Queries*, VII (1938), 97–103; (2) H. R. Schubert, *History of the British iron and steel industry from c. 450 B.C. to A.D. 1775* (London, 1957), 354–92.

following decade the new process spread into Kent and Surrey. Until about 1560 the Weald had a monopoly of iron production by blast furnace. In 1574, when the industry was at a peak, there were at least 51 blast furnaces and 58 forges operating in the Weald (Fig. 50); but there were only seven other blast furnaces throughout the country, three in the Midlands and four in South Wales and Monmouthshire. The traditional iron area of the Forest of Dean was slow in adopting the blast furnace in place of the bloomery, and the first blast furnace did not appear there until the 1590s. In the meantime, much Forest of Dean pig-iron was made for sale upstream along the Severn to the metalcraft centres of the west Midlands. When Leland visited Birmingham about 1540, he reported 'many smithes in the town', but it was not until later in the century that there is evidence not only of smithies, processing bar iron into tools and weapons in the town, but also of furnaces and forges, producing pig and bar iron. The first recorded blast furnaces in northern England were those at Heanor (Derbyshire), about 1576, and at Rievaulx (North Riding), about 1582. Gradually other blast furnaces and forges were built in the area extending along the eastern border of Derbyshire from Heanor to Sheffield.[1] The iron industry, both new and old, was responsible for the depletion of many woodlands and the deterioration of many roads. Neither complaints nor restrictive legislation, however, were able to check seriously the iron industry's expansion during the sixteenth century.[2]

Coalmining

A developing timber shortage during the later Middle Ages, due to increased industrial and domestic demands, encouraged the substitution of coal for wood wherever it was technically possible. The growth of the coal industry was both a cause and a consequence of parallel growth in other industries. Coal was widely worked on a small scale throughout the later Middle Ages, and came to be worked more intensively and on a larger scale during the second half of the sixteenth century. Except for the Tyne valley, there was no district from which, until after 1500, coal was regularly carried in quantities of more than a few hundred tons per annum for distances of more than a few miles from the outcrops. Small coal-workings added local pits to the landscape, the coal being used for lime-burning for agricultural and building purposes, for smith's work (although charcoal was preferred if obtainable) and for baking, brewing and salt-making.[3] During the fifteenth century, pits were sunk deeper than previously and precautions taken against flooding. A 'colepytte' at Kilmersdon (Somerset) was deep enough in 1437 to have an adit or drainage channel, and a pit at the same place was said in 1489 to be deep and dangerous: the 'wark' or spoil from these pits in the outcrop areas remains today in mounds of considerable size.[4] It was not until the late

[1] H. R. Schubert, 170–84; J. L. M. Gulley (1960), 275–7; L. T. Smith (ed.), *The Itinerary of John Leland*, III (London, 1910), 97; J. A. Langford, 'Birmingham at the time of Leland's visit', *Trans. Birmingham Archaeol. Soc.* (1882–3), 32–42; R. A. Pelham, 'The migration of the iron industry towards Birmingham during the sixteenth century', *Trans. Birmingham Archaeol. Soc.*, LXVI (1945–6), 147–9, and 'The establishment of the Willoughby ironworks in north Warwickshire in the sixteenth century', *Univ. Birmingham Hist. Jour.*, IV (1953), 18–29; W. H. B. Court, *The rise of the Midland industries, 1600–1838* (London, 1938), 33–44.

[2] State Papers Domestic, Elizabeth: Book 117, No. 39 — transcribed in D. and G. Mathew, 'Iron furnaces in south-eastern England and English ports and landing places, 1578', *Eng. Hist. Rev.*, XLVIII (1933), 91–9; B. Waites, 37; J. L. M. Gulley (1960), 198–200.

[3] J. U. Nef, *The rise of the British coal industry*, I (London, 1932), 8–9.

[4] J. A. Bulley, '"To Mendip for coal" — a study of the Somerset coalfield before 1830', *Proc. Som. Archaeol. and Nat. Hist. Soc.*, XCVII (1952), 46–78.

sixteenth century that deeper mines and mechanical pumping became common (although a pump worked by horse power had been used at Moorhouse, Co. Durham, in 1486).[1]

Coal mined in the Tyne valley was marketed beyond its immediate vicinity, and was exported from Newcastle by sea, in particular to London but also to other English ports, and to Flemish, Dutch and occasionally to French and German ports. Newcastle's coal trade declined during the fourteenth century, but became increasingly important when the port's trade in wool and hides collapsed at the end of the century. Before 1500, however, total annual shipments from the Tyne probably rarely exceeded 15,000 tons; by 1563–4 they reached 33,000 tons, and by 1597–8 as many as 163,000 tons. Much of this considerably increased production and export was a response to the rapidly growing demands of London. Mining from pits sunk within the manors of Whickham and Gateshead, to the south of the Tyne, was intensified; and, farther west, mining began at Winlaton, Stella and Ryton. Winlaton colliery alone produced more than 20,000 tons in 1581–2.[2]

During the late sixteenth century many new and deeper pits were sunk on most English coalfields. Production expanded as the market, both domestic and industrial, grew, and as supplies of wood fuel declined. Development was greater where there was access to water transport; the manor of Wollaton, for example, because of its proximity to the Trent and to abundant and accessible seams, was a favoured site in Nottinghamshire.[3] Many pits were still operated on a small scale and with primitive techniques, serving only local markets; such was the Earl of Shrewsbury's mine at Sheffield (Yorkshire) which between June 1579 and December 1582 had an annual output of about 1,200 tons.[4]

Other industries

The lead-smelting industry of the Peak District of Derbyshire saw an eastward migration away from its location on the limestone upland on to the gritstone edges to the east. This migration was largely the consequence

---

[1] 'The charters of endowment, inventories, and account rolls of the Priory of Finchale, in the County of Durham', *Publ. Surtees Soc*, VI (1837), cccxci: cited in L. F. Salzman, *English industries of the Middle Ages* (Oxford, 1923), 10.

[2] J. U. Nef, 9–26; C. M. Fraser, 'The north-east coal trade until 1421', *Trans. Archit. and Archaeol. Soc. Durham and Northumberland*, II (1962), 209–20.

[3] J. U. Nef, 57–109.

[4] L. Stone, 'An Elizabethan coalmine', *Econ. Hist. Rev.*, 2nd ser., III (1950–1), 97–106.

of a search for higher and draughtier sites for the hearths, and it resulted in a fundamental structural change in the industry towards the end of the fourteenth century. The smelting of lead had become separated from the other branches of the industry, creating a class of middlemen who bought dressed ore from the miners, smelted it, and resold it. An advantage of the new location of the hearths was their proximity to the main lead-marketing routes which led not only to Chesterfield, an important ex-change centre for the industry, but also to the Continent through the ports of Boston and, later, Hull. The new siting of the hearths had a considerable advantage over the old from the point of view of fuel supply. Resources of suitable timber on the upland had been virtually exhausted but there had been as yet little exploitation of the wood on the slopes of the gritstone edges. These new local sources of fuel were apparently sufficient to last until the latter half of the sixteenth century, when supplies of wood were imported into the region at a time when production of lead increased to meet the demand resulting from the wave of new building at home and abroad.[1] Lead mining in the Mendips also increased from the middle of the sixteenth century, and lead was exported from Bristol. Here, too, calamine, the ore of zinc, was discovered in 1566.[2]

The second half of the sixteenth century saw a similar intensification of activity in the exploitation of other minerals, often encouraged by royal patronage. Two great companies were chartered in 1568, the Mines Royal and the Mineral and Battery Works; these were the first companies to be formed in England for the manufacture of a product (copper and brass respectively) as distinct from companies formed for trading purposes.[3] German capitalists and workmen were invited to develop mineral in-dustries, notably the copper industry of the Lake District, centred on Keswick.[4] Much mineral exploitation towards the end of the sixteenth century was being checked by difficulties, especially that of flooding, but the period was also marked by technical experiments: shafts were sunk deeper; adits were pushed almost horizontally into hillsides; pumping and other machinery was introduced; and ores were smelted in furnaces

[1] J. P. Carr, 'The rise and fall of Peak District lead mining' in J. B. Whittow and P. D. Wood (eds.), *Essays in geography for Austin Miller* (Reading, 1965), 212–13. See also L. F. Salzman (1923), 41–68.

[2] J. W. Gough, *The mines of Mendip* (Oxford, 1930), 65–6, 82, 112.

[3] M. B. Donald, *Elizabethan copper* (London, 1955) and *Elizabethan monopolies* (London, 1961).

[4] F. J. Monkhouse, 'Some features of the historical geography of the German mining enterprise in Elizabethan Lakeland', *Geography*, XXVIII (1943), 107–13.

instead of open hearths.[1] Mineral production was extremely erratic, as that of tin in Devon and Cornwall illustrates. The introduction of shaft-mining in the late fifteenth century must have contributed to the sudden increase in the output of Devon tin round about 1500, while production fell at the end of the sixteenth century when production costs rose considerably and when drainage techniques were unable to cope with progressive flooding as the level of the mining sank deeper.[2]

The salt industry of Worcestershire and Cheshire continued to consume great quantities of turves and wood in the boiling of brine from salt springs.[3] Along the east coast, the last recorded evidence of salt-making at Fleet in Lincolnshire comes from 1455; here and elsewhere along the east coast, decayed salterns remain even today as relict features – clustered masses of irregularly shaped mounds, some of them 16 to 20 feet high.[4] In the north-east, along the coasts of Durham and Northumberland, coal began to be used for boiling brine in the sixteenth century, and resulted in considerable development, especially at South Shields at the mouth of the Tyne.[5]

The English salt industry underwent a 'commercial revolution' during this period. In 1334 England exported salt, but as the cloth industry developed the coastal salter had every incentive to leave salt-making for employment in the cloth industry. The rise in wages during the fourteenth century, in the absence of a general rise in the market price of salt, destroyed any cost advantages that English salt enjoyed in continental markets over the salt of northern Germany and the Low Countries. Furthermore, competition during the fourteenth century from cheaply produced salt in the Bay of Bourgneuf, to the south of the Loire estuary in France, helped to transform England from an exporter to an importer of coarse

[1] L. Stone, 98; F. J. Monkhouse, 'Pre-Elizabethan mining law, with special reference to Alston Moor', *Trans. Cumberland and Westmorland Antiq. and Archaeol. Soc.*, XLII (1942), 43–55; A. Raistrick, *Mines and miners of Swaledale* (Clapham, Yorks., 1955), 21–31.

[2] G. R. Lewis, *The Stannaries. A study of the English tin miner* (Cambridge, Mass., 1924), 39–64 and 252–65; A. R. Bridbury, 24–6; W. G. Hoskins, *Devon* (London, 1954), 133.

[3] H. J. Hewitt, *Mediaeval Cheshire* (Manchester, 1929), 110–14.

[4] E. H. Rudkin and D. M. Owen, 'The medieval salt industry in the Lindsey marshland', *Rep. and Papers, Lincs. Archit. and Archaeol. Soc.*, n.s. VIII (1960), 76–84; H. E. Hallam, 'Salt-making in the Lincolnshire Fenland during the Middle Ages', *Rep. and Papers, Lincs. Archit. and Archaeol. Soc.*, n.s., VIII (1960), 85–112.

[5] P. Pilbin, 'A geographical analysis of the sea-salt industry of north-east England', *Scot. Geog. Mag.*, LI (1935), 22–8; A. E. Smailes, *North England* (London and Edinburgh, 1960), 132–6.

salt. By 1364 more was imported than was exported, and exports almost entirely ceased by 1500. Bristol, Yarmouth (because of its connections with the herring industry) and London dominated the import of salt. With the dislocation of the salt trade by wars against France during the middle of the sixteenth century, and with increasing use of coal, the salt industry again became a large-scale enterprise, particularly at the mouths of the Wear and Tyne, where cheap coal was obtainable. Capital costs in the industry were high (one salt-works on the Wear in 1589 involved an investment of £4,000). The use of coal thus concentrated the industry and changed its character from domestic to capitalistic production.[1]

Another major industrial change was seen in the paper industry. Paper came increasingly to be used instead of parchment. When Caxton set up his printing office at Westminster in 1476, he used imported paper, and the overwhelming majority of books printed here in the sixteenth century continued to use imported paper. But the needs of printing, as well as of wrapping and writing, produced a rising demand for paper in the sixteenth century. A paper mill was at work near Hertford in 1495 but it had probably failed by 1507. In the 1540s there may have been no paper at all produced in England. Some of the mills established in the 1550s had short lives, and it was not until the latter decades of the century that the industry began to flourish. In 1558, a mill was established at Dartford (Kent) by a German, John Spilman, employing German workmen; and in 1589 Spilman was granted a monopoly for the making of white paper. The industry grew up along river valleys (water being needed as a raw material as well as for power in the pulping process) and in proximity to towns which provided supplies of rags (the principal raw material) as well as markets for the finished product. The growth of paper mills in the south-east owed much to the proximity of London: here was the largest English market for paper, and here was the largest English supply of linen rags (linen rags were much more important than woollen rags, and England imported most of her linens from France, Holland and Germany through London). The use of old cordage and sails as raw materials probably helped to attract brown (wrapping) paper mills to the proximity of such ports as Dover, Exeter and Southampton.[2]

[1] A. R. Bridbury, England and the salt trade in the later Middle Ages (Oxford, 1955), passim; E. Hughes, 'The English monopoly of salt in the years 1563–71', Eng. Hist. Rev., XL (1925), 334–50.

[2] D. C. Coleman, The British paper industry 1495–1860: a study in industrial growth (Oxford, 1958), 1–88; A. H. Shorter, Paper mills and paper makers in England,

TRANSPORT AND TRADE

Roads and rivers

There were minor, but no fundamental, changes in the network of internal communications between 1334 and 1600. Improvement of roads was highly localised, and concentrated especially on bridges. Maintenance of road conditions was very haphazard, and was the responsibility of local communities whose interests seldom transcended their own parish boundaries. The Highways Act of 1555 – significant as the first legislation ever passed applying to roads in general in England – achieved little more than the transference of responsibility from manorial officials to parochial surveyors of highways. It did little to improve the conditions of travel.[1]

River navigation remained important for the transport of bulky articles such as grain and timber, and a number of statutes were passed, more especially after 1500, to facilitate the removal of obstructions from rivers; the channel of the Lea, for example, so important for London, was improved between 1571 and 1581. After various attempts to improve the Exe below Exeter, the Exeter ship canal was dug in 1564–6 between Exeter and its outport of Topsham. It incorporated the earliest poundlocks in England, and has been called 'the first true canal of modern times in the British Isles'.[2] But the beginning of the 'canal age' was yet nearly two centuries away.

Although there was no fundamental improvement in the physical condition of the internal communications in England, there was, however, a concentration of inland trade into fewer centres; everywhere agricultural traffic, for example, tended to be drawn away from the smaller markets, ports, and fairs, and to be centred in the larger provincial towns such as Canterbury and Reading. There were far fewer market towns and villages in 1600 than in 1334, probably less than a third. In Norfolk, for example, where there had been 130 markets, there were only 31 by the sixteenth century; and the 53 in Gloucestershire had become 34. Moreover, the larger market towns began to specialise in particular types of products such as grain, cheese and butter, poultry, horses, sheep and leather products.[3]

1495–1800 (Hilversum, 1957), 22–50, and Fig. 1, p. 92; A. H. Shorter, *Paper making in the British Isles* (Newton Abbot, 1971), 13–19.

[1] H. J. Dyos and D. H. Alcroft, *British transport* (London, 1969), 30–1.
[2] *Ibid*, 36–8.
[3] A. Everitt, 'The marketing of agricultural produce', in J. Thirsk (ed.), *The agrarian history of England and Wales*, IV, 1500–1640 (Cambridge, 1967), 467–9 and 490–6.

Maritime trade: coastal and overseas

Coastwise traffic, so often overlooked, was an essential element in the economy of the country, and a very substantial amount of inter-regional trade was carried on by this means between the many harbours around the irregular coasts of the British Isles. One indication of this traffic was the growth of London, which drew its supplies from a very wide area, much of them by sea. It is difficult to see how London could have grown as it did during this period without the advantages of cheap seaborne traffic. It was in the sixteenth century that the Corporation of Trinity House received its first charter (in 1514) from Henry VIII. Later in the century it was granted authority to erect beacons and other marks for the guidance of mariners around the coasts of England.

There was also foreign traffic – in an ever-increasing quantity. In this connection, there were two important technical developments in ship-building: firstly, the transition from the one-masted to the three-masted ship, thus giving scope for a variety of sails with particular functions; secondly, the lengthening of a ship in relation to its beam, the transition from the 'round ship' to the 'long ship'. The sixteenth century in particular saw considerable expansion of the shipping industry, in response to the navy's growing requirements and to the expansion of trade (especially the east coast coal and the Atlantic fishing trades).[1]

Some idea of the scope of European commerce can be obtained for the fifteenth century from *The Libelle of Englyshe Polycye*, by an anonymous author, which appeared in 1436,[2] and from Sir John Fortescue's *Comodytes of England*, which was written at about the same time.[3] Trade in the spheres of the Baltic and North Sea was essentially in necessities – fish, salt, timber and forest products such as pitch, tar and potash. Export of wool to the Low Countries and northern France was from east coast ports of fundamental – but declining – significance (Fig. 51). By the end of the fifteenth century, on the other hand, the export of English cloth had

[1] H. D. Burwash, *English merchant shipping, 1460–1540* (Toronto, 1947); G. V. Scammell, 'English merchant shipping at the end of the Middle Ages', *Econ. Hist. Rev.*, 2nd ser., XIII (1960–1), 327–42, and 'Ship-owning in England circa 1450–1550', *Trans. Roy. Hist. Soc.*, 5th ser., XII (1962), 105–22; R. Davis, *The rise of the English shipping industry in the seventeenth and eighteenth centuries* (London, 1962), 1–8 and 44–6; G. J. Marcus, *A naval history of England. I. The formative centuries* (London, 1961), 1–67.

[2] G. Warner (ed.), *The Libelle of Englyshe Polycye* (Oxford, 1926).

[3] Printed in *The works of Sir John Fortescue*, 2 vols. (London, 1869).

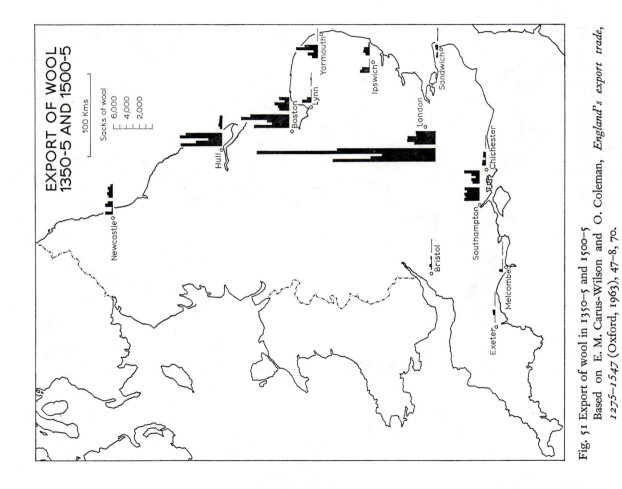

Fig. 51 Export of wool in 1350–5 and 1500–5
Based on E. M. Carus-Wilson and O. Coleman, *England's export trade,
1275–1547* (Oxford, 1963), 47–8, 70.

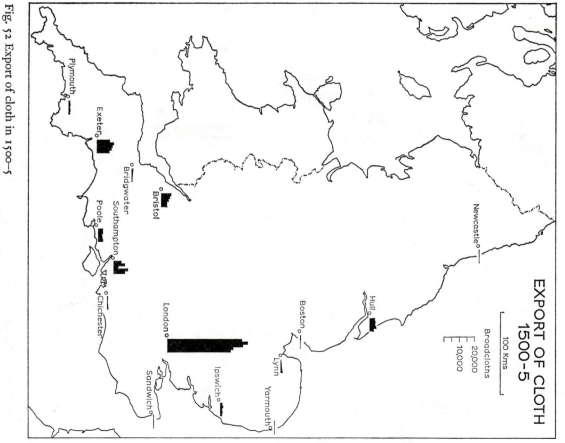

Fig. 52 Export of cloth in 1500–5
Based on E. M. Carus-Wilson and O. Coleman, *England's export trade, 1275–1547* (Oxford, 1963), 112.

become a very prominent element in the international trade of Europe. The export was primarily from London – as had been that of raw wool – and secondly from the ports of the south and south-west, in contrast to the earlier concentration on exporting of raw wool from the east coast ports (Fig. 52). From the west coast of France were imported salt and wine, the latter coming from south-west France through the ports of London, Southampton, Bristol and Hull.[1] Trade with Mediterranean countries and beyond was essentially in imported luxuries, such as spices, drugs, perfumes, sugar, precious stones, dyestuffs (indigo, madder, saffron), alum and carpets. The outward commodities were mainly woollens and linens together with some raw wool, metals and hides.

During the sixteenth century new trading companies extended the range of commercial activity, diversifying exports and more particularly imports. To the traditional northern and Mediterranean trades was added emergent Atlantic and Far Eastern trades. Thus, the main source of imported sugar seems already by about 1590 to have been shifting to Brazil and the West Indies. Furthermore, by this time refineries had been set up in England, and the finished product was being exported to continental markets.[2] From about 1566 onwards tobacco began to be imported from North America, and pipe smoking had become general by the end of the century. The formation of the East India Company of London in 1600 was another indication of the changing pattern of England's overseas trade.

TOWNS AND CITIES

The fortunes of English towns during this period were very varied and were characterised by great diversity both in time and place. Even so, some general changes can be discerned, such as the relative growth of many new industrial villages and small towns, the relative decline and later recovery of many of the old corporate towns, the close association of the condition of towns with the state of trade, and the increasing economic supremacy of towns in general and of London in particular. A comparison of the subsidies of 1334 and 1524 shows that urban wealth constituted a far larger proportion of total lay wealth at the end of this period than at the beginning. The fortunes of individual towns differed but collectively they strengthened their grip upon the national economy.[3]

[1] M. K. James, *Studies in the medieval wine trade* (Oxford, 1971).
[2] L. Stone, 'Elizabethan overseas trade', *Econ. Hist. Rev*, 2nd ser., II (1949–50), 30–58.   [3] A. R. Bridbury (1962), 77–82 and 111–13.

Industrial villages

A striking feature of the later Middle Ages was the growth of industrial villages, some of which developed (or were later to develop) into towns. The pattern of this new industrial settlement was markedly different from that of the old. Not confined within the walls or even the suburbs of the old towns, it was often a straggling growth integrated only by its market place and by the parish church, itself often wholly or partly rebuilt. Kersey and Long Melford, two cloth-making centres in Suffolk, still exhibit a linear pattern of unregulated growth along a main street.[1] The expansion of industries in the countryside is still exemplified at Castle Combe, in Wiltshire, where, during the first half of the fifteenth century, there was an impressive industrial growth alongside the stream, with the building of new and rebuilding of old houses, many of them in local stone. Among these new buildings at Castle Combe was a fifteenth-century church tower with its decorations based on cloth-working implements. Here, a growing class of craftsmen with no agricultural holdings lived in the valley at 'Nethercombe' while yeoman cultivators lived on the heights above the wooded valley at 'Overcombe'.[2] Industrial villages also emerged in the West Riding, such as Halifax, Leeds and Wakefield; and in the Midlands many small settlements developed important tanning industries.[3] In the mid-sixteenth century, Birmingham was more important for its tanning and clothing than for its metal industries, although the latter were beginning to assume increasing importance.[4]

But it was the cloth industry which, more than any other during the later Middle Ages, promoted small settlements into prosperous towns. Thus Totnes, a small village in Devon, blossomed suddenly during the fifteenth century into such a flourishing woollen manufacturing town that by 1523–7 it was among the twenty most prosperous provincial towns in England.[5] Involvement in cloth-making was enough to raise a

---

[1] M. W. Beresford and J. K. S. St Joseph, *Medieval England. An aerial survey* (Cambridge, 1958), 242–3 and 245–7.

[2] M. W. Beresford and J. K. St Joseph, 247–9; E. M. Carus-Wilson (1959–60), 197–204.

[3] H. Heaton, 54–79.

[4] W. H. B. Court, 33–43.

[5] E. M. Carus-Wilson, *The expansion of Exeter at the close of the Middle Ages* (Exeter, 1963), 17; J. Cornwall, 'English country towns in the fifteen-twenties', *Econ. Hist. Rev.*, 2nd ser., XV (1962–3), 54–69; W. G. Hoskins (1959), 177.

Table 5.1 *The ranking of towns, 1334–1525 (excluding the City of London, Westminster and Southwark)*

Town	1334	1524–5	Town	1334	1524–5
Bristol	1	2	Newbury	21	21
York	2	11	Plymouth	22	
Newcastle upon Tyne	3	a	Newark on Trent	23	
Boston	4	22	Peterborough (*cum membris*)	24	
Great Yarmouth	5	20	Nottingham	25	
Lincoln	6	15	Exeter	26	5
Norwich	7	1	Bury St Edmunds	27	13
Oxford	8	29	Stamford	28	30
Shrewsbury	9	26	Ely (*cum membris*)	29	
Lynn (King's and South)	10	8	Luton	30	
Salisbury	11	4	Reading		9
Coventry	12	3	Colchester		10
Ipswich	13	6	Lavenham		12
Hereford	14	19	Worcester		14
Canterbury	15	7	Totnes		16
Gloucester	16	17	Hull		18
Winchester	17		Hadleigh		23
Southampton	18	27	St Albans		24
Beverley	19		Leicester		25
Cambridge	20	28			

a Not taxed in 1524–5 but may have had second place.

place like Lavenham, in Suffolk, with scarcely a thousand inhabitants, also to a position among the first twenty towns. A comparison of the thirty most prosperous provincial towns in 1334 with those in 1524–5 shows clearly the rise of new towns as Lavenham, Hadleigh and Totnes and the relative decline of such old towns as York, Lincoln and Oxford.

The fate of the corporate towns

Many of the old-established towns had their populations drastically reduced by plague during the fourteenth century, and their economies checked during the malaise of the early fifteenth century. The physical spread of some seems to have been temporarily halted and there was even structural decay. During the later Middle Ages, plague appears to have become mainly an urban phenomenon, because of crowded and insanitary conditions.[1] In York, for example, there occurred a serious outbreak of bubonic plague accompanied by sweating sickness in 1550 and the summer of 1551.[2] Reduced populations resulted in derelict buildings. Numerous buildings fell into disuse and disrepair in Lincoln, for example, and at least 12 of its 46 parish churches seem to have decayed during the fourteenth and fifteenth centuries. In 1428, seventeen parishes in the city were returned as not having more than ten inhabitants each. Lincoln's suburbs were shrinking and so also were the back-streets of the city itself.[3] By the mid-sixteenth century many town castles were also in decay and town walls in ruins.[4] The age of private local wars had ended, and the decline of the town garrison had an impact upon its associated dependants. After the removal of the castle from Leicester, for example, a commission of enquiry into the state of house-property belonging to the Crown in the town disclosed, in the autumn of 1587, that there were no fewer than 235 tenements 'in great decay'.[5]

The prospects of both old and new towns, moreover, were closely associated with their involvement in industry and trade. The fortunes of Norwich, for example, were equally those of its worsted industry. Trade fluctuations checked any notable increase of population between the fourteenth and the early sixteenth century, by which time some of the city's buildings were in a bad state of repair. Many craftsmen avoided paying the city's charges, which included payments for street-paving, for the replacement of thatch with tiles and slates after disastrous fires, for the rebuilding of burnt houses and for the cleansing of rivers and streets. Both the worsted trade and the face of the city were given a lift during the late sixteenth century under the influence of Dutch and Walloon refugees who, by the 1580s, probably numbered one-third of its population.[6] In York, the growth of the cloth industry, trade, population and prosperity during the fourteenth century resulted in much building and rebuilding. On the other hand, the decline of the city's economy and numbers during the following century threw many houses into decay

[1] J. M. W. Bean (1962–3), 430.
[2] A. G. Dickens, 'Tudor York' in P. M. Tillott (ed.), *V.C.H. Yorkshire: The city of York* (1961), 120.
[3] J. W. F. Hill, *Medieval Lincoln* (Cambridge, 1948), 286–8.
[4] J. D. Mackie, *The earlier Tudors 1485–1558* (Oxford, 1952), 37–8.
[5] W. G. Hoskins, 'An Elizabethan provincial town: Leicester', in J. H. Plumb (ed.), *Studies in social history: a tribute to G. M. Trevelyan* (London, 1955), 33–67.
[6] K. J. Allison (1960), 73–4 and 79, and (1961), 61–2.

and dereliction.[1] Southampton also suffered a similar contrast of fortunes during the fifteenth and sixteenth centuries.[2] Towns intimately engaged in cloth production and marketing on a considerable scale suffered less than those without this economic backbone. Thus Winchester, lacking a staple industry, fell into serious decline, while Exeter rose as a centre for the manufacture and export of Devon kerseys and tin, particularly during the late fifteenth century. In addition to those industries common to every provincial town, such as milling and baking, Exeter acquired highly developed tanning, bell-making, cloth-making and cloth-finishing industries.[3]

Within towns, industrial quarters became more apparent. Much of Exeter's industrial activity was carried on beyond the city walls; thus below the west wall on Exe Island a multitude of leats, drawn from the Exe, were established to drive corn mills, fulling-mills and probably tanning mills.[4] There developed an increasing separation of the poor in the extra-mural suburbs and in the back-lanes and side-streets within the walled areas. Certain parishes in the old cities like Exeter, York and Leicester were almost entirely populated by labouring classes in the 1520s and other parishes were equally reserved for the wealthier.[5] Social and economic inequality in towns was aggravated during the later Middle Ages. W. G. Hoskins has calculated from the 1524 Lay Subsidy that in the larger English towns approximately half of the taxable population belonged to the wage-earning class.[6] In sixteen towns in Sussex, Buckinghamshire and Rutland the wage-earning class amounted to nearly 40% and yet it rarely owned as much as 10% of the wealth.[7] In the country as a whole, wealth was increasingly becoming concentrated in the capital.

London

Whereas in 1334 London had been not quite five times as wealthy as the richest provincial town (Bristol), in the 1520s the City of London alone was almost ten times as wealthy as Norwich, then the leading provincial

[1] J. N. Bartlett, 'The expansion and decline of York in the later Middle Ages', *Econ. Hist. Rev.*, 2nd ser, XII (1959–60), 17–33; E. Miller, 84–6.

[2] A. A. Ruddock, *Italian merchants and shipping in Southampton, 1270–1600* (Southampton, 1951).

[3] T. Atkinson, *Elizabethan Winchester* (London, 1963), 29–33; E. M. Carus-Wilson (1963), 5–16 and 22–4.

[4] E. M. Carus-Wilson (1963), 22.

[5] W. G. Hoskins (1955), 43–4, and 'English provincial towns in the early sixteenth century', *Trans, Roy. Hist. Soc.*, 5th ser, VI (1956), 19.

[6] W. G. Hoskins (1956), 17–19.   [7] J. Cornwall (1962–3), 63.

city, and more than fifteen times as wealthy as Bristol. In the subsidy of 1543–4, London paid thirty times as much tax as Norwich, and well over forty times as much as Bristol. Even the suburb of Southwark, across the river, paid more tax than Bristol. London contributed as much in 1543–4 as all the other English towns put together, from Norwich down to the smallest place that functioned as a local market centre.[1] The population of London increased threefold during the sixteenth century, to a large extent because of considerable immigration.[2] Its population in 1600 was probably about four to five times as great as it had been in 1334. Its prosperity was based upon an increasing share of England's trade, particularly of the cloth trade. A weekly cloth market was established in 1396 at Blackwell Hall, near the Guildhall, and the addition of several annexes in the course of the fifteenth century, and finally its destruction and rebuilding in 1588 as a 'new, strong and beautiful store-house', bore witness to the expanding textile trade and to London's control of the trade.[3]

Gradually London usurped the trading functions of many provincial towns. During the fifteenth century the distribution of cloth in York, for example, was taken out of the hands of the city's traders first by West Riding merchants and later by London merchants able to supply, in return for textiles, a large variety of imported goods hitherto unobtainable from local traders.[4] By the early sixteenth century, much cloth manufactured in Devon was being sent to London and thence exported to the Low Countries, rather than being exported to France from diverse West Country ports as it had been during the fifteenth century.[5] The ever-increasing influence of London's merchant class is clearly demonstrated in Southampton, whose prosperity during the fifteenth century was fostered by Italian merchants, nurtured by local merchants and ultimately killed by London merchants. From the middle of the fifteenth century, Londoners had taken a leading part in Southampton's commerce, using the town as an outport for trade between England and the Mediterranean. From trading with the Mediterranean, London merchants spread into every branch of commerce in Southampton, gradually swamping local merchants by their superior capital resources and large-scale ventures.

1 W. G. Hoskins (1955), 35.
2 P. Ramsey, *Tudor economic problems* (London, 1963), 110.
3 E. M. Carus-Wilson (1952), 420–1, and G. D. Ramsay, 25.
4 J. N. Bartlett, 30.
5 E. M. Carus-Wilson (1952), 28–9.

Southampton's role as London's outport was undermined when, during the first half of the sixteenth century, improvements in shipbuilding (especially in rigging) made the large square-rigger less cumbersome and less difficult to manoeuvre in the restricted channels amid the sandbanks of the Thames' estuary. And when, in 1514, the control of pilotage in the Thames was placed in the charge of a body of experienced lodesmen, the Brethren of Trinity House at Deptford Strand, navigational safety was considerably furthered. From 1540 onwards especially, London merchants forsook Southampton and increasing numbers of ships journeyed to the Thames rather than to Southampton Water. Southampton became moribund, with considerable structural decay.[1] Overland trade between Southampton and London was drastically reduced and intermediate towns like Salisbury were absorbed within the capital's economic hinterland as their cloth merchants exported more to the Low Countries via the Thames and less to France and Spain via Southampton Water.[2] Increasingly, London was dominating overseas trade as it came increasingly to dominate the English economy in general.

[1] A. A. Ruddock, 255–72.　　[2] G. D. Ramsay, 21–2.

Chapter 6

ENGLAND *circa* 1600

F. V. EMERY

One aspect of the flowering of Elizabethan England, drawing its strength from a variety of sources, was the beginning of a tradition of topographical writing.[1] John Leland's notes for a proposed 'Description of the realm of England' had not been published when he died in 1552 and, though used by many people, they were not printed until 1712. In the meantime, in 1577, there appeared William Harrison's *Historical description of the island of Britain* as an introduction to Holinshed's chronicles, and this was revised for a second edition in 1588, part of which was called *The description of England*. It was during these years that William Camden began the work that appeared in 1586 as the *Britannia*. Written in Latin, it was organised upon a county basis, and soon ran into a number of editions. The sixth edition in 1607 was illustrated by a series of county maps engraved from the surveys of Saxton, Norden and others. It was translated into English in 1610 by Philemon Holland, who called it a 'chorographical description.' Then in the following year came John Speed's *The theatre of the empire of Great Britaine* which aimed at 'presenting an exact geography' of the realm. It, too, was organised on a county basis; its text was abridged from Camden's *Britannia*, but there was a new set of maps, based likewise on those of Saxton, Norden and others.

The county maps that Camden and Speed were using constituted an innovation made possible by developments in surveying instruments and the art of surveying. Laurence Nowell in the 1560s had proposed to make maps of the English counties;[2] but what Nowell proposed, Christopher Saxton accomplished. He worked during the 1570s under the patronage of Thomas Seckford, lawyer and courtier, who obtained official support

---

[1] A. L. Rowse, 'The Elizabethan discovery of England', being ch. 2 (pp. 31–65) of *The England of Elizabeth* (London, 1951).

[2] R. Flower, Laurence Nowell and the discovery of England in Tudor times', *Proc. British Academy*, XXI (1935), 54.

for the enterprise. Saxton's atlas of England and Wales on a county basis appeared in 1579; it has been described as 'the first national atlas', [1] and it inaugurated a new era in cartography. Another surveyor of note was John Norden who aimed at producing a *Speculum Britanniae*, a series of county chorographies illustrated by maps. He succeeded in publishing only those for Middlesex (1593) and Hertfordshire (1598), but a number of his county maps appeared alone. These maps of Saxton and Norden provide much geographical information – not only about villages and market towns but also about woods, parks and hills; Norden's county maps were the first to show roads. John Speed's *Theatre* included on each county map an inset showing a plan or a bird's-eye view of the chief town of the shire, and so provides a unique collection of town plans.

The text of Camden and Speed's works is heavily weighted with references to antiquities. There are, it is true, many references to economic circumstances – to the ironworks of Sussex, to the grazing on Romney Marsh, to sheep on the Cotswolds, to the rich vales of Aylesbury and Belvoir, to the orchards of Kent and the saffron of Essex. But it is difficult to piece such fragments of information into a balanced picture of the national use of land. Camden, moreover, too often simply refers in a conventional way to the main kinds of land use as, for instance, in the Thames valley below Oxford, 'chequered with corn fields and green meadows, clothed on each side with groves'.

Parallel with the general accounts by Harrison, Camden and Speed, there were more detailed descriptions of particular counties. The county had for long been an important element in the administration of the realm, but now a new 'county self-consciousness', or provincial patriotism, was emerging. William Lambarde's *A perambulation of Kent* (1576) was the first, and was full of pride in the orchards, the deer parks and the countryside of Kent in general. He expressed the wish that 'some one able man in each shire' would 'describe his own country'. Richard Carew's *Survey of Cornwall* appeared in 1602, although he had been at work on it since the 1580s; it tells us much about tin-mining and fishing and about cattle and crops. Tristram Risdon's *The chorographicall description or survey of the county of Devon* was written in 1630, although it was not printed until 1714. It describes the six various combinations of soil and agriculture, ranging in quality from the South Hams, the very 'Garden of Devonshire',

[1] E. Lynam, *The mapmaker's art* (London, 1955), 63.

to Dartmoor 'barren and full of brakes and briars'.[1] These are only three of a number of county descriptions, some of which, like Risdon's *Devonshire*, did not see the light of print for many years. Some writers devoted themselves to towns and cities, and we are fortunate that among these was John Stow whose detailed *Survey of London* appeared in 1598.

Others wrote about their 'countries' on a more circumscribed scale than the county, as John Smith did in his profile of the hundred of Berkeley. 'In the body of this hundred', he said, 'are observed three steps or degrees', beginning with the vale alongside the Severn, 'which hath wealth without health'; farther upslope was the best land of all, and above it the Cotswold, 'which affordeth health in that sharpe aire, but lesse wealth'.[2] As we read these descriptions, and also those of foreign travellers, the emphasis varies. Sometimes it is upon tillage and the diligence of the husbandman, as with Camden, or upon the rich pastures and stores of cattle and sheep that so impressed the duke of Württemberg when he visited England in 1592.[3] All contributed to the geographical variations in what Michael Drayton was calling 'Albion's glorious isle'.[4]

POPULATION

Estimates of the total population of England about 1600 vary greatly. They can be based only upon such indirect evidence as that of parish registers, subsidy rolls, lists of communicants, and muster rolls of ablebodied men. There is also a variety of opinion about the multipliers that should be used to produce total populations from such figures. Most acceptable estimates vary around the figure of 4 million. William Harrison, about 1580, gave the number of 'able men for service' in the musters of 1574–5 as 1,172,674.[5] E. E. Rich, using the muster returns, with a multiplier of 4.0, suggested a total of 'something over four millions'.[6]

[1] F. V. Emery, 'English regional studies from Aubrey to Defoe', *Geog. Jour.*, cxxiv (1958), 315.

[2] *A description of the hundred of Berkeley... and of its in habitants*, by John Smith, being vol. 3 of J. Maclean (ed.) *The Berkeley Manuscripts* (Gloucester, 1885).

[3] W. B. Rye, *England as seen by foreigners in the days of Elizabeth and James I* (London, 1865), 30–1.

[4] M. Drayton, *Polyolbion, or a chorographicall description of Great Britaine* (London, 1612).

[5] W. Harrison, *Description of England (1577–1587)*, ed. G. Edelen (Cornell, 1968), 235.

[6] E. E. Rich, 'The population of Elizabethan England', *Econ. Hist. Rev.*, 2nd ser., II (1949), 247–65.

Whatever multiplier is used for 1574–5, the population some twenty-five years later was appreciably higher, but by 1600 there was a steadily decreasing rate of growth of population in most parts of England, making it a suitable point in time for a cross-sectional view.[1] The 'Liber Cleri' returns of 1603 put the number of communicants, recusants and non-conformists for the county as a whole at about 2 million, and using conventional multipliers this would yield a figure of about 4 million.[2] John Rickman, in the introduction to the 1841 Census, attempted to calculate the population of each county for a number of pre-Census years, i.e. before 1801. By applying certain assumptions to the figures for births, deaths and marriages, supplied by the parish clergy, he produced a total of 4.3 million for England in 1600 (Wales, including Monmouthshire, accounted for another 380,000). Subsequent research has shown how uncertain such estimates based upon parish registers must be, but, even so, Rickman's figure may be not far from the truth.

In view of the uncertainties, Fig. 53, based upon Rickman's tables, shows not densities per square mile, but variations, county by county, from the national density of 87.6 per square mile, and it may provide a basis for comparing different areas one with another. The most densely peopled part of the country comprised the metropolitan counties, a reflection of the intensive agriculture and varied industry in and around London. The high figures for Devon and Somerset also reflect a varied economy based not only on agriculture but on cloth manufacture, mining and commercial activities. Industrial Lancashire was beginning to emerge, and Worcestershire, with its vale of Evesham, its orchards, its salt industry and its metal-working in the north-east, also stood high. The figures for the counties of Northampton, Nottingham and Oxford may be thought surprisingly high, and that for Gloucestershire surprisingly low. The low figures for most northern counties are to be expected.

Fig. 53 is unsatisfactory not only because of the nature of the evidence upon which it is based but also because county averages are misleading. Thus southern Cambridgeshire was very much more populous than the northern half of the county with its undrained fen. Within each county,

[1] E. A. Wrigley (ed.), *An introduction to English historical demography* (London, 1966), 266.

[2] J. C. Russell, *British medieval population* (Albuquerque, New Mexico, 1948), 270–81; H. J. Habakkuk, 'The economic history of modern Britain', *Jour. Econ. Hist.*, XVIII (1958), 486–501; G. S. L. Tucker, 'English pre-industrial population trends', *Econ. Hist. Rev.*, 2nd ser., XVI (1963), 205–18.

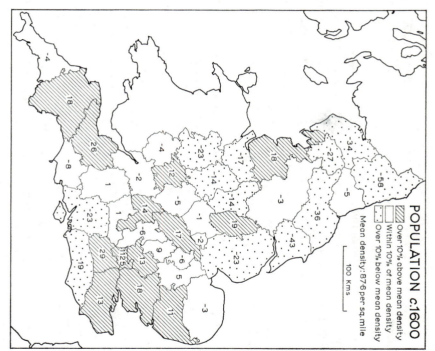

Fig. 53 Population *circa* 1600
Based on John Rickman's estimates in *Census of 1841: Enumeration Abstract*, 36 (P.P. 1843, xxii).

the ability of the land to support a farming population, and the intensity of various agrarian systems, were reflected in densities both higher and lower than the average for the county as a whole. Nowhere was this more true than in the uplands of the north. The justices of the peace for Westmorland claimed that their county was smaller, more barren, and more densely populated than any other county in England. Their view is substantiated by what is known about the concentration and compression of settlement in the dale or valley tracts within each fell parish.[1]

---

[1] J. Thirsk, 'Industries in the countryside', in F. J. Fisher (ed.) *Essays in the economic and social history of Tudor England* (Cambridge, 1961), 82.

We may take as a standard of measurement the average parish of southern England and the Midlands, say in the champion country of Leicestershire, with not more than 40 or 50 households accommodated on about 1,000 acres of land. Placed alongside this, the populations of 100, 200 and 300 families in the fell parishes of Westmorland, even though their acreages were ten times larger, must be regarded as dense.

Furthermore, the county totals include the populations of towns which held a significant and growing position in the realm. Any attempt to calculate the proportion of the total population of England that lived in towns is full of uncertainty, but a figure of about one-fifth may not be far wide of the mark.[1] If, however, London with some 250,000 people be excluded, the percentage drops to about 13 or so. It must be remembered, of course, that the line between country and town was at least as difficult to draw then as it is now. Many towns had open fields within their limits, and generally contained unmistakable reminders of the country in the form of barns, straw-ricks and hay-stacks, stalls, pens and penfolds for farm animals. On the other hand, in the countryside, farming pursuits were often combined with part-time occupations in rural industries, in cloth-making, weaving and spinning and in coalmining or iron-working.

An important feature of the population in general was its mobility. S. A. Peyton has shown from an examination of the subsidy rolls for Nottinghamshire that there was continuous movement.[2] When successive rolls are compared, 'names continually disappear, while new names occur, themselves in turn vanishing'. In a number of hundreds the majority of freeholders in 1612 appear to have held land for less than two generations. Some families whose names disappeared may have continued to reside in their parishes, although in poorer circumstances, but the evidence as a whole points to continuous and widespread migration. 'The population of the county of Nottingham between the years 1558 and 1641 was in a highly mobile condition.' The subsidy rolls for other counties point to similar movement – those for Bedfordshire,[3] Lincolnshire[4] and elsewhere.

[1] G. M. Trevelyan, *English social history* (London, 1942), 141.

[2] S. A. Peyton, 'The village population in Tudor Lay Subsidy rolls', *Eng. Hist. Rev.*, XXX (1915), 234–50.

[3] L. M. Marshall, 'The rural population of Bedfordshire, 1671–1921', *Beds. Hist. Rec. Soc.*, XVI (1934), 54 et seq.

[4] M. Campbell, *The English yeoman under Elizabeth and the early Stuarts* (London, 1942), 37–8.

Other evidence for Kent[1] and for Sussex[2] shows movement within the county and also from outside, sometimes from as far away as the north and west of England. E. E. Rich found that the Muster Rolls indicate the same mobility, and they sometimes include complaints about the continual 'remove' of men. 'Most of the Elizabethans who emerge from their backgrounds are members of families who have moved at least once in two generations. The Shakespear family moved from Snettisham to Warwick before William moved to London.'[3]

Some migrants went merely to neighbouring parishes; certainly this was so in Devonshire where it was unusual to leave the county unless one lived near the Somerset, Dorset or Cornish border. Dorset experienced a remarkable immigration of small farmers and gentry from Devon at this time.[4] Other migrants moved farther afield and frequently to the growing towns near and far. As at earlier periods, London was a great centre of attraction, but the numbers now involved must have been very considerable to account for its great growth during the sixteenth century to some 250,000 in spite of its high mortality rate.[5] Wherever they lived, Englishmen were still at risk when the harvest failed, and were likely to find themselves in the grip of starvation and sickness. The year 1600 itself was poised between a harvest pattern in the 1590s with more deficient, bad or dearth years ('the Great Famine') than in the following decade, which had six abundant or good years.[6]

THE COUNTRYSIDE

The man-made landscape of England in 1600, as at all other times, rested upon the basic physical division of the country into an upland zone and a lowland zone. The former, with its damp climate and with much of its

[1] P. Clark, 'The migrant in Kentish towns', being ch. 4 of P. Clark and P. Slack (eds.), *Crisis and order in English towns, 1500–1700* (London, 1971).

[2] J. Cornwall, 'Evidence of population mobility in the seventeenth century', *Bull. Inst. Hist. Research*, XL (1967), 143–52.

[3] E. E. Rich, 265; E. J. Buckatzsch, 'The constancy of local populations and migration in England before 1800', *Population Studies*, v (1951), 62–9.

[4] W. G. Hoskins, *Devon* (London, 1954), 173; C. J. Taylor, *Dorset* (London, 1970), 136.

[5] D. Cressey, 'Occupation, migration and literacy in East London, 1580–1640', *Local Population Studies*, v (1970), 53–60.

[6] W. G. Hoskins, 'Harvest fluctuations and English economic history, 1480–1619', *Agric. Hist. Rev.*, XII (1964), 28–46; C. J. Harrison, 'Grain price analysis and harvest qualities', *Agric. Hist. Rev.*, XIX (1971), 135–55.

land over 800 ft. above sea-level was largely a pastoral area. It included, it is true, lowland enclaves with tillage, but this was largely subordinate to pastoral pursuits – along the coastlands of Durham and Cumberland, in the Lancashire–Cheshire plain, and elsewhere. Most of this lowland was an enclosed countryside. Some of it had always been so, and had been taken in directly from the waste. Other parts of it had been enclosed from open fields, but open fields quite different in their functioning from those of the two- or three-field 'Midland' system; and the open fields that still remained were soon to disappear.

In the lowland zone, the main distinction was between enclosed country, wooded or hedged land, on the one hand, and open-field countryside, 'champaign' or 'champion' land on the other (Fig. 54). Leland had constantly referred to the presence of woodland in the enclosed districts and its absence in the open-field areas;[1] and Arthur Standish, in 1613, drew a contrast between enclosed country with timber, and the bare fields of 'champaign countries' where the countryside was open and fuel was scarce.[2] It was a distinction that could split a single county; so was it in Warwickshire between wooded Arden to the north and open Feldon to the south, or in Buckinghamshire between the open vale of Aylesbury and the wooded Chilterns. The open-field area was itself far from uniform. The Midland two- or three-field system was characterised, amongst other things, by the common pasturing of the village animals on the fallow land and on the stubble of the arable after harvest, hence the term 'common fields'. But there were many areas of open field where such common pasturing did not take place, and where there were also other differences. Lowland England included a variety of irregular open fields, and also areas where enclosures and open fields existed side by side.

One element in the English landscape to be found in upland and low-land zones alike comprised forests, woodland and the deer parks which were marked on the county maps of the time. The forests were used primarily for pleasure and for hunting, but, as well as producing venison, they sometimes provided grazing for animals and horses, and parts of some were even cultivated. They were not as important as they once had been in the geography of the realm. There were indeed still many deer parks, but some of them had started to disappear. Over the countryside in general much woodland remained, but complaints about its disappearance had begun to grow numerous.

[1] E. C. K. Gonner, *Common land and inclosure* (London, 1912), 322–3.
[2] A. Standish, *New directions of experience to the commons complaint* (London, 1613), 6.

F. V. EMERY

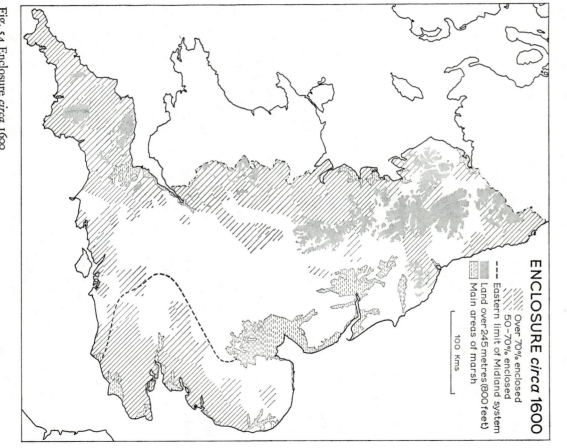

ENCLOSURE *circa* 1600

Over 70% enclosed
50-70% enclosed
Eastern limit of Midland system
Land over 245 metres (800 feet)
Main areas of marsh

100 Kms

Fig. 54 Enclosure *circa* 1600
Based on: (1) E. C. K. Gonner, *Common land and inclosure* (London, 1912),
map D; (2) H. L. Gray, *English field systems* (Cambridge, Mass., 1915).

While all this variety of arable and pasture and woodland cannot be portrayed accurately on maps, it is at any rate possible to make some broad generalisations; likewise about the distribution of types of rural settlement. Such generalisations must inevitably ignore the special features of this or that locality.

Rural settlement

The evidence does not permit us to discuss in any detail the relative distribution of the different types of settlement in which the population was disposed. But the basic contrast between nucleated and dispersed settlement was well appreciated by the men of the time. William Harrison about 1580 made the point clearly. In 'champaign' county, the houses stood 'all together by streets and joining one to another'; whereas 'in the woodland countries', they were 'dispersed here and there, each one upon the several grounds of their owners'.[1] It was essentially a contrast between open-field and enclosed districts, as could be seen in the historic division of Warwickshire into open-field Feldon with nucleated villages and wooded Arden with dispersed settlements. Even as Harrison wrote, dispersed houses were occasionally appearing in newly enclosed districts within open-field country. Francis Trigge, who strongly protested against enclosers, pointed out one of the results of their work in 1604: 'We may see many of their houses built alone like raven's nests, no birds building neere them.'[2] We also know from a variety of sources that the upland districts of England were characterised by small hamlets and scattered homesteads, although villages and small towns were to be found in the more fertile valleys and coastal plains.

The buildings that made up these settlements were still mostly of timber, and Harrison noted that 'as yet few of the houses of the commonalty' were 'made of stone'. He drew a contrast between the 'strong and well-timbered' houses of the 'woody soils' and those of 'the open country and champaign countries', having only a few upright and cross-posts with clay-covered panels between them.[3] But changes were already afoot, and 'the rebuilding of rural England' was well advanced.[4] What has been

---

[1] W. Harrison, 199 and 217.

[2] F. Trigge, *The humble petition of two sisters; the church and the commonwealth* (London, 1604).

[3] W. Harrison, 195.

[4] W. G. Hoskins, 'The rebuilding of rural England, 1570–1640', *Past and Present*, IV (1953), 44–59.

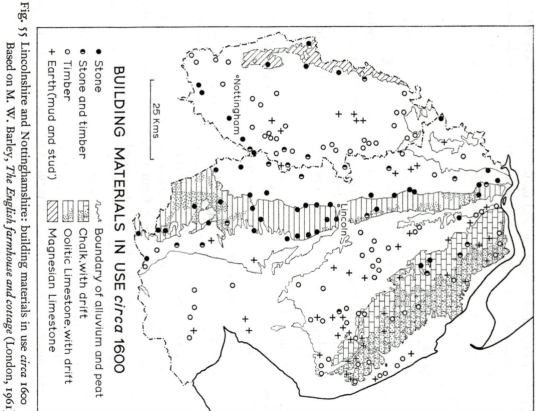

BUILDING MATERIALS IN USE *circa* 1600

- ● Stone
- ◉ Stone and timber
- ○ Timber
- + Earth (mud and stud)

- 〰 Boundary of alluvium and peat
- ▦ Chalk, with drift
- ▒ Oolitic Limestone, with drift
- ▨ Magnesian Limestone

°Nottingham

Lincoln

25 Kms

Fig. 55 Lincolnshire and Nottinghamshire: building materials in use *circa* 1600
Based on M. W. Barley, *The English farmhouse and cottage* (London, 1961), 82.

called the first phase of the housing revolution (1570–1615) saw the modernisation of medieval houses in the south-eastern counties and rather more new building farther away from London – among, for example, the prosperous yeomen farmers of Devonshire and the clothiers of the West Riding as well as among cottagers generally.

Stone was beginning to replace timber as the most usual building material, but only for the larger houses; it did not appear in the majority of farmhouses, even in stone country like the Cotswolds, until much later in the seventeenth century.[1] Brick came into general use later than stone. Even 'the ancient manors and houses' of gentlemen, so Harrison tells us, were for the most part of timber, except that those 'lately builded' were 'commonly either of brick or hard stone or both'. Newly built houses were to be seen on all sides, not only the grandiose piles like Holdenby in Northamptonshire and Wollaton in Nottinghamshire but in many manor-houses and farmsteads. That they still reflected regional characteristics may be seen from the details given in parsonage terriers of the sixteenth and seventeenth centuries (Fig. 55). One of the most striking was the 'magpie' half-timbered style of Lancashire, Cheshire and the west Midlands. Here, in a damper environment than that of counties like Essex, oak timbers were given a protective coating of tar or pitch, and plaster panels or brick nogging were whitewashed, making the contrast of black and white.[2]

Old men, wrote William Harrison about 1580, were noting a number of things 'to be marvelously altered in England within their sound remembrance'.[3] Many of the newly built mansions, such as Montacute House in Somerset, were set in gardens designed with the 'elaborate symmetry that delighted the Elizabethan eye'.[4] Chimneys and glazed windows were becoming more common; minor comforts, too, were increasing, such as feather beds instead of straw pallets, and pillows instead of wooden head rests, and platters and spoons made of metal instead of wood. In all such changes the south of England was in advance of the north.

The upland zone

Northern England. The Pennine uplands were largely lands of heath, moor and bog, and of rough grazing and poor pasture farming. The Border adjoining Scotland was moreover subject to raids from the Scots at least until 1603 and the union of the two Crowns. The main business of upland farmers was the breeding of sheep for wool and of young cattle for sale to lowland graziers for fattening. Transhumance was a feature of upland life, and stock was sent in early summer to hill pastures or 'grasshouses' on the better parts of the moorlands. Camden wrote of the men

[1] M. W. Barley, *The English farmhouse and cottage* (London, 1961), 101–3 and 123–5.
[2] A. Clifton-Taylor, *The pattern of English building* (London, 2nd ed., 1965), 32–3 and 46–51.
[3] W. Harrison, 200 *et seq.*
[4] E. J. M. Buxton, *Elizabethan taste* (London, 1963), 43–4.

of the northern Pennines 'who from the month of April lie out scattering and summering (as they term it) with their cattle in little cottages here and there, which they call sheals or sheilings'. Such outposts might also provide suitable temporary tillage plots, after which the land was allowed to revert to moor. Temporary intakes for cultivation were also to be found in the southern Pennines – in Rossendale, for example, and the Peak District. Fertile arable land was confined to the narrow valleys and dales that pierced the mountains and that provided a basis for a few small open fields and for meadow and permanent grass. Similar conditions characterised the Lake District and the North York Moors; here, too, cattle and sheep were the mainstays of the economy, together with some shifting cultivation.[1]

Larger expanses of more fertile land bordered the Pennine upland to the east and west. Along the eastern lowland much of the arable was both open and common, and lay in large fields. Such arrangements frequently resembled those of the Midlands in that a holder's strips were divided approximately between three fields; but they differed in having nearby large areas of pasture and waste that provided opportunities for a temporary expansion of the arable. In places there were arrangements similar to those of the infield–outfield system. This was a district transitional in field arrangements as well as in location between the Midlands and Scotland. By 1600, in many places, the arable open fields were in decay, and enclosure was spreading. The demands of the coalmining district around Newcastle, and those of the port itself, stimulated meat production supplemented by dairying, the rearing of oxen for draught, and horse breeding.[2]

Along the western lowlands, similar features could be encountered in Cumberland – some open fields, infield-outfield arrangements, and emphasis on cattle and sheep, with a tendency towards enclosure. Farther south, in Lancashire, much of the plain seems never to have been in open

[1] T. S. Willan and E. W. Crossley, *Three seventeenth century Yorkshire surveys* (Richmond, 1605; Middleham, 1605; Wensleydale, 1614), Yorks. Archaeol. Soc. Ser., 104 (1941), xx–xxiv; C. M. L. Bouch and G. P. Jones, *A short economic history of the Lake counties, 1500–1830* (Manchester, 1961), 94–101; G. Elliott, 'The decline of the woollen trade in Cumberland, Westmorland and Northumberland in the late sixteenth century', *Trans. Cumb. and West. Antiq. and Archaeol. Soc*, LXI (1961), 112–19.

[2] J. Thirsk (ed.), *The agrarian history of England and Wales*, vol. IV, 1500–1640 (Cambridge, 1967), 27. For further details of agrarian practice here and in the other districts surveyed below, see Mrs Thirsk's chapter 'The farming regions of England', *ibid.*, 1–112.

field but had been enclosed directly from wood-pasture and rough grazing. The small open fields, where they existed, were rapidly disappearing. In 1600, this was predominantly a pastoral country characterised by the rearing and fattening of cattle and sheep together with some dairying.[1]

The Welsh border counties. Enclosure in the counties bordering Wales had gone far by 1600. In many localities the land had been cleared from the waste as enclosed fields, held in severalty, with hedge or fence from the start. Elsewhere the small and irregular open fields were being enclosed, either in a piecemeal fashion or by more general agreement. Yet other areas remained in heath and waste and common. Some parts of these were ploughed up for a few years and then allowed to revert to rough grazing. It was predominantly a pasture-farming countryside with relatively little grain and with much cattle-rearing, dairying and pig-keeping.

There were variations dependent upon soil and other circumstances. The most important grain-growing area was in Herefordshire, but it was also a dairying county, and Camden's description of the county as 'fruitful for corn and cattle feeding' was echoed by other writers. Here, in the 'Golden Valley' of the Dore, Rowland Vaughan's experiments in watering meadows were taking place, and he was not the only man in the county interested in dairying. Herefordshire was also known for its orchards, so was Worcestershire and the adjoining parts of Gloucestershire; from the apples and pears cider and perry were made. Cheshire, on the other hand, had long been famous for its cheese, 'the best in Europe' according to Speed. Its small irregular open fields were quietly disappearing by agreement, many into pasture closes.[2]

The south-west. Much of the south-western peninsula consists of infertile upland country. Expanses of rough grazing on Dartmoor, Exmoor, Bodmin Moor and other hills were used as summer pastures for cattle and

[1] G. Elliott, 'The system of cultivation and evidence for enclosure in Cumberland open fields in the sixteenth century', *Trans. Cumb. and West. Antiq. and Archaeol. Soc.*, LIX (1959), 84–104; G. Youd, 'The common fields of Lancashire', *Trans. Hist. Soc. Lancs. and Cheshire*, CXIII (1961), 1–41; F. J. Singleton, 'The influence of geographical factors on the development of the common fields of Lancashire', *ibid.*, CXV (1963), 31–40; R. A. Butlin, 'Northumberland field systems', *Agric. Hist. Rev.*, XII (1964), 99–120.

[2] D. Sylvester, *The rural landscape of the Welsh borderland* (London, 1969), 230–1, 253–4, 262–70. See also T. Rowley, *The Shropshire landscape* (London, 1972).

heep. As in the north, transhumance was practised. There were likewise various forms of shifting cultivation and of infield–outfield cultivation. Plots selected for tillage were fertilised by paring off and burning the turf and mixing the ash with the soil ('denshiring').

Around the moors were the more fertile areas of the coastal plains and the vales. There were some open fields, but 'nothing to show that the tenants' holdings were normally distributed in equal parcels, one or more in each field'.[1] Moreover, in such a region, where most parishes had their own tracts of rough grazing, 'there was no absolute necessity to leave a half or a third of the ploughland under grass each year'.[2] In some areas, the open fields had undergone silent enclosure by agreement; other areas may have been enclosed directly from rough pasture.[3] Writers of the sixteenth century, in speaking of Cornwall and Devon, refer characteristically to enclosed fields surrounded by great hedges. Enclosed fields were also to be found over much of Somerset, west Dorset and south Gloucestershire. The south-west as a whole was predominantly a land of pastoral husbandry, with patches of mixed farming and grain growing on the coastlands and in the broader vales. John Norden in 1607 could described the vale of Taunton Deane as the paradise of England, with its fields of wheat, barley and oats and its dairies.[4] One corollary of enclosure was increasing local specialisation and production for the expanding markets. In Devonshire, men were enlarging their orchards and Richard Hooker, about 1600, referred to the careful management of orchards and apple gardens; cider was sold for the provisioning of ships.[5] East Somerset and west Dorset was a dairying countryside, and Yeovil a great cheese market. Domestic handicrafts (such as the making of cloth, gloves and bone-lace) supplemented the dairying occupations. To the north, the vales of Berkeley and of Gloucester were areas of mixed farming with dairying as an important element.

One of the most distinctive countrysides in the south-west was that of the Somerset Levels. Contemporary descriptions are similar to those of

1 W. G. Hoskins and H. P. R. Finberg, *Devonshire studies* (London, 1952), 283.

2 *Ibid*, 287.

3 P. J. Fowler and A. C. Thomas, 'Arable fields of the pre-Norman period at Gwithian', *Cornish Archaeology*, 1 (1962), 61–84; P. D. Wood, 'Open field strips at Forrabury Common, near Boscastle', *ibid*, 11 (1963), 26–33; E. M. Yates, 'Dark Age and medieval settlement on the edge of wastes and forests', *Field Studies*, 11 (1965), 133–53.

4 J. Norden, *The surveyor's dialogue* (London, 1607).

5 W. G. Hoskins (1954), 94.

the Fenland in eastern England – a watery spectacle with fowling, fishing and turf cutting, and with much grazing of cattle on pastures made fertile by winter flooding. Here, too, dairying was important, with an emphasis on butter and on Cheddar cheese. Some localities had been drained by 1600, but it is difficult to say how much of this was for arable and how much for pasture.

The lowland zone

The Midland area. The main physical characteristics of the Midland open-field system were the large open arable fields, generally two or three in number, each covering up to 400 acres. Under the three-field system, one was in winter-sown grain (wheat or rye), one in spring-grown grain (barley or oats) or pulse, and the third in fallow. Under the two-field system, one was in fallow and the other was divided between winter and spring grains. The fields were composed of 'furlongs' or blocks of land of varying size, and each furlong in turn was divided into strips, selions or ridges, usually of between one-quarter and one-third of an acre apiece. The arable land of a village may thus have comprised as many as two thousand separate strips or more. The holding attached to a farmstead consisted of a number of these strips, widely scattered and intermixed with the strips of other holdings. There were also common meadows allocated to each holding in small strips or 'doles'. Essential features in the working of such a system were: (1) the communal regulation of cultivation, with the strips in each field growing the same crop in any particular year; (2) the more or less equal division of the scattered strips of a holding between the two or three fields; and (3) the throwing open of the arable and meadow for common grazing in the fallow season and after the crops had been harvested. This was an invaluable right enjoyed by all cultivators in a village, and was the essential feature of the Midland system as opposed to other varieties of open field.[1]

By 1600 these arrangements still survived in their entirety on many Midland manors and partially on most. But it is clear that the system was

[1] H. L. Gray, *English field systems* (Cambridge, Mass., 1915), 39–49; C. S. and C. S. Orwin, *The open fields* (Oxford, 3rd ed., 1967), 59–66; M. W. Beresford and J. K. S. St Joseph, *Medieval England: an aerial survey* (Cambridge, 1958), 21–45; A. R. H. Baker, 'Howard Levi Gray and English field systems', *Agric. Hist*, xxxix (1964), 1–6. For comment on the terms 'open field' and 'common field', A. R. H. Baker, 'Some terminological problems in studies of British field systems', *Agric. Hist. Rev.*, xvii (1969), 138–40.

much more flexible than it was once thought to be. The manorial maps that become available in the latter part of the sixteenth century often reveal quite complex arrangements. They show that the unit of rotation was as likely to have been the furlong as the field. Moreover, individual furlongs could be put under grass although the rest of a field was in arable. Even individual strips could be left in grass by communal agreement; a farmer would then tether or hurdle his beasts during the growing season, and, after harvest, his strips would be thrown open, together with the stubble of his neighbours, for general grazing by the village animals. From such practices within the common fields sprang the incentive to consolidate strips, to enclose open land and to embark upon the convertible husbandry of alternate crops and grass. This was so especially in the densely settled claylands where mixed farming prevailed and where the possibilities of heavier crops and more livestock were realised. Innovations and variations became increasingly frequent; thus, at Harwell in Berkshire the rotation in two common fields was extended by grazing a fodder crop of vetches on the fallow field. Or again, in some villages there were blocks of enclosed land in the midst of fields which were still open. In Oxfordshire 19% of the townships were enclosed, although with variations in different parts of the county; on the Redlands around Banbury enclosure had touched only 13% of the townships, but 35% of those on the Chilterns were already enclosed.[1]

It was under such circumstances that the enclosures of the fifteenth and sixteenth centuries had taken place and had frequently involved the conversion of arable to pasture. In nine Midland counties, and especially in those of Leicester, Northampton, Rutland and south-east Warwick, over 8% of the total area had been enclosed between 1455 and 1607 (Fig. 47). Here, the new fields within which the flocks and herds were so carefully managed by the graziers were often as large as 50 to 100 acres, their strong fences and hedges sometimes following the rectilinear outlines of former furlongs, at other times transgressing them and cutting across the pattern of ridge and furrow.[2]

[1] R. H. Hilton, 'Medieval agrarian history', V.C.H. Leicestershire, II (1954), 145–98, 161, 197; M. A. Havinden, 'Agricultural progress in open-field Oxfordshire', Agric. Hist. Rev., IX (1961), 73–83; W. G. Hoskins, The Midland peasant (London, 1957), 67–70, 95; G. E. Fussell, Robert Loder's farm accounts, 1610–1620, Camden Society, 3rd ser., 53 (London, 1936).

[2] M. W. Beresford and J. K. S. St Joseph, 114–20; K. J. Allison et al., The deserted villages of Oxfordshire (Leicestershire, 1965), 8.

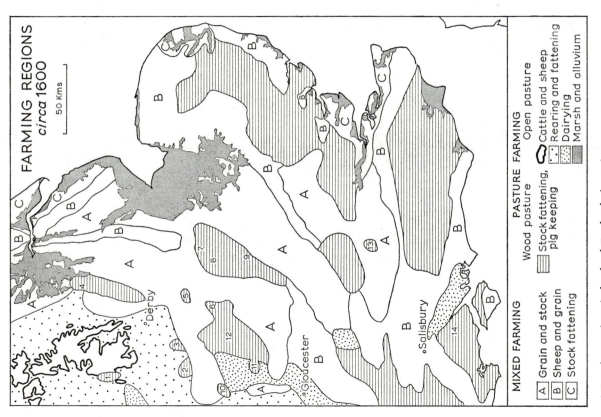

FARMING REGIONS
circa 1600

50 Kms

MIXED FARMING

A | Grain and stock
B | Sheep and grain
C | Stock fattening

PASTURE FARMING

Wood pasture   Open pasture

Stock fattening,   Cattle and sheep
pig keeping

                      Rearing and fattening

                      Dairying

                      Marsh and alluvium

Fig. 56 Farming regions in lowland England *circa* 1600
Based on J. Thirsk, *The agrarian history of England and Wales*, vol. IV, 1500–
1640 (Cambridge, 1967), 4.
Forests are numbered as follows: 1, Macclesfield; 2, Cannock; 3, Needwood;
4, Sherwood; 5, Charnwood; 6, Leicester; 7, Leighfield; 8, Rockingham;
9, Whittlewood and Salcey; 10, Kinver; 11, Feckenham; 12, Arden; 13,
Windsor; 14, New Forest.

In the mixed farming of the Midland area, grain and livestock were complementary to one another. Crops (including peas, beans and vetches as well as grain) were fed to the fattening animals; plough oxen and horses were needed to work the arable land; sheep were folded and cattle grazed on the fields to manure them. Various combinations of crops and stock were widespread through the clay lowlands from the vale of York southwards. On the clay plains around Oxford, for example, farmers grew and sold the staple grains, nearly half the sown land being under barley for malting and a quarter of it under wheat; much of their produce went down the Thames to London. Cattle were reared on most farms, fattened or added to dairy herds; smaller numbers of sheep and a few horses and pigs were also kept. In north Oxfordshire, the crops were fed to fat cattle which could walk to market and, with larger flocks of sheep, the farmer made most of his money from wool, beef and mutton. Graziers on the grand scale, like the Spencers at Wormleighton in south Warwickshire, kept as many as 14,000 sheep, and they also sold fat cattle to London; their 'stock ranching on great enclosed pastures' was made possible both by the grass and by the feeding-stuffs grown on their arable land.[1]

A second kind of mixed farming was the sheep and grain husbandry practised on the light soils of the chalk and Oolitic limestone outcrops – on the Yorkshire and Lincolnshire Wolds, the Lincolnshire Heath belt, the Cotswolds and on the chalk downs that radiate in all directions from Salisbury Plain. Here the farmer's mainstay was grain, first barley and then wheat. They were grown in the open fields, the fertility and soil texture of which were kept in good order by the large flocks of sheep folded on them. The primary purpose of keeping sheep was not for their wool but for the dung of the fold. They were fed in the daytime on the sheepdown, and then brought in at night to be folded on the arable. In Wiltshire, for example, about one half of the chalklands were sheepdown, about three-eighths in arable and the remainder in permanent grass. Meadow grass and the smaller arable crops (oats, peas, vetches) were used to supplement the downland grazing. The sheep of some villages numbered many thousands, even up to 10,000 and more. On thinner soils the inferior sheepdown merged into rabbit warrens. Parts of

[1] H. Thorpe, 'The lord and the landscape', *Jour. Birm. Archaeol. Soc.*, LXXX (1962), 38–77; J. A. Yelling, 'Common land and enclosure in east Worcestershire, 1540–1870', *Trans. and Papers, Inst. Brit. Geog.*, XLV (1968), 157–68; J. A. Yelling, 'The combination and rotation of crops in east Worcestershire, 1540–1660', *Agric. Hist. Rev.*, XVII (1969), 24–43.

the downland were occasionally broken up, tilled for one or more years and laid to sheep pasture. On the wolds of Yorkshire and Lincolnshire arrangements akin to the infield–outfield system were sometimes to be found.[1]

There were other variants in the Midland area (Fig. 56). The ancient royal forests were often characterised by pasture farming and by irregular field systems that had passed into enclosure by 1600. With large commons, it was not necessary to keep so much arable in fallow each year; moreover, assarts from the waste had not necessarily been incorporated into the open-field system. Thus the Northamptonshire forests of Rockingham, Salcey and Whittlewood, on heavy Boulder Clay, had some large commons and many old enclosures. Forested districts on light soils were characterised by temporary cultivation, and arrangements reminiscent of the infield–outfield system were also to be found – on the Bunter Sands of Sherwood Forest where 'breaks or temporary enclosures' were kept in tillage for five or six years and then allowed to revert to pasture; on the Bunter Sands of the Cannock Chase area and also in the Forest of Arden. The fertility of the arable on the light soils in and around the New Forest was maintained by the folding of sheep that grazed on the heaths and commons by day. This, too, was a pasture-farming area with pigs, cattle, and horses that also fed on the heaths; much of the area had passed into enclosure by 1600. Yet other variants were to be found in villages with substantial amounts of meadow along such rivers as the Thames, the Warwickshire Avon and the Wiltshire Avon; here again there were often irregular field arrangements and enclosure by 1600.[2] To sum up: the area of the traditional two- and three-field Midland system had many variants that seem to have been associated with differences in soil and geographical circumstance.

The Fenland. This was a distinctive area covering some 1,300 square miles and including two distinct types of country – siltlands towards the sea and peatlands inland.[3] The silt area, bordering the Wash, was quite as wealthy and prosperous as the uplands around, in places even more so. If open fields had ever existed here, they had certainly long disappeared. In any case, the amount of arable was small, and the main feature of the economy was its emphasis upon grazing, and the rearing and fattening of animals

[1] E. Kerridge, *The agricultural revolution* (London, 1967), 61–2, 105–7.
[2] H. L. Gray, 83, 107.
[3] H. C. Darby, *The draining of the Fens* (Cambridge, 2nd ed., 1956); J. Thirsk, *English peasant farming* (London, 1957).

for meat and hides. Sheep were also important, and Camden referred to the great pasture of Tilney Smeeth, in Norfolk, which supported 30,000 sheep. Seawards, towards the Wash, successive areas of marsh had been reclaimed, and beyond the outermost bank there was salt marsh liable to be flooded at spring tides.

On the landward side of the silt belt lay the peat area marked, as Camden wrote, by 'foule and flabby quavemires', and varied by stretches of water – the meres of Whittlesey, Ramsey, Soham and the like. Projecting above the general level of the peat surface were 'islands' where settlement was possible and where the cultivation of the arable followed a three-field system; the largest of the islands was that of Ely with a dozen or so villages. The characteristic occupations of the area were fishing and fowling, turf-cutting and reed-gathering, and, not least, the pasturing of animals, cattle, sheep, horses. Winter floods produced excellent summer pastures and a whole array of complicated regulations controlled these economic activities and also the waterways that flowed through the district. During the closing decades of the sixteenth century there had been attempts to drain various localities, and the idea of a large-scale project of draining took shape. At last, in 1600, came 'An Act for the recovering of many hundred thousand acres of marshes.' Of the main stretches of marsh in the kingdom that of the Fenland itself promised the most spectacular transformation, but it was not to come for thirty years or so.

To the north of the Fenland there was a smaller tract of marsh that straddled the Lincolnshire–Yorkshire–Nottinghamshire boundary, and that included the Isle of Axholme and the Hatfield Levels. Here, conditions resembled those of the Fenland, with fishing, fowling and pasturing, and with settlements sited on the islands.

East Anglia. The open-field system extended on to the light and medium soils of East Anglia, but with differences. One difference was that the strips constituting a holding did not lie widely scattered over two or three large open fields, but were concentrated in one particular part of the arable, maybe in adjacent furlongs. Thus a 1583 survey of Castle Acre in north-west Norfolk shows that it had three fields – East, Middle and West; but there the resemblance to a Midland village ceased, for 80% of one holding lay in Middle Field, about 75% of each of two holdings lay in West Field, and nearly 70% of a fourth in East Field.[1] Clearly the term 'field' was merely a topographical designation and had no functional

[1] H. L. Gray, 314–15.

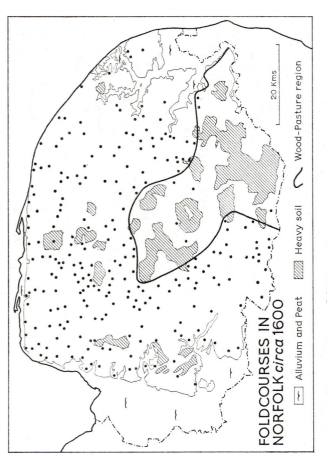

FOLDCOURSES IN
NORFOLK *circa* 1600

20 Kms

|⌐⌐| Wood-Pasture region

▨ Heavy soil

▨ Alluvium and Peat (for soil map).

Fig. 57 Fold-courses in Norfolk *circa* 1600

Based on: (1) K. J. Allison, 'The sheep-corn husbandry of Norfolk in the sixteenth and seventeenth centuries', *Agric. Hist. Rev.*, v (1957), 14; (2) *An economic survey of agriculture in the eastern counties of England* (Heffer, Cambridge, 1932), viii (for soil map).

significance. There was also considerable flexibility; adjoining strips in the same field were not necessarily under the same crop, and there were even strips of grass amidst growing grain.

Another difference lay in the fact that East Anglian villages usually contained two or three manors, and the flock of each manorial land had a 'fold-course' extending over arable and heath within clearly defined limits (Fig. 57).[1] Within the arable of each fold-course some areas must have been under winter crops, some under spring crops, while others lay fallow. During the autumn and winter both sheep and cattle grazed generally over the stubble of the unsown fields and over the heath; but for the rest of the year (from, say, March to October) access to the fallow arable was strictly reserved for the manorial flocks, each within its

[1] K. J. Allison, 'The sheep-corn husbandry of Norfolk in the sixteenth and seventeenth centuries', *Agric. Hist. Rev.*, v (1957), 12–30; A. Simpson, 'The East Anglian fold-course: some queries', *ibid.*, vi (1958), 87–96.

fold-course. Although this necessitated the labour of moving hurdles, it did secure an intensive manuring of the light soil. 'Some of the thriftless convenience of the midland system may have been sacrificed, but superior agricultural method and profitable sheep-raising were compensations.'[1]

A variant of this East Anglian open-field system was an arrangement by which the arable of a village was divided into two parts – a small intensively cropped infield near the settlement, and a much larger outfield on the heath or waste; a portion of the latter was broken up each year, cropped for a season or so and then allowed to revert to heath.[2] It was a system that was later to give its name to the Breckland, the area of the lightest and most hungry soil in East Anglia. One feature of the heathlands was the large number of rabbit warrens, and rabbit skins formed an important element in the exports from Blakeney on the north coast of Norfolk.[3]

The sheep-corn husbandry did not characteristically extend on to the heavier soils of central Norfolk. Here, pasture and meadow were much more important than arable; fold-courses were rare (Fig. 57); more land was enclosed; and cattle-rearing and dairying together with pig-rearing were important activities producing butter, cheese and bacon. It was a district that had much in common with the 'wood pasture district' of Suffolk.

South-east England. A large part of south-eastern England was almost entirely enclosed in 1600. Essex and Kent had been described in 1549 as counties 'which be most enclosed'. To them could be added much of Suffolk, Hertfordshire, Surrey and Sussex (Fig. 54). Some of this enclosed land may have been taken directly from the waste and wood; others parts had been enclosed, sometimes recently, from irregular open fields by agreement in a piecemeal fashion. Unenclosed parcels of strips were still to be found lying within enclosed fields. They were not the relics of a two- or three-field system; they were not subject to common grazing by the livestock of all the tenants; and they were tilled on a variety of rotations.[4]

[1] H. L. Gray, 329.
[2] J. Saltmarsh and H. C. Darby, 'The infield-outfield system on a Norfolk manor', *Econ. Hist.*, III (1935), 30–44; M. R. Postgate, 'The field system of Breckland', *Agric. Hist. Rev.*, X (1962), 80–101.
[3] B. Cozens-Hardy, 'The maritime trade of the port of Blakeney, Norfolk, 1587–90', *Norfolk Rec. Soc.*, VIII (1936), 19.
[4] A. R. H. Baker, 'The field systems of an East Kent parish (Deal)', *Archaeologia Cantiana*, LXXVIII (1963), 96–117; 'Open fields and parible inheritance on a Kent manor (Gillingham)', *Econ. Hist. Rev.*, 2nd ser., XVII (1964), 1–23; 'Field patterns in

Between this area of almost complete enclosure and the Midland common-field area there was a belt of country with irregular field arrangements and some enclosure – in the Chiltern parts of Hertfordshire, Buckinghamshire and Oxfordshire, and in Middlesex, east Berkshire and Surrey. In some parishes there were no traces of open field; in others, as well as closes, there were up to a dozen or so small open fields with strips lying in only a few of these, sometimes in only one field. As in the more fully enclosed area, such arrangements were very different from those of the two- or three-field systems, although in the Chilterns there were rights of common grazing over the fallow.[1]

An important element in the agriculture of the south-east was the demand of the London food market, and the large amount of enclosed land facilitated the development of individual practices and an intensified commercial agriculture. Access was provided by the many roads that converged upon the city and also by river and coastwise routes. Local farmers dealt with drovers and merchants who frequented the central London markets. Grain was a considerable crop in the Chilterns and in Essex and Kent; the soil was kept fertile by London's refuse and dung, and by sheep grazed on downs and waste and folded at night on the arable. A convertible husbandry was frequently practised. On most farms of the enclosed district of central Suffolk sheep were outnumbered by cattle, and some of the largest dairy herds in the country were to be found here. Cheese and butter were marketed at places like Woodbridge and Ipswich, and were dispatched by sea to London.

There was, in addition, a variety of land use associated with differing geographical circumstances. On the numerous heaths and commons there were extensive sheep-walks – on Blackheath and Hounslow Heath, for example, and on the Suffolk Sandlings. Parts of the heathland were enclosed for temporary tillage, and parts were occupied by numerous rabbit warrens and cattle commons. The heathy upland of the High Weald also carried many sheep, but was especially noted as a nursery of cattle. Along the coast the marshes were mostly under grass for cattle and sheep brought from elsewhere – on Romney Marsh, on the Pevensey Levels and on the marshes along the Thames estuary. In Kent, there were

seventeenth-century Kent', *Geography*, L (1965), 18–30; 'Field systems in the vale of Holmesdale', *Agric. Hist. Rev.*, XIV (1966), 1–24.
[1] D. Roden and A. R. H. Baker, 'Field systems of the Chiltern Hills and of parts of Kent from the late thirteenth to the early seventeenth century', *Trans. and Papers, Inst. Brit. Geog*, XXXVIII (1966), 76 and 85.

numerous cherry and apple orchards, especially in the Maidstone district and to the north between Rainham and Blean. Hopfields, too, were frequent here and elsewhere – in Surrey, Essex and Suffolk. Among other specialised crops were saffron which gave its name to Saffron Walden in Essex; weld, 'the dyer's weed' used for making yellow dye, which flourished on the chalklands near Canterbury and Wye; and carrots which were characteristic of the Suffolk Sandlings. In Middlesex there were dairy farms producing milk, cheese and butter, and from many places nearby large quantities of hay were brought to the Haymarket for the many horses in the city.

One of the most visible signs of the presence of London was the spread of market gardening, maintained to a great extent by immigrants from the Low Countries. Closes were being transformed into market gardens at nearby places such as Lambeth, Fulham, Putney, Whitechapel, Stepney and Greenwich, all within easy access of the city's manure. Among the produce were cabbages, carrots, parsnips, turnips and cauliflowers, some for the table and some to support the stall-fed cows that helped the milk supply of the city. That the London Gardeners Company received its charter in 1605 is indicative of this new element in the scene around the city.

Forests, woodland and parks

The acreage once subject to forest law had greatly diminished as the result of successive disafforestations. Some forests had completely disappeared in the sense that their territories had been relieved of forest jurisdiction and of control by an army of royal officials, the verderers, the regarders, the bow-bearers and others. In other forests a piecemeal nibbling had taken place; large or small tracts had been enclosed in return for payment, and were now in tillage or pasture. Even so, the amount of land under forest law was still considerable (Fig. 56), and Camden was prompted to reflect upon the waste involved. 'It is incredible', he wrote, 'how much ground the kings of England have suffered every-where to lie waste and have set apart and enclosed for deer.'[1] The red deer, the fallow deer and the roe deer were plentiful and they provided venison for the royal house-hold and for the king's friends. Other beasts of the chase were the fox and the marten. The wild boar was still to be found in Lancashire, Durham and Staffordshire, but the wolf had disappeared from England, though not from Scotland.

[1] W. Camden, *Britain, or a chorographicall description* (London, 1610), 293.

In spite of Camden's strictures, some good had resulted from the severity of the game laws in that much timber had been saved from the destruction involved in clearing land for agriculture and for the needs of industry. This was now available to meet the growing demand for timber for shipbuilding as a result of Tudor maritime expansion. The navy drew its timber from a number of royal forests, but its three main sources were the Forest of Dean, the New Forest and Alice Holt Forest in north-east Hampshire. The royal forests in general were far from being effective sources of supply, and contemporary surveys of the royal forests, such as that of 1608,[1] show large numbers of decaying trees as well as those fit for naval timber. Much of the supply came from private estates and from the woodlands of the nobility and landed gentry. An oak tree was at its best for shipbuilding when it was about one hundred years old, and there was much complaint about the premature cutting of half-grown oaks to meet unusual financial demands. The sale of timber provided the easiest way of obtaining ready money in the days of entailed estates.

During the sixteenth century, a number of Acts of Parliament had tried to restrict the cutting of timber for industrial purposes, and by 1600 there were many complaints about the loss of wood as a result of such activities as iron-making, glass-making and the provision of bark for tanning. William Harrison about 1580 had pointed to the fact that a man could often ride ten or twenty miles and find very little wood, or even none at all, 'except where the inhabitants have planted a few elms, oaks, hazels or ashes about their dwellings' as a protection 'from the rough winds'.[2] Arthur Standish, in 1611, pointed to the destruction that had taken place during the preceding twenty or thirty years, and urged that trees should be replanted.[3] Many of these complaints may well have been exaggerated. There were still great woods of beech and hazel as in the Chilterns and eastern Berkshire, and of oak as in the Weald. There were also many other districts characterised by a mixture of pasture and variegated forest, groves and glades, with stands of oak, ash, hazel, hawthorn, elm, birch, beech and holly. The fact was that not only had the woodland been reduced by clearance, but that the demand for timber had grown greater in pace with an increasing population and a general quickening of economic life. The real shortage was yet to come.

[1] Reprinted in H. C. Darby (ed.), *An historical geography of England before A.D. 1800* (Cambridge, 1936), 398–9.

[2] W. Harrison, 275.

[3] A. Standish, *The commons complaint* (London, 1611).

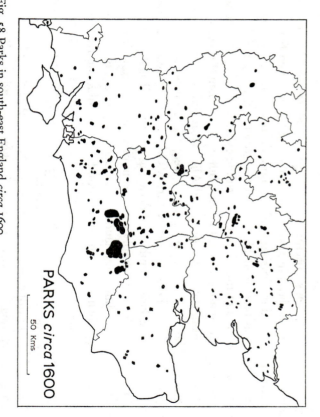

Fig. 58 Parks in south-east England *circa* 1600
Based on the county maps in John Speed, *The theatre of the empire of Great Britaine* (London, 1611).

There were, moreover, some landowners who refused to sell their trees and who preferred to keep them for ornamental purposes. This was so in the many deer parks that had come into existence around, or near to, the country houses of the realm. There were many royal deer parks; and, moreover, established or aspiring landowners, rich merchants and wealthy wool-staplers among them, made parks to signify their status. The parks were generally enclosed, wrote William Harrison, 'with strong pale made of oak' to prevent the deer from escaping.[1] The fences were sometimes raised on earth banks, often with a circumference 'of four or five miles, and sometimes more or less'. They are marked on the county maps of the time (Fig. 58); and they number nearly 800 in England, with another 30 or so in Wales.[2] But already many were disappearing. Richard Carew in his *Survey of Cornwall* (1602) pointed to a number that had disappeared within living memory and said that some owners had abolished the deer 'to give the bullockes place'. It was more profitable to

[1] W. Harrison, 254–5.

[2] H. C. Prince, *Parks in England* (Shalfleet, I.O.W., 1967), 1–2; E. P. Shirley, *Some account of English deer parks* (London, 1867).

breed bullocks than to graze deer. Even when parks were retained or created, they were reduced in area to provide land for grazing or tillage. Although deer parks were still numerous in 1600, they had begun to decline.

INDUSTRY

Manufacturing industry did not figure very prominently in the life of the English towns, apart from London and such clothing centres as Norwich and Colchester, together with a few other centres of growing industry. The occupational structure of most towns was varied, with no special emphasis on manufactures, and reflected the regional character of farming. Thus at Leicester, according to probate inventories made between 1560 and 1599, the principal occupations were as follows:[1]

	Per cent
Leather crafts (tanners, glovers, etc.)	19
Textiles (weavers, etc.)	19
Husbandry (graziers, etc.)	19
Victualling crafts (butchers, etc.)	14
Housing (carpenters, joiners, glaziers, etc.)	9
Retail trades (mercers, grocers)	6
Miscellaneous	14

A similar pattern was found in such provincial towns as Gloucester, Cirencester and Tewkesbury; they were not manufacturing centres, but handled the finishing processes and served as markets and sources of supply for the countryside.[2]

It follows that industries were to be found in rural areas, where they were very widely dispersed. The reasons for this are not difficult to see. In the first place, mechanical processes were generally carried out by means of water power, with which the country was well and generally supplied. Streams turned water-wheels for smelting tin in Cornwall, for draining collieries in Nottinghamshire, for forging iron in Sussex, and for fulling cloth in Lancashire. Likewise an important source of fuel was still charcoal, especially abundant in the wooded areas of late settlement in the Midlands or the Weald. Secondly, not only were the major raw materials like wool or iron well distributed, but so too was the population

[1] E. W. Kerridge in R. A. McKinley (ed.), *V.C.H. Leicestershire*, IV (1959), 178.
[2] A. J. and R. H. Tawney, 'An occupational census of the seventeenth century', *Econ. Hist. Rev*, V (1934), 37–8; L. A. Clarkson, 'The leather crafts in Tudor and Stuart England', *Agric. Hist. Rev*., XIV (1966), 25–39.

and the domestic market, London, as a great centre of population and trade, was exceptional in attracting to itself a variety of industry. One example must suffice: the copper and brass works sited on the Thames between Isleworth and Worton in Middlesex and shown on Norden's map as 'Coppermills'.

Some weight should also be given to the social reasons suggested by Joan Thirsk in her study of the handicraft industries that developed in communities already engaged in farming.[1] There was a coincidence of these industries with a populous society of small farmers, often mainly free-holders pursuing a pastoral economy such as dairying. This was typical of places as varied as north Wiltshire, central Suffolk, the Kentish Weald, Westmorland, the Fens, and the Yorkshire dales. The population of these areas was sometimes increased by immigrants attracted by their ample commons and weak manorial organisation. It was often maintained by the partible inheritance of land, which kept people in their native places and gave rise to large and immobile populations. There were thus many men whose farming left them time to engage in the subsidiary occupation of a handicraft industry; dairying, for instance, did not require as much hand labour as arable farming.

It is possible to show fairly exactly the importance of rural industry in Gloucestershire, a county in which manufacturing was important. Although agriculture was preponderant, farming in fact occupied only some 50% of the adult male population; the proportion varied so much in different parts of Gloucestershire that it might be safer to acknowledge that 'agriculture and industry were inextricably intertwined. Not only corn and cattle, but corn, wool, cloth, and, in some districts, even corn, coal, and iron, were almost joint products.'[2]

With these general considerations in mind, the long-standing industries of cloth-making and iron-working must be examined, together with the more recently developed coal industry and also a variety of other industrial activities.

The textile industries

Woollen cloth was England's main export, and its manufacture the largest industry in the realm. The cloth was of widely different character, depending upon the nature of the wool used and the method of manufacture. In a general way it was of three kinds; broadcloths usually

[1] J. Thirsk in F. J. Fisher (1961), 70–88.   [2] A. J. and R. H. Tawney, 42.

of fine quality; kerseys which were lighter, cheaper and maybe coarser; and worsteds which, unlike the other two types, did not need fulling. A bewildering variety of names were used for various local types and products – arras, callimancoes, carrells, frisadoes, minikins, pomettes, stamelles and very many others. The general distribution of the industry was still largely that of the latter part of the Middle Ages; cloth-making, although widespread, was associated mainly with three areas, the West Country, East Anglia and Yorkshire and Lancashire (Fig. 49).

The West Country included two cloth-making districts. Broadcloth was made over an area comprising the Cotswolds and extending into Wiltshire, Oxfordshire and Somerset. Kerseys, on the other hand, were made in west Somerset and Devon. In the broadcloth area Gloucestershire held a foremost place. Of all the spring-fed streams flowing across the Cotswold scarp, those of the Frome, the Cam and the Little Avon were the most constant and furnished the most reliable supply of power for water-wheels. The water was also as its softest here, best suited to the cleansing, scouring and fulling processes. Another natural advantage lay in the deeply incised valleys with their floors on Upper Liassic Clay; they were well suited to the making of reservoirs for large fulling mills, and, later, for the dye-houses for coloured cloth that supplanted the unfinished broadcloth. We are fortunate in being able to plot the distribution of the Gloucestershire industry for the year 1608 from John Smith's census of occupations (Fig. 59).[1] In the villages around Stroud in the Frome valley, around Dursley in the Cam valley and around Wotton-under-Edge, as many as 40 to 50% and over of the able-bodied men were employed solely in cloth-making. Such densities of clothiers, weavers, fullers and dyers had emerged primarily because of local advantages in water for power and washing.[2]

The location of cloth-making was not usually influenced by special resources of raw materials, because wool, fuller's earth and dyestuffs were generally available and few areas were decisively well off in them. From Gloucestershire the industry extended into the adjoining parts of Oxfordshire, where there were fulling mills at Witney and other places along the Windrush, and into Wiltshire where the Avon and its tributaries provided

[1] *The names and surnames of all the able and sufficient men in body…in 1608…* compiled by John Smith (London, 1902).

[2] R. Perry, 'The Gloucestershire woollen industry, 1100–1690', *Trans. Bristol and Gloucs. Archaeol. Soc.*, LXVI (1945), 49–137; J. Tann, 'Some problems of water power: a study of mill siting in Gloucestershire', *ibid*, LXXXIV (1965), 53–77.

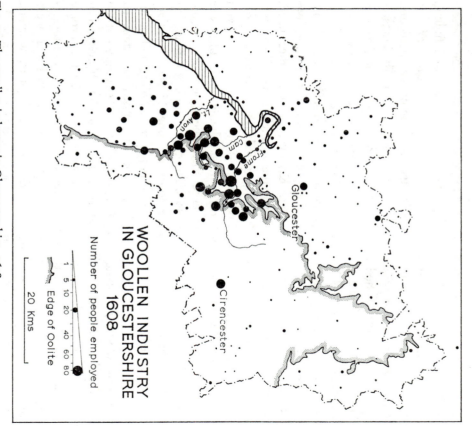

WOOLLEN INDUSTRY IN GLOUCESTERSHIRE 1608

Number of people employed

1 5 10 20 40 60 80

Edge of Oolite

20 Kms

Gloucester

Cirencester

Lt. Avon

Frome

Cam

Fig. 59 The woollen industry in Gloucestershire, 1608
Based on *The names and surnames of all the able and sufficient men in body...in
1608...compiled by John Smith* (London, 1902).

power for places extending from Malmesbury to Bradford on Avon and
so to Bath. Across the border the industry continued into Somerset,
where places such as Frome, Wells and Shepton Mallet developed
specialities.[1]

Westwards in Somerset, at Taunton and Wellington, kerseys were
made, and they were also a characteristic product of Devonshire with its

[1] E. M. Carus-Wilson and J. de L. Mann in E. Crittall (ed.), *V.C.H. Wiltshire*, IV
(1959), 133, 139, 141, 147, 150–1; E. M. Hewitt in W. Page, *V.C.H. Somerset*, II
(1911), 408, 411.

numerous streams suitable for fulling mills. Exeter, Crediton, Tiverton, Totnes and a large number of other places developed a degree of specialisation as between spinning and weaving. By 1600, the 'new draperies' of East Anglia had scarcely appeared in Devonshire except in and around Barnstaple in the north, but soon kerseys were to be supplemented and supplanted by serges or 'perpetuanos', so called for their long-lasting quality.[1]

In East Anglia, two separate districts were outstanding for cloth-making. In the north, Norwich was the centre of a group of towns and villages engaged in producing worsted for which the fulling mill and water power were not required; a few weavers of other kinds of textiles worked in Norwich itself (Fig. 60). In the south, along the Suffolk–Essex border, a number of villages made coarse cloths such as baize and kerseys. They included places such as Clare, Long Melford, Lavenham, Sudbury, Hadleigh and others situated along the south-eastward flowing rivers; this area also supplied yarn for the Norfolk weavers, and the manufacture of linen was gaining in importance. At the middle of the sixteenth century the industries of both East Anglian areas were in a depressed condition until all was changed as a result of the admission of Dutch and Walloon weavers fleeing from religious persecution. A few came to Norwich in 1554, and more in 1565. Others came to Colchester about 1570 and to other places. Their numbers increased in the years that followed. The manufactures they introduced included a varied range of cloth under a variety of names and spellings–bays, sayes, barracans, rashes, shalloons, bombazines and 'other outlandish commodities'. These were some of the lighter and finer fabrics of the so-called 'new draperies' and they did not involve fulling. By 1600 the East Anglian cloth areas were once more in a flourishing condition.[2]

The third main area for cloth-making was in the north. Here, the older 'clothing' towns, such as York, Beverley and Selby, had lost much of their industry; but other centres were rising, favoured by the abundance of water power for fulling mills. In the West Riding, a vigorous manufacture of coarse kersey, some coloured, was carried on at Halifax, Wakefield, Leeds and other places. The flourishing condition of the inhabitants, wrote Camden, 'confirms the truth of that old observation

[1] W. G. Hoskins (1954), 127; P. J. Bowden, *The wool trade in Tudor and Stuart England* (London, 1962), 50.

[2] G. Unwin in W. Page (ed.), *V.C.H. Suffolk*, II (1907), 249, 255, 257–8, 262, 267; D. C. Coleman, 'An innovation and its diffusion: the new draperies', *Econ. Hist. Rev.*, 2nd ser, XXII (1969), 417–29.

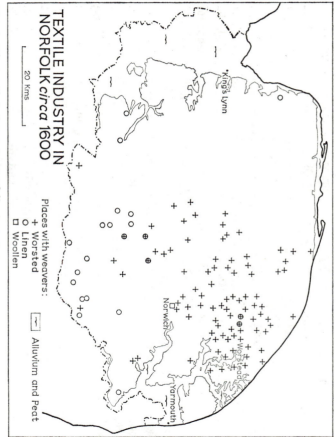

TEXTILE INDUSTRY IN NORFOLK *circa* 1600

20 Kms

Places with weavers:
+ Worsted
○ Linen
□ Woollen

⌐~¬ Alluvium and Peat

King's Lynn

Norwich

Worstead

Yarmouth

Fig. 60 The textile industry in Norfolk *circa* 1600 Based on K. J. Allison, 'The Norfolk worsted industry in the sixteenth and seventeenth centuries: 1. The traditional industry', *Yorks. Bull. Econ. and Soc. Research*, XII (1960), 75, and on additional information provided by Dr Allison.

that a barren country is a great whet to the industry of the natives'.[1] On the other side of the Pennines, the manufacture of coarse cloth was also growing around Manchester, Bolton and Rochdale, towns with few gilds and few special restrictions on trade and therefore well suited to be centres for the finishing and distributive sides of the industry; most of the weavers lived in the countryside around. By 1600 the making of kerseys had been supplemented by that of bays, one of the new draperies. A more important innovation was the making of fustian with a linen warp and a cotton weft; this seems to have been one of the new draperies that never flourished in East Anglia. This early introduction of cotton was to be of the greatest consequence for Lancashire, although all-cotton goods were not much made there for some years after 1600.[2]

[1] Quoted in H. C. Darby (1936), 374; M. Sellers in W. Page (ed.), *V.C.H. Yorkshire*, II (1912), 413–16.

[2] W. H. Chaloner and A. E. Musson, *Industry and technology* (London, 1963), 23;

The iron industry

Coal was not yet used in the iron-smelting industry, the chief centre of which was the Weald in south-eastern England (Fig. 61). There was plenty of ironstone in the Weald Clay and the Hastings Beds (principally in the Wadhurst Clay and the Ashdown Sands); the upper courses of the rivers, such as the Rother, Cuckmere and Ouse in Sussex, had deeply incised valleys and were therefore easily dammed, giving a steady head of water power for the bellows and hammers; finally, fuel was abundant locally in the form of charcoal burned from the brushwood of this wooded region. The Weald was also reasonably well placed to supply the London market. The blast furnaces were usually built of stone with buttresses and a bracing framework of wooden beams, about 18 ft. high. They were worked by two powerful pairs of bellows, in turn operated by a water-wheel; and they produced sows and pigs of cast-iron. These were re-heated and hammered at the finery, forge, or ironmill, thus giving blooms and bars of malleable wrought-iron which were suitable for general use by the smiths. Castings could also be made direct from the furnaces, including cannon and shot (for which the Wealden ironworks were famous) and domestic articles like iron firebacks.[1]

The supremacy of the Wealden area was still unchallenged in 1600, when it had 49 furnaces out of a total of 73 in England. There were also two furnaces not far away in Hampshire. An annual output of between 100 and 200 tons per furnace was usual. By then there were eleven in the west Midlands with its local Coal Measure ironstone and with charcoal from Wyre Forest. The 'free miners' of the Forest of Dean were slow in adopting the blast furnace, and there were only three in this area by 1620.[2] They nevertheless made much pig-iron for sale upstream along the Severn to the metalcraft centres of the west Midlands. South Staffordshire and north-east Worcestershire constituted the largest nail-producing area in England. Other products included such things as stirrups and spurs for horses, locks, fire-arms, cutlery and edged tools and a variety of ironmongery. We must envisage this as 'a countryside in

[1] M. C. Delany, *The historical geography of the Wealden iron industry* (London, 1921), 8–9, 19–21, 27–9, 32–3, 36–42; E. Straker, *Wealden iron* (London, 1931), 101–40; D. W. Crossley, 'The management of a sixteenth-century iron works', *Econ. Hist. Rev.*, 2nd ser., xix (1966), 273–88.

[2] C. Hart, *The industrial history of Dean* (Newton Abbot, 1971), 8.

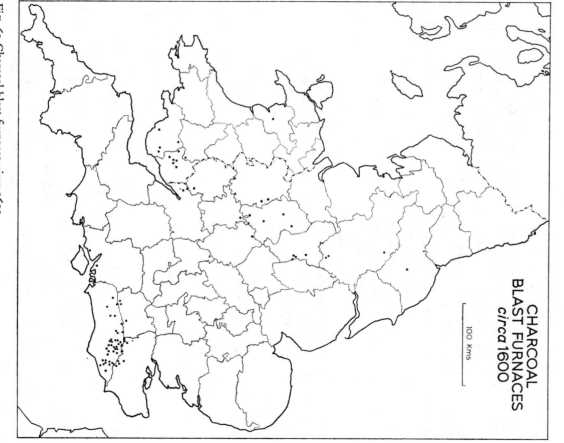

CHARCOAL
BLAST FURNACES
circa 1600

100 Kms

Fig. 61 Charcoal blast furnaces *circa* 1600
Based on H. R. Schubert, *History of the British iron and steel industry from*
c. 450 B.C. to A.D. 1775 (London, 1957), 354–92.

course of becoming industrialised; more and more a strung-out web of iron-working villages, market towns next door to collieries, heaths and wastes gradually and very slowly being covered by the cottages of the nailers and other persons carrying on industrial occupations in rural surroundings'.[1] But populous centres were beginning to emerge – places like Stourbridge, Dudley and Walsall, and especially Birmingham. Camden found Birmingham 'swarming with inhabitants, and echoing with the noise of anvils, for here are great numbers of smiths', and its tanneries and fulling mills were being displaced by forges and blade-mills. It may then have had a population of not more than about 2,000.[2]

There was another group of blast furnaces in an area stretching from east Derbyshire into south Yorkshire, also using Coal Measure ironstone. Associated with iron locally produced there was a variety of metal trades – nails, guns, ironmongery and, in particular, cutlery. The cutlers of Sheffield already had an established reputation throughout the country.

Coalmining

Coalmining had grown phenomenally during the sixteenth century, and by 1600 all the fields, apart from those of Kent, were being worked to some extent. Many of the monastic lands in northern and Midland counties which had passed into lay hands by 1540 were rich in coal, and their new owners had a thoroughly enterprising attitude to mining the seams. Long leases encouraged the heavy investment of capital in ambitious works; the Willoughbys of Wollaton in Nottinghamshire, for example, were putting nearly £20,000 a year into collieries and ironworks by 1600, installing rag-and-chain force-pumps to help draw the water from the deeper pits.[3] But it would be a mistake to assume that large-scale workings of this kind were typical of the industry as a whole. Far more characteristic was the colliery worked for the earl of Shrewsbury in Sheffield Park. Five men were generally employed in pits no deeper than 90 feet, producing between 1,200 and 1,400 tons a year; it was sold chiefly to the metal trades at Sheffield at a net profit of no more than £50 a year. Water was drained

[1] W. H. B. Court, *The rise of the midland industries, 1600–1838* (Oxford, 1938), 22; H. R. Schubert, *History of the British iron and steel industry from circa 450 B.C. to A.D. 1775* (London, 1957), 174–5 and 179–82.
[2] C. R. Elrington, P. M. Tillott and D. E. C. Eversley in W. B. Stephens (ed.), *V.C.H. Warwickshire*, VII (1964), 6–7, 81–3.
[3] W. H. Chaloner and A. E. Musson, 23; J. U. Nef, *The rise of the British coal industry*, I (London, 1932), 324.

away by adit or sough, but there was little machinery at pit-head or underground; much timber was used, being obtained freely from the earl's park. Output fluctuated with the high degree of absenteeism, for not only did the colliers observe a large number of feasts and fairs as holidays, they also followed other employment, probably on the land. 'The unmechanized, cheap, unenterprising little concern such as that at Sheffield', bringing in a regular income, was very common in the English coalfields.[1]

Coal was used not only in metal work at the forge but in a variety of other industries — in salt-making, soap-boiling, sugar-refining, glass-making, dyeing and brewing. It was also used in place of wood as domestic fuel, and a radical change in the method of heating houses had taken place over a short period. It is true that even late in Queen Elizabeth's reign some people objected to the smell of coal fires, but improved fireplaces and chimneys were helping to make coal the general fuel of the country. Moreover, not only was the consumption per head rising but the total population was increasing.

The cost of land carriage prohibited the haulage of coal for long distances, but much was transported along waterways such as the Trent and the Severn, and also around the coast by sea. In this respect the Northumberland–Durham field was well placed and its development had been particularly outstanding. Some of its mines were so situated that coal could be loaded direct into ships and the export from Newcastle alone reached about 160,000 tons by 1600. About a fifth of this went to foreign countries, to France, the Netherlands and Germany. The remainder went to ports along the south and south-east coasts, and especially to London, which, with about 250,000 inhabitants, provided a vast market in the context of the time. The average size of a cargo of coal increased from 56 tons in 1592 to 73 tons in 1606. A prodigious number of vessels bearing coal crowded the river below London Bridge. The number of lightermen and seamen employed was considerable, while the demand for coal-carrying ships was revolutionising the shipbuilding industry.[2]

Other industries

Other 'furnace industries', and the mines that kept them going, were well represented in the upland counties of northern and western England. In

[1] L. Stone, 'An Elizabethan coal mine', *Econ. Hist. Rev.*, 2nd ser., III (1950–1), 105.
[2] J. U. Nef, 390; J. Simmons, *Transport: visual history of modern Britain* (London, 1962), 17.

Derbyshire, the local justices of the peace felt that farmers were in danger of being outnumbered by the industrial workers: 'Many thousands live in work at lead mines, coal mines, stone pits, and iron works', becoming increasingly dependent for their food on Danzig corn brought from Hull along the Trent.[1] Derbyshire lead was transported downstream to Hull, the Peak being a main producer. The demand for lead had greatly increased; at home, the wave of new building meant that more lead was needed for roofing and plumbing, while (with tin) it was used by the pewterers; abroad, lead was one of England's traditional exports.[2] Shallow mines were thickly aligned along the veins of lead ore in the Carboniferous Limestone, often on moorlands up to 1,000 ft. above sea-level, between the rivers Dove and Derwent. The returns of the local justices showed that there were few parishes without lead mining, 'whereof our Hundred of High Peak hath much employment'. Around Wirksworth there was 'so great a multitude of poor miners' that the justices were compelled to tolerate a large number of ale-houses.[3] Lead was also mined in the northern Pennines, in the district around Alston, and again to some extent in the Lake District. Another lead-producing district lay in the Mendips, which, said Camden, were rich in lead mines.[4]

The West Country was also renowned for its tin mines, some in Devon but mostly in western Cornwall. Mine-shafts driven into the lodes (some of them worked by German miners) were replacing the streaming of placer deposits as traditionally practised in the eastern parts of Cornwall. The lodes of metallic ores were most abundant and at their richest in the narrow zone of contact between the slates and each of the granite masses, especially the western blocks of St Just, Carnmenellis, and Hensbarrow. Thus the western coinage towns of Helston and Truro, where the blocks of white tin received the Duchy seal as a guarantee of quality, were handling eight times as much tin as Liskeard and Lostwithiel in the east. The Cornish miners formed a distinct community, regarded somewhat unsympathetically at the time as 'ten thousand or twelve thousand of the roughest and most mutinous men in England'.[5] Carew described how they were 'let down and taken up in a stirrup by two men who wind

[1] J. Thirsk in F. J. Fisher (1961), 73.
[2] L. Stone, 'Elizabethan overseas trade', *Econ. Hist. Rev*, 2nd ser., II (1949), 45.
[3] J. H. Lander in W. Page (ed.), *V.C.H. Derbyshire*, II (1907), 177–8, 331.
[4] W. Camden, 230. See J. W. Gough, *The mines of Mendip* (Oxford, 1930), 112.
[5] A. L. Rowse, *Tudor Cornwall* (London, 1941), 54–62; W. S. Lewis, *The West of England tin mining* (Exeter, 1923), 5–6, 10, 39.

the rope', even where some of the workings were 300 feet deep. They toiled by candle light with pick-axe and wedges, draining their mines by means of 'sundry devices, as adits, pumps, and wheels driven by a stream'. Water power was as essential here as in the clothing districts: it worked the stamping-mills and crazing-mills that pulverised the ore; it washed the black tin sufficiently clean for the blowing-houses, and this was 'melted with charcoal fire, blown by a great pair of bellows moved with a water-wheel'.[1]

There was also a deliberate policy of trying to reduce England's dependence on imported copper, which brought new industries into being. Helped by German capital and hundreds of skilled workmen from Saxony, copper ore was mined and smelted at Brigham, near Keswick in Cumberland, under the auspices of the Society of Mines Royal, incorporated in 1568.[2] At the same time calamine, the ore of zinc, was found and worked in the Mendips, thus furnishing the means of making brass (from copper and zinc). Brass foundries and batteries were set up by the Society of Mineral and Battery Works.

The manufacture of salt was marked by the increasing use of coal for boiling. From early times there had been salt-pans at a large number of places around the coast, but in 1600 salt-making was especially flourishing at the mouths of the Tyne and Wear where coal was easily accessible and cheap. There were also important sources inland, some were in Cheshire where 'great store of white salt' was made at Nantwich, Northwich, and Middlewich; the brine was taken from the pits in wooden troughs to the wich-houses, where it was 'seethed in lead cauldrons'.[3] Charcoal was still the main fuel here and also at the Droitwich saltworks in Worcestershire; 'what a prodigious quantity of wood these salt works consume', complained Camden, 'though men be silent, yet Feckenham-forest, once very thick with trees…by their thinness declare'. William Harrison devoted a whole chapter to salt-making.[4]

A variety of other industries were also prospering in 1600, and were able to do so largely because of the substitution of coal for wood as a fuel. Glass-making, which had benefited from the French methods

[1] F. E. Halliday (ed.), *Richard Carew of Antony: The survey of Cornwall* (London, 1953), 92–4.
[2] C. M. L. Bouch and G. P. Jones, 118–27.
[3] W. Smith, *The particular description of England, 1588*, ed. by H. B. Wheatley and E. W. Ashbee (London, 1879), 9.
[4] W. Harrison, 375–8.

introduced in the 1550s, was shortly to be transformed by the cheaper process using an improved crucible and coal as its fuel. Glassworks were sited widely, from London to the New Forest, from the Weald to Staffordshire. Alum, much in demand as a fixing agent in dyeing and in curing skins, was mined and boiled at Whitby and in the Isle of Wight, again in an attempt to foster home production. Coal was also increasingly used in such activities as sugar-refining, soap-boiling and candle-making. Moreover, the transport of coal by sea 'probably accounts more than any other single factor for the growth of the English shipping industry in the late sixteenth and seventeenth centuries'.[1] The building of ships was already 'very much in practice' at Ipswich.[2] Some miscellaneous industries had been established with the aid of foreign immigrants. Many of them brought new skills which, in Kent alone, resulted in silk-weaving in Canterbury, a paper mill at Dartford, an iron-plate mill at Crayford, and in Sheppey 'a certain Brabanter' used pyrites to make copperas (iron sulphate), used as a mordant or fixing agent in dyeing cloth. The pull of the metropolitan market may also be seen in the siting of many paper mills along the Wye valley near High Wycombe.[3]

TRANSPORT AND TRADE

Roads and rivers

Tudor England inherited a well-developed road system based upon London, and incorporating many roads of Roman origin such as Watling Street. There were also cross-country highways such as that from Bristol to Gloucester and so by the Severn valley to Chester. These were supplemented by a network of many minor roads, most frequent where the market towns were most numerous. The extent to which all these various roads were used is only now being realised. Most of Southampton's imports were distributed by carriers' services to London, Bristol, and even as far afield as Manchester or Kendal. It has been said that the transport of goods and passengers before 1500 'could be easily and efficiently undertaken by road throughout southern England and the Midlands';[4] but it

[1] J. U. Nef, 325.

[2] F. Hervey (ed.), *Suffolk in the XVIIth century. The breviary of Robert Reyce, 1618* (London, 1902), 97–8.

[3] D. C. Coleman, *The British paper industry, 1495–1860* (Oxford, 1958), 49–50; A. H. Shorter, *Paper mills and paper makers in England, 1495–1800* (Hilversum, 1957), 22–50, 92; F. W. Jessup, *A history of Kent* (London, 1958), 98, 105.

[4] J. Simmons, 6–8 and 11.

must be remembered that the roads were unmetalled and unfit for heavy traffic.

The pattern of main roads in England by 1600 is outlined in Fig. 62, from which the focal position of London can be appreciated. Major post-roads went from it to the extremities of the kingdom, the longest making its way via Stamford, York, and Newcastle to the Scottish border at Berwick. Another ran to Coventry and Chester, with extensions to Carlisle and to North Wales. Other roads led to Bristol or to Gloucester and South Wales. The great road to Land's End made its way through Salisbury and Exeter. A post-road went through Canterbury to Dover, by no means the only significant Channel port, as Rye was 'the chiefest for passage betwixt England and France'.[1] Most English regions were within easy access of a main road. The chief exception was in the north, where the network of intersecting roads stopped short at the trans-Pennine connection from York to Chester. North of this cross-country line only the western and eastern post-roads were recorded. The pattern seems to be sparse beyond Lincoln, but there was a much-used ferry into the East Riding at Barton-upon-Humber, and Camden noted a road from Doncaster crossing the Trent below Gainsborough: 'it is a great road for pack-horses, which travel from the west of Yorkshire, to Lincoln, Lynn, and Norwich'.

London's primacy in the national pattern was repeated on a lesser scale in those cities which acted as provincial capitals: William Smith in 1588 entitled his summary of the English roads 'The highways from any notable town in England to the city of London, and likewise from one notable town to another'. The influence of York or Coventry may be deduced from their road connections not only with London, but with half-a-dozen major towns besides. Places such as Exeter, Bristol, Salisbury and Gloucester each had four or five main roads converging on it. For natural reasons, however, stretches of these roads could become impassable in the winter: the Exeter–Bristol road had a short-cut from Bridgwater to Axbridge through the marshes, 'but no man can travel it well except it be in summertime, or else when it is a great frost'. River-crossings do not seem to have been a deterrent, for by using no fewer than six ferries it was possible to follow a road from Southampton to Helford in Cornwall 'all along the sea coast'.[2]

Another feature of the road pattern was the advantage shared by counties immediately to the north and west of London, in the angle

1 W. Smith (1588, ed. 1879), 9.

2 Ibid, 69–72.

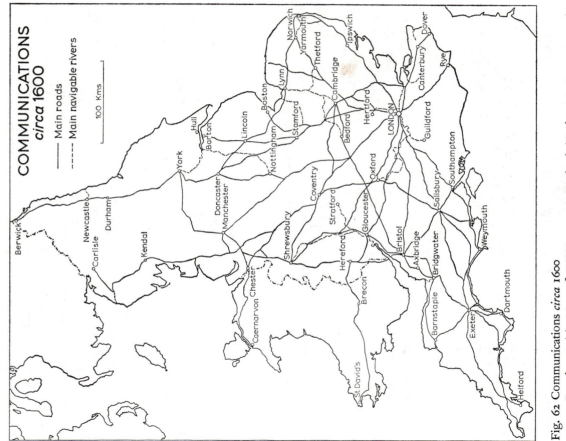

COMMUNICATIONS
circa 1600

—— Main roads
----- Main navigable rivers

100 Kms

Fig. 62 Communications *circa* 1600

Based on: (1) *Map of XVII century England* (Ordnance Survey, 1930);
(2) T. S. Willan, *River navigation in England, 1600–1750* (Oxford, 1936),
map 1.

DHG

formed by the roads to York and Bristol. These counties had alternative means of marketing in London, and they could supply the through traffic passing between London and more distant parts. Norden saw this very clearly in Hertfordshire, which was 'much benefited by through-fares to and from London northwards': its market towns were 'the better furnished with such necessaries as are requisite for inns, for the entertainment of travellers'. John Norden's appreciation of the importance of roads led him to mark them on his county maps; thus the 'principal highways from London through Middlesex, and towards what especial places in England they lead' were listed in his text and numbered on his maps.[1] Unfortunately the four variants of his manuscript map of Essex reveal that his depiction of the whole road pattern must be accepted with caution.[2] His road-maps nevertheless were symptomatic of a growing interest in roads, and in his *An intended guyde for English travailers* (1625) he included 'a new invention' of triangular distances between main towns.

Many roads were difficult to use because of excessively steep gradients or poor surface conditions, especially in clay country. When Camden travelled along Watling Street where it crossed the Gault Clay in Bedford-shire, he saw the improvements in what was 'heretofore a dirty (but now a very good) road', extremely troublesome to travellers in winter-time'.[3] By the Highways Act (1555) the responsibility for maintaining roads in good condition was placed upon the parishes through which they ran. Partly because parishes sometimes could not, or would not, do much about this, partly because of the pressing demands placed on the roads by an expanding economy, there had to be other regulations to keep the roads in working order. Some parishes were very conservative, and forbade, or tried to forbid, loads of more than one ton. Others were realistic, as for instance the Kent justices of the peace who in 1604 ordered the owners of carts carrying more than one ton of goods to pay five shillings for road repairs. Loads were becoming larger, some reaching to at least two-and-a-half tons, 'whereby the highway from Canterbury to Sittingbourne' (a stretch of the Dover Road on clay) 'is spoiled to the great annoyance of all travellers'. Some coal was being moved on the Midland roads, taken

[1] J. Norden, *Speculum Britanniae: An historicall and chorographicall description of Middlesex* (London, 1593).
[2] F. G. Emmison and R. A. Skelton, 'The description of Essex, by John Norden, 1594', *Geog. Jour.*, CXXIII (1957), 39.
[3] H. C. Darby (1936), 361.

by pack-horse trains along Watling Street from the Shropshire and Warwickshire coalfields, and (in 1581) carried 40 miles from Coleorton in Leicestershire. Wagons took heavy goods of all description: Ipswich, for example, had a thrice-weekly carriers' service to London. By the time John Taylor's *Carriers Cosmography* appeared in 1637, 'an organised system of goods transport' was in being.[1]

Increasing use was also made of navigable rivers. The only artificial canal (built with pound locks in 1564–6) bypassed obstructions in the estuary of the Exe, and joined Exeter with the sea at Topsham.[2] There were, however, a number of attempts to improve river courses themselves. Eight Acts of Parliament were passed for this purpose in the sixteenth century, such as that for improving the river Lea by a new cut (1571); the Lea supplied London with cheap corn and malt, but brought the bargemen into sharp conflict with millers, fishermen, and road-carriers who were content to see the river remain as it was.[3] For bulky loads such as grain, wool, coal, and building materials, river transport was far cheaper than road transport, especially on long journeys in the big trows and barges sailing the Severn or the Thames. Coal moved down the Severn and the Trent from the coalfields, and upstream from east-coast ports to areas served by the Thames, the Great Ouse, and the Welland.[4] The movement of grain, cheese, butter, and timber was contrary to this, going down the Ouse to King's Lynn, for example, or upstream to Nottingham on the Trent. There were at least 685 miles of navigable rivers, giving the country 'the foundation of its future system of internal water communications' (Fig. 62), but there were still 'great tracts of the country which lay more than 15 miles or one day's carriage by land from the sea or from a navigable river'.[5]

Maritime trade: coastal and overseas

Rivers either collectively or separately were not a self-contained transport system, because they were inseparable from the coasting trade (and, for that matter, the foreign trade) that passed through the English seaports.

[1] J. Simmons, 16–18.

[2] W. B. Stephens, 'The Exeter lighter canal, 1566–1698', *Jour. Transport History*, III (1957), 1.

[3] W. T. Jackman, *The development of modern transportation in England*, I (Cambridge, 1916), 165–8; G. B. G. Bull, 'Elizabethan maps of the lower Lea valley', *Geog. Jour.*, CXXIV (1958), 357–8.

[4] T. S. Willan, 'Yorkshire river navigation, 1600–1750', *Geography*, XXII (1937), 189.

[5] T. S. Willan, *River navigation in England, 1600–1750* (Oxford, 1936), 133.

The evaluation of inland navigation 'must take into account both the rivers and the sea'. The sea, so to speak, was 'merely a river round England, a river with peculiar dangers, peculiar conditions and peculiar advantages'.[1] A score of head ports and their many members were situated fairly regularly around the coastline, and few regions were unable to enjoy the very cheap transport provided by coasters.[2] Above all, the eastern ports were influenced by the London market, with its gigantic appetite for grain and coal. Newcastle sent each year, salt, glass and over 70,000 tons of coal. Hull shipped lead, iron and much else primarily to London; from King's Lynn went a constant flow of grain, malt, and butter. London itself shipped all kinds of goods to, and received cargoes from, almost every other port of any size in England and Wales.[3]

In foreign trade, furthermore, London's share 'was greater than that of all the provincial ports put together', and comprised between two-thirds and three-quarters of the nation's trade abroad.[4] Its superiority overshadowed Newcastle's shipments of 27,000 tons of coal a year to Emden, Hamburg, and other north European ports, or Bristol's trade with Ireland, France, and Spain (yielding customs dues of £2,112 in 1598–9). Of the other significant provincial ports, the ranking in 1594–5 (based on customs revenue for both imports and exports, but excluding prize cargoes) was Exeter (with its members Dartmouth and Barnstaple), Sandwich (and Dover), Hull, Ipswich, and Southampton. The London and provincial trades alike rested on the export of a few staple products, especially cloth, and the import of a wide range of raw materials, manufactured goods, foodstuffs and wines.[5] By 1600 the most striking development was the growing share of eastern ports in the export of grain, to the order of about 75% of the national whole. Another feature was the concentration of trade in fewer ports, as in King's Lynn and Yarmouth with their handling of 80% of the malt and barley, rye and wheat shipped from East Anglia.[6]

1 T. S. Willan (1936), 5.
2 W. Smith (1588, ed. 1879), 18, 50.
3 T. S. Willan, *The English coasting trade, 1600–1750* (1938), 111, 120, 126, 167, 192.
4 T. S. Willan, *Studies in Elizabethan foreign trade* (Manchester, 1959), 65; F. J. Fisher, 'London's export trade in the early seventeenth century', *Econ. Hist. Rev.*, 2nd ser., III (1950), 151–61; L. Stone, 'Elizabethan overseas trade', *Econ. Hist. Rev.*, 2nd ser., II (1949), 30–58.
5 T. S. Willan (1959), 90.
6 A. Everitt in J. Thirsk (1967), 525–6.

TOWNS AND CITIES

John Speed in *The theatre of the empire of Great Britaine* (1611), besides bringing together fifty town plans, marked on his county maps the market towns 'fit for buying and selling and other affairs of commerce'. They numbered 605, but a recent estimate has put the total at 760 for England with another 50 for Wales.[1] They varied greatly in size downwards from London, which may have had a population of about 250,000. Chief among the provincial centres was Norwich with about 15,000 people, and there were four other regional capitals each with a population of about 10,000 or over – York, Bristol, Newcastle upon Tyne and Exeter. Below these came another ten or so substantial towns each with a population of 5,000 or more. They were followed by about thirty towns with over 3,000 inhabitants apiece, and these were succeeded in turn by a variety of smaller towns, often with under 2,000, and sometimes with only 1,000 people or less.

A contemporary estimate placed the number of walled towns at about 100.[2] The walls had, for the most part, ceased to be defensible, but on the Scottish Border and along the coastline, walls had played a part in the anti-invasion plans of Elizabeth. Thus at Berwick 'the utmost town in England and the strongest hold', the earlier walls had been replaced by elaborate defences complete with Italianate fortifications and gun-emplacements.[3] Some towns were still confined within their ancient walls as at Nottingham and Northampton, but Speed's maps show that quite a number had expanded, frequently in the form of linear suburbs along main highways leading from the town gates. Such spread was also to be seen leading from unwalled towns such as Reading, where a sprawling built-up area was of long standing, and where the increasing population of Elizabethan times was accommodated by the subdivision of houses into tenements.[4]

In varying degrees, the towns were centres of trade and craftsmanship, being intimately bound to their hinterlands in the countryside and

[1] *Ibid.*, 467.   [2] W. Smith (1588, ed. 1879), 2.

[3] J. Speed, *The theatre of the empire of Great Britaine* (London, 1611), 89; M. W. Beresford and J. K. S. St Joseph, 177–9; I. MacIvor, 'The Elizabethan fortifications at Berwick-upon-Tweed', *Antiq. Jour.*, 45 (1965), 64–96.

[4] C. F. Slade, 'Reading', in M. D. Lobel (ed.), *Historic towns*, 1 (London and Oxford, 1969), 5–6; Banbury, Gloucester, Hereford, Nottingham and Salisbury are also included in this volume.

dependent upon good road and waterborne communications. But most of them had not yet lost their rural character. In Maitland's phrase, those who would study the history of towns 'have fields and pastures on their hands'.[1] Leading tradesmen were often farmers or graziers. At Leicester, 'pigs and cows went their way about the town, though ringed and herded, and as late as 1610 it was necessary to forbid winnowing in the streets'.[2]

Major provincial towns

Norwich seems to have had a population of about 15,000 or so. Virtually the whole of the town was contained within its medieval walls which enclosed an area of nearly one square mile, although as much as a quarter of this was not continuously built over. The city had suffered grievously in the sixteenth century from repeated fires and epidemics of plagues. In spite of these reverses it had begun to grow again when, in 1565, Dutch and Walloon refugees from religious persecution were invited to settle in the city and there weave their 'outlandish commodities', the so-called new draperies. At the end of the century they numbered about a quarter of the total population, and were reviving the depressed textile industry. 'By their means', so we read in a document of 1575, 'the city is well inhabited, and decayed houses re-edified and repaired that were in ruins', although initially the refugees had accentuated the poverty of overcrowded parishes such as St Paul's and All Saints.[3] Moreover, the central core of the city around the castle was improving its amenities and changing its appearance. River water was piped from New Mills to the Market Hall and Cross (1582); all new-built roofs were required to be of tile, slate or lead, and not of thatch, so as to reduce the risk of fire (1583); and two of the principal bridges across the Wensum were rebuilt in stone in place of wood (1591). These changes were consonant with Norwich's many functions as a regional capital, concerned with distributive and service industries as well as with the textile trades.[4]

York, with a population of about 12,000, was the most prominent city

---

[1] F. W. Maitland, *Township and borough* (Cambridge, 1898), 9.

[2] E. W. J. Kerridge in R. A. McKinley (ed.), *V.C.H. Leicestershire*, IV (1958), 99.

[3] Quoted in E. Lipson, *The history of the woollen and worsted industries* (1921), 23; J. F. Pound, 'An Elizabethan census of the poor', *Univ. Birm. Hist. Jour.*, VIII (1962), 140–4 and 150.

[4] J. F. Pound, 'The social and trade structure of Norwich, 1525–1575', *Past and Present*, XXXIV (1966), 49–69; Anon, *The history of the city and county of Norwich* (Norwich, 1768), 225–8, 242–4, 246.

in the north. It was the seat of an archbishopric and an important admini-strative centre; it was well served by inland waterways; and it had escaped the worst ravages of plague during the sixteenth century. Its walls enclosed some 143 acres, and linear suburbs extended from it on all sides. But much of its medieval glory was departing. It had lost a great deal of its clothing industry to other Yorkshire towns and villages; and it was losing its position as a port to places nearer the sea where larger sea-going ships could be accommodated. Hull gained much of what York lost.[1]

Bristol also may have contained about 12,000 people, and it had long grown beyond the circle of its medieval walls. 'For trade of merchandize', it was 'a second London, and for beauty and account next unto York', wrote Speed.[2] Set between the Frome and the Avon with their sheltered tidal harbours, it was well placed for trade. It was the port for one of the most vigorous cloth-making areas in England. It had connections with the Midlands and South Wales by means of the Severn and the Bristol Channel. Beyond, it traded with Ireland, Gascony and the Iberian peninsula, and its mariners and merchants were among the first English-men to take advantage of wider opportunities beyond Europe. Around the ancient centre had grown up a large industrial suburb and a fashionable residential area. It was far more salubrious than most of the towns of the realm. As William Smith could say in 1588, 'there is no dunghill in all the city', and Speed could point to its underground sewers removing 'all noysome filth and uncleaness'.[3]

Newcastle upon Tyne was a city of some 10,000 people. It lay where the main east-coast road to Scotland crossed the Tyne. Ships could berth at the foot of its Castle Hill, and the quayside quarters lay within the walls which enclosed about 150 acres, not all built over until the nineteenth century. The sixteenth century had seen a great increase in the use of coal in England generally, and the easily worked outcrops on the banks of the Tyne were being rapidly exploited. Shipments of coal from Newcastle increased fivefold between 1560 and 1600, and this activity was reflected in the prosperity of the town. It was able to spend a large sum of money to obtain a new and extended charter from the Crown in 1600. A bridge across the Tyne joined the town to Gateshead, but this was a borough in its own right, and attempts to annex it, during the sixteenth century, had failed.[4]

---

[1] A. G. Dickens in P. M. Tillott (ed.), *V.C.H. The city of York* (1961), 121–2, 129–30.

[2] J. Speed, 47.   [3] W. Smith (1588, ed. 1879), 34.   [4] J. Speed, 89.

Exeter was a city of about 9,000 people in spite of severe epidemics in the 1590s. Its ancient walls enclosed only 93 acres, but ribbon-like suburbs had grown out along the main roads, especially towards London, and these may have housed as much as a quarter of its population. It had its own clothing industry, and was also the outlet for the products of the clothing centres around. Direct access to the sea had long ceased, but the growth of its outport at Topsham, four miles down the estuary of the Exe, reflected the prosperity of the city itself.[1]

Lesser provincial towns

Of the lesser provincial towns of the realm, some were losing ground like the old cloth-making centres of Coventry and Salisbury with about 6,000 inhabitants each.[2] Canterbury, with its cathedral, had about 5,000 people.[3] The majority of county towns rarely exceeded 3,000 to 4,000 inhabitants apiece. Some, like Worcester and Hereford, had long linear extensions along each of half a dozen highways leading from their town gates.[4] Others, like Northampton and Nottingham, had virtually no suburbs. Their layouts were very varied. Of the smaller towns some occupied strong points, like Durham and Buckingham, around their castles. Others were roughly star-shaped where roads converged, as at Hertford and Bedford. Yet others, with their houses, shops and public buildings were merely strung along highways, as at Huntingdon and Kendal.

Market towns of all sizes, with weekly markets, were widely distributed over the countryside, and they often specialised in one or more products such as grain, wool, leather or cheese. Annual fairs also specialised more particularly in cattle, sheep or horses, and these attracted people from 50 miles and more. A characteristic meeting place was not only the market or the fair but the inn. A census of inns and ale-houses of 1577 enumerated over 1,600 inns in 25 English counties; those at Romford in Essex were the Blue Boar, the Swan and the White Horse.[5] A number of inns offered

---

[1] R. Pickard, *The population and epidemics of Exeter in pre-Census times* (Exeter, 1947), 14–15; W. T. MacCaffrey, *Exeter, 1540–1640* (Cambridge, Mass., 1958), 12–13; W. G. Hoskins, 'The Elizabethan merchants of Exeter', in S. T. Bindoff *et al.* (eds.), *Elizabethan government and society* (London, 1961), 164.

[2] W. G. Hoskins, *Provincial England: essays in social and economic history* (London, 1963), 72; Anon, *The history and antiquities of the city of Coventry* (Coventry, 1810), 51, 55, 71.

[3] W. K. Jordan, *Social institutions in Kent, 1480–1660* (Kent Archaeol. Soc., 75, 1961), 3.

[4] M. D. Lobel, 'Hereford' in M. D. Lobel (ed.), 9.

[5] A. Everitt in J. Thirsk (1967), 559.

special facilities to traders and their wares, and became centres for private bargaining.

Some markets towns were very small. William Harrison, about 1580, thought that the common run of market towns rarely had more than 60 households or 200 to 300 communicants which would imply total populations of 500 or less; but within their limited spheres such small towns played a vital role.[1] Of quite a different character were those places with mineral springs. Bath, after long neglect, began once more to attract visitors, but its heyday was still a long way off in 1600. The springs at Buxton and Matlock, as Camden noted, were being increasingly frequented. But none of these English spas could compare with the more famous continental resorts.[2]

A number of the lesser provincial towns in 1600 were prospering as a result of recent industrial development. Colchester, with perhaps 5,000 people, had almost as many houses standing 'without the walls' as within. Dutch immigrants, who began to arrive about 1570, gave its cloth-making industry a new lease of life. By 1609 it had become 'so populous...that there was not one house to be had at any rate'; by this time, the immigrants numbered 1,300.[3] Nearby in East Anglia, other clothing-making towns were also flourishing – Hadleigh, Lavenham, Long Melford and Sudbury, together with a host of smaller centres. In the clothing area of the West Country there were also prosperous sizeable towns such as Tiverton with 4,000 people, Barnstaple with 3,500, Crediton and others.[4] In the north, Halifax, Leeds and Wakefield were beginning to stand out as centres of the clothing trade. Other industrial towns were also beginning to emerge: Sheffield had about 2,200 people, apart from the inhabitants of various villages in its large parish.

Along the coast there was another group of prospering towns. Plymouth had more than doubled its numbers in the last decades of the sixteenth century until in 1600 it included about 7,800 people. New streets were built all round Sutton Pool in response to the town's role as the principal naval base in the Spanish war.[5] Southampton, on the other hand,

[1] W. Harrison, 217; W. G. Hoskins, 'Provincial life', in A. Nicol (ed.), *Shakespeare in his own age* (Cambridge, 1964), 13–20.
[2] J. A. Patmore, 'The spa towns of Britain', in R. P. Beckinsale and J. M. Houston (eds.), *Urbanization and its problems* (Oxford, 1968), 51–2.
[3] J. Speed, 31; P. Morant, *The history and antiquities of the county of Essex*, 1 (London, 1768), 72, 75–8, 105–39.
[4] W. G. Hoskins (1954), 113, 455; W. Smith (1588, ed. 1879), 44.
[5] W. G. Hoskins, *Local history in England* (London, 1959), 143–4.

was in decay largely because it had lost its trade to London.[1] In the Thames estuary, Greenwich, Rochester and Chatham had about 3,000 inhabitants each, so had Sandwich in Kent.[2] In East Anglia, Ipswich with about 5,500 people, was 'one of the most famous towns in England at this present for traffic and other respects'. Speed's plan shows a town much expanded beyond its walls, with suburbs along all the roads leading from the town, and a quay on the Orwell stretching as far as Stoke Bridge; and he saw it as 'full of streets plenteously inhabited'.[3] Farther north along the east coast were Great Yarmouth, Lynn, Boston and Hull, all reflecting the agricultural and industrial prosperity of their hinterlands. In the north-west Chester was a flourishing city 'with very fair and large suburbs'. It was still the post-town for Ireland, but the Dee was silting up and the town had 'lost the advantage of a harbour'. It was relying upon outports, such as Neston and Heswall, and it was being supplanted by Liverpool, in whose grain market the Lancashire dealers stood on one side and the Cheshire dealers on the other.[4]

London

London stood far above all other urban areas, being about sixteen times or so the size of Norwich, the largest provincial city. By 1600, 'Greater London' (including Westminster and Southwark) probably had a population of about 250,000.[5] The rapid growth of the sixteenth century could have been sustained only by immigration from the rest of England. There was also a foreign element that added a cosmopolitan character to its life. A return for 1573 gave a total of 4,287 for the city – 3,160 Dutch, 440 French, 423 Burgundians, 137 Italians, 58 Spaniards and 32 Scots, to name the chief elements.[6] There were others in Westminster, Southwark and the suburbs. The numerical preponderance of the newly growing suburbs is clear, and became more pronounced because accommodation within the old walled city could be increased only 'by turning great houses into tenements, and by building upon a few gardens'.[7]

[1] A. A. Ruddock, *Italian merchants and shipping in Southampton, 1270–1600* (Southampton, 1960), 258–62.

[2] W. Smith (1588, ed. 1879), 7; W. K. Jordan, 4; J. W. Jessup, 100–104; C. W. Chalklin, *Seventeenth century Kent* (London, 1965), 23–6.

[3] W. G. Hoskins (1959), 177; J. Speed, 76.

[4] A. Everitt in J. Thirsk (1967), 481.

[5] N. G. Brett-James, *The growth of Stuart London* (London, 1935), 495–8.

[6] E. E. Rich, 263.

[7] N. G. Brett-James, 499, quoting John Graunt in 1662.

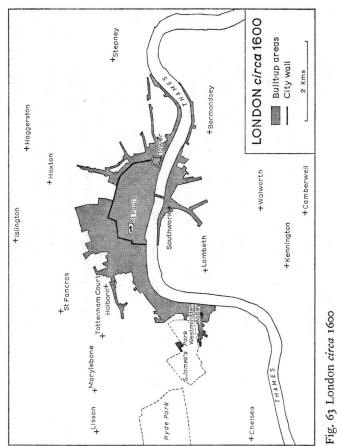

Fig. 63 London *circa* 1600
Based on N. G. Brett-James, *The growth of Stuart London* (London, 1935), map opposite p. 78.

The outline of Greater London may be reconstructed from Norden's plan of 1593 and Stow's unique description published in 1598 (Fig. 63). Only London Bridge spanned the Thames, which was unembanked and broader than it is today. At the southern end of the bridge lay the borough of Southwark which had been annexed to the city in 1550, and a continuous line of houses ran along the south bank where there were theatres and places for bull-baiting and bear-baiting. Below the bridge-head, sea-going ships anchored in mid-river, in the 'Pool', and they were served by barges, lighters and wherries which carried goods and passengers to the riverside wharves, warehouses and landing-stages.[1]

As viewed from the river, the boldest feature in the city's profile was the square tower of Old St Paul's, high above the steeples of the other 124 churches, and overlooking a warren of congested streets without a central focus. For decades new houses, gardens, and yards had covered any scrap of available land or made ground – in the city ditch, over Walbrook and the Fleet, on London Bridge, within the churchyards,

[1] *Ibid.*, 31.

encroaching on the streets not only at ground level but by the extension of the upper storeys of houses on brackets. No fewer than 80 houses had to be demolished, in 1565–70, to provide a site for the Royal Exchange, between Cornhill and Threadneedle Street. There had been widespread development of the 23 religious houses dissolved in the 'Great Pillage'. Most of the conventual churches were retained to serve parishes, but the monastic buildings made fine dwelling-houses for nobles or merchants, or they could be converted into tenements, business premises, even tennis courts. This kept to the city the aristocrats, courtiers, and officials who later in the seventeenth century gravitated towards Westminster. With the greatly increased demands being made upon them, London's supplies of water from the rivers, wells, and springs diminished in volume and deteriorated in quality. After 1582 came the new 'artificial forciers' worked by tide-mills, which pumped water from the Thames along pipes to the conduits; it was then distributed by water-carriers or, in a few cases, by quills direct to rich men's houses. This system was eclipsed by the New River Company in 1613, when a regular supply was brought to Finsbury from springs over 38 miles away at Amwell in Hertfordshire.[1]

From the Tower around to Shoreditch in the east, mean ribbon-like suburbs already reached along the Thames frontage to Limehouse and down the main roads, in Stow's view 'no small blemish to so famous a city to have so unsavoury and unseemly an entrance'.[2] The northern and western suburbs were more compact and had houses of quality, as at Chancery Lane and Holborn Hill; here, too, Camden tells us were 'some Inns for the study of common law'. Farther west, the suburbs forged a first physical link between London and Westminster. This went along the Strand, with a long line of former bishops' palaces beside the river, now transformed into 'large and goodly houses'. After 1603 the palace and abbey of Westminster drew to themselves permanently the seats of court and government, although the image lingered of 'twin-sister cities as joined by one street'.[3]

There were many attempts to curb the growth of Greater London following the proclamation in 1580 that prohibited the building of new houses within three miles of the city, and sub-letting within it. But it was difficult to check the dispersal of industry to the suburbs, including Southwark, maintained as it was by journeymen and apprentices who had

[1] W. H. Chaloner and A. E. Musson, 26.
[2] N. G. Brett-James, 59.
[3] N. G. Brett-James, 61, quoting Thomas Heywood in 1635.

broken away from the craft gilds within the city; by industries seeking lower rents and others (like tanning or soap-boiling) which were obnoxious in a residential setting; and by foreign immigrants who settled in suburban tenements, plying their trades in competition with the gilds.[1] The suburbs continued to grow, and to a fastidious Venetian observer it seemed as if they were 'inhabited by an inept population of the lowest description'.[2] Even so, Camden rejoiced in what to many was the dismaying growth of London. To him it was 'the epitome of all Britain, the seat of the British Empire'.

[1] H. J. Dyos, *Victorian suburb: a study of the growth of Camberwell* (Leicester, 1961), 34; 'The growth of a pre-Victorian suburb: south London, 1580–1836', *Town Planning Rev.*, xxv (1954), 67.

[2] Horatio Busino, 1618, quoted by H. and P. Massingham, *The London anthology* (London, 1950), 455.

Chapter 7

THE AGE OF THE IMPROVER: 1600-1800

H. C. DARBY

When the Tudor dynasty came to an end in 1603, the changes that were to transform the medieval landscape had already begun. Not only had they begun, but information about them was becoming increasingly abundant with the dissemination of the printed word. The tradition of English topographical writing inaugurated by Leland, Camden and Speed was soon to grow to great dimensions. As the seventeenth and eighteenth centuries progressed, there emerged a clearer picture of the differences in soil to be found in the various English counties. Compare, for instance, the description of Dorset by John Coker early in the seventeenth century with that by John Hutchins in 1774. The former referred generally to the difference between clayland and chalkland, but the latter drew much more definite pictures of the three divisions of the county — 'the down, the vale and the heath'.[1] This is not surprising in view of the increasing attention given to agricultural improvement. The Tudor husbandries of Fitzherbert and Tusser were the precursors of a large number of works concerned with the advancement of agriculture. The flow of treatises continued to increase throughout the eighteenth century when the name of Arthur Young came to dominate all others.[2]

Just as in agriculture the impulse to improve was widespread, so in industry was the impulse to contrive and invent — especially in the iron industry, in the textile industry and in coalmining. Reports and pamphlets began to multiply, and they dealt not only with machines but also with roads and canals. During these years, too, the first detailed maps of England were made. From 1759 until 1809 the Royal Society of Arts (founded in 1754) offered awards for county maps on a scale of one inch

[1] H. C. Darby, 'Some early ideas on the agricultural regions of England', *Agric. Hist. Rev.*, II (1954), 30–47; F. V. Emery, 'English regional studies from Aubrey to Defoe', *Geog. Jour.*, CXXIV (1958), 315–25.

[2] G. E. Fussell: (1) *The old English farming books from Fitzherbert to Tull, 1523 to 1730* (London, 1947); (2) *More old English farming books from Tull to the Board of Agriculture, 1731 to 1793* (London, 1950).

to a mile, and so stimulated the production of maps based upon original surveys and often showing details of land utilisation.[1] In the meantime the Ordnance Survey had been constituted in 1791, and its first one-inch sheet of Kent and part of Essex appeared on 1 January 1801.[2]

There was also a whole new body of writing about the features and problems of the changing economy of the time. Political economy and 'political arithmetic' emerged as distinct enquiries, and were exemplified in the works of such men as John Graunt (1620–74), Sir William Petty (1623–87) and Gregory King (1648–1712), some of whom were members of the Royal Society founded in 1660. Gregory King attempted estimates of the categories both of population and of land use. Towards the end of the following century there appeared Adam Smith's *The wealth of nations* in 1776 and T. R. Malthus's *Essay on . . . population* in 1798. The England of 1800 was not only two hundred years older than that of 1600. It was a different world with new landscapes and new problems.

These developments took place in the context of more general changes. In the middle of the seventeenth century, English life had been disrupted by the Civil War of 1642–8, but thereafter freedom from domestic warfare provided a background for continuous development. The political changes following the Civil War culminated in the Restoration in 1660, which, it has been said, 'has a better claim than most dates to be regarded as the economic exit from medievalism'.[3] Moreover, the realm was strengthened by union with Scotland. The two crowns had been worn by the same king since 1603, but legislative union was not realised until 1707 when the United Kingdom of Great Britain was created to the advantage of both countries. There were also wider changes. The eviction of the Hanse in 1598 and the formation of the East India Company in 1600 were 'symptomatic of the changing character of English trade at the end of the sixteenth century'.[4] Furthermore, tentative overseas expansion was accelerated by the maritime wars of the eighteenth century which extended British interests into the markets of the world. Not only the impulse to improve and invent, but also the impulse to expand was a powerful force transforming the geography of the kingdom.

[1] J. B. Harley, 'The Society of Arts and the surveys of English counties 1759–1809', *Jour. Roy. Soc. Arts*, CXII (1963–4), 43–6, 119–24, 269–75, 538–43.

[2] C. Close, *The early years of the Ordnance Survey* (Chatham, 1926); reprinted with an introduction by J. B. Harley (Newton Abbot, 1969).

[3] C. H. Wilson, *England's apprenticeship, 1603–1763* (London, 1965), 236.

[4] W. E. Minchinton (ed.), *The growth of English overseas trade in the seventeenth and eighteenth centuries* (London, 1969), 2.

POPULATION

A number of attempts have been made to estimate the population of England and Wales about 1600, based upon such sources as parish registers, muster rolls and hearth tax returns. Apart from a low figure of 2·5 million produced by Thorold Rogers, estimates range between 3·6 million and 5·0 million. A reasonably acceptable figure might lie somewhere between 4·0 and 4·5 million. Nearly one hundred years later, a contemporary estimate by Gregory King, in 1695, put the total at 5·5 million, and this has been more or less supported by D. V. Glass who would, however, reduce it to 5·2 million.[1] John Rickman, who was in charge of the first Census of 1801, attempted to argue backwards on the basis of the births, deaths and marriages entered in parish registers. For 1700, he produced a final figure of 6·0 million; for 1750 it was 6·5 million which Deane and Cole revised to 6·1 million; and for 1780 it was 7·5 million. With the first Census itself we are on firmer ground, and for 1780 it was 7·5 million, but Rickman himself later revised this to 9·2 million (including 0·6 million for Wales).[2] In looking at these figures, it must be remembered that subsequent research has shown how uncertain any estimates based upon parish registers must be; compulsory registration by Act of Parliament did not come until 1838.[3] Even so, we may reasonably conclude that the rate of growth was slow before 1750; that it rose during the next 30 years; and that it became rapid after 1780. Here was a demographic revolution, a 'population explosion'.

Traditionally, the main cause of the increase was assumed to be a rapidly falling death rate, largely the result of advances in medical knowledge.[4] But the emphasis on medical advance has been challenged by two medical

---

[1] D. V. Glass, 'Gregory King's estimate of the population of England and Wales, 1695', *Population Studies*, III (1950), 358.

[2] P. Deane and W. A. Cole, *British economic growth, 1688–1959* (Cambridge, 2nd ed., 1967), 98 et seq. For a summary of various estimates see B. R. Mitchell with P. Deane, *Abstract of British historical statistics* (Cambridge, 1962), 5.

[3] For general summaries of the problem, see M. W. Flinn, *British population growth, 1700–1850* (London, 1969). See also P. Deane, 'The demographic revolution', being ch. 2 (pp. 20–35) of *The first industrial revolution* (Cambridge, 1965); D.V. Glass and D. E. C. Eversley (eds.), *Population in history* (London, 1965). A number of the key papers are reprinted with an introduction in M. Drake (ed.), *Population in industrialization* (London, 1969).

[4] E.g. G. T. Griffith, *Population problems of the age of Malthus* (Cambridge, 1926); M. C. Buer, *Health, wealth and population in the early days of the industrial revolution* (London, 1926).

historians who point to the lack of new drugs, to the high death rate in hospitals, to the continued dangers of midwifery and surgery before the days of anaesthetics and antiseptics, and to the unlikelihood that inoculation against small-pox had much effect.[1] P. E. Razzell, however, has thrown doubt on this view as far as inoculation for small-pox was concerned. Inoculation was first introduced into England in 1721 and mass inoculations became common after about 1760; this alone, says Razzell, 'could theoretically explain the whole of the increase in population'.[2] There are also other possible explanations for a decline in the death rate. Part of the answer might lie in the natural history of endemic and epidemic disease. Why, for example, did plague, 'the greatest single agent of mortality', disappear from England after the outbreak of 1665–7?[3] Or again part of the answer might lie in a generally improved standard of living, including a better and more varied diet associated with improvements in agriculture and transport. Certainly life seemed to have become less precarious when Gilbert White, looking back over ten years or so, wrote in 1774, 'Such a run of wet seasons a century or two ago would, I am persuaded, have occasioned a famine.'[4]

Some scholars, however, have doubted whether the death rate did substantially fall during the later decades of the eighteenth century, and they have concluded that a rising birth rate was 'the major cause' of population growth at this time.[5] It could have been associated with an earlier age of marriage and with greater economic opportunity, but both these possibilities are matters of controversy. The relation of birth rate and death rate, and the dynamics of population growth in general, have prompted much ingenious discussion. Given both a falling death rate and a rising

[1] T. McKeown and R. G. Brown, 'Medical evidence related to English population changes in the eighteenth century', *Population Studies*, IX (1956), 119–41.

[2] P. E. Razzell, 'Population change in eighteenth-century England. A reinterpretation', *Econ. Hist. Rev.*, 2nd ser., XVIII (1965), 312–32.

[3] K. F. Helleiner, 'The population of Europe from the Black Death to the eve of the vital revolution', being ch. I of E. E. Rich and C. H. Wilson (eds.), *The Cambridge economic history of Europe*, IV (1967), 95.

[4] Gilbert White, *The natural history of Selborne* (Everyman's Library), 151. See E. L. Jones, *Seasons and prices. The role of the weather in English agricultural history* (London, 1964), 55, 109 and 146.

[5] J. T. Krause, 'Changes in English fertility and mortality, 1781–1850', *Econ. Hist. Rev.*, 2nd ser., XI (1958), 52–70. See also: (1) K. H. Connell, 'Some unsettled questions in English and Irish population history, 1750–1845', *Irish Historical Studies*, VII (1951), 225–34; (2) H. J. Habakkuk, 'English population in the eighteenth century', *Econ. Hist. Rev.*, 2nd ser., VI (1953), 117–33; and (3) P. Deane and W. A. Cole, 122 et seq.

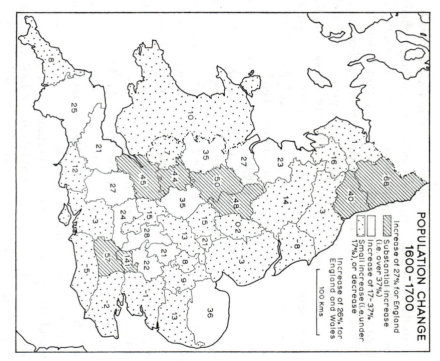

Fig. 64 Population change, 1600–1700
Based on John Rickman's estimates in *Census of 1841: Enumeration Abstract*, 36–7 (P.P. 1843, xxii).

birth rate, the discussion has then centred around which was the more effective in inaugurating a sharp and continued increase in population, and also around the precise period in which this happened. Finally, we must add that it is not possible to say to what extent the movement of people into and out of the country affected the rate of increase. There was certainly immigration from Ireland and Scotland, and also a stream of newcomers who came from Europe in connection with the new industries such as the manufacture of paper, glass and silk. Against this must be set the movement of Englishmen overseas, but we do not know to what extent one balanced the other.

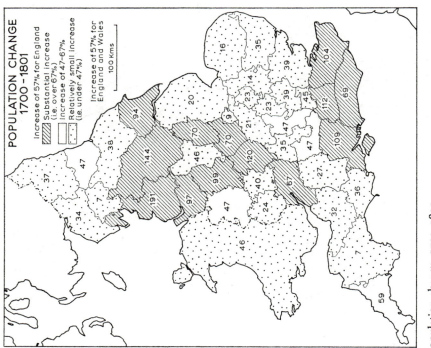

POPULATION CHANGE
1700–1801

Increase of 57% for England
▨ Substantial increase
(i.e. over 67%)
⬚ Increase of 47–67%
· Relatively small increase
(i.e. under 47%)

Increase of 57% for
England and Wales

100 Kms

Fig. 65 Population change, 1700–1801
Based on: (1) John Rickman's estimates in *Census of 1841: Enumeration Abstract*, 37 (P.P. 1843, xxii); (2) P. Deane and W. A. Cole, *British economic growth, 1688–1959* (Cambridge, 2nd ed, 1969), 103.

Whatever be the doubts about the beginnings and causes of the increase of population, we know that it did not take place evenly over the kingdom, as may perhaps be seen from Fig. 64 (for 1600–1700) and Fig. 65 (for 1700–1801), based upon Rickman's figures modified (for 1700–1801) by Deane and Cole who point to the inevitable limitations of the evidence.[1] Any interpretation of the figures can only be unsatisfactory, and all one can hope for is a very general picture.

[1] P. Deane and W. A. Cole, 101–6. For an earlier attempt at mapping, see E. C. K. Gonner, 'The population of England in the eighteenth century', *Jour. Roy. Stat. Soc.*, LXXVI (1913), 261–303.

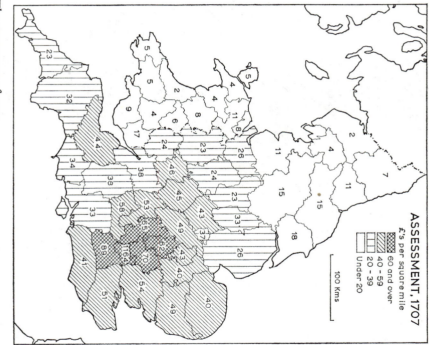

Fig. 66 The assessment of 1707
Based upon A. Browning, *English historical documents, 1660–1714* (London, 1953), 317–21.

As an alternative one can construct maps showing the distribution of wealth on the basis of tax assessments, but these, again, have many limitations.[1] Such a map is Fig. 66 for the 1707 assessment,[2] and, at any rate, it bears out Defoe's statement that England south of the Trent was 'the most populous part of the country, and infinitely fuller of great towns, of people, and of trade' than the northern part of the realm.[3] By the first

[1] Examples of maps appear in: (1) A. Browning, *English historical documents, 1660–1714* (London, 1953), 458–9; (2) J. H. Andrews, 'Some statistical maps of Defoe's England', *Geog. Studies*, III (1956), 44.

[2] Based on the figures in A. Browning, 317–21.

[3] Daniel Defoe, *A tour through England and Wales* (Everyman's Library), I, 253.

Census of 1801 we are on more certain ground. Fig. 67 shows that there had been great changes. Except for Middlesex and Surrey, which reflected the growth of London, the most populous county was Lancashire. Considerably below it came the West Riding, Warwickshire and Staffordshire, the areas of growing industry. Not far below these, were Gloucestershire (with the port of Bristol) and Worcestershire (with saltworks at Droitwich and ironworks in the north). But the more rural counties had also shared in the general increase to varying degrees.

To what extent was the emergence of areas of dense population the result of natural increase or of immigration into them? Deane and Cole have attempted tentatively to distinguish between natural increase and net immigration for each county, again on the basis of Rickman's figures.[1] Considering the evidence as a whole, two points emerge. In the first place, a large number of migrants went to the London area, particularly to Middlesex and Surrey, but also to Essex and Kent; immigration rather than natural increase accounted for their increasing populations, and so for the growth of London. In the second place, immigration into the rising industrial areas played a surprisingly small part in their growth. Lancashire even showed a net loss of migrants during 1751–81, and Warwickshire during 1781–1801. Staffordshire and the West Riding showed a consistent loss throughout the century. It was not until after 1800 that net immigration into these counties became important. What gave these four counties their respective accumulations of population before 1800 were birth rates considerably above the national average. Unlike the London area, they had grown largely through natural increase, not only by attracting immigrants. The table on p. 310 includes the period 1800–30 in order to show how the trends continued.

We must ask why natural increase was a more important element in the growing populations of Staffordshire and the West Riding than in those of Lancashire and Warwickshire. The answer may lie in the more rapid development of large towns in the latter counties. Liverpool and Manchester between them accounted for nearly 25% of the population of Lancashire in 1801. Birmingham and Coventry accounted for about 40% of that of Warwickshire. On the other hand, Leeds and Sheffield accounted for only about 14% of the population of the West Riding (and

[1] P. Deane and W. A. Cole, 106–22. For a criticism see: (1) M. W. Flinn, 28–9; (2) L. Neal, 'Deane and Cole on industrialization and population change in the eighteenth century', *Econ. Hist. Rev.*, 2nd ser., XXIV (1971), 643–7; and (3) W. A. Cole, 'Rejoinder', *ibid.*, 648–51.

Table 7.1 *Annual rates of migration and natural increase, 1701–1830 (rates per 1,000 of population)*

	Net migration				Natural increase			
	1701–50	1751–80	1781–1800	1801–30	1701–50	1751–80	1781–1800	1801–30
Lancashire	+3.0	−1.2	+11.6	+6.2	+2.7	+10.6	+13.4	+16.0
Warwickshire	+2.2	+6.5	−3.8	+3.4	+2.8	+5.3	+13.3	+11.9
Staffordshire	−4.4	−4.3	−0.9	+0.3	+6.9	+14.4	+13.8	+17.1
West Riding	−3.4	−0.8	−0.6	+1.6	+8.1	+14.6	+12.6	+15.9
Middlesex	+15.8	+14.0	+12.2	+12.2	−15.6	−9.3	−1.2	+4.1
Surrey	+11.2	+18.1	+13.4	+10.3	−10.8	−4.8	+3.3	+8.8

Source: P. Deane and W. A. Cole. *British economic growth, 1688–1959* (Cambridge, 2nd ed, 1967), 115.

less than 10% if sixteen of the townships within the large parish of Leeds be excluded). Wolverhampton (the town, not the parish) and Stoke on Trent together had only about 12% of the population of Staffordshire. There is, moreover, an accidental element in any figures for migration based upon counties, because the migrants ignored county boundaries. Birmingham, on the margins of Warwickshire, drew many of its immigrants not only from its own county but from neighbouring Staffordshire and, to a less extent, from Worcestershire and Shropshire;[1] so must Manchester from the West Riding; whereas Leeds, centrally placed in the West Riding, was less likely to attract immigrants from outside its county. Whatever the complications and imperfections of these figures, it would seem that industrial expansion as such was associated with high rates of natural increase and that urban expansion as such (especially in

[1] R. A. Pelham, 'The immigrant population of Birmingham, 1686–1726', *Trans. Birmingham Archaeol. Soc.*, LXI (1937), 45–80. See also W. H. B. Court, *The rise of the Midland industries, 1600–1838* (Oxford, 1938), 48–50.

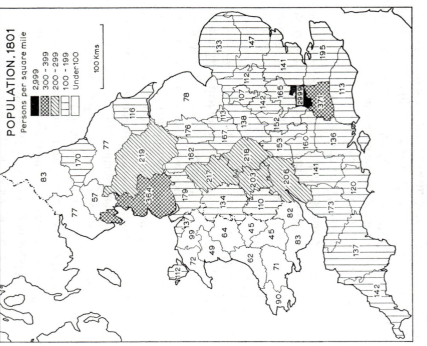

POPULATION.1801
Persons per square mile

- 2,999
- 300 - 399
- 200 - 299
- 100 - 199
- Under 100

100 Kms

Fig. 67 Population in 1801 (by counties)
Based on *Census of 1801: Enumeration*, 451, 496 (P.P. 1802, vii).
For the distribution by registration districts, see Fig. 83 on p. 393 below.

view of high urban death rates) was associated with migration. Contemporary writers, such as J. Massie in 1758, pointed to the continual movement of population from the country districts and smaller towns to the larger towns;[1] and Arthur Young in 1774 exclaimed: 'Let any person go to Glasgow and its neighbourhood, to Birmingham, to Sheffield, or to Manchester...how then have they increased their people? Why, by emigrations from the country.'[2] Most of the Sheffield apprentices, for example between 1624 and 1799, 'sons of yeomen, farmers and labourers',

[1] J. Massie, *A plan for the establishment of charity houses* (London, 1758), 99.
[2] A. Young, *Political arithmetic* (London, 1774), 63.

came from villages within fifteen miles of the town.[1] The Board of Agriculture report on Warwickshire (1793) in speaking of the effects of enclosure, said that 'the hardy yeomanry of the country villages have been driven for employment into Birmingham, Coventry and other manufacturing towns'.[2] The Leicestershire report also spoke of depopulation for the same reason and of families having to 'migrate into towns'.[3] Other county reports contain similar statements.[4] It must be remembered, however, that the relationship between enclosure and population change was much debated at the time and has also been much debated since then.[5] There was, moreover, considerable short-distance mobility 'in local populations in several areas in the south and east of England', that is outside the growing industrial centres.[6]

It is not surprising that the phenomena of the time were being discussed by contemporaries. The first edition of T. R. Malthus's *Essay on the principle of population* appeared in 1798 and it explained the growth of population as a response to economic opportunity. But looking backwards, the relationship between the two is seen to be extremely complicated. Can we be sure that the rise in birth rate did not precede the period of rapid industrial growth? H. J. Habakkuk has formulated a number of pertinent questions. 'Did the Industrial Revolution create its own labour force? Or did the vagaries of disease and weather produce an additional population that either stimulated an Industrial Revolution or had the luck to coincide with one independently generated?'[7] Attempts to answer such questions are impeded by uncertainty about the chronology of population growth in relation to economic growth. Moreover, the continuous interaction between the two makes it difficult to separate cause from effect. We may be clearer about these matters when there are more local studies

[1] E. J. Buckatzsch, 'Places of origin of a group of immigrants into Sheffield, 1624–1799', *Econ. Hist. Rev.*, 2nd ser., II (1950), 306.

[2] J. Wedge, *General view of the agriculture of the county of Warwick* (London, 1794), 40.

[3] W. Pitt, *General view of the agriculture of the county of Leicester* (London, 1809), 16.

[4] T. Davis, *General view of the agriculture of the county of Wiltshire* (London, 1794), 88; J. Holt and R. W. Dickson, *General view of the agriculture of the county of Lancashire* (London, 1815), 393.

[5] J. D. Chambers, 'Enclosure and labour supply in the industrial revolution', *Econ. Hist. Rev.*, 2nd ser., V (1953), 319–43.

[6] E. J. Buckatzsch, 'The constancy of local populations and migration in England before 1800', *Population Studies*, V (1951–2), 62–9.

[7] H. J. Habakkuk, 'The economic history of modern Britain', *Jour. Econ. Hist.*, XVII (1954), 500.

along the lines of those by J. D. Chambers, D. E. C. Eversley, E. A. Wrigley and others.[1]

THE COUNTRYSIDE

The new husbandry

Convertible husbandry. The eighteenth century has often been hailed as the great century of agricultural improvement, but many of its ideas and practices had long been anticipated. Not only were these ideas described by the numerous agricultural writers of the seventeenth century, but documentary evidence shows that they, or many of them, were being put into practice in a large number of localities throughout the country. Indeed it has been claimed that 'the agricultural revolution took place in England in the sixteenth and seventeenth centuries and not in the eighteenth and nineteenth'.[2] An important feature of the agriculture of these earlier centuries was the introduction of what has been variously described as convertible, alternate, field-grass or up-and-down, husbandry, that is what a later age was to call ley farming. Under this system, arable land was put under grass for a period in order to rest, and pasture land was ploughed up. Here was a breakthrough in farming techniques. The advantages of this for those with enclosed land had been described by Fitzherbert as early as 1523,[3] and they were set out at some length by Walter Blith about 1650.[4] By this time the practice had become widespread.

The impact of these new ideas upon open-field agriculture was also considerable because the routine of unenclosed farming was much more flexible than writers of the nineteenth and twentieth centuries sometimes supposed it to have been. In the sixteenth century there were Midland parishes where some strips in the open fields were put under grass for a few years, and where an intermixture of arable and ley could be en-

[1] J. D. Chambers, 'The vale of Trent, 1670–1800: A regional study of economic change', *Econ. Hist. Rev*, Supplement, No. 3, 1957; D. E. C. Eversley, 'A survey of population in an area of Worcestershire from 1660 to 1850 on the basis of parish registers', *Population Studies*, x (1957), 230–53; E. A. Wrigley, 'Family limitation in pre-industrial England', *Econ. Hist. Rev.*, 2nd ser., xix (1966), 82–109.

[2] E. Kerridge, *The agricultural revolution* (London, 1967), 15. See also E. L. Jones (ed.), *Agriculture and economic growth in England, 1650–1815* (London, 1967) For a convenient review with a useful bibliography, see D. Woodward, 'Agricultural revolution in England, 1500–1900: a survey', *Local Historian*, ix (1971), 323–37.

[3] J. Fitzherbert, *The boke of husbandrye* (London, 1523).

[4] W. Blith: (1) *The English improver* (London, 1649); (2) *The English improver improved* (London, 1652).

countered in each of three fields.[1] Different farmers grew different crops in the same field, and the introduction of leys enabled a more efficient use of fallow. On parts of some fallow fields, catch or 'hitch' crops were sometimes grown, leaving the fields bare for only part of the year; such hitch crops usually comprised peas, vetches and lentils. Thus Robert Loder of Harwell in Berkshire, in 1610–20, 'hitched' about a quarter of his fallow field every year.[2] The right of common grazing on the fallow fields did not prevent innovation, and village agreements indicate how local arrangements were devised to meet particular opinions and needs. Studies of such counties as Leicestershire,[3] Lincolnshire[4] and Oxfordshire[5] show how frequent were such arrangements by the early seventeenth century. In seventeenth-century Leicestershire, for example, 'within certain broad limits, the open-field farmer could do what he liked with his own strips'.[6] Such possibilities were aided by the exchange and consolidation of strips. It is true that on the eve of enclosure in the eighteenth century there were villages where, in William Marshall's phrase 'the spirit of improvement' lagged behind,[7] and where a fairly rigid open-field system persisted un-altered from medieval times; moreover, fallowing was still necessary on many heavy wet clays.[8] But there were also many open-field villages where rotations were flexible, where ley farming was important and where change and progress were very evident. Marshall's account of the Midland counties in 1790 set out clearly the advantages of current practice 'in keeping the land in grass and corn alternately'.[9]

New crops. The advance of convertible husbandry in the seventeenth and eighteenth centuries was bound up with the increasing range of field crops available to the farmer. Two main groups of crops provided possibilities of change – roots, on the one hand, and clover and grasses on the other.

[1] E. Kerridge (1967), 194 *et seq*; J. Thirsk, *The agrarian history of England and Wales*, vol. IV, 1500–1640 (Cambridge, 1967), 178.
[2] G. E. Fussell (ed.), *Robert Loder's farm accounts, 1610–1620*, Camden Soc., 3rd ser., 53 (London, 1936), 11, 41, 91–8, 103–4.
[3] W. G. Hoskins, *The Midland peasant* (London, 1957), 152 *et seq*.
[4] J. Thirsk, *English peasant farming* (London, 1957), 99 *et seq*.
[5] M. A. Havinden, 'Agricultural progress in open-field Oxfordshire', *Agric. Hist. Rev.*, IX (1961), 73–83.
[6] W. G. Hoskins, 'The Leicestershire farmer in the seventeenth century', *Agricultural History*, XXV (1951), 9–20.
[7] W. Marshall, *The rural economy of the West of England*, I (London, 1796), 106–7.
[8] E. Kerridge (1967), 27–8.
[9] W. Marshall, *The rural economy of the Midland counties*, I (London, 1790), 184, 187.

Both groups of crops had been mentioned by Barnaby Googe in a book (1577) which was largely a translation of a work by Conrad Heresbach of the Low Countries, where the most advanced farming in Europe was to be found.[1] The new field crops enabled a greater number of stock to be carried through the year, an important fact in an age without efficient fertilizers. The number of sheep, in particular, greatly increased. More animals meant more manure for the more flexible rotations. Root crops were of value to the land as clearing crops; and clover and the leguminous grasses added nitrogen to the soil and so further increased its fertility.

Carrots and turnips had already been grown in gardens in England, and there are references to both as garden crops grown by Dutch immigrants after 1565 outside the city of Norwich.[2] About 1600 there are indications that carrots were grown in fields in the Suffolk Sandlings; and, after about 1670, another centre of carrot growing developed in the vale of Taunton Deane in Somerset.[3] Carrots also spread elsewhere but not to the same degree. Turnips, on the other hand, became widespread over much of south-eastern England. They appeared as a field crop relatively suddenly in High Suffolk in the middle of the seventeenth century.[4] They extended into Norfolk and into the Chilterns long before 1700; and in 1724 Daniel Defoe could speak of turnip cultivation as being 'spread over most of the south and east parts of England'.[5] By the middle of the eighteenth century the turnip had reached the Midlands and the north, although it was not extensively cultivated. On heavy soils, dwarf rape or cabbages formed an alternative to turnips; they also had appeared as field crops before 1700. Another alternative was swedes, but not until well into the eighteenth century.

In the meantime, about 1650, Sir Richard Weston of Sutton Court in Surrey, who had travelled in the Low Countries, was advocating the field cultivation not only of turnips but of clover and grasses.[6] Weston's

---

[1] Barnabe Googe, *Foure bookes of husbandrie by M. Conradus Heresbachius....Newly Englished and increased by B. Googe* (London, 1577).

[2] K. J. Allison, 'The sheep-corn husbandry of Norfolk in the sixteenth and seventeenth centuries', *Agric. Hist. Rev.*, v (1957), 27.

[3] E. Kerridge (1967), 268–9.

[4] E. Kerridge: (1) 'Turnip husbandry in High Suffolk', *Econ. Hist. Rev.*, 2nd ser., VIII (1955–6), 390–2.

[5] Daniel Defoe, *Tour*, I, 58.

[6] R. Weston, *A discours of husbandrie used in Brabant and Flanders, shewing wonderful improvement of land there* (London, 1605). The imprint date seems to be an error, maybe for 1650.

advocacy was continued by Andrew Yarranton of Worcestershire, famous as a political economist. He wrote in 1663 on 'the great improvement of lands by clover', and said that since Weston's day some clover was sown 'in most counties', but that it had not become more popular owing to ignorance and mismanagement.[1] Even so, mixtures of varieties of clover with perennial rye-grass and other grasses became increasingly common as the seventeenth century passed into the eighteenth. A specialised treatise of 1671 on one of the grasses – sainfoin – was attributed to Sir John Pettus, and was described on its title page as 'being useful for all ingenious men'.[2] The advantage of turnips, clover and the grasses were set out by a variety of agricultural writers, by, for example, John Worlidge in a book first published in 1669 and reprinted with additions many times up to 1716.[3] The Royal Society of Arts, in the 1760s, attempted to encourage the improvement of clean grass seed by offering prizes;[4] and in 1790 William Curtis, a prominent botanist, wrote an account of British grasses which ran into several editions.[5]

The cultivation of other crops was also advocated by the writers of the time.[6] Among them were flax, hemp, woad, madder, saffron, liquorice, weld, rape, and they were grown locally to a greater or less degree. One new crop deserves special mention – the potato.[7] First introduced from the New World into Ireland in the 1580s, it was at first grown only in gardens. John Forster set out its many advantages in 1664;[8] it appeared as a field crop in rotations on the mosslands of Lancashire before 1700, but its spread beyond the northern parts of the realm was slow. By 1800 it seems to have been little cultivated in the south, except in gardens – in the market gardens of Middlesex and Essex, for example. This 'valuable

---

[1] A. Yarranton, *The improvement improved, by a second edition of the great improvement of lands by clover* (London, 1663). For this edition see G. E. Fussell (1947), 75.

[2] Anon, *St Foine improved: a discourse showing the utility and benefit which England hath and may receive by the grasse called St Foine* (London, 1671). No author's name appears on the title page, but the book has been attributed to Sir John Pettus.

[3] J. Worlidge, *Systema agriculturae* (London, 1669).

[4] H. T. Wood, *A history of the Royal Society of Arts* (London, 1913), 119.

[5] W. Curtis, *Practical observations on the British grasses* (London, 1790).

[6] J. Thirsk, 'Seventeenth-century agriculture and social change', *Agric. Hist. Rev.*, XVIII (Supplement, 1970), 148–77.

[7] R. Salaman, *The history and social influence of the potato* (Cambridge, 1949), 434 et seq.

[8] J. Forster, *Englands happiness increased...by a plantation of the roots called potatoes* (London, 1664).

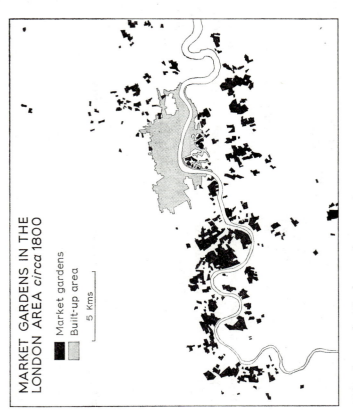

MARKET GARDENS IN THE
LONDON AREA *circa* 1800

▉ Market gardens
▒ Built-up area

|_____| 5 Kms

Fig. 68 Market gardens in the London area *circa* 1800
Based on T. Milne, *Plan of the cities of London and Westminster, circumjacent towns and parishes etc, laid down from a trigonometrical survey taken in the years 1795–1799* (London, 1800).

root', as William Marshall called it, had to wait until the nineteenth century for its wider cultivation.

Around London, in particular, agriculture was beginning to take on a suburban character, and market gardening spread to help feed the growing population. According to Samuel Hartlib, it 'began to creep into England' from Holland and Flanders around about 1600. In 1651 old men could still remember the first gardeners who came to Fulham and other places to grow a variety of vegetables – cabbages, cauliflowers, turnips, carrots and parsnips among them.[1] At the end of the eighteenth century, the Board of Agriculture reports on Middlesex and Surrey gave interesting accounts of the market gardens enriched by plentiful supplies of manure from London, and employed in raising vegetables for the

[1] S. Hartlib, *Legacie, or an enlargement of the discourse of husbandry used in Flanders and Brabant* (London, 1651), 9.

London market,[1] and Thomas Milne's map of 1798 shows how widespread they had become (Fig. 68).

Fertilisers. The spread of new methods and new crops was not paralleled by the discovery of new fertilisers. Writers, and the farmers they wrote about, were well aware of the importance of manuring the soil, but the substances they advocated and used were very largely the traditional ones. Most important was the dung of the sheepfold and the farmyard, but other manures were also in use, including oxblood, soot, peat, offal, bones, seaweed, soap ashes with the growth of soap-boiling, sea-shells, sheep's trotters, furriers' clippings and all kinds of waste and refuse including rags and old leather, but only rarely night-soil. Sir Hugh Plat in 1594 enumerated a large number of fertilising agents;[2] a second edition appeared in 1653, and many other writers of the seventeenth and eighteenth centuries also discussed the properties of various manures.

With the use of manures may be grouped the improvement of soils by the application of marl and of lime and chalk. Thin and sandy soils were improved by a dressing of clay marl while clay soils gained from a dressing of lime. These were not new practices, but they appear to have become increasingly current in the seventeenth century, and there was much discussion about methods in the literature of the time. Men like John Evelyn in 1675 were distinguishing between different kinds of soil and the treatment appropriate for each.[3] The progress that had been made by the end of the eighteenth century may be seen from the county reports of the Board of Agriculture, each of which, in a chapter headed 'Improvements', devoted a section to the problems of manuring, marling and liming.

But in spite of this discussion, and although in the 1790s bones were being ground instead of being crushed before being applied to the land, there was no fundamental advance in methods of fertilising because the limits to such an advance were set by contemporary knowledge of chemistry. As early as 1756 Francis Home had emphasised the dependence of agriculture upon chemistry in a book that was reprinted in 1759 and

[1] J. Middleton, *A view of the agriculture of Middlesex* (London, 1807), 328–38; W. Stevenson, *General view of agriculture of the county of Surrey* (London, 1809), 414–19. For a 'general view of the former and present state of market gardens' within twelve miles of London, see D. Lysons, *The environs of London*, 5 vols. (London, 1795–1810), IV, 573–6; V, 446–8.

[2] H. Plat, *The jewel house of art and nature* (London, 1594).

[3] J. Evelyn, *Terra: a philosophical discourse of earth* (London, 1675).

in 1776.[1] George Fordyce's book on the same subject was published in 1765[2] and its fifth edition appeared in 1796, the year in which the newly formed Board of Agriculture published his account of different kinds of manure.[3]

New implements. The spirit of improvement that was abroad in agriculture in the seventeenth and eighteenth centuries did not greatly manifest itself in the invention of new tools and machines. As with fertilisers so with implements, the pace was slow. A wide variety of traditional ploughs remained in use, but only here and there can improvement be discerned. About the year 1730 the so-called Rotherham plough, based upon Dutch designs, was developed at Rotherham in Yorkshire, and came into use in the north and east. Then in 1784 there appeared James Small's treatise on ploughs, one of a number of such treatises in the seventies and eighties; it described an improved design and remained a standard work for the next generation or so.[4] But such improvements, and there were others, came only slowly into general use. It was not until the closing years of the eighteenth century that iron ploughs were replacing wooden ones.[5]

The common method of sowing seed for long remained that of broad-casting. Various suggestions were made for 'setting,' or drilling seed at regular intervals, in order both to avoid waste and to benefit the ensuing crop – by Sir Hugh Plat in 1600,[6] by Edward Maxey in 1601,[7] by Gabriel Plattes in 1639[8] and by Josiah Worlidge in 1669.[9] There were various patents for mechanical sowers but they too came to little or nothing. Then about 1701 Jethro Tull invented a drill to sow seeds in evenly spaced rows sufficiently wide apart to allow the soil between them to be pulverished and cleared of weeds; but an account of his work was not published until 1731, after which there were many editions.[10] Tull thought that a combination of drilling and weed control would do away with the

[1] F. Home, *The principles of agriculture and vegetation* (Edinburgh, 1756).

[2] G. Fordyce, *Elements of agriculture and vegetation* (Edinburgh, 1765).

[3] G. Fordyce, *Plan for ascertaining the effects of different sorts of manures* (London, 1796).

[4] J. Small, *Treatise of ploughs and wheel carriages* (Edinburgh, 1784).

[5] C. Culpin, *Farm machinery* (London, 1938), 59–60.

[6] H. Plat, *The new and admirable arte of setting of corne* (London, 1600).

[7] E. Maxey, *New instruction of plowing and setting of corne* (London, 1601).

[8] G. Plattes, *A discovery of infinite treasure* (London, 1639).

[9] J. Worlidge, *Systema agriculturae; the mystery of husbandry discovered* (London, 1669).

[10] J. Tull, *The new horse-houghing husbandry* (London, 1731).

necessity of using a manure, and his work was followed by much controversy; but, whatever may be said, the fact remains that many have regarded his drill as the most important agricultural invention of the eighteenth century. Originally designed for grass and other light seeds, it came to be of especial value for sowing turnips. Improved varieties of drills followed, James Cooke's drill of 1783 being the most important;[1] but, if we may judge from the Board of Agriculture county reports, drills, although widely known, were not in general use even by the end of the century.

Improvements in threshing machines were suggested at various times in the eighteenth century; one of the most successful was Andrew Meikle's design of 1786, but it was not widely in use in the south until after 1800. In the meantime, with the quickening temper of the last two decades of the eighteenth century, numerous patents were taken out not only for sowing and threshing machines, but for harrows, winnowers, scarifiers, reapers, chaff-cutters, turnip-slicers and a variety of other mechanical aids.[2] But the fruits of this enterprise were not commonly available in 1800.

The improvement of livestock. The new fodder crops made possible the carrying of greater numbers of breeding stock than hitherto, and it is clear that in the seventeenth century many farmers were attempting to improve the quality of their sheep and cattle by importing rams and bulls from elsewhere. Although they worked with empirical methods, the care taken in selective breeding may be seen from the activities of men like Henry Best about 1640[3] and John Franklin about 1670.[4] By the early 1700s there were a number of well-known successful breeders of livestock, and in the 1720s Daniel Defoe could speak of the rich graziers of Leicestershire and of the Midlands generally.[5]

It is against this background that the work of Robert Bakewell (1725–95) must be viewed. He began his experiments at Dishley in Leicestershire

[1] J. Cooke, *Drill husbandry perfected* (Manchester, 1784); J. Horn, *The description and use of the new invented patent universal sowing machine, for broadcasting or drilling every kind of grain, pulse and seed* (Canterbury, 1786).

[2] R. E. Prothero, 'Landmarks in British farming', *Jour. Roy. Agric. Soc.* 3rd ser, III (1892), 27–8.

[3] C. B. Robinson (ed.), *Rural economy in Yorkshire in 1641, being the farming and account books of Henry Best of Elmswell* (Surtees Society, 1875), 1, 12, 20, 27–8, 75–6.

[4] T. B. Franklin, *British grasslands* (London, 1953), 90.

[5] Daniel Defoe, *Tour*, 11, 89.

in 1745. Hitherto, improvers of livestock had tried to obtain the qualities they wanted by outcrosses; Bakewell's method was to inbreed. His cattle became famous and his sheep – the New Leicester breed – even more so. He also experimented with the improvement of pigs and farm-horses. Dishley soon became a centre that attracted agriculturalists from all over the country. Although he left no account of his methods, his name became one of the best known in the history of British agriculture; and before his death others were applying the new methods of selective breeding to the various regional breeds of livestock in Britain.[1]

The strides made by animal husbandry, and indeed by agriculture in general, was reflected in cattle shows, sheep-shearings and ploughing matches. The Bath and West of England Agricultural Society was founded in 1777, and a large number of counties started experimental farms and societies for improvement; that many were short-lived does not alter the fact that their foundation bore witness to the new energy in the closing years of the eighteenth century.[2] In 1793, the Board of Agriculture was established, and its county reports published between 1793 and 1815 enable us to obtain an idea of recent improvement and of the state to which British agriculture had reached.[3]

Enclosing the arable

The enclosure that aroused so much opposition in the fifteenth and sixteenth centuries continued into the seventeenth.[4] There was an enquiry in 1607, into enclosure and depopulation within Leicestershire, Northamptonshire and five nearby counties.[5] The year 1630 saw another enquiry covering five of the Midland counties, and commissions were again appointed in 1632, 1635 and 1636 to restrict enclosure and to halt depopulation. Unrest was also reflected in the pamphlet literature of the time – in such tracts as Robert Powell's *Depopulation arraigned* (1636) and John Moore's *The crying sin of England* (1653) which was none other than 'such as doth unpeople towns and uncorn fields'. But the arguments

[1] R. Trow Smith, *A history of British livestock husbandry, 1700–1900* (London, 1959), 45–69.
[2] Lord Ernle, *English farming past and present* (4th ed., 1927), 209.
[3] R. Mitchison, 'The old Board of Agriculture (1793–1822), *Eng. Hist. Rev.*, LXXIV (1959), 41–69.
[4] E. M. Leonard, 'The inclosure of common fields in the seventeenth century', *Trans. Roy. Hist. Soc*, XIX (London, 1905), 101–46.
[5] E. F. Gay, 'The Midland revolt and the inquisitions of depopulation of 1607', *Trans. Roy. Hist. Soc*, n.s., XVIII (1904), 195–244.

against 'the decay of tillage' and 'depopulating enclosures' began to lose their force in the latter half of the seventeenth century. Friendly arrangements for enclosure became more frequent and its advantages more clear. The last anti-enclosure bill was rejected by the House of Commons in 1656, and in the same year Joseph Lee was showing how many Leicestershire villages had been enclosed without decay of tillage and without depopulation.[1] Writers of the new age began to find virtues in enclosed land. Samuel Fortrey in 1663 argued that only by enclosure could land be put to the use for which it was best suited, and other arguments in favour of enclosure were set out by such writers as John Worlidge (1669), John Mortimer (1707) and John Lawrence (1726).[2]

The enclosure that took place during these years was by private agreement under which the body of commoners in a village appointed commissioners and surveyors to allot the new holdings. Confirmation of the agreement was then sought from the Court of Chancery or Court of Exchequer. Such a procedure required the unanimous consent of all commoners concerned. It was sometimes possible to buy out individuals, but not always; and to overcome this difficulty, recourse was made to private Acts of Parliament, which became increasingly numerous from the 1720s onwards. These differed from decrees in Chancery in that they could initiate enclosure and compel dissentients to agree.[3]

At this point in time, on the verge of the great changes brought about by parliamentary enclosure, we may pause to ask what had been the effect upon the landscape of the enclosing activity of the years since, say, 1450. In view of the outcry about depopulation in the Midland counties, and in view of the flood of pamphlets, the sustained legislative activity and the many deserted villages, it may seem strange to find that, after all, the Midlands was the main area of parliamentary enclosure. The counties which had produced such a volume of complaint in Tudor times were the very ones in which open fields flourished triumphantly right on into the eighteenth and even into the nineteenth century. How is the paradox to be explained?

It is to be explained by the fact that much of England outside the Midlands had already been enclosed before the fifteenth and sixteenth

[1] J. Lee, *A vindication of a regulated enclosure* (London, 1656).
[2] S. Fortrey, *England's interest and improvement* (Cambridge, 1663); John Worlidge, *Systema agriculturae* (London, 1669); John Mortimer, *The whole art of husbandry* (London, 1707); John Laurence, *A new system of agriculture* (London, 1726).
[3] See E. C. K. Gonner, *Common land and inclosure* (London, 1912), 51 *et seq.*; W. E. Tate, *The English village community and the enclosure movements* (London, 1967), 46 *et seq.*

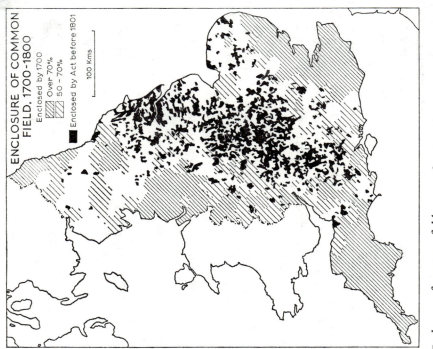

Fig. 69. Enclosure of common field, 1700–1800
Based on: (1) E. C. K. Gonner, *Common land and inclosure* (London, 1912),
map C; (2) G. Slater, *The English peasantry and the enclosure of common fields*
(London, 1907), 73.

centuries, and that these 'old enclosures', and the economy they expressed,
often went back to much earlier times. The agrarian changes of 1450–
1650 thus brought little or no depopulation to such counties as Essex,
Kent, Devon, Hereford and Worcester. That these changes, on the other
hand, brought 'serious suffering and disturbance' to the grain counties of
the Midlands did not mean that all the Midland open fields had disappeared.[1]
It was in fact far otherwise. A variety of contemporary evidence shows
that much, indeed most, of the Midlands was still in open field at the
beginning of the eighteenth century. John Morton, the historian of

[1] R. H. Tawney, *The agrarian problem in the sixteenth century* (London, 1912), 265.

11-2

Table 7.2 *Parliamentary enclosure of open fields*

Before 1750	64	1800–09	574
1750–59	87	1810–19	422
1760–69	304	1820–29	101
1770–79	472	1830–39	75
1780–89	150	1840–49	84
1790–99	398	1850 and after	106

The number of acts for each decade is calculated from G. Slater, *The English peasantry and the enclosure of common fields* (London, 1907), 268–313.

Northamptonshire, for example, wrote in 1712 that 'the enclosures lie dispersedly up and down in the county . . . Yet far the greatest part of the county is still open.'[1] The same could be said of other counties. It was not until well into the eighteenth century that enclosure began to transform the Midland landscape on a broad scale (Fig. 69).

The general pace of open-field enclosure, decade by decade, can be seen from Gilbert Slater's figures which are certainly underestimates.[2] There was a total of at least 2,800 such acts, and about one half of these had been passed by 1800. An attempt has been made to show that the intensity of enclosing activity was associated with variations in interest rates and agricultural prices; the 1780s, for example, was largely a decade of high interest rates and difficult conditions of borrowing and also one of relatively little enclosure.[3] But such correlations must not be pressed too closely.[4]

The physical operation of enclosure must often have been a most complicated one. To survey one, two or three thousand interlocking places, noting the varying title of each; to value and then to redistribute; to set out roads; to reorganise the drainage of the fields; to settle conflicting claims; and finally to proclaim an award – all involved long effort that sometimes stretched out over five or six years or more.[5] It was,

[1] J. Morton, *The natural history of Northamptonshire* (London, 1712), 15.
[2] W. E. Tate, 88.
[3] T. S. Ashton, *An economic history of England: the eighteenth century* (London, 1955), 40–1.
[4] J. D. Chambers and G. E. Mingay, *The agricultural revolution, 1750–1880* (London, 1966), 82–4; R. A. C. Parker, *Enclosures in the eighteenth century* (Hist. Assoc., London, 1960), 5.
[5] M. W. Beresford, 'Commissioners of enclosure', *Econ. Hist. Rev.*, XVI (1946), 137.

moreover, an expensive operation.[1] What the enclosure commissioners did for a village was to lay out its territory anew and so recreate much of its geography. The effect upon the scenery of much of England may be summed up under two main headings – the conversion of arable to pasture and the spread of the hedgerow.

Not all enclosures meant conversion to pasture, but many did. The changes in one locality as compared with another depended upon a variety of local circumstances and upon local initiative, but it was enclosure that provided opportunities for land to be used in the way best suited to its soil and climate.[2] The transformation that took place in the vale of Belvoir, in Leicestershire, is instructive. Gabriel Plattes, in 1639, had described it as the best corn land in Europe,[3] yet in 1809 the Board of Agriculture county report could say: 'the richest land in the vale, formerly tillage, has been laid to grass; and the poorer land up the hills, and the skirtings of the vale, formerly a sheep-walk, have been brought into tillage'.[4] The enclosure of twelve parishes in the vale had taken place between 1766 and 1792, and their stiff Lias clay soils were now proving most profitable under grass, while the light Marlstone soils of the bordering upland were taking on a new value with the agricultural advances of the time. What was true of the vale of Belvoir was true of Leicestershire as a whole. It was, wrote William Marshall in 1790, 'not long ago an open arable county including a proportion of cows and rearing cattle; now a continued sheet of greensward'.[5] What was true of Leicestershire was true of the Midland counties in general. As Arthur Young wrote in 1774:

The fact is this; in the central counties of the kingdom, particularly Northamptonshire, Leicestershire and parts of Warwic, Huntingdon and Buckinghamshires, there have been with 30 years large tracts of the open field arable under that vile course, 1 fallow, 2 wheat, 3 spring corn, inclosed and laid down to grass, being much more suited to the wetness of the soil than corn.[6]

Much of England came to look greener than before.

[1] J. M. Martin, 'The cost of parliamentary enclosure in Warwickshire', *Univ. of Birmingham Hist. Jour.*, IX (1964), 144–57. Reprinted in E. L. Jones (ed.), *Agriculture and economic growth in England, 1650–1815* (London, 1967).

[2] E. C. K. Gonner, 329; see also 236–7.

[3] G. Plattes, *A discovery of infinite treasure, hidden since the world's beginning* (London, 1639).

[4] W. Pitt, *General view Leicester* (1809), 14–15.

[5] W. Marshall, *Rural economy of the Midland counties*, 2 vols. (London, 1790), I, 193.

[6] A. Young, *Political arithmetic* (London, 1774), 148.

The Board of Agriculture reports made many references to these changes. The Norfolk report, after referring to the new husbandry on the light soils, went on to speak of 'the wonderful improvements' also to be found on clays and loams which sometimes became twice as valuable in pasture as they could ever be in arable; such improvements, the report added, were especially striking in the Midland counties.[1] The Northamptonshire report also referred to the new order; when enclosed, the lighter lands of the county were 'kept more in tillage' but the strong, heavy lands were 'generally laid down to permanent pasture'.[2] Or again, 'many of the open fields' of Warwickshire had been enclosed and 'converted into pasture', thus needing 'much fewer hands to manage them than they did in their former open state'.[3] Other county reports also mentioned depopulation.[4] There was much debate at the time about depopulation and the social effects of enclosure, and modern enquirers have continued the argument at length.[5]

Enclosure changed the appearance of the Midland landscape in yet another way. To our eyes the unenclosed countryside would have looked bare; there was little or nothing to break the sweep of the wind across the ploughed earth. With enclosure came the need for permanent fences, and the most convenient form of fence was the hedgerow made by planting quick or live cuttings usually of hawthorn. The enclosure acts laid down that new allotments of land were to be, as one act put it, 'well and sufficiently hedged and ditched' within an appointed time. The details varied from place to place, and the county reports of the Board of Agriculture, around about 1800, frequently devoted a good deal of space to the methods and problems of fencing. One result of the planting of hedgerows was to give the countryside the appearance of being much more wooded than it really was. As William Marshall put it in 1787, 'the eye seems ever on the verge of a forest, which is, as it were by enchantment, continually changing into inclosures and hedgerows'.[6] In some localities, in the Cotswolds or

[1] N. Kent, *General view of the agriculture of the county of Norfolk* (London, 1794), 73–4.

[2] J. Donaldson, *General view of the agriculture of the county of Northampton* (London, 1794), 5.

[3] J. Wedge, *General view Warwick* (1794), 20.   [4] See p. 312 above.

[5] See, for example: (1) J. D. Chambers (1953), 319–43; (2) J. D. Chambers and G. E. Mingay, *The agricultural revolution, 1750–1880* (London, 1966), 77–105. For a convenient summary, with bibliography, see G. E. Mingay, *Enclosure and the small farmer in the age of the industrial revolution* (London, 1968).

[6] W. Marshall, *The rural economy of Norfolk*, 1 (London, 1787), 4.

the Isle of Purbeck for example, hedgerows were replaced by stone walls built from the readily available material.

Sixteenth- and seventeenth-century topographers had drawn a distinction between the 'woodland' and the open or 'champaign' parts of England, between the old enclosed counties such as Essex, Kent, Devon, Worcestershire and the open unhedged counties of the Midlands. This distinction was now being obliterated by parliamentary enclosure, but not completely so because the layout of the new hedgerows, and of the fields they bounded, often presented a somewhat different aspect from that of the lands of old enclosure. In the latter were to be found small circular or irregular shaped fields. In the former, on the other hand, the fields were usually larger and were often bounded by straight lines. They bore the mark of conscious planning and of the surveyor with his chain. It is interesting to reflect what the English landscape would look like had wire fencing been available during the age of parliamentary enclosure.

Changes in the woodlands

The clearing that had taken place in the Middle Ages still left England with much woodland, but its extent was getting smaller, and the demands upon it were multiplying with the increase in population and the quickening of economic activity.[1] One great devourer of trees was industry. The lead and copper mines of the Lake District and the tin mines of Cornwall, for example, needed pit props for mining and charcoal for smelting. Glass-making, too, involved 'a continuale spoile of woods'; so did tanning, salt-making and a host of other industrial activities. Perhaps the greatest demand was that made by the iron industry of the Weald, the west Midlands and the Forest of Dean. There were many attempts during the seventeenth century to use coke instead of charcoal for smelting; and, after the success of Abraham Darby at Coalbrookdale in Shropshire about 1709, the iron industry began to free itself from dependence on wood.

Other demands on wood, however, were increasing. The expansion of England's mercantile marine, and the development of the navy, depended upon an adequate supply of oaks for the hulls of ships; fir trees for masts had to be imported from the Baltic and also from New England.[2] Of the

[1] H. C. Darby, 'The clearing of the English woodlands', Geography, XXXVI (1951), 71–83; H. G. Richardson, 'Some remarks on British forest history', Trans. Roy. Scottish Arboricultural Soc., XXXV (1921), 157–67; XXXVI (1922), 174–97.

[2] R. G. Albion, Forests and sea power (Cambridge, Mass., 1926), especially ch. 3, 'England's diminishing woodlands' (pp. 95–138).

many royal forests, the three main ones upon which the navy relied were the Forest of Dean, the New Forest and Alice Holt Forest in north-eastern Hampshire. The woods of private landowners were at times important suppliers, but they formed a somewhat uncertain source. The expansion of arable and pasture constituted another demand; and, about 1650, Walter Blith could speak of former woodlands 'which now inclosed are grown as gallant cornfields as be in England', and he mentioned western Warwickshire, northern Worcestershire together with Staffordshire, Shropshire, Derbyshire and Yorkshire.[1]

A number of Acts of Parliament in the sixteenth century had already tried to restrict the cutting of timber; but they do not seem to have been very effective, and by 1600 there were many complaints about the loss of woodland. Arthur Standish, in 1611, complained about the destruction during the preceding twenty or thirty years and about the fact that no replanting was taking place.[2] Many of these complaints may well have been exaggerated, and it has been argued that 'the universal timber and fuel crisis cannot have appeared very formidable over much of England by 1600 or even 1630'.[3] The situation seems to have grown worse in the decades that followed, and the outbreak of civil war in 1642 served to encourage indiscriminate destruction; those royalists who were fined had frequently to sell their oaks to meet the penalties; and woods on sequestrated estates were often sold to settle claims.[4] Camden's *Britannia*, in 1610, described the hills of Oxfordshire as clad with woods, but when Edmund Gibson edited that work a century or so later, in 1695, he added: this 'is so much alter'd by the late Civil Wars, that few places (except the Chiltern-country) can answer that character at present'.[5] It is clear that by the Restoration in 1660 the amount of woodland in England had been much reduced; that the reduction was continuing may be seen from the fact that between 1660 and 1667 over 39,000 trees were cut down in the Forest of Dean to feed the iron furnaces.[6]

[1] W. Blith (1649), 83.

[2] A. Standish, *The commons complaint, wherein is contained two special grievances: the first the general destruction and waste of woods in this kingdom . . .* (London, 1611), p. 1.

[3] G. Hammersley, 'The crown lands and their exploitation in the sixteenth and seventeenth centuries', *Bull. Inst. Hist. Research*, XXX (1957), 155. See p. 363 below.

[4] C. H. Firth (ed.), *Life of William Cavendish, Duke of Newcastle* (London, 1886), 149.

[5] W. Camden, *Britannia*, ed. by E. Gibson (London, 1695), 257.

[6] Evidence given in *Journals of the House of Commons*, XLIII (1787–8), 564–5; see R. G. Albion, 132.

The Navy Board, in their alarm over the timber shortage, consulted the Royal Society which had been founded in 1660; and this in turn asked a member of its council, John Evelyn, to examine the problem. His report, which appeared in 1664, surveyed the destruction of wood due to the demands of shipping, industry and tillage; and it appealed to the landed gentry to plant trees.[1] The book ran into many editions, and in that of 1679 Evelyn went so far as to claim that millions of trees had been planted as a result of his advocacy. To what extent Evelyn's claim was true has been debated, but it is clear that the trees planted during these years came to maturity in time to sustain the British navy through the wars of the eighteenth century, but the problem always remained acute. At the end of the Seven Years' War (1756–63) a Liverpool shipwright named Roger Fisher painted a very gloomy picture; many counties, he wrote, had only a quarter or even a tenth of the woods they had possessed forty years earlier.[2] Fir trees for masts were increasingly imported from the Baltic and North America. Concern for naval timber continued throughout the eighteenth century. In 1787, commissioners were appointed to enquire into the state of the royal forests and their seventeen reports were issued between 1787 and 1793. Most of them dealt with separate forests but some were of a general character, especially the eleventh (1792).[3] Its conclusion was that cultivation was crowding out woodland, that the woods were approaching general exhaustion, and that the government should plant in the royal forests.

In the middle of the Seven Years' War, in 1758, the Royal Society of Arts had begun to offer gold and silver medals to encourage the planting of timber.[4] The first recipient was the duke of Beaufort who was awarded a gold medal for sowing 23 acres in Gloucestershire with acorns. Awards continued to be made up to 1835, and, as a result, it has been estimated that well over 50 million trees were planted. One feature of the planting of the time was the widespread use of new species. The traditional timbers of England were the hardwoods – oak, ash, beech, elm – but, now, softwoods were introduced, although not to everyone's liking – the Scots pine, the larch, the spruce, and varieties of fir from abroad. Some of this planting was not for utility but for the adornment of the landscape parks

[1] J. Evelyn, *Sylva: or a discourse of forest trees, and the propagation of timber in His Majesty's dominions* (London, 1664).

[2] R. Fisher, *Heart of oak, the British bulwark* (London, 1763).

[3] R. G. Albion, 135–6 and 439.

[4] H. T. Wood, *A history of the Royal Society of Arts* (London, 1913), 145–50.

that were becoming such a feature of the eighteenth-century landscape. Very often, ornament and utility were combined. A general picture of what Arthur Young called 'the modern spirit of planting'[1] at the end of the century may be obtained from the sections dealing with 'Woods and Plantations' that appeared in the Board of Agriculture county reports. In Northumberland, for example, we are told that 'plantations, on an extensive scale', were 'rising in every part of the county', adding greatly to its 'ornament and improvement'.[2] It was the same in the south of the realm, and in Gloucestershire there were numerous planters 'who have skreened the bleak spots of the Cotswolds, and have improved the general face of the county'.[3]

Many of the county reports that tell of this new planting, tell also of the grubbing up of trees for cultivation. The author of the Gloucestershire report noted that 'in every year many acres of beech woods are destroyed and given up to the plough'.[4] Arthur Young had witnessed similar operations in Suffolk and thought that corn and grass were of much more value than timber.[5] Thomas Preston of Suffolk was equally emphatic: 'The scarcity of timber ought never to be regretted, for it is a certain proof of national improvement.'[6]

The making of water-meadows

Meadowland had always formed an important element in the economy of the countryside, and an acre of meadow was usually much more valuable than an acre of arable. That riverside meadows received much benefit from the overflowing of their streams was generally recognised; but in the late sixteenth and early seventeenth centuries an innovation in the management of meadows greatly increased the benefits of such inundation. This was the artificial flooding of meadows at appropriate times by irrigation channels so as to stimulate the growth of grass. The resulting crop of hay

[1] A. Young, *General view of the agriculture of the county of Norfolk* (London, 1804), 381.

[2] J. Bailey and G. Culley, *General view of the agriculture of the county of Northumberland* (London, 1794), 15.

[3] T. Rudge, *General view of the agriculture of the county of Gloucester* (London, 1807), 243.

[4] *Ibid.*

[5] A. Young, *General view of the agriculture of the county of Suffolk* (London, 1804), 166.

[6] *Journals of the House of Commons*, XLVII (1792), 343; see R. G. Albion, 119.

supported a large number of sheep, and these in turn provided substantial amounts of manure.[1]

Long known in Italy, artificial water-meadows were said to have been made in England by that curious Elizabethan figure, Sir Horatio Palla-vicino, who had bought an estate at Babraham to the south of Cambridge.[2] Early in the following century Rowland Vaughan published an account of the water-meadows he had made over some years in the Golden Valley in Herefordshire. The sight of 'a spring breaking out of a mole-hill with the grass very green where it ran' gave him, so he said, the idea that con-trolled flooding would be very good for grass, and he advocated the general adoption of the practice.[3] Vaughan's book was dedicated to the earl of Pembroke, and it may be more than chance that it was soon adopted on the Pembroke estates in Wiltshire.[4] The seventeenth-century topo-grapher of Wiltshire, John Aubrey, said that water-meadows had been constructed in the valleys of the Wylye and the Ebble about 1635, and in those of the Kennet and the Avon about 1646.[5] The technique spread to other downland valleys, not only in Wiltshire but in the nearby counties of Berkshire, Dorset and Hampshire. It became an essential feature of chalkland farming, and was integrated into the 'sheep and barley' hus-bandry of the district. Fed by day on the meadow, the sheep were folded at night on the arable and greatly enriched it with their manure.

Water-meadows were also to be found elsewhere in the first half of the seventeenth century. About 1638 Sir Richard Weston of Sutton in Surrey described how he had obtained great crops of hay from irrigated meadows.[6] Other writers, too, were advocating the new system. Walter Blith, about 1650, set out instructions for improving land by 'floating or watering'.[7] In 1669 John Worlidge of Petersfield in Hampshire could say that the practice had become 'one of the most universal and advantageous

[1] E. Kerridge (1) 'The sheepfold in Wiltshire and the floating of the water-meadows', *Econ. Hist. Rev.*, 2nd ser., VI (1954), 282–9.

[2] W. Gooch, *General view of the agriculture of the county of Cambridge* (London, 1813), 258.

[3] R. Vaughan, *Most approved and long experienced waterworkes, containing the manner of winter and summer drowning of meadow and pasture* (London, 1610).

[4] J. Thirsk (1967), 182.

[5] John Britton (ed.), *The natural history of Wiltshire by John Aubrey, F.R.S.* (written between 1656 and 1691) (London, 1947), 93 and 104.

[6] R. Weston, *A discours of husbandrie used in Brabant and Flanders, showing wonder-ful improvement of land there* (London, 1605). The imprint date seems to be in error, maybe for 1650.   [7] W. Blith (1649), 2.

improvements in England within these few years'.[1] We certainly hear of water-meadows in such counties as Worcester,[2] Warwick and Leicester where Robert Bakewell constructed some at Dishley,[3] but they were never very common in the east or the north of the country.

From the seventeenth right through the eighteenth century, water-meadows formed an important element in the agriculture of many areas, especially in that of the chalk country. Towards the end of the eighteenth century a number of treatises on 'the art of floating land' appeared,[4] and in the last decade or so an account of it was included under the heading of 'improvements' in many of the county reports of the Board of Agriculture. The outstanding description was in Thomas Davis's report on Wiltshire. As he said: 'none but those who have seen this kind of husbandry, can form a just idea of the value of the fold of a flock of ewes and lambs, coming immediately with bellies full of young quick grass from a good watermeadow, and particularly how much it will increase the quantity and quality of a crop of barley.'[5]

The writers of the time distinguished two main types of water-meadow – the flowing or 'floated' meadow and the catchwork meadow. The former type was characteristic of alluvial valleys across which channels were made to carry water from the main stream and then back to it at a lower level, the circulation being controlled by a system of hatches and weirs. The system varied in complexity according to the lie of the land, but Fig. 70 shows the kind of network encountered in the valley of the Avon below Salisbury.[6] Thus it was that many valleys, particularly those of the chalkland, became criss-crossed by a network of channels. The catchwork meadow was to be found on hillsides. Water from a spring or small stream was led by a new cut with but a slight fall along the side of

---

[1] J. Worlidge (1669), 16–17.

[2] R. C. Gaut, *A history of Worcestershire agriculture and rural evolution* (Worcester, 1939), 123–4.

[3] W. Marshall (1790), 284–6.

[4] G. Boswell, *An account of the advantage and method of watering meadows by art as practised in the county of Gloucester* (Cirencester, 1789; also 1790); T. Wright, *The art of floating land as it is practised in the county of Gloucester* (London, 1799); T. Wright, *On the formation and management of floated meadows* (Northampton, 1808); W. Smith, *Observations on the utility, form and management of water meadows* (Norwich, 1806).

[5] T. Davies, *General view Wiltshire* (1794), 30–1.

[6] H. P. Moon and F. H. W. Green, 'Water meadows in southern England' in F. H. W. Green, *Hampshire* in 'The land of Britain', ed. by L. D. Stamp, pt. 89 (London, 1940).

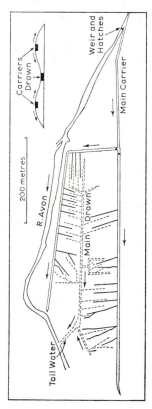

Fig. 70 Water-meadows along the River Avon (Hampshire)

Based on H. P. Moon and F. H. W. Green, 'Water meadows in southern England' in F. H. W. Green, *Hampshire* (L. of B., pt. 89, London, 1940), 375. Meadows are flooded by means of 'carriers' and then drained by means of 'drawns'.

a hill, and then allowed to escape, thus watering the slope before rejoining the stream below.

The draining of the marshes

The General Draining Act of 1600, 'for the recovering of many hundred thousand acres of marshes', not only marked the end of some decades of experiment, but was, in a sense, a promise of changes to come. Of all the many stretches of marsh in the kingdom, that of the peatland of the southern Fenland saw the most spectacular transformation.[1] At first, there was great opposition from those with vested interests in the fen pastures and in the fen streams. Many proposals were made in the early years of the seventeenth century, but nothing effective was done until in 1630–1 the fourth earl of Bedford, together with thirteen 'co-adventurers' contracted to drain the peat areas of the southern Fenland. They secured the services of the Dutch engineer, Cornelius Vermuyden, who had been at work upon the marshes of Hatfield Chase and Axholme to the north.[2] Under his direction, cuts, drains and sluices were made, the most important works being the two Bedford rivers that form so prominent a feature of the countryside today. Their activities were interrupted by disputes and by the Civil War, so that it was not until 1652 that the area now known as the Bedford Level was finally declared to be drained.

[1] H. C. Darby, *The draining of the Fens* (Cambridge, 3rd ed., 1968); L. E. Harris, *Vermuyden and the Fens* (London, 1953); R. L. Hills, *Machines, mills and uncountable costly necessities* (Norwich, 1967).

[2] J. Thirsk, 'The Isle of Axholme before Vermuyden', *Agric. Hist. Rev.*, I (1953), 16–28; L. E. Harris, 41–53.

The completion of the work was followed by great success, and on the newly reclaimed land there was not only 'all sorts of corne and grasse' but also such crops as rape, onions, flax, hemp, mustard, chicory and woad. But in spite of this 'deluge and inundation of plenty',[2] it was evident before the end of the century that all was not well. Neither Vermuyden nor his associates seemed to have realised that, as soon as it was drained, the surface of the peat would rapidly become lower, due partly to shrinkage and partly to the wasting of the peat owing to bacterial action. The result was that the peat surface was soon below the levels of the channels into which it was supposed to drain. Complaints about flooding became frequent; what had seemed a promising enterprise in 1652 had become a tragedy by 1700. The solution lay in the introduction of windmills for pumping, and it was this that saved the situation as the seventeenth century passed into the eighteenth. Windmills became increasingly frequent; and in the eighteenth century they, in their hundreds, gave a distinctive character to the fen landscape.

But the windmill was to prove unsatisfactory. It was at the mercy of gale and frost and calm, and as the peat surface became increasingly drier and lower, windmill drainage became more and more ineffective. The irony was that improved drainage only caused the peat surface to sink more rapidly, and therefore become increasingly liable to flooding. There was also difficulty over the outfalls of the fenland rivers. Observers at the end of the eighteenth century were shocked at the 'misery and desolation' before them. Large stretches of country were relapsing into 'waste and water' as may be seen on Vancouver's map of 1794.[3] A solution was not to appear until the use of steam-driven pumps about 1820.

In the meantime there had been much cutting of drains and embanking of land in other parts of England, but success was often only partial and was frequently followed by deterioration. A brief view of conditions in some other marshland areas must serve to give an idea of the state of affairs by 1800. There had been various improvements in the Fenland to the north of the Bedford Level but some had been only temporary; and, in spite of much discussion and activity especially after about 1760, East, West and Wildmore Fens in Lincolnshire were still stretches of marsh

[1] Sir William Dugdale in 1657; quoted in H. C. Darby (1968), 280.
[2] T. Fuller, *History of the University of Cambridge* (London, 1655), section v.
[3] C. Vancouver, *General view of the agriculture of the county of Cambridge* (London, 1794).

and water at the beginning of the nineteenth century.[1] Holderness and the Hull valley, again in spite of much activity after about 1760, remained subject to persistent flooding.[2] The Somerset Levels, after various vicissitudes, had been greatly improved during the last decades of the eighteenth century, as we may read in John Billingsley's county report for 1798, but his map shows that stretches of undrained bogs still remained,[3] and the area as a whole was for long subject to inundation.[4] Although there had been much piecemeal activity in Lancashire, a large-scale map of the county in 1786 could still indicate considerable stretches of coastal marsh and mossland.[5] But changes were afoot; thus in 1780 Rainford Moss was reclaimed and planted with potatoes; and other mosslands were soon to disappear under the terms of various enclosure acts. The draining of Trafford Moss was almost over by 1800, but that of the larger expanse of Chat Moss still remained undrained in 1800.[6] Along the coast, Martin Mere in the Fylde had been drained (not for the first time) in the 1780s, only to be reflooded in 1789.[7] Marton Mere to the east of Blackpool was also still a lake, but an act for its reclamation was passed in 1798. The meres of northern Shropshire, which were such a feature of John Rocque's map of 1752,[8] were also soon to disappear (Fig. 71). Joseph Plymley, in 1803, could report that much progress had taken place over the preceding twenty years,[9] and in 1801 had come an act for 'inclosing, draining and improving', yet other Shropshire moors.[10]

On the other side of the country the area of the Ouse–Trent lowland is of particular interest because not only draining but another operation

[1] W. H. Wheeler, A history of the fens of south Lincolnshire (Boston and London, 2nd ed. 1896).

[2] J. A. Sheppard: (1) The draining of the Hull valley (York, 1958); (2) The draining of the marshlands of south Holderness and the Vale of York (York, 1966).

[3] J. Billingsley, General view of the agriculture of the county of Somerset (London, 1798).

[4] M. Williams, The draining of the Somerset Levels (Cambridge, 1970).

[5] J. B. Harley (ed.), William Yates's map of Lancashire, 1786 (Historic Society of Lancs. and Cheshire, 1968).

[6] J. Holt, General view of the agriculture of the county of Lancaster (London, 1794), 86–103.

[7] H. Brodrick, 'Martin Mere', Eighth report of the Southport Society of Natural Science, 1902–1903 (Southport, 1903), 5–18.

[8] J. Rocque, Actual survey of the county of Salop (London, 1752).

[9] J. Plymley, General view of the agriculture of the county of Shropshire (London, 1803), 223.

[10] E. J. Howell, Shropshire (L. of B., pt. 66, London, 1941), 288–9.

— that of warping — was employed to make new land. The West Riding report, in 1799, said that warping had been tried to the west of Goole about fifty years earlier but that it had not been common for 'more than 20 or 25 years'.[1] Arthur Young, in the same year, described the process in some detail, and extolled the 'vast fertility' of the soil yielded by the silt-laden waters of the Humber.[2] Warping involved the controlled flooding of embanked areas at high tides; after depositing its silt the water was allowed to run off at low tides. When the salt had been leached out, the new soil yielded rich crops of wheat, beans and potatoes. The practice of artificial warping, Young said, was peculiar to this part of Britain, and could be extended with enormous profit to other districts.

This brief catalogue must stand for all the stretches of marsh in England where changes were afoot. We can generalise and say that the earlier half of the seventeenth century had seen a great burst of activity of which the Bedford Level was the chief triumph. The latter half of the eighteenth

---

[1] R. Brown, *General view of the agriculture of the West Riding of Yorkshire* (Edinburgh, 1799), 167.

[2] A. Young, *General view of the agriculture of the county of Lincoln* (London, 1799), 276–88. See (1) R. Creyke, 'Some account of the process of warping', *J.R.A.S.*, v (1844), 398–405; (2) J. Thirsk (1957), 230–1, 289–90.

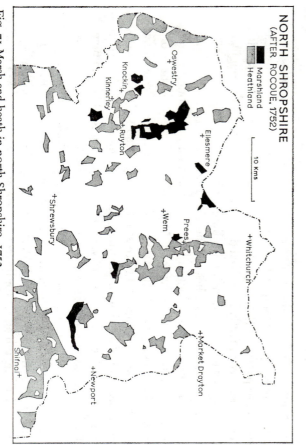

NORTH SHROPSHIRE
(AFTER ROCQUE, 1752)

Marshland
Heathland

10 Kms

Oswestry
Knockin
Kinnerley
Ruyton
Ellesmere
Wem
Prees
Whitchurch
Shrewsbury
Market Drayton
Newport
Shifnal

Fig. 71 Marsh and heath in north Shropshire, 1752
Based on John Rocque, *Actual survey of the county of Salop* (London, 1752).

century saw a quickening of intensive local activity, but this effort, too, was limited by the technical expertise of the time.

Under-draining the claylands

For the greater part of their history, the claylands of England were ill-drained, and water could often be seen standing upon their impervious stiff soils in winter (Fig. 143). Some drainage of surface water was secured by the furrows of the ridged-up fields,[1] but the general impression conveyed by Arthur Young's *Tours*, and by much writing of the eighteenth century, is that of countrysides water-logged in winter, and parched and cracked in summer.

The possibility of transforming this state of affairs, and of giving the claylands a new value – whether as arable or as pasture – was provided by under-draining.[2] By this practice, parallel trenches were cut across fields, bushes or stubble or loose stones were placed along the bottoms of the trenches which were then filled up with earth. The date at which such under-draining began to be practised is uncertain. It was not mentioned in Thomas Tusser's *Five hundreth pointes of good husbandry* in 1573, but in 1649 Walter Blith described how to make covered drains with the aid of stones and bushes.[3] Richard Bradley, in 1727, gave an account of a similar method of 'hollow ditching or draining', which he described as a recent invention 'chiefly practised in Essex'.[4] Arthur Young, in 1769, said that he had found the practice 'scarce any where but in Essex and Suffolk',[5] but it is clear that before 1750, and even earlier, it was also to be found at least in the nearby counties of Hertford and Norfolk.[6]

During the latter part of the eighteenth century, from about 1764 onwards, Joseph Elkington, a Warwickshire farmer, drained his sloping fields by means of trenches up to 5 feet below the surface of the

[1] E. Kerridge, 'A reconsideration of some farmer husbandry practices', *Agric. Hist. Rev.*, III (1955), 26–40.
[2] H. C. Darby, 'The draining of the English claylands', *Geographische Zeitschrift*, LIV (1964), 190–201.
[3] W. Blith (1649), 23–4.
[4] R. Bradley, *A complete body of husbandry* (London, 1727), 133–4.
[5] A. Young, *A six weeks tour through the southern counties of England and Wales* (London, 1768), 209; see also 67.
[6] P. Pusey et al., 'Evidence on the antiquity, cheapness, and efficacy of thorough draining, or land-ditching, as practised throughout the counties of Suffolk, Hertford, Essex and Norfolk', *J.R.A.S.*, IV (1843), 23–49.

ground;[1] and in 1797 he was awarded a grant of £1,000 for his work by Parliament. Even so, the principles of underground drainage were far from being understood, particularly in near-level fields of impervious clay.

An alternative method of draining heavy clay was by means of 'mole drains' which Stephen Switzer had described how to make as early as 1724.[2] By this method a trench was dug some 18 inches deep, and a rod of wood or a length of rope was laid along it; the trench was filled with clay, and the rod or rope was then withdrawn to leave an underground channel along which water could run. An improved way of making such channels was by means of a mole plough, and various types of mole ploughs were produced towards the end of the eighteenth century. These were drawn by horses or by means of a windlass and cable, and their purpose was to cut a channel below the surface of the ground by means of a sharp metal plug or 'mole' which had been inserted into a hole dug at the commencement of each drain.[3]

Drainage was very much in the minds of agriculturalists by the end of the century. The county reports of the Board of Agriculture included sections on 'Draining' in their chapter on 'Improvements'. The topics discussed included the proper depth of the drains, the distance between them, the most suitable material for filling them, and the various types of mole ploughs. The general tenor of the accounts is that while much had been done, still more remained to be done. As the Warwickshire report said: 'There is no improvement that can be made on land, productive of more salutary effects than that of draining.'[4]

The reclamation of the heathlands

The new crops and new methods of husbandry that were becoming common in the seventeenth century were of especial importance to the inherently infertile light soils of many parts of England. Turnips fed on the ground to sheep enabled the barren sands to be manured and also consolidated by treading; marling further improved the loose soil. Among the questions asked in 1664 by the 'Georgical Committee' of the newly formed

[1] J. Johnstone, *An account of the most approved mode of draining land: according to the system practised by Mr Joseph Elkington* (Edinburgh, 1797).

[2] S. Switzer, *The practical fruit-gardiner* (London, 1724), 25–6.

[3] A. Young, *General view of the agriculture of the county of Essex*, II (London, 1807), 194–201.

[4] A. Murray, *General view of the agriculture of the county of Warwick* (London, 1813), 147.

Royal Society was: 'And who they are if there be any in your County that have reduced *Heaths* into profitable Lands?'[1] No quantitative measure of the amount of reclamation that took place is possible, but we can obtain a broad picture.

Turnip husbandry appeared on the Norfolk heathlands during the years 1670–80, and by this time sainfoin and clover had also become common there.[2] By 1700 they were to be found in a variety of places, and references to the new husbandry became frequent in the topographical literature of the time. In Cambridgshire in 1701 Robert Morden was noting how sainfoin 'does wonderfully enrich the Dry and Barren Grounds of that county', presumably on the heathy chalk belt that stretches across the south of the county from Royston to Newmarket.[3] The same feature was to be found on other downlands.[4] In 1724 Daniel Defoe referred more than once to the change that had taken place in Hampshire, Wiltshire and Dorset. It was remarkable, he wrote, 'how a great part of these downs comes by a new method of husbandry, to be not only made arable, which they never were in former days, but to bear excellent wheat, and great crops too', all of which was done, he added elsewhere, by the folding of sheep upon the ploughed land.[5] As Thomas Davis wrote of Wiltshire at the end of the eighteenth century: 'The first and principal purpose of keeping sheep is undoubtedly the *dung of the sheepfold*.'[6]

The changes in Norfolk have received more attention than those elsewhere, due in part to the writings of Arthur Young. Here were the estates of Viscount Townshend (1674–1738) at Raynham, of Sir Robert Walpole (1676–1745) at Houghton,[7] and of Thomas Coke (1697–1755) at Holkham;[8] the last-name is not to be confused with that of his kinsman

---

[1] R. Lennard, 'English agriculture under Charles II: Evidence of the Royal Society's "Enquiries"', *Econ. Hist. Rev*, IV (1932), 25.

[2] E. Kerridge (1967), 273, 279, 283.

[3] R. Morden, *The new description and state of England* (London, 1701), 13.

[4] E. L. Jones, 'Eighteenth-century changes in Hampshire chalkland farming', *Agric. Hist. Rev*, VIII (1960), 5–19.

[5] Daniel Defoe, *Tour*, I, 187, 282, 285.

[6] T. Davis, *General view Wiltshire* (1794), 20.

[7] J. H. Plumb: (1) 'Sir Robert Walpole and Norfolk husbandry', *Econ. Hist. Rev.*, 2nd ser, v (1952), 86–9; (2) *Sir Robert Walpole* (London, 1956).

[8] C. W. James, *Chief Justice Coke and his descendants at Holkham* (London, 1929); R. Parker, 'Coke of Norfolk and the agricultural revolution', *Econ. Hist. Rev.*, 2nd ser, VIII (1955–6), 156–66.

Thomas William who succeeded to the estate in 1776. We must now recognise these not so much as innovators but as powerful advocates of 'the agriculture to which the general epithet of Norfolk husbandry peculiarly belongs'.[1] The name of 'Turnip Townshend' is particularly associated with the development of the so-called 'Norfolk Four Course', in which wheat, turnips, barley and clover followed one another in succession; but this was only one of a number of variant systems that involved turnips and seeds.[2]

The new rotations coupled with increased marling produced spectacular results. Arthur Young, in 1768, was loud in his praise:

All the country from Holkham to Houghton was a wild sheep-walk before the spirit of improvement seized the inhabitants; and this glorious spirit has wrought amazing effects; for instead of boundless wilds, and uncultivated wastes, inhabited by scarce anything but sheep, the county is all cut into inclosures, cultivated in a most husband-like manner.[3]

Allowance must be made for the exaggeration of enthusiasm, but other evidence also bears witness to great changes. Some idea of these can be gained from the diary of the François de la Rochefoucauld who visited East Anglia in 1784. At Massingham, reclamation had taken place, 'within the last thirty or forty years', and the owner had made 'immense sums'. At Dunton there was a prosperous farm of 1,600 acres which fifty years earlier had lain all uncultivated; the same was true of other farms, and la Rochefoucauld could only exclaim: 'The fertility of this land is wholly artificial.'[4] To this north-western area of Norfolk, Arthur Young gave the name of 'Good Sand', and this was adopted by many as a convenient regional designation (Fig. 72). It was 'good' in contrast to the infertile sandy area to the south, which was later known as the Breckland, and which stretches from Norfolk into Suffolk.[5]

Striking changes had also taken place during the eighteenth century on the light soils of Lincolnshire (Fig. 74). A picture of the transformation on the chalk wolds is given in the agricultural reports of Thomas Stone

1 A. Young, *General view Norfolk* (1804), 3.
2 N. Riches, *The agricultural revolution in Norfolk* (Chapel Hill, N.C., 1937), 76 *et seq.*
3 A. Young (1768), 21.
4 J. Marchand and S. C. Roberts (eds.), *A Frenchman in England, 1784* (Cambridge, 1933), 221, 228, 230.
5 The word 'Breckland' was coined by W. G. Clarke in 1894 – see W. G. Clarke, *In Breckland wilds* (Heffer, Cambridge, 2nd ed., 1937), 1 and 174.

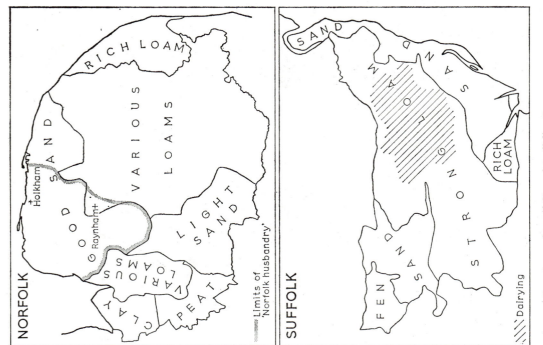

NORFOLK

RICH LOAM

SAND

+Holkham

⊙ Raynham+

GOOD

VARIOUS LOAMS

LIGHT SAND

VARIOUS LOAMS

CLAY

PEAT

Limits of
Norfolk husbandry.

SUFFOLK

SAND

LOAM

SAND

STRONG

RICH LOAM

FEN SAND

Dairying

Figs. 72 and 73 Soil regions of Norfolk and Suffolk, 1804
Based on: (1) A. Young, *General view of the agriculture of the county of Norfolk*
(London, 1804), map; (2) *General view of the agriculture of the county of Suffolk*
(London, 1804), map and (for dairies) p. 199.
It is impossible to provide an accurate scale.

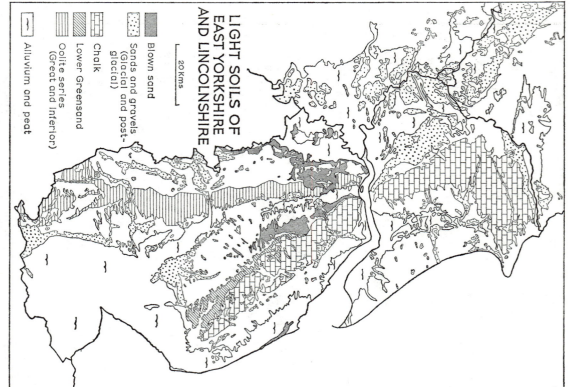

LIGHT SOILS OF
EAST YORKSHIRE
AND LINCOLNSHIRE

20 Kms

Blown sand

Sands and gravels
(Glacial and post-
glacial)

Chalk

Lower Greensand

Oolite series
(Great and Inferior)

Alluvium and peat

Fig. 74 Light soils of east Yorkshire and Lincolnshire
Based on *Geological Survey Quarter-Inch Sheets* 8, 12; *One-Inch Sheets* (*New
Series*), 31–5, 39, 40–4, 50–5, 59, 60, 62–5, 67–73, 76–8, 85–7, 100; *One-Inch
Sheets* (*Old Series*), 70, 83, 85, 86.

in 1794 and Arthur Young in 1799. 'Forty years ago', wrote Young, 'it was all warren from Spilsby to beyond Caistor; and by means of turnips and seeds, there are now at least twenty sheep kept to one before.' Elsewhere, he spoke of 'the immense and rapid progress turnips had made' in transforming 'the bleak wolds and heaths'. Even so, as he said, much warren still remained.[1] The oolitic belt that ran for forty miles north and south of Lincoln had also been changed. In the 1770s it was a tract of heath and gorse given over largely to rabbits; by 1799 it was a countryside of 'profitable arable farms' with turnips and seeds much in evidence,[2] although Young's account must be tempered by that of Thomas Stone who pointed time and again to poor management.[3] To the north, there were also changes on the Yorkshire Wolds, made especially by Sir Christopher Sykes who died in 1801. One local historian of 1798 was able to report that 'the land which formerly presented to the eye a dreary and uncultivated waste, now wears the appearance of an hospitable region'.[4]

Many other stretches of light soil elsewhere in the kingdom were also transformed in the eighteenth century. Samuel Rudder, the historian of Gloucestershire, could describe in 1779 how 'within the last forty years prodigious improvements' had taken place on the Cotswolds;[5] and Thomas Rudge summed up the changes of the century by describing how 'many thousand acres' of 'little more than furze and a few scanty blades of grass' had been turned into arable with the aid of turnips and seeds together with the 'treading and excrements of the sheep' folded upon them.[6] A similar story could be told, for example, of patches of glacial sand in northern Shropshire (Fig. 71), of the 'turnip soils' of Northumberland, and of parts of the Bunter and Keuper sandstone outcrops in the Midlands. Much of the 'Sandlings' in eastern Suffolk (Fig. 73) had become arable by 1784 with the help not only of turnips but also of carrots, for which the district became well-known.[7]

But there were limits to the magic of the new husbandry, and many tracts of barren hungry land remained unreclaimed. Such, for instance, were the Breckland with its blowing sands, the Dorset heathlands around

[1] A. Young, *General view Lincoln* (1799), 6, 225, 115, 382, 390, 393, 224, 12.

[2] *Ibid.*, 78, 136.

[3] T. Stone, *General view of the agriculture of the county of Lincoln* (London, 1794), 15, 32, 38, 46.

[4] T. Hinderwell, *The history and antiquities of Scarborough* (York, 1798), 265.

[5] S. Rudder, *A new history of Gloucestershire* (Cirencester, 1779), 21.

[6] T. Rudge, *General view Gloucester* (1807), 89, 133.

[7] J. Marchand and S. C. Roberts, 179–83.

Poole harbour, parts of the Suffolk Sandlings, stretches of the sandy
outcrops in Kent, Sussex and Surrey, and many other like areas, all
'calling loudly for improvement', as one county report of the Board of
Agriculture put it.[1] Even near London there were considerable stretches
of heath, rough grazing and commons. Nathaniel Kent could say in 1775
that 'within thirty miles of the *capital*, there is not less than 200,000 acres
of waste land'.[2] Hounslow Heath and Finchley Common, notorious
haunts of highwaymen, were only two of such tracts marked on William
Faden's map of 1802.[3]

Landscape parks and gardens

In so far as they owed their character to human effort, most of the landscapes of England, as elsewhere, were but the incidental by-products of
economic activity. But in some ages and in some localities people con-
sciously aimed at producing certain kinds of scenery. Such were the
ornamental gardens of the seventeenth and eighteenth centuries.[4] Gardens,
it is true, were not new in England. The Elizabethan period had seen
a considerable increase in the study and practice of making gardens with
flowers, fruit herbs and mazes, and early in the next century (1625) Francis
Bacon's essay 'Of Gardens' described the ideal form of a great garden
which should cover not less than thirty acres.[5] With the Restoration in
1660 this interest was not only greatly quickened but given a new accent.

When, in the 1690s, Celia Fiennes toured England on horseback, she
found a large number of newly built houses surrounded by gardens and
parks. A park with 'fine rows of trees' was coming to be regarded as
a necessary element in the dignity of a county seat.[6] Stimulated by John
Evelyn's *Sylva* (1664) people were planting not only for profit but for
ornament. Early in the next century, Daniel Defoe was struck by this new

---

[1] A. Young, *General view of the agriculture of the county of Sussex* (London, 1793), 8.

[2] N. Kent, *Hints to gentlemen of landed property* (London, 1775), 101.

[3] For this and John Rocque's map of 1754, see E. C. Willatts, *Middlesex and the London Region* (L. of B., pt. 79, London, 1937), 283 *et seq.*

[4] General accounts include: (1) H. F. Clark, *The English landscape garden* (London, 1948); (2) R. Dutton, *The English garden* (London, 2nd ed, 1945); (3) M. Hadfield, *Gardening in England* (London, 1960); (4) C. Hussey, *English gardens and landscapes, 1700–1750* (London, 1967); (5) H. C. Prince, *Parks in England* (Shalfleet, I.O.W., 1967).

[5] Francis Bacon, *Essayes or Counsels* (Everyman's Library), 137–43.

[6] C. Morris (ed.), *The journeys of Celia Fiennes* (London, 1947), 55, 67, 68, 151, 233 *et al.*

addition to the variety of the English scene. 'The alteration is indeed wonderful thro' the whole kingdom', he wrote. The fine houses 'surrounded with gardens, walks, vistas, avenues' were giving 'a kind of character to the island of Great Britain in general'.[1]

The gardens that were coming into being were of a very formal character. André le Nôtre (1613–1700), who planned the layout at Versailles, had many followers in England, and the influence of his geometrical designs was very evident. Avenues and walks were laid out in straight lines, frequently radiating from one point, parterres were arranged in stiff and symmetrical patterns, trees were cut and clipped with precision, geometry triumphed over Nature. Some of these formal layouts are well seen in the engravings of Johannes Kip, Thomas Badeslade and others around 1700.

Early in the eighteenth century many were beginning to criticise this formality. Joseph Addison, in 1712, in *The Spectator* essays poured scorn upon gardens 'laid out by the rule and line'; he preferred an 'artificial rudeness' to 'neatness and elegancy'.[2] In 1728, Batty Langley could think of nothing 'more shocking than a stiff regular garden'.[3] Such criticism was symptomatic of a new inspiration in gardening design. The restrictions of the formal style were being broken down, and the irregular garden was in turn to develop into the landscape park. The new fashion owed much to the long procession of travellers who made the Grand Tour over the Alps to Italy. They began to look around through the eyes of the Italian school of landscape painters; and they admired the wild scenery painted by Salvator Rosa, and the classical landscapes of Claude Lorraine and Gaspard Poussin with their ruined temples and broken columns.[4]

It was, apparently, the poet William Shenstone (1714–63) who invented the term 'landscape gardener', and said that 'the landskip painter is the gardiner's best designer';[5] his own ornamental farm at The Leasowes in Worcestershire attracted many visitors.[6] Alexander Pope (1688–1744), an important advocate of the new style, said that 'all gardening is landscape

[1] Daniel Defoe, *Tour*, I, 167.

[2] J. Addison, *The Spectator*, Essay 414 (1712).

[3] B. Langley, *New principles of gardening* (London, 1728), iv.

[4] E. W. Manwaring, *Italian landscape in eighteenth century England* (New York, 1925).

[5] W. Shenstone, 'Unconnected thoughts on gardening', an essay printed in *The works in verse and prose of William Shenstone Esq.*, II (London, 1764), 129.

[6] A. R. Humphreys, *William Shenstone* (Cambridge, 1937), *passim*.

painting';[1] and, towards the end of the century, Horace Walpole put it clearly; 'an open country is but a canvass on which a landscape might be designed'.[2] The making of a landscape park thus became an exercise in which stretches of grass, clumps of trees and expanses of water were important elements. Something of this new vogue can be seen in the works of Sir John Vanbrugh (1664–1726) and Charles Bridgeman (d. 1738), but the new freedom was brought to its fullest expression in the works of three great gardeners of the eighteenth century, William Kent (1685–1748), Lancelot Brown (1715–83) and Humphry Repton (1752–1818).

William Kent had been a painter, had spent several years in Italy and was a collector of Italian landscape paintings.[3] He turned from painting and, according to William Mason, 'worked with the living hues that Nature lent, and realised his landscapes'.[4] He used Bridgeman's device of the 'ha ha' or sunk fence to abolish the separation of garden and park, so bringing the latter right up to the walls of a house. He began the destruction of avenues; he replaced formal canals by serpentine streams; and he relieved large stretches of open lawn and grass by the planting of trees in clumps; he has been described as the inventor of the clump which was to feature so prominently in the English scene.[5] He introduced temples, colonnades and obelisks, often to terminate a vista. But alongside this freedom, there was still an element of formality in Kent's designs. They represented not so much Nature herself but, to use Horace Walpole's phrase, Nature 'chastened or polished'.[6]

Elegant informality was carried a step further under the leadership of Lancelot Brown.[7] Some 188 major landscape parks have been attributed to him and they included such magnificent examples as those around the great houses of Blenheim, Chatsworth and Luton Hoo. When consulted, he had the habit of saying: 'I see great capability of improvement here',

[1] J. Spence, *Anecdotes, observations, and characters, of books and men* (London, 1820), 144.

[2] H. Walpole, *The history of the modern taste in gardening*, first appeared in 1771, reprinted frequently. The reprint used here appears in I. W. U. Chase, *Horace Walpole: gardenist* (Princeton, 1943), and is accompanied by an essay; the quotation is from p. 37.

[3] M. Jourdain, *The work of William Kent* (London, 1948).

[4] W. Mason, *The English garden: a poem in four books* (London, 1782).

[5] H. F. Clark, plate 3.

[6] I. W. U. Chase (ed.), 27.

[7] D. Stroud, *Capability Brown* (London, 1950).

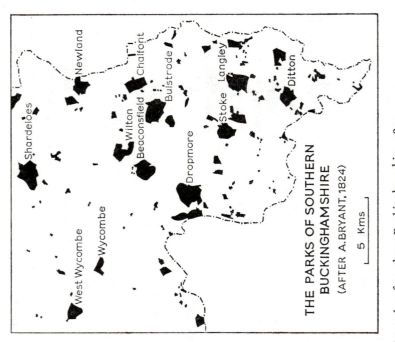

THE PARKS OF SOUTHERN
BUCKINGHAMSHIRE

(AFTER A.BRYANT, 1824)

5 Kms

Fig. 75 The parks of southern Buckinghamshire, 1824
Based on *Map of the county of Buckingham, from an actual survey by A. Bryant
in the year 1824* (London, 1825).
The scale of the original is 1½ inches to 1 mile.

and became known as 'Capability Brown'. Serpentine walks, stretches of
water, plantations and clumps of exotic as well as native trees now became
the fashion. The romantic style also appeared in other ways – in garden
temples, arches, obelisks, grottoes and follies of various kinds and in ruins
which, if they did not already exist, were constructed in a Gothic manner.
There was also another source of inspiration. Increasing contacts with the
East were reflected in the new vogue of chinoiserie, stimulated by Sir
William Chambers (1726–96) who wrote an account of oriental gardening in 1772.[1] Pagodas, Chinese temples and lattice-work bridges were
added to the embellishments of the English park. It was Chambers who
designed the ruined arch, the classical temple and the pagoda that can still

[1] W. Chambers, *A dissertation on oriental gardening* (London, 1772).

be seen in Kew Gardens; his Turkish mosque, Moorish alhambra and
Gothic church have disappeared. Here, too, must be mentioned the
introduction to the English scene of new shrubs, such as the magnolia
in 1688, the hydrangea in 1736, the buddleia in 1774, the fuchsia in 1788,
and varieties of rhododendron between 1736 and 1763.[1]

Capability Brown's style, his broad 'shaven lawns' and his repetitious
rounded clumps, provoked criticism, which became loud after his death
in 1783. Many people came to prefer not nature made elegant and 'beautiful', but what William Gilpin called the boldness and roughness of nature
herself;[2] and the new mode of the rugged and the 'picturesque'[3] was
advocated by such writers as Sir Uvedale Price (1747–1829)[4] and Richard
Payne Knight (1750–1824).[5] Amidst much contending discussion, the
third great gardener of the eighteenth century, Humphry Repton, tried
to steer a middle course between elegance and the picturesque.[6] His work
can be seen from his handsomely bound manuscript 'Red Books' which
discussed and showed landscapes before and after 'improvement'.[7] Over
some thirty years he 'improved' no fewer than 220 places, including
Regent's Park in London, and echoes of his work occur in Jane Austen's
Mansfield Park published in 1815.

Whatever the artistic differences among the landed gentry and professional gardeners of the eighteenth century, the consequences of their
work were very great for the English landscape. As Uvedale Price wrote
in 1794, these embellishments were giving 'a new and peculiar character
to the general face of the country'.[8] The effect was to be seen not only
around the great palaces of the realm – Blenheim, Chatsworth, Stowe and
the like, but also around what the county gazetteers were in the habit of
calling 'a neat mansion pleasantly situated in a park'. Small men as
well as great had followed Joseph Addison's advice and made 'a pretty

[1] A. M. Coats, *Garden shrubs and their histories* (London, 1963), *passim*.
[2] W. Gilpin, *Observations relative chiefly to picturesque beauty, made in the year 1772, on several parts of England* (London, 1786), I, xv.
[3] C. Hussey, *The picturesque: studies in a point of view* (London, 1927).
[4] U. Price, *An essay on the picturesque, as compared with the sublime and the beautiful* (London, 1794).
[5] R. P. Knight, *The landscape, a didactic poem* (London, 1794).
[6] H. Repton: (1) *Sketches and hints on landscape gardening* (London, 1795); (2) *The theory and practice of landscape gardening* (London, 1803); (3) *Fragments on the theory and practice of landscape gardening* (London, 1816).
[7] D. Stroud, *Humphry Repton* (London, 1962).
[8] U. Price, I.

landskip' of their possessions.[1] The county maps of the time, such as those by Bryant, Jeffreys, Faden and others, bear witness to the widespread distribution of these parks. In the Chilterns, for example, they were numerous (Fig. 75) but not more so than in some other districts (Fig. 90).

Mountains and moorlands

Topographical writers of the seventeenth century continued to echo William Camden's descriptions of the highland regions of England. Much of the Pennines, for example, was 'wild, solitary and unsightly', and Exmoor was 'a filthy barren ground'. Other adjectives included desolate, rugged, rocky and bleak. In 1696 Gregory King gave the following estimate of land use in England and Wales in millions of acres:[2]

Arable land	11
Pasture and meadow	10
Woods and coppices	3
Forests, parks, commons	3
Barren land etc.	10
Houses, gardens, churches etc.	1
Water and roads	1
	——
	39

The true area of England and Wales is 37.3 million acres and his total was therefore some 2 million too great; but it would seem that about one quarter of the country comprised what he described as 'heaths, moors, mountains and barren land'. We cannot apportion this between upland and lowland, but there is no doubt that by far the greater part of it consisted of upland country. Daniel Defoe in the 1720s made many references to waste, wild, barren, desolate uplands and frightful mountains.[3]

Fifty years or so later, Arthur Young drew attention to the many barren tracts in the realm, especially in the north:

From the most attentive consideration, and measuring on maps pretty accurately, I am clear there are at least 600,000 waste acres in the single county of Northumberland. In those of Cumberland and Westmoreland, there are as many

[1] J. Addison, *The Spectator*, Essay 414 (1712).
[2] Gregory King, *Natural and politicall observations and conclusions upon the state and condition of England* (1696). Reprinted in G. E. Barnett (ed.), *Two tracts by Gregory King* (Baltimore, 1936), 35.
[3] Daniel Defoe, *Tour*, e.g. I, 259, 263, 267 (Devon); II, 176 (Derbyshire), 185 (Yorkshire), 271 (Westmorland).

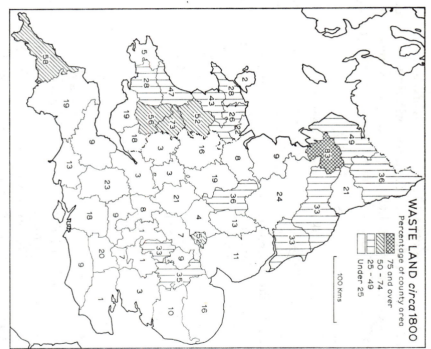

Fig. 76 Waste land *circa* 1800
Based on *General report on enclosures drawn up by order of the Board of Agriculture* (London, 1808), 139–41.

more. In the north and part (*sic*) of the west riding of Yorkshire, and the contiguous ones of Lancashire; and in the west part of Durham are yet greater tracts: you may draw a line from the north point of Derbyshire to the extremity of Northumberland, of 150 miles as the crow flies, which shall be entirely across waste lands; the exceptions of small cultivated spots, very trifing.[1]

He also found 'immense' tracts of waste in Devonshire and Cornwall.

During the eighteenth century, as earlier, there was much nibbling around the edges of the waste land, and old farms were being enlarged

[1] A. Young, *Observations on the present state of the waste lands of Great Britain* (London, 1773), 37.

Table 7.3 *Estimated enclosure of common pasture and waste*

Period	Acts	Acres (in 1,000s)
1700–60	56	75
1761–1801	521	752
1802–44	808	939
1845 and after	508	335
Total	1,893	2,101

or new ones created.[1] Here and there, improvement had taken place as a result of paring, burning and liming to produce good grass and even arable. Some 'taking in' of the waste was but temporary; tracts were tilled for a few years and then allowed to revert to rough pasture. Much of the moorland was used for grazing in summer, and came to be, in the latter part of the century, divided by miles and miles of drystone walls that brought a new element to upland landscapes.[2] Arthur Young, for example, described how a farmer at Dalton in the North Riding improved 'an extensive moor' during 1755–66. 'His first business was the inclosure, which he did by walling; the surface of the moor yielded, in some places, a sufficiency of stones for this work, but in many others pits were sunk for them.'[3] Or again in 1777, William Bray wrote of enclosure near Malham in the West Riding: 'Many of these pastures, which are of great extent, have been lately divided by stone walls.'[4] It seems that while enclosure of open fields was most rapid before 1800, enclosure of commons and waste was most rapid after 1800. Even so, Gilbert Slater's figures in table 7.3 show that at least some 800,000 acres or so of common waste land had been enclosed by 1801, mostly after 1760, and his figures are on the low side.[5]

[1] E.g. W. G. Hoskins, 'The reclamation of the waste in Devon, 1550–1800', *Econ. Hist. Rev.*, XIII (1943), 80–92; S. R. Eyre, 'The upward limit of enclosure on the East Moor of north Derbyshire', *Trans. and Papers, Inst. Brit. Geog.*, No. 23 (1957), 61–74.

[2] A. Raistrick, *The story of the Pennine walls* (Clapham, Lancaster, 1946).

[3] A. Young, *A six months tour through the north of England*, II (London, 1770), 433.

[4] W. Bray, *Sketch of a tour into Derbyshire and Yorkshire etc.* (London, 2nd ed., 1783), 308.

[5] For the figures in the table see: (1) G. Slater, *The English peasantry and the enclosure of common fields* (London, 1907), 267; (2) A. H. Johnson, *The disappearance of the small landowner* (London, 1909), 90; (3) W. E. Tate, *The English village community and the enclosure movements* (London, 1967), 88.

The Board of Agriculture, established in 1793, began at once to enquire into the condition of the waste lands, and each county report contained a chapter on 'Wastes'. Arthur Young summarised this information in 1795[1] and again in 1799.[2] These estimates were modified a little in a further report produced in 1808.[3] The total area of waste land was put at 7.8 million acres (6.2 in England and 1.6 in Wales), and its distribution among the counties is shown on Fig. 76. The overwhelming predominance of waste in those counties with much land over 800 feet above sea-level is very apparent. In the meantime, Sir John Sinclair had presented two reports in 1795 and 1797 on 'the means of promoting the cultivation and improvement of the waste, uninclosed, and unproductive lands of the kingdom'.[4] It was hoped that part at any rate might be turned into arable land, part into good grazing and part into woodland. But in spite of local changes, the uplands continued to be areas of rough grazing and peat bog. There were limits to 'the spirit of improvement' that so characterised the eighteenth century.

Although of little economic value, parts of the upland area were acquiring a new interest in the eyes of many travellers. The cult of the 'picturesque' which flourished in the latter part of the eighteenth century brought with it an appreciation of wild and rugged scenery. To the Lake District, in particular, came many writers and artists. The poet Thomas Gray had written a journal of his visit in 1769, and its publication in 1775 did much to make the district a fashionable tourist area. The first Guide to the Lakes, by Thomas West, appeared in 1778 and reached its seventh edition in 1799. By this time the district had been described by many others, including such well-known writers as William Gilpin, Thomas Pennant, Arthur Young and Mrs Ann Radcliffe.[5] But the writer who did most to give the Lake District a special place in English Literature was William Wordsworth, born in 1770 at Cockermouth nearby. His 'Guide to the Lakes' first appeared anonymously in 1810, and has been hailed by successive generations as a classic of geographical description.[6]

[1] A. Young, 'Waste lands, by estimation in Great Britain', Annals of agriculture, XXIV (1795), 10–17.

[2] A. Young, 'Waste lands', Annals of agriculture, XXXIII (1799), 12–59.

[3] General report on enclosures drawn up by order of the Board of Agriculture (London, 1808), 139–41.

[4] Reports from Committees of the House of Commons, IX (1803), 201–25.

[5] C. Hussey, 107, 126; E. W. Manwaring, 175, 181–3, 192–5; W. G. Collingwood, Lake district history (Kendal, 1928), 155–60.

[6] W. W. Merchant (ed.), A guide through the district of the Lakes . . . by William Wordsworth (London, 1951).

INDUSTRY

The industrial revolution

The term 'industrial revolution' was given currency by Arnold Toynbee, who placed the beginnings of the movement at about the year 1760.[1] Later writers, with thoughts of the continuity of history in mind, challenged the idea of any sudden and dramatic changes. Paul Mantoux, as early as 1906, held that 'in spite of the apparent rapidity of its development, the industrial revolution sprang from far-distant causes'.[2] The year 1660, rather than 1760, has been hailed as marking the turning point in economic growth and industrial expansion;[3] some writers have placed it even in the sixteenth century.[4] A number of studies of individual industries and in-dividual regions have shown the continuity of eighteenth-century changes and their relation to those of the seventeenth century.[5] It would seem, as Herbert Heaton said in 1932, that 'a revolution which continued for 150 years and had been in preparation for at least another 150 years may well seem to need a new label'.[6]

The application of statistical enquiry to historical generalisation has produced more exact estimates of the period of greatest change. W. G. Hoffmann, in calculating the rate of growth of total industrial output from 1700 onwards, came to the conclusion that the year 1780 was 'the approxi-mate date at which the annual percentage rate of industrial growth was first greater than 2, a level at which it remained for over a century'. The change was 'so definite that it clearly marks an epoch in the evolution of Britain's economy'.[7] T. S. Ashton also concluded that 'after 1782 almost every statistical series of production shows a sharp upward turn'.[8] W. W. Rostow advanced the view that the breakthrough, the

---

[1] A. Toynbee, *Lectures on the industrial revolution in England* (London, 1884).

[2] P. Mantoux, *The industrial revolution in the eighteenth century* (London, 1928), 25. English translation of French edition of 1906.

[3] C. Wilson, *England's apprenticeship, 1603–1763* (London, 1965), 185–205.

[4] E.g. J. U. Nef, 'The progress of technology and the growth of large-scale industry in Great Britain, 1540–1640', *Econ. Hist. Rev.*, v (1934), 3–24.

[5] E.g. A. P. Wadsworth and J. de L. Mann, *The cotton trade and industrial Lancashire, 1600–1780* (Manchester, 1931); W. H. B. Court, *The rise of the Midland industries* (Oxford, 1938).

[6] H. Heaton, 'The industrial revolution', *Encyclopedia of the Social Sciences*, VIII (New York, 1932), 5.

[7] W. G. Hoffmann, *British industry, 1700–1950* (Blackwell, Oxford, 1955), 30–2.

[8] T. S. Ashton, *An economic history of England: the 18th century* (London, 1955), 125.

DHG

'take-off into self-sustained growth', belonged in England to the years 1783–1802.[1] Others have viewed the acceleration as a two-phase process, beginning in the 1740s with a 'considerably sharper upward trend in the 1780s and 1790s'.[2] Yet others have suggested modifications of these dates.[3]

Not only the dates of 'the industrial revolution' but also its causes have produced a variety of opinion. They have been viewed as connected with agricultural change, with population growth, with the expansion of trade, with increasing demand, with the lowering of the rate of interest, with technological invention or with religious dissent. Clearly, the process of growth was complex, and the search for any single group of causes is fruitless. Many circumstances were involved, interrelated maybe but varying in importance at different times.[4]

Deane and Cole found 'little evidence of growth in the first four decades' of the eighteenth century.[5] To contemporaries it did not seem so. Daniel Defoe in the 1720s saw, or seemed to see, rapid economic develop-ment: 'New discoveries in metals, mines, minerals; new undertakings in trade; inventions, engines, manufactures, in a nation, pushing and improving as we are: these things open new scenes every day, and make England especially shew a new and different face in many places.'[6] To the early years of the century belong the Newcomen engine (1705–6), the smelting of iron by coke (1709), Kay's flying shuttle (1733), although one must distinguish between invention and its use in economic enterprise. Whatever the causes, the great period of technological change did not come until after 1760. In addition to the inventions of Hargreaves (1764), Arkwright (1769), Watt (1769), Crompton (1779) and Cort (1784), there were many other innovations by forgotten men who improved one

[1] W. W. Rostow, 'The take-off into self-sustained economic growth', *Economic Journal*, LXVI (1956), 25–48; *The stages of economic growth* (Cambridge, 1960); *The economics of take-off into sustained growth* (London, 1963).

[2] P. Deane and W. A. Cole, 58 and 280; P. Deane and H. J. Habakkuk, 'The take-off in Britain', being ch. 4 (63–82) of W. W. Rostow (1963).

[3] E.g. R. M. Hartwell, *The industrial revolution in England* (Hist. Assoc., London, 1966), 11.

[4] Convenient summaries may be found in the following: P. Deane, *The first industrial revolution* (Cambridge, 1965); C. Wilson, *England's apprenticeship, 1603–1763* (London, 1965); M. W. Flinn, *The origins of the industrial revolution* (London, 1966); R. M. Hartwell (ed.), *The causes of the industrial revolution in England* (London, 1967); R. M. Hartwell, *The industrial revolution and economic growth* (London, 1971).

[5] P. Deane and W. A. Cole, 280.

[6] Daniel Defoe, *Tour*, II, 133.

device or added to another. The number of patents sealed in each decade of the eighteenth century show the increase after 1760 and the acceleration after 1780.[1]

1700–9	22
1710–19	38
1720–9	89
1730–9	56
1740–9	82
1750–9	92
1760–9	205
1770–9	294
1780–9	477
1790–9	647

The interpretation of these figures, and the relationship between science and technology in general, have given rise to much discussion.[2] Many patents were of little or no value; others were followed by time-lags before they led to development; but at any rate, in T. S. Ashton's words, they 'may serve perhaps as a rough index of innovation'.[3]

Improvements and increased production were manifested in a whole range of miscellaneous industries – in the non-ferrous industries such as copper, brass, lead and tin; in the manufacture of sugar, candles, paper, soap and glass; in the making of pottery, when Josiah Wedgwood opened his works at Burslem in 1759; in brewing; in the chemical industries; and in salt-making, for rock salt had been discovered in 1670 during explorations for coal at a depth of about 100 feet near Northwich in Cheshire.[4] But the take-off into sustained growth between 1780 and 1800 was marked by outstanding developments in three sectors – the textile industry, the iron industry and the coal industry in association with the steam engine. Each of these three developments must be examined separately. Taken together they profoundly changed the material appearance of England.

The textile industries

Taking a broad view, the distribution of the woollen and worsted industries during the seventeenth and early eighteenth centuries was still

[1] B. R. Mitchell with P. Deane, 268.
[2] See the interesting collection of studies in A. E. Musson (ed.), *Science, technology, and economic growth in the eighteenth century* (London, 1972).
[3] T. S. Ashton, 'Some statistics of the industrial revolution in Britain', *The Manchester School*, XVI (1948), 214–34; the quotation is from the reprint of extracts in A. E. Musson (1972), 117.
[4] A. E. Musson and E. Robinson, *Science and technology in the industrial revolution* (Manchester, 1969); A. and N. L. Clow, *The chemical revolution* (London, 1952); L. Weatherill, *The pottery trade and north Staffordshire, 1660–1760* (Manchester, 1971).

very much the same as in the later Middle Ages – that is they were carried on in most parts of the country, but pre-eminently in three districts, East Anglia, the West Country and Yorkshire. A picture of this distribution in the 1720s may be obtained from Daniel Defoe's *Tour* (Fig. 77). In East Anglia there were two main textile districts – north-east Norfolk and the Suffolk–Essex border. In the former, the weavers of Norwich employed 'all the country round in spinning yarn for them'; likewise in the latter district, the weavers of Colchester, Sudbury and other places were sustained by the spinning wheels of the villagers nearby. In the West Country there were a large number of towns 'principally employ'd in the clothing trade'; and between them were 'innumerable villages, hamlets, and scattered houses, in which, generally speaking, the spinning work of all this manufacture is performed'. In Yorkshire, the industry was also dispersed over the countryside, several settlements merging into one another to give, in Defoe's words, the impression of 'one continued village' in the neighbourhood of Halifax.[1]

Some of the outlying textile centres declined during the seventeenth and eighteenth centuries. Salisbury and its woollen manufacture, for example, was said by Yarranton in 1677 to be 'much decayed of late years'; so was the industry of Worcester, Kidderminster and Bewdley,[2] and the textile workers of Reading were also reduced to 'a very small number' by 1695.[3] In Kent, Defoe tells us, the once 'very considerable cloathing trade' was 'quite decay'd'.[4] The same could be said of some other places. There were also changes within the three main textile districts themselves. The early product of the West Riding was a coarse cloth, but towards the end of the seventeenth century worsted cloth also began to be made there,[5] and the subsequent advance of the industry was reflected in the growth of Leeds, Bradford, Huddersfield, Wakefield and Halifax, all noted by Defoe.[6] By 1772, the value of the worsted cloth made in the West Riding equalled that made in Norwich and Norfolk generally.[7] Even so, the production of fine quality worsteds in the latter area was increasing most of the time, and absolute decline did not begin until the

[1] Daniel Defoe, *Tour*, I, 61, 17, 279–80; II, 194.
[2] A. Yarranton, *England's improvement by sea and land* (London, 1677), 207, 146, 162.
[3] W. Camden, *Britannia*, ed. by E. Gibson (London, 1695), 152.
[4] Daniel Defoe, *Tour*, I, 115.
[5] H. Heaton, *The Yorkshire woollen and worsted industries* (Oxford, 1920), 257, 263–8.
[6] Daniel Defoe, *Tour*, II, 185–204.
[7] J. James, *History of the worsted manufacture* (London, 1857), 285; E. Lipson, *The history of the woollen and worsted industries* (London, 1921), 241.

early decades of the nineteenth century.[1] To the south, however, the textile manufacture of the Suffolk–Essex towns and villages had definitely decreased before the middle of the eighteenth century. Philip Morant, the historian of Essex, said in 1748 that the trade of Colchester had 'removed in a great measure into the west and northern parts of this kingdom, where provisions are cheaper, the poor more easily satisfied, and coals are very plentiful'.[2] West Country industry was also beginning to run down in the eighteenth century. Its serges could not stand competition from the Norwich 'stuffs', and unemployment began to appear in a number of centres about 1750 or so.[3] Then towards the end of the century, competition from Yorkshire began to be felt. J. Collinson, for example, the historian of Somerset, writing in 1791, could point to the decline of woollen manufacture in many places – Milverton, Yeovil, Taunton and others.[4]

There had long been some small-scale textile manufacturing on the western side of the Pennines. Woollens were made in the upland valleys of east Lancashire and linens on the plain to the west; the latter were originally based on locally grown flax but were relying on Irish yarn by the sixteenth century. In the years around 1600 or so, the making of fustian became established between the woollen and the linen districts in an intermediate belt running through Bolton and Blackburn.[5] Fustian, made of linen warp and cotton weft, had been introduced from the Continent as part of the 'new draperies' in the sixteenth century, but had never flourished in East Anglia. In Lancashire, on the other hand, the fustian area extended at the expense of the woollen area. Moreover, cotton yarn began to be used increasingly in the making of other types of cloth until cotton manufacture soon became an industry in its own right, and Lancashire cottons began to challenge imported cotton fabrics from India. Progress was steady but slow until about the middle of the seventeenth century, after which the industry began to develop much more rapidly.

[1] J. H. Clapham, 'The transference of the worsted industry from Norfolk to the West Riding', *Econ. Jour.*, xx (1910), 195–210.

[2] P. Morant, *The history and antiquities of Colchester* (London, 1748), 75.

[3] W. G. Hoskins, *Devon* (London, 1954), 128–9; E. A. G. Clarke, *The ports of the Exe estuary, 1660–1860* (Exeter, 1960), 102 et seq.; J. de La Mann, *The cloth industry in the west of England from 1640 to 1880* (Oxford, 1971), 32 et seq.

[4] J. Collinson, *The history and antiquities of the county of Somerset*, I (Bath, 1791), 13, 204, 226.

[5] A. P. Wadsworth and J. de L. Mann, *The cotton trade and industrial Lancashire, 1600–1780* (Manchester, 1931), 11–25. See also G. W. Daniels, *The early English cotton industry* (Manchester, 1920).

An important element in this growth after 1750 or so was a series of inventions which transformed the making of textiles. These inventions were as relevant to wool as to cotton, but they were earlier and more readily adopted by the latter industry. They may be summarised as follows:

(1) The speed of weaving was greatly increased by the invention of John Kay's flying shuttle, patented in 1733, and widely adopted in the 1750s and 1760s. It upset the balance between weaving and spinning in that the capacity of the greatly improved weaving looms exceeded the amount of yarn that was being produced.

(2) Successive improvements in spinning came from the spinning jennies of Lewis Paul (1738) and James Hargreaves (1764) and especially from the water-frame of Richard Arkwright (1769). Arkwright's machine took its name from the fact that it could be operated by water power; and it hastened the transition from domestic to factory spinning. Samuel Crompton's spinning mule of 1779 marked a further advance, and the rate of cotton spinning now outstripped that of weaving. Subsequent modifications adapted the spinning machines for other fibres such as wool and flax.

(3) An attempt to redress the balance between spinning and weaving was made by Edmund Cartwright, whose powered weaving loom was patented in 1785. Various difficulties, however, prevented the successful use of power-driven looms until the early years of the following century. In 1789 Cartwright patented a wool-combing machine, but this likewise did not come into general use until after 1800.

Not only wool and cotton but two other branches of the textile industry were also involved in the changes of the eighteenth century. As early as 1589 William Lee of Calverton near Nottingham had invented a mechanical knitting frame which imitated the movements of hand-knitting in the making of stockings, but it met with opposition and Lee took refuge in France. After his death in about 1610, his companions returned to Nottingham with their frame-knitting machines. The industry spread into the adjoining counties of Derby and Leicester, and there was also a substantial development of frame-knitting in London.[1] The London industry began to decline, but that of the Midlands, aided by cheaper wage rates, expanded during the eighteenth century. There were also a number of innovations. In 1730, Nottingham was first making cotton stockings from imported Indian yarn.[2] Lee's frame was improved, especially by Jedediah Strutt of

[1] E. Lipson, *The economic history of England*, II (2nd ed, London, 1934), 104–9.
[2] J. D. Chambers, *Nottinghamshire in the eighteenth century* (London, 1932), 95, 114.

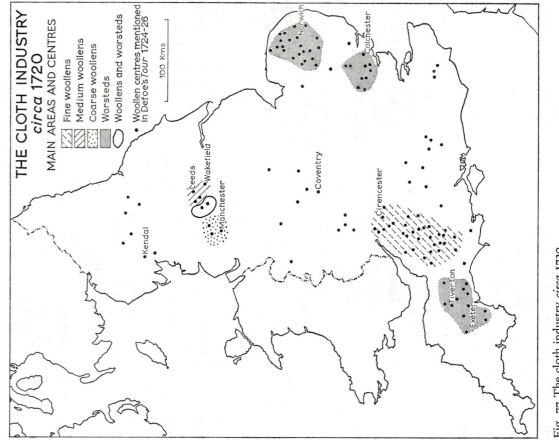

THE CLOTH INDUSTRY
circa 1720
MAIN AREAS AND CENTRES

Fine woollens
Medium woollens
Coarse woollens
Worsteds
Woollens and worsteds
Woollen centres mentioned
in Defoe's *Tour* 1724–26

100 Kms

Norwich
Colchester

Leeds
Wakefield
Manchester
Kendal

Coventry

Cirencester

Tiverton
Exeter

Fig. 77 The cloth industry *circa* 1720
Based on: (1) Daniel Defoe, *A tour through the whole island of Great Britain*
(London, 1724–7); (2) P. J. Bowden, *The wool trade in Tudor and Stuart
England* (London, 1962), 49.

Belper who made ribbed stockings in 1758; and there was also 'a multiplicity of smaller technical innovations'.[1] Richard Arkwright, coming from Lancashire in 1768, erected his first cotton-spinning mill, driven by horses, at Nottingham. Then, after patenting his water-frame in 1769, he started his first water-powered factory at Cromford in Derbyshire in 1771. Others were built a few miles away at Belper and Milford before his return to Lancashire in 1777. By this time, the London industry was surpassed by that of the Midlands where it was marked by a degree of specialisation – woollen stockings at Leicester, cotton at Nottingham and silk at Derby.

In the meantime, there had been developments in the silk industry. Its growth had been helped by Flemish weavers in the late sixteenth century and by the application of the knitting frame in the manufacture of silk hose. Its expansion was accelerated after the Restoration in 1660, and then through the skill of Huguenot refugees from France after the Revocation of the Edict of Nantes in 1685. A considerable colony of Huguenots settled at Spitalfields in the east end of London, but there were also others at Coventry, Macclesfield, Norwich and elsewhere. The manufacture was extended after 1717 when John and Thomas Lombe introduced water-driven silk-making machinery from Italy, and set up a factory on the banks of the Derwent at Derby. This large establishment has been regarded as the precursor of the factory system, and it anticipated Arkwright's use of water power. 'Thomas Lombe was to silk what Richard Arkwright half a century later was to cotton.'[2] But the market for silk was limited by its luxury nature and by the difficulty of exporting it to the older silk-manufacturing centres on the Continent and elsewhere. The great expansion of the textile industries was in the manufacture of woollens and, more especially, that of cotton.[3] Indian calicoes and muslins had long been in demand in Britain, and the home-produced version found a readymade market. In competition with other textiles such as linen and silk, the woollen and cotton markets soon widened to massive proportions.[4]

The various inventions, and the successive improvements upon them, had marked consequences not only for the growth of the textile industries but also for their geographical distribution. Spinning machines increasingly driven by water power were located along rivers, especially along the

[1] J. D. Chambers (1932), 35.
[2] E. Lipson, II (1934), 103.
[3] P. Deane (1965), 84–99.
[4] M. M. Edwards, *The growth of the British cotton trade, 1780–1815* (Manchester, 1967).

Table 7.4   *Water-driven spinning mills, 1788*

Lancashire	41	Hertfordshire	1
Derbyshire	22	Leicestershire	1
Nottinghamshire	17	Worcestershire	1
Yorkshire	11	Gloucestershire	1
Cheshire	8	Isle of Man	1
Staffordshire	7		
Westmorland	5	Flintshire	3
Berkshire	2	Pembrokeshire	1
Cumberland	1		
Surrey	1	Scotland	19

Source: Anon, *An important crisis in the callico and muslin manufactory of Great Britain, explained* (London, 1788).

steeply graded Pennine streams. As we have seen, the first was installed at Cromford in Derbyshire in 1771. A list of 1788 shows a total of 120 water-driven cotton-spinning mills in England, 4 in Wales and 19 in Scotland. There was a heavy concentration in Lancashire, with which may be grouped Cheshire, and there were appreciable numbers in Nottinghamshire, Derbyshire and Yorkshire.[1] It is not absolutely clear why the cotton industry should have been thus concentrated in Lancashire, although a number of reasons have been advanced – the linen–fustian background, the damp climate, the lime-free water and the availability of imports.

The water-power phase with its rural distribution was soon challenged by steam power.[2] The first steam engine to be used for cotton spinning seems to have been at Manchester in 1783 – at Richard Arkwright's cotton mill. A number of modified Savery and Newcomen engines were soon used to drive spinning mills in the 1780s and 1790s. In the meantime the first Boulton and Watt engine was used for spinning at Papplewick near Nottingham in 1785.[3] In the following year one was installed for the Cark Cotton Company in north Lancashire; the first in Manchester seems to have been installed in 1789, and in the following years its superiority began to be appreciated in spite of its expense. By 1800 there were 42

[1] *An important crisis in the callico and muslin manufactory of Great Britain, explained* (London, 1788). This anonymous and scarce pamphlet is discussed in P. Mantoux, *An industrial revolution in the eighteenth century* (London, 1928), 253–4; Mantoux, however, gives the table inaccurately.

[2] A. E. Musson and E. Robinson, 393–426.

[3] E. Baines, *History of the cotton manufacture in Great Britain* (London, 1835), 226.

Boulton and Watt engines employed in the Lancashire cotton industry.[1] But they had not yet succeeded in replacing the older types, and, moreover, many engineers pirated various features of Watt's patents to produce their own engines. Boulton and Watt engines probably numbered not more than one-third of the total steam engines in the county by 1800. 'Steam-powered mechanization was proceeding more rapidly in Lancashire in the late eighteenth century than has hitherto been supposed.'[2] The application of power to weaving, however, was slow. Edmund Cartwright patented a power loom in 1785. His small factory at Doncaster in 1787 was worked at first by animals, and his attempt to introduce steam power here (1789) and at Manchester (1791) was defeated largely owing to the opposition of the weavers.[3] It was not until the early years of the next century that power-driven weaving looms were successfully established.

The water-power phase in the woollen industry came later than in the cotton industry; and it was of shorter duration because by the time that machines were established in the 1790s, steam was the obvious source of power. Thus the way was open for the full development of the factory system, and for the close association of the textile industries with coalfields, particularly with those of Lancashire and Yorkshire (Figs. 91 and 92). The success of the Lancashire industry can be measured by the eightfold increase in the import of raw cotton during 1780–1800.[4] Within a quarter of a century or so, it had become one of the most important industries in the country, and in the early years of the nineteenth century, the annual value of its production was to outstrip that of the woollen industry. In 1795, the growth of the industry was described as 'perhaps absolutely unparalleled in the annals of trading nations'.[5] W. W. Rostow has viewed it as the 'original leading sector in the first take-off' for the industrial revolution.[6]

The iron industry

In 1600 the most important iron-producing area was the Weald with 49 blast furnaces, but there were also eleven in the west Midlands, another

---

[1] J. Lord, *Capital and steam power* (London, 1923), 167–71.
[2] A. E. Musson and E. Robinson, 426.
[3] P. Mantoux, 247–8; E. Lipson (1921), 166.
[4] P. Deane and W. A. Cole, 183.
[5] J. Aikin, *A description of the country from thirty to forty miles around Manchester* (London, 1795), 3.
[6] W. W. Rostow (1960), 53.

eight in Yorkshire and Derbyshire, and a few others elsewhere (Fig. 61).[1] Curiously enough, the traditional iron centre of the Forest of Dean lagged behind. The monopoly of the 'free miners' of the Forest had hindered the adoption of the blast furnaces, and the first was not erected until the 1590s; but others soon followed in the early decades of the next century.[2] In the meantime, here as elsewhere, the old-fashioned bloomeries were disappearing.

The available statistics are unreliable, but there seems to have been a decline in iron production between about 1620 and 1660, especially in the Weald. The traditional view was that this general decline continued for another hundred years,[3] but it has been argued that this was not so, judging by the erection of new furnaces and forges between 1660 and 1760 in areas almost entirely outside the Weald.[4] Even so, the growth in output must have been very limited, in view of the competition of cheap iron from abroad, particularly from Sweden. The older view also attributed the postulated decline to a scarcity of wood for charcoal,[5] but, again, it has been argued that the scarcity has been much exaggerated.[6] Ironworks were very selective in their use of timber; the best charcoal was made from coppice timber of 20 years or less, and systematic coppicing in the seventeenth and eighteenth centuries ensured the continuation of the industry.[7]

The distribution of the industry in the early eighteenth century can be seen from the list of furnaces and forges compiled by John Fuller in 1717.[8] His total of 60 furnaces and 114 forges is probably not complete, but it may well provide a fair picture (Fig. 78). Most furnaces were located

[1] H. R. Schubert, *History of the British iron and steel industry from c. 450 B.C. to A.D. 1775* (London, 1957), 175.

[2] *Ibid*, 183–8.

[3] T. S. Ashton, *Iron and steel in the industrial revolution* (2nd ed, London, 1951), 13.

[4] M. W. Flinn, 'The growth of the English iron industry', *Econ. Hist. Rev.*, 2nd ser., XI (1959), 144–7.

[5] T. S. Ashton (1951), 15; H. R. Schubert, 218–22.

[6] G. Hammersley, 'The Crown Lands and their exploitation in the sixteenth and seventeenth centuries', *Bull. Inst. Hist. Research*, XXX (1957), 136–61; see also M. W. Flinn (1959), 148–53.

[7] H. R. Schubert, 222.

[8] E. W. Hulme, 'Statistical history of the iron trade of England and Wales, 1717–1750', *Trans. Newcomen Soc.*, IX (1929), 12–35; B. L. C. Johnson, 'The charcoal iron industry in the early eighteenth century', *Geog. Jour.*, CXVII (1951), 167–77; H. G. Roepke, *Movements of the British iron and steel industry – 1720 to 1951* (Urbana, Illinois, 1956).

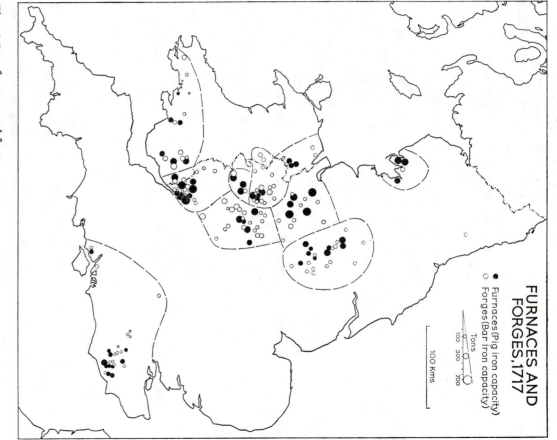

FURNACES AND
FORGES, 1717

● Furnaces (Pig iron capacity)
○ Forges (Bar iron capacity)

Tons
100
300
700

100 Kms

Fig. 78 Iron furnaces and forges, 1717
Based on: (1) E. W. Hulme, 'Statistical history of the iron trade of England
and Wales, 1717–1750', *Trans. Newcomen Soc*, IX (1930), 12–35; (2) B. L. C.
Johnson, 'The charcoal iron industry in the early eighteenth century', *Geog.
Jour*, CXVII (1951), 168.
The areas are those named on Fig. 79.

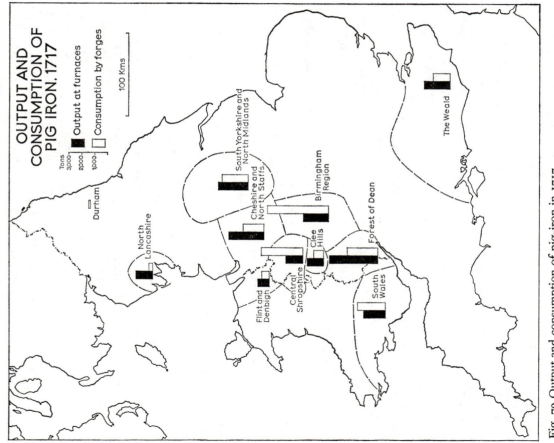

OUTPUT AND
CONSUMPTION OF
PIG IRON. 1717

◼ Output at furnaces
▢ Consumption by forges

Tons
3000
2000
1000

Durham

North Lancashire

Flint and Denbigh

Central Shropshire

Clee Hills

South Yorkshire and North Midlands

Cheshire and North Staffs

Birmingham Region

Forest of Dean

South Wales

The Weald

100 Kms

Fig. 79 Output and consumption of pig-iron in 1717
Based on B. L. C. Johnson, 'The charcoal iron industry in the early eighteenth
century', *Geog. Jour.*, CXVII (1951), 168.

fairly near supplies of iron ore, and, rivalling the Wealden ironstones, were now the Coal Measure ironstones of Yorkshire and Derbyshire, of Staffordshire and Shropshire, and of Monmouthshire and South Wales, and also the haematite areas of the Carboniferous Limestone in north Lancashire and the Forest of Dean itself. As well as ironstone and wood, water power for the furnace bellows and forge hammers was an important localising factor. Not all furnaces produced for local forges. Surpluses of high-quality pig-iron from the haematite areas of north Lancashire and the Forest of Dean were exported by sea and up the Severn to west Midland forges. The relative output and consumption of pig-iron, district by district, is shown on Fig. 79.

Many attempts were made to find an alternative to charcoal as a fuel in blast furnaces. The earliest patent for the use of coal in iron-making was granted as early as 1589. Others followed, among them being that granted in 1621 to Dud Dudley who claimed to have made iron of good quality with coal.[1] But no great advance was made until about 1709 when Abraham Darby was smelting with coke at Coalbrookdale in Shropshire. Not all types of coal were found to be suitable, and years of experiment followed before coke successfully and generally replaced charcoal. In 1760 there were only seventeen coke furnaces in blast in England and Wales (none in Scotland).[2] In the meantime another advance was made in the 1740s when Benjamin Huntsman, at Handsworth on the outskirts of Sheffield, devised a method of making good quality cast steel, but the great expansion in the use of steel had to await the inventions of the latter half of the nineteenth century.

It has been said that until 1775 the output of iron increased on average not more than one per cent per annum. 'The year 1775, in fact, marks more clearly than most dates selected as boundary-stones the end of one economic period and the beginning of another.'[3] About this time James Watt was developing an improved steam engine able to generate a strong blast and so increase the efficiency of smelting by coke.[4] The wider application of steam power not only opened up new possibilities for the manufacture of iron and ironwares, but, in turn, created a greater demand for them. Then, in 1783–4 Henry Cort patented processes which enabled high-grade wrought-iron to be made with coke. The forge-masters were now freed from their dependence upon wood, just as the furnace owners had been after 1709; not only was the charcoal-based industry finally outmoded,

[1] E. Lipson, II (1934), 159–60.
[2] H. R. Schubert, 332–3.
[3] T. S. Ashton (1951), 60.
[4] P. Deane (1965), 106.

but imported ore was no longer needed except for the manufacture of the best steel.[1]

In 1774 there were 31 coke furnaces in blast in Britain as a whole (including 5 Scottish). By 1790 there were 81 (including 12 Scottish) as compared with 25 charcoal furnaces (including 2 Scottish).[2] Cort's patent was confiscated in 1789, and the way was open for iron-masters to introduce improvements without having to pay royalties. Moreover, the continental wars from 1793 onwards provided a stimulus. By 1806 the coke furnaces in blast numbered 162 (including 18 Scottish) as compared with 11 charcoal furnaces 'still in use in different counties'.[3] Not only had the number of coke furnaces increased, but the output of each had become very much greater. The increase in the total annual output of pig-iron in tons between 1717 and 1806 was as follows, always remembering that the figures can only be approximate.[4]

	England and Wales	Scotland
1717	18,000	?
1788	61,000	7,000
1796	109,000	16,000
1806	235,000	23,000

The 1806 grand total of 258,000 included only 7,800 tons from charcoal furnaces. The transition from charcoal to coke was all but complete.

In the meantime there had also been profound geographical changes. The coke-based industry had become associated with the coalfields of the Midlands, Yorkshire, Derbyshire, South Wales and Scotland – those areas with Coal Measures that include ironstone as well as coal (Fig. 93). The coalfields of Northumberland and Durham and of south Lancashire, with little or no ironstone, on the other hand, had but few blast furnaces. The Wealden industry had all but disappeared; at Ashburnham the last furnace closed down about 1812 and the last forge about 1828.[5] The Forest of Dean industry survived longer, and the failing supply of

[1] T. S. Ashton (1951), 87 et seq.; A. Birch, The economic history of the British iron and steel industry, 1784–1879 (London, 1967), 22–44.

[2] H. Scrivenor, A comprehensive history of the iron trade (London, 1841), 359–61; H. R. Schubert, 333.

[3] H. Scrivenor, 97.

[4] H. Scrivenor, 57, 87, 95 and 97; W. E. Hulme, 22; T. S. Ashton (1951), 98.

[5] E. Straker, Wealden iron (London, 1931), 369; H. R. Schubert, 169, 366.

iron ore in the eighteenth century was made good by importation. Coke for smelting was not introduced until late in the eighteenth century. A few blast furnaces lingered on until the last was closed down about 1890.[1]

Even in the days of charcoal fuel, the coalfields had already attracted much iron manufacture. Coal Measure ironstone was available for the charcoal blast furnaces, and these produced pig-iron for the forges which in turn produced bar-iron for the smiths. Although not employed in smelting, coal was used by the smiths in the making of their wares. By the end of the seventeenth century the development of such activities was marked in three areas – the west Midlands, south Yorkshire and north Durham.

The west Midland smiths were already well known,[2] and in 1677 Andrew Yarranton described how the iron manufactures of Stourbridge, Dudley, Wolverhampton, Sedgley, Walsall and Birmingham were 'diffused all England over'.[3] Nails and edge-tools were of especial importance; guns and fire-arms increased the variety. The local iron was of low-grade quality, and the forges drew much of their supplies from elsewhere, especially from the Forest of Dean by way of the Severn;[4] the imbalance between furnaces and forges can be seen from Fig. 78. The introduction of brass in the eighteenth century gave promise of the remarkable diversification that came to be a feature of the metal trades of Birmingham – buttons, buckles, toys, jewellery and plated goods of all kinds.[5] The Soho Manufactory established by Matthew Boulton at Handsworth just outside Birmingham in 1761 soon achieved a national reputation as a 'nursery of ingenuity'.[6] After 1775, the partnership of Boulton and Watt at Soho resulted in the design and construction of steam engines which were to have so marked an effect upon the course of industrial change. To the west, in the Coalbrookdale area, the construction of the first iron bridge (across the Severn) in 1779 was a portent of things to come. A boat made of cast-iron plates was launched on the Severn in

[1] R. Jenkins, 'Iron-making in the Forest of Dean', *Trans. Newcomen Soc*, v (1926), 42–65; F. T. Baber, 'The historical geography of the iron industry of the Forest of Dean', *Geography*, XXXVII (1942), 54–62; C. Hart, *The industrial history of Dean* (Newton Abbot, 1971), 152.

[2] W. H. B. Court, 33 *et seq*.

[3] A. Yarranton, *England's improvement by sea and land* (London, 1677), 56–9.

[4] B. L. C. Johnson, 169.

[5] H. Hamilton, *The English brass and copper industries to 1800* (London, 2nd ed., 1967), 260–73.

[6] W. Hutton, *An history of Birmingham* (Birmingham, 1781), 271.

1787, and in the following year the Coalbrookdale works provided 40 miles of cast-iron pipes for the water supply of the city of Paris.[1]

A second important area of iron manufacturing was the south Yorkshire coalfield with local iron and wood and with power from the tributaries of the Don. Forges had long supplied iron for smiths and toolmakers in the neighbourhood of Sheffield, Rotherham and Barnsley. Continued progress is indicated by the Act of Parliament in 1624 which empowered the company of the Cutlers of Hallamshire to make bye-laws. Local iron was supplemented by imports, through Hull, from Sweden, and about 1724 Daniel Defoe could write of 'the continued smoke of the forges' at Sheffield which were always at work making 'all sorts of cutlery-ware'.[2] It was at Handsworth, just outside Sheffield, in the 1740s, that Benjamin Huntsman devised a method of making high-quality steel with coke, and nearby at Masborough another large ironworks was coming into existence.[3] Sheffield grew more slowly than Birmingham, but the variety of its products was almost as great, and included such articles as knives, shears, scissors and sickles.

A third iron manufacturing district – in north-east Durham – had different origins from those of the other two districts. There was hardly any ironstone in its Coal Measures, and no blast furnaces appear in lists of the eighteenth century. But there was wood and coal, and Newcastle was a centre for importing Swedish iron. It was at Sunderland in 1682 that Ambrose Crowley, a smith of Greenwich, established an ironworks with the aid of continental workmen; it was more economical to send iron to the north than to bring coal to the forges of the Weald.[4] In 1690, the factory was moved to Swalwell and Winlaton near Newcastle upon Tyne, and a variety of ironwork was produced in the eighteenth century –anchors, chains, nails, chisels, hammers, agricultural implements; Defoe in 1724 spoke of the considerable manufacture of hardware here 'lately erected after the manner of Sheffield'.[5] In 1770 Arthur Young

[1] P. Mantoux, 315.   [2] Daniel Defoe, *Tour*, II, 183.

[3] M. W. Flinn and A. Birch, 'The English steel industry before 1856, with special reference to the development of the Yorkshire steel industry', *Yorks. Bull. Econ. and Soc. Research*, VI (1954), 163–77; A. Raistrick and E. Allen, 'The south Yorkshire ironmasters (1690–1750)', *Econ. Hist. Rev.*, IX (1938–9), 168–85; A. Raistrick, 'The south Yorkshire iron industry, 1698–1756', *Trans. Newcomen Soc.*, XIX (1940), 51–86.

[4] M. W. Flinn: (1) 'Sir Ambrose Crowley, Ironmonger, 1658–1713', *Explorations in Entrepreneural history*, V (Cambridge, Mass., 1953), 162–80; (2) *Men of iron: the Crowleys in the early iron industry* (Edinburgh, 1962).

[5] Daniel Defoe, *Tour*, II, 252.

described the works as 'supposed to be the greatest manufactory of its kind in Europe'.[1] Whether this was true or not, it is clear that north-east Durham had become the scene of important and unusual activities.

Coalmining

The substitution of coal for wood as domestic fuel, begun in Tudor times, continued at an ever-increasing rate during the seventeenth century. Contemporary references bear witness to the widespread use of coal, and there was some working on almost every British coalfield. The total production in Britain in 1700 has been estimated at about 3 million tons, of which about 700,000 tons were produced in Scotland and Wales. By 1800, the total exceeded 10 million tons, including 2.4 million in Scotland and Wales.[2] Some coal from Northumberland and Durham was exported overseas, but the greater part went to swell the coastwise traffic to London and other ports of the east coast and the English Channel (Fig. 81). The smaller coalfields of the north-west and of Wales also shipped increasing quantities to Ireland and southwards to the English Channel and sometimes even to London and beyond.[3] The increasing volume of coal was distributed inland from the coastal ports by river, by the Thames, the Severn, the Trent and other navigable waterways, and also by carts and pack-horses.[4] Coal from the Midland fields also went by inland waterways and by road. The development of a network of canals after 1760 greatly facilitated its wider distribution and transformed the possibilities for wider industrial change (Fig. 80).

Coal was being used not only as a domestic fuel but more and more for industrial purposes. By 1600 it had become 'almost the universal fuel for the innumerable lime kilns' that produced lime for mortar and for agriculture.[5] It was also used in metal working of all kinds – iron, lead, copper and silver. It was taking the place of wood in the manufacture of salt, sugar and soap, in the making of glass, starch and candles, in the production of bricks and tiles, and in dyeing and brewing. After about 1709 it was used in the smelting, as well as in the forging, of iron. Later in the century,

[1] Arthur Young (1770), 13.
[2] J. U. Nef, *The rise of the British coal industry*, 2 vols. (London, 1932), I, 19–20.
[3] J. U. Nef, I, 90–1; T. S. Willan, *The English coasting trade, 1600–1750* (Manchester, 1938), 55–69.
[4] J. U. Nef, I, 95–108; T. S. Willan, *River navigation in England, 1600–1750* (Oxford, 1936), 123–5.
[5] J. U. Nef, I, 205.

after the inventions of Henry Cort in 1783–4, 'integrated iron and coal concerns sprang up on all the more important coalfields'.[1]

A revolution in coalmining began early in the eighteenth century. Until then, the extraction of coal had been mainly by means of bell-pits and adits, although there were some shafts in Northumberland as deep as 400 feet.[2] The main difficulties in the development of mining operations were proper ventilation and adequate drainage. Underground workings were easily flooded, and various devices were used for pumping; but a great advance came with the inventions by Thomas Savery and Thomas Newcomen of their steam engines worked by coal. Savery was a Cornishman familiar with the difficulties of mining copper. His first patent was taken out in 1698, and he described it in a pamphlet of 1707 entitled 'The Miner's Friend'. Newcomen's engine appear in 1705–6, and was another advance. By 1720 it had been improved and was soon widely in use not only for mining but for pumping purposes generally. The new steam engine not only made deeper mining possible but was itself a great consumer of coal.

A further advance took place in 1769 when James Watt patented a much improved steam engine which worked on a different principle and used less coal. The patent was extended in 1775 for another 25 years, and in 1781 Watt took out another patent for rotary motion by which his steam engine ceased to be merely a pump but could be adapted for driving machinery of all kinds. 'From that moment the whole field of industry was thrown open to it.'[3] The adoption of the Boulton–Watt engine for pumping in coalmines was slow because the saving of coal was of little advantage to colliery owners.[4] It has been estimated that by 1800 about 500 Boulton and Watt engines were in use in England and Wales,[5] but this can only have been a fraction of the total number of steam engines at work.[6] In the first place, the old Savery and Newcomen engines had been adapted by John Smeaton and others for many general uses as well as for pumping, and some were 'still in use' in the 1820s.[7] In the second

---

[1] T. S. Ashton and J. Sykes, *The coal industry of the eighteenth century* (Manchester, 1929), 6.

[2] *Ibid.*, 10.

[3] P. Mantoux, 340.

[4] T. S. Ashton and J. Sykes, 40.

[5] H. W. Dickinson, *A short history of the steam engine* (2nd ed., ed. A. E. Musson, London, 1963), 88.

[6] A. E. Musson and E. Robinson, 393–426.

[7] J. Farey, *A treatise on the steam engine* (London, 1827), 422.

place, there were 'pirate engines' constructed by rival engineers who infringed Watt's patents, and whose activities gave rise to much litigation in the 1790s. Altogether, it has been estimated that about 1,200 steam engines were at work by 1800.[1] They were used in a wide variety of industries – not only in the textile, iron and mining industries but in corn mills, potteries, glassworks, breweries and for pumping in connection with canals and waterworks.[2] The application of the steam engine to traction and haulage was soon to come.

In this way, the preliminaries for the development of large-scale industry were over, and the age of steam had begun. But it must be emphasised that the coalfields were already centres of industry – of textile manufacturing associated with water power, and of metal working associated with the presence of iron as well as coal. But coal, by means of the steam engine, was now providing a source of power that dwarfed all other sources. It was capable of a thousand applications far outside the coalfields, but it was on those fields themselves that its effects were most dramatically to be seen. As Arthur Young wrote in 1791, 'all the activity and industry of this kingdom is fast concentrating where there are coal pits'.[3]

TRANSPORT AND TRADE

Transport by road

An Act of Parliament in 1555 had made each parish responsible for the upkeep of the roads within its boundaries.[4] Various other acts in the early half of the seventeenth century attempted to make this system of maintenance more effective; but, in the absence of any central organisation, very little was achieved; moreover, the methods of making and repairing roads were very primitive. In the meantime, the growing volume of traffic increased the need for improvement, particularly for the improvement of the roads leading from London. A new step was taken in 1663

[1] J. R. Harris, 'The employment of steam power in the eighteenth century', *History*, LII (1967), 131–48.

[2] J. Lord, 175.

[3] A. Young, *Annals of agriculture*, XVI (London, 1791), 552.

[4] General accounts include: (1) S. and B. Webb, *English local government: the story of the king's highway* (London, 1913); (2) W. T. Jackman, *The development of transportation in modern England*, 2 vols. (Cambridge, 1916); W. Albert, *The turnpike road system in England, 1663–1840* (Cambridge, 1972). For a convenient short account, see B. F. Duckham, *The transport revolution, 1750–1830* (Hist. Assoc., London, 1967).

when an act enabled the justices of the three counties of Hertford, Huntingdon and Cambridge to levy tolls for the repair of the section of the Great North Road that passed through them; the first tollgate or turnpike was erected at Wadesmill to the north of Ware in Hertfordshire. The act was intended to continue in force for eleven years, and in 1664–5 this term was extended to twenty-one years for Hertfordshire only. Both acts, however, were allowed to expire and the tollgates were removed; before the end of the century the road had again become 'dangerous and impassible'.[1]

The setback was only temporary. Within a few years a new system of maintenance by which those who used roads were made to pay for their upkeep was introduced. Tollgates were again authorised in 1695 on a stretch of the London–Harwich road. The increasing awareness of the importance of roads in the national life had already been indicated by the publication in 1675 of John Ogilby's *Britannia*, which set out the principal roads of England and Wales in strips or bands, incorporating notes, on either side about the roads themselves and various features to be encountered along them. Its 102 plates were the result of a systematic measurement 'by the wheel', and it became the prototype for a large number of road books.

By 1700 only seven turnpike acts had been passed, but during the next fifty years they averaged nearly ten a year. From 1750 onwards, interest and activity greatly increased, and the average was soon 40 acts a year, reaching over 50 a year in the decade 1790–1800. The typical turnpike trust was small in scope, intended to meet a local need. It only supplemented, and did not replace, the parish system of maintenance which remained in force until 1835. The new turnpike trusts were particularly numerous in the rapidly growing industrial areas – in Lancashire and the West Riding, in Midland counties such as Warwickshire and Staffordshire, and in the counties around London. The trusts frequently met with opposition, and bands of raiders at night destroyed many tollgates and burned the houses of the toll collectors.

Daniel Defoe in the 1720s devoted much space to the new roads. Turnpikes, he said, were 'very great things'. Roads formerly dangerous and scarce passable in winter had become firm, safe and 'easy to travellers, and carriages as well as cattle'. The effect upon trade was frequently 'incredible'. Some roads were 'exceedingly throng'd' with 'a vast number of carriages' carrying malt, barley, grain or cheese, with 'infinite droves of black cattle, hogs and sheep', and with pack-horses, mail, and ordinary

[1] W. T. Jackman, I, 63.

'travellers on horseback'. But much remained to be done, and Defoe dwelt at length upon the horrors of 'the deep clays' of some Midland counties where chalk or stones were not available for repairs. Stretches of roads in the Weald were also very bad.[1]

The evidence for the latter part of the eighteenth century varies. In 1767, Henry Homer could say: 'It is probable that there is no one circumstance which will contribute to characterize the present age to posterity so much as the improvements which have been made in our public roads.'[2] On the other hand, there were often complaints about this or that stretch of road, especially about their condition in winter. Arthur Young, travelling in the 1760s and 1770s, described some stretches, even of turnpike roads, as terrible, infamous or execrable.[3] Many trusts gave poor service and they all suffered from the lack of a central administration.

It must be emphasised that from a technical point of view, turnpikes marked no advance. Roads were still made by piling up loose material, and there was no attempt to provide a firm well-drained basis for the road surface. In order to prevent the formation of deep ruts, various acts tried to enforce the use of broad wheels, and some of this legislation was consolidated in the General Highway Acts of 1766 and 1773. But by this time the technique of road-making was beginning to improve. John Metcalfe (1717–1810) was one of the first engineers to pay attention to solid foundations and to drainage in the making of roads. Thomas Telford (1757–1834) and J. L. McAdam (1756–1836) followed, and the name of the latter has been preserved in the word macadam. Better coach design helped, and the speed of coach travel was improved. The journey from London to York took four days in 1754 (as it had done in 1706) but this was reduced to two days by 1774; similar improvements took place on other routes. John Palmer in 1784 persuaded the government to send the royal mail by coach, at first between London and Bristol; and the new mail coaches served as incentives for improved passenger services. Bridge-building, too, increased the speed of movement. Before 1750, large rivers were usually crossed by fords and ferries, but the early civil engineers were also bridge-builders, and in this connection must be mentioned not

[1] Daniel Defoe, Tour, II, 117–32.

[2] H. Homer, An enquiry into the means of preserving and improving the publick roads (Oxford, 1767), 3.

[3] E.g. A. Young (1768), 72, 99, 111, 112, 211. For the variety of opinion, see the anthology of eighteenth-century extracts in C. W. Scott-Giles, The road goes on

only Telford but John Smeaton (1724–92), and John Rennie (1761–1821). The first iron bridge, built across the Severn by John Wilkinson and Abraham Darby in 1779, pointed the way to further possibilities.[1]

Transport by water

In spite of these improvements there were limits to the possibilities of moving heavy or bulky goods by road. It has been said that 'if Britain had had to depend on her roads to carry her heavy goods traffic the effective impact of the industrial revolution might well have been delayed until the railway age'.[2] There were, however, other means of transport available. It has been estimated that in the first half of the seventeenth century there were at least 685 miles of navigable rivers, including the great arteries of the Thames, the Severn, the Trent, the Yorkshire Ouse and the Great Ouse.[3] But rivers were often obstructed, they were subject to changes of level, and many stretches were unnavigable. The example of the Netherlands and the increasing traffic of bulky goods were in men's minds, and Acts of Parliament for the improvement of various stretches of river became increasingly frequent after 1660. Pamphleteers such as Andrew Yarranton advocated making 'rivers navigable in all places where art can possibly effect it'.[4] The Aire and Calder navigation was improved primarily for the Yorkshire woollen industry about 1700. Shortly after this, the Weaver was improved with the Cheshire salt trade in mind, and the Don was improved for the Sheffield steel manufacture.[5] Many other streams were improved during these years. T. S. Willan estimated that the mileage of navigable rivers increased from at least 685 miles in 1660 to at least 960 in 1700 and to about 1,160 miles in 1726 or so (Fig. 80).[6]

The improvement of river channels led naturally to the possibility of making new ones.[7] The idea of connecting the Severn and the Thames, and so facilitating the movement of coal from the Forest of Dean, was mooted several times in the seventeenth century, by Thomas Proctor in 1610, by John Taylor in 1641, and by others. Francis Mathew, who took up the idea in 1655, also envisaged the systematic joining of other rivers, of the Warwickshire Avon with the Welland, and of the Suffolk Waveney with the Trent and the Yorkshire Ouse by various links. In spite of many

[1] A. Raistrick, *Dynasty of iron founders* (London, 1953), 193–207.
[2] P. Deane (1965), 73.
[3] T. S. Willan (1936), 133; W. T. Jackman, I, chs. 3 and 5.
[4] A. Yarranton, *England's improvement by sea and land* (London, 1677), 7 and 64–6.
[5] T. S. Willan (1938), 136.   [6] *Ibid.*, 133.   [7] T. S. Willan (1936), 7–10.

Fig. 80 Canals and waterways, 1800–50
Based on C. Hadfield, *British canals: an illustrated history* (2nd ed., Newton Abbot, 1966).
See also Fig. 106, p. 505 below.

pamphlets and much parliamentary activity, work was confined to improving old channels rather than making new ones until well on into the eighteenth century.

The canal age is generally regarded as beginning in 1761 with the opening of a canal connecting the coalmines at Worsley with Manchester. It crossed the Irwell by an aqueduct forty feet high, and was made for the Duke of Bridgewater by the engineer James Brindley; the price of coal in Manchester was immediately halved. Other canals soon followed in the 1760s and 1770s, not only in Lancashire but in the Midlands and elsewhere. The rate of construction slowed down during the recession associated with the American War of Independence (1775–83), but it started up again in the 1780s only to become greater than ever. The years 1792–3 saw a 'canal mania'. Speculation was rife, and many schemes remained only on paper.

Within thirty years or so inland transportation in England had been transformed (Fig. 80). Birmingham became the centre of a network of waterways. The Mersey was linked to the Severn in 1772; the Trent to the Mersey in 1777; the Severn to the Thames in 1789; the Mersey to the Trent and the Thames in 1790. Canals crossed the Pennines along three different routes. Liverpool, Hull, Birmingham, Bristol and London were linked together. The connection between London and the Midlands was improved by the making of the Grand Junction Canal in 1793–1805. So far as can be calculated, the cost of transport by canal was one quarter to one half that by road.[1] The easy transport of coal was now possible and the way was clear for the development of widespread industrial changes. Inland navigation also played its part in the agricultural improvement of the age by making possible the cheap transport of lime, manure and grain. It also carried building materials to expanding towns.

There was also another form of water transport of vital importance in the expanding economy of the age – coastwise shipping. Adam Smith could point out in 1776 that over the same period six or eight men, 'by the help of water carriage, can carry and bring back the same quantity of goods between London and Edinburgh, as fifty broad-wheeled waggons, attended by a hundred men, and drawn by four hundred horses'.[2] The Port Books reveal how varied and how widespread was the coastal traffic of the age. The commodities included butter, cheese, wool, cloth, salt, iron, glass,

[1] W. T. Jackman, II, 724–9.
[2] Adam Smith, *An enquiry into the nature and causes of the wealth of nations* (London, 1776), Book I, ch. 3.

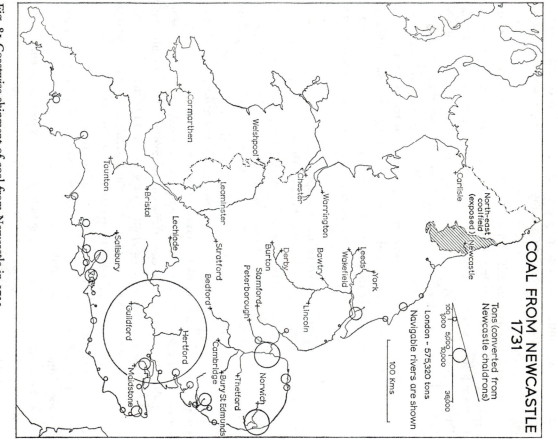

Fig. 81 Coastwise shipment of coal from Newcastle in 1731
Based on: (1) T. S. Willan, *The English coasting trade, 1600–1750* (Manchester,
1938), 211. The figures come from P.R.O. Exchequer K.R. Port Books 23617,
and are for the period Christmas 1730 to Christmas 1731; (2) T. S. Willan,
River navigation in England, 1600–1750 (Oxford, 1936), map 3.

COAL FROM NEWCASTLE
1731

Tons (converted from
Newcastle chaldrons)

100,000
50,000
10,000

5,000
1,000

36,000

London – 575,320 tons
Navigable rivers are shown

100 Kms

North-east
coalfield
(exposed) Newcastle

Carlisle

Welshpool
Chester
Carmarthen
Warrington
Leominster
Bawtry
Derby
Burton
Leeds
Wakefield
York
Lincoln
Taunton
Bristol
Lechlade
Stratford
Stamford
Peterborough
Bedford
Salisbury
Guildford
Hertford
Cambridge
Bury St Edmunds
Thetford
Norwich
Maidstone

timber and, above all, grain and coal. The London trade towered above that of any other port, but a multitude of substantial ports and small harbours also traded with one another as T. S. Willan's account so well shows.[1] Ships increased in size, and tonnage increased in amount. Cargoes of coal in three-masted square-rigged ships from Newcastle dominated the trade of the east-coast ports, and extended far along the south coast where they encountered far competition from South Wales (Fig. 81). The import of coal into London rose from about 74,000 tons in 1605 to nearly half a million by 1700 and to over 1.4 million tons by 1800.[2] During these years, civil engineers were at work transforming estuaries and river moorings into docks and harbours. England's first wet dock was opened at Rotherhithe in 1700; the second was opened at Liverpool in 1715, and the third at Bristol shortly after.[3] Others followed. John Rennie (1761–1821) alone was involved in over seventy harbour schemes.[4]

Overseas trade

The improvement of harbours and the growth of ports reflected, to a great extent, the increase in England overseas trade.[5] In 1600 English trade was almost entirely with Europe, and varieties of woollen cloth accounted for some 80% of the total exports by value. The main imports were wine from France, linen and metal-ware from central Europe via the Netherlands, together with flax, hemp, iron and timber from the Baltic lands, and cotton, raisins, dyestuffs and silks from those of the Mediterranean. After the Civil War, this pattern was transformed so much that the years 1660–1760 witnessed what has been called a 'commercial revolution', which took place in spite of, and in some ways as a result of, the many maritime wars of the age.

[1] T. S. Willan, *The English coasting trade, 1600–1750* (Manchester, 1938).

[2] J. U. Nef, II, 381–2; T. S. Ashton and J. Sykes (1929), 249–51.

[3] A. F. Williams, 'Bristol port plans and improvement schemes of the 18th century', *Trans. Bristol and Gloucs. Archaeol. Soc.*, LXXXI (1963).

[4] C. T. G. Boucher, *John Rennie, 1761–1821: the life and work of a great engineer* (Manchester, 1963).

[5] For general accounts see: (1) W. E. Minchinton (ed.), *The growth of English overseas trade in the seventeenth and eighteenth centuries* (London, 1969); (2) R. Davis, *The rise of the English shipping industry in the seventeenth and eighteenth centuries* (London, 1962). There is a convenient summary in R. Davis, *A commercial revolution* (Hist. Assoc., London, 1967). For statistics of trade, see G. N. Clark, *Guide to English commercial statistics, 1697–1782* (London, 1938), and E. B. Schumpeter, *British overseas trade statistics, 1697–1808* (Oxford, 1960).

Table 7.5   Foreign trade, 1700–98

(a) Totals in £000s

	England and Wales			Great Britain	
	1700–1	1750–1	1772–3	1772–3	1797–8
Imports	5,819	7,856	12,430	13,588	23,903
Domestic exports	4,468	9,123	9,739	10,195	18,301
Re-exports	2,123	3,428	5,802	6,932	11,802

(b) Geographical distribution (in percentages for each group)

	England and Wales			Great Britain	
	1700–1	1750–1	1772–3	1772–3	1797–8
Total imports from:					
Europe	61.5	46.5	34.6	34.2	29.3
Ireland etc.	4.9	8.8	10.5	10.6	13.1
North America	6.4	11.2	11.6	14.5	7.1
West Indies	13.5	18.9	24.8	23.7	25.0
East India	13.3	14.0	17.7	16.2	24.2
Africa	0.4	0.5	0.6	0.6	0.3
The Fisheries	—	0.1	0.2	0.2	1.0
Domestic exports to:					
Europe	82.1	69.4	39.9	39.2	21.1
Ireland etc.	3.2	7.6	9.4	9.9	9.0
North America	5.7	10.7	25.3	26.0	32.2
West Indies	4.6	4.9	12.0	12.0	25.2
East India	2.6	6.4	8.4	8.1	9.0
Africa	1.8	1.0	5.0	4.8	3.5
The Fisheries	—	—	—	—	—
Re-exports to:					
Europe	77.8	62.1	63.0	65.4	77.5
Ireland etc.	7.5	17.7	19.0	18.2	10.9
North America	5.0	11.2	9.0	8.7	3.1
West Indies	6.2	4.1	2.9	2.5	4.1
East India	0.5	2.0	1.2	1.0	0.6
Africa	3.0	2.9	4.9	4.1	3.7
The Fisheries	—	—	—	—	0.1

(Ireland etc. included Ireland, the Isle of Man and the Channel Islands. The Fisheries includes Greenland, Iceland and the Northern and Southern Fisheries; East India implies Asia.)

Source: Calculated from P. Deane and W. A. Cole, *British economic growth, 1688–1959* (Cambridge, 2nd ed., 1967), 87.

Not only did English overseas trade greatly expand but its character fundamentally changed. To Europe now went a more diversified range of goods, but in 1750 woollens still accounted for about 46% of English-produced exports by value. By the end of the eighteenth century this share had fallen to 29%, but the export of cotton yarn and fabrics had grown from almost nothing to 24%. In return, the various parts of Europe continued to supply their different traditional products.

The outstandingly new feature of the commercial revolution was the trade with tropical lands and the development of a great English re-export trade. From the West Indies came such products as sugar, cotton and rum; and from the Middle East and Far East came tea, coffee, spices, silks and calicoes. A large part of these arrived in Britain only to be exported. Woollen goods could hardly feature in the return trade to these warm lands. To West Africa went weapons, metal goods and spirits in exchange for slaves, ivory and gold. The slaves were sold in the West Indies in return for tropical products; the ivory and gold dust went to the East in return for its products that, from the mid-seventeenth century onwards, included tea. There was also a growing varied trade with the mainland of North America. Tobacco, rice, indigo, cotton, furs and timber came in return for slaves and manufactured goods.

Something of this growing, and increasingly intricate, international network of trade can be seen from Table 7.5. During the eighteenth century, re-exports came to form a large part of English overseas commerce to the great advantage of the English shipbuilding industry. This entrepôt trade lay behind the growth of Liverpool, Bristol and other ports and also behind the growth of London and its rise to be the financial centre of the world. The changes were also reflected in the general life of eighteenth-century England – in such things as the coffee-house, the drinking of tea and chocolate, the consumption of sugar and the making of mahogany furniture.

TOWNS AND CITIES

In spite of changes during the seventeenth century, the five largest towns (London apart) in the early years of the eighteenth century were still the same as in 1600 – Norwich, York, Bristol, Newcastle and Exeter; each seems to have had between ten and twenty thousand inhabitants. Daniel Defoe in the 1720s, however, could speak of some towns 'lately encreas'd in trade and navigation, wealth, and people, while their neighbours decay', and he had much to say about the rise or decline of individual towns.

Liverpool, he wrote, was 'one of the wonders of Britain' and was 'visibly' increasing in size. Manchester, too, was 'much encreased within these thirty or forty years'. Sheffield now had 'at least as many, if not more people in it than the city of York'; Coventry was 'a large and populous city'; so were Leeds and Leicester. Hull was a substantial centre of commerce, especially for the woollen goods of the West Riding. As well as Newcastle, Sunderland nearby, and Whitehaven[1] in the north-west had grown to be very considerable 'of late' by reason of the coal trade. Yarmouth, King's Lynn and Portsmouth were also thriving ports.[2] Defoe did not mention Birmingham but we know from other sources that by the 1720s it had grown tenfold since the mid-sixteenth century to over 15,000 people,[3] an increase that was helped by immigrants from the parishes around, particularly from those to the west.[4]

By the end of the eighteenth century, there had been greater and accelerating changes (see table 8.2, p. 459). Bristol alone kept its place among the first five. Although not serving a large industrial area, it had a wide agricultural hinterland and a variety of local manufactures such as soap, glass and sugar. Its coastwise trade was considerable, and its overseas connections were wide – with Ireland, Europe, the New World and Africa. Sugar, rum and tobacco became important imports and it also benefited from the slave trade. By 1801 its population had become over 60,000.[5] Norwich, it is true, had more than doubled in size since 1600 but it was no longer the largest provincial city.[6] Exeter and York had grown more slowly and now lagged behind with only about 16,000 to 17,000 inhabitants apiece. The old woollen towns of East Anglia and the West Country had been far outstripped.

The new rising centres were in the developing industrial areas. The population of Manchester-Salford had been about 12,000 or so in Defoe's day; it rose to about 30,000 by 1775 and to 84,000 in 1801. Around

1 J. E. Williams, 'Whitehaven in the eighteenth century', *Econ. Hist. Rev.*, 2nd ser., VIII (1955–6), 393–404.

2 Daniel Defoe, *Tour*, I, 43; II, 255–6, 261–2, 183, 83, 204, 88, 242, 273; I, 65–7, 73, 136–8.

3 M. J. Wise and B. L. C. Johnson in M. J. Wise (ed.), *Birmingham and its regional setting* (Birmingham, 1950), 174.   4 R. A. Pelham, 45–80.

5 W. E. Minchinton, 'Bristol – metropolis of the west in the eighteenth century', *Trans. Roy. Soc.*, 5th ser., IV (1954), 69–89.

6 P. Corfield, 'A provincial capital in the late seventeenth century: the case of Norwich', being ch. 8 (pp. 263–310) in P. Clark and P. Slack (eds.), *Crisis and order in English towns, 1500–1700* (London, 1972).

Manchester–Salford, smaller centres were also growing. Oldham, for example, in 1760 was a village of not more than 400 inhabitants. The earliest factories were built there about 1776–8, and by 1801 its population had grown to about 12,000. The story of places like Bolton, Bury, Stockport and Wigan was similar.[1] It is not easy to be precise about the size of some of these places because their very large parishes included a number of separate townships, but a number of townships themselves had populations of 10,000 and over by 1801; they were large industrial villages which had yet to become corporate boroughs. Nearby, the port of Liverpool, with under 10,000 inhabitants in the 1720s, had increased to 78,000 by 1801. Its growth reflected that of its hinterland. Its exports were coal, salt and manufactured goods; and it imported raw materials such as cotton and sugar. Like Bristol it had a considerable coastwise trade and connections with Ireland and Europe. It also benefited from the notorious 'triangular trade' – exchanging goods for slaves in West Africa, selling them in the West Indies, and returning home with raw materials.[2] By the end of the eighteenth century it was larger than Bristol, and had supplanted it as the main port of the west.

Across the Pennines, Defoe had mentioned the 'vast cloathing trade' of five towns in particular – Leeds, Halifax, Bradford, Huddersfield and Wakefield.[3] Easily the largest was Leeds with some 12,000 people, more than Manchester. Its further growth was relatively slow; it seems to have had only 17,000 in 1775, but in the 1790s its rate of growth accelerated to produce, in 1801, just over 31,000, or much more if the townships within its parish are included.[4] The other four towns still had under 10,000 inhabitants each, but this is misleading because with their subsidiary townships they formed even more than in Defoe's day 'one continued village'.[5] To the east, Hull, with its first tidal basin opened in 1778, had reached 30,000 by 1801.[6] To the south, Birmingham had passed the 70,000 mark, and nearby were the lesser centres of Dudley, with just over 10,000 inhabitants, and Wolverhampton with about 13,000; but again the precise population assigned to each depends upon what townships are included; all were

[1] P. Mantoux, 367–8.
[2] F. E. Hyde, 'The growth of Liverpool's trade, 1700–1950', in W. Smith (ed.), *A scientific survey of Merseyside* (Liverpool, 1953), 148–63.
[3] Daniel Defoe, *Tour*, II, 187.
[4] P. Mantoux, 369.
[5] Daniel Defoe, *Tour*, II, 194.
[6] W. G. East, 'The port of Kingston-upon-Hull during the industrial revolution', *Economica*, XI (1931), 190–212.

joined by a web of iron-working villages into a large built-up area. In the same way the adjoining parishes and townships of the Potteries (which much later became the borough of Stoke-on-Trent) accounted for a population of about 28,000. There were also other growing centres becoming as obvious in the England of 1801 as they were important in its economy. Nottingham, with silk and hosiery, and Sheffield with cutlery, had each grown to about 30,000 inhabitants. Coventry and Leicester had over 16,000 apiece. The north-east, in spite of its long history of coalmining, was still very rural in character. Apart from Newcastle with about 28,000, Gateshead on the opposite bank of the Tyne had another 9,000; to the south, at the mouth of the Wear, was Sunderland with about 12,000 people.

Along the south coast, many small ports had long histories of coastwise and cross-Channel trade; and, in the eighteenth century, some benefited greatly from the growing trade with North America and the West Indies. Plymouth and Portsmouth also became naval centres of critical importance during the maritime wars with Holland and France in the seventeenth and eighteenth centuries. With their harbours and arsenals and dockyards they grew by 1801 to places among the ten largest provincial towns in the kingdom – Plymouth with 43,000 people and Portsmouth with 32,000. There were also naval centres along the Thames estuary, and, among these, Chatham, with its rope-walk, mast-yard and anchor works, was the largest in 1801 with 11,000. At this time, Dover, well placed as a packet station for the Continent, had grown to 15,000 people.

Centres of growth were not limited to ports and industrial areas. Many market and county towns showed a modest growth, especially towards the end of the eighteenth century, as a result partly of improving communications or partly of some local advantage, but there were not many with populations above 5,000 by 1801. One category of growing centres must be mentioned because it constituted a new element in the urban scene, and had considerable potentiality for the future – that of inland spas and seaside resorts.[1] Doctors in the sixteenth and seventeenth century were beginning to emphasise the use of waters for medicinal and curative purposes. After the Restoration in 1660, places with mineral springs began

[1] E. W. Gilbert, 'The growth of inland and seaside health resorts in England', Scot. Geog. Mag., LV (1939), 16–35; E. W. Gilbert, Brighton, old ocean's bauble (London, 1954); J. A. Patmore, 'The spa towns of Britain', being ch. 2 (pp. 47–69) of R. P. Beckinsale and J. M. Houston (eds.), Urbanization and its problems (Oxford, 1968).

to assume a more prominent position, not only as places of medical treatment but as centres of fashion. The springs of Bath had been known in Roman times, and they began to attract attention once more in Tudor and Stuart times. Then, after the Restoration, Charles II and his court went there in 1663 and the way was prepared for the glory of the eighteenth century. The physician Dr Oliver, the architects John Wood (father and son) and the man of fashion Richard (or Beau) Nash, between them raised Bath to its pre-eminent position. Other inland resorts with springs were soon following this example – Harrogate, Epsom, Buxton, Cheltenham, Tunbridge Wells and others. Springs had been discovered at Scarborough in 1620 but its future was to be mainly bound up with sea-bathing. The practice was already common when, in 1750, Dr Richard Russell published his famous treatise on the curative properties of sea water. Soon, the inland spas were being rivalled by the coastal resorts. The Prince of Wales (afterwards George IV) began to visit Brighton in 1783; and George III began to visit Weymouth in 1789. Other seaside places were also attracting visitors – Margate and Worthing among them. All, whether inland or coastal, began to exhibit certain common architectural features such as crescents, terraces, promenades and assembly rooms. By the end of the century, Bath with its surrounding parishes, included some 32,000 people. Brighton and Scarborough had about 7,000 inhabitants each. The others were smaller, but they were to grow greatly in the nineteenth century.

In a category by itself stood London, so much larger than any other town in the country. Its exact size is a matter for argument, and any figures can convey only orders of magnitude. By 1600 it had expanded beyond the city walls and may have comprised some 250,000 people.[1] It was thus about sixteen times larger than Norwich, the greatest provincial city. It continued to grow until at the time of the Restoration in 1660 it may have included 460,000 people. The built-up area was now extending westwards over Covent Garden and Lincoln's Inn Fields. The plague of 1665 claimed perhaps as many as 100,000 victims, and, in the following year the Great Fire turned three-quarters of the old city within the walls into an expanse of smoking rubble. But these were only temporary checks. The speed of rebuilding was as remarkable as it was beneficial.[2] Houses were now of brick and stone; streets were wider; paving and drainage were better; and St Paul's Cathedral was completed in 1710. In the meantime, the

---

[1] N. G. Brett-James, *The growth of Stuart London* (London, 1935), 496–512.
[2] T. F. Reddaway, *The rebuilding of London after the Great Fire* (London, 1940).

13   D H G

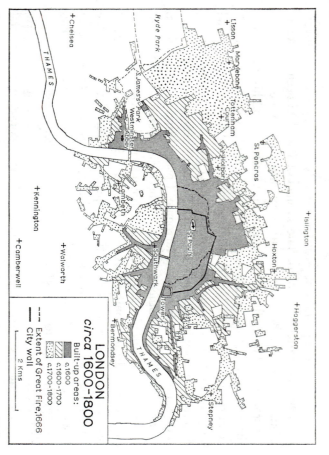

Fig. 82 London, 1600–1800
Based on: (1) N. G. Brett-James, *The growth of Stuart London* (London, 1935), maps opposite pp. 78 and 494; (2) T. Milne, *Plan of the cities of London and Westminster, circumjacent towns and parishes etc, laid down from a trigonometrical survey taken in the years 1795–1799* (London, 1800).

expansion westward had continued; the built-up areas of London and of Westminster were no longer connected only by the Strand along the north bank of the Thames, but also by the built-up area of the 'West End' around Piccadilly and St James's. The total population of this enlarged area in 1700 was over 600,000; and, as deaths greatly exceeded births, it had grown largely as a result of immigration.[1] London in its widest sense had outstripped Paris and had become the most populous unit in Europe.

To Daniel Defoe, in the 1720s, it was 'a prodigy of buildings' with 'new squares and new streets rising up every day'. Villages that had formerly stood in the country were now joined together by 'continued buildings'.[2] By 1750 its population had reached about 675,000. Within

[1] M. D. George, *London life in the eighteenth century* (London, 1930), 24 and 329–30; P. E. Jones and A. V. Judges, 'London population in the late seventeenth century', *Econ. Hist. Rev*, VI (1936), 45–63. See also J. Summerson, *Georgian London* (London, revised ed, 1970).
[2] Daniel Defoe, *Tour*, I, 314–15.

this extensive built-up area there were many separate parts and a fairly marked differentiation into business, legal, industrial, residential and government areas.[1] Apart from ferries, London Bridge itself provided the only crossing to the south bank until Westminster Bridge was opened in 1750 and Blackfriars Bridge in 1769, after which the borough of Southwark began to grow rapidly (Fig. 82). By the time of the first Census in 1801, the population of this built-up area amounted to nearly 960,000, of which about 750,000 lived to the north of the river, and the other 210,000 or so to the south, in the counties of Kent and Surrey.[2]

Nearly one Englishman in ten was now a Londoner, and the supply of this vast and growing agglomeration constituted one of the major elements in the economic geography of seventeenth- and eighteenth-century England.[3] It had gathered a variety of industries including watch-making, silk-weaving, boot-making, brewing and distilling, and also the manufacture of soap, sugar and of a variety of items, such as anchors, associated with boat-builders' yards. The countryside around was marked by the development of market gardening and by agricultural specialisation such as dairying and the production of hay for horses.[4] But the influence of London was felt far beyond its immediate surroundings. Defoe noted more than once 'how every county in England furnish'd something of its produce towards the supply of the city of London',[5] and the general truth of this is borne out by the widespread distribution of places sending carriers to London.[6] There was also the immense volume of coastwise shipping in which the bulky imports of coal and grain stood out prominently.[7] Furthermore, after the Peace of Paris in 1763, British overseas

[1] O. H. K. Spate, 'The growth of London', A.D. 1660–1800', in H. C. Darby (ed.), *An historical geography of England before A.D. 1800* (Cambridge, 1936), 529–48.

[2] *Census of 1851: Population Tables, I*, vol. I under London (P.P. 1852–3, lxxxv). This contains a retrospective summary for 1801.

[3] F. J. Fisher, 'The development of the London food market, 1540–1640', *Econ. Hist. Rev.*, v (1935), 46–64; F. J. Fisher, 'The development of London as a centre of conspicuous consumption in the sixteenth and seventeenth centuries', *Trans. Roy. Hist. Soc.*, 4th ser, xxx (1948), 37–50; E. A. Wrigley, 'A simple model of London's importance in changing English society and economy, 1650–1750', *Past and Present*, No. 37 (1967), 44–70.

[4] G. B. G. Bull, 'Thomas Milne's land utilization map of the London area in 1800', *Geog. Jour.*, cxxii (1956), 25–30. See p. 317 above.

[5] Daniel Defoe, *Tour*, I, 265.

[6] J. H. Andrews, 'Some statistical maps of Defoe's England', *Geog. Studies*, III (1956), 33–45.

[7] T. S. Willan (1938), 141–5.

possessions and influence were world-wide, and to the port of London came a rich harvest of trade as varied as it was considerable. The river below London Bridge was crowded with vessels discharging cargoes into lighters to be landed at the quays and wharfs along the river banks. The congestion produced much discussion which led to a parliamentary committee in 1796 and to the construction in 1800–2 of the West India Dock across the neck of the Isle of Dogs.[1] It was the beginning of a new phase in the history of the port. About this time the first Census could describe London as 'the Metropolis of England, at once the Seat of Government and the greatest Emporium in the known world'.

[1] J. H. Bird, *The geography of the port of London* (London, 1957), 44–9, 76.

Chapter 8

ENGLAND *circa* 1800

H. C. PRINCE

A traveller returning to England in 1800 after many years abroad, or a Frenchman or a German seeing the country for the first time, would have been struck by one new and pervasive quality of the English countryside – its neatness. Kent, the Garden of England, a county familiar to continental visitors, exhibited the most assiduously manicured scenery, with coppiced woodlands, well-pruned orchards, elaborately trained hopgrounds, verdant water-cress beds, the smoothest downs and fields of crops that were the envy of foreign observers and the pride of returning expatriates. In many other districts the soil was cultivated to the pitch of perfection, seeds were drilled in straight rows, plants regularly spaced and clean weeded, grass closely shorn and kept uniformly green and, in the newly enclosed Midlands, fields were square and bounded by neatly cut and laid hawthorn hedges. Augustan villas which half a century earlier had been sited in the midst of barren wastes, now looked over acres of idyllic pastoral scenery, embellished with artificial lakes, rustic temples and mock heroic monuments. The distinctive characteristic of the rural landscape at that time was that it was contrived so as to give an impression of orderliness and opulence. Even so, squalor had not been banished, and some of the most wretched farms lingered in highly improved regions. But pastoral elegance was no longer a poetic ideal, and it was the ambition of many practical husbandmen to create their own *ferme ornée*.

Nor had towns escaped the tidying hand of Georgian planners. On the ashes of the City of London, over the charred remains of Blandford in Dorset, and on many other devastated and derelict sites, splendid stone buildings had arisen. London's West End terraces, malls and squares marched in measured steps across the fields of Middlesex, Bath's crescents ascended the slopes of the Avon valley, and new shopping streets and residential quarters, studiously planned and proportioned according to the classical rules, imposed a sense of order and dignity on English towns such as they had not experienced before and were not to experience again. The whole nation appeared to be engaged in reorganising its landscape, and ever-growing numbers of people were coming forward to occupy new farms, to move into new town houses and to open new businesses.

At the same time, we must remember that England in 1800 was a nation at war. It is true that in this age war did not call for the enormous organisation of twentieth-century conflicts, but it brought dislocation and the absorption of manpower and, with but two brief intervals, it continued from 1793 to 1815. In spite of high prices and financial burdens, the geographical consequences of war were not entirely negative. For agriculture, the war provided an incentive to improve methods of farming and to extend the land under cultivation; the attack on wasteland became more vigorous. For industry it meant a stimulus to shipbuilding, to iron manufacturing in the form of guns, chains and anchors, and to woollen manufacturing for clothing soldiers and sailors. Moreover, after the battle of Trafalgar in 1805, England's supremacy at sea was virtually complete, and Napoleon's attempts at blockade came to naught. Continental competitors in overseas markets were crippled and English trade with the New World flourished in spite of many vicissitudes. London's role as a financial centre was increased at the expense of Amsterdam and Paris.

POPULATION AND SETTLEMENT

On the tenth of March 1801 in every parish, township or place, enumerators were instructed to make a house-to-house enquiry to ascertain the number of inhabited houses, the number of families, the number of male and female persons, the number employed in agriculture, in trade, manufactures or handicrafts, and in other occupations. The census recorded 8,331,434 inhabitants in England and 541,546 in Wales; but these figures were defective,[1] and subsequent adjustments made retrospectively in later censuses brought the figures up to 8.6 and 0.6 million respectively.[2] In any case, the census of 1801 was incomplete. No count was taken in the Isles of Scilly, in the Channel Islands nor in the Isle of Man. For many other places the enumeration was inaccurate because the boundaries of parishes and townships were not known, and for some places no returns had been received by the time the abstract was compiled. Moreover, a large part of the population was away from home on census day. No less than 469,188 were serving in the army, navy and merchant fleet; another 1,410

1 *Census of 1801: Enumeration*, 497 (P.P. 1802, vii).
2 *Census of 1851: Population Tables*, I, vol., 1, xxviii (P.P. 1852–3, lxxxv).

were convicts 'on board the hulks'; and an untold number of homeless vagrants, deserters and fugitives escaped the attention of the enumerators. Despite its defects and omissions the first census provided a fuller and more accurate record than the most reliable previous calculations. It marked the beginning of a statistical age.

The population was predominantly rural, and only 19% of the people in England lived in towns of over 20,000 inhabitants. If the built-up area of London, north and south of the river, be excluded from the calculation the figure becomes just over 7%. Wales had no town with over 20,000 inhabitants, and so the corresponding figures for England and Wales as a whole are reduced to $17\frac{1}{2}$% and just under 7%. Even so, England had a higher proportion of town-dwellers than any other European country, and almost twice as many people lived in London, its imperial capital and chief seaport, as in either Paris or Constantinople. But beyond London no town had more than 100,000 inhabitants, whereas in France, outside Paris, both Marseilles and Lyons exceeded that number. Not only had England few large towns but few townsmen were street-bred. Some had become townsmen as their rural birthplaces were surrounded by new buildings, while some were migrants from elsewhere. A Bolton cotton operative asked by the Factory Commission to say whence spinners were recruited at the beginning of the nineteenth century replied: 'A good many from the agricultural parts; a many (*sic*) from Wales; a many (*sic*) from Ireland and from Scotland.'[1] At Preston and other places the building of a cotton mill might create a 'sudden and great call and temptation for hands from the country, of this county and others, and many distant parts'.[2] It was, above all, from the neighbouring countryside that the factories drew their 'herds of Lancashire boors'.[3] Among Sheffield apprentices, 'sons of yeomen, farmers and labourers made up about 45% of the immigrants', most of whom came from villages within fifteen miles of the town.[4]

To an observer in 1801 the evidence for the concentration of people in northern England, for the ascendancy of industry over agriculture, and

[1] *Factories Inquiry Commission: Supplementary Report, 1834*, Pt I, 169 (P.P. 1834, xix).

[2] A. Young (ed.), *Annals of agriculture*, xv (1791), 564. Between 1784–6 the place of publication was Bury St Edmunds; between 1786 and 1815 (when it ceased), the place was London.

[3] *The callico printer's assistant* (C. O'Brien, London, 1789). B.M. 1420. b. 7.

[4] E. J. Buckatzsch, 'Places of origin of a group of immigrants into Sheffield, 1624–1799', *Econ. Hist. Rev.*, 2nd ser., II (1950), 306.

for the size of the urban population were novel and impressive findings of the census. To a modern observer the most striking fact is the smallness of the population. To recall what it was like to live in England at that time we must remove three out of every four persons in the present population. We must reduce London to a centre of about 960,000 inhabitants, shrink every provincial town to less than one-tenth of that number, depopulate large towns in the coalfield manufacturing districts, and wipe out almost all dormitory suburbs. Everywhere in 1800 we should have been aware of the sights, sounds and smells of the countryside. From St Paul's Cathedral we should have been able to view London in its entirety, surrounded on every side by sylvan landscapes of meadows, orchards, deer parks, woods, heaths and rutted lanes. The streets of London and other towns were jammed with horses, carts and farmers. On the outskirts of the capital were inns that catered for the traffic of country wagoners, Welsh drovers and Irish haymakers.

At the other extreme, farmers had relentlessly pushed back the edge of cultivation across heaths and up hills. Thousands of new homesteads clung precariously to footholds beyond the present limits of settlement. In 1801 many places in England and Wales had more people than they were to have in the twentieth century; they were situated in mountains and moorlands, in the Lake District and fells of the north country, in remote Pennine dales, in central Wales, in the south-west peninsula. In 1801, mining supported many upland communities which have since decayed. Tin mining in Cornwall was declining, but the fortunes of the Cornish mining towns were revived by the exploitation of copper at greater depths.[1] At the beginning of the nineteenth century Cornwall and Devon together produced more than two-thirds of the world's copper and the copper mine at Parys Mountain in Anglesey was the largest in Europe. On the Carboniferous Limestones of the Peak District and of Alston Moor, lead production had not yet reached its zenith.[2] In the Mendip Hills veins of lead were nearly exhausted but men were busy digging for zinc at a hundred mines.[3] Low-grade iron ore was still being worked in Sussex and south Lancashire, and coal was dug in Pembrokeshire and the North Riding of Yorkshire. Among the mountains of Westmorland or wild Wales a traveller was rarely out of sight of human

278–9.

1 W. J. Rowe, *Cornwall in the age of the industrial revolution* (Liverpool, 1953), 175.
2 A. E. Smailes, *North England* (London and Edinburgh, 1960), 148–51, 185–6,
3 J. W. Gough, *The mines of the Mendips* (Oxford, 1930), 175–80, 226–30.

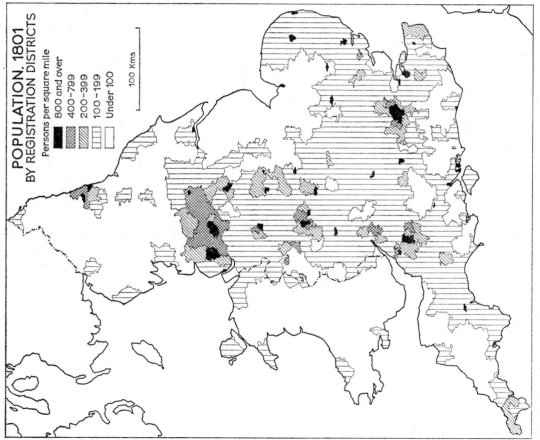

POPULATION. 1801
BY REGISTRATION DISTRICTS

Persons per square mile

800 and over
400–799
200–399
100–199
Under 100

100 Kms

Fig. 83 Population in 1801
Based on the retrospective evidence as given in *Census of 1851: Population
Tables*, *I*, 2 vols. (P.P. 1852–3, lxxxv, lxxxvi).

habitations and from the edges of wolds or downs in the south country dozens of villages might be viewed at a glance.

In 1801, compared with later periods, population was evenly spread (Fig. 83). More than half the surface of England carried a density of between 100 and 200 people per square mile. The farming counties in southern and eastern England, where recently light land had been brought into cultivation, had attained average densities of population. These districts also possessed few uninhabited houses, and many villages had more families than houses to accommodate them. Towns, too, were often crowded. It was true that in Birmingham, which had grown without plan or control, every workman, so a witness said before a Select Committee, 'has a house of his own'; but in most towns, houses occupied by more than one family were common.[1] In London, an average household included a family and one or two lodgers or servants, the rookeries of St Giles and Whitechapel being more than twice as congested as the rest of the metropolis. Crowding was especially intense at spas and seaports. In many seaports every second house was shared by two families, while in Newcastle, Sunderland, Chatham, Falmouth and Plymouth there were two families for every available house. Places as far apart in distance and in character as Manchester, Nottingham, Oxford, Carlisle and Exeter were hardly less crowded.

Rural settlement

Foreigners observed many farms and solitary dwellings standing far apart, but most people lived in compact villages, in places that had been named in Domesday Book. Homes and farm buildings clustered around village greens, abutted on to village streets, stood close to churches, castles, bridges or harbours. In shape and size each village differed somewhat from every other, and was continually changing as houses were built, repaired, demolished and rebuilt. A zone in which villages predominated extended from Durham to Devon. In the north of England, memories of the 1745 campaign were still fresh and the layout of villages recalled a long history of wars and border raids. Some larger villages in east Durham, a few in Northumberland and also a few in west Cumberland were closely built around broad greens. By the beginning of the nineteenth century these open spaces were no longer used as refuges for villagers' stock but as places to hold cattle sales, race meetings and holiday festivities, 'affording a proper place for the sports of childhood,

[1] *Report of the Lords Committees on the Poor Laws, 1817,* p. 180 (P.P. 1818, v).

the recreation of youth, and even the relaxation of old age'.[1] Green villages were also common in districts far beyond the reach of marauders, in the east Midlands and close to London. Over most of the Midlands, from the Pennines to the Chilterns, it was still somewhat unusual to see a farm or cottage outside a village. Certainly, wherever open fields survived, villages were compact and isolated farms absent.

Villages varied greatly in size, in layout and in distance from one another. Considerations of shelter, security, access to water supplies, the quality of soils and differences in farming practice were all, to some extent, reflected in their siting. Southward into the chalk country, lines of small villages picked out the base of the chalk outcrop or followed the courses of streams. In Somerset, linear villages were arrayed along river banks, ridges and dunes, while circular villages ringed dry points, hilltops, and contacts of two geological formations around springs.[2] Every part of midland England displayed similar contrasts in the plans of its villages.[3]

The number of buildings in a village was sometimes controlled deliberately by the owners and occupiers of the land. Parishes in the hands of only one or two landowners were often small communities of large farmers, some of the smallest being 'closed' townships where the Poor Law was strictly enforced to prevent indigent newcomers from taking up residence and becoming charges on the rates. 'Open' townships were large rambling villages crowded with small farmers, shopkeepers, artisans and landless labourers. In the eastern counties, open villages furnished the earliest gangs employed by large farmers for stone-picking, potato-setting, singling and lifting turnips, haymaking and harvesting.[4]

West of the Severn the transition from a landscape of villages to one of hamlets was discontinuous. Outlying groups of villages appeared among hamlets and dispersed farmsteads in Herefordshire. Farther north, villages

[1] E. Mackenzie, *An historical, topographical and descriptive view of the county palatine of Durham*, 2 vols. (Newcastle, 1834), II, 182; H. Thorpe, 'The green villages of County Durham', *Trans. and Papers, Inst. Brit. Geog.*, XV (1949), 155–80.

[2] B. M. Swainson, 'Rural settlement in Somerset', *Geography*, XX (1935), 112–24.

[3] E.g. H. M. Keating, 'Village types of Nottinghamshire', *Geography*, XX (1935), 283–94.

[4] W. Hasbach, *A history of the English agricultural labourer* (London, 1908), 134n., 195–8; D. R. Mills, 'The poor laws and the distribution of population c. 1600–1800, with special reference to Lincolnshire', *Trans. and Papers, Inst. Brit. Geog.*, XXVI (1959), 185–95.

clustered on flat arable lands between Shrewsbury and Stafford, on both sides of the Dee and Mersey estuaries, in the Fylde and around Morecambe Bay; but the most remarkable group of villages stood over 1,000 feet above sea-level in the Peak District. They were the highest permanent settlements in England, preserving in stone the distinctive features of Midland villages. Their farms crowded closely along their main streets and beyond lay furlong upon furlong of narrow curving strip fields enclosed, small as they were, by dry stone walls.

Westward, from the Cotswolds, towards Devon, villages became smaller, more irregular in arrangement and hamlets appeared more frequently. Devonshire, in the main, was a countryside of hamlets, of small groups of three or four farmhouses with their yards and cottages, some with an inn or a smithy, some dignified by the presence of a parish church. The area was threaded with narrow winding lanes and deeply rutted tracks leading to a multitude of hamlets. In the west and north, compact villages were by no means rare among the hamlets. Villages in the West Country, unlike those in the Midlands, were surrounded by dispersed hamlets and isolated farms, and the degree of dispersion increased west of the Exe.[1] In west Cornwall the only compact settlements were fishing villages, mining centres and a few inflated hamlets surrounded by strip fields.[2] The settlement pattern was further complicated by the presence of large numbers of solitary farms, substantial bartons in Devon, remote trevs in Cornwall, and squatters' cottages high on the moors.

In the Lake District, and over most of the Pennine uplands, few settlements possessed more than twenty houses. Chapels, bridges, mills or rows of cottages with an inn, served as centres for extensive parishes. In remote dales the largest habitations were farmsteads, massive stone buildings nestling in sheltered nooks on valley sides. Some were crumbling monastic granges, some were fortified manor houses, converted castles or peles, some were substantial and imposing farms standing alone at the edge of the moors. In many areas the characteristic dwellings were humble cottages, narrow windowed, squat chimneyed, slate roofed, built of rough unhewn stone, with lichens, mosses, ferns and flowers growing in their crannied walls. In 1810, William Wordsworth's *A guide through the district of the Lakes* describes the long, low cottages:

[1] W. G. Hoskins, *Devonshire studies* (London, 1952), 289–333.
[2] W. G. V. Balchin, *Cornwall* (London, 1954), 35–9, 72–3, 88–92.

Cluster'd like stars some few, but single most,
And lurking dimly in their shy retreats.[1]

But these unassuming dwellings were already disappearing fast as strangers were attracted to the mountains. Wealthy purchasers, 'if they wish to become residents, erect new mansions out of the ruins of the ancient cottages',[2] but old-established landowners, not to be outdone, rebuilt their family seats and laid out model villages such as Lowther.

Hamlets and isolated farms were characteristic not only of the uplands but also of formerly wooded districts in the English lowlands. Hamlets predominated in Cannock Chase, in Charnwood, in the Arden district of Warwickshire, in the Forest of Dean and in the well-wooded areas of southern and eastern England. In the Weald, hamlets or straggling villages, in mid-Suffolk and the Hampshire Basin scattered farmsteads, were numerous. Where timber had once been plentiful, timber-framed or half-timbered buildings were characteristic, sometimes weatherboarded, sometimes hung with wooden shingles. Many old houses also had brick chimneys and gable ends, and some occupied moated sites. In old-enclosed country north of the Thames in Middlesex, in the Chilterns, in Essex, and in parts of East Anglia, farms were dispersed and hamlets, loosely ranged around spacious commons, at the edges of heaths or along roads, bore such names as '-end' and '-green'.[3]

In many parts of England, particularly in the Midlands, new farms were arising among old villages. In distant Northumberland, where neither villages nor isolated houses had stood before, large farms were laid out with huts in rows to accommodate labourers.[4] Between small villages in the Solway lowlands and in the vale of Eden, single massive courtyard farms were not uncommon. East of the Pennines, where reclamation had advanced up to the edges of the moors and over the wolds, a rash of neat pantiled-roofed farms with parlours and sash windows had broken out in newly enclosed fields.[5] South of the Humber, on the Lincolnshire

[1] W. M. Merchant (ed.), *A guide through the district of the Lakes...by William Wordsworth* (London, 1951), 96.

[2] *Ibid*, 127.

[3] H. Thorpe, 'Rural settlement', being ch. 19 (pp. 358–79) of J. W. Watson and J. B. Sissons (eds.), *The British Isles: a systematic geography* (London and Edinburgh, 1964).

[4] W. Marshall, *The review and abstract of the county reports to the Board of Agriculture: Northern Department* (York, 1808), 40.

[5] A. E. Smailes, 156.

Wolds, on the sands of Nottinghamshire, Norfolk, Suffolk and Berkshire similar bright brick-built farmhouses and field barns were springing up in newly enclosed or improved districts.[1] Most were no more than thirty years old. Farms dating from the mid-eighteenth century were also to be found in a number of other areas – e.g. in the vale of York, in the vale of Trent, on the drift-covered wolds of Leicestershire, Northamptonshire and in the Feldon district of Warwickshire. The names of lodges, granges and farms commemorated, among other events, the colonisation of America and India, the reigning house of Hanover and the battles in the wars with France.[2] Farm building was not confined to arable districts. The Board of Agriculture reporters noted that in mid-Cheshire modern dairy farms were as spacious and well-designed as any in the kingdom; and in the dairy land of north-west Wiltshire they were built 'on a much superior plan to those in many other counties', while in Middlesex they were asserted to be 'perfect models of their kind'.[3] Enclosure and improvement were imposing a new pattern of settlement on the old.

No part of England, least of all the south-east, entirely escaped the transforming hand of Georgian architects and planners. On the great roads out of London government officials and citizens of London built themselves comfortable brick boxes. Old halls and courts were replaced by porticoed country houses, and at the side of the road the paint was still fresh on new gate-lodges, toll-houses and renovated posting inns. New brickwork and stucco was to be seen far beyond a half day's journey from the City. On many estates landowners were pulling down dark hovels and putting up well proportioned cottages, but in Gloucestershire and other counties it was reported that 'the popular complaint against the dilapidation of cottages is but too well founded'.[4] In making landscape gardens, whole villages had been swept away and built anew at Nuneham Courtenay, Milton Abbas, Ickworth and other places. In the new

[1] A. Harris, *The rural landscape of the East Riding of Yorkshire, 1700–1850* (Oxford, 1961), 70–2; M. B. Gleave, 'Dispersed and nucleated settlement in the Yorkshire Wolds, 1770–1850', *Trans. and Papers, Inst. Brit. Geog.*, xxx (1962), 105–18; N. Riches, *The agricultural revolution in Norfolk* (Chapel Hill, N.C., 1937), 144–6.

[2] W. G. Hoskins, *The making of the English landscape* (London, 1955), 157–9.

[3] H. Holland, *General view of the agriculture of Cheshire* (London, 1808), 82; T. Davis, *General view of the agriculture of the county of Wiltshire* (London, 1811), 169; J. Middleton, *A view of the agriculture of Middlesex* (2nd ed., London, 1807), 45.

[4] T. Rudge, *General view of the agriculture of the county of Gloucester* (London, 1807), 47.

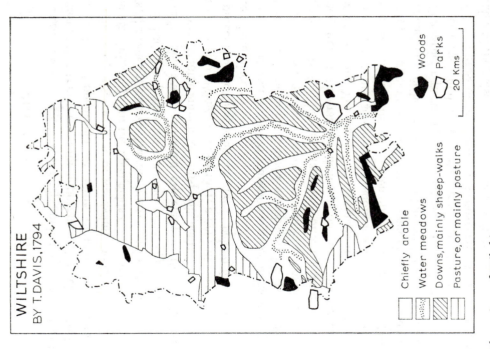

WILTSHIRE
BY T.DAVIS,1794

Chiefly arable
Water meadows
Downs, mainly sheep-walks
Pasture, or mainly pasture

Woods
Parks
20 Kms

Fig. 84 The regions of Wiltshire, 1794
Based on T. Davis, *General view of the agriculture of Wiltshire* (London, 1794).

manufacturing districts factory owners commissioned model villages for their work-people at Etruria, Cromford, Styal and Mellor, and new villages were built as inland navigations extended into the Midlands. Apart from medieval churches and a few other enduring architectural monuments, most buildings and some entire villages in old England were no older than those in New England beyond the Atlantic.

THE COUNTRYSIDE

In 1800 England was more or less at midpoint in the process of parliamentary enclosure that transformed the appearance of much of the countryside. The traditional open-field agriculture of some 1,500 or so villages, largely in the Midlands, had been, or was being abolished by private Acts of Parliament (Fig. 69); but the open fields of an even greater number of villages still awaited their acts. Another kind of enclosure, that of commonable waste, had also proceeded a considerable way by private act, but, again, much open waste remained especially in the northern counties (Fig. 76). A substantial amount of non-parliamentary enclosure by mutual agreement was also taking place. In 1801, a general enclosure act simplified the proceedings leading to an act, and so helped forward the changes that were taking place.[1]

A picture of the countryside in transition is given in the county reports of the Board of Agriculture, a body established by Act of Parliament in 1793. It was not a government department in the modern sense of the term, but a society for the encouragement of agriculture, supported by government grants. Its first president was Sir John Sinclair and its first secretary was Arthur Young, and it continued until 1822, when it was dissolved. One of its first actions was to sponsor a series of 'General views' of the agriculture of each county, written upon a uniform plan. There were two editions, an earlier quarto edition for private circulation and comment, and a later and more elaborate octavo edition usually written by a different author. The earlier reports appeared between 1795 and 1815. Taken together, they provide not only a view of the progress of enclosure but a conspectus of English agriculture in general, and they also contain a wealth of information on topics other than agriculture.[2]

Each report was normally accompanied by what was usually called a 'map of the soil', but very often the maps were based not so much on soil as on land use or on relief or on a combination of all three.[3] William

[1] G. Slater, *The English peasantry and the enclosure of common fields* (London, 1907), 140, 268–313; W. E. Tate, *The English village community and the enclosure movements* (London, 1967), 88, 130.

[2] G. East, 'Land utilization in England at the end of the eighteenth century', *Geog. Jour.*, LXXXIX (1937), 156–72.

[3] H. C. Darby, 'Some early ideas on the agricultural regions of England', *Agric. Hist. Rev.*, II (1954), 30–47.

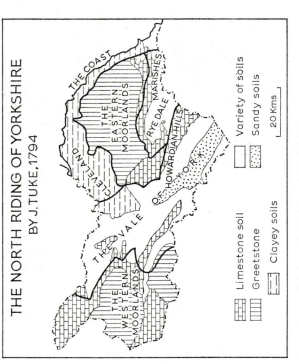

THE NORTH RIDING OF YORKSHIRE
BY J. TUKE, 1794

THE COAST
CLEVELAND
THE EASTERN MOORLANDS
RYE DALE
MARISHES
THE VALE OF YORK
HOWARDIAN HILLS
THE WESTERN MOORLANDS

Limestone soil
Greetstone
Clayey soils
Variety of soils
Sandy soils

20 Kms

Fig. 85 The regions of the North Riding of Yorkshire, 1794
Based on J. Tuke, *General view of the agriculture of the North Riding of Yorkshire*
(London, 1794).

Smith, the so-called 'father of English geology' had produced a geological map of the country around Bath in 1799 and a map of the geology of England in 1801. Some of the authors of the Board's reports may have consulted William Smith or his records, but others carried out their own surveys. A number of authors confessed to the difficulty of delineating complex soil variations accurately or in detail, and they sometimes solved this problem by indicating on their maps areas of 'miscellaneous' or 'various' soils. Examples of the maps are those for Norfolk (Fig. 72), Suffolk (Fig. 73) and Wiltshire (Fig. 84). The divisions of the first two counties were based on differences in soil texture; that of Wiltshire upon differences in land use. One of the most interesting was John Tuke's map of the North Riding (Fig. 85). On a division according to soil he super-imposed boundaries of districts 'each remarkable either for its climate, soil, surface or minerals (i.e. rocks)'. Each district was distinguished 'by the name by which it is usually known', or, if none existed, by some appro-priate descriptive name. All the districts included more than one type of soil. Only in east and south-east England did agricultural regions cor-respond at all closely with soil divisions.

During these years there was another attempt to delineate the agricultural variety of England. William Marshall (1745–1818) was strongly critical of the Board of Agriculture's plan for describing what he called 'the rural economy of England' in terms of counties, and he pointed to the fact that 'agricultural districts' frequently ignored county boundaries. Thus the dairy district of northern Wiltshire extended into Berkshire and Gloucestershire, and the cider country or 'fruit liquor district, of the Wye and the Severn', included parts of Herefordshire, Gloucestershire and Worcestershire; in the same way, the Fenland needed 'six county reports to treat of it'. After producing studies of various parts of England, Marshall, between 1808 and 1817 proceeded to 'review and abstract' the Board of Agriculture reports, county by county, and the five volumes of this review were re-issued as a whole in 1818.[1] These, together with Marshall's work in general, provide a valuable supplement to the Board of Agriculture's picture of England in the years around 1800.

Arable land

In 1801 estimates of acreages of land in different states of cultivation for every county in England and Wales were abstracted from the reports of the Board of Agriculture and from parliamentary accounts by Benjamin Capper,[2] and these may be compared with the estimates of W. T. Comber a few years later.[3] The total area of arable land in 1801 was estimated to be between eleven and twelve million acres. A still larger total would have been arrived at if all land in temporary grass had been added, and if the true extent of the surface area of certain Welsh and west Midland counties had been known. In constructing Fig. 86 estimates from the later Board of Agriculture reports have been used in preference to Capper's where they approximate more closely to the total surface area of a county. Whatever the defects of the figures they probably indicate the relative proportions of land devoted to different uses.

Most districts in eastern England were continuously in crops, but some western districts rarely saw a newly turned furrow, and along the western border of the West Riding it was observed that 'nearly the whole of the good land is under the grazing system, while corn is raised upon the

---

[1] For a summary of William Marshall's writings, see D. McDonald, *Agricultural writers, 1200–1800* (London, 1908), 215–16.

[2] B. P. Capper, *A statistical account of the population and cultivation, produce and consumption, of England and Wales* (London, 1801), 66–73.

[3] W. T. Comber, *An enquiry into the state of national subsistence* (London, 1808), 52.

Table 8.1 *Land use in England and Wales in 1801 and 1808*
(in acres)

	1801 B. P. Capper	1808 W. T. Comber
Arable land	11,350,501	11,575,000
Pasture and meadow	16,796,458	17,495,000
Total cultivated area	28,146,959	29,070,000
Woods and coppices	} 5,664,156	1,641,000
Commons and wastes		6,473,000
Buildings, roads, water etc.	3,454,740	1,316,000
Total uncultivated area	9,118,896	9,430,000
Grand total	37,265,855	38,500,000

The correct total area of England and Wales is 37,325,000 statute acres.

inferior or moorish soils'.[1] Proportions of arable and grass varied from county to county. The leading arable county, Norfolk, was about two-thirds under crops, and the leading grazing county, Leicestershire, was about three-quarters grass, but at this time there was no marked distinction, as in later times, between corn counties in the east and grazing counties in the west. In the west, tillage occupied over one half of Herefordshire, almost a half of Worcestershire and one-third of Cornwall, while in the east almost two-thirds of Surrey, Suffolk and Lincolnshire were under grass.

No less important were differences in farming enterprise and in the manner of cultivating the land. Some agricultural products took their names from and gave a unique character to particular localities. By 1800, mustard from the silt fens of Norfolk, Pontefract liquorice, Aylesbury ducks, Stilton cheese and Burton ale had achieved nation-wide renown and had entered the export market. On the other hand, neither orcharding nor hop-growing were localised. Hops were to be found in counties as far apart as Shropshire, Nottinghamshire, Cornwall and Suffolk as well as in Kent and Worcestershire.[2] Around the towns, farmers used a variety of urban refuse as manure, and supplied urban markets with perishable food-stuffs and industrial crops. The Board of Agriculture reports described in

[1] G. B. Rennie *et al*, *General view of the agriculture of the West Riding of Yorkshire* (London, 1794), 77.
[2] D. C. D. Pocock, 'England's diminishing hop acreage', *Geography*, XLIV (1959), 14–21.

some detail the market gardens around London, dependent upon ample supplies of manure, and employed in raising vegetables for the London market.[1] Thomas Milne's map shows their extent in about 1800 (Fig. 68). Near cloth manufacturing centres in Yorkshire, East Anglia and the West Country, in addition to food and fodder crops, flax, hemp, teasles, weld, woad and saffron were grown for industrial uses. Part-time spinners and weavers in these districts brought into cultivation land yielding too little to support a family, and such part-time farmers were less attentive than full-time farmers to the finer points of husbandry. They considered farming not as a business but 'only as a matter of convenience; speak of spinning-jennies, and mills, and carding machines, they will talk for days with you.'[2] 'Never inquire about the cultivation of land or its produce within ten or twelve miles of Manchester', wrote another; 'the people know nothing about it.'[3]

The quality of farming varied greatly from one district to another, independently of the area devoted to particular farming types. Arable farming was considered 'the great object of the Hertfordshire husbandry', whereas in neighbouring Buckinghamshire, also a predominantly arable county, grasslands were praised as 'the inestimable treasures which give character to the husbandry'.[4] In Suffolk, where extensive areas were devoted to dairying and sheep farming, arable farming had achieved a high pitch of perfection while in Lincolnshire the management of arable land left much to be desired, particularly on wet clay soils, and in Shropshire large tracts of arable were wretchedly cultivated.[5] The best arable farming was to be found in Norfolk, in Kent and in parts of Hertfordshire, while the finest grassland husbandry was practised in Middlesex and Leicestershire.

The leading enterprise of most arable farms was the production of bread grain, and to that end was directed the greatest effort at improvement. Between 1780 and 1801 the price of wheat trebled.[6] Population increased

---

[1] J. Middleton, *View Middlesex* (1807), 328–38; W. Stevenson, *General view of the agriculture of the county of Surrey* (London, 1809), 414–19.

[2] R. Brown, *General view of the agriculture of the West Riding of Yorkshire* (Edinburgh, 1799), 77, 225.

[3] J. Holt, *General view of the agriculture of the county of Lancaster* (London, 1795), 211n.

[4] A. Young, *General view of the agriculture of Hertfordshire* (London, 1804), 55.

[5] A. Young, *General view of the agriculture of the county of Suffolk* (London, 1804), 46–8; A. Young, *General view of the agriculture of the county of Lincoln* (London, 1799), 92; J. Plymley, *General view of the agriculture of Shropshire* (London, 1803), 161, 348.

[6] Lord Ernle, *English farming past and present* (4th ed., London, 1927), 441.

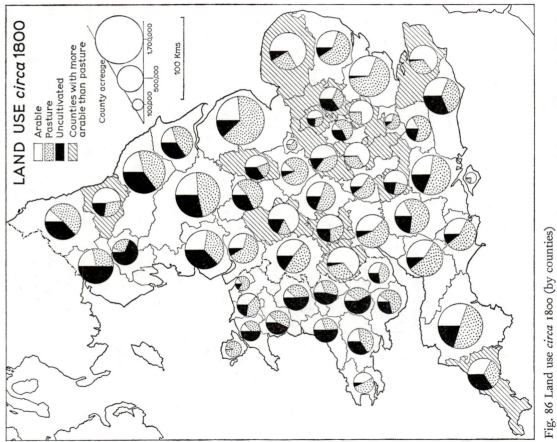

LAND USE *circa* 1800

Arable
Pasture
Uncultivated
Counties with more
arable than pasture

County acreage

1,700,000

500,000

100,000

100 Kms

Fig. 86 Land use *circa* 1800 (by counties)
Based on B. P. Capper, *A statistical account of the population and cultivation,
produce and consumption, of England and Wales* (London, 1801), 66–73; supplemented by the county *General views* of the Board of Agriculture.

rapidly, imports from enemy-held territory were restricted, and crop failures between 1794 and 1800 created an almost continuous apprehension of impending famine.[1] In 1795, wheat was severely damaged by winter frost, but it was not a bad year for spring-sown barley. In 1799 a cold wet summer was followed by an exceptionally bad harvest and in 1800, the wheat crop was again about 25% below average.[2] After 1795 several attempts were made to estimate how much grain was grown, and in 1801 the Home Office called for returns from every parish in England and Wales of the acreages growing wheat, barley, oats, potatoes, peas, beans and turnips or rape.[3] The crop returns, although they were incomplete and often inaccurate, not only provide a rough measure of the relative importance of different crops in different areas, but they enable us to locate these differences with greater precision than is possible from the Board of Agriculture reports. At a parochial scale we may identify areas where the acreage of wheat exceeded that of barley or oats or rye and also areas where other cereals were dominant. Fig. 87 provides an indication of the variations to be encountered in Leicestershire.

The total acreage growing crops in 1801 may only be guessed but there can be little doubt that over half the land in crops was sown with wheat, barley, oats and rye. The area occupied by grain crops was probably not less than six million and possibly nearer seven million acres, of which wheat perhaps occupied about three million acres, oats nearly as many, barley no more than one million and rye less than one-quarter of a million acres. Wheat not only occupied a large and increasing area; but the amount of grain yielded by each acre was also increasing more rapidly than at any previous time.[4] By the end of the eighteenth century wheat had become the almost universal bread grain of all classes and occupations.[5] The Trent, which separated wheat-eaters from the rest at the beginning of the eighteenth century, no longer marked such a dietetic boundary. In the new manufacturing districts in the north labouring families ate wheat

[1] W. F. Galpin, *The grain supply of England during the Napoleonic period* (New York, 1925), 213–19

[2] A. Young (ed.), *Annals of agriculture*, XXXIII (1799), 129–53, 194–218, 253–74, 346–65, 518–31; *ibid.*, XXXIV (1800), 95–8.

[3] H. C. K. Henderson, 'Agriculture in England and Wales in 1801', *Geog. Jour.*, CXVIII (1952), 338–45; W. E. Minchinton, 'Agricultural returns and the government during the Napoleonic wars', *Agric. Hist. Rev.*, I (1953), 29–43.

[4] M. K. Bennett, 'British wheat yields per acre for seven centuries', *Econ. Hist.*, III (1935), 27–8.

[5] W. J. Ashley, *The bread of our forefathers* (Oxford, 1928), 1–2.

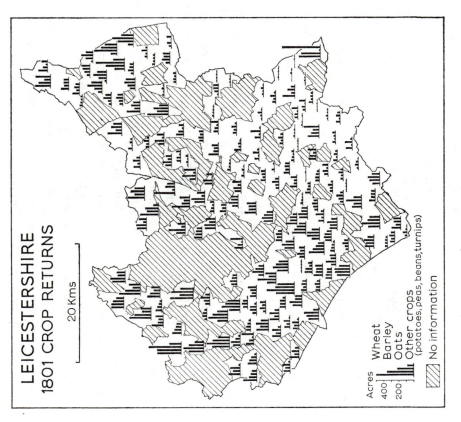

LEICESTERSHIRE
1801 CROP RETURNS

20 Kms

Acres
400
200

Wheat
Barley
Oats
Other crops
(potatoes, peas, beans, turnips)

No information

Fig. 87 Crop returns for Leicestershire, 1801
Based on W. G. Hoskins, 'The Leicestershire crop returns of 1801', being
pp. 127–53 of W. G. Hoskins (ed.), *Studies in Leicestershire agrarian history*
(Leicester, 1949); this constituted vol. XXIV of *Trans. Leicestershire Archaeol.
Soc.*

bread. In the West Riding, where in 1799 the wheat acreage was twice
that of barley, the vicar of Wakefield observed, 'the prodigious Number
of Tradesmen, Mechanics and Husbandmen, who twenty Years back
subsisted on Oat and Barley Cakes, as their favourite Diet, now consume
none but the *best* Wheaten Flour'.[1] The Board of Agriculture reports
repeatedly described wheat as a modern production. In Lancashire it had

[1] Quoted in B. P. Capper (1801), 59.

been introduced so recently that in 1775 it had 'scarcely yet acquired the name of corn', which in general is applied only to barley, oats and rye'.[1] Rye was still grown widely in Durham, but only in southern Northumberland was it spoken of as 'the most general bread of the labouring people'.[2] In 1808 the amount of rye grown in Bedfordshire was reported to be 'much decreased of late years'; the falling demand for it as a bread grain rendered its production unprofitable on 'any soil of moderate fertility'.[3] On light sands in south-west Norfolk and on the forest sands of Nottinghamshire some rye was grown but 'scarce at all used for bread'.[4] Here, and in southern England, it was raised as spring fodder for sheep, while on the very poorest soils in Suffolk buckwheat was regarded as 'a very valuable crop'.[5] On the Welsh Border, where both rye and maslin, a mixture of rye and wheat sown together, were once widely cultivated, rye was sown 'but sparingly' and 'good wheat' was grown instead.[6] The extension of wheat growing accompanied the improvement of soils by marling and liming and the introduction of a regular course of cropping. In south-west England, where little rye was grown, the wheat acreage expanded at the expense of barley. In Cornwall most people ate barley bread. In Cheshire, in 1794 agricultural labourers still subsisted on barley bread but by 1808 wheat was the leading bread grain and little barley was cultivated.[7]

In the north, wheat was grown less extensively than oats. Even so, oats were not generally used for making bread except in the Lake District, in Craven, in the Yorkshire dales and in north Lancashire.[8] In the vale of Pickering and in north Staffordshire oatbread was baked only in times of scarcity.[9] Throughout the northern counties, however, oatmeal appeared

[1] T. Percival, 'Observations on the state of population in Manchester, and other adjacent places, concluded', *Philosophical Transactions*, LXV (1775), 327.
[2] J. Bailey and G. Culley, *General view of the agriculture of the county of Northumberland* (London, 1794), 80.
[3] T. Batchelor, *General view of the agriculture of the county of Bedford* (London, 1808), 386.
[4] A. Young, *General view of the agriculture of the county of Norfolk* (London, 1804), 304; R. Lowe, *General view of the agriculture of the county of Nottingham* (London, 1798), 46.
[5] A. Young, *General view of Suffolk* (1804), 304.
[6] J. Duncumb, *General view of the agriculture of the county of Hereford* (London, 1805), 66; J. Plymley, *General view of Shropshire* (1803), 173.
[7] H. Holland, *General view of Cheshire* (1808), 298.   [8] H. C. K. Henderson, 342.
[9] J. Tuke, *General view of the agriculture of the North Riding of Yorkshire* (London, 1800), 127; W. Pitt, *General view of the agriculture of the county of Stafford* (London, 1796), 163.

in a variety of dishes such as porridge, oatcakes, crowdies or hasty puddings. In southern and eastern counties scarcely any oatmeal entered the diet, and in the south Midlands the area devoted to oats was very small. In the west, where oats were widely grown, they were used as fodder for livestock.

Beans were cultivated as a fodder crop on heavy land, particularly in the south-east. About one-fifth of the arable land in Middlesex was devoted to beans, and they were tended 'in the most clean and perfect manner'.[1] They were also extensively grown in Essex and Kent and on the strongest soils in Surrey. North and west from London they were characteristic of common-field husbandry on claylands, and were often cultivated in a slovenly manner. In Cheshire and in the north-east their acreage was increasing.

The turnip was described by Arthur Young as 'the crop which, in Norfolk, is made the basis of all others', and the advance of agricultural improvement in the north and west of the county was marked by the spread of turnip cultivation. By 1804 they were sown 'indiscriminately on all soils in Norfolk'.[2] In Suffolk they had 'changed the face of the poorer soils', and in Hertfordshire they were sown 'wherever turnips can be sown'.[3] Their cultivation had continued to spread during the closing years of the eighteenth century. In 1796 John Boys stated that in Kent, 'thirty years ago, hardly one farmer in a hundred grew any; and now there are few especially in the upland parts that do not sow some every year'.[4] In Surrey it was reported that their cultivation had 'certainly extended considerably' since 1790.[5] In Lincolnshire, wrote Arthur Young in 1799, 'thousands of acres' flourished where about thirty years earlier 'there was scarcely a turnip to be seen'.[6] On light soils in Nottinghamshire, on the Yorkshire Wolds and as far west as Cornwall they were now regularly cultivated. Fig. 88, based on information derived largely from the Board of Agriculture reports, attempts to show where turnips were grown in rotation at least once in five years. The map gives a very incomplete picture because the available information varies greatly from county to county. R. Parkinson's survey of Rutland records the course of cropping

[1] J. Middleton, *View Middlesex* (1807), 241.
[2] A. Young, *General view Norfolk* (1804), 219.
[3] A. Young, *General view Suffolk* (1804), 95; A. Young, *General view Hertfordshire* (1804), 61.   [4] J. Boys, *General view Kent* (1796), 92.
[5] W. Stevenson, *General view Surrey* (1809), 243.
[6] A. Young, *General view Lincoln* (1799), 138.

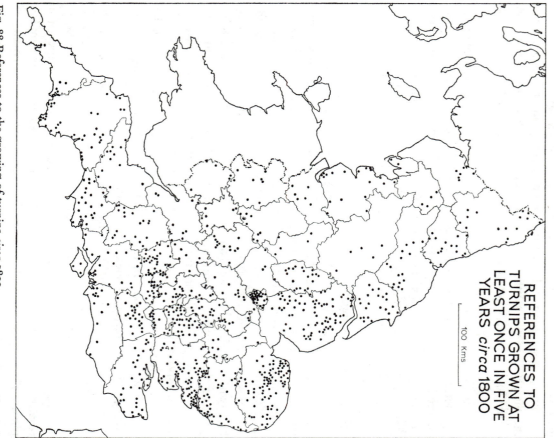

REFERENCES TO
TURNIPS GROWN AT
LEAST ONCE IN FIVE
YEARS *circa* 1800

100 Kms

Fig. 88 References to the growing of turnips *circa* 1800
Based on: (1) the county *General views* of the Board of Agriculture; (2) Arthur
Young (ed.), *Annals of agriculture*, 45 vols. (Bury St Edmunds, 1784–6;
London, 1786–1815). (3) and the writings of William Marshall.

for each parish,[1] G. B. Worgan names only a few places in Cornwall to illustrate general observations,[2] while some localities are reported in accounts published in Arthur Young's *Annals of agriculture* (1784–1815) and in the writings of William Marshall. Notwithstanding the deficiencies of the map, it is apparent that turnip husbandry had become firmly established in East Anglia and the counties to the north-west of London.

In Essex, and in Northumberland, for example, some land appears to have already become tired of turnips and clover, while many farmers in the Midlands and in the north-west considered turnips unsuited to all but dry soils, a view shared by few observers acquainted with Norfolk practice.[3] In the Midlands, so William Marshall observed, 'not one acre in a hundred, taking the district throughout, is subjected to the turnip culture'.[4] In 1795 Edward Harries, touring through Shropshire, described a tract of arable land in the Severn valley as 'a fine sound country, chiefly arable' but with 'scarce a turnip fallow', though no land can be better adapted to that crop'.[5] In Cheshire, William Marshall complained, 'even in 1800 – *no Turnips*! – not even on the Turnip Lands, with which the County abounds'.[6] On light soils in these areas change was imminent. By 1803 turnips had appeared on the sands of eastern Shropshire, and in 1808 Henry Holland welcomed their recent arrival in the central and southern dairy districts of Cheshire;[7] but they were late in coming to some counties such as Dorset.[8]

Swedes thrived in some counties where turnips made little progress. Their introduction into the Midlands and into southern England followed, and was associated with, the replacement of grassland sheep by arable breeds.[9]

[1] R. Parkinson, *General view of the agriculture of the county of Rutland* (London, 1808), 45–9.

[2] G. B. Worgan, *General view of the agriculture of the county of Cornwall* (London, 1811), 55–7, 67–71.

[3] A. Young, *General view of the agriculture of the county of Essex*, II (London, 1807), 12; J. Bailey and G. Culley, *General view Northumberland* (1794), 184.

[4] W. Marshall, *The rural economy of the Midland counties*, 2 vols. (London, 1790), I, 253.   [5] A. Young (ed.), *Annals of agriculture*, XXIV (1795), 379–80.

[6] W. Marshall, *The review and abstract of the county reports to the Board of Agriculture: Western Department* (York, 1809), 7.

[7] J. Plymley, *General view Shropshire* (1803), 173; H. Holland, *General view Cheshire* (1808), 156.

[8] W. Stevenson, *General view of the agriculture of the county of Dorset* (London, 1812), 251.

[9] E. L. Jones, 'Eighteenth-century changes in Hampshire chalkland farming', *Agric. Hist. Rev.*, viii (1960), 17.

In Staffordshire it was said that 'the most material alteration that has taken place in the cultivation of this county since 1796 has been the pretty general introduction of the Swedish turnip'.[1] They were particularly successful in the Soar valley in Leicestershire, and were also advancing in Devon and Cornwall. With few exceptions they were shunned in eastern districts and in the south Midlands; and in drier localities in the east, rape and carrots were often substituted for turnips. Rape was gaining ground in the east Midlands, while carrots were also favoured especially in the south and east of the country. Cabbages were also successfully brought into the root course in places as far apart as Durham, Cheshire and Suffolk.

Among the new crops none was of more importance than the potato, because it yielded large quantities of human food, serving to supplement available supplies of bread. Potatoes were raised on all well-manured dry light soils in Lancashire and Cheshire.[2] They were pre-eminent in the Fylde and their acreage was expanding in the northern counties generally. In Cornwall and Devon they produced 'certain and prolific' yields, supplying much of England from their surplus.[3] On rich sandy loams in Somerset and in Gloucestershire they were eagerly taken up, and in Herefordshire they were reported to be 'gaining ground every year: near towns in particular'.[4] In Shropshire they were still not commonly cultivated in 1803 but their acreage was increasing annually. In the highly cultivated districts of eastern and south-eastern England, they were grown only in gardens or for feeding pigs, and in the south Midlands generally they were still hardly to be seen. Although they were rarely grown more than five miles from the metropolis they were one of the leading crops in the gardens of its immediate neighbourhood.[5]

At the beginning of the nineteenth century arable farming was prospering. Both landowners and farmers were spending large sums of money to bring more land into cultivation, to improve the soil, to increase productivity, to experiment with new techniques, and to venture into new enterprises. They were applying a wide range of manures in such quantities as to alter the character of soils. William Stevenson, describing the

[1] W. Pitt, *General view Stafford* (2nd ed., 1813), 72.

[2] A. Young (ed.), *Annals of agriculture*, XIX (1793), 332–4; *ibid.*, XXIV (1795), 568–72.

[3] G. B. Worgan, *General view Cornwall* (1811), 72; C. Vancouver, *General view of the agriculture of the county of Devon* (London, 1808), 197–200.

[4] J. Duncumb, *General view Hereford* (1805), 66.

[5] W. W. Glenny in W. Page and J. H. Round (eds.), *V.C.H. Essex*, II (1907), 474–7.

soils of Surrey, asserted that 'in the strict and chemical sense of the term *clay*, no soil which has long been under cultivation has a just claim to it'.[1] And the transformation of extensive areas of light soils was no less spectacular than the conversion of clays in Surrey. From rabbit warrens and sheep-walks in west Norfolk the prosperous arable district of Good Sand had been called into existence, 'and, by dint of management', wrote Arthur Young in 1804, 'what was thus gained has been preserved and improved, even to the present moment'.[2] Other extensive agricultural regions were similarly won from from the waste. In south-east England only the most obdurate, villainous tracts of heath remained for William Cobbett to reproach.

Near towns, market gardens and fields were heavily dressed with organic manures from hearths, stables, dairies, slaughter houses, fish markets and the sweepings from streets, courtyards and rubbish dumps. John Middleton estimated that Middlesex alone probably received 250,000 cart loads of London manure each year and that as much again went to neighbouring counties.[3] From manufacturing districts in the north came reports of experiments evaluating the fertilising properties of bones, blood, hair, oil cake, soap ashes, soot, rags and various composts. The great value of bones was discovered in Yorkshire but it was not until 1794 that an enterprising farmer in Lancashire began grinding them at a mill.[4] Green manuring was practised on some sandy soils with varying success, and the value of paring and burning in many different localities was keenly debated.

The texture and acidity of many soils were profoundly modified by the application of mineral manures. Enormous quantities of peat, sand, clay, limestone and marl were spread and mixed with soils of opposing characteristics. William Marshall termed marl 'the grand fossil manure of Norfolk', and manuring, chiefly with marl, was praised by Arthur Young as 'the most important branch of the Norfolk improvements, and that which has had the happy effect of converting many warrens and sheep-walks into some of the finest corn districts in the kingdom'.[5] In Lancashire, marl was declared 'the great article of fertilization, and the

1 W. Stevenson, *General view Surrey* (1809), 20.
2 A. Young, *General view Norfolk* (1804), 3.
3 J. Middleton, *View Middlesex* (1807), 374.
4 Lord Ernle, 218n.
5 W. Marshall, *The rural economy of Norfolk*, 2 vols. (London, 1787), I, 16; A. Young, *General view Norfolk* (1804), 402.

foundation of the improvements'; it was also claimed as 'unquestionably one of the most important of the Cheshire manures'.[1] It was held in high esteem in Essex, in Hertfordshire, in the Chilterns and in some localities in south and south-west England. Chalk was extensively used on the Lincolnshire Wolds, and in 1812 it was reported to be 'coming into use' in the East Riding of Yorkshire,[2] Light drifting sands and sandy loams benefited greatly from judicious additions of marl but heavy soils were often injured by excessive applications which increased rather than relieved their tenacity.

In many districts, burnt lime was found to be cheaper and more effective than marl. In Warwickshire the effects of lime had been found 'from experience so infinitely superior to those of marl, that the last will in a short time cease to be used at all'.[3] Leicestershire farmers had learned that marl was unnecessary for their grass, and in heavily marled Lancashire it was lime that was considered 'the best manure for grass'.[4] In Derbyshire, lime was considered essential in the coal district and marling was being abandoned.[5] In Staffordshire, William Pitt thought that lime was used in a more extensive way than elsewhere and its use was still increasing.[6] In the North Riding of Yorkshire and further north, lime was superseding marl.

Burnt lime was used in many districts where marling had not been practised previously. In the Mendips it was described as 'the great article of modern improvement of these hills', and in the vicinity of Cheddington in Dorset, where no lime had been used six years earlier, it was reported in 1812 to be 'now in general use'.[7] High in the dales and on the moors of north Yorkshire, where rye had been the leading grain crop, William Marshall noted that 'the alteration of the soil by liming' was such that wheat had become a more prevalent crop.[8] In the Weald of Kent some

[1] J. Holt, General view Lancaster (1795), 111; H. Holland, General view Cheshire (1808), 221.

[2] H. E. Strickland, General view of the agriculture of the East Riding of Yorkshire (York, 1812), 212.

[3] A. Murray, General view of the agriculture of the county of Warwick (London, 1813), 150.

[4] W. Pitt, General view of the agriculture of the county of Leicester (London, 1809), 188; J. Holt, General view Lancaster (1795), 124, 128.

[5] J. Farey, General view of the agriculture of Derbyshire, 3 vols. (London, 1811–17), I, 148.

[6] W. Pitt, General view of the agriculture of Stafford (1796), 126.

[7] J. Billingsley, General view of the agriculture of the county of Somerset (Bath, 1798), 105; W. Stevenson, General view Dorset (1812), 352.

[8] W. Marshall, The rural economy of Yorkshire, 2 vols. (London, 1788), I, 312.

farmers used marl in 1785, but in the Surrey Weald in 1813 lime was 'almost universally used'.[1] The efficacy of liming depended upon adequate soil drainage. On waterlogged land lime was worse than useless.

In 1800, in most arable districts both marl and lime were used in smaller quantities and more frequently than hitherto. Fig. 89 shows places where applications of marl or lime were mentioned in contemporary agricultural writings; the map gives a very incomplete picture, because the information varies from county to county. Even so, the distribution illustrates some preferences of farmers largely ignorant of the chemical properties of soils and manures. Before Humphry Davy lectured on agricultural chemistry in 1803 no scientific principles governed the practice of manuring.[2] Farmers followed local traditions and learned lessons from practical experience.

About 1800 great progress began to be made in the invention, manufacture and utilisation of implements for arable farming. Lighter and more efficient harrows, horse-hoes, horse-rakes, scarifiers, chaff-cutters, turnip-slicers, drills, drill-rollers as well as reaping, mowing and haymaking machines were patented in large numbers. Norfolk led the country in the number and variety of its farm implements. The Norfolk plough, modified in detail, was adopted with enthusiasm by farmers in north-western counties. Threshing machines, widely used in Norfolk and Suffolk before 1800, were now general in northern Northumberland and were 'becoming very prevalent' in Devon.[3] Between 1794 and 1804 twenty new threshing mills were erected in Suffolk, and in 1813 the first steam engine to be used solely for agricultural purposes was installed at a farm at Haydon in Norfolk. In northern England, drill husbandry was an innovation; but in Suffolk, where drills had long been in use for sowing roots and pulses on light soils, they were now used for sowing grain on heavy land. The most rapid advances in the development of agricultural machinery were made in eastern districts, particularly for cultivating light soils, but wet lands derived no small benefit from mechanisation in speeding the sowing of seed and in the gathering of crops. In northern counties iron was employed not only in the manufacture of agricultural machinery but also

[1] A. Young (ed.), *Annals of agriculture*, II (1785), 64; W. Stevenson, *General view Surrey* (1809), 498.

[2] The substance of the lectures was published ten years later as *Elements of agricultural chemistry* (London, 1813).

[3] J. Bailey and G. Culley, *General view Northumberland* (1794), 49; C. Vancouver, *General view Devon* (1808), 121.

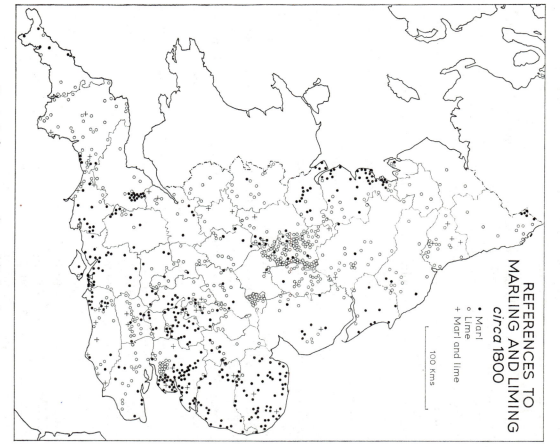

REFERENCES TO
MARLING AND LIMING
circa 1800

- Marl
○ Lime
+ Marl and lime

100 Kms

Fig. 89 References to marling and liming *circa* 1800
Sources as for Fig. 88.

in building strong gates and fences. By 1800 northern and some western districts were beginning to catch up with Norfolk in the efficiency of their arable farming.

The advance of improvement left large areas of arable land untouched. No fewer than two million acres lay fallow each summer. A large proportion of the arable land in a belt extending from Durham to Dorset was fallowed one year out of three, in some localities as often as one out of two. In such areas as the grain growing district of the West Riding bare fallows prevailed 'to a much greater extent than necessary', and in the vale of Gloucester the continuance of fallows was attributed to 'prejudice and an attachment to the practice of their forefathers'.[1] Common arable fields were not confined to the Midland belt. In Middlesex 11,000 acres out of a total of 14,000 acres of arable land remained in common fields in 1807, and fallowing was universal before the introduction of green and root crops.[2] On the heavy loam soils of north Surrey, from Carshalton to Runnymede, more than 8,000 acres lay in common fields in 1809.[3] But the area of common arable fields in both Middlesex and Surrey had been greatly reduced since 1795).

Fallowing was also widely practised on enclosed land. Kent had no common arable fields, but fallowing was generally pursued on cold stiff soils. In one large district in Essex, Arthur Young reported 'half the arable being under a dead summer fallow', and 'on all but sound dry turnip soils, it is universal'.[4] In Lancashire few open fields remained but fallowing was widespread, while in Northumberland, except where turnips were grown, 'naked fallows' still persisted, 'with an almost universal opinion that it is absolutely necessary to the fertility of the land'.[5] But there was much discussion generally among contemporaries about the value of fallowing.[6]

Grassland

In 1800 England was a grassy country, much greener than France. For every acre of arable land there were one and a half acres of grass, and the grassland acreage was expanding faster than the arable. 'The predilection

[1] T. Rudge, *General view Gloucester* (1807), 104.
[2] J. Middleton, *View Middlesex* (1807), 158.
[3] W. Stevenson, *General view Surrey* (1809), 470–7.
[4] A. Young, *General view Essex*, I (1807), 201.
[5] J. Holt, *General view Lancaster* (1795), 43; J. Bailey and G. Culley, *General view Northumberland* (1794), 62.
[6] J. C. Loudon, *An encyclopedia of agriculture* (London, 1815), 320–1, 740–2.

14

for *pasture* land which many years past seems to have been universally manifested', was attributed by John Billingsley to a growing demand for meat and dairy produce in London and other centres of population.[1] In the Midlands the value of many open fields when enclosed and converted into pasture was doubled; and in old enclosed districts great grain growing areas, such as the Fylde, were laid down to grass, a process which, throughout Lancashire, 'seems yearly increasing'.[2]

Both Lancashire and Cheshire were counted among the more productive grassland districts in the country; rich grazing lands were the glory of Lincolnshire; Middlesex produced some of the most valuable hay in the world; Romney Marsh fattened sheep and cattle more rapidly than any other district, and breeders and graziers in Leicestershire 'surpassed every other county in the kingdom, and I suppose every other country in the universe'.[3] In the west, each locality excelled in some branch of grassland management. Dairying dominated the broad vales of the Mersey, Dee, Severn and Avon. William Marshall observed that the western parts of the realm, apart from the hill country, were 'almost wholly applied to the produce of the DAIRY: Cheeses of different qualities being its common (or prevailing) production'.[4]

The greenest patches in the pastoral scene were picked out by ribbons of riverside meadows whose fertility was regularly restored by the silt spread over them during floods. Such were the fine grasslands of the Thames and Lea valleys and the vale of St Albans. The extensive natural meadows of Staffordshire and Derbyshire were enriched by the lime-impregnated waters of the Derwent and of the Dove, while those of Shropshire, flooded by the Severn, were 'constantly mown without any other manure being bestowed upon them'.[5] There, as in other counties, artificial channels were cut to divert silt-laden waters to places beyond the reach of natural floods. On the warplands of Lincolnshire and the Humber estuary, the silt was impounded by regulating the tidal flow. On the water-meadows of the chalk country of southern England muddy water from arable fields was floated down during the autumn. They were hot beds for grass, raising an early bite for sheep in March and April. The

[1] J. Billingsley, *General view Somerset* (1798), 154.
[2] J. Holt, *General view Lancaster* (1795), 71; Leicestershire and Northamptonshire are referred to in N. Kent, *General view of the agriculture of the county of Norfolk* (London, 1796), 73–4.
[3] W. Pitt, *General view Leicester* (1809), 216.
[4] W. Marshall, *Western Department* (1809), 1–2.
[5] J. Plymley, *General view Shropshire* (1803), 180.

irrigation of 20,000 acres of meadow in south Wiltshire was described as 'the greatest and most valuable of all improvements'.[1] In a few localities in Rutland, in Buckinghamshire and elsewhere, water-meadows produced only coarse rank unpalatable herbage. In many other parts of England, from the Yorkshire dales to Cornwall, meadows were watered by catch-work channels.

Almost all river valleys possessed some excellent natural and artificial meadows, but extensive areas of pasture in the grazing districts in the north and west called loudly for improvement. Even in such counties as Lancashire, Cheshire, Shropshire and Herefordshire, celebrated for their fat-stock and dairy produce, much pasture land was neglected and badly managed. In the dry eastern counties many pastures were 'certainly of inferior consideration and merit to the arable lands', and without manure and cultivation grasslands produced little or no hay.[2] Arthur Young thought the condition of grassland in Suffolk could 'scarcely be worse', but Norfolk appeared to be yet more depressing.[3] 'No where', he declared, 'are meadows and pastures worse managed.' In north-west Norfolk some irrigation had been attempted by Lord Walpole and Thomas William Coke,[4] but the practice was not widely imitated in the arable districts of England.

On heavy clays in the south Midlands, immense tracts of grass were ill-drained and infested with reeds and sedges. Draining was rudimentary or completely lacking in most areas. In Cheshire it was estimated that the value of meadows might be doubled by the provision of adequate drainage. In several Midland counties some wet and swampy land was already drained by open ditches but much remained to be done. In the Welsh borderland ditching had hardly begun; and in the south-west, the art of draining was neither widely practised nor properly understood. On all claylands the greatest need was for effective underdraining.[5] Large areas of the Midlands, once under the plough, benefited from the ridge and furrow beneath the turf, and ridging was still the most widely practised method of improving drainage on flat or gently sloping surfaces. Mole draining was successful in Huntingdonshire, Cambridgeshire and Bedford-

[1] T. Davis, *General view Wiltshire* (1811), 116.
[2] A. Young, *General view Essex*, II (1807), 94.
[3] A. Young, *General view Suffolk* (1804), 159; A. Young, *General view Norfolk* (1804), 370.
[4] A. Young (ed.), *Annals of agriculture*, XXXVII (1801), 510–12.
[5] H. C. Darby, 'The draining of the English clay-lands', *Geographische Zeitschrift*, LII (1964), 193.

shire, and was attempted further west in the vale of Oxford and in the vale of Gloucester. Hollow-draining, with stones or bushes or with bricks and tiles, was general on heavy soils in Suffolk, but there, as in mid-Norfolk and north Essex, it was carried out mostly on arable land. In 1795, hollow-draining was introduced into Lincolnshire, and in 1797 it was stated to have 'lately made its way into Northumberland'.[1] But in western districts it was still practically unknown. By 1797 Joseph Elkington's method of tapping water at its source in springs and sloughs was employed at a number of places in the Midlands. Each district perfected its own draining techniques and the results were patchy.[2]

In addition to land under permanent grass a considerable area was sown with artificial grasses in leys of more than one year. Clover and ryegrass were the two most widely grown ley crops. In the eastern counties the cultivation of clover was long established and well understood, but the rich loams of Suffolk and south Norfolk were already tired of clover, its place being taken by ryegrass, tares and vetches.[3] In Hertfordshire, on the other hand, clover had been cultivated without ill-effect, 'probably as long, or longer, than in any part of the kingdom'.[4] Clover and ryegrass were grown for one or two years in arable rotations in Lincolnshire, and the counties to the north. They were sown for short leys, but not extensively, in the Cotswolds, in the west Midlands and, to a smaller extent, in Lancashire and Cheshire. They had reached few places west of the Severn or north of the Ribble.

Lucerne and sainfoin were newly introduced as hay crops in many localities. Lucerne was not widely cultivated north of the Thames, but sainfoin was considered the most valuable of all cultivated grasses over much of the chalk country of southern England from Wiltshire to the Chilterns. It was also grown extensively on the light soils of Cambridgeshire and on the sandy soils of west Suffolk, where it was described as 'this noble plant, the most profitable of all others'.[5] On the Cotswolds artificial grasses were reckoned to be 'necessary to the very existence' of

[1] A. Young, *General view Lincoln* (1799), 242; J. Bailey and G. Culley, *General view Northumberland* (1794), 111.

[2] A. Young (ed.), *Annals of agriculture*, XVI (1791), 542–4, 550; J. Johnstone, *An account of the most approved mode of draining land: according to the system practised by Mr Joseph Elkington* (Edinburgh, 1797).

[3] A. Young, *General view Norfolk* (1804), 257; A. Young, *General view Suffolk* (1804), 104.

[4] A. Young, *General view Hertfordshire* (1804), 115.

[5] A. Young, *General view Suffolk* (1804), 106.

a farm, and sainfoin was cultivated 'with great success'.[1] Its cultivation spread to the well managed leys of the vale of York, but had not been successful in the Midlands.

In many districts, old grassland was never broken up. Landowners and farmers either thought it unprofitable or lacked the tools to do it. On some farms ploughing up was expressly forbidden by the terms of leases. In 1800, to promote the conversion of old pastures into tillage, the Board of Agriculture offered substantial prizes for essays describing the best methods to be followed.[2] The response was disappointing, and, apart from small patches broken up for potato growing, the area of permanent grass continued to expand.

A large part of the grassland of England and Wales, together with a considerable proportion of the arable and most of the rough grazing land, was devoted to feeding sheep. The area of land carrying or supporting sheep at different times during the year may have been as much as three-quarters of the whole surface area of the country. About 1800 the number of sheep was variously estimated, but the most detailed calculation was that made by John Luccock, a Leeds wool-stapler.[3] County by county he examined the numbers of long- and short-woolled sheep of fourteen distinct types or breeds. The resulting total, including the number of the lamb crop and of the annual slaughter, added up to 26,150,463, or almost three times the human population. This compared well with an estimate of 25,589,214 made by Arthur Young in 1779.[4] On unimproved hill grazings there was an average of under one sheep per three acres, and some Pennine moors were overstocked at that density. At the opposite extreme, the fine pastures of Romney Marsh fattened more than five sheep per acre. On pastures of intermediate quality, on downland or on Cotswold pasture an acre carried about one sheep.

The leading enterprise for most sheep farmers was wool production. The hardy mountain sheep of northern England and Wales were bred mostly for their fine short wools, but long-woolled varieties were established on the fells of Northumberland, on Exmoor and in other moorland areas. Hill sheep were successfully crossed with improved lowland strains to produce mutton as well as wool. On the limestone hills of central

---

[1] T. Rudge, *General view Gloucester* (1807), 174–5.

[2] *Communications to the Board of Agriculture*, III (London, 1802). The volume contains fourteen essays on the conversion of grassland into tillage.

[3] J. Luccock, *The nature and properties of wool* (Leeds, 1805), 338.

[4] A. Young (ed.), *Political arithmetick, Part II* (London, 1779), 28.

England, on the Lincolnshire Wolds and on the moors of north Yorkshire and Durham, the new crosses were well established. On the chalklands of southern England the old long-horned breeds were being displaced by fat South Downs and New Leicesters so rapidly that it seemed the old stock might 'soon be extinct'.[1] The long-woolled Lincolnshire breed held its own on the marshlands but New Leicesters were reported to be 'spreading very rapidly over the county'.[2] In Norfolk, South Downs had taken possession of all the Good Sand region, the native sheep retaining a dwindling territory in the Breckland.[3] Suffolk was now the home of the Norfolk sheep, but South Downs and New Leicesters were advancing steadily.[4] In all parts of the Midlands the New Leicesters were ousting other breeds but in the West Country they made little headway.

Apart from Romney Marsh, the best grazing lands were appropriated to cattle, but little is known about the total area used to feed them or about the numbers of stock kept. Contemporary estimates put the number at between three and a little over four million.[5] A continually changing element in the cattle population was the number of imported livestock. Each year thousands were driven on the hoof across country from Wales and from Scotland, and in 1801 no fewer than 31,543 Irish cattle were landed at English ports.[6] Some made their way to extensive fattening pastures in the vale of Trent, and on Humberside, but it was not only northern districts that received these droves. They also went to arable farms in the east Midlands and East Anglia where they outnumbered native beef-stock.

While breeders spent more energy and took greater pains to perfect strains of fat-stock than to improving the milking qualities of dairy cattle, dairying remained the most profitable enterprise for a great majority of cattle farmers. Every important livestock region produced cheese, butter, cream or fresh milk. In the north, where the emphasis was on rearing, the vale of Eden, and the Craven and Kendal lowlands sold large surpluses of butter throughout the country. In the husbandry of Lancashire the

---

[1] W. Marshall, *Rural economy of the southern counties of England*, 2 vols. (London, 1798), I, 347.
[2] A. Young, *General view Lincolnshire* (1799), 371.
[3] A. Young, *General view Norfolk* (1804), XV, 449.
[4] A. Young, *General view Suffolk* (1804), 209–16.
[5] A. Young, *Political arithmetick, Part II* (1779), 28, 31; G. E. Fussell, 'Animal husbandry in eighteenth century England, Part I, Cattle', *Agric. Hist.*, XI (1937), 102.
[6] G. R. Porter, *The progress of the nation* (London, 1847), 345.

dairy was the principal source of income. It not only supplied fresh milk to the populous textile manufacturing centres around Manchester and Leeds but also marketed large quantities of cheese. Cheshire, with 92,000 cows in milk, was also one of the leading dairy districts in England.[1] The claylands and clay loams of the Midlands were largely devoted to dairying, while in the vicinity of London some 8,500 cows were kept solely to supply the metropolis with fresh milk.[2] No attention was paid to the breed of the cow as long as it was a good milk producer. 'All round London,' wrote Thomas Baird, 'but particularly near Hackney, Islington, and for several miles thereabouts the cowkeepers engross every inch of land they can procure.'[3] But even in summer when grass was most abundant it was necessary to supplement the feed with grains.

Draught oxen were still to be seen ploughing strong land in the West Country. In Cornwall they were used for all kinds of work; in north Devon they were the premier draught animals, and in north Somerset they were preferred to horses for ploughing. They were also favoured on heavy soils in Herefordshire and in Shropshire. Some were kept in Gloucestershire, in Wiltshire and in Oxfordshire, but they were virtually extinct in eastern England.

Horses were being used in increasing numbers in most parts of the country. John Middleton probably over-estimated the total number at 1,800,000; but it is unlikely that there were fewer than one million.[4] In 1791, London and its environs alone used more than 31,000 horses for non-agricultural purposes. Perhaps as many as ten times that number were working in other parts of the country along canals, in wagon trains, in stage coaches, in hackney carriages. In addition, possibly a quarter of a million were kept for riding, for cavalry, for racing and for private carriages. Those working on farms were yet more numerous. Derbyshire and Leicestershire were the leading horse-breeding counties raising the finest horses for draymen and carters. The North Riding was equally famous for its saddle and coach horses, while many other localities bred farm horses. On the eve of the invention of the steam locomotive horses provided almost the sole source of motive power. Indeed, the capacity of the new machines was assessed in terms of their horse power.

[1] H. Holland, *General view Cheshire* (1808), 252.
[2] J. Middleton, *General view Middlesex* (1807), 417.
[3] A. Young (ed.), *Annals of agriculture*, XXI (1793), 112.
[4] J. Middleton, *General view Middlesex* (1807), 639.

Other livestock were relatively unimportant in England. Geese, ducks and chickens picked their way across the stubble and scratched among the winnowings in the farmyard. The pig was the universal scavenger and a usual accompaniment of the cottage garden. The population of the manufacturing districts in Lancashire and in the Midlands had little taste for pork, and in the West Riding Robert Brown complained that there were too many pigs. Only at the distilleries on the south side of London were pigs fattened in large numbers, being brought from as far afield as Berkshire, Shropshire and the East Riding. Hardly less important than domestic pigs and poultry as sources of food were the deer, rabbits and wildfowl of forest, heath and marsh.

Woods, plantations and parks

In 1800 probably no more than two million acres of woodland remained in England and Wales, now one of the least wooded of all north European nations. Most of the country within twenty miles of a seaport had been stripped of shipbuilding timber. The oakwoods along the valleys of the Tyne, Wear and Tees had been denuded not only of ship timber but also of small wood for pit props. The coastal plain of County Durham had also been cleared of its wood, so that the shipbuilders of Tynemouth, Sunderland and Middlesbrough now looked to the 25,500 acres of woods in the North Riding for their supplies. But much of that extensive area had already been cut over, and the shipyards of Whitby and Scarborough had consumed most of the available timber.[1] All that remained were slender saplings raised from the stools of timber trees.

The situation on the west coast was no better. Cheshire suffered a severe shortage of oak timber, and could not even find enough oak bark for her tanneries. Neither there nor in Lancashire had any considerable tract of ancient woodland been spared. The Lake counties possessed remnants of native woods but they were seriously deficient in oak. In Gloucestershire, the Forest of Dean, once a great storehouse of naval timber, furnished barely sufficient mature trees to repair the ships that put in at Bristol. The woods of Worcestershire had been reduced to spinneys and copses producing substantial quantities of hop poles and billet wood but little else of value.

The naval dockyards along the Channel coast were desperately short of suitable materials. The uplands of Devon and Cornwall were bare of trees, and the few oaks growing at the heads of valleys were 'wasting in

[1] J. Tuke, *General view North Riding* (1800), 187.

a most alarming manner'.[1] Timber trees had largely disappeared from the oakwoods of Dorset and Hampshire, and were no longer plentiful even in the depths of the Wealden forests. The south-east produced immense quantities of wood for hop poles, hurdles, faggots, barrel staves, charcoal and material for firing kilns and making gunpowder, but little sound timber. Essex and Middlesex had extensive areas of forest and parkland but no unexploited reserves of fully grown timber. In Norfolk and Suffolk, thousands of acres of oak, elm and ash were regularly cut for poles and billets but their mature woods 'hardly deserved mentioning'.[2]

Some of the least wooded counties were well stocked with hedgerow trees. To the west, in Devon, Somerset and Herefordshire; to the east, in Essex, Kent, Surrey and Sussex, hedgerows were both numerous and heavily timbered. The hedges of Devon were deep and rambling, whereas those of Sussex were kept 'in a state of garden cleanness'.[3] In both districts free growing oaks in field and hedge spread their branches to give shade to cattle and, when they were felled, to yield valuable pieces of curved and crooked timber. About two-thirds of the English oak used in building a man-of-war had to be curved or crooked.[4] A few hedgerow trees were specially trained to produce timber of exceptional size and shape required for the frame of a ship. Such trees were worth at least twice as much as ordinary oaks.[5]

In the centre of England were to be found the densest stands of timber. William Marshall remarked that the Midland counties, 'with little latitude, may be said to contain all the ship timber now growing in the kingdom', but as he wrote canals were 'taking off the produce of the interior'.[6] Bristol shipbuilders cleared the great timber from the lower Severn valley, made inroads into the oakwoods of Herefordshire and, working their way upstream, now drew their main supplies from 'the very fine woods of oak' left standing in Shropshire.[7] The Forest of Arden in Warwickshire still had considerable tracts of oak woods; and the forests, parks and chases

[1] G. B. Worgan, *General view Cornwall* (1811), 98; C. Vancouver, *General view Devon* (1808), 457.

[2] A. Young, *General view Suffolk* (1804), 165.

[3] A. Young (ed.), *Annals of agriculture*, xx (1793), 289.

[4] W. Marshall, *On planting and rural ornament*, 2 vols. (3rd ed., London, 1803), I, 49.

[5] R. G. Albion, *Forests and sea power* (Cambridge, Mass., 1926), 7–9.

[6] W. Marshall, *The review and abstract of the county reports to the Board of Agriculture: Midland Department* (York, 1818), 381.

[7] J. Plymley, *General view Shropshire* (1803), 212.

in Northamptonshire were abundantly wooded. Thirty years after the opening of the Trent and Mersey Canal, its industries consuming increasing quantities of wood each year, Staffordshire was 'well stocked with all kinds of timber, notwithstanding the immense quantities that have been cut down of late years'.[1] The east Midlands, on the other hand, were unremarkable for their timber resources.

Landowners planted not only oak, ash, beech and elm, but many other native, exotic, deciduous and evergreen trees. It scarcely needed the repeated exhortations of John Evelyn's *Sylva*, revised and considerably enlarged by John Hunter in 1801, or practical hints from William Marshall, or the generous prizes of the Royal Society of Arts, to induce gentlemen to 'adorn their goodly mansions and demesnes with trees of venerable shade and profitable timber'.[2] During their lifetime they had little to gain but they might expect handsome profits to accrue to their grandchildren and great-grandchildren. At Longleat, an acre of fully grown oaks and other trees was worth £1,500, and in many parts of the country plantations bearing the names of victorious admirals anticipated future glories at sea and high prices at the shipyards.[3] For the present, landowners contented themselves with the pleasure of planting; their efforts contributed more to the adornment of parks and to the mantling of bare wastes than to a solution of the immediate shortage of naval timber.

The largest estates were the most steadfast planters and the leading suppliers of merchantable timber. At Bowood in Wiltshire the earl of Shelburne planted 150,000 trees each year.[4] Much planting was of oak but the duke of Portland's estate at Welbeck also raised great quantities of beech, larch, Spanish chestnut, Weymouth pine and many other conifers.[5] On neighbouring estates in the Dukeries 'the spirit of planting' was reported to have 'prevailed much', conifers being nurtured with as much pride as oaks.[6] On the light soils of Norfolk millions of trees were planted. At Holkham alone more than two million were planted between 1781 and 1801, of which fewer than one in six were oaks. Among 48 other varieties, cherries, chestnuts, hazel, poplars, evergreen oaks, Scots pine, spruce and larch figured prominently.[7] In Lincolnshire and on the Yorkshire

[1] W. Pitt, *General view Stafford* (1796), 92.
[2] J. Hunter (ed.), *John Evelyn's Sylva*, 2 vols. (York, 1801), II, 303.
[3] *Ibid.*, II, 299; T. Davis, *General view Wiltshire* (1811), 191.
[4] A. Young (ed.), *Annals of agriculture*, VIII (1787), 76.
[5] J. Hunter (ed.), I, 89.
[6] R. Lowe, *General view Nottingham* (London, 1798), 53, 89.
[7] A. Young, *General view Norfolk* (1804), 382–3.

Wolds great mixtures of trees were propagated by Sir Joseph Banks, the duke of Ancaster and Sir Christopher Sykes, but on light soils farther north more larch was planted than any other species.

In many new plantations conifers were introduced as nurse trees. In south-east England, Scots pine and larch were sparsely disseminated among oaks and intermixed with seedlings of tender native and exotic trees in clumps and ornamental plantations. In the west and north they occupied a large proportion of the ground. In Shropshire, Joseph Plymley reported, 'many modern plantations of various sorts of firs and pine, generally mixed with different deciduous trees'.[1] In Cheshire, extensive plantations on large estates were mostly coniferous, and in 1795 the Royal Society of Arts awarded a gold medal for the planting of half a million conifers there. On high ground near Ambleside, the bishop of Llandaff attempted to establish 100 acres of mixed trees, but like most sites in Westmorland it was found to be 'too cold for any sort of wood except the fir and larch', the larch flourishing where all else failed.[2] Oak, ash, alder, birch, hazel were coppiced for billet wood and charcoal but rarely grew into tall trees. The clothing of rocky slopes with fast growing conifers was welcomed as a means of utilising and improving poor soils. The Board of Agriculture report for Cumberland, for example, referred to 'a large plantation of larches thriving exceedingly well, on the steep edge of the west side of Skiddaw'.[3]

Some writers objected to the appearance of conifers. William Wordsworth thought the scenery of the Lake District was being deformed by 'the small patches and large tracts of larch plantations that are overrunning the hill-sides'.[4] But Bailey and Culley had nothing but praise for the plantations rising in every part of northern England. In Northumberland larch more than pine or spruce was adding 'greatly to the ornament of the country',[5] In the North Riding larch grew rapidly and produced sound timber, but in southern England Scots pine was preferred. It brought about a complete change in the appearance of the Bagshot Sands in Surrey, of the Greensand outcrops from Bedfordshire to Wiltshire, and of the sands of Hampshire and Dorset. In Cornwall and other western counties

[1] J. Plymley, *General view Shropshire* (1803), 212.
[2] A. Pringle, *General view of the agriculture of the county of Westmorland* (Edinburgh, 1797), 278.
[3] J. Bailey and G. Culley, *General view of the agriculture of the county of Cumberland* (London, 1794), 202. [4] W. M. Merchant (ed.), 120.
[5] J. Bailey and G. Culley, *General view Northumberland* (1794), 109.

many different kinds of trees grew equally well, lending varied colours and textures to the scene, breaking the harsh baldness of the skyline.

Within thirty miles of London, from 'the rich blue prospects of Kent, to the Thames-watered views in Berkshire', the landscapes created by Charles Bridgeman, William Kent, Lancelot Brown and Humphry Repton had transformed the face of the countryside.[1] Hundreds of parks lined the roads out of London; hilltops were crowned with ornamental clumps of trees; streams tumbled over artificial cascades to spread out in broad lakes; towers, follies, ruins and temples caught the eye at the end of each studiously contrived vista. Grounds laid out and planted in the early eighteenth century were now fashionable showplaces visited by admiring tourists. Thomas Whateley's *Observations on modern gardening*, reprinted in 1801, commended Claremont, Esher Place, Blenheim, Caversham, the Leasowes, Woburn Farm, Painshill, Hagley, Persfield and, above all, Stowe. Stowe epitomised all the splendour and magnificence of the early enthusiasm for landscape gardening, but Horace Walpole's romantic taste inclined to the subtleties of Rousham or to the alpine savagery of Painshill, where 'all is great, foreign, and rude'.[2] By 1800 the new rage for picturesque scenery, described in William Gilpin's *Tours*, had reached its height, and many travellers set out to discover rugged grandeur in the Welsh borderland, beckoned by Richard Payne Knight 'to some neglected vale'.[3]

In lowland England, thousands of acres were landscaped in the smooth manner perfected by Lancelot Brown (Fig. 90). The elegant compositions of shaven lawn, calm water, rounded clumps, winding drives were repeated at many other places by Humphry Repton. Some were early landscape gardens remodelled in detail or extended to take in fresh prospects but some were newly created from farmland or from waste. It was, above all, on tracts of poor soil that landscaping made its greatest impact. Englishmen were slowly losing their fear and hatred of uncultivated land. Much had been tamed and cultivated, but even where cultivation was a forlorn hope it was possible to take pleasure in picturesque disorder and perhaps to improve the scenery by judicious planting.[4]

1 H. Walpole, 'History of the modern taste in gardening', in W. Marshall, *On planting and rural ornament*, I, 242.
2 *Ibid*, 237.
3 R. P. Knight, *The landscape, a didactic poem* (London, 1794), Book 3, line 235.
4 H. C. Prince, *Parks in England* (Shalfleet, I.O.W., 1967).

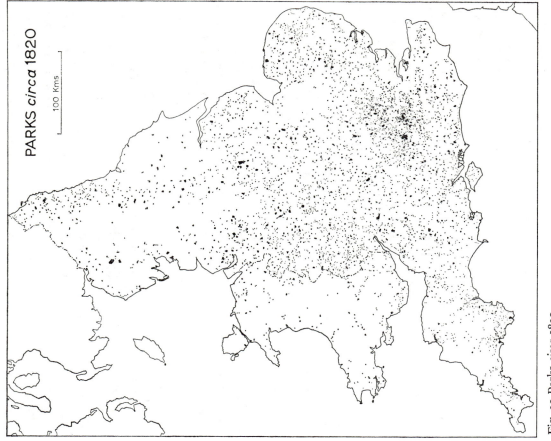

PARKS *circa* 1820

100 Kms

Fig. 90 Parks *circa* 1820
Based on the county maps of A. Bryant (1822–35), C. & J. Greenwood
(1817–33) and R. G. Baker (1821).

Heath, moor and marshland

If England and Wales had somewhat less wood in 1800 than at present, there was probably more unimproved and waste land. It is difficult to obtain an accurate assessment of the area of unimproved land partly because of the inherent difficulty of drawing a line between managed woodland and tumbled-down scrub, or of distinguishing downland pasture from rough hill grazing or meadow liable to flood from undrained marsh. A report of 1808 estimated the waste land at 7.8 million acres (6.2 in England and 1.6 in Wales).[1] This was higher than W. T. Comber's figure of 6.5 million for 'commons and wastes'.[2]

As well as wastes, commons and rough grazing, much cut-over wood-land remained derelict awaiting improvement. Thus Delamere Forest contained nearly 10,000 acres of waste land and in well-timbered Stafford-shire 9,220 acres of Needwood Forest lay 'in a state of nature'.[3] Between 1787 and 1793 the reports of the Middleton Commission, appointed to enquire into the state of the Crown forests, painted a gloomy picture of neglect. A large part of the royal forests lay treeless and idle. Vast areas were reported to be totally unproductive, yet little was being done to improve them.[4] William Gilpin's *Remarks on forest scenery*, written while he was vicar of Boldre in the New Forest, describes that picturesque waste as 'consisting of heathy land and carpet lawns interspersed with woods'.[5] On the other side of Hampshire Gilbert White described Woolmer Forest in 1789 as consisting 'entirely of sand covered with heath and fern' and without 'one standing tree in the whole extent'.[6] There were occasional improvements; thus in 1798 Robert Lowe reported that at least 20,000 acres of the forest district of western Nottinghamshire had been privately enclosed, some reclaimed for agriculture, some planted with trees.[7] Vast stretches of desolate heath lay within a few miles of London. Surrey alone had 70,000 acres of wastes and commons, mostly on the dry

[1] *General report on enclosures drawn up by order of the Board of Agriculture* (London, 1808), 139–41.
[2] See p. 403 above.
[3] H. Holland, *General view Cheshire* (1808), 208. W. Pitt, *General view Stafford* (1796), 102.
[4] R. G. Albion, 135–6.
[5] W. Gilpin, *Remarks on forest scenery* (London, 1791), ed. T. D. Lander (London, 1834), I, 125.
[6] Gilbert White, *A natural history of Selborne* (Everyman's Library), 17.
[7] R. Lowe, *General view Nottingham* (1798), 51–98, 150.

sands in the western half of the county; and in Middlesex open heaths at Hounslow and Hampstead almost reached the edge of the built-up area.[1] Or, again, the improved arable husbandry of Norfolk and Suffolk scarcely touched the Breckland. In 1804 in the 16 miles from Newmarket to Thetford there were 'immense wastes', and another 18 miles as far as Swaffham deserved 'to be called a desert'.[2] There were smaller but no less dreary stretches surrendered to rabbit warrens and sheep-walks in the Sandlings of east Suffolk and in north Norfolk. In Lincolnshire much light land had already been reclaimed, but furze and broom still encumbered thousands of acres on the Wolds and along the Heath belt north and south of Lincoln. The sandy heaths of Cannock Chase and Sutton Coldfield in Staffordshire occupied more than 30,000 acres, and almost as large an area was waste in pastoral Cheshire.

A small part of the heathland was occasionally brought into cultivation. In Lincolnshire and the East Riding, 'the greater part of the Wold townships which remain open have a large quantity of *outfield* in ley land'.[3] About 1800, similar practices were reported in Norfolk, in Nottinghamshire, in Staffordshire, in Lancashire, in Shropshire, on the Mendips and as far west as Devon and Cornwall.[4]

Open moors over 800 feet above sea-level were rarely if ever broken up. About one-third of northern England was uncultivated mountain and moor. The economy of hill farms in these areas was almost exclusively pastoral. In the Lake District the practice of transhumance survived where cattle were kept, but when sheep replaced cattle, seasonal movement ceased and summer shielings were abandoned.[5] In the southern Pennines, hill farms bred and reared store cattle during the summer. Autumn sales provided the fattening pastures of the Midlands with one of their main sources of store animals.[6] But over most of the mountain and moorland areas of England and Wales cattle were far less numerous than sheep, and their rough grazing was devoted mostly to sheep raising.

[1] W. Stevenson, *General view Surrey* (1809), 456–70; J. Middleton, *General view Middlesex* (1807), 112–120.

[2] A. Young, *General view Suffolk* (1804), 170; A. Young, *General view Norfolk* (1804), 385.

[3] I. Leatham, *General view of the agriculture of the East Riding of Yorkshire* (London, 1794), 42.

[4] W. Smith, *An economic geography of Great Britain* (London, 1949), 17–18.

[5] A. E. Smailes, 65.

[6] R. W. Sturgess, 'A study of agricultural change in the Staffordshire moorlands, 1780–1850', *North. Staffs. Jour. Field Studies*, 1 (1961), 77–85.

At the edges of the waste, steady nibbling by cottagers and small farmers added thousands of tiny enclosures to the cultivated area. Near the woollen manufacturing districts in the West Riding of Yorkshire, on the edge of Rossendale and in Devonshire piecemeal inning was particularly active.[1] In northern Pennine dales, on the sides of the North York Moors and on the East Moor of Derbyshire, larger intakes of ten, twenty or even fifty acres were added to a few farms.[2] Patches of moorland were occasionally brought into cultivation by paring and burning the turf, by draining and by dressing the surface with lime; and a succession of oat crops was grown before the land reverted to grass. The practice was widespread in Devon and Cornwall and was also prevalent in Somerset, in parts of the Midlands and in Lancashire. But on the drier uplands of eastern England it was considered harmful.

Some of the largest remaining tracts of waste lay in the coastal marshes and peat-filled basins around the Humber, in the Fens, in the Somerset Levels and on the mosses of south Lancashire. The peatlands of eastern England had already been extensively embanked and drained; the 'cars' and 'marshes' of the vale of Pickering, the 'cars' of the Hull valley and the warplands of the vale of York were largely protected.[3] Ague had almost been eliminated from the Trent and Humber lowlands and, after the draining of Walling Fen, that country was now 'full of new-built houses, and highly improved'.[4] In Lincolnshire over a century of unremitting effort had brought 150,000 acres of marsh and fen into cultivation, to become 'one of the richest tracts in the kingdom'. The remaining 66,000 acres of wet land was fast disappearing, and the draining of Deeping Fen, completed in 1799, represented a 'very capital improvement'. To the north, the watery expanses of East Fen, West Fen and Wildmore Fen were drained within the next ten years.[5] In Huntingdonshire, in Suffolk

[1] E.g. G. H. Tupling, *The economic history of Rossendale* (Manchester, 1927), 42–69; W. G. Hoskins, 'The reclamation of the waste in Devon, 1550–1800', *Econ. Hist. Rev.*, XIII (1943), 80–92.

[2] J. Chapman, 'Changing agriculture and the moorland edge in the North York Moors', unpublished M.A. thesis, University of London, 1961; S. R. Eyre, 'The upward limit of enclosure on the East Moor of north Derbyshire', *Trans. and Papers, Inst. Brit. Geog.*, XXIII (1957), 61–74.

[3] J. A. Sheppard: (1) *The draining of the Hull valley* (East Yorks. Local Hist. Soc., No. 8, York, 1958); (2) *The draining of the marshlands of south Holderness and the Vale of York* (*Ibid.*, No. 20, York, 1966).

[4] A. Young, *General view Lincoln* (1799), 6.

[5] H. C. Darby, *The draining of the Fens* (2nd ed, Cambridge, 1956).

and in Cambridgeshire new tracts of peatlands were being reclaimed and sown with rape and clover. Among the largest, nearly 14,000 acres of Burnt Fen in Suffolk were now cultivated.[1]

By 1800 the condition of the Bedford Level, drained in the seventeenth century, was fast deteriorating. Hundreds of windmills failed to clear the water from miles of drains, dykes and lodes, and, as the peat surface shrank, windmills became less and less effective. In July 1805, surveying the country between Whittlesey and March, Arthur Young was shocked at the sight 'in all which tract of ten miles, usually under great crops of cole, oats and wheat, there was nothing to be seen but desolation, with here and there a crop of oats or barley, sown so late they can come to nothing'. He could only conclude: 'The Fens are now in a moment of balancing their fate; should a great flood come within two or three years for want of an improved outfall, the whole country, fertile as it naturally is, will be abandoned.'[2]

In Oxfordshire, Otmoor remained an undrained bog of 4,000 acres, 'the most considerable, and at the same time the most valuable waste in this county'.[3] In 1800 no part of the Somerset Levels was without a drainage system, however rudimentary, but the greatest progress had been made in the Brue and King's Sedgemoor while little had yet been done in the southern levels.[4] Draining was carried out in a piecemeal fashion and the newly reclaimed lands were inadequately protected against floods. In Devon and Cornwall small patches of coastal marsh were walled in and improved. New walls were raised along the coasts of Kent, Essex and around Morecambe Bay. In south Lancashire large areas of Trafford Moss, Rainford Moss, Bolton Moss and Bootle Moss were already growing potatoes with clover, vetches and barley sown to prepare the ground for more potatoes. The largest stretches, such as Chat Moss, remained neglected, and Cheshire still had 18,000 acres in unimproved bogs and mosses.

INDUSTRY

In 1800 the beginning of the industrial revolution, associated particularly with the invention of the steam engine and the improvement of textile machinery, left few scars on the face of England. Contemporaries marvel-

[1] A. Young (ed.), *Annals of agriculture*, XVI (1791), 463–76.
[2] A. Young (ed.), *Annals of agriculture*, XLIII (1805), 545–7.
[3] R. Davis, *General view of the agriculture of the county of Oxford* (London, 1794), 22.
[4] M. Williams, *The draining of the Somerset Levels* (Cambridge, 1970), 131–52.

led at the power of steam to drive mighty instruments of mass-production but few at that time visualised the shapes of the infernal landscapes to come. The maps of the time show no closely packed streets of houses, and no densely built towns outside Manchester, Salford, Leeds and Bradford. The manufactured products of England came from hundreds of mills, mines and small workshops scattered over the countryside.

Rural industries

Districts which, at the present time, are largely agricultural, in 1800 possessed a great variety of manufacturing activities. In addition to flour mills, bakeries, maltings, breweries, sawmills, slaughterhouses and tanneries (processing the produce of farms and forests), rural brickyards, tileries, joineries, wheelwrights' shops, cooperages, saddleries, village smithies and foundries supplied the nation with a large part of its building materials, implements and tools. The countryside teemed with craftsmen making many articles that we now do without or obtain from factories.

The Board of Agriculture reports, mentioning only the most specialised and highly localised industries, indicate how diverse those activities were.[1] Bedfordshire made osier baskets, reed mats, pillow lace; hemp spinning was nearly extinct but straw plaiting was widely practised not only there but also in Cambridgeshire, Hertfordshire and Buckinghamshire. In many villages and small towns in Suffolk, in Norfolk and in Lincolnshire, sacking, coarse linen and hempen cloth were woven, in addition to large quantities of woollen and worsted thread spun in cottages. Kent, the Garden of England, had not only oasthouses but dyeworks, saltworks, copperas works, paper mills, gunpowder mills, linen and calico printing shops. Surrey made paper, oil, snuff, leather, parchment, ironware and printed cloth. Both Sussex and Hampshire were predominantly agricultural counties but the dying charcoal iron industry of the Weald and Henry Cort's rolling mill at Fontley were situated within their boundaries. Berkshire, Oxfordshire, Gloucestershire, Wiltshire, Dorset, Somerset and Devon produced a bewildering variety of yarns and fabrics from sailcloth to silk ribbons. In most counties, a number of villages specialised in the making of pottery, glassware, hats, gloves, boots, buttons, pins, iron and brass ware, tin-plate, soap and candles. Over the whole countryside these industries were widely diffused.

In a large number of villages on the borders of Devon, Somerset and Dorset, in west Wiltshire and in the Gloucestershire Cotswolds, as much

[1] Lord Ernle, 308–12.

or more than half the population was reported in the 1801 Census to be engaged in manufacturing. These were districts where, in addition to the usual rural crafts, cloth-making was carried on in a majority of homes. Men and boys plied their looms, while women and children spun yarn. There was also a wide scatter of manufacturing households in the worsted-spinning districts of north Essex and East Anglia. But the highest pro-portion of the population engaged in rural and domestic textile industries outside south Lancashire and the West Riding was in the east Midlands, in southern Nottinghamshire, southern Derbyshire and western Leicester-shire. Moreover the wire-drawing, nailmaking, ironmongery and hard-ware trades were dispersed throughout the country from the Forest of Dean through Worcestershire and Staffordshire to the Lake District, and individual firms put out work over a very wide area. Warrington file makers distributed materials to be made up by outworkers in a number of villages in south Lancashire and Cheshire; while Dudley nailmakers employed domestic workers as far afield as the Stour and Tame valleys.[1]

Although most industrial activities were highly scattered, important branches of some leading industries were rapidly concentrating in a few localities. Gunsmiths, locksmiths and button makers were increasing in hundreds of small workshops in Birmingham, watchmakers in Coventry and in Clerkenwell, and cutlers in Sheffield.[2] Salisbury cutlers still pro-duced blades superior in workmanship to any in the country; Stafford, too, manufactured large quantities of cutlery and held a share in overseas markets, Birmingham sword-makers prospered during the wars, but the steel industry of Woodstock in Oxfordshire had lately succumbed to competition from Sheffield and Birmingham.

Industries that were growing were spreading most rapidly in one or two localities. There was an immense increase in the production of salt but most of that increase was produced in Cheshire and on Tyneside. By 1800 the Weaver navigation was carrying nearly 140,000 tons of salt a year down to the Mersey and bringing 100,000 tons of coal up to the saltworks. Cheshire salt was exported all over the world for preserving fish and meat. It was also shipped to the Potteries for glazing, and some soda was

[1] T. S. Ashton, *An eighteenth-century industrialist: Peter Stubbs of Warrington, 1756–1806* (Manchester, 1939), 9–22; S. Timmins, *The resources, products and industrial history of Birmingham and the Midland hardware district* (London, 1866), 86.

[2] J. H. Clapham, *An economic history of modern Britain: the early railway age, 1820–1850* (2nd ed, Cambridge, 1930); M. J. Wise, 'On the evolution of the jewel-lery and gun quarters in Birmingham', *Trans. and Papers, Inst. Brit. Geog*, xv (1949), 57–72.

extracted.[1] About four-fifths of the salt output was exported for half a million pounds.

Glass production increased greatly as the standards of lighting in homes and workshops were improved. In 1800 its value was estimated at about one and a half million pounds.[2] A score of provincial towns and ports continued to make crown and bottle glass on a small scale. London and Southwark were still by far the largest producers of most types of glass, but the increasing demand for thick plate glass was met solely by a large new works at Ravenhead near St Helens. Large new glasshouses were rising on Tyneside to supply window glass; and at Stourbridge, already a prospering glass-making town, 'a number of very lofty and spacious glasshouses' were added at the end of the eighteenth century.[3]

The manufacture of pottery, which a century earlier had been carried on in every town in England, was fast becoming concentrated on a district of some twelve square miles containing five small towns identified simply by the name 'The Potteries'. Although the five towns retained their separate names, they had become joined together so closely by 1795 that they struck a traveller 'as but one town'.[4] Josiah Wedgwood's porcelain was more highly valued than the fine china of Derby, Chelsea, Worcester and the Delft ware of Mortlake. Cratefuls of earthenware plates and tea-cups shipped from a hundred other Staffordshire potteries were cheaper than those made in Gateshead or Barnstaple or Caughley. A large part of the two million pounds' worth of pottery manufactured in England and Wales came from this district. The great concentration of production had been achieved by a labour force of fewer than 10,000, largely without the assistance of labour-saving machines, relying upon roads and canals for the transport of the bulk of its raw materials, including much of its clay, lead and salt. By 1800 waterways linked the Potteries with Liverpool, Hull and Bristol serving distant markets.[5] In their search for economical methods of using materials, north Staffordshire manufacturers had perfected new glazes, new methods of printing, and of preparing bone paste, and in 1793 had built a steam mill to crush flints. But the scale of operations

[1] H. Holland, General view Cheshire (1808), 12–73, 315–24.
[2] W. Smart, Economic annals of the nineteenth century, 1801–20 (London, 1910), 21.
[3] W. Pitt, General view Stafford (1796), 168.
[4] J. Aikin, A description of the country from thirty to forty miles around Manchester (London, 1795), 516.
[5] H. A. Moisley, 'The industrial and urban development of the north Staffordshire conurbation', Trans. and Papers, Inst. Brit. Geog., XVII (1951), 149–65.

was never large. Josiah Wedgwood employed fewer workers at Etruria than William Reynolds at a new china manufactory opened in 1797 at Coalport in Shropshire which gave work to 400 people.[1]

Large commercial enterprises were taking an increasing share of the processing industries, and were flourishing in the metropolis and in other large centres of population. According to one estimate, brewing ranked third in value of output following woollen cloth and leather goods.[2] Enormous quantities of small beer were brewed at home for private consumption, and for sale at thousands of brewing victuallers' houses, but in 1800 twelve leading London firms brewed over one million barrels of porter, nearly one-quarter of all the strong beer consumed in England and Wales.[3] The great London breweries supplied most of the ale houses in Middlesex and large numbers in Kent and Surrey. They and a group of breweries on the Trent at Burton, Nottingham and Newark were the largest exporters of beer.[4] Distilling was also highly localised, more than half the spirits sold and paying duty came from London. London produced almost all the gin drunk in the country.[5]

In 1800 leather goods manufactured in England and Wales were worth more than ten and a half million pounds, their value exceeded only by that of woollen cloth. Almost every town from Berwick to Penzance had its tanners and curriers, leather being used for many articles now made from rubber, plastics and cloth. The daily wear that men and horses gave their boots, belts, saddles and straps was a good deal harder than at present. By the end of the eighteenth century many women and children of labouring families were wearing leather shoes, but many others still went barefoot or wore wooden clogs. A multitude of independent cobblers and saddlers made most of the shoes and harness worn in the country but London shops were supplied with shoes made in Stafford, Northampton, Kettering and Wellingborough. Curriers in inner north-east London and in the West End supplied glovers, bookbinders, coach trimmers and strap makers. The greatest concentration of tanneries in the country lay to the south of the Thames. About one-eighth of the nation's tanning business was transacted in Bermondsey, and a number of large tanneries were

[1] J. H. Clapham, 185; J. Plymley, *General view Shropshire* (1803), 341.
[2] W. Smart, 22–3.
[3] J. Middleton, *View of Middlesex* (1807), 583; G. R. Porter, 572.
[4] P. Mathias, 'Industrial revolution in brewing', *Explorations in entrepreneurial history*, v (1952–3), 208–24.
[5] D. George, *London life in the eighteenth century* (3rd ed., London, 1951), 40.

situated nearby in the Wandle valley from Mitcham to Croydon. Liverpool alone had a tannery larger than any in London, and hides imported from Ireland and the New World were processed in south Lancashire and Cheshire.[1] The home market consumed almost the entire output of the industry but English leather was highly valued abroad for its suppleness.

The woodworking crafts used little machinery and almost no mechanical power. A circular saw, invented in 1790, was used only in naval dockyards, and turning machines were speeding the making of chairs at High Wycombe. Shoreditch was the most important centre of furniture making. In 1800 most counties possessed one or two paper mills but the industry was concentrating on chalk streams near London, in Hertfordshire, Kent, Berkshire and Hampshire.

The woollen and worsted industries

The manufacture of woollen and worsted cloth had for long been 'supposed the sacred staple and foundation of all our wealth'.[2] It employed more workers than any other industry. The value of fabric produced for home consumption amounted to eleven million pounds, and exports were worth another eight million pounds. In 1806 a Select Committee on the woollen manufacture of England reported that production 'has been gradually increasing in almost all the various parts of England in which it is carried on; in some of them very rapidly'.[3] Exports continued to increase while the demand for wool steadily outpaced home production.

But the benefits of rising productivity were not evenly distributed. The industry brought prosperity to some areas while others were depressed. Labour-saving machines had caused widespread unemployment among spinners of woollen and worsted yarn. While spinning declined in almost all the old-established centres of the industry it boomed in the West Riding (Fig. 91). Not unnaturally, commentators blamed the new factories for depriving the rest of the country of its livelihood. In 1791, a west of England manufacturer asserted that 'Yorkshire, by dint of such machines and engines, not only use all their wool, but send down into the west country and buy it up out of the very mouths of the wool dealers

---

[1] J. Statham, 'The location and development of London's leather manufacturing industry since the early nineteenth century', unpublished M.A. thesis, University of London, 1965, 70, 81, 85.

[2] A. Young, *The farmers' letters to the people of England* (London, 1767), 22.

[3] *S.C. on the woollen manufacture in England, 1806, Report*, 3 (P.P. 1806, iii).

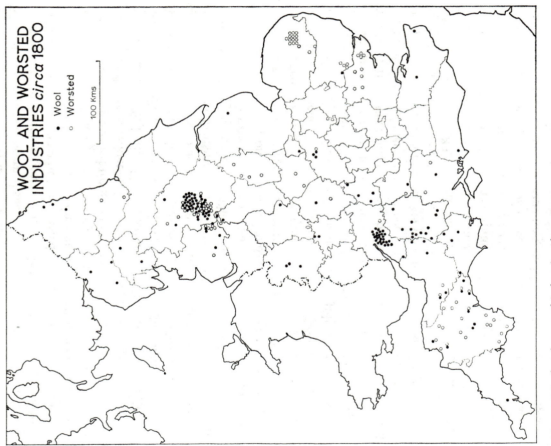

WOOL AND WORSTED
INDUSTRIES *circa* 1800

• Wool
○ Worsted

100 Kms

Fig. 91 Wool and worsted industries *circa* 1800
Based on the manufacturers mentioned in: (1) *S.C. on the woollen manufacture
in England, 1806, Report and Minutes of evidence* (P.P. 1806, iii); (2) the
county *General views* of the Board of Agriculture; (3) Arthur Young (ed.),
Annals of agriculture, 45 vols. (Bury St Edmunds, 1784–6; London,
1786–1815).

and clothiers, and thereby take our trade with it'.[1] As far as the spinning of woollen yarn was concerned such complaints contained a small element of truth, but the fortunes of woollen weaving varied widely from district to district, while the state of the worsted trade was yet more complex.

The introduction of machines for spinning woollen yarn brought distress to much of Wiltshire and to neighbouring Gloucestershire. In 1796 Sir Frederick Eden learned from a clothier at Seend in Wiltshire that 'the poor, from the great reduction in the price of spinning, scarcely have the heart to earn the little that is obtained by it'.[2] Because of the decline in spinning, Stroud and other Cotswold villages, according to one observer, 'fell into decay, and almost wholly into beggary', whence colonies of mendicants poured into adjacent towns. Even where trade was active, as at Chippenham in Wiltshire in 1796, Arthur Young lamented that there were 'many begging children'. The use of spinning jennies in the home meant that families were made redundant for everyone that obtained work.

Cotswold weavers, on the other hand, successfully resisted attempts made in 1792 to install flying shuttles at Trowbridge in Wiltshire; and in south Gloucestershire few weavers were out of work, and after 1796 unemployment practically ceased. The setting up of gig-mills to dress woollen cloth appeared as a fresh threat, and in July 1802 serious riots broke out in Wiltshire. The prospect of further outrages did nothing to encourage manufacturers to adopt new machines, such as shearing frames. A witness from Bradford on Avon before the Select Committee of 1806 stated: 'if that machinery had not been instituted there would not be so many boys running about the streets without shoes or stockings on, and nearly half-starved'.[3]

Gloucestershire clothiers were not modernising their methods, but their order books were full, and unemployment was negligible in comparison with that of neighbouring counties. In Hampshire, the decay of the serge industry was attributed to the war and was optimistically 'expected to revive again on the return of peace'.[4] In Berkshire, clothiers abandoned hope of regaining the trade they had lost, while in Oxfordshire, they had resigned themselves to a 'very depressing poverty'. Even at Witney woollen manufacturing had shrunk to a point where there was

[1] H. Wansey, *Wool encouraged without exportation; or practical observations on wool and the woollen manufacture* (London, 1791), 69.
[2] F. M. Eden, *The state of the poor*, 3 vols. (London, 1797), III, 796.
[3] S.C. *woollen manufacture, 1806, Minutes of evidence*, 308 (P.P. 1806, iii).
[4] C. Vancouver, *General view of the agriculture of Hampshire* (London, 1810), 404.

'very little expectation of its ever reviving'.[1] It had also diminished in the Midlands; the country around Kettering no longer employed more than 3,000 spinners and most districts had taken up other domestic crafts.

In south-western England the decline of spinning was widespread, and severe hardship afflicted places where weaving had also declined. In north Devonshire many women were out of work, and John Collinson painted a gloomy picture of the Somerset industry in 1791. At Milverton the manufacture of serges and druggets had much declined; in the country around Keynsham they were 'now entirely dropt'; and at Pensford the industry was 'dreadfully decayed', and 'bereft of the benefit of trade, many of the houses are fallen into ruins'.[2] At Taunton, once the foremost manufacturing town in Somerset, the industry had been so reduced that it seemed unlikely to recover, and only at Wiveliscombe and Wellington were looms still active.[3] The loss of Spanish and Mediterranean markets delivered a death blow to the serge industry in south Devon, but its life was prolonged for a few years by contracts from the East India Company. As late as 1796 in Exeter itself, despite the use of many spinning jennies, there was little unemployment: 'on the contrary, they are not able to get spinners'.[4]

The most depressed textile villages in the country were in north Essex and in mid-Suffolk where not only had the spinning of woollen and worsted yarn declined but the production of Colchester baize had fallen to a quarter of what it had been before the outbreak of war in 1793. Woollen manufacturing had once been the leading industry of Essex, 'but from its long continued dwindling condition' so it was said in 1806, 'it is uncertain whether it will many years remain so'.[5] By 1803 no fewer than 38,337 persons, or 17% of the population of Essex, received poor relief.[6] Nowhere in England was so large a proportion of the population reduced to poverty as along the borders of Suffolk and Essex.

By contrast, woollen manufacturing in the West Riding was booming. Evidence submitted to the Select Committee of 1806 showed that possibly 65,000 Yorkshire people were engaged in the industry, most of whom were domestic workers living 'in villages and detached houses, covering

---

[1] A. Young, *General view of the agriculture of Oxfordshire* (London, 1809), 325, 328.
[2] J. Collinson, *The history and antiquities of the county of Somerset*, 3 vols. (Bath, 1791), II, 400, 429; III, 13.
[3] J. Billingsley, *General view Somerset* (1798), 296.
[4] A. Young (ed.), *Annals of agriculture*, XXVIII (1797), 634.
[5] A. Young, *General view Essex*, II (1807), 390.
[6] *Ibid*, II, 414.

the whole face of a district of from 20 to 30 miles in length, and from 12 to 15 in breadth'.[1] The opinions of witnesses were divided about the extent to which factories had fostered the expansion of trade, but two facts are abundantly clear. Firstly, that in the West Riding factories manufactured only about one-sixteenth of all the woollen cloth made in the country.[2] Secondly, that the prosperity of the industry was shared by factory and domestic workers alike. In Armley, for example, the number of domestic clothiers doubled between 1786 and 1806, and comparable increases were reported in other villages around Leeds.[3] On the other hand, superfine broadcloth produced in factories was acknowledged to be both cheaper and better in quality than domestic stuff, and factory owners paid higher wages than domestic masters. Factories raised the standards for their rivals largely by the superiority of their organisation. They possessed no power looms and few mechanical aids. Power-driven machinery was first applied to the initial processes of preparing wool for spinning – to scribbling, carding and slubbing – before it was applied to spinning. But as soon as steam engines were installed, the number of spinning mules increased rapidly. In 1796 Arthur Young was informed that six or seven steam engines were working in Leeds, and by 1800 there were probably as many as twenty in the town.[4]

In 1800 much less worsted than woollen cloth was woven in England but no branch of the worsted industry was depressed. Both in Norwich and in the country around Halifax and Bradford production was rising; and elsewhere, clothiers turned from the uncertainties of the woollen trade to an assurance of modest earnings in worsteds. Mechanisation took over slowly, and not until 1800 was a steam engine used to drive a worsted spinning mill at Bradford.[5] By this time the West Riding had already surpassed Norwich in the value of its worsted manufacture. Both districts however were still increasing their output, and it is clear that Norwich produced most of the finest cloth, and that the industry gave employment directly or indirectly to about 100,000 workers in Norfolk.[6] In 1805 the stranglehold of northern competition was felt when Norwich began to buy machine-spun yarn from Yorkshire mills. By 1808 some 300 woolcombers

[1] S.C. woollen manufacture, 1806, Minutes of evidence, 9, Report, 9 (P.P. 1806, iii).
[2] Ibid., Minutes of evidence, 89.
[3] Ibid., Minutes of evidence, 16, 94, 158, 444.
[4] A. Young (ed.), Annals of agriculture, XXVII (1796), 310.
[5] J. James, History of the worsted manufacture in England (London, 1857), 592–3.
[6] J. K. Edwards, 'The decline of the Norwich textile industry', Yorks. Bull. Econ. and Soc. Research, XVI (1964), 31–41.

had departed for the West Riding. As competition stiffened, Norwich diversified its range of products, using cotton, silk, alpaca and mohair warps in an increasing proportion of its cloth. Norwich turned to the production of mixed cloths much later than most other districts.

The cotton industry

Lancashire had long been weaving fustians, half-worsteds and mixed woollen and linen cloths in addition to pure woollens, while many other districts changed from producing one type of fabric to another. Along the Bristol Avon, for example, a number of mills were converted from woollens to mixed cloths to cottons, or from cottons to worsteds.[1] In Rossendale the first steps in mechanisation were taken in the manufacture of woollens. Most water mills were spinning woollen thread and towns such as Oldham and Rochdale still produced considerable quantities of woollen goods.[2] In Lancashire and the West Riding corn mills were taken over by carders, combers and spinners, while in Derbyshire some early cotton mills were built alongside iron forges. There was a great deal of flexibility in the use to which mills were put and a diversity in the types of fabric produced, but the growing number of cotton mills in south Lancashire made a deep impression on a traveller in 1791. He reported: 'there is scarcely a stream that will turn a wheel through the north of England that has not a cotton mill upon it.'[3]

The production of cotton cloth increased more rapidly than that of other textiles. Imports of raw cotton doubled between 1790 and 1800 and more than doubled between 1800 and 1810. In 1802 the value of cotton yarn and cotton cloth exported exceeded that of woollen goods, and cotton goods became the leading export. During its first phase of expansion the industry dispersed to remote water-power sites in upland areas. About 1790 it was more widely dispersed than at any later period. A number of counties had one or more spinning mills, but the greatest numbers were located in two districts: the first comprising south Derbyshire, Nottinghamshire and parts of Leicestershire and Staffordshire; the second comprising south Lancashire, north Cheshire and north Derbyshire (Fig. 92). The east Midlands was the birthplace of the cotton factory.[4] As long

[1] S. J. Jones, 'The cotton industry in Bristol', *Trans. and Papers, Inst. Brit. Geog.*, XIII (1947), 63–7.

[2] G. H. Tupling, 193–206.   [3] *European magazine*, XX (1791), 140.

[4] D. M. Smith, 'The cotton industry in the east Midlands', *Geography*, XLVII (1962), 256–69.

as water power dominated the industry the number of mills in Derbyshire increased. By 1807 there were 51 mills in south Derbyshire and almost as many again in Nottinghamshire, north Leicestershire and east Staffordshire. The cotton spinners of Nottingham supplied twist to domestic hosiers but the spinners multiplied their output while the hosiery trade expanded very slowly. The application of steam power to cotton spinning gave Nottinghamshire a new lead. The first Boulton and Watt engine installed in a cotton mill was at Papplewick in 1785, and by 1800 Nottinghamshire cotton mills had fifteen such engines, Leicestershire had two while Staffordshire and Derbyshire each had only one. Both Middlesex with four and Durham with three had surpassed Derbyshire in the application of Boulton and Watt engines to cotton manufacture.

It was Lancashire, however, that held an undisputed lead in cotton manufacturing. It had achieved its ascendancy very rapidly. When Arkwright founded his water-powered spinning factory at Cromford in Derbyshire in 1771, the cotton industry in Lancashire was poor and struggling, completely overshadowed by a long-established woollen industry. But by 1788 nearly twice as many water-driven spinning mills were situated in Lancashire as in Derbyshire;[1] and Lancashire advanced more rapidly than Nottinghamshire in equipping its mills with steam engines. The first Lancashire steam-powered factories were set up at Manchester, Bury, Preston, Oldham and Chorley. In 1795 Sir Robert Peel owned factories in twelve different places and employed 15,000 workers, including most of the workers in Bury. The Horrocks opened three factories in Preston, while Oldham underwent an 'extraordinary change from the scale of a mere village to that of one of the most populous towns in the kingdom'.[2] At Stockport, nearby in Cheshire, Samuel Oldknow built a steam cotton mill in 1790, and by 1795 there were twenty-three factories in the town, of which four were driven by steam.[3] But the most spectacular growth took place in Manchester. By 1795 it was preeminent among cotton towns. John Aikin prefaced his Description of the country from thirty to forty miles around Manchester with the observation that: 'The centre we have chosen is that of the cotton manufacture; a branch of commerce, the rapid and prodigious increase of which is, perhaps, absolutely unparalleled in the annals of trading nations'.[4] Describing the

[1] Anon, An important crisis in the callico and muslin manufactory of Great Britain explained (London, 1788). See table on p. 361 above.

[2] E. Butterworth, Historical sketches of Oldham (Oldham, 1849), 117–18.

[3] J. Aikin (1795), 445–6.

[4] Ibid, 3.

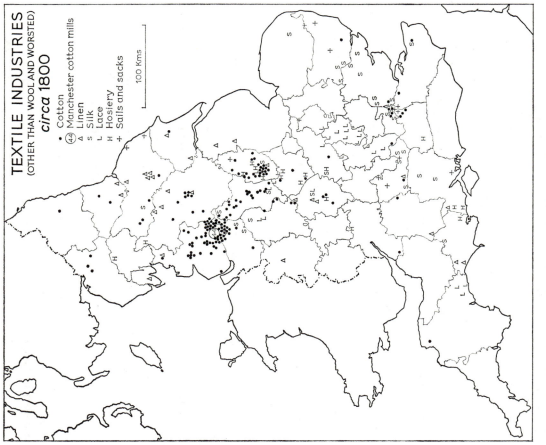

TEXTILE INDUSTRIES
(OTHER THAN WOOL AND WORSTED)
circa 1800

100 Kms

• Cotton
㊹ Manchester cotton mills
△ Linen
s Silk
L Lace
H Hosiery
+ Sails and sacks

Fig. 92 Textile industries (other than wool and worsted) *circa* 1800
Based on the places mentioned in: (1) G. W. Daniels, 'Samuel Crompton's
census of the cotton industry in 1811', *Econ. Hist.*, II (1930), 107–16; (2) the
county *General views of the Board of Agriculture*; (3) Arthur Young (ed.),
Annals of agriculture, 45 vols. (Bury St Edmunds, 1784–6; London, 1786–
1815).

process by which Manchester outstripped its neighbours, William Radcliffe wrote in 1828: 'there was not a village within thirty miles of Manchester, on the Cheshire and Derbyshire side, in which some of us were not putting out cotton warps and taking in goods, employing all the weavers of woollen and linen goods who were declining those fabrics as the cotton trade increased; in short we employed every person in cotton weaving who could be induced to learn the trade'.[1]

The application of steam power first to spinning, and later to weaving, set the seal on the supremacy of Lancashire, and above all on Manchester. In 1800 Lancashire possessed 42 out of 84 Boulton and Watt steam engines working in cotton-spinning mills in England; and these probably constituted not more than one-third of the total number of steam engines at work in the cotton industry of the country at this time.[2] Power-loom weaving, however, was still virtually unknown. A steam-driven loom had been set up at Manchester in 1791, but was destroyed by the hand-weavers, and it was not until 1803 that a satisfactory steam-driven loom was built at Stockport. It was improved by successive patents and was soon in use in several Lancashire towns.[3]

The other textile industries were of minor importance (Fig. 92). The fortunes of the linen industry had been submerged by the rise of cotton. Lancashire had almost entirely gone over to producing cotton goods. Leeds still produced linen thread, and linen spinning lingered in Somerset and Dorset, but hardly any progress had been made in the mechanisation of linen manufacture. The hosiery industry grew sluggishly. Between 1782 and 1812 the number of stocking frames increased from about 20,000 to 29,590, almost 85% of which were located in well-established centres, especially in Leicestershire and Nottinghamshire.[4] The cap-making and knitwear industries elsewhere stagnated, and the woollen spinners supplying them with yarn suffered much hardship. In London also, stocking frames were fast disappearing. The industry contracted at the edges and became concentrated in the east

[1] W. Radcliffe, *Origin of the new system of manufacture, commonly called power loom weaving* (Stockport, 1828), 12.

[2] A. E. Musson and E. Robinson, *Science and technology in the industrial revolution* (Manchester, 1969), 426. See p. 453 below.

[3] E. Baines, *History of the cotton manufacture in Great Britain* (London, 1835), 234; J. Wheeler, *Manchester, its political, commercial and social history, ancient and modern* (London, 1836), 107; C. Hardinck, *History of the borough of Preston and its environs* (Preston, 1857), 375.

[4] G. R. Porter, 206.

Midlands.[1] Lace-making was still a rural handicraft industry, especially in the villages of Bedfordshire and Buckinghamshire. The silk industry had moved out of Spitalfields in two directions: north-east to the depressed woollen centres of Essex and East Anglia; and also north-west to Macclesfield and Congleton in Cheshire and to Derbyshire where water-driven silk-throwing mills were established on fast flowing streams.

The iron industry

The phenomenal growth of the iron industry around 1800 clearly foreshadowed the changes that were to take place in almost every branch of industry during the nineteenth century. In 1800, iron manufacturing was of vital importance to the nation's war effort, its products were diffused, 'by numerous meandering streams, into every department of civil life', its surpluses found their way into new export markets.[2] In its various branches iron manufacturing gave direct employment to about 200,000 workers, and possibly as many more were engaged in ancillary activities such as mining ore and coal, or in engineering and distributive trades.

Not only had the total output of pig-iron doubled and doubled again in each of the last two decades of the eighteenth century but the locations of furnaces shifted during the same period. The early seats of charcoal smelting in the Forest of Dean, in the Weald and in Furness were abandoned. In 1794 it was reported that the last remaining Wealden furnace (at Ashburnham) 'blows six months in the year, and is considered as an auxiliary to the limeworks'.[3] It seems to have closed down in 1812; and the last forge, also at Ashburnham, ceased to work in 1828.[4] The Weald continued to manufacture considerable quantities of lime and charcoal.

From being widely scattered in wooded spots accessible to supplies of charcoal, the smelting of iron concentrated upon five areas: in Shropshire, in the Black Country, in south Yorkshire, in Durham and Northumberland, and, towards the end of the century, in South Wales and Monmouthshire (Fig. 93). In 1802, some forty furnaces were being built or had been

[1] D. M. Smith, 'The British hosiery industry at the middle of the nineteenth century: an historical study in economic geography', *Trans. and Papers, Inst. Brit. Geog.,* XXXII (1963), 125–42.

[2] From a speech in the House of Commons against the proposed duty on iron in 1806, cited in H. Scrivenor, *A comprehensive history of the iron trade* (London, 1841), 102.

[3] A. Young (ed.), *Annals of agriculture,* XXII (1794), 269.

[4] E. Straker, *Wealden iron* (London, 1931), 369.

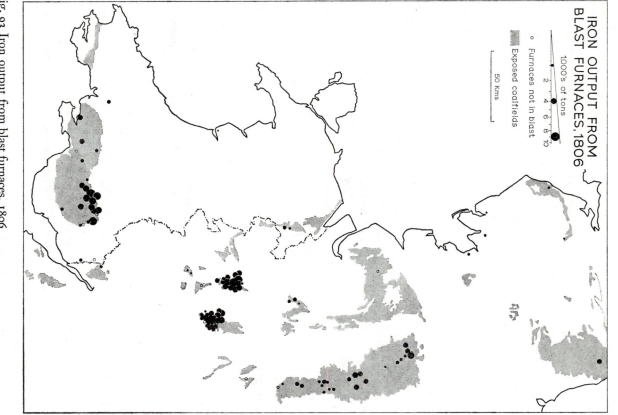

IRON OUTPUT FROM
BLAST FURNACES, 1806

1000's of tons

2
4
6
8
10

○ Furnaces not in blast

Exposed coalfields

50 Kms

Fig. 93 Iron output from blast furnaces, 1806
Based on E. Buckley, 'Number of furnaces and make of iron in England,
Scotland and Wales in the year 1806', *Report of the Commissioner on the State
of Population in the Mining Districts*, Appendix, p. 25 (P.P. 1849, xxii).

completed since 1796 at places in these areas. Between 1802 and 1806 another sixty were built in the same localities.[1] In 1785 Coalbrookdale had been described by Arthur Young as 'a very romantic spot', notwithstanding 'the noise of the forges, mills, etc. with all their vast machinery, the flames bursting from the furnaces with the burning of the coal and the smoak of the lime kilns, are altogether sublime'.[2] In 1803, according to a contemporary, the number of blast furnaces in the seven miles between Ketley and Willey, in eastern Shropshire, exceeded 'any within the same space in the kingdom'.[3] The works employed some 6,000 hands and consumed 260,000 tons of coal each year. Both the Coalbrookdale Company and John Wilkinson at Bradley built hundreds of new houses to accommodate incoming workers, but the furnaces grew so fast and so thickly that there was a persistent shortage of labour.[4] A Shropshire ironmaster giving evidence before a Select Committee on the coal trade in December 1800 summed up his predicament: 'There are new furnaces for the smelting of iron constantly erecting; and it is now difficult to get pitmen to work the coals.'[5]

In the Black Country and in South Wales, furnaces were multiplying even more rapidly than in Shropshire, and by 1806 the output of pig-iron from South Wales was greater than that either from Staffordshire or from Shropshire. Attracted by high wages offered in the Welsh valleys, ironworkers journeyed from Stourbridge and other centres in the west Midlands.[6] South Wales ironmasters were quick to take advantage of the inventions of Henry Cort and Peter Onions, and forged their iron economically by puddling and rolling instead of by hammering.

In the ironworks of Shropshire and Staffordshire, Boulton and Watt steam engines were first applied to purposes other than pumping. The first engine to blow a furnace had been installed by John Wilkinson at New Willey in Shropshire, and in 1800 nearly two-thirds of the Boulton and Watt engines used in ironworks were located either in Shropshire or in Staffordshire. The need for steam coal, as well as for coke, drew furnaces towards the coal measures, but water-power sites remained important for forges and for smithies. On the Stour and its tributaries, wrote William Pitt, 'are a number of very considerable iron works, where pig iron from

[1] H. Scrivenor, 96–8.

[2] A. Young (ed.), *Annals of agriculture*, IV (1785), 168.

[3] J. Plymley, *General view Shropshire* (1803), 340.

[4] T. S. Ashton, *Iron and steel in the industrial revolution* (2nd ed., London, 1951), 199.

[5] *Ibid.*, 144.

[6] *Ibid.*, 199.

15

Staffordshire and Shropshire foundries, and elsewhere, is rendered malleable, and worked into bars, rods and sheet iron'.[1] Around Sheffield, the fast flowing streams were dammed to form an almost continuous succession of mill ponds for cutlers and edge-tool makers, while smelting and steel-making concentrated near the coal pits. In east Durham, the iron industry, although much smaller, made a distinctive contribution to the local economy, and produced such articles as anchors, chains and edge-tools.

In 1806 no more than one ton of pig-iron in every thirty was made in the few surviving charcoal furnaces in Cumberland, Lancashire and the Forest of Dean, and more than nine-tenths of the nation's output was smelted on five coalfields.[2] The demand for castings seemed insatiable. Cast-iron rails were fitted to tramways at Coalbrookdale, Sheffield and on Tyneside. Cast-iron beams and girders were used in factory and mine-shaft construction. Cast-iron bridges spanned the Severn and the Wear. Miles of cast-iron water pipes were laid in London and cast-iron barges sailed on the Severn. Castings also began to enter in small quantities into the construction of machinery. Wheels and pistons were cast and machines for textile manufacturing, nailmaking and for furnace bellows were fabricated to an increasing extent from iron. A mechanical engineering industry was called into existence.

Coal and steam

Advances in the techniques of mining and utilising coal had been less far-reaching than those in iron-working. The substitution of coal for wood as a domestic fuel proceeded slowly. Houses had first to be provided with adequately ventilated fireplaces and chimneys to draw off the toxic fumes; and many districts where firewood was scarce lay beyond the reach of coal. Very little was hauled overland more than five or ten miles to iron-works or steam engines. Household coal burned in London and along the coast from the Tweed to the Exe was almost entirely shipped by sea from the Tyne and the Wear. About four-fifths of the large output of North-umberland and Durham was sent away by sea. Coal was also shipped across the Irish Sea mainly from Cumberland, but also from Lancashire,

[1] W. Pitt, *General view of the agriculture of the county of Worcester* (London, 1810), 279.

[2] E. Buckley, 'Number of furnaces and make of iron in England, Scotland and Wales in the year 1806', *Report of the Commissioner on the State of Population in the Mining Districts*, Appendix, p. 25 (P.P. 1849, xxii).

from lowland Scotland and from South Wales. In the meantime, canals were extending the markets of the inland coalfields.[1]

In 1800, the age of steam had dawned but the sky was not yet darkened with the smoke of factory chimneys. Probably no more than 1,200 steam engines were at work in England and Wales.[2] Coalmining and copper mining were the industries most dependent on steam for pumping. In most manufacturing districts, however, steam engines still contributed little, if anything, to the total amount of power being used. In Sheffield in 1794, only five cutlers' wheels were driven by steam while 111 were water-powered.[3] The Soho factory in Birmingham, established by Boulton in 1761, was dependent for some years on water power, and the competition for available water was keen in the Midlands.[4] Most engineers were engaged not in building steam engines, but in constructing and improving water wheels and water-powered machines. Indeed, it has been said that 'advances in water power deserve stressing because they have been unduly overshadowed by the early development of steam engines'.[5] In almost all localities water was a greater source of power than steam, and for some types of work, such as shipping and fen draining, wind power was more important than both. Experiments with steam traction had begun on water, but on the road horses reigned supreme. The number of horses kept in London and Middlesex alone far exceeded the aggregate horsepower of all steam engines at work in England and Wales.

The location of Boulton and Watt engines (Fig. 94) broadly indicates where investment in power-driven machinery was heaviest. They made a disproportionately large contribution to driving machines in the technically most advanced factories.[6] They were most numerous in cotton mills in Lancashire, Nottinghamshire, Yorkshire and Cheshire. Ironworks in

[1] T. S. Ashton and J. Sykes, *The coal industry of the eighteenth century* (Manchester, 1929), 226–39.

[2] J. R. Harris, 'The employment of steam power in the eighteenth century', *History*, LII (1967), 131–48.

[3] A. Allison, 'Water power as the foundation of Sheffield's industries', *Trans. Newcomen Soc.*, XXVII (1949–51), 221–4.

[4] P. Mantoux, *The industrial revolution in the eighteenth century* (London, 1928), 332–3; W. H. B. Court, *The rise of the Midland industries, 1600–1838* (Oxford, 1938), 249–52; R. A. Pelham, 'The water-power crisis in Birmingham in the eighteenth century', *Univ. of Birmingham Hist. Jour.*, IX (1963), 64–91.

[5] A. E. Musson and E. Robinson (1969), 71.

[6] J. Lord, *Capital and steam power* (London, 1923), 166; for a revaluation of Lord's view see A. E. Musson and E. Robinson, 'The early growth of steam', *Econ. Hist. Rev*, n.s., XI (1959), 418–59.

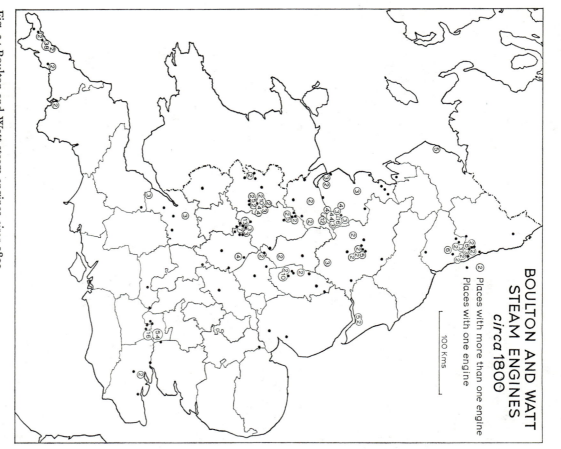

Fig. 94 Boulton and Watt steam engines *circa* 1800
Based on the Boulton and Watt Engine Book in the Boulton and Watt
Collection, Birmingham Public Library.

Shropshire and Staffordshire were well equipped with them, as were the deepest coal pits. But by far the largest number of pumping engines employed in coal mines were atmospheric engines of the Savery or Newcomen type, improved by John Smeaton, James Brindley and others. Boulton and Watt engines were economical in their consumption of fuel and were preferred to atmospheric engines in districts where coal had to be imported. They were widely used in the tin and copper mines of Cornwall. The metropolitan area, also a long distance from coalfields, used Boulton and Watt engines for pumping at waterworks, breweries and distilleries. Middlesex had steam-powered corn mills, sugar mills, oil mills, cotton mills, bleacheries and dyeworks. Middlesex and Surrey together had about as many Boulton and Watt engines as Lancashire, but Boulton and Watt probably built no more than one-third of the steam engines used in Lancashire.[1] In all colliery districts Boulton and Watt engines were outnumbered, and also in such centres of manufacturing as Leeds and Birmingham. In 1802 William Reynolds reported that in Shropshire 180 steam engines were at work,[2] but at that time the firm's Engine Book records that only 43 Boulton and Watt engines had been installed in the county.

Boulton and Watt held exclusive patent rights but other manufacturers successfully pirated their inventions. The ironmaster, John Wilkinson, closely associated with the partnership, built his own engines infringing the patents. In 1800 some of the early engines were worn out, rebuilt or removed, and some were performing work different from that for which they had been intended. Pumping engines were blowing furnaces, and engines installed to grind flour were used to crush oil seeds. The Boulton and Watt Engine Book records some of the removals and changes of use. For steam engines made by other firms no comparable lists exist.

TRANSPORT AND TRADE

Transport by road

One of the greatest obstacles to the growth of manufacturing in inland districts was the high cost of transport by road.[3] Rates differed for different kinds of goods. For coal, iron ore, lime and other minerals the profitable

[1] A. E. Musson and E. Robinson (1969), 426.
[2] J. Plymley, *General view Shropshire* (1803), 340n.
[3] For a convenient short account, see B. F. Duckham, *The transport revolution, 1750–1830* (Hist. Assoc., London, 1967).

range of overland movement was rarely more than ten to twenty miles from a pit. The city of York, situated within sixteen miles of productive coal measures, obtained shipments of coal from Newcastle upon Tyne, over two hundred miles away, and the coastlands of North Wales imported coal more cheaply from South Wales than from the landlocked Flint and Denbigh coalfields.

Arthur Young and other travellers frequently remarked on the execrable condition of the roads, but these were not uniformly bad nor were parochial highways invariably worse than turnpike roads. The worst roads were often those carrying the heaviest traffic. Worst of all were the well-trodden roads between large towns, churned by herds of iron-shod cattle, by waddling geese, by lumbering wagons, by trains of mules laden with panniers of coal and by the travellers who complained. The roads that approached London were described as muddy rutted tracks, deep in filth, insufferably bad and disgraceful. In Middlesex the turnpike roads were worse than most parish roads, and in Surrey the direct roads leading to London were often impassable while the cross roads were generally good. The industrial districts of south Lancashire and the West Riding had the densest network of roadways in the country but the condition of some defied description. The turnpike roads of Lancashire were notoriously bad, rivalling those leading to the mining districts of Shropshire which, 'particularly in the clay parts of the county, are almost impassable to any but the inhabitants'.[1]

The nature of the terrain contributed to the poor state of many roads. In Surrey, 'on the clays of the Weald and on the sands as well as in the low parts of the county near the Thames', the roads were 'very indifferent'.[2] In the Weald of Kent the turnpikes were 'as bad as can be imagined' but common roads, although neglected, were not so bad.[3] Steep inclines were often more difficult to negotiate than miry claylands. Even a short hill such as Highgate Hill took wagons several hours to climb, and accidents often occurred when teams of horses lost their footing. Many steep ascents and descents were the principal handicaps of roads in Devonshire, while in northern England the difficulties of crossing hill country were increased by the hazards of snow, ice, poor visibility and by waterlogged

[1] J. Plymley, General view Shropshire (1803), 273.
[2] W. Stevenson, General view Surrey (1809), 547.
[3] J. Boys, General view Kent (1796), 168–9. See G. J. Fuller, 'The development of roads in the Surrey–Sussex Weald and coastlands between 1700 and 1900', Trans. and Papers, Inst. Brit. Geog., xix (1953), 37–49.

mosses. Rivers were also major obstacles, especially in thinly peopled districts where bridges were far apart.

Even rugged country had some tolerably good roads where traffic was light. Turnpikes were reported to be excellent in the Lake District, and those of Northumberland and Durham were in good repair although most township roads in the north were much neglected. The rural areas of eastern England, not traversed by through traffic, were better served by roads than any other part of the country. In Norfolk, Arthur Young said, in 1804, that they had greatly improved during the preceding twenty years; in Suffolk he thought they were 'uncommonly good'; and in Essex he found it 'impossible to say too much in praise of the roads of most districts'.[1] He might have had reservations about crossing the Lea valley, but Mile End Road was by no means the worst approach to London.

The area around London and the leading industrial districts were given more attention in road improvement acts than other parts of the country, and by 1800 substantial advances had been made in Nottinghamshire, Staffordshire and Durham. Before he died in 1810 the blind John Metcalfe had improved over 200 miles of roadway in Yorkshire. Thomas Telford held the appointment of Surveyor of Public Works in the industrial county of Shropshire, while John Loudon Macadam perfected his method of giving roads a hard metalled surface near Bristol and extended his practice to thirty-four other districts. In 1808, the roads in Cheshire were 'in a state of progressive improvement', stone being brought from Mow Cop, from Flintshire and from Anglesey to lay on their surfaces.[2] Three roads out of London to Canterbury, to Bath and to Portsmouth were greatly improved and kept in excellent repair. Industry itself helped to remedy the problem it had created. Iron girders were used in the construction of new bridges designed by John Smeaton, Thomas Telford and John Rennie, while slag and cinders were incorporated in road metal. Even so, the roads of 1800 were inadequate to carry the increasing burden industry imposed upon them.

Canal and coastwise traffic

Adam Smith observed that water transport was cheaper and more efficient than land carriage as a means of opening distant markets for manufactured goods. He reasoned that 'it is upon the sea-coast, and along the banks of

[1] A. Young, *General view Norfolk* (1804), 489; A. Young, *General view Suffolk* (1804), 227; A. Young, *General view Essex* (1807), II, 384.
[2] H. Holland, *General view Cheshire* (1808), 302.

navigable rivers, that industry of every kind naturally begins to subdivide and improve itself'.[1] The mines of Tyneside produced more coal than those of any other district because it was profitable to ship Newcastle coal coastwise to London and to the south coast as far west as Plymouth. The inland coalfields of the Midlands were at a disadvantage until they were provided with waterways. Many inland navigations were created by dredging, widening and embanking existing streams, but the most spectacular results were achieved by cutting canals (Figs. 80 and 106). In Manchester the price of coal had been halved since the opening of the Worsley Canal in 1761, and the Birmingham Canal (opened in 1772) halved the price of coal in Birmingham. With the coming of the Soar and Erewash navigations and their southward extension to Leicester in 1794, Leicestershire coal owners lost between 30% and 40% of their trade to Derbyshire competitors.[2] The making of the Grand Junction canal in 1793–1805 opened the way to London, but high freight rates prevented Midland coal from flooding that lucrative market. Even so, coal was easily the largest single commodity carried by canals.

In Cheshire, canals gave a new lease of life to the salt industry, and they were also of great value to farmers as a cheap means of conveying marl, lime and fuel. Benjamin Capper believed that the cutting of canals also provided fat-stock producers with an efficient means of carrying animals to market and so helped the conversion of arable land to grass in the Midlands.[3] The rapid growth of Birmingham, the Black Country and the Potteries as manufacturing centres was 'much promoted by canal conveyance'.[4] Not only the trade in coal but that in iron ore, limestone, lime and manufactured goods was considerably extended. Neighbouring Shropshire was well behind Staffordshire in canal construction but made rapid progress after 1800.[5] By contrast with coalmining County Durham, which as late as 1810 had no artificial canals, some rural areas were well served by the great canals that passed through them.

As well as the new movement along inland waterways, coastwise shipping continued increasingly to integrate the economic activities of various parts of the British Isles. Even agriculture lost some of its

[1] Adam Smith, *An enquiry into the nature and causes of the wealth of nations* (London, 1776), Book I, ch. 3.
[2] T. J. Chandler, 'Communications and a coalfield; a study in the Leicestershire and south Derbyshire coalfield', *Trans. and Papers, Inst. Brit. Geog*, XXIII (1957), 165.
[3] B. P. Capper (1801), 37. [4] W. Pitt, *General view Stafford* (1796), 165.
[5] J. Plymley, *General view Shropshire* (1803), 307.

parochial exclusiveness. Cast-iron rollers from the Carron works in Lanark-shire and iron-framed ploughs from Rotherham in Yorkshire were shipped to farms in Norfolk. Haymakers and harvesters crossed from Ireland and from Scotland and travelled by inland waterways to farms in the English Midlands. But agriculture was not the economic activity most deeply affected by improved coastwise transportation. Facilities for moving minerals, particularly coal, greatly expanded as ships increased in size and tonnage. More shipping was engaged in carrying coal than any other commodity, but lead, tin, bricks, building stone, slates, china clay, hard-ware and brass also bulked large.

Overseas trade

In registered tonnage of shipping, Newcastle ranked second only to London. Liverpool held third place, followed closely by the coal ports of Sunderland and Whitehaven. Bristol's fleet ceased to grow, while ships from Exeter and from King's Lynn carried diminishing shares of England's maritime commerce. Most of the tonnage was engaged in coastwise traffic. The sheer bulk of the coastal trade – particularly in coal – dwarfed that overseas. 'Foreign trade', wrote J. H. Clapham, 'had not yet com-pletely lost its primitive characteristic – the exchange of precious things.'[1] But it was increasing, and the growing industrialisation of the country was reflected in its exports.[2] Textiles were important, but with cotton goods rivalling those of wool; the products of metal manufacture, especially of iron, were becoming significant. Among imports, timber alone constituted a bulk cargo, and the import of fir from northern Europe was a consider-able item in this trade. In addition, small quantities of softwood were received from Canada, and increasing amounts of teak, mahogany and tropical hardwoods were arriving from the south Atlantic and the Indian Ocean. The growing demand for tropical products was associated with a radical reorientation of British overseas trade towards greater dependence on the New World and the Far East, and away from traditional links with Europe and the Mediterranean. Imports of raw cotton, sugar, tobacco, coffee and tea increased dramatically year by year.

More ships, of larger size and sturdier construction, were built to carry this long-distance trade. Shipyards, as a consequence, suffered from chronic shortages both of manpower and materials. The strain was increased by the demands of the navy, called upon to protect extended

[1] J. H. Clapham, 237.
[2] E. B. Schumpeter, *British overseas trade statistics, 1697–1808* (Oxford, 1960).

lines of communication, to blockade the European shores and to defeat the hostile fleets united under Napoleon's command. The navy not only maintained four dockyards on the lower Thames and the Medway (at Deptford, Woolwich, Sheerness and Chatham) but also held bases on the English Channel at Portsmouth and Plymouth. To a large extent, the fortunes of war were decided along the south coast.

In the Irish Sea and along the shores of the North Sea, fishing, coastal transport and, in times of peace, trade with the Baltic and with the Low Countries, flourished. New docks were opened at Hull, but the most spectacular transformations were those taking place in London and Liverpool. The digging of deep-water dock basins and the building of gigantic warehouses were tangible expressions of the opening of a new sphere of commerce. The ships whose masts crowded the Thames and the Mersey included some that had voyaged from distant ports throughout the world, including the Pacific.[1] Compared with the volume of overseas trade conducted at the end of the Elizabethan age, the quantity and variety of goods passing through English ports at the beginning of the nineteenth century was prodigious, and the rate of expansion was accelerating. England had become the most prosperous mercantile nation in the world.

TOWNS AND CITIES

In 1801 the legal and political status of towns bore little relation to the numbers of their inhabitants, to the size of their built-up areas, or to their social and economic functions. Some cathedral cities and many ancient corporate towns had fewer citizens than many villages. Uninhabited Old Sarum returned a member to parliament while the 31,000 parishioners of Sheffield were unrepresented. Of the provincial towns enumerated in the census, five had each a population of over 50,000; another eight had each a population of over 20,000; and yet another thirty had each a population of over 10,000 (see table 8.2 below). The combined population of these towns with over 10,000 inhabitants was about one million. Taking all towns together (whether with more or less than 10,000 inhabitants each), four categories may conveniently be distinguished: county and market towns, ports, manufacturing centres and resorts.

The county and market towns were almost entirely built around market places, serving extensive rural areas. Tiny Queenborough (545 inhabitants)

[1] A. M'Konochie, *A summary view of the statistics and existing commerce of the principal shores of the Pacific Ocean* (London, 1818).

Table 8.2 *Chief provincial towns in 1801*

The table includes all those places in the 1801 Census Tables designated as 'towns' as distinguished from 'parishes', 'townships' or 'hamlets'. The populations of London, Westminster and neighbouring towns, such as Southwark and Greenwich, are excluded from this list. Figures are given to the nearest thousand.

Over 50,000		10,000–20,000 *(cont.)*	
Manchester/Salford[a]	84,000	Stockport	15,000
Liverpool	78,000	Shrewsbury	15,000
Birmingham	74,000	Wolverhampton	13,000
Bristol[b]	64,000	Bolton	13,000
Leeds[c]	53,000	Sunderland	12,000
		Oldham	12,000
20,000–50,000		Blackburn	12,000
Plymouth	43,000	Preston	12,000
Norwich	37,000	Oxford	12,000
Bath	32,000	Colchester	12,000
Portsmouth/Portsea[d]	32,000	Worcester	11,000
Sheffield	31,000	Ipswich	11,000
Hull	30,000	Wigan	11,000
Nottingham	29,000	Derby	11,000
Newcastle upon Tyne	28,000	Hundersfield[e]	11,000
		Quick[f]	11,000
10,000–20,000		Warrington	11,000
Exeter	17,000	Chatham	11,000
Leicester	17,000	Carlisle	10,000
York	16,000	Dudley	10,000
Coventry	16,000	King's Lynn	10,000
Chester	15,000	Cambridge	10,000
Dover	15,000	Reading	10,000
Great Yarmouth	15,000		

[a] Manchester, 70,000; Salford, 14,000.
[b] Bristol city, 41,000; Barton Regis hundred, 23,000.
[c] Including 17 townships within the Liberty of Leeds.
[d] Portsmouth, 8,000; Portsea, 24,000.
[e] Now in Rochdale (Lancashire).
[f] Now in Saddleworth (West Riding).

on the Isle of Sheppey held two markets each week, the city of Worcester held three, while Manchester had only one. Some towns were cathedral cities, some had medieval castles, some held assizes, some had large schools, but many had changed little in size or shape since 1600. While much new civic building was to be seen, no Georgian cathedrals had been built, no new fortresses raised, and no new universities founded. The city of York had been surpassed in population by Exeter, which had expanded its manufacturing activities, but many other cathedral cities (such as Gloucester, Lincoln and Salisbury) had fewer than 10,000 people. Much the largest cathedral city was Norwich. Almost all the built-up area of the city was contained in the square mile or so within its walls, with much open space around the castle and the cathedral. The walls were washed on the southern side by the floods of the Wensum, and sea-going wherries reached the quays of Thorpe nearby. Norwich had now reached the summit of its importance as a centre of worsted manufacturing, while its banks, its corn exchange and its posting inns served travellers from an extensive prosperous agricultural region. Its annual cattle fair, its weekly corn and cattle markets and its three market days each week set it above all other market centres in Norfolk.[1] In East Anglia, as throughout the country, there was a well-ordered hierarchy of market centres. Places of similar size were regularly spaced, provided similar services, and were very similar in their appearance and compactness.

All the leading seaports had experienced a remarkable growth in trade, in volume of shipping and in population. Bristol had grown in size but had been overtaken by Liverpool, which not only had a lion's share of the sugar and slave trades but handled most of the country's raw cotton imports. This was a very recent development. In spite of the incidents of war, the shipping using the port increased greatly in the years around 1800, and cotton began to figure prominently in imports from the West Indies. Liverpool was becoming the main supply port for the expanding industrial areas of south Lancashire and the Midlands.[2] The first docks had been dug but the waterfront itself was constantly crowded with ships, and the streets leading from the quays were among the most densely crowded in England. On the east coast, Newcastle was a much smaller town than Liverpool, but its port had a larger fleet and handled more cargo. The country's demand for coal was insatiable, and the Tyne, its leading supplier, increased production and grew wealthy. In 1801 on Tyneside nearly 40,000 people were engaged in mining and in shipping coal, while the combined population of Newcastle and neighbouring Gateshead numbered 37,000.[3] The coal trade had created more than black dirt in

[1] R. E. Dickinson, 'The distribution and function of the smaller urban settlements of East Anglia', Geography, XVII (1932), 22–4, 28.

[2] F. E. Hyde, 'The growth of Liverpool's trade, 1700–1950', in W. Smith (ed.), A scientific survey of Merseyside (Liverpool, 1953), 154–5.

[3] Anon., A picture of Newcastle-upon-Tyne (Newcastle upon Tyne, 1807).

Newcastle. In 1788, a Theatre Royal was opened and the Assembly Rooms were more spacious than any but those at Bath. Downstream, beyond Sandgate towards Wallsend, there was a growing industrial suburb with a multitude of glasshouses, iron foundries, lead refineries, potteries, soap-works and steam-driven flour mills.

Farther south along the east coast, the port of Hull boomed as inland navigations drew traffic from the Trent valley and the Midlands to the Humber; fishing, flour milling and the coal trade expanded.[1] The east-coast ports of Sunderland and Yarmouth, which also shared in the coast-wise coal trade, were now larger than Lynn and Boston, whose approaches were too shallow to admit large ships. A most remarkable growth in population had also taken place in the ports along the English Channel where a vast apparatus of naval armament was concentrated. Plymouth, now the sixth largest provincial town, had more than twice as many in-habitants as the city of Exeter, while Portsmouth had more than four times the population of the ancient port of Southampton. In Kent, the dock-yard towns of Greenwich and Chatham and the packet station of Dover were each larger than the city of Canterbury (with 9,000 inhabitants).

Among the five provincial towns with populations of more than 50,000 were the great manufacturing centres of Manchester–Salford and Birming-ham. In 1794, *Scholes's Manchester and Salford Directory* listed 600 streets of which 61 'were laid out but not built upon'. There was space in the town for 'handsome country houses on every hill, elegantly furnished and surrounded by as elegant pleasure grounds'.[2] But land close to the mills was crowded with tightly packed rows of cottages and with three-storey tenements whose cellars accommodated some of Manchester's 5,000 immigrant Irish. A dozen other textile manufacturing centres in Lancashire and Yorkshire each had a population of more than 10,000, but only a handful exercised the powers and privileges of incorporated towns. In the West Riding, Bradford, Huddersfield and Halifax were still only parishes, but Leeds, the first place to build a steam-driven woollen spinning mill (1972), had achieved the dignity of urban status. The textile manu-facturing conurbations of Lancashire and Yorkshire were thus only fore-shadowed on contemporary maps. The emergent Black Country, however, was recognised as 'the most populous vicinity in the kingdom out of the metropolis', employing great numbers in the manufacture of hardware,

[1] W. G. East, 'The port of Kingston-upon-Hull during the industrial revolution', *Economica*, XI (1931), 190–212.

[2] *European Magazine*, XX (1791), 216.

toys, nails, glass, iron, coke and lime. William Pitt estimated 'that a figure of 200 square miles might be marked out, having the town of Dudley near its center', which contained 'upwards of two hundred thousand people'.[1] In Birmingham, 8,000 of its 12,000 inhabited houses had been built since the coming of the canal in 1765;[2] but the road between Wolverhampton and Birmingham was still far from being continuously built-up. The Soho works were situated in green fields, and the iron furnaces between Dudley and Wednesbury stood far apart along the canal banks. The degree of dispersion was even wider in the parish of Sheffield. Unlike the west Midland centres it remained the largest and one of the most scattered villages in England, inaccessible to navigable waterways. The five fast flowing streams which plunged from the edge of the Pennines were no longer relevant to its industrial development.[3] In 1770 some 229 waterwheels were turning while steam was driving 132 grinding wheels, but in 1794 only 123 waterwheels had driven tilt hammers and grinding wheels, mostly near the centre of Sheffield.[4] Not only was the cutlery industry concentrating upon the coal seams but Huntsman's steelworks, the first modern plant of its kind, was producing steel on a large scale. Sheffield plate and Britannia plate, produced in hundreds of small workshops near the valley bottom, were competing successfully with silver ware and pewter, and contributing to the grime and congestion of the village centre.

Among the growing settlements of England were the inland spas and watering places, but only a few were, as yet, corporate towns. Bath was no longer at the height of fashion, although it remained a spacious city of tall houses, many in multiple occupation as lodgings.[5] Its broad shopping thoroughfare, Milsom Street, and the graceful crescents climbing the hillsides, commanded extensive views over the Avon valley and open country. Other inland resorts such as Tunbridge Wells, Epsom and Buxton were thriving, while Cheltenham, Leamington and Harrogate were still quiet genteel rural retreats.[6] The most novel development was

[1] A. Young (ed.), *Annals of agriculture*, VII (1787), 463.

[2] J. A. Langford, *A century of Birmingham life; or a chronicle of local events from 1741 to 1841*, 2 vols. (Birmingham, 1868), I, 444.

[3] R. N. Rudmose Brown, 'Sheffield: its rise and growth', *Geography*, XXI (1936), 175–84.

[4] G. I. H. Lloyd, *The cutlery trades* (London, 1913), 157, 443.

[5] R. A. L. Smith, *Bath* (London, 1944), 92.

[6] E. W. Gilbert, 'The growth of inland and seaside health resorts in England', *Scottish Geog. Mag.*, LV (1939), 22; J. A. Patmore, 'The spa towns of Britain', being

the creation of seaside resorts, following the model of Scarborough which first achieved fame as a spa. Charles Vancouver's report on Hampshire describes Lymington, Christchurch and Southampton as 'much frequented in the summer for sea-bathing'. At Southampton, the company was 'much more select than usual at such places'; and although the accommodation was dear, it was 'generally of the best sort and truly elegant'.[1] At Lowestoft and Margate, the bathing machines of London shopkeepers mingled with fishermen's boats. The king patronised Weymouth and Lyme Regis but the most extravagant of the new resorts was Brighton, frequented by the Prince of Wales. The building of the Royal Pavilion had begun in 1786, and was to continue intermittently until 1822. Of the 1,300 houses in Brighton in 1801, as many as '211 were let solely as lodging houses, and in another 208 lodgings were to be obtained'.[2] Such development did not meet with universal approval, and Arthur Young firmly disapproved of those who idled away their summer months 'in this sort of inglorious obscurity' that 'London dissipation carries in its train'.[3]

London was unique among cities, a metropolis of bewildering complexity (Fig. 82). The population of its built-up area amounted to about 960,000, of which 750,000 lived to the north of the river, and the other 210,000 to the south, in the counties of Kent and Surrey.[4] One in ten of the inhabitants of England was a Londoner. J. P. Malcolm in 1803 thought it easier 'to attempt to describe the varying form of a summer cloud, than to trace from year to year the outline of London'.[5] On the north-west, the frontier of building had reached the 'New Road', constructed in 1756–7 (now named Marylebone, Euston and Pentonville Roads). Most of Westminster and Holborn were covered with broad streets and spacious squares, and the view from Queen Square (in Holborn) to Hampstead was 'hidden by the majestic houses adorned with Tuscan pillars' which had lately been put up in Guildford Street.[6] To the south of the Thames, the roads from Westminster and Blackfriars bridges converged in St

ch. 2 (pp. 47–69) of R. P. Beckinsale and J. M. Houston (eds.), *Urbanization and its problems* (Blackwell, Oxford, 1968).

[1] C. Vancouver, *General view Hampshire* (1810), 420, 427.

[2] E. W. Gilbert, *Brighton, old ocean's bauble* (London, 1954), 96–7.

[3] A. Young (ed.), *Annals of agriculture*, xxvii (1797), 118.

[4] *Census of 1851: Population Tables*, I, vol. 1, London (P.P. 1852–3, lxxxv). This contains a retrospective summary for 1801.

[5] J. P. Malcolm, *Londinium redivivum*, 4 vols. (London, 1803–7), 1, 5.

[6] *Ibid.*, 1, 6.

George's Fields; and here, in 1814, it was said that 'with very little exception, the whole line of each road is now skirted on both sides with houses and other buildings'.[1]

This growing London consumed a large part of the country's produce. London used more bricks and tiles than any other part of the kingdom, and by night a zone of brick kilns was described by Henry Hunter as forming 'a ring of fire' and pungent smoke around the city.[2] Beyond the zone of brick kilns were the meadows that provided hay and grazing for thousands of horses and dairy cows kept by Londoners, and farther afield lay the orchards and market gardens that provided the fruit and fresh vegetables brought daily to the London markets (Fig. 68).[3] London not only had the largest fruit and vegetable markets in the country, but Smithfield was the largest cattle market, Billingsgate the largest fish market, and London merchants dominated the trade in leather, wool, precious metals and a multitude of other commodities.[4]

The years around 1800 saw a new stage in the development of the port of London. Between 1796 and 1800 a series of parliamentary committees enquired into the congestion of shipping in the Thames and made recommendations for the improvement of the port. The total quay accommodation at London was little greater than at Bristol, but the tonnages of goods entering and leaving were very much greater. Valuable and perishable cargoes, delayed in the river for weeks, were being lost through pilferage and decay. The West India Docks, across the Isle of Dogs, opened in 1802, removed hundreds of ships from the river and relieved the pressure on warehouse space in the City and Shadwell. With the opening of Commercial Road in 1803, Limehouse and Poplar became dockland settlements. At this time the masts and flags of naval artificers waved proudly above the trees in Limehouse, and the large mansions of opulent merchants looked out on fields that were being covered with miles of new streets.[5] To its supremacy in Britain, London was adding a new domination as the capital of a vast empire and the commercial and financial centre of the world.

[1] E. W. Brayley, *London and Middlesex*, 2 vols. (London, 1814), II, 86.
[2] H. Hunter, *The history of London*, 2 vols. (London, 1811), II, 2; T. Baird, 'London brick fields', in A. Young (ed.), *Annals of agriculture*, XXI (1793), 150.
[3] G. B. G. Bull, 'Thomas Milne's land utilization map of the London area in 1800', *Geog. Jour.*, CXXII (1956), 25–30.
[4] R. Westerfield, *Middlemen in English business, particularly between 1660 and 1760* (New Haven, 1915), 420.
[5] J. P. Malcolm, II, 85.

Chapter 9

CHANGES IN THE EARLY RAILWAY AGE: 1800–1850

ALAN HARRIS

Some parts of rural England still wore in 1800 an outwardly medieval aspect that had yet to be transformed by enclosure, land drainage and improved systems of farming. Even by 1850 the transformation was not complete, but James Caird's England was much nearer that of the twentieth century than the England of only fifty years earlier described by the Board of Agriculture reporters. This was also a half-century of accelerated change in the towns, many of which, in assuming new functions, acquired new dimensions, new townscapes and problems which, in scale, if not in kind, were also new. Although steam power had already proved itself a potent agent of industrial change by 1800, its greatest effects were then still to be experienced. Half a century later, despite the survival of numerous relics of an earlier order, a new industrial geography, in which communication by rail played an increasingly important part, had been largely created. In demographic history also, these were remarkable years. As G. Kitson Clark has written, 'If nothing else had happened in the first half of the nineteenth century, the growth of population alone would have secured that Victorian England was decisively different...from the Georgian England of 1800.'[1]

POPULATION

Not only were there twice as many people living in England and Wales in 1851 as half a century earlier but they showed, as the century advanced, an increasing tendency to accumulate 'in great and well-defined masses'.[2] The proportion of the total population living in towns rose progressively as the flow of labour from the rural areas gathered momentum, increased, from the 1820s onwards, by a stream of migrants from Ireland. The

[1] G. K. Clark, *The making of Victorian England* (Oxford, 1962), 66.
[2] P. Gaskell, *The manufacturing population of England* (London, 1833), 9.

Table 9.1 *Population of England and Wales, 1801–51*
(Figures in millions; population for England in brackets)

	Total	Decennial increase %
1801	8.9 (8.3)	—
1811	10.2 (9.5)	14.00
1821	12.0 (11.2)	18.06
1831	13.9 (13.0)	15.80
1841	15.9 (14.9)	14.27
1851	17.9 (16.9)	12.65

Source: Census of 1851: Population Tables, I, vol. 1, p. xxxiii, (P.P. 1852–3, lxxxv.)

townward movement affected first the rate of growth, and then the absolute size, of rural communities, many of which in 1851 were smaller than they had been twenty years earlier.[1] This was true, for example, of about one hundred parishes in Leicestershire and of some sixty in Dorset.[2] Many rural communities in Lancashire and Cheshire were also shrinking in size at this period, although the combined population of the two counties increased by 185% during 1801–51.[3]

In spite of local decreases, the only English county to experience a net loss of population during 1801–51 was Wiltshire, although three counties in upland Wales had also suffered in this way.[4] The rural population, still widely sustained by local crafts and industries, continued to increase in the country at large, although at a consistently lower rate than that of the great towns and manufacturing districts, especially after 1820.[5] The overall changes in England as a whole may logically be considered, first in the

[1] J. Saville, *Rural depopulation in England and Wales, 1851–1951* (London, 1957), 5.

[2] *Census of 1851: Population Tables*, I, vol. II, Leicestershire, 12–26 (P.P. 1852–3, lxxxvi); *ibid*, vol. 1, Dorsetshire, 28–40 (P.P. 1852–3, lxxxv).

[3] R. Lawton, 'Population trends in Lancashire and Cheshire from 1801', *Trans. Hist. Soc. Lancs and Cheshire*, CXIV (1963), 193; J. T. Danson and T. A. Welton, 'On the population of Lancashire and Cheshire, and its local distribution during the fifty years 1801–51', *ibid.*, X (1858), 19, 21.

[4] R. P. Williams, 'On the increase of population in England and Wales', *Jour. Roy. Stat. Soc.*, XLIII (1880), 484–5.

[5] W. Farr, 'Population', in J. R. McCulloch, *A statistical account of the British Empire* (London, 1837), 410; R. P. Williams, 466–7, 470; R. Lawton, 'Rural depopulation in nineteenth-century England', in R. W. Steel and R. Lawton (eds.), *Liverpool Essays in Geography* (London, 1967), 227–8.

context of the demographic revolution of the time, and secondly in terms of the internal redistribution of population by which it was accompanied.

The growth shown in table 9.1 underlines one of the most remarkable features of the period. The rate of increase between 1811 and 1821, even allowing for the presence of many ex-servicemen, is much the highest on record for England and Wales. Such a vast increase of population was certainly without precedent, and it is likely that the addition to the population of England and Wales in the first three decades of the nineteenth century was greater than that which had occurred between the Restoration and the first census.[1]

The study of demographic change during this period raises difficult and controversial questions.[2] The major issue is whether the more important influence on the growth of population was a rise in the birth rate or a decline in mortality. The crude death rate, according to Brownlee, fell from 28.6 per thousand in 1781–90 to 21.1 per thousand in 1811–20.[3] Brownlee's estimates have not found universal acceptance. Far from falling, J. T. Krause suggested that the death rate may actually have risen between 1780 and 1820.[4] In a re-examination of the evidence, Deane and Cole emphasised the importance of regional variations in demographic experience at this period, and concluded that while a fall in mortality was most marked in London and in the rural areas, population growth in districts that were undergoing rapid industrialisation was 'much more clearly due to an increased birth rate'.[5] Whether this resulted from earlier marriages, greater fertility, or an increase in illegitimacy is uncertain.

After 1837, when official statistics of births, deaths and marriages were first available, there is less room for dispute. From 20.9 per thousand in 1845, the lowest figure recorded since 1838, the crude death rate climbed to 25.1 per thousand in 1849, before falling sharply in 1850.[6] Both the

[1] W. H. B. Court, *A concise economic history of Britain from 1750 to recent times* (Cambridge, 1964), 5.

[2] M. W. Flinn, *British population growth 1700–1850* (London, 1970); M. Drake, *Population in industrialization* (London, 1969), 1–10.

[3] J. Brownlee, 'The history of the birth and death rates in England and Wales . . . from 1570 to the present time', *Public Health*, XXIX (1916), 232.

[4] J. T. Krause, 'Changes in English fertility and mortality, 1781–1850', *Econ. Hist. Rev.*, 2nd ser, XI (1958), 56–7, 69.

[5] P. Deane and W. A. Cole, *British economic growth, 1688–1959* (Cambridge, 2nd ed., 1967), 133.

[6] B. R. Mitchell and P. Deane, *Abstract of British historical statistics* (Cambridge, 1962), 36.

infant mortality rate and the birth rate continued at a high level. A high rate of mortality in great towns, it is widely believed, pushed up the national death rate this time.[1] Already the subject of official concern in the first *Annual Report of the Registrar-General* (1837–8),[2] the health of towns was soon to attract widespread and unfavourable comment with the publication in 1842 of Edwin Chadwick's monumental *Report on the sanitary condition of the labouring population*. At their worst, the great towns were enormous reservoirs of disease and infection:

Over half the deaths were caused by infectious diseases alone. Cruel over-crowding and malnutrition gave respiratory and other forms of tuberculosis a grotesque predominance among the fatal infections, and enabled typhus to take a steady annual toll. Infant diseases, product of dirt, ignorance, bad feeding, and overcrowding, swept one in two of all children born in towns out of life before the age of five. Other infections – smallpox, scarlet fever, and the like – lingered among the overcrowded poor, occasionally bursting into epidemic fury. And from the human and animal excrements which over-flowed from cesspools, littered streets and tainted streams...came more filth diseases like typhoid and summer diarrhoea, that clung endemic in the towns, or like Asiatic Cholera which in 1832 had killed 16,437 in its first ravaging swoop on England and Wales.[3]

William Farr's sombre figures tell a similar story. Whereas the expectation of life at birth in 1841 averaged 41.16 years for England and Wales, it was 26 years in Liverpool. Two years later the expectation of life at birth was 24.2 years in Manchester compared with 40.2 years in the country at large.[4] A comparatively high birth rate and a favourable balance of migration, however, ensured that most towns continued to grow in size.[5] Growth was most marked in the coastal resorts and the centres of mining and manufacturing industry; it was slowest in the county towns.[6] Some small towns which failed to attract railways, or whose industries were

[1] T. H. Marshall, 'The population problem during the industrial revolution', *Econ. Hist.* (1929), 429–56; reprinted in E. M. Carus-Wilson (ed.), *Essays in economic history* (London, 1954); the reference is to pp. 327–9.

[2] *Annual Report of the Registrar-General for 1837–8* (1839), 76–8 (P.P. 1839, xvi).

[3] R. Lambert, *Sir John Simon 1816–1904, and English social administration* (London, 1963), 59.

[4] W. Farr, *Vital statistics* (ed. W. A. Humphreys, 1885), quoted in W. Ashworth, *The genesis of modern British town planning* (London, 1954), 59.

[5] A. K. Cairncross, 'Internal migration in Victorian England', *The Manchester School*, XVII (1949), 86–7.

[6] *Census of 1851: Population Tables*, I, vol. I, p. xlix (P.P. 1852–3, lxxxv).

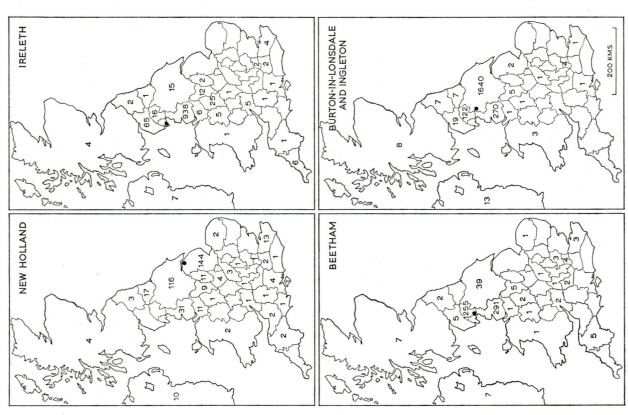

Fig. 95 County of birth of the inhabitants of four centres in 1851
Based on Public Record Office, H.O. 107/2118, 2275, 2442 and 2277.
These refer respectively to New Holland, Ireleth, Beetham, and Burton-
in-Lonsdale with Ingleton.

declining, grew little or not at all. By 1841, 29% of the population were living in towns of 20,000 inhabitants or more, compared with 18% in such places in 1811.[1]

Weavers from Bowland[2] and the linen villages near Knaresborough,[3] cloth workers from Wiltshire and Gloucestershire,[4] lead miners from Derbyshire and Alston,[5] together with agricultural workers from many counties, helped to swell the size of the towns. Much migration was local in character although, as both J. W. House and R. Lawton have observed, long-distance migration appears to have been more important than has sometimes been suggested.[6] Fig. 95 indicates that experience in this respect varied considerably from place to place. Short-distance migration to the nearest growing town and, within the large towns themselves, a movement from centre to periphery, probably accounted for a high proportion of migration at this period, just as it did later in the century.[7] The ranks of the long-distance migrants may have been recruited particularly from those who possessed some specialist skill, such as the paper-makers from Kent, the puddlers and rollers of iron from Staffordshire and Shropshire and the mine captains from Cornwall who made their appearance by 1851 in the enumerators' returns for Lake District parishes.[8]

Not all areas were equally affected by the drift to the towns. On the Yorkshire Wolds, for example, the scene of much agricultural improvement after 1800, demand for labour was rising in the 1830s, and loss of

47.

[1] A. F. Weber, *The growth of cities in the nineteenth century* (New York, 1899), lxxxvi).

[2] *Census of 1851: Population Tables, I*, vol. II, Lancashire, 46–9 (P.P. 1852–3, lxxxvi).

[3] S.C. on *Manufactures, Commerce, and Shipping*, 601 (P.P. 1833, vi); W. G. Rimmer, *Marshalls of Leeds, flax-spinners, 1788–1886* (Cambridge, 1960), 134, 163.

[4] *Census of 1851: Population Tables, I*, vol. I, Wiltshire, 22–5 (P.P. 1852–3, lxxxv); *ibid*, Gloucestershire, 18–19, 22–5.

[5] J. P. Kay, 'Report on the migration of labourers', in *First Ann. Rep. of Poor Law Commissioners for England and Wales*, 185 (P.P. 1835, xxxv). R.C. on *Poor Laws*, Part I, Rural Queries, Appendix B, p. 98 (P.P. 1834, xxx).

[6] J. W. House, *North-eastern England: population movements and the landscape since the early 19th century* (Newcastle upon Tyne, 1954), 12; R. Lawton, 'The population of Liverpool in the mid-nineteenth century', *Trans. Hist. Soc. Lancs. and Cheshire*, CVII (1955), 108.

[7] R. Lawton, 'Population changes in England and Wales in the later nineteenth century: an analysis of trends by registration districts', *Trans. and Papers, Inst. Brit. Geog.*, XLIV (1968), 68.

[8] Public Record Office, H.O. 107/2275 and 2442 (1851); A. Harris, *Cumberland iron: the story of Hodbarrow Mine 1855–1968* (Truro, 1970), 19.

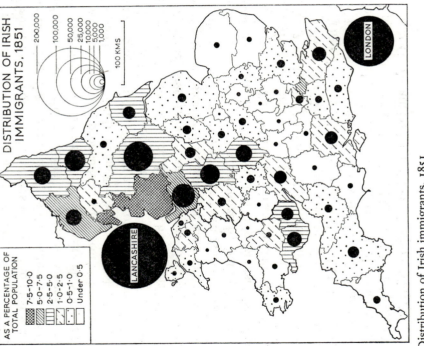

DISTRIBUTION OF IRISH
IMMIGRANTS. 1851

200,000
100,000
50,000
25,000
10,000
5,000
1,000

100 KMS

LONDON

LANCASHIRE

AS A PERCENTAGE OF
TOTAL POPULATION

7·5–10·0
5·0–7·5
2·5–5·0
1·0–2·5
0·5–1·0
Under 0·5

Fig. 96 Distribution of Irish immigrants, 1851
Based on *Census of 1851: Population Tables, II*, vol. I, pp. ccxc–ccxcvi
(P.P. 1852–3, lxxxviii, pt I).

population appears to have been slight.[1] On the nearby claylands, on the other hand, where improvement was less marked and the effects of agricultural depression more severe, the exodus of rural labourers was considerable.[2] In several parts of the country, enclosure and land reclamation were followed by local gains in population.[3]

[1] S.C. on Agric. 2nd Rep. and Mins. of Evidence, 57–8 (P.P. 1836, viii, pt I);
Census of 1851: Population Tables, I, vol. II, Yorkshire, East Riding, 48–66 (P.P. 1852–3, lxxxvi).
[2] S.C. on Agric. 2nd Rep. and Mins. of Evidence, 49–53 (P.P. 1836, viii, pt I);
Census of 1851: Population Tables, I, vol. II, Yorkshire, East Riding, 48–66 (P.P. 1852–3, lxxxvi).
[3] A. Redford, *Labour migration in England, 1800–1850* (Manchester, 1926), 65–9.

Many recruits to the urban population were of Irish origin. To the already familiar bands of Irish harvesters who worked their way through the grain counties,[1] there were added, as the century advanced, tens of thousands of poverty-stricken Irish immigrants, whose 'roaming and restless habits appear to have carried them to every place where there was any prospect of obtaining profitable employment'.[2] Between 1841 and 1851 the number of Irish-born in England and Wales increased from 291,000 to 520,000 or from 1.8% to 2.9% of the total population.[3] They were to be found in many rural areas, in the mining districts and 'in every manufacturing or commercial town',[4] but especially in and around Liverpool, Manchester, London and Glasgow (Fig. 96). So numerous were they in Manchester by the 1830s that their countrymen on arrival felt 'as if they were coming to an Irish town'.[5] Here, as elsewhere, the Irish formed closed communities, separated by both religious affiliation and degree of poverty from most of their neighbours. They also tended to form colonies, 'Irish' quarters,[6] which acquired the reputation of being 'the lowest, dampest, dirtiest, most unhealthy and ruinous' in their respective districts.[7] Their inhabitants served the needs of many different trades, but mostly those in which skill either counted for little or was readily acquired. They found work as navvies and as dock labourers and as 'hodmen' in the building trade.[8] They were also to be found in the sugar-houses and chemical works on Merseyside and in the ironworks and coal mines of the north-east.[9] Only a few, like Daniel Dudgeon, Controller of Customs at Goole, attained a position of responsibility.[10]

[1] B. M. Kerr, 'Irish seasonal migration to Great Britain, 1800–1838', *Irish Hist. Studies*, III (1943), 372–3.

[2] *Report on the state of the Irish poor in Great Britain*, 433 (P.P. 1836, xxxiv).

[3] R. Lawton, 'Irish immigration to England and Wales in the mid-nineteenth century', *Irish Geography*, IV (1959), 38.

[4] *Rep. on Irish Poor*, 433 (P.P. 1836, xxxiv).   [5] *Ibid.*, 453.

[6] *Relief of distress of destitute poor (Ireland), Papers relating to the immigration of Irish paupers in Liverpool*, 17 (P.P. 1847, liv); R. Lawton (1955), 104; F. Beckwith, 'The population of Leeds during the industrial revolution', *Thoresby Miscellany*, XLI (1948), 173; H. Cooper, 'On the cholera mortality in Hull during the epidemic of 1849', *Jour. Roy. Stat. Soc*, XVI (1853), 351.

[7] *Rep. on Irish Poor*, 437 (P.P. 1836, xxxiv).

[8] *S.C. on Railway Labourers*, 501 (P.P. 1846, xxxiv); *Rep. on Irish Poor*, 429, 431, 433–5 (P.P. 1836, xxxiv).

[9] J. W. House, 38, 40; T. C. Barker and J. R. Harris, *A Merseyside town in the industrial revolution: St Helens, 1750–1900* (Liverpool, 1954), 281–3.

[10] P.R.O., H.O. 107/2350 (1851), 15.

The net result of half a century of population change is shown on Fig. 97. Between 1801 and 1851 the northern industrial districts, several metropolitan counties and the west Midlands increased their share of the total population, whereas eastern, southern and north-western England declined in relative importance. A pattern that was to be familiar for many years to come had been established.

THE COUNTRYSIDE

As William Howitt remarked in 1838, with evident satisfaction, there were many corners of rural England in the early railway age where 'primitive living and primitive habits' lingered.[1] Yet even Howitt, whose interests led him to emphasise the quaint and the picturesque, felt obliged to include in *The rural life of England* a chapter on scientific farming, together with some comments on the spread of manufacturing industry to hitherto rural areas. Even as he wrote, many parts of the countryside were in a state of swift transition. Common arable fields were fast disappearing and would soon become a curiosity. Extensive areas of moor and heath were in process of enclosure and improvement. Works of land drainage were under way, and 'high farming' was becoming more general. Its influence was more evident in some districts than others, but the spirit of change was generally abroad in rural England during the first half of the nineteenth century.

In 1800 much land still awaited enclosure – both open-field arable and open common pasture and waste. Parliamentary proceedings were facilitated by a General Enclosure Act in 1801, and still further by other general acts in 1836 and 1845. It would seem that after 1800 over 1,200 acts enclosed about 1.8 million acres of arable (see table on p. 324), and another 1,300 or so acts enclosed about 1.3 million acres of common pasture and waste (see table on p. 351). But these figures are probably on the low side, and they exclude non-parliamentary enclosure.[2] Figs. 98 and 99 show, for arable and for commons and waste, the total effect of parliamentary enclosure during the eighteenth and nineteenth centuries, i.e. from the 1720s to 1870 or so.

There were also other changes of the greatest importance for the land.

[1] W. Howitt, *The rural life of England* (London, 3rd ed., 1844), 100. The first edition appeared in 1838.

[2] W. E. Tate, *The English village community and the enclosure movements* (London, 1967), 88.

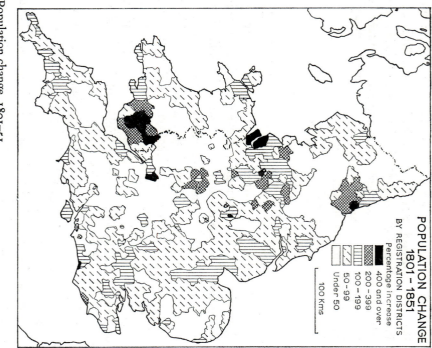

POPULATION CHANGE
1801–1851
BY REGISTRATION DISTRICTS

Percentage increase

400 and over
200–399
100–199
50–99
Under 50

100 Kms

Fig. 97 Population change, 1801–51
Based on R. Lawton in J. W. Watson and J. B. Sissons (eds.), *The British Isles: a systematic geography* (London and Edinburgh, 1964), 232.

The Board of Agriculture in 1803 commissioned Humphry Davy to lecture on the connections between chemistry and plant physiology, and the substance of these lectures was published in 1813 under the title *Elements of agricultural chemistry*. Then in 1840 came an English edition by Lyon Playfair of Justus von Liebig's *Chemistry in its application to agriculture and physiology*. The age of chemical fertilisers was inaugurated, and nitrate of soda, Peruvian guano and superphosphate of lime were now added to the traditional manures. There were also improvements in the implements of tillage – in ploughs, harrows, drills, haymakers, chaff-cutters and other machines. A number of attempts had been made to produce satisfactory reaping machines, and Patrick Bell's reaper in 1828 marked a great advance. Steam power was used for threshing and for other agricultural purposes in the early years of the century, but it was not applied to ploughing and cultivating until after 1850. The new spirit in agriculture could be seen in such events as the foundation of the Royal Agricultural Society in 1838, the establishment of Rothamsted Experimental Station in 1843, and the institution of the Royal College of Agriculture in 1845. Legislation also benefited agriculture; the Tithe Commutation Act was passed in 1836 and an act for the repeal of the Corn Laws in 1846. By the time of the Great Exhibition in 1851, much yet remained to be done, but agriculture had already begun to reflect the scientific and industrial changes of the century.

Enclosure of the arable

In many English counties the process of eliminating the common arable fields, which was already well advanced by 1800, had been largely completed by the end of the Napoleonic Wars. But, even so, arable land held in common still accounted in 1820 for nearly 10% of the total area of Cambridgeshire, for just over 8% of Oxfordshire, and for about 4 to 5% of Bedfordshire, Buckinghamshire and Huntingdonshire.[1] Even where the proportion of common arable land was very small, the tidying-up process sometimes involved the enclosure of many thousands of acres. In Wiltshire, for instance, about 64,000 acres, including much open arable land, were enclosed between 1815 and 1850;[2] the comparable figure for the East Riding was 45,000 acres.[3] By 1850, certainly by 1870, open fields had disappeared from all but a few rare villages (Fig. 98).

Contemporary writers were well aware of the local importance of enclosure in the years after 1815. J. A. Clarke, writing of Lincolnshire in 1851, numbered enclosure among the agricultural achievements of the preceding thirty years.[4] To C. S. Read, in 1854, the greatest improvements that had occurred in Oxfordshire farming since the appearance of Arthur Young's General view of that county in 1809, were 'those produced by

[1] E. C. K. Gonner, Common land and inclosure (London, 1912), 279–81.

[2] Acreage calculated from awards at the Wiltshire Record Office, Trowbridge, and from W. E. Tate, 'A hand list of Wiltshire enclosure acts and awards', Wilts. Archaeol. and Nat. Hist. Mag., LI (1945). I am grateful to Miss T. E. Vernon for assistance with the Wiltshire awards.

[3] Acreage calculated from awards in the East Riding Registry of Deeds and County Record Office, Beverley.

[4] J. A. Clarke, 'On the farming of Lincolnshire', Jour. Roy. Agric. Soc, XII (1851), 330.

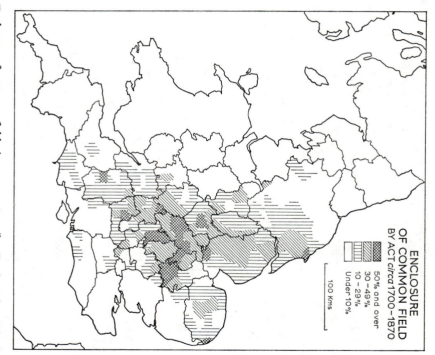

Fig. 98 Enclosure of common field *circa* 1700–1870 (by Registration Districts)
Based on E. C. K. Gonner, *Common land and inclosure* (London, 1912), map A.

the extension of enclosures'.[1] And of Cambridgeshire, where common arable fields had survived in large numbers to annoy William Cobbett in the 1820s,[2] Samuel Jonas declared in 1846, 'few counties, if any, have improved more in cultivation...all the common fields have been enclosed with the exception of five or six parishes'.[3]

In these counties, as elsewhere, the incidence of enclosure differed greatly from one district to the next. In Wiltshire nearly 90% of the area enclosed between 1815 and 1850 lay south of the escarpment of the

[1] C. S. Read, 'On the farming of Oxfordshire', *J.R.A.S.*, xv (1854), 248.
[2] William Cobbett, *Rural rides* (Everyman's Library), I, 79–80.
[3] S. Jonas, 'On the farming of Cambridgeshire', *J.R.A.S.*, vii (1846), 38.

Wiltshire Downs.[1] Or again in Lincolnshire, where 49 parishes were enclosed by Act of Parliament between 1815 and 1845, no less than 28 were in the coastal marshland of Lindsey.[2] These examples serve as a reminder of the complexity of the factors that influenced the pace of enclosure. In some districts it was the light-soiled uplands which, despite the work of earlier improvers, still awaited attention in the nineteenth century. Elsewhere it was on the claylands that the task remained largely unfinished. Open fields of high potential value in relation to the costs of enclosure sometimes survived long after similar land nearby had been enclosed. There is no simple explanation of these facts, as J. Thirsk and H. G. Hunt, among others, have demonstrated.[3] The progress of enclosure at this period, as in the past, was influenced by the distribution of landownership as well as by the nature of the soil, by local population pressure as well as by legislation which facilitated or retarded enclosure, by the extent of agricultural improvement within the framework of the open fields as well as by changing economic conditions.[4]

Enclosure of arable land was followed in some districts, particularly on the heavy claylands, by the laying of arable down to grass due to the increased profitability of heavy land for livestock as compared with grain. By 1850, the broad distinction between the arable and the grazing had been drawn, and the result was summed up in James Caird's generalisation of 1850.[5] His map showed the contrast between what he called 'the chief corn districts' of the south and east, and 'the principal grazing, green crop, and dairy districts' of the Midlands and the west (Fig. 112). In both areas market gardening continued to spread near the large towns, and particularly around London where it was fostered, as William Cobbett said in 1821, by 'the demand for crude vegetables and repayment in manure'.[6]

[1] Calculated from enclosure awards and from W. E. Tate (1945).

[2] J. Thirsk, *English peasant farming* (London, 1957), 292.

[3] J. Thirsk (1957), 237–40, 292–4; H. G. Hunt, 'The chronology of parliamentary enclosure in Leicestershire', *Econ. Hist. Rev.*, 2nd ser., x (1957), 265–72.

[4] M. A. Havinden, 'Agricultural progress in open-field Oxfordshire', *Ag. Hist. Rev.*, ix (1961), 73; A. Harris, *The rural landscape of the East Riding of Yorkshire, 1700–1850* (Oxford, 1961), 64–5.

[5] J. Caird, *English agriculture in 1850 and 1851* (London, 1852), frontispiece.

[6] William Cobbett, *Rural rides* (Everyman's Library), i, 47.

Reclamation and improvement

By 1850, enclosure of common arable fields was practically at an end. On the other hand, the age-old task of enclosing and improving the upland commons remained far from complete. It is perhaps hardly surprising that William Dickinson's enthusiasm for what had been achieved in Cumberland since 1815 should have been tempered with a note of regret since, according to his own estimate in 1852, at least 40,000 acres of reclaimable land in that county were still burdened with common rights.[1] In other western counties, too, much enclosure in the uplands awaited the years of 'high farming' after 1850. Even so, progress had been made by mid-century; particularly during the immediate aftermath of the Napoleonic Wars, when awards arising out of the great surge of war-time enclosure activity were still numerous; and again after 1836, though on a reduced scale, as economic conditions slowly improved and enclosure procedure was simplified and reduced in cost.

The upland counties which figure prominently on Fig. 99 do so by virtue of their unenclosed pastures. In Cumberland, for instance, the attack on the waste between 1815 and 1850 was marked by the passing of twenty private acts solely for the enclosure of commons and waste, and by a dozen enclosures under the general acts of 1836 and 1845;[2] Elsewhere in northern England, further inroads were made into the commons of Westmorland, the North and West Ridings of Yorkshire, Northumberland and Durham. In the Pennines and moors of the North Riding, for example, about 50,000 acres were enclosed between 1815 and 1850;[3] in the uplands of Durham about 23,000 acres[4] and in Northumberland 13,000 acres.[5] Farther south, the act of 1815 enclosing the enormous Forest of Exmoor

[1] W. Dickinson, 'On the farming of Cumberland', *J.R.A.S.*, XIII (1852), 289.

[2] W. E. Tate, 'A hand list of English enclosure acts and awards: Cumberland', *Trans. Cumberland and Westmorland Antiq. and Archaeol. Soc.*, n.s., XLIII (1943), 175–98.

[3] Calculated from a card-index of enclosure acts and awards relating to the North Riding, compiled by C. K. C. Andrew and kept at the County Hall, Northallerton.

[4] Calculated from a typescript list of enclosure awards at the County Record Office, Durham, and from W. E. Tate, 'Durham field systems and enclosure movements', *Proc. Soc. of Antiquaries of Newcastle upon Tyne*, 4th ser., X, no. 3 (1943), 119–40.

[5] Calculated from W. E. Tate, 'A hand list of English enclosure acts and awards: Northumberland', *Proc. Soc. of Antiquaries of Newcastle upon Tyne*, 4th ser., X, No. 1 (1942), 39–52.

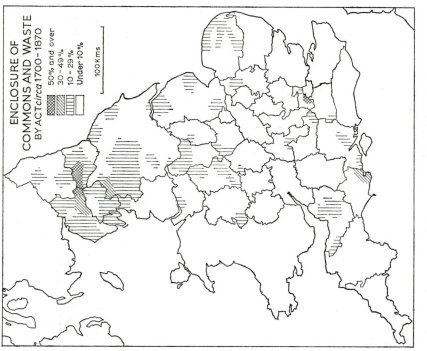

ENCLOSURE OF
COMMONS AND WASTE
BY ACT *circa* 1700–1870

50% and over
30 – 49%
10 – 29%
Under 10%

100 KMs

Fig. 99 Enclosure of commons and waste *circa* 1700–1870 (by Registration Districts) Based on E. C. K. Gonner, *Common land and inclosure* (London, 1912), map B.

covered an area of about 20,000 acres mostly in Somerset.[1] The partial reclamation of this land by the Knight family and their tenants after 1820 was in itself an achievement on the grand scale.

The new enclosures extended in some areas well beyond the economic limit of cultivation, and the making of the award was not necessarily followed by a major change in land use. Title was established, common rights were extinguished or regulated, and miles of division fence, of stone or earth, were laboriously erected; but the land remained under pasture, which was sometimes 'improved' only in being freed from

[1] C. S. Orwin, *The reclamation of Exmoor Forest* (Oxford, 1929), 20–1; W. G. Hoskins and H. P. R. Finberg, *Devonshire studies* (London, 1952), 332.

common rights.[1] Under favourable circumstances, as in parts of Cumberland for example, oats were occasionally grown at elevations of 900 to
1,000 feet 'by way of renewing and improving the land for pasturage'.[2]
Hard-won intakes continued to be made along the moorland edge
without parliamentary sanction, though their total extent is difficult to
determine. According to John Watson, writing in 1845, it was usual in
Cumbria and the Pennines of west Yorkshire for small landowners with
large families and with little employment for one half of the year, to
reclaim 'small patches of craggy ground from off their larger sheep walks'.[3]
For this purpose, as Watson explained in a passage which has about it an
almost timeless quality, it was necessary to hack, dig and trench, 'making
use of the stones for fences, drains and roads, or otherwise stacking them
up in corners, or upon the worst parts of the land; and nibbling out all
and every patch that is considered worth the labour'.[4] The impression
that methods of reclaiming the uplands had changed little over the years is
reinforced by Stephen Glover's statement that, in upland Derbyshire,
improvement involved the 'removal of surface stones, stubbing, paring,
burning, draining, and the free use of Peak lime'.[5] Piecemeal enclosure
was also slowly transforming the appearance of the mining districts of
west Cornwall. Here, as Richard Thomas reported in 1819, 'thousands
of acres of downs, commons, and wastes' were 'enclosed and improved
by the miners and others on a small scale'.[6] That this activity continued
locally well into the nineteenth century is borne out by descriptions in the
eighteen-forties.[7] Industry and agricultural improvement were also closely
associated in Cumberland and west Durham, where the London Lead
Company was engaged after 1815 in draining, liming and afforesting its
high moorland estate near Alston.[8] In the lead districts of Yorkshire and

[1] See B. Loughbrough, *Some geographical aspects of the enclosure of the vale of Pickering in the eighteenth and nineteenth centuries*, unpublished M.A. thesis, University of Hull, 1960, 106ff.

[2] W. Dickinson, 217.

[3] J. Watson, 'On reclaiming heath land', *J.R.A.S.*, VI (1845), 95.

[4] *Ibid.*

[5] S. Glover (ed. by T. Noble), *The history and gazetteer of the county of Derby*, 2 vols. (Derby, 1831–3), I, 162.

[6] R. Thomas, *Report on a survey of the mining district of Cornwall from Chasewater to Camborne* (London, 1819), 14.

[7] C. Redding, *An illustrated itinerary of the county of Cornwall* (London, 1842), 143; W. F. Karkeek, 'On the farming of Cornwall', *J.R.A.S.*, VI (1845), 411, 445.

[8] A. Raistrick, *Two centuries of industrial welfare; the London (Quaker) Lead Company* (London, 1938), 32.

the northern Pennines, miner-farmers were carving small intakes from the moors;[1] and in the uplands of Rossendale, the improvement of hitherto unprofitable moorland was stimulated by the presence nearby of an expanding manufacturing district.[2]

The conversion of much downland, warren and heath into arable on the sandlands and the chalk and limestone soils of lowland England continued a trend that was already under way in 1800, and that marked a further stage in the shift of cereal production from the heavy to the light lands. Enclosure, paring and burning, marling and chalking, all played a part in this transformation, but no less important was the use that was made of green crops, of 'light' manures such as bones, guano and superphosphates, of oil-cake which yielded a rich return in the form of dung, and of the closer integration of livestock and arable husbandry. To Samuel Sidney in 1848, the wolds of north Lincolnshire presented 'an unbroken succession of large farms', composed, almost 'entirely of arable land'.[3] He saw 'hundreds of sheep feeding off turnips in the fields' and 'scores of young stock feeding off cake and hay, and treading down straw in the yard'.[4] Philip Pusey, who had travelled over Lincoln Heath a few years earlier, was equally impressed by what he saw. 'Every stubble-field was clean and bright; all the hedges kept low, and neatly trimmed; every farmhouse well built... and surrounded by rows of high, long, saddle-backed ricks.'[5] Fifty years earlier both these districts had contained large expanses of rabbit warren and sheep-walk. On the chalklands of Yorkshire, Wiltshire, Hampshire and Dorset too, the acreage under the plough increased appreciably during the first half of the nineteenth century,[6] but not all

[1] R. T. Clough, *The lead smelting mills of the Yorkshire Dales* (Leeds, 1962), 21; A. Raistrick, *Mines and miners of Swaledale* (Clapham, 1955), 83–4; C. J. Hunt, *The lead miners of the northern Pennines* (Manchester, 1970), 152–9.

[2] G. H. Tupling, *The economic history of Rossendale* (Chetham Society, Manchester, 1927), 229.

[3] S. Sidney, *Railways and agriculture in north Lincolnshire: rough notes of a ride over the track of the Manchester, Sheffield, Lincolnshire, and other railways* (London, 1848), 72.

[4] *Ibid*, 74.

[5] P. Pusey, 'On the agricultural improvements of Lincolnshire', *J.R.A.S.*, IV (1843), 287; H. C. Darby, 'The Lincolnshire Wolds', *The Lincolnshire Historian*, No. 9 (1952), 315–24.

[6] C. Howard, *A general view of the agriculture of the East Riding of Yorkshire* (London, 1835), 111; E. Kerridge in E. Crittall (ed.), *V.C.H. Wiltshire*, IV (1959), 73; J. Wilkinson, 'The farming of Hampshire', *J.R.A.S.*, XXII (1861), 290, L. H. Ruegg, 'Farming of Dorsetshire', *J.R.A.S.*, XV (1854), 437.

these districts attained the same high standards of husbandry as Lincoln-shire.[1]

Both agriculture and industry encroached upon the heathlands which, island-like, survived in many parts of the lowlands. On the Forest Sands of Nottinghamshire, on the Keuper Sandstones of mid-Cheshire, on the sandy wastes of Breckland and in the southern vale of York, enclosure, tree-planting, bone-tillage and turnip husbandry made steady progress.[2] And as the Cannock coalfield was developed, heathland there disappeared beneath sprawling mining settlements.[3] The preservation of rough land for sporting purposes, however, meant that an alternative use was not always possible.[4] Attempts to check the enclosure of commons in the vicinity of populous places also acted as a barrier to improve-ment.[5]

'As for draining,' Philip Pusey wrote in 1842, 'there is not a county, nor any large proportion of parishes, or even of farms, in which it ought not to be done.'[6] At the district level, a new drainage era had in fact opened some years earlier, in 1819–20, with the installation of a steam pump near Littleport in the Isle of Ely.[7] During the following decade steam pumping was adopted in other parts of the Fenland basin, and by 1848 it was being used 'extensively and with vast advantage' both there and in several other low-lying districts.[8] But, as J. A. Clarke ad-mitted, though considerable progress had been made by the 1840s, the task of draining this area remained unfinished, not least because more

[1] T. Walkden, 'On the advantages of ploughing-up down-land', *J.R.A.S.*, IV (1843), 83–5.

[2] R. W. Corringham, 'Agriculture of Nottinghamshire', *J.R.A.S.*, VI (1845), 2–7; W. Palin, 'The farming of Cheshire', *J.R.A.S.*, V (1844), 73; R. N. Bacon, *The report on the agriculture of Norfolk* (London, 1844), 19, 92; G. Legard, 'Farming of the East Riding of Yorkshire', *J.R.A.S.*, IX (1848), 92.

[3] M. J. Wise (ed.), *Birmingham and its regional setting* (Birmingham, 1950), 274.

[4] S. Glover, II (1833), 94, 103.

[5] F. E. Hyde, 'Utilitarian town planning, 1825–1845', *Town Planning Rev.*, XIX (1947), 157–8.

[6] P. Pusey, 'On the progress of agricultural knowledge during the last four years', *J.R.A.S.*, III (1842), 185.

[7] H. C. Darby, *The draining of the Fens* (Cambridge, 2nd ed, 1956), 222; C. T. G. Boucher, *John Rennie, 1761–1821* (Manchester, 1963), 136.

[8] *The rural cyclopedia*, II (Edinburgh, 1848), 81; W. H. Wheeler, *History of the fens of South Lincolnshire* (Boston and London, 1868), 35, 132; M. Williams, *The draining of the Somerset Levels* (Cambridge, 1970), 144, 160–1; J. Thirsk (1957), 286.

efficient drainage inevitably produced further shrinkage on the peat fens.[1] Improvement schemes were frequently accompanied by renewed efforts to rectify the deficiencies of the local rivers. In the Fenland, for instance, the Eau Brink Cut (1821) removed a great bend along the lower Ouse near King's Lynn, and a new outfall on the Nene (1830) replaced a shifting and frequently silted channel.

While steam pumps in some districts helped to bring new land into cultivation and prevented the old from reverting to fen, in others their use was still unknown in 1850. Along the Hull valley, partly drained by that year, drainage was still carried out by windmill, and ague, though rare, had not yet disappeared.[2] Or again, across the Pennines, Martin Mere, 'the Great Fen of Lancashire', received the benefit of steam drainage only in 1850.[3] Although the technical means were now available, effective drainage was too often delayed by the survival of antiquated and unco-ordinated administrative machinery, and by disagreement between those who wished to drain the land and those who, for various reasons, were opposed to change.[4] Many years were to pass in some districts before these obstacles were overcome. Outside the great levels reclamation was also proceeding, although on a less ambitious scale. Chat Moss, in Lancashire already diminished in size by drainage works, received further attention from the improvers, and by 1833 good crops of oats and potatoes were growing there on newly recovered mossland.[5] Further north, in the Wyre valley, a combination of draining and marling was turning the local moss-lands into arable land of high quality.[6] Warping and land drainage were performing a similar service for parts of the Humberhead marshes.[7]

'It is upon the lighter lands – the sandy, the loamy, the peaty soils – that the main expenditure of skill has hitherto taken place', a writer in the

[1] J. A. Clarke, 'On the Great Level of the Fens', *J.R.A.S.*, VIII (1847), 110, 112; *ibid*, 'On trunk drainage', *J.R.A.S.*, XV (1854), 1–73.
[2] J. A. Sheppard, *The draining of the Hull valley* (East Yorkshire Local History Society, York, 1958), 21; H. Cooper, 'On the relative prevalence of diseases in Hull, and the effects of season upon disease', *Jour. Roy. Stat. Soc.*, XVI (1853), 355.
[3] H. White, 'A detailed report of the drainage by steam-power of a portion of Martin Mere, Lancashire', *J.R.A.S.*, XIV (1853), 165.
[4] J. Henderson, 'Report upon the Rye and Derwent drainage', *J.R.A.S.*, XIV (1853), 136, 140; J. A. Sheppard, 15, 18.
[5] *S.C. on Agriculture*, 187–8 (P.P. 1833, v).
[6] *Lonsdale Magazine*, I (1820), 199–204; W. J. Garnett, 'Farming of Lancashire', *J.R.A.S.*, X (1849), 24–8.
[7] R. Creyke, 'Some account of the process of warping', *J.R.A.S.*, V (1844), 398; W. Edwards, 'On dry warping at Hatfield Chase', *J.R.A.S.*, XI (1850), 180–3.

Edinburgh Review observed in 1845, and then went on to point out that 'the first great stride which England has to make in the cultivation of her arable lands, is in the adaptation of her clay soils to the alternate husbandry'.[1] The heavy clays, which without the benefit of adequate drainage were unkind to root crops and folded sheep, and which in many areas were managed on a costly and relatively inflexible system of husbandry, had not only failed to keep pace with the light lands in agricultural progress but had also become problem areas.[2] Field-drainage did something to redress the balance, and the draining methods of James Smith of Deanston in Perthshire did much to improve the heavier soils.[3] On some clayland estates draining was tackled energetically in an attempt to mitigate the worst effects of agricultural depression.[4] The strong clays of Huntingdonshire were said to be partly tile-drained in 1836 and, though erratic, some progress had by that time been made elsewhere.[5] Field-drainage was greatly accelerated during the 1840s, however, by the use of machinery which enabled first ordinary drainage tiles and later pipe-tiles to be mass-produced at low cost.[6] In some areas at least under-draining was soon followed by the extension of turnip husbandry on to land which a few years earlier had been considered too wet for it.[7] But in 1850 it was still possible to assert without serious fear of contradiction that 'if there be any land which requires improvement, it is our real heavy clays'.[8]

The changes which have been indicated were accompanied by others. Thus territorial boundaries of the various breeds of livestock continued to alter. Moorland breeds of sheep were in retreat, in upland Derbyshire and elsewhere, in the face of enclosure and reclamation,[9] and agricultural

[1] *Edinburgh Review*, LXXXI (1845), 94–5.

[2] E. L. Jones, *Seasons and prices: the role of the weather in English agricultural history* (London, 1964), ch. 9.

[3] J. Smith, *Remarks on thorough draining and deep ploughing* (Stirling, 1831).

[4] F. M. L. Thompson, *English landed society in the nineteenth century* (London, 1963), 245–55.

[5] *S.C. on Agriculture, First Report and Mins. of Evidence*, 60, 168; *Second Report etc.*, 73, 173 (P.P. 1836, viii).

[6] J. Parkes, 'Report on drain-tiles and drainage', *J.R.A.S.*, IV (1843), 372–3; H. C. Darby, 'The draining of the English clay-lands', *Geographische Zeitschrift*, LII (1964), 190–201; G. E. Fussell, *The farmer's tools, 1500–1900* (London, 1952), ch. I.

[7] J. Grey, 'A view of the past and present state of agriculture in Northumberland', *J.R.A.S.*, II (1841), 182.

[8] P. Pusey, 'On the progress of agricultural knowledge during the last eight years', *J.R.A.S.*, XI (1850), 406.

[9] S. Glover, I (1831), 162.

improvement led to the replacement of the restless Norfolk, admirable as heath sheep, by the Southdown.[1] The Shorthorn breed of cattle had displaced the Longhorn from most parts of Westmorland by 1845 and from all Lancashire by 1849.[2] Many similar examples could be found.

Two further changes invite comment. As land was enclosed and reclaimed, many of the gaps that remained to be filled in the pattern of rural settlement were occupied by outlying dwellings. On Exmoor, in 1850, Thomas Acland found cottages and farmsteads where not long before there had been unimproved and uninhabited moorland;[3] and in the textile district of the West Riding new farmsteads, built, like their predecessors in the same area, of native gritstone, now made their appearance high up on the valley sides.[4] Locally, as on the Crown estate of Sunk Island in the East Riding, the building of farmsteads followed coastal reclamation. Other new farmsteads were the result of the enclosure of arable fields; a number of these appeared some time after the initial enclosure, as large allotments were gradually organised into new farm units. Villages also changed. Many grew substantially in size during the first three or four decades of the nineteenth century, and under this stimulus farmsteads were subdivided and rows of cottages added wherever space was available, which meant that gaps in the old village plan were now often closed by a process of infilling. Tiny cottages for landless labourers appeared side by side with elegant town-style houses which owed little or nothing to the vernacular tradition of the district.[5]

Although the age of country-house building on a grand scale was slowly drawing to a close, the eighteenth-century taste for architectural ostentation continued well into the new century. Landed aristocrats like the 5th duke of Rutland at Belvoir in Leicestershire, and the 15th earl of Shrewsbury at Alton Towers in Staffordshire, were not alone in building or refashioning great houses in the heart of the countryside, sometimes in the Classical tradition, but more frequently, as the century advanced,

[1] W. Youatt, *Sheep*, London (1837), 307–10; H. Raynbird, 'On the farming of Suffolk', *J.R.A.S.*, VIII (1847), 308.

[2] W. Youatt, *Cattle* (London, 1834), 200; F. W. Garnett, *Westmorland agriculture, 1800–1900* (Kendal, 1912), 184; W. J. Garnett, 39.

[3] T. D. Acland, 'On the farming of Somerset', *J.R.A.S.*, XI (1850), 688ff.

[4] J. C. R. Camm, *Industrial settlement in the Colne and Holme valleys, 1750–1960*, unpublished M.Sc. thesis, University of Hull, 1963, pp. 115–17.

[5] R. B. Wood-Jones, *Traditional domestic architecture of the Banbury region* (Manchester, 1963), 197–9, 288–90; W. G. Hoskins, *The Midland peasant* (London, 1957), 272–3.

in the Gothic style.[1] There were also many new settlers — retired merchants and industrialists in the Lake District afford a well-known example — who were also active in building country residences.[2] As at Edensor, on the Chatsworth estate in Derbyshire, 'improvement' sometimes involved the remodelling of a complete village;[3] or, more frequently, the landscaping of a park after the manner of Repton or in the gardenesque style of John Claudius Loudon.[4] But, as the *Quarterly Review* noted, by the middle of the nineteenth century few attempts were being made to 'improve on the extensive scale that was adopted by Brown and his school'.[5]

Apart from the planting of trees for ornament and for fox covers or game preserves, very little afforestation took place. The English navy increasingly relied upon imported timber. It is true that the shortage during the French wars had stimulated the management of the Crown forests and the planting of oak, but not very effectively.[6] In any case, before the oaks so planted had come to maturity, the era of the wooden ship was over. Rural England, particularly lowland England, had become a countryside of grain and grass, and so it remained.

If the emphasis has been placed here on the changing elements in the rural scene that is not to deny the existence of other more permanent features. Many thousands of acres of old enclosed country must have altered little in outward appearance, though here and there hedgerows were removed and old pastures ploughed out as the burden of tithe was lifted;[7] some of these areas, denied the stimulus of enclosure, remained agriculturally backward.[8] And against the factory-farms of Northumberland with their steam-driven equipment may be set those of Middlesex where, so it was alleged in 1836, standards of land management were but little improved on those of 'our forefathers'.[9]

[1] H. M. Colvin, *A biographical dictionary of English architects, 1660–1840* (London, 1954), 730; J. C. Loudon, *An encyclopaedia of gardening* (London, 1859), 256–63.

[2] William Wordsworth, *A guide through the district of the lakes* (Malvern, Facsimile edition, 1949), 62.

[3] H. M. Colvin (1954), 506. I am indebted to the Librarian and the Trustees of the Chatsworth Settled Estates for permission to see plans of Edensor at Chatsworth.

[4] D. Clifford, *A history of garden design* (London, 1962), 173, 184; H. C. Prince, 'The changing landscape of Panshanger', *Trans. East Herts. Archaeol. Soc.*, XIV (1959), 42–58.

[5] *Quarterly Review*, XCVIII (1855), 215.

[6] R. G. Albion, *Forests and sea power* (Cambridge, Mass., 1926), 137–8, 399.

[7] G. Buckland, 'On the farming of Kent', *J.R.A.S.*, VI (1845), 301; H. Evershed, 'On the farming of Surrey', *J.R.A.S.*, XIV (1853), 412.

[8] J. Thirsk (1957), 263.

[9] S.C. on Agriculture, First Report etc., 179 (P.P. 1836, viii).

INDUSTRY

Throughout the first half of the nineteenth century agriculture retained its place as the most important British industry judged in terms of employment, although its share both of the total occupied population and of the national income slowly declined as manufacturing industry and mining increased in importance.[1] Their growth was accompanied by a flood of literature, varying in character from Andrew Ure's massive *Dictionary of arts, manufacturers, and mines* (London, 1848). More of the 'well authenticated facts' about the country's economy called for by G. R. Porter in *The progress of the nation* (London, 1836) were gathered on a systematic basis, even though the full statistical age had not yet arrived.

Coalmining

Coal output mounted quickly between 1815 and 1850 as the demands of industry, of domestic users and of transport grew more insistent, and it may have risen from 16 million tons in 1816 to 49 million in 1850.[2] Much of the increase occurred between 1820 and 1840, when many new mines were 'sunk and worked with great rapidity and to a great extent in various parts of England'.[3]

Mining was extended in both area and depth. In Northumberland and Durham, mining activity spread into hitherto untouched or little worked districts. Slowly at first, and then with great speed as the railway network was evolved, new colliery districts were opened up in Northumberland, in south-west Durham, and in the concealed coalfield of east Durham.[4] In these areas, as elsewhere, the advent of the safety lamp and of improved techniques of winding, pumping and ventilation enabled deeper seams to be won, and by 1835 the shafts of Monkwearmouth colliery, the deepest mine in the north-east, had reached 1,590 feet.[5] Expansion of output in the

[1] C. Booth, 'Occupations of the people of the United Kingdom, 1801–81', *Jour. Roy. Stat. Soc.*, XLIX (1886), Appendix A; P. Deane and W. A. Cole, ch. 5.
[2] *R.C. on Coal*, 861, 883 (P.P. 1871, xviii); see P. Deane and W. A. Cole, 216. The figures relate to the United Kingdom.   [3] *Westminster Review*, XL (1843), 418.
[4] T. Y. Hall, 'The extent and probable duration of the northern coal field', *Trans. North of England Inst. of Mining Engineers*, II (1854), *passim*; A. E. Smailes, 'Population changes in the colliery districts of Northumberland and Durham', *Geog. Jour.*, XCI (1938), 222–4.
[5] [J. Holland], *The history and description of fossil fuel, the collieries and coal trade of Great Britain*, London (1835), 188–9.

north-east, from about 4.8 million tons in 1816 to about 10.5 million in 1851,[1] was accompanied by the rise of new coal ports at Seaham Harbour, at Hartlepool and West Hartlepool and on Tees-side, and by a great increase in the volume of coal shipments. By the middle of the century, however, the traffic in seaborne coal from the north-eastern ports, which for so long had dominated the London market, was entering upon a new and competitive phase. The change had been marked, in 1845, by the arrival in London of the first consignments of railborne coal. These amounted to little more than 8,000 tons in that year, but by 1851 they had reached nearly 248,000 tons.[2] The 1840s was a critical decade in the north-east.[3]

Although the northern coalfield remained in 1850 the most productive in the country, other coalfields had been growing in importance during the previous half century. Shipments of coal from west Cumberland more than doubled between 1819 and 1849, for example, and new mining districts were developed there, particularly after the construction of the Maryport and Carlisle and other railways during the 1840s.[4] Rising demand from the alkali, glass and copper industries of St Helens combined, with the Cheshire salt industry and a growing market for steam coal, to stimulate mining in south-west Lancashire.[5] Many new mines were opened to win both coal and ironstones in the Potteries,[6] in south Staffordshire,[7] along the north crop of the South Wales coalfield,[8] and in central Scotland where the era of heavy industry had begun.[9]

[1] J. W. House, 61.

[2] R.C. on Coal, 865–6 (P.P. 1871, xviii).

[3] A. J. Taylor, 'The third marquis of Londonderry and the north-eastern coal trade', Durham University Journal, n.s., XVII (1955), 22; P. M. Sweezy, Monopoly and competition in the English coal trade 1550–1850 (Cambridge, Mass., 1938), especially ch. 10.

[4] M. Dunn, An historical, geological and descriptive view of the coal trade of the north of England (Newcastle upon Tyne, 1844), 133; O. Wood, The development of the coal, iron and shipbuilding industries of west Cumberland, 1750–1914, unpublished Ph.D. thesis, University of London (1952), 97ff., 125.

[5] T. C. Barker and J. R. Harris, ch. 15.

[6] J. Hedley, 'Mines and mining in the north Staffordshire coal field', Trans. North of England Inst. of Mining Engineers, II (1854), passim; M. W. Greenslade in J. G. Jenkins (ed.), V.C.H. Staffordshire, VIII (1963), 101–3, 169, 222.

[7] M. J. Wise (ed.), 232–8.

[8] J. H. Morris and L. J. Williams, The South Wales coal industry 1841–1875 (Cardiff, 1958), 8–12; E. G. Bowen (ed.), Wales (London, 1957), 209–11.

[9] J. B. S. Gilfillan and H. A. Moisley in R. Miller and J. Tivy (eds.), The Glasgow region (Glasgow, 1958), 169–72.

The iron and steel industries

'Ingenuity furnishes endless occasion for fresh demand, while at the same time equal industry is apparent in the corresponding exertions which create the supply.'[1] Thus wrote Scrivenor of the iron trade in 1841, after several decades of erratic expansion had carried pig-iron production in England and Wales from about 220,000 tons in 1806 to 1,155,000 tons in 1840.[2] Seven years later the figure was to exceed 1,450,000 tons. Behind this progress lay the expanded gas and water undertakings of the country and new railway, engineering and shipbuilding industries. Amongst the forces influencing supply were improvements at both the furnace and the forge, which resulted in great economies in the consumption of fuel, in larger blast furnaces and in a more efficient method of puddling iron.[3]

The manufacture of iron advanced most rapidly on the coalfields of South Wales and Scotland, which together accounted for 63% of the total pig-iron output of Great Britain in 1847, compared with 40% in 1806.[4] During this period the west Midlands was surpassed by each of these districts. Yet pig-iron production multiplied more than sixfold in Staffordshire between 1806 and 1847, and in Shropshire by some 60%.[5] From south Staffordshire, where more than fifty new blast furnaces appeared between 1823 and 1847,[6] there flowed an ever-increasing variety of foundry and forge pig-iron, castings, puddled iron and finished metal goods; and the coalfield, which within living memory had retained a rural aspect, was transformed by 1846 into 'a continuous city of fire-belching furnaces and smoke-vomiting chimneys'.[7] On the Shropshire coalfield, so Thomas Smith said in 1836, 'the smoke and blackness of furnaces, forges, and foundries, give a tone to every part of the prospect'.[8] Similar changes were taking place, although on a smaller scale, in Derbyshire and

[1] H. Scrivenor, *A comprehensive history of the iron trade* (London, 1841), vi.

[2] B. R. Mitchell and P. Deane, 131.

[3] H. R. Schubert, 'The extraction and production of metals: iron and steel', in C. Singer *et al.*, *A history of technology*, IV (Oxford, 1958), 109–14; B. R. Mitchell, 'The coming of the railway and United Kingdom economic growth', *Jour. Econ. Hist.*, XXIV (1964), 326–8.

[4] I. L. Bell, *The iron trade of the United Kingdom* (London, 1886), 9.

[5] B. R. Mitchell and P. Deane, 131. A comparatively small output in north Staffordshire is included for 1806.

[6] H. Scrivenor (1854 ed.), 135, 295.

[7] H. Miller, *First impressions of England and its people* (Edinburgh, ed. 1857), 49.

[8] T. Smith, *The miner's guide* (Birmingham, 1836), 114.

Yorkshire.[1] In steel production, however, Yorkshire was pre-eminent: the output of blister and cast steel from Sheffield rose swiftly after 1830 as new markets for steel goods were found outside the local cutlery trades.[2] The coalfields of Warwickshire and Lancashire, which were poorly endowed with both ironstones and coal of high coking quality, failed to share in the general growth of the iron industry.[3] Its development was tardy and of limited extent also in Northumberland and Durham.[4] By mid-century, however, the iron industry of the north-east was about to pass through a period of accelerated change which ultimately affected the metal trades of districts far removed from the local region. The haematite ores of Cumberland and Furness were exported mainly to Scotland and South Wales; there was very little local iron manufacturing owing to the unsuitability of the local coal for coking.[5]

The textile industries

The continued growth of the cotton textile industry was one of the marvels of the age. Its 'rapid growth and prodigious magnitude', the *Edinburgh Review* commented in 1827, were 'the most extraordinary phenomena in the history of industry'.[6] The output of cotton textiles increased steadily from the middle of the 1820s, and by 1835 at least 185,000 mill hands and perhaps 200,000 hand-loom weavers were engaged in the industry.[7] Three

[1] K. Warren, 'The Derbyshire iron industry since 1780', *East Midland Geographer*, II, No. 16 (1961), 19–21; J. F. W. Johnston, 'The economy of a coal-field', *Proc. Geological and Polytechnic Society of the West Riding of Yorkshire*, I (1849), 50.

[2] M. W. Flinn and A. Birch, 'The English steel industry before 1856 with special reference to the development of the Yorkshire steel industry', *Yorks. Bull. Econ. and Soc. Research*, VI (1954), 173–4; J. C. Carr and W. Taplin, *A history of the British steel industry* (Oxford, 1962), II.

[3] M. J. Wise (ed.), 295–6; S. H. Beaver, 'Coke manufacture in Great Britain: A study in industrial geography', *Trans. and Papers, Inst. Brit. Geog.*, XVII (1952), 135, 138; A. Birch, 'The Haigh Ironworks, 1789–1856: a nobleman's enterprise during the industrial revolution', *Bull. John Rylands Library*, XXXV (1952–3), *passim*.

[4] I. L. Bell, 'On the manufacture of iron in connection with the Northumberland and Durham coal-field', *Trans. North of England Inst. of Mining Engineers*, XIII (1864), 111–24.

[5] J. D. Kendall, 'Notes on the history of mining in Cumberland and north Lancashire', *Trans. North of England Inst. of Mining and Mechanical Engineers*, XXXIV (1885), 92–4; A. E. Smailes, *North England* (London and Edinburgh, 1960), 183–4.

[6] *Edinburgh Review*, XLVI (1827), I.

[7] E. Baines, *History of the cotton manufacture in Great Britain* (London, 1835), 383, 394; R. C. O. Matthews, *A study in trade-cycle history: economic fluctuations in Great Britain, 1833–1842* (Cambridge, 1954), especially 127–8 and ch. 9.

years later the number of mill workers had risen to almost 220,000 and by 1850 to 292,000, though the number of hand-loom weavers meanwhile had fallen sharply.[1]

The face of the principal cotton-manufacturing areas quickly changed. The altered appearance of the district following the building of numerous cotton mills during the twenties and thirties formed the subject of much contemporary comment in south Lancashire.[2] The number of cotton mills at work within the township of Manchester rose from 44 in 1820 to 63 in 1826,[3] and in Heaton Norris, nearby, from about a dozen in 1825 to 20 in 1836.[4] Mills and printworks were crowded into the valleys of the Irwell and its tributaries, which in the vicinity of Manchester quickly became little more than open sewers. For a time new mills also continued to appear in outlying centres of the industry such as Carlisle, Lancaster and Bristol.[5] In Hull, a port without any tradition of textile working but in control of nearly 70% of the country's export trade in cotton twist and yarn, the industry gained for itself new territory in 1838.[6]

Although most of the new growth took the form of steam-driven factories, the older water-powered mills were slow to disappear (Figs. 100 and 101). Thus more than half the horse-power employed in the Derbyshire cotton industry in 1850 was derived from water-wheels.[7] Even in Lancashire, where conversion to steam was far advanced by the 1830s, water remained an important source of power in the valleys of Rossendale

[1] *Factory Returns, 1838,* 59, 205, 295 (P.P. 1839, xlii); *Factory Returns, 1850,* 470 (P.P. 1850, xlii); G. A. Wood, 'The statistics of wages in the nineteenth century – XIX, The cotton industry', *Jour. Roy. Stat. Soc.,* LXXIII (1910), 594–6.

[2] J. Butterworth, *A history and description of the towns and parishes of Stockport, Ashton-under-Lyne, Mottram-Longden-Dale, and Glossop* (Manchester, 1827), 60–1.

[3] E. Baines, 395.

[4] E. Butterworth, *A statistical sketch of the County Palatine of Lancaster* (London, 1841), 90.

[5] W. Parson and W. White, *History, directory and gazetteer of Cumberland and Westmorland* (Leeds, 1829), 118, 152; M. M. Schofield, *Outlines of an economic history of Lancaster, Part II 1680–1860* (Lancaster, 1951), 111–15; S. J. Jones, 'The cotton industry in Bristol', *Trans. and Papers, Inst. Brit. Geog.,* XIII (1947), 73–4.

[6] J. M. Bellamy, 'Cotton manufacture in Kingston upon Hull', *Business History,* IV (1962), 92–5.

[7] A. J. Taylor, 'Concentration and specialization in the Lancashire cotton industry, 1825–1850', *Econ. Hist. Rev,* 2nd ser., I (1949), 115; D. M. Smith, 'The cotton industry in the east Midlands', *Geography,* XLVII (1962), 264–5.

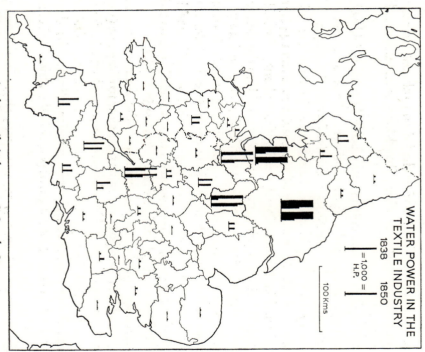

Fig. 100 Water power in the textile industry, 1838 and 1850. Based on: (1) *Factory Returns, 1838* (P.P. 1839, xlii); (2) *Factory Returns, 1850* (P.P. 1850, xlii).

and along the flanks of the uplands further south.[1] Some country mills, of which those near Ingleton in Yorkshire are an example, were able to obtain coal almost literally from underfoot.[2] Without ready access to coal, however, many remote mills were placed at a disadvantage in an age of steam, and by 1850 they had either disappeared or else had been converted to other uses.[3] The weaving of cotton by hand disappeared even more

[1] G. North, 'Industrial development in the Rossendale valley', *Jour. Manchester Geog. Soc.*, LVIII (1961–2), 17; H. B. Rodgers, 'The Lancashire cotton industry in 1840', *Trans. and Papers, Inst. Brit. Geog.*, XXVIII (1960), 138–9.
[2] A. Harris, 'The Ingleton coalfield', *Industrial Archaeology*, V (1968), 318.
[3] J. D. Marshall, *Furness and the industrial revolution* (Barrow-in-Furness, 1958), 54; J. D. Marshall and M. Davies-Shiel, *The Lake District at work, past and present* (Newton Abbot, 1971), 18, 21.

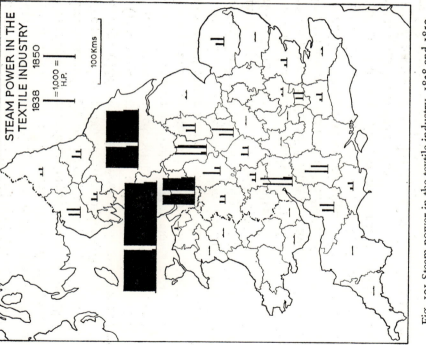

STEAM POWER IN THE
TEXTILE INDUSTRY

1838   1850

=1,000 = H.P.

100 Kms

Fig. 101 Steam power in the textile industry, 1838 and 1850.
Sources as for Fig. 100.

rapidly. About Oldham, Royton and Crompton, Wheeler reported in 1836, 'loom-shops have been deserted, looms sold or broken-up, and whole families have gone to the mills for employ'.[1] The number of power looms at work increased almost threefold in Lancashire between 1835 and 1850,[2] by which date the specialised power-weaving district in the north of the county was just beginning to emerge.[3]

With approximately 69% of the mill hands of the cotton industry in England and Wales in 1838, Lancashire contained by far the largest group

[1] J. Wheeler, *Manchester: its political, social and commercial history* (London and Manchester, 1836), 233.

[2] G. R. Porter, *The progress of the nation* (London, 1851 ed.), 200.

[3] H. B. Rodgers, 145–51.

of cotton workers in the country; Cheshire, with between 16% and 17%, and Yorkshire with about 6%, lagged far behind, as did the Midlands with 6.5%.[1] By 1850 geographical concentration had advanced a stage further (Fig. 102). In that year Lancashire, with 70% of the cotton mills in England and Wales, contained 74% of the employees, 73% of the spindles and 79% of the power looms engaged in the industry.[2]

The changes which overtook the worsted industry were hardly less striking (Fig. 103). Between 1838 and 1850 employment in worsted factories more than doubled, while hand-weaving of worsteds began rapidly to diminish.[3] The use of cotton warps, alpaca and mohair 'imparted a new character to the worsted industry',[4] enabling its products to compete with those of cotton in the market for cheap, light fabrics.[5] As the industry expanded, concentration within the West Riding became more marked: Yorkshire's share of the country's worsted workers rose from approximately 85% in 1838 to 90% in 1850.[6] Lancashire, Leicestershire and Norfolk, the only other counties with a substantial worsted industry, could then claim between them only some 10% of the country's worsted workers.[7] Nowhere was the quickening of activity within the Yorkshire worsted trade more apparent than in Bradford, where many new worsted mills were added between 1800 and 1830, and where the town grew rapidly and untidily.[8] Attracted by an expanding industry, worsted merchants transferred their activities to Bradford, making it the undisputed centre of the English worsted trade.[9]

The transformation of woollen manufacturing from a domestic to a factory industry proceeded relatively slowly. According to Baines, writing in 1859, even then there were as many woollen workers employed

[1] *Factory Returns, 1838* (P.P. 1839, xlii). The Midlands here include Derbyshire, Nottinghamshire, Staffordshire, Leicestershire and Warwickshire.

[2] *Factory Returns, 1850, 456–7* (P.P. 1850, xlii).

[3] *Factory Returns, 1838* (P.P. 1839, xlii) and *1850* (P.P. 1850, xlii); E. M. Sigsworth, 'Bradford', in C. R. Fay, *Round about industrial Britain, 1830–1860* (Toronto, 1952), 119–20, 125–7.

[4] J. James, *History of the worsted manufacture in England* (London, 1857), 470–1.

[5] E. M. Sigsworth, *Black Dyke mills* (Liverpool, 1958), 43–5.

[6] *Factory Returns, 1838* (P.P. 1839, xlii) and *1850* (P.P. 1850, xlii). But see J. K. Edwards, 'The decline of the Norwich textiles industry', *Yorks. Bull. Econ. and Soc. Research*, XVI (1964), 37–8.

[7] *Factory Returns, 1850* (P.P. 1850, xlii).

[8] J. James (1857), 605.

[9] E. M. Sigsworth, 'Fosters of Queensbury and Geyer of Lodz, 1848–1862', *Yorks. Bull. Econ. and Soc. Research*, III (1951), 67.

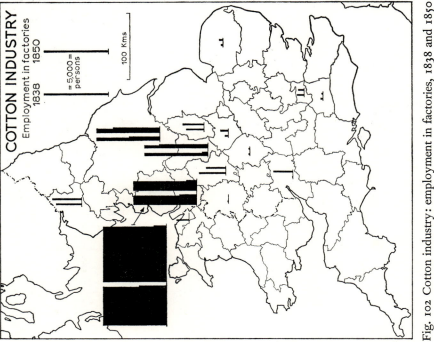

COTTON INDUSTRY

Employment in factories

1838   1850

■ = 5,000 = persons

100 Kms

Fig. 102 Cotton industry: employment in factories, 1838 and 1850
Sources as for Fig. 100.

outside the factories as in them.[1] Mechanisation of the weaving process was particularly slow, and the number of power looms working woollen goods remained small until the fifties. The Yorkshire factories were largely steam-powered by 1838, but the process of conversion was still incomplete in 1850 (Figs. 100 and 101). In several other districts, too, water power retained its hold on the industry. In the West Country, only Wiltshire adopted steam power on an extensive scale: there steam engines generated 64% of the power in 1838 and 77% in 1850.[2] But in Gloucestershire and Somerset steam was responsible at both dates for less than half

[1] E. Baines, 'On the woollen manufacture of England', *Jour. Roy. Stat. Soc.*, XXII (1859), 9.

[2] *Factory Returns, 1838* (P.P. 1839, xlii) and *1850* (P.P. 1850, xlii).

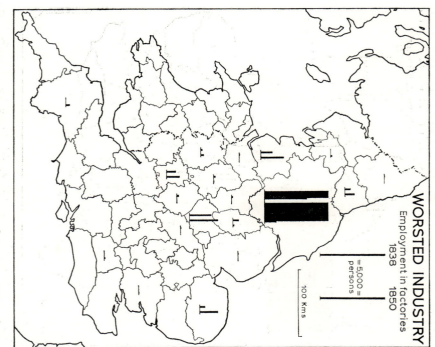

Fig. 103 Worsted industry: employment in factories, 1838 and 1850
Sources as for Fig. 100.

the power generated. The Devonshire serge manufacturers showed even
less inclination to convert to steam; the mills returned there in 1838 and
1850 were powered almost entirely by water-wheels. Yorkshire's share
of the factory workers increased markedly during this period as the pace
of investment in the local woollen industry gathered momentum (Fig. 104).
The rate of new building in the 1830s was said to be 'immense, enough
to astonish anybody',[1] and from 1833 to 1838 the number of woollen
mills in the county increased from an estimated 129 to a reported 606.[2]

---

[1] *Report from the S.C. on Manufactures (1833)*, quoted in F. J. Glover, 'The rise
of the heavy woollen trade of the West Riding of Yorkshire in the nineteenth century',
Business History, IV (1961), 10.   [2] F. J. Glover, 10.

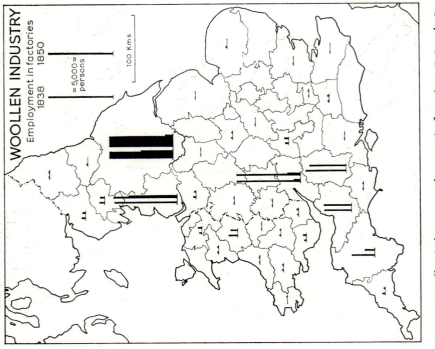

WOOLLEN INDUSTRY
Employment in factories
1838   1850

☰ = 5,000 =
persons

100 Kms

Fig. 104 Woollen industry: employment in factories, 1838 and 1850
Sources as for Fig. 100.

By 1850 the figure was 880.[1] Small valley settlements such as Slaithwaite and Holmfirth expanded into towns.[2] And as the shoddy manufacture emerged from its 'limited, rude and tentative' origins after 1813, the population of Batley steadily increased, and between 1831 and 1851 almost doubled itself.[3] Sustained by weaving, mining and quarrying, however, many of the old upland hamlets remained small but thriving communities.[4]

The old also persisted alongside the new in other branches of the textile industry, some of which changed but little in character during the first

[1] *Factory Returns, 1850* (P.P. 1850, xlii).   [2] J. C. R. Camm, 103–13.
[3] S. Jubb, *The history of the shoddy-trade* (London, 1860), 100.
[4] J. C. R. Camm, 114–31.

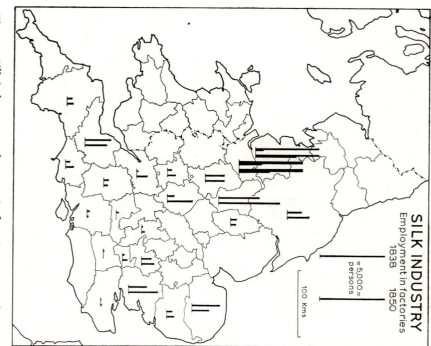

Fig. 105 Silk industry: employment in factories, 1838 and 1850
Sources as for Fig. 100.

half of the nineteenth century. With the development of machinery,
Spitalfields lost its position as a leading producer of silk to Cheshire and
Derbyshire and to the depressed woollen centres of East Anglia and Essex.
The introduction of steam power accelerated the process. The first steam
mill for spinning silk was started in Manchester in 1819–20, and the next
ten years saw a rapid increase of silk manufacture in Lancashire which had
'hardly been reckoned a silk county at all thirty years earlier'.[1] The older
silk areas in Derbyshire and Cheshire also benefited, and the latter had
a larger number of silk-factory workers by 1838 than any other county.

[1] J. H. Clapham, *An economic history of modern Britain: free trade and steel,
1850–1886* (Cambridge, 1932), 28.

New silk-spinning factories were also to be found in the traditional textile areas of East Anglia, Essex and the West Country (Fig. 105). But the weaving of silk often still remained a hand-loom domestic industry. The weaving of plain ribbons by steam power was already advanced in Derby, Leek and Congleton at a time when Coventry's fancy ribbon trade was still conducted on a domestic basis cheek by jowl with shops, warehouses and slaughter-houses.[1] Likewise in Nottingham, Leicester and Derby, hosiery and lace-making, in spite of the invention of machines, were still domestic industries, and had undergone no general technical transformation. Though the first steam-powered hosiery factories appeared in the 1840s, framework knitting remained in 1850 largely a domestic industry.[2] The introduction of steam-powered flax-spinning machines into Leeds about 1820 made it the centre of the English linen industry, and factory weaving had also made much progress by 1850. But with developments in Scotland, linen was ceasing to be an important English industry.

Other industries

Several extractive industries other than coal and iron also contributed significantly to the changing scene: quarrying, in such widely differing areas as the Lake District and the industrial Pennines;[3] and the working of china clay in Cornwall[4] and of brick clays in scores of townships, both rural and urban.[5] Mines and subsidence flashes were becoming increasingly numerous on the Cheshire saltfield,[6] while lead mining was responsible for considerable improvements in the road system of the northern Pennines.[7] The output of tin in Cornwall increased as deeper mining

[1] *Rep. Assistant Hand-loom Weavers' Commissioners*, pt 4, pp. 67, 353, 360, 363 (P.P. 1840, xxiv); J. Prest, *The industrial revolution in Coventry* (Oxford, 1960), 24, 93–6.

[2] L. A. Parker in W. G. Hoskins and R. A. McKinley (eds.), *V.C.H. Leicestershire*, III (1955), 15–16; D. M. Smith, 'The British hosiery industry at the middle of the nineteenth century, an historical study in economic geography', *Trans. and Papers, Inst. Brit. Geog.*, XXXII (1963), 141.

[3] J. D. Marshall, 45–7; J. C. R. Camm, 119.

[4] J. Rowe, *Cornwall in the age of the industrial revolution* (Liverpool, 1953), 117n.

[5] *Eighteenth Report of the Commissioners of Excise Inquiry* (on bricks), Appendix B, 166 (P.P. 1836, xxvi); H. A. Shannon, 'Bricks – a trade index, 1785–1849', *Economica*, n.s., I (1934), 301; M. Robbins, *Middlesex* (London, 1953), 48–50.

[6] *Chambers's Edinburgh Journal*, XI (1849), 181–3; K. L. Wallwork, 'The mid-Cheshire salt industry', *Geography*, XLIV (1959), 174.

[7] T. Sopwith, *An account of the mining districts of Alston Moor, Weardale, and Teesdale* (Alnwick, 1833), 8–9; L. C. Coombes, 'Lead mining in east and west Allendale', *Archaeologia Aeliana*, 4th ser., XXXVI (1958), 258–60.

became more general. Perhaps most striking of all, however, were the changes which accompanied the continued growth of the copper industry in the south-west. The output of copper ore from Cornwall and Devon soared from about 78,000 tons in 1815 to more than 150,000 in 1850,[1] a remarkable achievement which for a time made the district the world's leading producer of copper.[2] The effects of this and other local mining activities were not confined to the mineralised districts, where engine-houses and 'dead-heaps' yearly grew more numerous. For, as mineral railways were constructed to link the mines with the coast, tiny places such as Devoran and Point Quay rose to prominence as shipping points for ores and coal.[3] It was during this period, too, that the foundations were laid of a mining engineering industry which by 1850 was famous far beyond the south-west.[4] These developments were inevitably accom-panied by population changes. The population of Gwennap, for example, where 30% of British copper originated between 1823 and 1832, grew dramatically during these years;[5] so did that of Tavistock between 1841 and 1851, the decade which saw the opening of Devon Great Consols mine.[6]

The development of the great staple industries on the coalfields of the Midlands and north naturally attracted much contemporary comment; but, on the other hand, the continuing importance of London as a centre of manufacturing industry was not always fully recognised.[7] Yet, sus-tained in large measure by 'the consumption and vast commerce' of the metropolis,[8] London's industries retained throughout the first half of the century both their collective importance and rich variety. Like the watch, clock and jewellery trades of the Clerkenwell district, many had behind them a long tradition of skilled craftsmanship. Others were still com-

1 R. Hunt, *British mining* (London, 2nd ed., 1887), 892.

2 D. B. Barton, *A history of copper mining in Cornwall and Devon* (Truro, 1961), 91.

3 D. B. Barton, *The Redruth and Chasewater railway, 1824–1915* (Truro, 1960).

4 A. K. H. Jenkin, *The Cornish miner* (3rd ed., 1962, London), 174–7.

5 C. C. James, *History of Gwennap* (Privately printed, Penzance, n.d. [1952]), 129, 195.

6 D. B. Barton, *A historical survey of the mines and mineral railways of east Cornwall and west Devon* (Truro, 1964), 71–3; *Census of 1851: Population Tables*, I, vol. I, Devonshire, 48–9 (P.P. 1852–3, lxxxv).

7 P. G. Hall, 'The location of the clothing trades in London, 1861–1951', *Trans. and Papers, Inst. Brit. Geog.* XXVIII (1960), 155.

8 Census of 1831, quoted by J. H. Clapham, *An economic history of modern Britain: the early railway age, 1820–1850* (2nd ed., Cambridge, 1930), 68.

paratively new in 1850. Iron shipbuilding, for example, was added to the already heterogeneous group of industries which clustered along the banks of the Thames below London Bridge; and the manufacture of cheap ready-made clothing established a foothold in the East End.[1] For a time, too, London was pre-eminent in the new and fast expanding field of mechanical engineering.[2]

TRANSPORT AND TRADE

Transport by road

The extension of the turnpike system and new techniques of road making produced very great changes in internal communications. About 18,200 miles of English road had been turnpiked by 1821.[3] By 1848, with the brief golden age of the roads already in the past, the figure stood at 19,900 miles.[4] J. L. McAdam's method of surfacing roads with small stones broken to size on a well-drained foundation was widely adopted on many of the trunk roads, first in the neighbourhood of Bristol and Bath, and later (although according to McAdam's own testimony before 1820) in 'almost every county in the south of England'.[5] By that year, he claimed, his methods were also finding favour in other parts of the country. The regional distribution of Road Acts, as calculated by W. T. Jackman, indicates that the most progressive districts lay within the Home Counties and the industrial Midlands and north.[6] By 1830, as appears from C. and J. Greenwood's *Atlas of the counties of England* (1834), the great manufacturing areas of the country and much of the area around London had been 'covered with an elaborate network of turnpike roads linking together every place that could be called a town'.[7]

[1] O. H. K. Spate, 'Geographical aspects of the industrial evolution of London till 1850', *Geog. Jour.*, XCII (1938), 426-7; J. Thomas, *A history of the Leeds clothing industry* (Yorks. Bull. Econ. and Soc. Research, Occasional Paper No. 1, 1955), 12; P. G. Hall, 165-7.

[2] A. E. Musson, 'James Nasmyth and the early growth of mechanical engineering', *Econ. Hist. Rev*, 2nd ser., x (1957), 121-2.

[3] Anon., 'Turnpike roads in England and Wales', *Jour. Roy. Stat. Soc.*, I (1839), 542; *Abstract of Returns relative to the Expense and Maintenance of the Highways of England and Wales*, 258 (P.P. 1818, xvi).   [4] *Accounts and Papers*, 413 (P.P. 1847-8, lx).

[5] *S.C. on Turnpike Roads and Highways in England and Wales*, 315 (P.P. 1820, ii).

[6] W. T. Jackman, *The development of transportation in modern England* (London, 2nd ed, 1962), 743.

[7] G. H. Tupling, 'The turnpike trusts of Lancashire', *Manchester Literary and Philosophical Society Memoirs*, XCIV (1953), 43.

The Greenwoods' *Atlas* of 1834 also serves as a reminder that a century of turnpike activity had left many gaps in the system. Lancashire south of the Ribble, for instance, was the only part of that county with a close network of turnpike roads, which even there failed to extend to the moss-lands along the Cheshire boundary; and in east Leicestershire and the East Riding of Yorkshire, both predominantly rural and agricultural, the pattern of turnpike roads remained skeletal.[1] In fact, most counties which possessed an impressive mileage of turnpike road possessed also a very much greater mileage of other highways. Thus, although in 1848 the Yorkshire trusts controlled 1,737 miles of road, this represented little more than 18% of the total length of highways in the county.[2] The proportion was still lower in Devon, and lowest of all in East Anglia and Cornwall, where less than 10% of the mileage of highways had been turnpiked.[3] In the country as a whole, turnpike roads represented in 1838 about one-fifth of the mileage of all highways.[4]

Detailed study of almost any district will yield evidence of changes in the pattern and character of the roads at this period. Sometimes the changes were due to the Commissioners of Enclosure, 'those merciless annihilators of rural scenery',[5] who laid out many new roads and straightened countless old ones during the last phases of parliamentary enclosure. Frequently, however, the changes were on a larger scale and were of considerable regional significance. As the trans-Wealden routes between London and the coastal resorts grew in importance after 1800, for example, several new turnpike roads were constructed and a number of old ones re-aligned.[6] There were similar changes within south Lancashire where most of the roads were old highways, which had been turnpiked and repaired; others, including those between Bolton and Bury and between Bolton and Chorley were, either wholly or in part, new roads, less tortuous and more easily graded than the old routes between the same towns. In scores of places, including many along the line of Telford's improved Holyhead road, and on the main routes across the Pennines,

[1] P. Russell in W. G. Hoskins and R. A. McKinley (eds.), *V.C.H. Leicestershire*, III (1955), 80; A. Bryant, *A map of the East Riding of Yorkshire* (London, 1829).

[2] *Accounts and Papers*, 413 (P.P. 1847–8, lx); *Report of the Commissioners for Inquiry into the State of the Roads in England and Wales*, 631 (P.P. 1840, xxvii).

[3] *Rep. Commissioners on Roads*, 630–1 (P.P. 1840, xxvii).

[4] *Ibid*, 630–1. The calculation is for England alone.

[5] *Quarterly Review*, XXIII (1820), 102.

[6] G. J. Fuller, 'The development of roads in the Surrey–Sussex Weald and coastlands between 1700 and 1900', *Trans. and Papers, Inst. Brit. Geog.*, XIX (1954), 46.

narrow lengths of highway were widened and 'angular turnings and un-necessary hills' eliminated.[1]

An immediate consequence of these improvements was a reduction in the journey times of many long-distance stage coaches: by the 1830s these were maintaining average speeds of nine and ten miles an hour, including stops, over such routes as those between London and Glasgow, and London and Bristol.[2] To many places, and particularly to those which lay, as did Hounslow, Newbury and Kendal, along great thoroughfares and at important staging-points, the coaches brought an air of bustling prosperity. Thirty-four coaches daily passed through Newbury in the heyday of the coaching era.[3] An improvement in the performance and frequency of coaches along the great arterial roads of the country was accompanied by the emergence, within the industrialised areas, of complex networks of coach services, no less highly organised than those of the open roads and frequently supporting a greater density of traffic.[4] Equally important were the new public transport facilities within the great towns. As these grew larger, suburb and centre were linked by means of a variety of short-stage horse-drawn vehicles. Thus a close network of omnibus routes had appeared in London by the middle of the 1830s, and embryonic networks of a similar kind were emerging elsewhere.[5]

Moreover, by the 1820s, if not earlier, the whole country was covered by a network of carrier services by wagon and van. In the larger centres of population the number of firms engaged in such activities ran into scores; in 1824 ninety-four different firms operated carrier services from Manchester alone.[6] As the volume of traffic using the roads gradually increased, acute problems of traffic congestion became apparent. They were probably most severe in central London where, despite street improvements, the difficulty of moving freely by road remained a subject

[1] E. Mogg, *Paterson's roads* (London, 1829), 179–93; W. B. Crump, *Huddersfield highways down the ages* (Huddersfield, 1949), 82–8.

[2] H. W. Hart, 'Some notes on coach travel, 1750–1848', *Jour. Transport History*, IV (1959–60), 148–9.

[3] *R.C. on Municipal Corporations*, 230 (P.P. 1835, xxiii).

[4] G. C. Dickinson, 'Stage-coach services in the West Riding of Yorkshire between 1830 and 1840', *Jour. Transport History*, IV (1959–60), 1–11.

[5] T. C. Barker and M. Robbins, *A history of London transport*, I (London, 1963), 14–40; H. J. Dyos, 'The growth of a pre-Victorian suburb: south London, 1580–1836', *Town Planning Rev.*, XXV (1954–5), 69–70; G. C. Dickinson, 'The development of suburban road passenger transport in Leeds, 1840–95', *Jour. of Transport History*, IV (1959–60), 214–15.   [6] G. H. Tupling, 54.

of constant complaint.[1] The problem was not confined to the metropolis. Thus it was said in 1829 that in Carlisle the city's Saturday market filled 'all the principal streets and many of the lanes' with a press of farmers' cars and other traffic.[2] So bad were conditions in some provincial towns that the removal of markets and fairs from the principal streets formed a major consideration in any scheme for improving local amenities.[3]

Canal and coastwise traffic

The first twenty years of the nineteenth century saw some notable changes in the pattern of English waterways. In southern England, water communication in the country between Thames and Severn was both extended and improved in 1810 with the opening of the Wilts & Berks Canal for narrow boats and the Kennet & Avon Canal for barges. After 1819, by which time the North Wilts Canal had become available, through traffic, following the older Thames & Severn Canal, could, if necessary, avoid the shallows of the upper Thames below Lechlade by using for part of the journey the Wilts & Berks and North Wilts Canals.[4] More useful as an alternative to a difficult section of river navigation, however, was the Gloucester & Berkeley Canal, which was opened in 1827, and which enabled vessels of up to 700 tons to reach Gloucester without first navigating the shoals of the Severn between that city and Sharpness.[5]

In the Midlands, the network of waterways was strengthened by the addition of the Grand Union Canal which linked the Leicestershire & Northamptonshire Union Canal at Gumley in Leicestershire with the recently finished Grand Junction Canal at Long Buckby in Northamptonshire (Fig. 106). Thus, in 1814, was completed 'the great line of canals which extended from the Thames to the Humber',[6] for the Leicestershire & Northamptonshire Union in turn provided access to the waterways of the Soar and Trent valleys. To the already complex network of canals

[1] S.C. on Metropolis Improvements, 34 (P.P. 1836, xx); T. C. Barker and M. Robbins, 10–14.

[2] W. Parson and W. White, 148.

[3] See, for example, K. J. Allison in P. M. Tillott (ed.), V.C.H., The city of York, 488–9.

[4] C. Hadfield, British canals: an illustrated history (London, 1950), 78–82.

[5] W. G. East, 'The Severn waterway in the eighteenth and nineteenth centuries' in L. D. Stamp and S. W. Wooldridge (eds.), London essays in geography (London, 1951), 108–10.

[6] Prospectus of the Grand Union Canal Co. quoted by A. T. Patterson in W. G. Hoskins and R. A. McKinley (eds.), V.C.H. Leicestershire, III, 102.

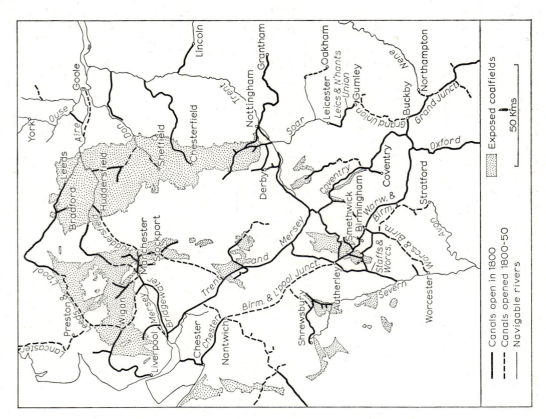

Fig. 106 Canals and waterways of the Midlands and the north, 1800–50
Based on C. Hadfield (ed.), *The canals of the British Isles*, 6 vols. (Newton
Abbot, 1966).

linking the Birmingham region with the Severn was added in 1816 a connection with the Worcester & Birmingham Canal, whose flight of thirty locks at Tardebigge lifted it through a vertical height of 217 feet at a gradient of about 1 in 50.[1] In the north, two canals, the Huddersfield Narrow and the Leeds & Liverpool, completed in 1811 and 1816 respectively, provided new water routes across the Pennines.

Standing apart from all these, both in time and character, was the Birmingham & Liverpool Junction Canal, which by means of numerous cuttings and embankments achieved in 1835 a more or less direct route between the Staffordshire & Worcestershire Canal at Autherley and the Chester Canal at Nantwich.[2] As an act of faith in improved canal navigation, the Birmingham & Liverpool Junction was magnificent; but it was, nevertheless, the last of the great trunk waterways to be constructed in this country during the canal age. It was cut at a time when efforts were being made to increase the competitive efficiency of canals in various ways. These involved both the introduction on a number of routes of express services for passengers and goods, and also alterations to the canals themselves. According to Sir George Head, writing in 1835, an effective opposition to the coaches was maintained in his day by the passage-boats 'Water Witch' and 'Swiftsure', which ran a daily service along the Lancaster Canal between Preston and Kendal.[3] Much more enduring in their results, however, were engineering works such as those which shortened the length of the Oxford Canal by 13½ miles; replaced the summit locks of the Birmingham Canal at Smethwick by a deep cutting; doubled the Harecastle Tunnel; and increased the capacity of the reservoirs which served the summit locks of the Grand Junction at Tring.

A substantial amount of traffic was soon moving along some of the newly completed waterways. The Gloucester & Berkeley carried 107,000 tons in its first year of operation (1827), and more than 399,000 tons in 1837.[4] The Kennet & Avon handled nearly 342,000 tons of traffic in 1838, and 360,000 tons ten years later.[5] Still more intensively used was the Regent's Canal, in London (1820), which was carrying almost half a

---

[1] J. H. Appleton, *The geography of communications in Great Britain* (Oxford, 1962), 6.　　[2] L. T. C. Rolt, *Thomas Telford* (London, 1958), 173–86.

[3] G. Head, *A home tour through the manufacturing districts of England in the summer of 1835* (London, 1836), 45.

[4] *R.C. on Canals and Inland Navigations of the United Kingdom, First Report*, 1, Part II, Appendix 14, p. 43 (P.P. 1906, xxxii).

[5] *Return relating to Inland Navigation and Canal Companies in England and Wales*, 68ff. (P.P. 1870, lvi).

million tons of traffic in 1828 and more than one million tons in 1848.[1] By contrast, the Bude Canal (1825), which supplied an agricultural district with sea sand for use as fertiliser, carried little more than 52,000 tons of traffic in 1848.[2] Of the other new canals at least one, the Sheffield (1819), had by 1850 attracted to its banks much heavy industry[3] and another, the Carlisle, was instrumental after 1823 in furthering the development in Caldewgate, Carlisle, of an industrial quarter near the canal basin.[4]

The most important of the new waterways formed an integral part of an already elaborate network of canals and navigable rivers, and played a vital role in the economy at large. The list of commodities handled by the Grand Junction, which carried about one million tons of traffic in 1838, reads like an inventory of industrial England. Besides coal, iron and building materials, it carried Cheshire salt, Stourbridge glass, Staffordshire pottery, Manchester textiles, a variety of metal goods, foodstuffs, agricultural produce, and much else.[5] Waterborne coal was of great importance in the industrial growth of Leicester.[6] Until the trade was disrupted by the opening of the Leicester & Swannington Railway in 1832, Leicester was supplied with coal from the Nottinghamshire–Derbyshire coalfield by way of the Erewash valley and the Loughborough and Leicester Navigations.[7] The importance of waterborne coal to a large manufacturing town may be illustrated more precisely in the case of Manchester. In 1834, 463,000 tons of coal reached Manchester by way of the Ashton, Bolton, Bridgewater and Rochdale Canals, compared with 247,000 tons by road and a mere 21,000 tons by the new Liverpool & Manchester railway.[8] The Kennet & Avon Canal acted as an important outlet for Somerset coal; by means of its feeder, the Somersetshire Coal Canal, it linked the Somerset

[1] Ibid, under Regent's Canal.
[2] Ibid, under Bude Canal; C. Hadfield, 'James Green as canal engineer', Jour. Transport History, I (1953–4), 48.
[3] A. W. Goodfellow and A. J. Hunt in D. L. Linton (ed.), Sheffield and its region (Sheffield, 1956), 165, 234, 236.
[4] W. Parson and W. White, 148; A. E. Smailes, North England (Edinburgh and London, 1960), 255.
[5] C. Hadfield, 'The Grand Junction Canal', Jour. Transport History, IV (1959–60), 104–5.
[6] A. T. Patterson, Radical Leicester: a history of Leicester 1780–1850 (Leicester, 1954), 39, 40, 260.
[7] J. E. Williams, The Derbyshire miners: a study in industrial and social history (London, 1962), 41–2.
[8] R.C. on Coal: Appendix to Report of Committee E, Table 55, p. 1161 (P.P. 1871, xviii).

coalfield with markets as far east as Reading, beyond which Somerset coal competed with seaborne coal from London and coal from the west Midlands carried by way of the Oxford Canal.[1] Even on the Driffield Navigation, a minor Yorkshire waterway, coal was an important item of traffic.[2] Of all the waterways for which tonnage figures are available, none carried more than those of the Birmingham Canal system in 1838, and almost 4,700,000 tons a decade later. This chiefly represented, as Joseph Priestley had observed a few years earlier, a vast flow of coal, iron ore and limestone to local ironworks, many of which occupied canal-side locations, and a movement in transit of finished and semi-finished iron goods.[3]

Writing in 1837, and thus towards the end of the period of major canal development in Britain, J. R. McCulloch claimed for England an extent of canal navigation 'unparalleled in any other country', with the exception of Holland';[4] and he devoted more attention to canals than to either roads or railways, the last of which he dismissed in a page or two. Yet, such was the pace of change, that three years later Francis Whishaw was writing at length about some of the consequences of living 'in the times of railways', not the least of which was the virtual cessation of canal building.[5]

In emphasising the role of the canal age in the developing economy of the nation, we must not forget the continuing importance of another form of water transport – coastwise traffic. The outstanding element in this was the enormous shipment of coal (some two million tons in 1830) to London, the most important consumer and chief distributor; a large number of ships worked on regular beats from north-east ports to those of the Thames estuary and elsewhere along the east and south coasts. Furthermore, there was an enormous miscellaneous traffic entering and leaving all the many harbours, great and small, around the coast. Amidst the variety of ships, a new type was beginning to make its appearance – the steamboat.

---

[1] *Ibid*, 877–8; J. A. Bulley, 'To Mendip for coal – a study of the Somerset coalfield before 1830', *Proc. Somerset. Archaeol. and Nat. Hist. Soc*, XCVII (1953), 55–6.

[2] *S.C. on the State of the Coal Trade*, 500 (P.P. 1830, viii); J. Priestley, *Historical account of the navigable rivers, canals, and railways, throughout Great Britain* (London, 1831), 220.

[3] J. Priestley, 228.

[4] J. R. McCulloch, 188.

[5] F. Whishaw, *Railways of Great Britain and Ireland practically described and illustrated* (London, 1840).

It is easy to lose sight of the fact that locally the established order of things had been changed by the steamboat before the advent of railways. Steamer excursions on the Thames to Richmond had assumed sufficient social importance by 1833 to find a place in the deliberations of the Select Committee on Public Walks.[1] More important still was the growth of steamer traffic downstream to Gravesend, Margate and Ramsgate.[2] Passenger traffic on the Thames steamboats was not confined to the summer months, and in 1837 it could be claimed that Gravesend owed its popularity as a residential town to the steamer services which linked it with London.[3] The effects of the steamers were felt over a wider area with the development of steam-packet services round the coast and to continental ports. Thus, innkeepers along the Great North Road complained in 1837 that, for long journeys, travellers preferred the cheaper steam packets to the coaches.[4] Since the travelling public would go out of their way 'to get to the steam', however, coach traffic on some cross routes actually benefited from the change.[5] In at least one district the introduction of steamboats led to changes of far-reaching geographical importance, as Richard Edmonds' account of the Cornish parish of Madron showed.[6] He explained in 1839 that until recently 'the early vegetables from Penzance and Mount's Bay have been principally consumed in Cornwall; but the facility of conveyance by steam-boats is now so great, that during the spring of 1838 a very considerable portion of them was sent to London and other distant markets'. Edmonds went on to predict that 'for the future Mount's Bay will most probably be the great spring-garden, not merely of Cornwall, but of all England'.

The growth of railways

Railways had their origin in the horse tramways that were in use before 1700 on many coalfields. By 1800, especially in some mining districts, they had become a well established part of both local and regional transport systems.[7] Further progress awaited Richard Trevithick's high-pressure

[1] *S.C. on Public Walks*, 362 (P.P. 1833, xv).

[2] T. C. Barker and M. Robbins, 40–3.

[3] *Report on Municipal Corporation Boundaries*, 45 (P.P. 1837, xxvii).

[4] *S.C. on Internal Communication Taxation*, 304, 309, 332 (P.P. 1837, xx).

[5] *Ibid.*, 332.

[6] R. Edmonds, 'A statistical account of the parish of Madron', *Jour. Roy. Stat. Soc.*, II (1839), 207.

[7] F. Atkinson, *The great northern coalfield, 1700–1900* (London, 1968), 48–52; M. J. T. Lewis, *Early wooden railways* (London, 1970), *passim*.

steam engine after 1801, and the improved design of locomotive engines, notably by George Stephenson after 1814. In 1825 the Stockton & Darlington Railway was opened, and was soon carrying passengers as well as minerals and goods. Five years later, on 15 September 1830, the Liverpool & Manchester Railway was opened, and achieved a success that surprised even its promoters. The railway age may be said to have begun.[1]

Railways multiplied after 1830, somewhat slowly at first and then with great rapidity in the years that followed the two bursts of speculative activity, from 1836 to 1837 and from 1845 to 1847, which together laid the foundations of the future railway system. Up to the end of 1838 the length of line open to public traffic was about 500 miles.[2] By July 1841 this had increased to some 1,400 miles.[3] Railway openings sanctioned during the following two years brought the total in June 1843 to about 1,700 miles.[4] In 1848 alone, more than 740 miles of new railway were opened, and at the end of that year the network of lines extended over 3,900 route miles.[5] By the close of 1849 more than 4,600 miles of railway were in operation in England and Wales, out of a total of 5,996 miles in the United Kingdom.[6] In little more than a generation the railway network of the country had evolved from a series of highly localised mineral tramways into a national system of lines carrying both goods and passengers (Fig. 122).

By the end of 1844 the outlines of a railway system had begun to emerge. 'The Great North-Western artery', 240 miles in length,

[1] The following sources have been consulted in the compilation of this section: W. M. Acworth, *The railways of England* (London, 5th ed, 1900); H. G. Lewin, *Early British railways* (London, 1925); E. Cleveland-Stevens, *English railways: their development and their relation to the state* (London, 1915); C. E. R. Sherrington, *The economics of rail transport in Great Britain*, I (London, 1928); E. T. MacDermot and C. R. Clinker, *History of the Great Western Railway* (London, 1964); H. P. White, *A regional history of the railways of Great Britain*, III *Greater London* (London, 1963); J. H. Clapham, *An economic history of modern Britain: the early railway age, 1820–1850* (Cambridge, 1930); J. H. Appleton, *The geography of communications in Great Britain* (Oxford, 1962).

[2] *R.C. on Railways*, Pt I, p. ix (P.P. 1867, xxxviii). Unless otherwise stated, all figures refer to England and Wales. Dates appended to the name of a railway refer to the opening of the line.

[3] Anon., 'List of the railways in the United Kingdom', *Jour. Roy. Stat. Soc*, IV (1841), 176–7.

[4] *S.C. on Railways*, 600–1 (P.P. 1844, xi).

[5] *Report of the Railway Commissioners for 1848*, 277 (P.P. 1849, xxvii).

[6] *Report of the Railway Commissioners for 1849*, 5 (P.P. 1850, xxxi).

connected London with Lancaster by means of the London & Birmingham (1838) and Grand Junction (1837) Railways and the several continuations of the latter beyond Warrington, where this north–south route met and crossed the Liverpool & Manchester Railway (1830). A steamer service linked the new town and port of Fleetwood, at the seaward end of the Preston & Wyre Railway (1840), with Ardrossan. Passengers from London could reach Darlington by using the London & Birmingham Railway as far as Rugby, and then the lines of the newly consolidated Midland Railway to Normanton, and thereafter those of the York & North Midland (1840) and Great North of England (1841) companies.

The completion of the Bristol & Exeter Railway in May 1844 extended rail communication by the broad gauge of the Great Western as far as Exeter. Lines to Folkestone, Brighton and Dover were already open, as was the South Western Railway to Southampton and Gosport. The luck-less Eastern Counties line had reached, and stopped at, Colchester, while the Northern & Eastern, even more impecunious, was struggling to reach Cambridge. Elsewhere, a number of cross-country routes had been developed: from Newcastle to Carlisle (1838), from Leeds to Selby (1834), from Selby to Hull (1840), from Manchester to Normanton (and Leeds, 1841), and from Manchester across the Pennines to Sheffield via Woodhead (1845).

No national plan lay behind this growth. 'The railways of England grew up piecemeal and haphazard in short, unconnected lengths.'[1] It will be apparent from the examples cited, however, that several trunk lines did in fact emerge at an early date; an original main line was often extended by arrangement with other companies in anything but a haphazard manner, and a process of amalgamation was already consolidating local companies into larger units. Many new lines were added to the railway map following the promotions of 1845–7; and by 1850 Scotland and Eng-land were linked by rail both by way of Carlisle and Berwick. Considerable progress had also been made in constructing lines in East Anglia and Lincolnshire; and work was under way on the Great Northern beyond Peterborough on the route between King's Cross and York. Competing lines had appeared, 'blocking lines' had been built to exclude rivals, and amalgamations had produced, *inter alia*, the London & North Western, the Lancashire & Yorkshire, the Manchester, Sheffield & Lincolnshire, and the London, Brighton & South Coast Companies.

The effects of railways were soon apparent. Coaching services along main roads swiftly declined, though some survived until 1850, and even later, between places which were not served by rail. But by 1848, when the last of the ordinary mail and stage coaches ceased to run out of London, the days of this traffic were numbered.[1] With it went the posting trade. 'Many inns and public houses, once in full business, have disappeared', Henry Tremenheere wrote of Brentford in 1843;[2] and the same was true of other places throughout the country. For a time, however, many cross routes leading to the new railway stations were busier than ever, much to the discomfort of turnpike trustees whose gates, designed to catch a hitherto predictable flow of traffic, now proved to be badly situated.[3] The railways stimulated the growth of local passenger and carrier services, and, since these tended to use a limited number of routes within built-up areas, congestion frequently occurred.[4] In 1846 it was claimed that the rail journey between Reading and Paddington could be made in the time taken by an omnibus in travelling between Paddington and the City.[5] London's traffic problem was aggravated by the peripheral situation of the main-line railway termini north of the Thames. Excluded from the City by the Corporation of London, these generated a large volume of cross-town traffic which included heavy goods and merchandise as well as passengers. Pickford's leviathan wagons lumbered through the narrow City streets between the railway terminals and the receiving-houses of the railway com-panies, often situated in old coaching inns. ' One cart comes perhaps with two or three chests of tea, a waggon with a couple of tons of iron, another with a bale of goods.' The result was that many streets were completely blocked.[6]

The full weight of railway competition had fallen upon many canals by 1850, and though tonnage figures frequently showed an improvement over earlier years, receipts were falling due to the reduction of tolls. By 1847, most of those canals that were destined to be transferred to railways had already changed hands.[7] River navigations were also affected: receipts from tolls on the Yorkshire Ouse, for instance, fell from £5,108 in 1837

[1] H. W. Hart, 146.

[2] H. Tremenheere, 'Agricultural and educational statistics of several parishes in the county of Middlesex', Jour. Roy. Stat. Soc., VI (1843), 126–7.

[3] S.C. on Turnpike Trusts, 411, 422 (P.P. 1839, ix).

[4] T. C. Barker and M. Robbins, 64–8.

[5] Metropolitan Railway Commission, 233 (P.P. 1846, xvii).

[6] Metropolitan Railway Commission, 177 (P.P. 1846, xvii). See J. R. Kellett, The impact of railways on Victorian cities (London, 1969), 35–40.

[7] E. Cleveland-Stevens, 91; C. Hadfield (1950), 191.

to £1,540 in 1850.[1] These, however, were only some of the more obvious effects of the coming of railways. As early as 1843, the G.W.R. was undermining the privileged position held by Middlesex farmers in the London market.[2] Henry Evershed, writing in 1853 of Surrey, noted that a new trade in milk and vegetables had been developed in the county since the completion of the South Western Railway.[3] A year earlier William Bearn had noticed fat beasts being sent from Northamptonshire to Smith-field by rail instead of, as formerly, on the hoof.[4] New coke ovens were built at Clay Cross, Camden Town and elsewhere by the railway com-panies for the supply of their locomotives.[5] the 'touch of the South Eastern Railway' was said to have given new life to Folkestone.[6] But the railways could also deprive. Thus, opening of railways was said to have damaged the trade of the Ingleton coalfield, which hitherto had served a large rural area on the borders of Lancashire, Westmorland and Yorkshire; 'railway coal' from south Lancashire and the main Yorkshire coalfield was in future to compete there with the local product.[7] This corner of England was no more than one of many, however, in which by 1850 'steam, the great magician of the nineteenth century', had been at work and, through railways, made its presence felt.[8]

By 1850 the railway had become a familiar part of the English scene. On every side the land had been 'bridged and cut and tunnelled',[9] and these manifestations of railway activity, together with railway inns, railway docks, railway streets and a handful of railway towns, bore ample testimony to 'the altered appearance of the country, produced by the formation of railways'.[10] In some places, notably perhaps at Euston, Chester and Newcastle, the railway station, architecturally splendid and functionally efficient, was an imposing symbol of the new age.[11] The

[1] B. F. Duckham, 'Inland waterways: some sources for their history', *The Amateur Historian*, VI (1963), 9.   [2] H. Tremenheere, 122.

[3] H. Evershed, 'On the farming of Surrey', *J.R.A.S.*, XIV (1853), 402.

[4] W. Bearn, 'On the farming of Northamptonshire' *J.R.A.S.*, XIII (1852), 47.

[5] F. Whishaw, 233, 298, 347, 423.

[6] *Illustrated London News*, XVII (1850), p. 48.

[7] *Lancaster Gazette*, 12 May 1866; *Lancaster Guardian*, 20 July 1872.

[8] S. Sidney, 21.   [9] J. H. Clapham (1930), 389.

[10] W. White, *History, gazetteer, and directory of the county of Essex* (Sheffield, 1848), 239.

[11] C. L. V. Meeks, *The railway station: an architectural history* (London and New Haven, 1957), especially 4, 35, 39; A. A. Arschavir, 'The inception of the English railway station', *Architectural History*, IV (1961), *passim*.

country station, less monumental but infinitely varied, had brought to many rural areas both a new settlement form and a new architecture.[1]

Overseas trade

In the merchant fleet, however, steam power made slow progress, and much of England's large and increasingly important overseas trade continued to be handled by sailing ships.[2] Among the imports, manufactured goods were of little importance. But from India, the West Indies and, above all, from the U.S.A. came great quantities of raw cotton; and from Germany, Spain and Australia came an increasing volume of wool. The import trade in raw materials was swollen by flax and hemp from the Baltic and by large amounts of timber from both colonial and foreign sources. In 1840, raw materials and semi-manufactured goods together were responsible for 56% by value of all imports,[3] An extensive trade in such commodities as sugar, coffee, tea, tobacco, grains and spirits – some of which were eventually re-exported – accounted for a further 40% of the total in that year.

By contrast, the export trade was dominated by manufactured goods, of which by far the most important were textiles. The greatly expanded cotton industry supplied the major share, accounting for at least 40% of the declared value of domestic exports during most of the period 1815–50.[4] In the 1840s, as in the 1820s, Europe and North America absorbed most of the country's exports, though other outlets were developed, particularly in Asia, Africa and South America.[5] New docks and warehouses in London, Liverpool, Hull and elsewhere, helped to meet the demand for more and better port facilities, while the reduction and eventual removal of tariffs during the 1840s prepared the way for free trade.

[1] J. Simmons, *The railways of Britain: an historical introduction* (London, 1961), ch. 3; M. Robbins, *The railway age* (London, 1962), ch. 7.

[2] For a general review of overseas trade see J. H. Clapham, *An economic history of modern Britain: the early railway age, 1820–1850* (Cambridge, 2nd ed., 1930), especially chs. 6 and 12. Statistics of trade and their interpretation are discussed in B. R. Mitchell and P. Deane, *Abstract of British historical statistics* (Cambridge, 1962), 274–337, and in W. Schlote, *British overseas trade from 1700 to the 1930s*, trans. and edited W. O. Henderson and W. H. Chaloner (Oxford, 1952).

[3] P. Deane and W. A. Cole, *British economic growth, 1688–1959* (Cambridge, 1964), 33. The figures refer to the United Kingdom.

[4] J. Potter, 'Atlantic economy, 1815–60: the U.S.A. and the industrial revolution in Britain', in L. S. Presnell (ed.), *Studies in the industrial revolution* (London, 1960), 259; P. Deane and W. A. Cole, 295.

[5] J. D. Chambers, *The workshop of the world* (London, 1961), 99.

TOWN AND CITIES

Not only were the large towns almost everywhere becoming substantially larger, but they were doing so at a rate that was prodigious. Thus the population of Liverpool grew by over 30% between 1811 and 1821, and by more than 40% in each of the following two decades.[1] Birmingham's population increased by more than 40% between 1821 and 1831, as did that of Manchester, Leeds and Sheffield, all of which attained their maximum rate of increase during this decade.[2] Some well-established urban communities, smaller than these but still large enough to rank among the 'large towns and populous districts', developed even faster. The population of Bradford, for example, increased by more than 60% between 1821 and 1831, making it the second most rapidly growing English city during the decade; the first place over the same period belonged to Brighton with an increase of just over 66%.[3]

In the Census of 1801, the population of London appeared as just under one million. Around it were many growing suburbs, and the metropolitan area (to be defined by the Metropolis Management Act of 1855) covered about 117 square miles and, in 1851, contained some 2.4 million people. This area corresponded fairly closely with the Metropolitan Police District which had been instituted in 1829 (and, incidentally, with the county of London to be created in 1888). But in 1839, the Metropolitan Police District was greatly extended to cover an area within 15 miles or so from Charing Cross (Fig. 141).[4] It included the whole of Middlesex and parts of Essex, Hertfordshire, Surrey and Kent; it covered nearly 693 square miles, and included a total of nearly 2.7 million people (see table on p. 673). This area within the 1839 limits came to be regarded as constituting 'Greater London'.

In 1801 no provincial city had as many as 100,000 inhabitants, but by 1851 there were seven. As in the case of London, it is difficult to define their limits in a geographical as opposed to an administrative sense, but the municipal boroughs of Liverpool and Manchester each had over 300,000 inhabitants. The other municipal boroughs with over 100,000

---

[1] R. P. Williams, 468–9.

[2] *Ibid.*, 486–7.

[3] E. W. Gilbert: (1) *Brighton, old ocean's bauble* (London, 1954), 97; (2) 'The growth of Brighton', *Geog. Jour.*, XCIV (1949), 38.

[4] J. F. Moylan, *Scotland Yard and the metropolitan police* (2nd ed., London, 1934), 82–3.

people were, in descending order, Birmingham, Leeds, Bristol, Sheffield and Bradford. There were thirteen other municipal boroughs each with over 50,000 people. Furthermore, nearly one million other town-dwellers lived in places with between 20,000 and 50,000. In this last group were towns as different in origin, function and appearance as Cambridge, Walsall, Wigan and York, yet all formed part of a rapidly expanding urban environment which by 1851 had come to contain about one half of the total population of England.[1] As Robert Vaughan rightly observed in 1843, the early railway age was also an 'age of great cities'.[2]

The spreading built-up areas. Such a vast and sudden accession of population could hardly fail to be accompanied by profound changes in both the extent and the character of the urban landscape. 'The increase of London since the commencement of the present century has exceeded...that of the last in celerity and extent', it was said in 1842, 'and is visible on all sides'.[3] The growth of the metropolis, a source of both pride and annoyance, amazed contemporaries, not least on account of its continuing character. Already in 1826 a vast 'province of bricks', London was growing at a rate that appeared to admit of no check.[4] 'Year by year the map of the metropolis takes new forms, and juts out in every direction fresh angles...houses are not built singly, but by wholesale.'[5] Many of the additions, as the *Illustrated London News* remarked, were big enough in themselves to rank as 'so many towns'.[6] By the late 1840s, the built-up area, that extraordinary 'admixture of the beautiful and the mean',[7] sprawled irregularly and, in places, tenuously from Hammersmith to Stratford and from Holloway to Camberwell; its salients thrust out along a host of roads, and its expanding periphery engulfed both swollen villages and open country (Fig. 140).[8] Along the river, the opening of the West India Dock in 1802 was followed by the construction of others with their associated warehouses, factories and dwellings. By 1828 there had

1 *Census of 1851: Population Tables, I*, Vol. I, Table XXVIII, p. 1 (P.P. 1852–3, lxxxv).
2 R. Vaughan, *The age of great cities* (London, 1843), 1.
3 S. Lewis, *A topographical dictionary of England*, III (5th ed., London, 1842), 123.
4 *Quarterly Review*, XXXIV (1826), 192.
5 *Chambers's Edinburgh Journal*, XII (1850), 141.
6 *Illustrated London News, Supplement*, January 1845, 141.
7 C. Knight, *London pictorially illustrated*, I (London, 1841), 16.
8 *Illustrated London News, Supplement*, January 1845, p. vii; H. C. Prince, 'North-west London, 1814–1863', in J. T. Coppock and H. C. Prince (eds.), *Greater London* (London, 1964), 80–117.

come into existence the London Docks, the East India Dock, the St Katharine's Docks and the Surrey Commercial Docks.

The built-up area of Manchester also expanded steadily. J. Aston's map of Manchester and Salford, dated 1804, shows a compact built-up area that was still virtually surrounded by open country.[1] But the lines of streets, laid out but not yet fully developed, indicated the shape of things to come. Although the open countryside remained for some time within easy walking distance of much of the town,[2] the rural fringe was pushed steadily outwards, and by 1850 Greater Manchester covered about seven square miles.[3] The urban area on Merseyside, which was still relatively small and confined to the Liverpool shore of the Mersey at the beginning of the nineteenth century, received a considerable accession after 1820 with the rise of Birkenhead, that other 'glory on the Mersey's side',[4] and by the 1840s the land on both sides of the river was undergoing rapid change.[5]

An uncompromising scenic transformation was frequently effected within the space of a few years. The valley of the lower Medlock, for instance, which in 1804 presented a fringe of open country to the southern outskirts of Manchester, had become by 1836 the site of a substantial industrial suburb, 'created within a few years by the erection of factories'[6] and displaying 'forests of chimneys, clouds of smoke and volumes of vapour, like the seething of some stupendous cauldron'.[7] Fields and gardens in and around Birmingham were quickly engulfed by buildings between 1820 and 1850.[8] And in Sheffield, where a period of rapid physical expansion culminated in a building mania during 1835 and 1836, no fewer than 156 new streets were built or projected between 1831 and 1836.[9] At such times the physical results of town growth could be measured in weeks rather than years.

[1] J. Aston, *Manchester guide*, Manchester (1804).
[2] *First Report of the Commissioners on the State of Large Towns and Populous Districts*, 572 (P.P. 1844, xvii); A. Redford, *The history of local government in Manchester*, II (London, 1940), 213–15.
[3] H. B. Rodgers, 'The suburban growth of Victorian Manchester', *Jour. Manchester Geog. Soc.*, LVIII (1961–2), 4.
[4] *Illustrated London News*, X (1847), 228.
[5] R. Lawton, 'The genesis of population', in W. Smith (ed.), *Merseyside: a scientific survey* (Liverpool, 1953), 122.   [6] J. Wheeler (1836), 269.
[7] C. Redding, *An illustrated itinerary of the county of Lancaster* (London, 1842), 8.
[8] C. Gill, *History of Birmingham*, I (Oxford, 1952), 363–5; M. J. Wise (ed.), 213.
[9] G. C. Holland, *The vital statistics of Sheffield* (London, 1843), 29, 53.

Inland and seaside resorts. It was not only large and well-established towns which, in Sir John Clapham's graphic phrase, 'were growing like toad-stools'.[1] Indeed, one of the outstanding characteristics of town growth at this period was the rise to prominence of many small or hitherto insignificant places.[2] Cheltenham was one such place. In 1806 the town consisted principally of the High Street and, nearby, the first of the 'rows of white tenements', with green balconies' which were later to catch Cobbett's unfriendly eye.[3] Immediately beyond these lay open country, some of which was still cultivated in common fields. As a spa Cheltenham had enjoyed fame for some time, but its population in 1801 was barely 3,000. The discovery of new mineral springs and their skilful exploitation led to rapid growth, so that by 1831 the population had risen to 23,000. On the north side of the town, the extensive suburb of Pittville was laid out on the estate of Joseph Pitt, an M.P. who had acquired the land at the enclosure of the town fields in 1806.[4] Still unfinished when H. S. Merrett's plan of the town appeared in 1834, Pittville was the counterpart on the north of the earlier estate around the Montpellier pump room of Henry Thompson on the other side of the town.[5] Cheltenham was ceasing to be simply a 'drinking spa' and, like Bath at a somewhat earlier date, it was also becoming a fashionable residential town.[6] The growth of Leamington was even more remarkable. From 543 in 1811 the population of the parish of Leamington Priors soared to 2,183 within ten years. By the late 1820s Leamington had become 'a rich and elegant town' with stuccoed hotels, numerous boarding-houses and fashionable shops.[7] Twenty years later, by which time the population had risen to over 13,000, every road leading into the town had been 'seized upon and flanked with buildings'.[8]

Rapid and sustained growth was more common, however, in the rising seaside resorts than among the inland spas. The population of Torquay, for example, rose from 838 in 1801 to 5,982 in 1841, and to 11,474 in 1851.

[1] J. H. and M. M. Clapham, 'Life in the new towns', in G. M. Young (ed.), *Early Victorian England 1830–1865*, I (London, 1934), 227.

[2] W. Ashworth, 8.   [3] William Cobbett, I, 33.

[4] Hunt & Co, *City of Gloucester and Cheltenham directory and court guide* (London, 1847), 7; H. P. R. Finberg, *Gloucestershire* (London, 1955), 89.

[5] H. S. Merrett, *Plan of Cheltenham and its vicinity* (Cheltenham, 1834); scale approx. 16 in. to one mile.

[6] A. B. Granville, *The spas of England and principal sea-bathing places*, 2 vols. (London, 1841), II, 309–11.

[7] S. Lewis, III, 40; J. A. R. Pimlott, *The Englishman's holiday: a social history* (London, 1947), 99.   [8] A. B. Granville, II, 223.

The town was recommended by doctors as a winter retreat for their consumptive patients, and its growing fame was reflected in rows of elegant houses and a number of detached villas.[1] Farther north, the sea-bathing resorts along the Lancashire coast were also growing vigorously. Edward Baines estimated in 1825 that at least half the buildings in Southport had been erected within the previous four years, and observed that the resort was still expanding.[2] A few years later William Thornber, commenting on the changing face of Blackpool, noted that it was fast assuming 'the air and importance of a town'.[3]

Railway and canal towns. The fashionable seaside resorts built in the style of Belgravia-by-the-sea, and the not-so-fashionable with their variety of styles, formed an important element in the changing urban scene, but there were other, and very different, new towns, which were no less characteristic of the age. Like Middlesbrough, several towns were called into existence by new lines of communication.[4] In Yorkshire, Goole was created by the Aire and Calder Navigation Company at the seaward end of its new canal, the Knottingley & Goole completed in 1826. In the same year the first docks were opened, and on the nearby dock estate a town rose 'in point of elegance and uniformity . . . the handsomest in the north of England'.[5] Twenty years later, by arrangement with the Wakefield, Pontefract & Goole Railway Company, whose line reached the town in 1848, the present Railway Dock was built and further growth took place.[6] Goole had a population of 4,700 in 1851 and, in little more than twenty years, had risen from a small village 'to the dignity and importance of a considerable shipping port'.[7] The early growth of Goole was bound up with both canal and railway links, but more characteristic of the period were the new railway towns. These were created to meet the special needs of the railway companies and, like Wolverton in Buckinghamshire, were

[1] A. B. Granville, II, 472; W. G. Hoskins, *Devon* (London, 1954), 500.

[2] E. Baines, *History, directory, and gazetteer of the County Palatine of Lancaster*, II (Liverpool, 1825), 552; F. A. Bailey, *A history of Southport* (Southport, 1955), 59.

[3] W. Thornber, *An historical and descriptive account of Blackpool and its neighbourhood* (Poulton, 1837), 226, 230.

[4] A. Briggs, *Victorian cities* (London, 1963), ch. 7.

[5] *Hull Advertiser*, 21 July 1826.

[6] G. F. Copley, *An historical and descriptive guide to the Wakefield, Pontefract and Goole Railway* (Pontefract, 1848), 58.

[7] G. Head, 222; J. Bird, *The major seaports of the United Kingdom* (London, 1963), 145–6.

situated strategically in relation to the railway system which each was intended to serve.[1] 'Wolverton forms a remarkable example of what railway enterprise may effect', the *Railway Chronicle* wrote in 1844.[2] 'A few years since it exhibited nothing but farms and uplands; it is now the centre of a flourishing community . . . a colony of engineers and handicraftmen', which had grown up since 1838 around the locomotive works of the London & Birmingham Railway Company. A 'neat, brick-built, clean little town of eight or ten streets', and with a population of about 1,500, had appeared there by 1844.[3] A similar 'mechanical settlement in an agricultural district' developed after 1843 at Crewe, following the removal there from Edgehill, near Liverpool, of the workshops of the Grand Junction Railway Company.[4] But not all railway colonies were planted, as these were, in the heart of the countryside. New Swindon, for example, grew up next to the old market town of Swindon. When this happened, the physical contrast between the old and new parts of the settlement might assume considerable social as well as geographical significance.[5]

Urban order and disorder. Apart from the railway colonies, examples of ordered town development at this period were comparatively rare.[6] Where controlled development did occur, however, the results were often immediately apparent. When, for example, in 1822 the Aire & Calder Navigation Company was preparing to build houses on its Goole estate, plans and specifications were drawn up for the contractors.[7] 'The Proprietors of the Canal have laid down a plan and elevation . . . according to which all the buildings are to be erected of fine brick or stone, and covered with blue slates.'[8] The uniformity of appearance that resulted was much praised. But as Goole's trade prospered and the town was enlarged, rows

[1] B. J. Turton, 'The railway town, a problem in industrial planning', *Town Planning Review*, XXII (1961), 100; H. Perkin, *The age of the railway* (London, 1970), 127–31.

[2] *Railway Chronicle*, 1 June 1844, p. 165.

[3] *Ibid.* 165; [F. B. Head], *Stokers and pokers* (London, 4th ed., 1849), 82.

[4] *Chambers's Edinburgh Journal*, XIII (1850), 392; W. H. Chaloner, *The social and economic development of Crewe, 1780–1923* (Manchester, 1950), 42ff.

[5] K. Hudson, 'The early years of the railway community in Swindon', *Transport History*, I (1968), 146–7, 150.

[6] W. Ashworth, ch. 2.

[7] ACN 1 (19), 16 December 1822, 4 August 1823 (British Transport Record Office, London).

[8] *Hull Advertiser*, 21 July 1826.

of houses of various styles appeared in close proximity to the company's estate, which formed the heart of the port. At Birkenhead, the effect of planning, though incompletely realised, was even more striking as James Law's plan of 1844 indicates.[1] On this, Paxton's Birkenhead Park, opened three years later, interposes a green barrier between the regularly formed streets of the central area and the spacious villa-dotted suburbs.[2] To the earl of Stamford, Ashton-under-Lyne owed its wide streets and good drains; Edgbaston, wholly villas in 1842, owed its character to the policies of Lord Calthorpe, the principal landowner.[3]

But for every Edgbaston there were a score of places where growth was controlled only by the prevailing state of the market for houses and by the inadequate clauses of local improvement and building acts. The well-known Blue Books of the 1840s contain a wealth of descriptive material, from which the effects of unco-ordinated and rapid town growth emerge in vivid detail. Many working-class industrial suburbs of the north were developing features similar to those of Chorlton upon Medlock where, the Health of Towns Commissioners were informed, in 1844, of houses which were built standing back to back.[4] In Hull, where standards of building were falling rapidly, there were by 1840 many congested 'courts of a very peculiar construction, a court within a court, and then another court within that'.[5] The building of working-class houses so as to 'economize the land' was not, however, peculiar to the north.[6] Extreme conditions of congestion were reached in Nottingham, where the outward expansion of the town was hampered until 1845 by the presence of common fields.[7] And filthy, overcrowded courts and alleys, some of them old but many built or rebuilt since 1800, could be found within a few minutes' walk of the main thoroughfares of London, Brighton and other towns. Indeed, one of the

[1] J. Law, *Plan of the township or chapelry of Birkenhead* (London, 1844).

[2] W. Ashworth, 40; H. R. Hitchcock, *Early Victorian architecture in Britain*, I (London and New Haven, 1954), 450-3.

[3] *Second Report on the State of Large Towns and Populous Districts*, Appendix II, pp. 311, 323 (P.P. 1845, xviii); *S.C. on Buildings Regulation and Improvement of Boroughs*, 292 (P.P. 1842, x).

[4] *First Report of Commissioners on Large Towns, Appendix*, 60 (P.P. 1844, xvii).

[5] *S.C. on the Health of Towns*, 439 (P.P. 1840, xi).

[6] The process is effectively illustrated by M. W. Beresford, *Time and place: an inaugural lecture* (Leeds, 1961).

[7] J. D. Chambers, 'Nottingham in the early nineteenth century', *Trans. Thoroton Society*, XLVI (1943), 28-31; K. C. Edwards, 'The geographical development of Nottingham', in K. C. Edwards (ed.), *Nottingham and its region* (Nottingham, 1966), 370-1.

objects of Nash's Regent Street project (1817–23) had been to separate untidy, squalid Soho from the fashionable estates of the West End.[1]

The appearance of towns was affected also by improvement schemes, carried out either by local authorities or by private individuals.[2] The cutting of New Oxford Street through the slums of St Giles's, for example, between 1845 and 1847 swept away dilapidated courts and alleys; but many people who were displaced by this improvement went to swell the numbers in already crowded districts nearby, which in consequence deteriorated still further in social status.[3] At about this time, too, Manchester and Salford opened their first public parks[4] while the former town prohibited the building of houses back to back in 1844.[5] Elsewhere, old streets were widened and new ones cut, markets rehoused and paving and lighting extended by local improvement schemes; but their effect was seldom other than piecemeal. Redevelopment on the scale of that carried out in central Newcastle by Grainger and Dobson during the 1830s – a scheme at once so ambitious and in its original form so aesthetically satisfying that it has been regarded as the culmination of Georgian urbanism – was exceptional in its scope and vision.[6]

The suburbs. As the towns grew larger, so did suburbia. It was not a new phenomenon, but during the first half of the nineteenth century the effects of suburban growth were becoming increasingly obvious in the vicinity of the great towns. Those who could afford to live at some distance from their place of work were encouraged to desert the inner areas of towns, as a contemporary account of Manchester explains, because of the 'annoyance of smoke, the noise and bustle of business, and perhaps also the growing value of building land, for shops and warehouses in the central parts'.[7] Aided by better communications, Manchester merchants forsook the central areas for Pendleton, Cheetham Hill, Higher Broughton and

[1] J. Summerson, *Georgian London* (London, 1945), 168–71; H. J. Dyos, 'Urban transformation. A note on the objects of street improvement in Regency and early Victorian London', *International Review of Social History*, II (1957), 261.

[2] B. Keith-Lucas, 'Some influences affecting the development of sanitary legislation in England', *Econ. Hist. Rev.*, 2nd ser., VI (1954), 294–5; F. Clifford, *The history of private bill legislation*, II (London, 1887), 291.

[3] H. R. Hitchcock, I, 378–80.

[4] *Illustrated London News*, IX (1846) 12, 114. F. E. Hyde, 156.

[5] A. Redford (1940), 86.

[6] A. E. Smailes (1960), 169; H. R. Hitchcock, I, 374.

[7] Quoted in T. S. Ashton, *Economic and social investigations in Manchester, 1833–1933* (London, 1934), 37. The quotation refers to c. 1840.

beyond.[1] Successful cutlers sought out more attractive quarters on the western outskirts of Sheffield,[2] and Hull merchants removed to villages along the flanks of the Yorkshire Wolds, some miles outside the town.[3] Around London, suburbia advanced across the fields of Camberwell,[4] studded Lewisham with the residences of the wealthy[5] and added genteel fringes to Stepney and Hackney.[6]

The exodus of the well-to-do and the prosperous was accompanied by changes in the character of many of the older residential districts. As was said of London in 1826, 'the shopkeeper has discovered it to be most profitable in every sense to remove his family out of town; he places his stock in trade in the apartments they occupied, and employs the warehouse rent thus saved in hiring a "pretty tenement" at Islington, Knightsbridge, or Newington, where his children thrive in a purer air'.[7] In Sheffield, houses were converted into cutlery workshops;[8] and in Manchester, where the central area was said in 1842 to be 'in a state of constant alteration', buildings were modified in a variety of ways, 'sometimes to divide, sometimes to enlarge, to apply them to other uses – to convert houses into warehouses...and warehouses into workshops'.[9] In central Liverpool, as elsewhere, the coming of the railway also led to the removal of houses.[10] Still other houses in once fashionable districts were subdivided into tenements and became slums. As one class moved out, it was explained in 1846, 'a second grade go in' and eventually only 'very low parties' were attracted to such areas.[11] Many of the newcomers were themselves compelled to move in the course of time owing to the insistent pressure of commercial expansion, and by 1850 the inner areas of London

[1] E. Butterworth, 83, 87–8; H. B. Rodgers (1961–2), 5–6; T. W. Freeman, 'The Manchester conurbation', in C. F. Carter (ed.), *Manchester and its region* (Manchester, 1962), 54.

[2] S. Pollard, *A history of labour in Sheffield* (Liverpool, 1959), 6; A. J. Hunt, 'The morphology and growth of Sheffield', in D. L. Linton (1956), 228–37.

[3] G. A. Cooke, *Topographical and statistical description of the county of York* (London, n.d., c. 1820), 284.

[4] H. J. Dyos, *Victorian suburb: a study of Camberwell* (Leicester, 1961), 33.

[5] S. Lewis, III, 72.

[6] M. Rose, *The East End of London* (London, 1951), ch. 12.

[7] *Quarterly Review*, xxxiv (1826), 195.   [8] S. Pollard, 5.

[9] S.C. on Buildings Regulation etc., 256 (P.P. 1842, x).

[10] T. W. Freeman, *The conurbations of Great Britain* (Manchester, 1959), 109; H. J. Dyos, 'Railways and housing in Victorian London', *Jour. Transport History*, II (1955–6), 12.

[11] *Metropolitan Railway Commission*, 214 (P.P. 1846, xvii).

and the larger provincial towns were slowly being drained of their residential population.[1] 'People do not live in the city of London', William Tite declared in 1846 with magnificent exaggeration.[2] Friedrich Engels, more careful in his choice of words, found the commercial heart of Manchester without 'permanent residents' and 'deserted at night'.[3]

Public health. Of the many problems that faced the towns none was more urgent than an adequate and unpolluted supply of water. Most towns obtained their water supply from nearby wells, springs and rivers (Fig. 107). London and Plymouth, which for many years had drawn supplies of water from distant sources, were in this respect unusual. During the first half of the nineteenth century, however, several industrial towns in northern England in turn established waterworks within the Pennines and Rossendale. Reservoirs were constructed and water was conveyed by gravity to distribution points near the towns. At Bolton the nearby uplands at Belmont provided a convenient site for waterworks,[4] but the extensive works of Liverpool and Manchester, which were initiated but not completed before 1850, were situated far from the points of consumption, at Rivington and Longdendale respectively.[5] Under the 'new system of gathering-grounds', as it was called in 1850, distant uplands sometimes acquired significance as the 'water-farms' of the expanding towns.[6]

The provision of improved supplies of drinking water was only one aspect of the sanitary problem facing the towns, as Edwin Chadwick recognised. His concern with the environmental causes of disease led him to emphasise the importance of water as a means of removing town refuse quickly and cheaply.[7] In Chadwick's view, the engineering solution to the sanitary problem lay in the provision of a constant supply of water to every house, and, he argued, that waste matter could thus be flushed from the houses through self-scouring sewers and then discharged harmlessly

1 T. W. Freeman (1959), 5, 7, 32-7, 82, 109, 138, 172.

2 *Metropolitan Railway Commission*, 64 (P.P. 1846, xvii).

3 F. Engels, *The condition of the working-class in England in 1844*, trans. and ed. by W. O. Henderson and W. H. Chaloner (Oxford, 1958), 54.

4 J. Black, *A medico-topographical, geological, and statistical sketch of Bolton and its neighbourhood* (Bolton, n.d., 1836), 38; *Centenary of the Bolton corporation waterworks undertaking, 1847-1947* (Bolton, 1947).

5 B. D. White, *A history of the corporation of Liverpool, 1835-1914* (Liverpool, 1951), 56-7; A. Redford II (1940), 181-5.

6 *Quarterly Review*, LXXXVII (1850), 498.

7 E. Chadwick, *Report on the sanitary condition of the labouring population of Great Britain* (London, 1842), 370.

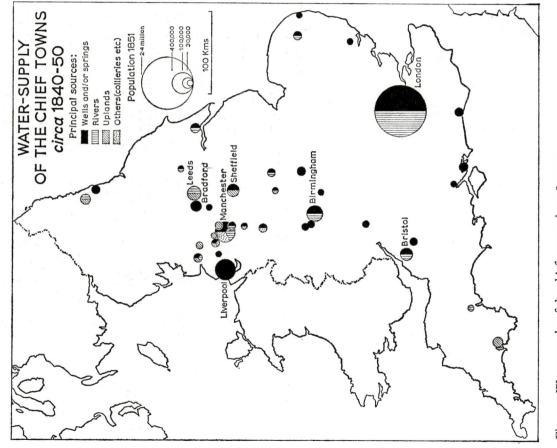

Fig. 107 Water supply of the chief towns *circa* 1840–50
Based on a variety of contemporary sources, e.g. Parliamentary Papers and
local reports.
The chief towns are those with over 30,000 inhabitants in 1851.

and profitably on the outskirts of towns.[1] In the urban, as in the rural, areas there was thus a growing awareness of the importance of land drainage. And in the person of James Smith, of Deanston, whose ideas on the drainage of cities and the construction of sewers were quoted with approval by Chadwick, skills developed in the one environment were applied to the solution of problems in the other.[2] The municipal corporations established by the Municipal Corporations Act of 1835 provided a basis for more effective local government, and by the late 1840s the need for drastic remedial measures to improve the health of the urban population was widely recognised; the Public Health Act was passed in 1848 and the sanitary reformers were hard at work.[3] But many years were to elapse before the new era of public health was reflected unmistakably in the appearance of the towns.

[1] R. Lambert, 61–2; W. H. G. Armytage, *A social history of engineering* (London, 1961), 140.

[2] E. Chadwick, 379. I am grateful to Professor John Saville for drawing my attention to this aspect of James Smith's work.

[3] R. A. Lewis, *Edwin Chadwick and the public health movement, 1832–1854* (London, 1952), especially ch. 8.

Chapter 10

ENGLAND *circa* 1850

J. B. HARLEY

The England of 1850, in the view of Sir John Clapham, 'had turned her face towards the new industry – the wheels of iron and the shriek of the escaping steam'.[1] In contemporary eyes she was now the 'workshop of the world', and the Great Exhibition of 1851 offered to all comers an un-equivocal proof of her supremacy as a manufacturing nation. But amidst so much industrial reality, Léonce de Lavergne diagnosed a measure of ambivalence, a reluctance to come of age as an industrial and urban state. He felt that Englishmen harboured 'a repugnance to being shut up within the walls of the towns', and that the taste of the wealthier part of the nation was for a country life. Nor, in his opinion, was an explanation hard to find because 'in travelling through England, one cannot help being constantly impressed with the contrast between town and country . . . The largest towns, like Birmingham, Manchester, Sheffield, or Leeds, are in-habited only by workmen and shopkeepers, and the parts of the town occupied by their dwellings have a poor and melancholy appearance'.[2] If this Frenchman's short acquaintance with England led him into over simplification, he had touched correctly upon a deep dichotomy in its geography. Never, previously, had town and industry occupied so much of England, nor yet had they presented so contrasted a face to that of the countryside. The extremes had been heightened by the expansion of urban squalor and industrial dereliction on the one hand, and, on the other, by the coming to maturity of the work of the Georgian improvers, enclosers and landscape artists. Indeed, the sharply contrasted landscapes of early Victorian England mirrored vividly the deep cleavages in contemporary society: the 'two nations' were at no time more clearly portrayed than in the human geography of 1850.

[1] J. H. Clapham, *An economic history of modern Britain: free trade and steel, 1850–1866* (Cambridge, 1932), 22.

[2] L. de Lavergne, *The rural economy of England, Scotland, and Ireland* (Edinburgh and London, 1855), 120–31.

527]

POPULATION

The distribution of population

In 1851 England was inhabited by some 16.9 million people, and Wales by just over one million. Their distribution reflected the marked quickening in the differentiation of commercial, manufacturing and mining areas from the rest of the country (Fig. 108). There were four main areas in which the average population density exceeded 800 persons per square mile. First, London was still the most crowded centre of population; when the English Registration Districts were arranged in order of magnitude of population density, the first 15 places were filled by the small but teeming London districts (a dozen of which had over 100,000 people per square mile).[1] Even the extra-metropolitan districts of Middlesex were more densely peopled than any other part of England except Lancashire. Secondly, the great coalfield concentration of people embracing east Lancashire and west Yorkshire formed the largest area of England where the population exceeded 800 persons per square mile: in aggregate it contained more inhabitants than all of London, Middlesex, Surrey and Kent.[2] The third outstanding populous area in 1851 was the west Midlands, particularly the industrial areas of Birmingham, the Black Country and the Potteries, but with smaller populations in Coventry and the east Warwickshire coalfield. Fourthly, such densities were matched, but over a lesser area, in the industrial district of Durham county. Elsewhere in England, islands of higher population density reflected smaller coalfields such as those of Bristol and Coalbrookdale, mining districts such as that of west Cornwall, and a few especially prosperous centres of rural industry.

These concentrations were directly related to migration. In 1851 'the towns, in the mass, were mainly inhabited by immigrants...Out of 3,336,000 people, of 20 years of age and upwards, living in London and 61 other English and Welsh towns in 1851, only 1,337,000 had been born in the town of their residence.'[3] The percentage varied according to local circumstances. In London there were roughly 46% native-born, in Leeds 58%, but in Manchester-Salford and Bradford this fell to just above 25% and in Liverpool below 25%. Not only the large industrial centres were thronged with strangers; Brighton had one of the largest proportions of

---

[1] *Census of 1851: Population Tables, I,* vol. 1, p. cxi (P.P. 1852–3, lxxxv).
[2] A. Redford, *Labour migration in England, 1800–50* (Manchester, 1926), 13.
[3] J. H. Clapham, *An economic history of modern Britain: the early railway age, 1820–1850* (Cambridge, 1926), 536.

immigrants (about 80%) for a town of its size,[1] and even Harrogate, a more staid northern spa, had 59%.[2] Many of the newcomers had come from the countryside, especially from areas adjacent to the main industrial regions; in Lancashire, for example, the largest numbers of English immigrants recorded in 1851 had originated in Cheshire, Derbyshire and Shropshire. In total, however, the most substantial influx into the towns had been from Ireland; the year 1851 marked the climax of the post-famine arrival of the Irish. They now numbered 727,000 in England, and were most numerous in the north-west,[3] where they dominated the immigrant communities of many Lancashire towns – not only the larger such as Liverpool and Manchester, but the smaller such as Ashton-under-Lyne and Wigan (Fig. 96). Frequently they had segregated into distinctive colonies; in Liverpool, behind the docks and business district; in Manchester, in areas such as 'Little Ireland' where Engels had found them in 1844; and in London in parishes such as St Giles, Shadwell and Whitechapel.[4]

These crowded immigrant quarters were only one aspect of the overall unhealthiness of urban life. The facts of medical geography in the towns often make grisly reading: at mid-century the expectation of life was low; the rates of mortality, disease and general sickness were high; and epidemics of cholera, smallpox and typhus not uncommon. Pioneer investigations of health and disease in England (in progress at mid-century) serve to underline these spatial variations in the national mortality; not only was the normal death rate higher in the towns, but also the epidemic death rate, which, as at the time of the cholera of 1848–9 and 1853–4, struck heaviest at the urban dwellers.[5] Indeed, the primary purpose of the Public Health Bill of 1848 was to improve those towns where the crude death rate exceeded 23 per thousand (the national average crude death rate stood at 21.8 per thousand in 1851–2). It is a measure of the magnitude of the problem, that, by the end of 1853, no less than 182 local health boards had been set up, some voluntarily, others compulsorily.[6] Few of the larger

[1] *Ibid.*, 537.   [2] J. A. Patmore, *An atlas of Harrogate* (Harrogate, 1963), 19.
[3] A. Redford, 137; R. Lawton, 'Irish immigration to England in the mid-nineteenth century', *Irish Geography*, IV (1959), 35–54.
[4] R. Lawton, 'The population of Liverpool in the mid-nineteenth century', *Trans. Hist. Soc. Lancs. and Cheshire*, CVII (1955), 89–120; T. W. Freeman, *Pre-famine Ireland* (Manchester, 1957), 43–7; J. H. Clapham (1926), 60.
[5] E. W. Gilbert, 'Pioneer maps of health and disease in England', *Geog. Jour.*, CXXIV (1958), 172–83.
[6] G. Slater, *The making of modern England* (London, 1919), 168.

J. B. HARLEY

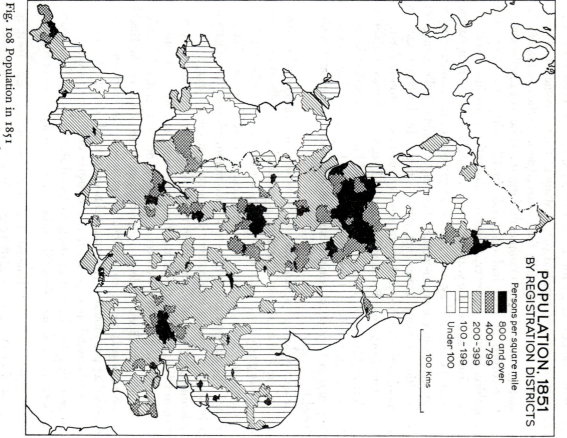

Fig. 108 Population in 1851
Based on *Census of 1851: Population Tables, I*, vol. I, pp. cxi–cxiii (P.P.
1852–3, lxxxv).

towns were absent from this list, but smaller places could be just as un-healthy; Salisbury, for example, in the decade 1841–50 attained a mean annual crude death rate as high as 28 per thousand.[1] On the other hand, the rural areas with their lower death rates were still regarded as 'the healthy reservoirs from whence the ravages amongst the town population were repaired'.[2] Population densities lessened sharply away from the towns and coalfields, but the countryside, too, was more crowded with people than ever before, and large areas had densities of over 200 persons per square mile.

Population and administrative areas

The inadequacy of an ancient system of administrative areas for the needs of the contemporary distribution of population was much in evidence by 1850. The authors of the report on the Census of 1851 were clear in their condemnation: 'The old division of the country into parishes, townships, and counties, is open to many...objections...Parishes are, in many instances, almost inextricably intermingled; and they vary in population from single families to tens of thousands of families; in extent, from a few hundreds of acres to many thousands of acres. The counties are also irregu-larly and unequally constituted.'[3] The evidence in the 1851 Census for the size of the ancient English parish (Fig. 109) reveals the nature of this problem more precisely. There is little to support a concept of a 'normal' sized rural parish, and there are striking regional inequalities in parish size. Particularly sharp is the contrast between the small parishes of parts of southern and eastern England – averaging below 3,000 acres – and the large parishes of north-western England – averaging over 12,000 acres. But many smaller-scale variations were superimposed on this general pattern. Wealden and Fenland parishes, for example, were well above the average acreage for lowland England, while in highland counties, such as those of the Lake District, the coastal parishes were generally much smaller than those of the interior.

These facts hardly mirrored the realities of local government in 1850, as they might have done at the time of the creation of the ecclesiastical parishes. The large parishes of northern England, with their constituent townships, had long proved too unwieldy to administer functions such

[1] R. A. Waterhouse in R. B. Pugh and E. Crittall (eds.), *V.C.H. Wiltshire*, V (1957), 325.

[2] W. Smith, *An economic geography of Great Britain* (London, 1949), 131.

[3] *Census of 1851: Population Tables, I*, vol. 1, p. lxxix (P.P. 1852–3, lxxxv).

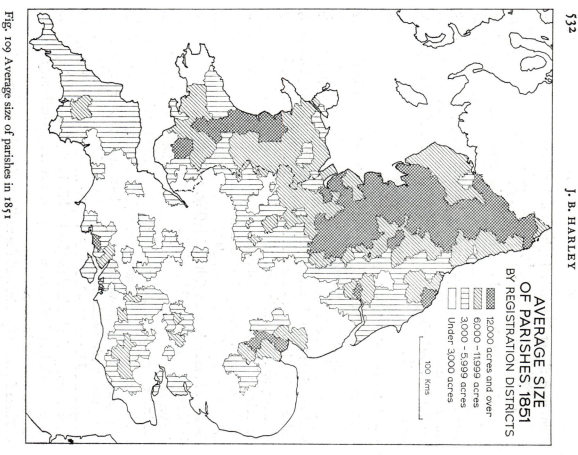

Fig. 109 Average size of parishes in 1851
Based on *Census of 1851: Population Tables, I* (P.P. 1852–3, lxxxv, lxxxvi)

as those of the poor law system; and, prior to the formation of the Poor
Law Unions in 1834, this responsibility had frequently been vested in the
township. The decline of the ecclesiastical parish as the smallest effective
cell of local government had gone so far, that, by 1851, about a third of all
the 16,000 'parishes or places' in England and Wales were separately

administered townships.[1] The stronghold of these multiple-township parishes lay in the northern counties; the parish of Wigan, for example, 28,433 acres in area, contained 13 townships.[2] In southern England, although the parish was more frequently co-terminous with the township, it had not been immune from subdivision into subordinate areas such as tithings and hamlets.

Three other considerations added complexity to this chaotic patchwork of small administrative areas. First, there were still extensive areas not contained within the boundaries of any parish (they were estimated to number nearly 600), and, until an Act of 1857, these extra-parochial tracts (including areas of old royal forest and some reclaimed fens) enjoyed virtual exemption from taxation, from maintaining the poor, from the militia laws, and from repairing highways. Secondly, there were many fragmented parishes. Prior to the Divided Parishes Acts of the last quarter of the century, the number of parishes in two or more parts was nearly 1,300.[3] Thirdly, civil and ecclesiastical parishes had often ceased to be co-extensive. Boundary changes made for civil purposes had not affected ecclesiastical parishes, and, similarly, subdivisions or amalgamations of ecclesiastical parishes had been made independently by the church authorities. Such ecclesiastical areas had proliferated. Many chapelries had 'acquired boundaries as definite and generally recognised as those of the parent Parish', although their number was 'not exactly ascertainable'; furthermore, since the Church Building Acts (beginning in 1818), fragmentation had gone further with the creation around new churches of 'Ecclesiastical Districts', the boundaries of which, even to the clergy, 'unprovided with maps or plans' were uncertain.[4] Not least, the growth of non-established religions (in 1851 they claimed almost half of all church-goers),[5] with their own administrative areas, meant that the ancient parish was not even the sole territorial unit of religious life.

The consequences of these obsolete areas were most deleterious in the industrial and urban areas. The size of settlements had far outgrown the inherited administrative framework, and at mid-century there were still

[1] *Census of 1851: Population Tables, I*, vol. I, p. cxvi (P.P. 1852–3, lxxxv); V. D. Lipman, *Local government areas, 1834–1945* (Oxford, 1949), 26.

[2] *Census of 1851: Population Tables, I*, vol. II, Lancashire, 34 (P.P. 1852–3, lxxxvi).

[3] V. D. Lipman, 27, 69–72; *Guides to official sources, No. 2; Census reports of Great Britain 1801–1931* (H.M.S.O., 1951), 97.

[4] *Census of 1851: Population Tables, I*, vol. I, pp. lxxi–lxxii (P.P. 1852–3, lxxxv).

[5] *Census of 1851: Religious Worship, Report and Tables* (P.P. 1852–3, lxxxix).

places like the town of Bury in Lancashire, with over 39,000 inhabitants, which possessed 'no more municipal organisation than that of a rural village'.[1] This state of affairs illustrates the fact that the Municipal Corporations Act (1835) was only a starting point in the creation of valid units of administration in harmony with the social geography of mid-Victorian England. In 1851, although 196 reformed boroughs had been created, places as populous as the parliamentary boroughs of London (the Tower Hamlets alone contained over half a million people), as Stoke-on-Trent, as Brighton or Huddersfield, were still without charters of incorporation.[2] And the existence of a charter did not necessarily confer the will or the capacity to cope with the problems of urban growth; at York, for example, the new corporation had done little for its citizens, although the population of the municipal borough had increased by over 10,000 between 1831 and 1851.[3] By 1850 the pattern of administrative areas was still very inchoate. The period around mid-century was characterised by the creation of a variety of authorities for special purposes – for the administration of the Poor Laws, of health and sanitation, of the highways and of education[4] – but the multi-purpose local authority areas were yet to be put on the map in the second half of the century.[5] Thus, although the ancient parish was in rapid decay as an effective cell of local government, the creation of suitable areas to replace it was still in its infancy.

THE COUNTRYSIDE

The occupation tables of the 1851 Census[6] showed that agriculture still held a position of great importance, employing no less than one-quarter of the adult males aged 20 years and over in England and Wales – a total exceeding that in the greatest of the manufacturing industries. Fig. 110 depicts the regional contrasts between the 'counties of maximum rusticity', as Clapham termed them,[7] and those where the growth of industry

[1] *Second Report of Commissioners for inquiring into the State of Large Towns and Populous Districts*, Appendix, Part II, 21 (P.P. 1845, xviii).

[2] *Census of 1851: Population Tables, I*, vol. I, pp. lxviii–lxix (P.P. 1852–3, lxxxv).

[3] E. M. Sigsworth in P. M. Tillott (ed.), *V.C.H., The city of York* (1961), 254–5.

[4] V. D. Lipman, 34–76.

[5] K. B. Smellie, *A history of local government* (London, 4th ed, 1968), 41–70.

[6] *Census of 1851: Population Tables, II*, vols. I and 2 (P.P. 1852–3, lxxxviii, pts I and 2).

[7] J. H. Clapham (1932), 252.

had transformed the economy. Some groups of registration districts in East Anglia, in southern England, in Devon and in parts of Yorkshire had over 60% of their adult male population engaged in agriculture. On the other hand, in the registration districts surrounding London, and in the mining and manufacturing regions, below 40% (and sometimes less than 25%) were recorded as employed in agriculture. Such areas formed a broad belt stretching from Lancashire and the West Riding southwards into the west Midlands; other important outliers where the agricultural population had been eclipsed by 1851 were in Durham, west Cumberland and Cornwall.

A consideration of the actual (rather than the percentage) distribution of those employed in farming presents a somewhat different pattern (Fig. 111). The lowest densities of males per square mile employed in agriculture lay predictably upon the mountains and moorlands of northern England, and, to a less extent, in the south-west. High densities occurred, again as we might anticipate, in south-eastern England – particularly in parts of Essex, Hertfordshire, Kent and Surrey – where the rural population as a whole was most numerous. On the other hand, the continued importance of agricultural occupations – indeed their apparent intensification – upon some of the coalfields was marked. As Thomas Welton, a contemporary statistician, observed: 'The numbers employed upon the land in the midst of collieries and manufacturing towns were really surprising;...perhaps most of all in...the South Staffordshire district, where the number of agriculturalists per square mile was nearly as high as in Hertfordshire.'[1] The textile areas of Lancashire and the West Riding, which had more agriculturalists per square mile than much of rural England, were in a similar category. No doubt this was to be explained partly by ambiguities in the census classification,[2] and partly by the fact that many smallholders also worked in a factory or mine. But the conclusion is statistically inescapable, that amid furnaces and factories, men still clung tenaciously to work on the land.

By 1850, the disappearance of common-field husbandry had divested the township of its former importance as a unit of local agricultural organisation. The statistics of farm size contained in the 1851 Census are, therefore, a timely index of contemporary contrasts in the organisation of

[1] T. A. Welton, *Statistical papers on the census of England of Wales, 1851* (London, privately printed, 1860), 50.

[2] *Census of 1851: Population Tables, II*, vol. I, pp. lxxviii–lxxix (P.P. 1852–3, lxxxviii, pt I).

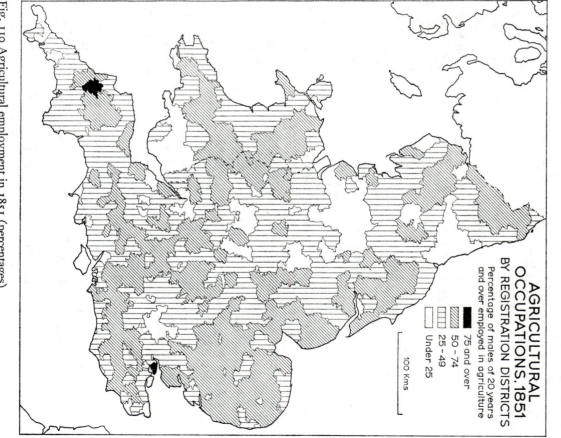

AGRICULTURAL
OCCUPATIONS. 1851
BY REGISTRATION DISTRICTS
Percentage of males of 20 years
and over employed in agriculture

75 and over
50 - 74
25 - 49
Under 25

100 Kms

Fig. 110 Agricultural employment in 1851 (percentages)
Based on *Census of 1851: Population Tables, II* (P.P. 1852–3, lxxxviii, pts. 1
and 2).

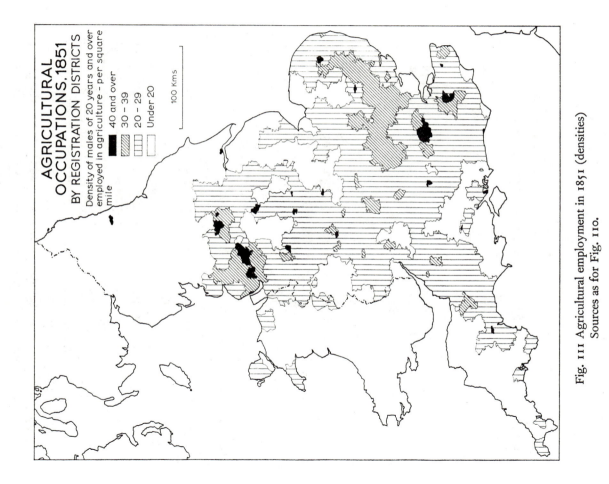

AGRICULTURAL
OCCUPATIONS. 1851
BY REGISTRATION DISTRICTS

Density of males of 20 years and over
employed in agriculture - per square
mile

40 and over

30 – 39

20 – 29

Under 20

100 Kms

Fig. 111 Agricultural employment in 1851 (densities)
Sources as for Fig. 110.

farming. Large farms of 500 acres and over were most numerous in eastern and southern England and in Northumberland; and small farms of below 100 acres were most important in the north Midland counties, in the north-west and the south-west (although we may note that the census excluded the acreage of hill pastures). Within these general divisions there were, however, internal contrasts: in lowland England, for example, smaller farms occurred on the claylands and in the Fens, while large farms were to be found on the light-soil areas of the chalk, limestone and sandstone outcrops.[1] Around the larger towns, and particularly around London, arable farming had in places given way to market gardening to supply an increasing demand for vegetables and 'all sorts of garden produce'.[2]

Among the more ubiquitous features in the countryside of 1850, land-scape parks continued to grow in number as the railways opened fresh areas such as the Chilterns to the homes of London merchants.[3] The older creations of Lancelot Brown and Humphry Repton were sometimes re-modelled to suit the taste of the age, but they continued to exhibit a style which often transcended that of any single region. In general, however, the rural landscape is most easily described in terms of its infinite variety, and the contrasts in farming measured by the census were amply sub-stantiated by the descriptions of contemporary writers. James Caird, for example, broadly divided the English countryside into the 'Corn' and the 'Grazing' counties (Fig. 112). 'The chief commodity of the western farmer', he wrote, 'is the produce of his dairy, his cattle and his flock. The large eastern farmer looks principally to his wheat and barley.'[4] If the general validity of his distinction is clear, a twofold division is, however, inadequate for a fuller account of the English countryside.

We should be clearer about the face of England at mid-century if the tithe surveys of the 1840s had been plotted. They were made on a parish basis and cover about three-quarters of the country. 'They provide a record of the use of land; whether cultivated or uncultivated, arable or grass, orchard or hop ground, heath or marsh, wood or agriculturally un-productive.'[5] This information has been plotted for a number of districts

[1] D. B. Grigg, 'Small and large farms in England and Wales: their size and distri-bution', *Geography*, XLVIII (1963), 268–9.

[2] J. C. Clutterbuck, 'The farming of Middlesex', *Jour. Roy. Agric. Soc.*, V (1869), 18.

[3] H. C. Prince, *Parks in England* (Shalfleet, I.O.W., 1967), 10.

[4] J. Caird, *English agriculture in 1850–51* (London, 1852), 482.

[5] H. C. Prince, 'The tithe surveys of the mid-nineteenth century', *Agric. Hist. Rev.*, VII (1959), 25.

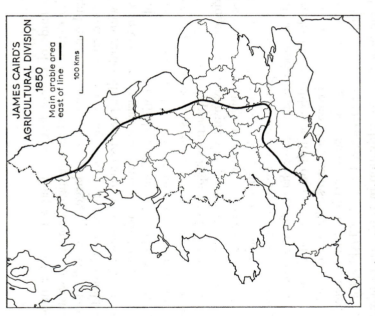

JAMES CAIRD'S
AGRICULTURAL DIVISION
1850

Main arable area
east of line

100 Kms

Fig. 112 Corn and grazing counties in 1850
Based on J. Caird, *English agriculture in 1850–51* (London, 1852).

and sample parishes, e.g. for the Chilterns (Figs. 113 and 114); but much work remains to be done before the full contribution of the surveys has been extracted for the country as a whole.[1]

Both Caird and the Royal Agricultural Society (in its series of prize essays on the agriculture of England) had used the county as a unit of description and, within the county, a scheme of subdivisions based usually on either geology or soil texture.[2] From these divisions within individual counties there can be built a larger system of agricultural regions in England as a whole. Broadly speaking there were five types of region – the sandlands, the fens, the south-east chalk and limestone areas, the claylands and the uplands.

[1] Examples of the use of the material are to be found in many of the 92 county reports of the Land Utilization Survey – L. D. Stamp (ed.), 'The land of Britain' (London, 1936–46).

[2] H. C. Darby, 'Some early ideas on the agricultural regions of England', *Agric. Hist. Rev.*, II (1954), 41–7.

The sand districts

The light lands of eastern England, long a model for improvers, had emerged by 1850 as districts of mature high farming. In western Norfolk, Caird found much to admire 'in the large, open, well cultivated fields, divided from each other by straight lines of closely-trimmed thorn hedges, and tilled with garden-like precision and cleanliness'. It was still a source of wonder to contemporaries that such abundant crops 'were growing, in many places, on an entirely artificial soil'. The Norfolk farmer was regarded as without equal in the permanent improvement of the soil by claying and marling, which, as earlier, was still widely favoured.[1] But not only in Norfolk had such practices helped to fashion a new countryside. In Cambridgeshire, for example, the 'poor and hungry lands' would have remained sterile but for high farming and enclosure which had greatly enhanced their productivity.[2] And in the western 'sand district' of Nottinghamshire comparable achievements could be reviewed; no district in England had 'undergone a greater change for the better'.[3] Large flocks of sheep replaced the rabbits which once browsed the forest lands, and fine crops of turnips and corn were harvested from large fields divided by quickset hedges. Such grand transformations had attracted substantial capital investment. Part of Delamere Forest in Cheshire, for example, disafforested and brought into cultivation in 1856, was reclaimed with the help of a light tramway to transport marl to the new fields.[4] Optimism about these areas still ran high: at least half of Cannock Chase appeared capable of cultivation to James Caird,[5] and it was regarded as out of place that it should remain untilled in a densely peopled county. On the other hand, some reclamation already embraced terrain beyond even the skill of the improver to improve. It was reported from western Suffolk that much cultivated land was 'apt to blow, that is to be driven by the wind', and it was ranked 'amongst the worst soils'; while from the contiguous district

[1] J. Caird, 163; B. Almack, 'On the agriculture of Norfolk', *J.R.A.S.* V (1844), 307; H. C. Prince, 'The origin of pits and depressions in Norfolk', *Geography*, XLIX (1964), 15–32.

[2] J. Caird, 151; S. Jonas, 'On the farming of Cambridgeshire', *J.R.A.S.*, VII (1846), 43.

[3] R. W. Corringham, 'Agriculture of Nottinghamshire', *J.R.A.S.*, VI (1845), 2–3.

[4] R. B. Grantham, 'A description of the works for reclaiming and marling parts of the late Forest of Delamere, in the county of Cheshire', *J.R.A.S.*, XXV (1864), 369–80.

[5] J. Caird, 243.

of Norfolk, C. S. Read concluded that, though improved, it 'must ever remain poor and comparatively barren'.[1]

Arable farming dominated the improved sandlands by 1850. In the Good Sand region of Norfolk, the tithe surveys reveal almost continuous cultivation in some areas. But Norfolk was, in any case, an arable county and, by mid-century, it had some 63% of its area under the plough.[2] Here, as over much of England, the arable acreage had reached its maximum extent. Similarly, on the less sterile areas of Bagshot Heath 60% was under the plough; in the upper Wey Basin in Surrey it was again the lighter soils based on the Bargate Beds, the Brick Earth and the Upper Greensand which were most cultivated (67% to 91%). The Hastings Beds, although partly clay, and the Upper Greensand of Sussex were likewise predominantly arable. Even within the 'grazing' counties, the lighter soils derived from the Triassic and Old Red sandstones and from fluvio-glacial sands and gravels, stood out at mid-century as islands of arable farming – in Warwickshire and Worcestershire and in the North Riding of Yorkshire.[3]

Increased productivity was closely linked to the better organisation of farming. Light-land farms were, in the main, relatively large and were tenanted by men of capital and of an improving zeal. Investment was reflected in the face of the countryside: in the bones, guano, nitrates and superphosphate which increased yields; in the new machinery – ploughs, drills, scarifiers, clod-crushers and steam-driven threshing machines;[4] and, here and there, in new and substantial brick buildings which were gradually replacing 'the inconvenient, ill-arranged hovels, the rickety wood and thatch barns and sheds', which Caird and others found to be widespread.[5] On the other hand, high farming had not dispensed with the armies of field labourers – male and female – who still undertook many of the seasonal tasks of farming; the abundance of cheap labour, especially in

[1] S. Lewis, *A topographical dictionary of England*, IV (London, 7th ed., 1848), 259–60; C. S. Read, 'Recent improvements in Norfolk farming', *J.R.A.S.*, XIX (1858), 265–6.   [2] J. E. G. Mosby, *Norfolk* (L. of B., pt. 70, London, 1938), 142–9.

[3] E. C. Willatts, *Middlesex and the London region* (L. of B., pt. 79, London, 1937), 268–70; D. W. Shave in L. D. Stamp and E. C. Willatts, *Surrey* (L. of B., pt. 81, London, 1941), 394; E. W. H. Briault, *Sussex* (L. of B., pts. 83–4, London, 1942), 498–505; A. W. McPherson, *Warwickshire* (L. of B., pt. 62, London, 1946), 789–840; K. M. Buchanan, *Worcestershire* (L. of B., pt. 68, London, 1944), 521–664; M. M. Milburn, 'On the farming of the North Riding of Yorkshire', *J.R.A.S.*, IX (1848), 499–511.

[4] C. S. Orwin and E. H. Whetham, *History of British agriculture, 1846–1914* (London, 1964), 7–10; F. M. L. Thompson, 'The second agricultural revolution, 1815–1880', *Econ. Hist. Rev.*, XXI (1968), 62–77.   [5] J. Caird, 490.

southern England, lessened the economic advantages of mechanisation. The debate continues both about the degree of diffusion of new agricultural technologies by 1850 and about the regional variations in the labour productivity of the English farm worker.[1]

As in 1800 the basis of light-land farming was a grain and sheep system with the Norfolk four-course rotation integrating its branches. Its principles were widely disseminated; it was the rotation 'most generally approved'; and examples of its adoption (often as a condition of lease) occurred from Cornwall to Northumberland. But equally, by 1850, a reaction had set in against too dogmatic an adherence to its tenets. Even in Norfolk, although the four-course was still hallowed at Holkham Park, many scientific farmers had deviated from it with success.[2] In any case, on poorer soils and in the less climatically favoured north of England, oats and rye entered the rotations; and everywhere local practice diversified the field crops from a long list of clovers, grasses and roots. Moreover, uniformity of rotation did not necessarily go hand-in-hand with identical farm economies. In Norfolk, there was an emphasis on the production of malting barley which was then shipped to London. But in many four-course districts, fat lambs and wool, and even some beef, were raised as well as cereal crops; in Cheshire, with its 'Sand-Land Dairy Farms', and in parts of Staffordshire and the West Riding, a four-course rotation was associated with dairy farming. Elsewhere, the sandy soils carried market gardens. In Bedfordshire they occupied the sandy heaths of the Lower Greensand outcrop; the sandy loams of the Lea valley were 'appropriated to market-gardening and nurseries'; and, on the blown sands behind the shore of northern Wirral, horticulture also flourished.[3]

Fenland and marsh

The Fens, despite the different challenges to improvement they had offered, may, by 1850, be grouped with the sandlands as lowland districts where men could look back on a period of remarkable progress. The Isle of Ely improvements have rendered advisable', J.R.A.S., xxi (1860), 258–66.

[1] E. J. T. Collins, 'Harvest technology and labour supply in Britain, 1790–1870', Econ. Hist. Rev., XXII (1969), 453–73.

[2] P. D. Tuckett, 'On the modifications of the four-course rotation, which modern improvements have rendered advisable', J.R.A.S., xxi (1860), 258–66.

[3] W. Palin, 'The farming of Cheshire', J.R.A.S., v (1844), 59–62; F. Beavington, 'The change to more extensive methods in market gardening in Bedfordshire', Trans. and Papers, Inst. Brit. Geog., XXXIII (1963), 89; also W. Bennett, 'The farming of Bedfordshire', J.R.A.S., XVIII (1857), 18; H. Evershed, 'Agriculture of Hertfordshire', J.R.A.S., xxv (1864), 270; J. Caird, 261–3.

to Samuel Jonas was 'a wonderfully fine district, and one in which more improvement had taken place within a few years than any other'. He could see the changes wrought by the application of the steam engine to draining, and by the claying and enclosure of lands which were still in progress.[1] The countryside was already 'traversed by excellent roads and railways and . . . mostly freed from the overflow of floods'.[2] The present landscape was nearly fashioned. It was then, as now, an arable region specialising in the cultivation of grain crops, potatoes and vegetables. Its fertility was becoming an agricultural legend, and J. A. Clarke, reviewing its agriculture in 1848, could write of 'the abundance and luxuriance' of the fenland crops, as he looked across the 'immense plain of dark arable fields'.[3] In addition, other formerly boggy morasses, such as the Isle of Axholme, King's Sedgemoor, the Holderness carrs and the mosslands west of the Pennines, although still presenting problems,[4] were systematically becoming, with the light lands, the new granaries of nineteenth-century England.

The south-east chalk and limestone areas

James Caird caught the essence of much of the chalkland when he wrote of a bare and undulating outline, of lofty downs still untouched by the plough, of white flinty roads, of coppices and widely spaced farmsteads, and, in the valleys, of the grass of the irrigated water-meadows. There was, of course, regional variety. On the Hampshire chalklands the country was 'rather warmer' than in Dorset and Wiltshire; and the soil variations in a chalkland district were so manifold in Berkshire as to require a 'lengthened description'. The history of enclosure had also influenced the countryside: in Yorkshire, where improvement of the Wolds had begun earlier, the landscape was 'all enclosed, generally by thorn hedges, and plantations everywhere grouped over its surface'; Salisbury Plain, on the other hand, was still raw, 'the face of the country . . . bare and un-sheltered, with no fence dividing field from field'.[5]

Farms on the chalk were amongst the largest in England. On Salisbury

[1] S. Jonas, 62; H. C. Darby, *The draining of the Fens* (Cambridge, 2nd ed., 1956), 237–46.   [2] J. Caird, 179.

[3] J. A. Clarke, 'On the Great Level of the Fens, including the Fens of south Lincolnshire', *J.R.A.S.*, VIII (1847), 132.

[4] M. Williams, *The draining of the Somerset Levels* (Cambridge, 1970), 209–29.

[5] J. Caird, 57, 79, 92, 310; J. B. Spearing, 'On the agriculture of Berkshire', *J.R.A.S.*, XXI (1860), 9–10; E. W. H. Briault, 501; J. Thirsk, *English peasant farming* (London, 1957), 259.

Plain they ranged from 800 even up to 5,000 acres, and cultivated fields often lay two miles distant from a homestead; in Wiltshire, chalk farms ranged from 400 to 1,000 acres.; in Hampshire and in Yorkshire, too, the holdings were above the average for the county.[1] They had attracted tenants with capital, who could pay high rents and invest in high farming. New barns, yards, cottages and wold ponds had appeared on the hillsides; some farmers spent a pound an acre every year on fertilisers in Lincolnshire.[2] And yet, even in these areas there were corners of backwardness. On the Sussex Downs, Caird found ancient wooden ploughs drawn by teams of six oxen; in some areas the dilapidated farm buildings did not measure up to high farming, and labourers' cottages (notwithstanding the model dwellings described here and there with such pride) were more often than not squalid and insanitary.[3] As in many agricultural regions there were numerous instances of highly progressive and grossly antiquated farming continuing virtually side by side.[4]

The chalklands were part of arable England. Of 19 parish tithe surveys in the Yorkshire Wolds 13 revealed an arable acreage equivalent to two-thirds or more of the total farmed area, and in only one survey did the arable area fall below a half. Some parishes on the Surrey Downs had up to three-quarters of their area under the plough,[5] although in Wiltshire more extensive tracts remained un-enclosed. As on the other light lands, the mid-century marked the end of an era in land-use history, in so far as the maximum arable acreages were reached – a major consequence of the preceding agrarian revolution. A variable acreage of unimproved pasture remained as sheep-walk on most farms; and, in addition, especially in the south-western chalklands, irrigated water-meadows were still universally important. In Wiltshire, for example, these meadows covered about 20,000 acres, and had been elaborated wherever possible.[6] The variants of a staple

[1] J. Caird, 80–1, 8; 310; E. Little, 'Farming of Wiltshire', *J.R.A.S*, V (1844), 62.
[2] D. B. Grigg, 'Changing regional values during the agricultural revolution in south Lincolnshire', *Trans. and Papers, Inst. Brit. Geog.* (1962), 91–103; J. B. Spearing, 33: J. Thirsk (1957), 259.
[3] J. Caird, 127–8; L. H. Ruegg, 441–5; *Report of Special Assistant Poor Law Commissioners on the Employment of Women and Children in Agriculture*, 15–25 (P.P. 1843, xii).
[4] J. D. Chambers and G. E. Mingay, *The agricultural revolution, 1750–1880* (London, 1966), 172.
[5] A. Harris, *The rural landscape of the East Riding of Yorkshire, 1700–1850* (Oxford, 1961), 102; D. W. Shave, 393.
[6] E. Kerridge in E. Crittall (ed.), *V.C.H. Wiltshire*, IV (1959), 72.

grain and sheep husbandry, as on the sandlands, prevailed. In Buckinghamshire, Dorset and Wiltshire sheep were bred and folded to enrich the soil for crops of wheat and barley: they were the 'manure carriers' of a specialised cereal farming, and a maxim of the husbandry was 'the greater the number of sheep, the greater the quantity of corn'. On the Hampshire and Berkshire downs, however, the sheep themselves received more attention and were 'a source of profit as well as manure'. The Sussex chalk farmer also relied on his breeding flocks of Southdown ewes.[1] Sheep were indeed the dominant livestock along the length of the chalk outcrops. Stocking of up to one sheep an acre was common, and such counties as Lincolnshire (with 1,000,000 sheep) and Wiltshire (with about 600,000 sheep) had particularly dense sheep populations.[2] The principles of the Norfolk four-course rotation were widely but not rigidly embodied. Chalkland rotations were adapted to local soils. In Wiltshire, for example, on the flint and chalk loams the ordinary four-course was used; on the light flinty soils, fertilisers had enabled a longer course with an extra grass or rape crop to be practised; on the heavy lands – often on the level hill tops – where turnips grew with difficulty, a modified three-course was employed 'in which wheat and green crops predominated; on the sand outcrops along the edge of the downs a four-course was again followed; while on land that had become 'clover sick' or 'turnip tired' new green crops were substituted.[3]

Finally, of the light lands, the Jurassic Limestone shared many of the characteristics of chalk farming. James Caird had noted this similarity when he remarked that the heath farming of Lincolnshire much resembled that of the wolds. The whole outcrop there was a fairly continuous arable farming region. Contemporary descriptions of the Cotswolds (where the arable terrains were still named 'downsy' land because they were remembered as sheep-walk), of the 'Stonebrash district' of Oxfordshire, and of the 'Red Stony Soil' of Northamptonshire, confirm that the grain–sheep economy had colonised much of the limestone belt. And beyond, from Northamptonshire to Durham, on the narrow strip of the Magnesian Limestone, similar conditions were continued into the more pastoral counties.[4]

---

[1] J. Caird, 59, 93, 128; J. B. Spearing, 12; J. Farncombe, 'On the farming of Sussex', *J.R.A.S.*, XI (1850), 76–80.

[2] L. H. Ruegg, 407, 432; A. Harris, 105; J. H. Clapham (1932), 276; E. Kerridge in E. Crittall (ed.), *V.C.H. Wiltshire*, IV (1959), 72.

[3] E. Little, 162–6; E. Kerridge in E. Crittall (ed.), *V.C.H. Wiltshire*, IV (1959), 73.

[4] J. Caird, 190; J. Bravendar, 'Farming of Gloucestershire', *J.R.A.S.*, XI (1850), 132–45; C. S. Read, 'On the farming of Oxfordshire', *J.R.A.S.*, XV (1854), 189–276;

18

The claylands

Clayland landscapes provided a sharp contrast to the spacious appearance of the light lands. A closely knit countryside of small fields and high hedgerows was the characteristic which frequently caught the eye of contemporaries. Edward Little, writing of north Wiltshire, captured this essential difference: 'Instead of open down country...the whole consists of enclosures, some of which are very small; and in many places the hedgerows are so thickly stocked with trees as to give the appearance of an extensive plantation when viewed from a distance.'[1] But the differences were not merely visual; they were economic and technical. There were still many obstacles to high farming on these claylands, amongst them, some of the features which gave the countryside its distinctiveness. The plentiful hedgerow timber, for example, was one impediment to good farming; 'the luxuriant foliage of summer', Caird explained, must overshadow the surface, and draw from the soil much of that nutriment which fields would otherwise yield to the farmer's stock'.[2] Many agricultural writers were in agreement with him.[3] The magnitude of the problem was illustrated in Devon. Here, in 1844, ten parishes totalling 36,976 acres contained no less than 7,997 fields whose 1,651 miles of hedge occupied 7% of the area.[4] Hedge-grubbing became an approved part of high farming.[5] Only occasionally did a more conservative farmer defend his massive hedges as providing shelter for his stock, although William Johnston, on aesthetic grounds, regretted their destruction and the fact that land was regarded 'as nothing else than a manufactory of agricultural produce'.[6] Ridge and furrow was also regarded as a hindrance to improvement. In

[1] E. Little, 172.   [2] J. Caird, 40.

[3] J. Grigor, 'On fences', *J.R.A.S.*, VI (1845), 194–228; W. Cambridge, 'On the advantage of reducing the size and number of hedges', *ibid*, 333–43; J. H. Turner, 'On the necessity for the reduction or abolition of hedges', *ibid*, 479–88; R. Baker, 'On the farming of Essex', *J.R.A.S.*, V (1844), 39; J. Farncombe, 85; J. Bravendar, 128–9.

[4] J. Grant, 'A few remarks on the large hedges and small enclosures of Devonshire and the adjoining counties', *J.R.A.S.*, V (1844), 420–9.

[5] J. H. Clapham (1932), 267.

[6] W. Johnston, *England as it is, political, social and industrial*, 3 vols, I (London, 1851), 4.

W. Bearn, 'On the farming of Northamptonshire', *J.R.A.S.*, XIII (1852), 44–113; S. H. Beaver, *Yorkshire; West Riding* (L. of B., pt. 46, London, 1941), 180–1; J. H. Charnock, 'On the farming of the West Riding of Yorkshire', *J.R.A.S.*, IX (1848), 289–90.

some districts it was a hallmark of agricultural inertia; in Northampton-shire, the ridge and furrow lands had remained under grass, fossilising the old arable strips 'in the same state as when they were sown down at the time of enclosure'. Elsewhere, as in Cheshire, the 'butts' and 'reins' were still a customary, albeit obsolete, method of land drainage. At mid-century farmers were still reluctant to disturb their 'consecrated form' and thereby lay bare the less fertile tops of the ridges, although, as in Worcestershire, they hindered the use of new machinery. Samuel Jonas was over-optimistic when he looked forward (in western Cambridgeshire) to a not too distant day when the high backs would be 'gradually ploughed down and sized into straight uniform lands'.[1]

In 1850, poor drainage was still a major stumbling block to progress on the claylands, despite the availability of government grants for drainage, and of mass-produced tile drains from new machines.[2] In many counties, however, a start had been made. Even in Surrey, condemned by Caird for its rural backwardness, another reporter found the London Clay 'naturally stubborn to cultivate', but made productive by draining. And in Berkshire, Bedfordshire and Huntingdonshire much clayland drainage had been successfully executed.[3] On the other hand, there were many areas where drainage was urgently recommended; Philip Pusey had believed in 1842 that one-third of England required draining. The heavier the land, the slower the progress. From Devonshire to Durham, 'cold clay' farming came in for criticism as being backward or defective – its soils described variously as wet, stubborn, exhausted and sterile. A description of the Lias clays of the vale of Marshwood in Dorset as being 'a terrible rough country' probably fitted many such districts. Even in Huntingdonshire, the essayist who in one sentence had praised its farmers as managers of strong land, in the next censured 'the large extent of poor, undrained, unproductive grassland, which yet remains unimproved'.[4] Although the

[1] W. Bearn, 80; W. Palin, 62–3, 77–8; C. W. Hoskyns, *Talpa: or the chronicles o, a clay farm* (London, 1854), 29; C. Cadle, 'The agriculture of Worcestershire', *J.R.A.S.*, 2nd ser., III (1867), 449; S. Jonas, 55.

[2] H. C. Darby, 'The draining of the English clay-lands', *Geographische Zeitschrift*, LII (1964), 190–201.

[3] J. Caird, 118; H. Evershed, 'On the farming of Surrey', *J.R.A.S.*, XIV (1853), 402; J. B. Spearing, 23; W. Bennett, 5; G. Murray, 'On the farming of Huntingdon', *J.R.A.S.*, 2nd ser., IV (1868), 266.

[4] P. Pusey, 'On progress of agricultural knowledge during the last four years', *J.R.A.S.*, III (1842), 170; H. Tanner, 'The farming of Devonshire', *J.R.A.S.*, IX (1848), 460; J. Caird, 334–6; L. H. Ruegg, 240; G. Murray, 266.

statistical evidence for the areas drained by 1850 – as later – is uncertain, it is clear that the technical mastery of the clay soils was far from complete.[1] In contrast to the light lands the heavy lands were still relatively backward and the 'agricultural revolution on the English clays', debated by some historians, was only partly under way.[2] The small size of many clayland farms, tenanted by uneducated men of insufficient capital to effect improvement, was a contributory factor in this lack of progress.[3]

Two groups of clayland enterprise can be identified: those where grassland was characteristic; and those where grass and arable were more equally intermixed. Grassland farms dominated the heavy clay vales and lowlands of the grazing counties. In southern Cheshire, for example, the tithe surveys showed a proportion of two-thirds grass to one-third (or even less) arable, and in eastern Leicestershire permanent grass sward also occupied up to two-thirds of the farm area.[4] The pasture varied from the carefully managed 'prime old grazing lands' of the vale of Aylesbury to parts of Leicestershire with 'wet, spongy, hassocky, ant-hill-covered pasture', and one critic asserted that a large proportion of English grassland needed improvement.[5] In 1850 the large towns still obtained their milk supplies mainly from suburban farms or from town cowsheds[6] and the farmers in the specialist dairying regions still manufactured either cheese – a Cheshire, a Stilton, a Gloucestershire – or butter as in Northamptonshire and western Warwickshire. In other districts, such as Corve Dale in Derbyshire and South Hams in Devonshire, the emphasis was upon the rearing and feeding of beef cattle, but a diversified livestock region such as the vale of Aylesbury relied on its dairies, its fat cattle and its flocks of ewes.[7]

[1] A. D. M. Phillips, 'Underdraining and the English claylands, 1850–80: a review', *Agric. Hist. Rev.,* XVII (1969), 44–55.

[2] R. W. Sturgess, 'The agricultural revolution on the English clays', *Agric. Hist. Rev.,* XIV (1966), 104–21; E. J. T. Collins and E. L. Jones, 'Sectoral advance in English agriculture', *Agric. Hist. Rev.,* XV (1967), 65–81; R. W. Sturgess, 'The agricultural revolution on the English clays: a rejoinder', *Agric. Hist. Rev.,* XV (1967), 82–7.

[3] W. J. Moscrop, 'A report on the farming of Leicestershire', *J.R.A.S.,* 2nd ser., II (1866), 293; H. Evershed (1853), 412; E. Little, 173; R. W. Corringham, 30; J. Farncombe, 81.

[4] C. S. Davies, *The agricultural history of Cheshire 1750–1850* (Manchester, 1960), 128; W. J. Moscrop, 292.

[5] J. Caird, 3; W. J. Moscrop, 337; R. Smith, 'The management of grassland', *J.R.A.S.,* X (1849), 2.

[6] E. H. Whetham, 'The London milk trade, 1860–1900', *Econ. Hist. Rev.,* XVII (1964), 369.

[7] J. Caird, 3.

The second group of farming types on the heavier soils had a higher proportion of arable. The clays, prior to the agricultural revolution on the light lands, had been the principal granaries, but by 1850, some clay districts such as the Fylde in Lancashire were rapidly changing to a grassland husbandry.[1] Many other districts retained a strong interest in arable farming despite its higher costs of production. Thus in the 1840s some 60% of the Boulder Clay area of south-east Hertfordshire was cultivated. The Chilterns with its cover of clay-with-flints was also an arable area, with stretches of grassland confined largely to its valleys and its landscape parks, as may be seen from the surveys mapped by F. D. Hartley (Figs. 113 and 114). It was matched by the extensive tracts of clayland under the plough on the Weald Clay of Surrey and Sussex, and on the London Clay of Essex.[2] Even away from south-east England, on the claylands of the grazing counties, many arable fields survived in 1850. They formed the basis of farm economies which combined some grain growing with the management of livestock, but, in these localities, arable farming had remained at its most backward. There were farms where green crops had only been at most partially introduced into the rotation; the old three-fold course – fallow, wheat, beans, as on the clays of the vale of York[3] – was widely practised, and crop yields were well below the national average. In the claylands the enclosure of the common arable fields had been far from a universal prelude to improvement.

The uplands

The mountains and moorlands of England (although the scene of such grand transformations as the reclamation of Exmoor)[4] were not subjected, even in the heyday of high farming, to the thorough-going improvement of the downs and heaths of lowland England. As contemporaries noted, a harsher environment had imposed certain absolute boundaries to human

[1] W. Smith, 'Agrarian evolution since the eighteenth century', in A. Grime (ed.), *A scientific survey of Blackpool and district* (London, 1936), 44–7.
[2] L. G. Cameron, *Hertfordshire* (L. of B., pt. 80, London, 1941), 329–32; D. W. Shave, 393; H. C. K. Henderson, 'Our changing agriculture: the Adur Basin, Sussex', *Jour. Min. of Agric.*, XLIII (1936), 630, 632–3; E. A. Cox, 'An agricultural geography of Essex c. 1840', unpublished London University M.A. thesis, 1963; E. A. Cox and B. R. Dittmer, 'The tithe files of the mid-nineteenth century', *Agric. Hist. Rev.*, XIII (1965), 12.
[3] G. Legard, 'Farming of the East Riding of Yorkshire', *J.R.A.S.*, IX (1848), 99.
[4] T. D. Acland, 'On the farming of Somersetshire', *J.R.A.S.*, XI (1850), 688–93; C. S. Orwin, *The reclamation of Exmoor Forest* (Oxford, 1929).

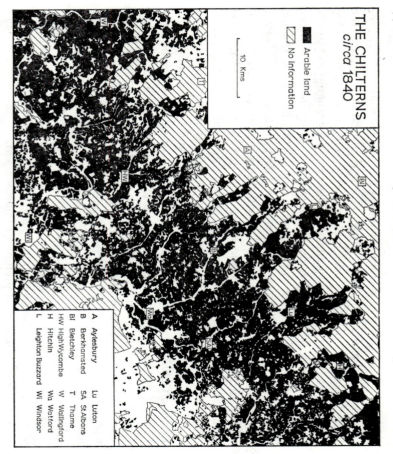

THE CHILTERNS
circa 1840

■ Arable land

▨ No Information

10 Kms

A	Aylesbury	Lu	Luton
B	Berkhamsted	SA	St Albans
Bl	Bletchley	T	Thame
HW	High Wycombe	W	Wallingford
H	Hitchin	Wa	Watford
L	Leighton Buzzard	Wi	Windsor

Fig. 113 The Chilterns: arable land *circa* 1840
Based on the tithe returns as plotted in F. D. Hartley, 'The agricultural geography of the Chilterns *c.* 1840', unpublished M.A. thesis, University of London, 1953.

endeavour. Cultivation was limited by altitude. On the Mendip Hills, where much of the land was between 800 feet and 900 feet above sea-level, although there were substantial patches of land cultivated intelligently by recently established farmers, T. D. Acland noted in 1851 that 'a large part of the hill must be reclaimed over again before it can be properly farmed'.[1] And farther north in Cumberland William Dickinson explained that 600 feet was the upper limit for wheat, although oats succeeded up to 800 feet and were occasionally sown at higher altitudes when land was broken up to improve the pasture. But optimism was often tempered with caution. Thus, although J. J. Rowley praised the efforts in Derbyshire to enclose

[1] T. D. Acland, 729. See M. Williams, 'The enclosure and reclamation of the Mendip Hills, 1770–1870', *Agric. Hist. Rev.*, XIX (1971), 65–81.

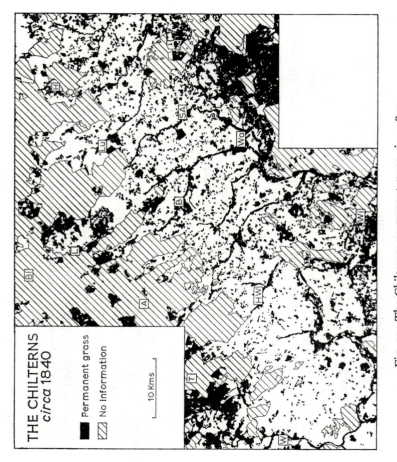

THE CHILTERNS
circa 1840

■ Permanent grass
▨ No information

10 Kms

Fig. 114 The Chilterns: permanent grass *circa* 1840
Source as for Fig. 113.
For the key to the initials see Fig. 113.

the Millstone Grit moorland, he had to admit that there were many 'difficulties and drawbacks; the seasons are critical; the winters long'. Moreover, much upland grazing remained as unstinted common, and its associated abuses impeded fencing and draining. The result, as the first Ordnance Survey maps for northern England show, was that the upper edge of cultivation was ragged, with many single enclosures detached, here and there, from the body of the improved land. J. H. Charnock, writing of the heather moors of the West Riding, noted that each year saw additional examples of moorland enclosure, but he had to admit to only partial and superficial improvement. In the lead dales of Durham and Northumberland – where grass fields reached up to 1,500 feet and even 1,700 feet – the agricultural holdings were small and the moors 'not half stocked'. Already, at mid-century, the farmers of Westmorland had before them the disheartening spectacle of land cultivated in the Napoleonic Wars, but subsequently 'permanently depreciated'.[1]

Thus, the progress of moorland improvement, although much had been accomplished, must not be overestimated. Arable fields, especially on the Millstone Grits, were few and scattered as the tithe evidence for Derbyshire suggests.[2] The best pastures lay on the Carboniferous Limestone. Craven was esteemed for its 'high feeding qualities', while the 'rich' grazing valleys of the North Riding – such as Wensleydale and Swaledale – were noted for their productiveness.[3] Here, the pastoral economy of the hill farm was most prosperous and, like English farming as a whole, it depended on the populous industrial regions. The products of hill farming were summed up by Webster when he wrote: 'so long as the tall chimneys of Yorkshire and Lancashire smoke, so long will the Westmorland farmer have a never-failing demand for all his produce – beef, mutton, butter, cheese and wool'.[4]

Other forms of upland utilisation were also well established by mid-century. Reservoirs had been created in the valleys of the Pennine upland. Afforestation with larch and fir had continued in Cumbria, stimulated by the demand for pit-props; and heather moorlands were rented for shooting. Tourists came to the hills in increasing numbers, and, in the Lake District, there was a new class of competitors for the ownership of the soil, 'the merchant princes of the manufacturing districts, who eagerly buy up any nook where they may escape from their own smoke, and enjoy pure air and bracing breezes, with shooting and fishing'.[5]

INDUSTRY

The Census of 1851 included the first scientific attempt to classify occupations,[6] and it has stimulated several studies of the distribution of

[1] W. Dickinson, 'On the farming of Cumberland', *J.R.A.S.*, XIII (1852), 289; J. J. Rowley, 'The farming of Derbyshire', *J.R.A.S.*, XIV (1853), 49; J. H. Clapham (1932), 259; J. H. Charnock, 299; D. Macrae, 'The improvement of waste lands', *J.R.A.S.*, 2nd ser., IV (1868), 321; T. G. Bell, 'A report upon the agriculture of the county of Durham', *J.R.A.S.*, XVII (1856), 92; C. Webster, 'On the farming of Westmorland', *J.R.A.S.*, 2nd ser., IV (1868), 7, 17.
[2] H. C. K. Henderson, 'Changes in land utilization in Derbyshire', in A. H. Harris, *Derbyshire* (L. of B., pt. 63, London, 1941), 71–4.
[3] J. H. Charnock, 300; M. M. Milburn, 514–16.
[4] C. Webster, 16.   [5] *Ibid*, 8.
[6] *Guides to official sources*, No. 2 (H.M.S.O., 1951), 30.

industry in the mid-nineteenth century. Augustus Petermann's pioneer distribution map of occupations compiled to illustrate the Census Report, had stood the test of time sufficiently to form the basis of J. H. Clapham's analysis of the 'localisation of manufacture about the end of the early railway age', and of R. Lawton's recent study which has re-mapped Petermann's data in a more assimilable form.[1] Other accounts, such as those of C. Day and T. A. Welton, have also used this census to comment on the distribution of industry.[2]

It is clear that the boundaries of industrial England (despite the survival of much loosely distributed rural industry) were more narrowly drawn than ever before. In particular, the coalfields had consolidated their position as the premier seats of industrial activity – especially in the textile and metal manufacturing industries. But the census also revealed the danger of oversimplifying the industrial structure. In spite of the well-known specialisation between coalfields, the major industrial regions were seldom a simple amalgam of coalmining and one staple manufacture. South Lancashire, for example, not only contained concentrations of workers in the major industries such as coalmining, the metal industries and the manufacture of textiles, but also in a variety of other manufactures, such as paper making, glass and earthenware (Fig. 121). On the coalfield of north Staffordshire, the district of the Potteries was of growing importance.

The coal and iron industries

Coal and iron ore, in this classic age of the blast furnace, were the twin pillars of a number of English industrial regions. The *Mineral statistics* enumerate some 1,704 collieries which produced an unprecedented total of 47 million tons of coal.[3] The collieries varied enormously in size and layout and, moreover, many small shafts, as a scrutiny of the map evidence confirms, had no place in the early editions of the *Mineral statistics*: the figures of coal production may therefore be an underestimate. The regional totals of iron-ore production are even harder to compile. The main feature of iron-ore mining was its concentration (up to 95%) upon the Coal-Measure ores (Fig. 115); and coal and iron sometimes came from the same

[1] J. H. Clapham (1932), 22–46; R. Lawton, 'Historical geography; the industrial revolution', in J. W. Watson and J. B. Sissons (eds.), *The British Isles: a systematic geography* (London and Edinburgh, 1964), 229–38.

[2] C. Day, 'The distribution of industrial occupations in England', *Trans. Connecticut Academy of Arts and Sciences*, XXVIII (1927), 79–235; T. A. Welton, *passim.*

[3] R. Hunt, *Mineral statistics of the United Kingdom for 1853 and 1854* (Mem. Geol. Survey, 1855), 50.

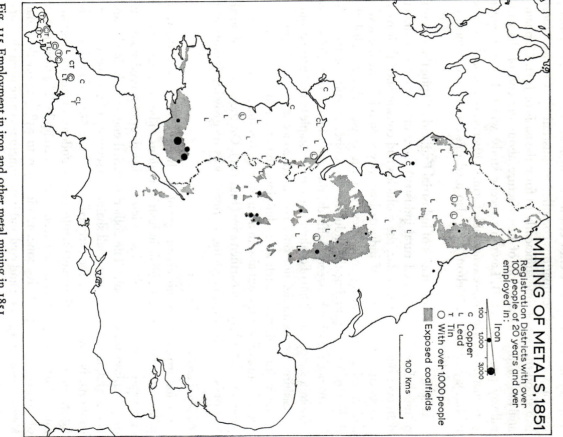

Fig. 115 Employment in iron and other metal mining in 1851
Based on *Census of 1851: Population Tables, II* (P.P. 1852–3, lxxxviii,
pts. 1 and 2).

pit, so that iron mines were not listed separately in the *Mineral statistics*. The confused topography of such an area was illustrated to the south-west of Bradford. What the Ordnance surveyors called 'coal and iron pits', were intermingled with 'collierys' and 'coal pits' in a complex mining landscape. The coal and iron pits fed a suite of blast furnaces directly by means of a tramway, the whole site providing a small epitome of the marriage of coal and iron which had virtually completed the relocation of the iron industry by 1850.[1] In Derbyshire and south Yorkshire, local coal was also used to smelt local iron. Only in Furness did a few charcoal-iron furnaces remain in production.[2]

Of the English regions where coal and iron occurred in bulk, the coke-iron industry of the Black Country outstripped all others. It ranked first in the English output of iron ore and pig-iron (1.2 and 0.75 million tons respectively) and third in coal production.[3] It is hard to visualise the intensity of heavy industry in this relatively small district – at most only 16 by 17 miles. The accounts of eye-witnesses suggest that perhaps nowhere else in England had the countryside been so transformed. A landscape was mapped out 'interspersed with blazing furnaces, heaps of burning coal in process of coking, piles of iron-ore calcining, forges, pit-banks, and engine chimneys; the country being besides intersected with canals, crossing each other at various levels; and the small remaining fields of grass or corn, intermingled with heaps of refuse of mines or of slag from the blast furnaces'.[4] Although many of its collieries were small, shallow and ill-equipped, and the scale of production of its blast furnaces was small, the Black Country was perhaps the most remarkable heavy industrial region of 1850.[5] The early industrial revolution had almost run its full course: the district was soon to reach its peak in mineral production (the productive thirty-foot coal seam was already substantially exhausted)[6] at its point of greatest national importance.

The north-eastern coalfield region, unlike the Black Country, stood on

[1] Yorkshire O.S. 6", First Edition, Sheet 216 (1852).
[2] R. Hunt, 'The present state of the mining industries of the United Kingdom', *Jour. Roy. Stat. Soc*, xix (1856), 317.
[3] B. R. Mitchell and P. Deane, *Abstract of British historical statistics* (Cambridge, 1962), 115–31; G. C. Allen, *The industrial development of Birmingham and the Black Country 1860–1927* (London, 1929), 84–96.
[4] *First Report of the Midland Mining Commission, South Staffordshire*, iv (P.P. 1843, xiii).
[5] *Ibid*., civ; W. Smith (1949), 124; G. C. Allen, 145.
[6] B. C. L. Johnson and M. J. Wise, 'The Black Country, 1800–1850', in M. J. Wise (ed.), *Birmingham and its regional setting* (Birmingham, 1950), 232–4.

the threshold rather than at the climax of an industrial boom. A large-scale iron industry around Middlesbrough based on the Jurassic ores of the Cleveland Hills was being set up. Although the Skinningrove and Eston seams were newly opened only in 1850, by 1854, 59 modern furnaces were already in blast, to make the region second in English pig-iron production. Coalmining, on the other hand, had done no more than retain its ancient supremacy (Fig. 116). Northumberland and Durham had the greatest production (15.4 million tons) of any coalfield – over a quarter of the English output.[1] Moreover, many of its collieries were modern and well equipped: the pits were deeper than elsewhere, the pit engines of larger capacity, and the most modern shaft ventilation had been introduced. The results of a rapid spread of mining across the coalfield can be assessed by 1850. The expanding railway network had provided a crucial stimulus, for few collieries were, by then, without a rail link.[2] Many collieries had become the nucleus of self-contained pit-head villages planted in the open countryside. West Cramlington, on a branch of the North Eastern Railway, was not untypical of the newly colonised areas. The terraces of miners' cottages formed three sides of a square; the colliery works and railway sidings lay in the centre; and the Blue Bell Inn, the Primitive Methodist Chapel and the Mechanics Institute completed this small community. The names of the terraces of houses: 'Crank Row', 'Long Row', 'Smoky Row', 'Pump Row', 'Brick Row', 'Foreman Row', prosaic and mechanical, were resonant of the new industrial age which called them into being.[3] These hastily built rows of brick or stone cottages, sometimes with blue-slate roofs, were becoming ubiquitous in the industrial regions of north England, not only in Northumberland and Durham, but in Cleveland, where they 'appeared suddenly, like fungi' on the bleak open plateau.[4] Likewise in west Cumberland they straggled around pit-head and blast furnace in the industrial hinterland of Maryport, Workington and White-haven. But in these northern industrial areas the climax of heavy industry lay in the future.

The secondary metal-using trades, as the 1851 Census reveals, were more widely dispersed than the heavy iron industry (Fig. 117). Nonetheless, the outstanding concentrations (with the exception of London and south Lancashire) were associated with important areas of coal and iron

[1] R. Hunt (1855), 41, 51.
[2] A. E. Smailes, *North England* (London and Edinburgh, 1960), 165.
[3] Northumberland O.S. 6″ First Edition, Sheets 80 (1864) and 88 (1864); and Durham O.S. 6″ First Edition, Sheet 20 (1861).
[4] A. E. Smailes, 230.

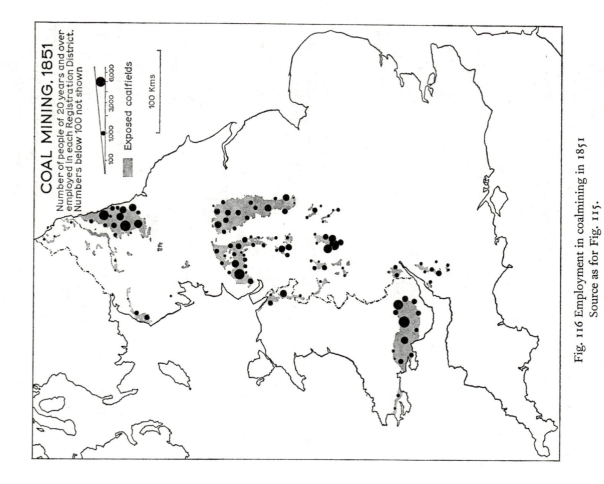

COAL MINING, 1851

Number of people of 20 years and over
employed in each Registration District.
Numbers below 100 not shown

Exposed coalfields

6,000
3,000
1,000
100

100 Kms

Fig. 116 Employment in coalmining in 1851
Source as for Fig. 115.

production. Many Black Country towns had developed specialist 'hard-ware' trades: at Wolverhampton, tinplate, japanned and holloware articles; at Darlaston, nuts and bolts; at Halesowen, Dudley and Rowley, nails; and at Walsall, saddlers' ironmongery[1] – to cite but a few. And in Birmingham, the metropolis of this metal-manufacturing region, the trades were 'so numerous and so varied … that no one could be found who would undertake to describe them all';[2] although the staple brass, gun, button and jewellery industries were located within distinctive quarters of the town.[3] The Sheffield district, coming second to the west Midlands in the metal trades, was not dissimilar in industrial organisation for it, too, lacked the large factory. Within the staple cutlery, file-making and tool-making trades, there were many firms occupying small work-shops. On the eve of the introduction of the Bessemer process (when over four-fifths of the nation's small output of perhaps 60,000 tons of steel came from Sheffield) there were no less than 135 small steel-making firms in the town.[4] In the metal-using (as in most) industries, old and new processes thrived side by side; technological innovation had been sporadic in its impact; 'this was not the machine age' as W. H. B. Court put it, 'but the machine age had begun'.[5]

The textile industries

By 1850 the machine age had become most mature in the textile districts on the coalfields of east Lancashire and west Yorkshire. James Caird found that 'the evidences of a busy population everywhere present them-selves'.[6] But south Lancashire was no Black Country. It was still to one observer 'like a vast city scattered amongst meads and pastures'.[7] Even in intensively mined areas, such as that around Wigan, collieries were often small by Northumberland standards; the pits were scattered and, beneath the spidery arms of small mineral railways, the ancient field patterns were

[1] G. C. Allen, 65–83.

[2] S. Timmins, *The resources, products and industrial history of Birmingham and the Midland hardware district* (London, 1866), 208.

[3] M. J. Wise, 'On the evolution of the jewellery and gun quarters in Birmingham', *Trans. and Papers, Inst. Brit. Geog.*, XV (1951), 59–72.

[4] C. Singer *et al.* (eds.), *A history of technology*, V (Oxford, 1958), 53; G. P. Jones, 'Industrial evolution' in D. L. Linton (ed.), *Sheffield and its region* (Sheffield, 1956), 156.

[5] W. H. B. Court, *A concise economic history of Britain from 1750 to recent times* (Cambridge, 1958), 178.

[6] J. Caird, 264.

[7] S. Bamford, *Walks in south Lancashire and on its borders* (Blackley, 1844), 10.

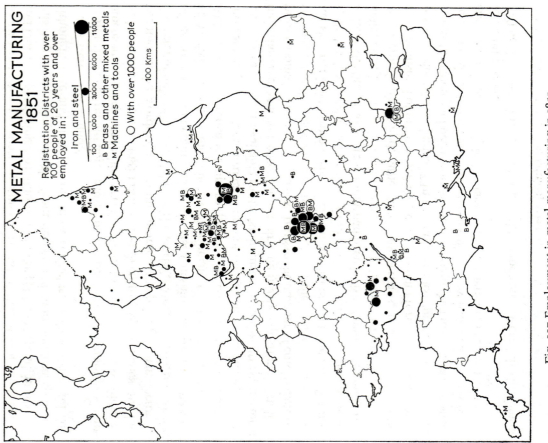

METAL MANUFACTURING
1851

Registration Districts with over
100 people of 20 years and over
employed in:

Iron and steel

11,000
6,000
3,000
1,000
100

B Brass and other mixed metals
M Machines and tools

○ With over 1,000 people

100 Kms

Fig. 117 Employment in metal manufacturing in 1851
Source as for Fig. 115.

not yet erased.[1] Coalmining was, however, an important strand in the south Lancashire economy. It ranked second to that of Northumberland and Durham, producing 9.9 million tons in 1854 (Fig. 116); and, moreover, this output came increasingly from deeper and more modern collieries, although, here and there, as on most coalfields, shallow drift and adit mines were to be found.[2]

But the most remarkable feature of the economic geography of mid-nineteenth century Lancashire was the extent to which the cotton industry had concentrated there (Fig. 118). Through cotton, south Lancashire had become, in the words of one contemporary, 'the principal manufacturing district, not only of England, but of the world'.[3] Moreover, the boundaries of the cotton region embraced north-east Cheshire and the extreme west of Yorkshire. It was the fullest embodiment of the new factory age. Although there remained up to perhaps 59,000 hand-loom weavers, mechanisation had gone further in the cotton industry than in any other.[4] Of the steam power available to British industry at mid-century, the most important regional concentration was driving nearly a quarter of a million power looms in the cotton industry. In 1850 the number of cotton factories in England and Wales was nearly 1,800,[5] and their buildings dominated the skyline; it was from their tall chimneys that at least one traveller now took his landmarks rather than from 'church spires and monumental columns'.[6] But art as well as money had been lavished upon the new cotton mills; even in Salford, it was reported that 'some of them have a good architectural effect', and, rather fancifully perhaps, that 'were they built of stone instead of brick, when they cease to vomit forth smoke they might pass for triumphal columns'.[7] In the multi-storied spinning mills of south Lancashire there had been more scope for design than in the weaving sheds of the north, although even in remote valleys they had retained 'a forceful architectural character', in the face of the accelerating change from water to steam power.[8]

The isolated cotton mill was still important. Certainly in remote

[1] Lancashire O.S. 6", First Edition, Sheet 93 (1849).
[2] B. R. Mitchell and P. Deane, 115; A. J. Taylor, 'The Wigan coalfield in 1851', Trans. Hist. Soc. Lancs. and Ches., CVI (1954), 119.
[3] T. A. Welton, 61.
[4] J. H. Clapham (1932), 119.
[5] Factory Returns, 1850, 16 (P.P. 1850, xlii).
[6] C. Redding, The pictorial history of the county of Lancaster (London, 1844).
[7] Ibid, 33.
[8] J. M. Richards, The functional tradition in early industrial buildings (London, 1958), 75-7.

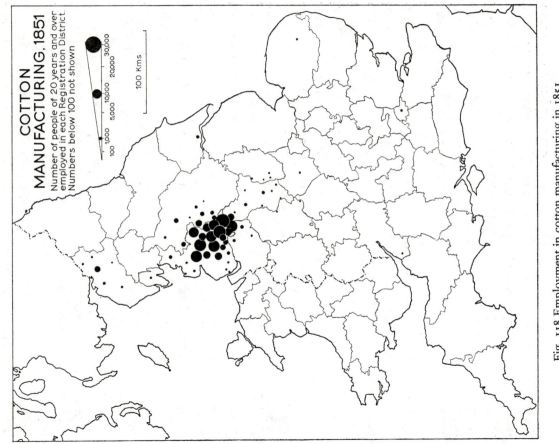

COTTON
MANUFACTURING, 1851

Number of people of 20 years and over
employed in each Registration District.
Numbers below 100 not shown

100 Kms

100   1,000   5,000   10,000   20,000   30,000

Fig. 118 Employment in cotton manufacturing in 1851
Source as for Fig. 115.

moorland valleys small water-powered mills had remained in production; while numerous bleach and dye works, printworks, and 'tenter grounds', where cloth was stretched and dried, are delineated in streamside locations on the First Edition of the six-inch Ordnance Survey Map in Lancashire. There were also steam-powered mills situated away from the stream sides in fairly open country. Hurst Cotton Mills, for example, north-east of Ashton-under-Lyne, lay within the perimeter of a landscaped park, together with a small coalpit, a terrace of cottages, a Methodist chapel and a school. Such factory villages – the equivalent, in the cotton province, of the north-eastern colliery settlements – had become a characteristic element in the industrial landscape.[1] But many erstwhile industrial villages were, by 1850, the new cotton towns. Contemporary maps captured them at a time of expansion; to the north-east of Bury, for example, around Chesham Field Cotton Mill, we can glimpse, in the half-finished rows of terraced houses and in the roads mapped out but not built along, these new units of workplace and home.[2] In many locations the paramount importance of coal and communications in the determination of site is demonstrated: factories frequently abutted on to canals, turnpike roads and railways, and were seldom far from a coalpit. Indeed, it has been argued that a cotton firm 'incurred a substantial penalty in siting a mill more than a few miles from the nearest supply, especially since many of its competitors had coal almost literally on their doorsteps'.[3] In general, therefore, with the notable exception of Preston, cotton manufacture had become concentrated on the coalfield; at the same time, the well-known geographical specialisation in the Lancashire cotton industry was largely shaped by 1850: not only was the separation of the spinning and weaving branches established, but 'every cotton town had acquired a distinctive industrial personality'.[4] Away from Lancastria the cotton industry was in relative decline and poorly adapted to the age of steam. In the east Midlands, for example, although there were 110 cotton-spinning mills in 1850, they represented only 6.5% of the total cotton mills in England (as compared with the region's 39% at the end of the eighteenth century); many were still water-powered.[5]

[1] Lancashire O.S. 6", First Edition, Sheet 105 (1848); S. Pollard, 'The factory village in the industrial revolution', *Eng. Hist. Rev*, LXXIX (1964), 513–31.
[2] Lancashire O.S. 6", First Edition, Sheet 88 (1851).
[3] H. B. Rodgers, 'The Lancashire cotton industry in 1840', *Trans. and Papers, Inst. Brit. Geog.*, XXVIII (1960), 140.
[4] *Ibid*, 146.
[5] D. M. Smith, 'The cotton industry in the east Midlands', *Geography*, XLVII (1962), 256–69.

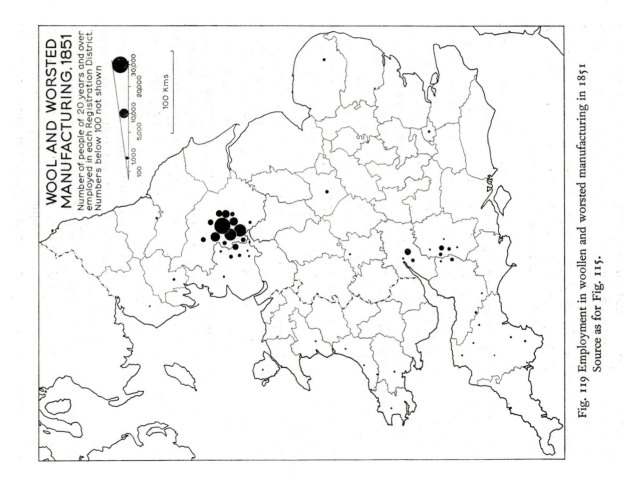

WOOL AND WORSTED
MANUFACTURING. 1851

Number of people of 20 years and over
employed in each Registration District.
Numbers below 100 not shown

100 Kms

100   1,000   5,000 10,000 20,000 30,000

Fig. 119 Employment in woollen and worsted manufacturing in 1851
Source as for Fig. 115.

A traveller across the Pennines in 1850 would have been struck by the similarity of the industrial landscape as he passed eastwards from the Lancashire zone of cotton, to the west Yorkshire zone of worsted and woollen manufacture (Fig. 119). These districts had much in common. In the first place, the locations had achieved stability. In the English worsted industry the decay of the Norfolk manufacture and the rise of west Yorkshire was such that, by 1850, some 87% of the spindles and 95% of the looms were installed in the latter district. In the woollen industry, although Yorkshire's share of the national manufacture was less (from 40% to 50%), the industry had likewise 'almost reached the pitch of concentration which it showed in the twentieth century'.[1] Secondly, by 1850, internal geographical specialisation in west Yorkshire matched that of Lancashire. Worsted manufacture, concentrated in the north-west, was more of an urban factory industry, prominent in large towns such as Bradford and Halifax. Woollen manufacture, in the south-east of the region, apart from the dominance of Leeds, was carried on, contrariwise, in smaller mills with riverine sites in the lesser towns or industrial villages, and the conversion to steam power was less complete. Thirdly, however, both types of factory – worsted and woollen – had spread as in Lancashire, away from the streams, and were often sited near to coalpits.

West Yorkshire, although fully in the factory age, was still a curious hybrid scene of agricultural and industrial activity. Even allowing for artist's licence, there was apparently some reality in the contemporary paintings of textile mills with rustic figures and scenes hard by the factory gate.[2] It was in this district, after all, that Caird had discovered a competent class of smallholders producing butter and milk for the local market.[3] On some parts of the major coalfields, industry had blackened and blighted the landscape; but on others, factories and collieries were set against a rural backcloth.

The dominance of cotton and woollen manufacture should not lead us to overlook the contribution of other textiles to the industrial economy of 1850. If flax had ceased to be a major English industry – although sailcloth was still made in a number of ports – silk retained third place in the textile industries in terms of the numbers employed.[4] There were silkworkers in 1851 not only in the main textile provinces of Lancastria and

[1] W. Smith (1949), 113; J. H. Clapham (1932), 27.
[2] E. M. Sigsworth, Black Dyke mills (Liverpool, 1958), 176–8.
[3] J. Caird, 286–8.
[4] J. H. Clapham (1932), 29.

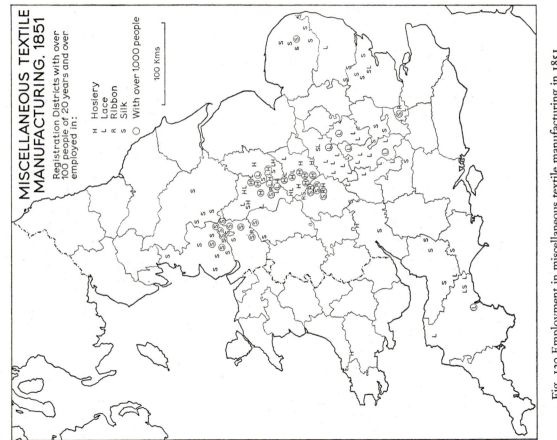

MISCELLANEOUS TEXTILE
MANUFACTURING, 1851

Registration Districts with over
100 people of 20 years and over
employed in:

H   Hosiery
L   Lace
R   Ribbon
S   Silk
◯   With over 1,000 people

100 Kms

Fig. 120 Employment in miscellaneous textile manufacturing in 1851
(Hosiery, lace, ribbon, silk)
Source as for Fig. 115.

Yorkshire, but also in the east Midlands and Warwickshire, in Norfolk and Essex, in London and Hertfordshire, and in the south-western counties (Fig. 120). Fully mechanised silk factories could be found in all the major textile areas, but the industry was also characterised by a widespread survival of the out-worker system and, as in other textile trades where the factory system was only partly developed, by a distinctive domestic architecture. Separate loomshops were housed in an upper storey, in extra ground-floor rooms or in out-buildings; in places there had developed an almost regional style of working-class housing.[1]

Tin, copper and lead mining

The non-ferrous mining industry, by 1851, was of great relative importance in parts of Devon, Cornwall and the northern Pennines. These districts were uniquely dependent on their mines (Fig. 115). In Devon and Cornwall, Thomas Welton calculated that mining employed 28% or over of the adult male population in ten registration districts; this figure rose to 53.9% in the Redruth District. In the north Pennine area as a whole mining was of comparable importance (28.8%) rising to 45.9% in Weardale Registration District, to 51.5% in that of Reeth and to 59.6% in that of Alston.[2] These statistics reflect both the absence of other industries and the poverty of agriculture; mining had spread over inferile moorlands to provide the only inducement to a sporadic agricultural colonisation. Such areas were also distinctive amongst industrial regions because they had attained, in general, the peak of their production (iron-ore and coal output, on the other hand, continued to rise sharply after 1850). Thus the 'Cornish Copper Kingdom', with Redruth as its capital, was soon to reach its maximum output in 1856, and the lead dales of the northern Pennines were also near their zenith. Indeed, in the latter area, some of the richest deposits were already exhausted, and in Cornwall a legacy of derelict mines remained in 1850 from earlier speculative booms.[3]

In the south-western region, different types of mine were intricately

---

[1] O. Ashmore, *The industrial archaeology of Lancashire* (Newton Abbot, 1969), 27–37; D. M. Smith, *The industrial archaeology of the east Midlands* (Dawlish, 1965), 34–50; W. J. Smith, 'The architecture of the domestic system in south-east Lancashire and the adjoining Pennines', in S. D. Chapman (ed.), *The history of working-class housing: a symposium* (Newton Abbot, 1971), 249–75.

[2] T. A. Welton.

[3] W. J. Rowe, *Cornwall in the age of the industrial revolution* (Liverpool, 1953), 305, 323–32; J. W. House, *North-eastern England: population movements and landscape since the early 19th century* (Newcastle upon Tyne, 1954), 32.

mixed, but the salient features of the mining landscape, the engine house with its tall chimney, the ore-dressing floors and smelting houses, the waste heaps, the mineral railways, the adits and watercourses, and the groups of miners' cottages – sometimes mere hovels of cob and thatch[1] – were common to all districts. Mining was closely concentrated upon the fringe of mineralised rocks around the edge of the granite; the stream working of tin still survived, but most of the ore came from deep mines. Copper production had, however, outstripped that of tin, and the copper workings employed three-quarters of the mining population in 1851 (Fig. 115). Lead took third place in the mining activities of south-west England.

The centre of lead mining lay in Pennine England (Fig. 115) and the lead dales of Northumberland and Durham clearly surpassed those of Derbyshire and Yorkshire. The mineralised zone was richest near the head of the dales, and as a result mining had spread upwards to nearly 2,000 feet, although smelting sometimes took place down-dale.[2] The first edition of the Ordnance Survey 6" Map, published for this area at mid-century, captured the detail of these landscapes in their economic heyday: the derelict and active pits, the old levels, the washing places for lead, the chimneys and the smelting mills – all framed by great rectangular intakes into the moorland edge. At Allenhead, for example, a fairly extensive lead-mining colony had grown up at 1,300 feet above sea-level, and mines and old workings scarred the hillside: administrative buildings, workshops and cottages, flanked by plantations, nestled below in the valley floor.[3]

Manufacturing industries outside the coalfields

Away from the coalfields there was a great variety of manufacturing in-dustry in 1850, but it only rarely made such a spectacular contribution to the landscape as did mining, metal smelting and textile manufacture. More often than not these industries took second place to other activities: in the countryside to agriculture, in the ports, and apparently in London, to commerce and trade. And yet, although the non-coalfield industry was not outstanding in the industrial topography of the steam age, it employed, as J. D. Chambers has reminded us, many more people than did the 'great industry' of the coalfields. In 1851, there were still more shoemakers than coalminers, and more handicraft blacksmiths than men tending iron

[1] W. J. Rowe, 152; A. H. Shorter *et al.*, *Southwest England* (London, 1969), 164–8.
[2] A. E. Smailes, 274.
[3] Northumberland O.S. 6", First Edition, Sheets 93 (1865) and 111 (1865).

furnaces; and those employed in the non-mechanised industries still out-numbered those in the mechanised industries by three to one.[1]

In rural England, many 'industries' were still really handicraft trades; production was for the local market so that the industrial workers were widely disseminated relative to the total population in such trades as milling, smithing, shoemaking and tailoring. As a group the handicraft industries were of most importance within two districts: in the east Midlands — stretching from Nottinghamshire south-east into Bedfordshire — and in the West Country. The first, the east Midlands, provided several examples of industries which had not migrated to steam-powered factories, and, although the countryside might be semi-industrialised, the domestic system had altered the rural landscape but little (Fig. 120). Such was the hosiery industry, surviving only in a condition of stagnation. There were, it is true, a number of hosiery factories in towns such as Derby, Leicester and Nottingham, but the majority of the framework knitters were still out-workers in converted farm buildings and cottages, where the appli-cation of power was almost unknown, and where the frame itself had re-mained practically unaltered for a hundred years or more.[2] Domestic lace manufacture was important in an area stretching from Derbyshire to London (Fig. 120); and in Bedfordshire and West Hertfordshire the craft of straw plaiting to make hats and bonnets was also organised as a domestic industry (Fig. 121). The Northamptonshire boot and shoe industry, although it had taken on a modern location, was likewise a small-scale handicraft trade, resisting mechanisation, and its workers were con-trolled by 'manufacturers' in Leicester and Northampton. In the West Country too, the industrial geography was mostly static. The Wiltshire woollen industry was exceptional in that steam was used far more than in Gloucestershire or Somerset.[3] And in Devon the woollen industry still relied on the water-wheel, and the making of gloves, lace and silk was a cottage industry. In many small country towns and villages the industrial legacy of the pre-railway era lay scarcely disturbed.

[1] J. D. Chambers, *The workshop of the world* (London, 1961), 21–2; J. H. Clapham (1932), 23–5.

[2] D. M. Smith, 'The British hosiery industry at the middle of the nineteenth cen-tury: an historical study in economic geography', *Trans. and Papers, Inst. Brit. Geog*, XXXII (1963), 125–42; F. A. Wells, *The British hosiery and knitwear industry* (Newton Abbot, 2nd ed., 1972), 106–17; L. A. Parker in W. G. Hoskins and R. A. McKinley (eds.), *V.C.H. Leicestershire*, III (1955), 2–23; W. G. Hoskins, *Leicestershire* (London, 1957), 81.

[3] J. de L. Mann in E. Crittall (ed.), *V.C.H. Wiltshire*, IV (1959), 173.

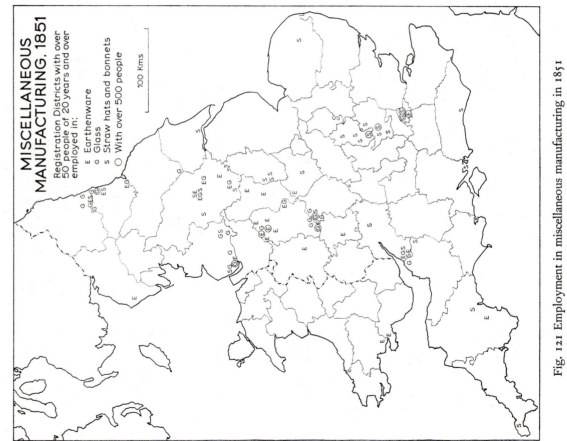

MISCELLANEOUS
MANUFACTURING, 1851

Registration Districts with over
50 people of 20 years and over
employed in:

E Earthenware
G Glass
s Straw hats and bonnets
◯ With over 500 people

100 Kms

Fig. 121 Employment in miscellaneous manufacturing in 1851
(Earthenware, glass, straw hats)
Source as for Fig. 115.

In contrast, the industries of the rapidly growing ports foreshadowed one of the more important industrial growth points after 1850, and dock-side industries, processing imported raw materials, were becoming more prominent in the waterfront scene. At Bristol, a report of 1823 advocated the establishment of a cotton industry, and a factory was built in 1837;[1] in Liverpool, in addition to oil-seed milling, sugar refining, soap and tobacco manufacturing, there were some 30 grain mills in 1845 (a quarter of them driven by steam), and more were constructed in the next decade;[2] and at Hull, too, oil-seed milling was an established industry.[3] If the formation of dockside industrial zones was well under way, the foundations of a modern shipbuilding industry were only just discernible in 1850. Ship-yards building iron vessels were in production at Birkenhead on the Mersey, at Millwall on the Thames, and, in a small way, on the Tyne. But such coastal seats of industry, in comparison with those of the coalfields, were as yet in their infancy.

London was, as it had always been, the exception. It constituted by any yardstick the most important single concentration of economic activity in the whole country. Some of its metal, textile and miscellaneous industries were, in terms of numbers employed in 1851, of national pre-eminence (Figs. 117, 120 and 121). In both highly specialised trades and in the more staple – including clothing, silk and printing – it had the greatest single labour force. Much of this industry was crammed into a zone in central and inner-suburban London, especially in the East End and Southwark, and in the West End.[4] It was comprised then, as now, of a chain of distinctive industrial quarters, each link housing its own trades: the silk-workers in Bethnal Green and Whitechapel (containing the parish of Spitalfields); the instrument-makers, goldsmiths and watchmakers in the parishes of Clerkenwell and St Luke; the printing trades in a quarter of the City and Finsbury; and the clothing trades, ranking as the largest metropolitan industry, in centres both in the West End (in Westminster) and in the East End (in Stepney).[5] But if no list can convey the number

1 S. J. Jones, 'The growth of Bristol', Trans. and Papers, Inst. Brit. Geog., xi (1946), 77.
2 W. Smith (1949), 548–9.
3 J. H. Bird, The major seaports of the United Kingdom (London, 1963), 128.
4 P. G. Hall, The industries of London since 1861 (London, 1962), 28–9, 37; P. G. Hall, 'Industrial London: a general view' in J. T. Coppock and H. C. Prince (eds.), Greater London (London, 1964), 226–35.
5 Ibid, 37–70; P. G. Hall, 'The location of the clothing trades in London, 1861–1951', Trans. and Papers, Inst. Brit. Geog., XXVIII (1960), 158–78.

and complexity of these trades, we can at least point to their common characteristics. In organisation they were an urban variant of the domestic system – a complex web of small masters, apprentices and out-workers; the division of labour was accomplished by splitting the manufacturing process into a series of stages – each performed by a different group; hence, the typical unit of production was small, little mechanised, and requiring none of the space or paraphernalia of large-scale industry. These enterprises blended unobtrusively into the urban landscape; the small workshop, often a converted house or shop, was the most typical industrial building of inner London. Perhaps, because of this, the importance of their production came to be overlooked by contemporaries.[1] Indeed, public attention was so sharply focused on the great industries of the coalfields, that, in the year of the Great Exhibition, John Weale concluded that London could be regarded as a 'vast trading and commercial, rather than a manufacturing town'.[2]

TRANSPORT AND TRADE

Road, canal, rail and coastwise transport

In 1850 we can review both the developed network of improved roads and inland waterways and, in the railways, describe a brand-new system of internal communications, challenging both turnpike road and canal. But we must not allow the more dramatic advance of the railways to overshadow the continuing importance of the older forms of transport. As the heyday of coaching receded, the symptoms of decline were indeed written large and clear. One by one the famous stage coaches ceased to run along the main roads of the country; by 1850 the coaches of the Sleepy Leeds, the Peak Ranger, the Red Rover had all passed into the history of transport; some ignominiously to end their days as summer-houses.[3] The revenues of the turnpike trusts were likewise plunging sharply,[4] yet, at the same time, the total mileage of turnpike roads in 1848 was greater than at any earlier date in the century. It is true that there were marked regional differences in the extent of turnpiking (varying from about a third of the total roads of Middlesex to under 10% in Cornwall, Cumberland and

[1] P. G. Hall (1960), 158.
[2] J. Weale, *London and its vicinity exhibited in 1851* (London, 1851), 220.
[3] G. M. Young (ed.), *Early Victorian England 1830–1865*, 2 vols. (London, 1934), II, 294.
[4] H. J. Dyos and D. H. Aldcroft, *British transport: an economic survey from the seventeenth century to the twentieth* (Leicester, 1969), 223.

Norfolk),[1] but on a national as on a local scale the physical legacy of new roads was permanent. The organisation which created them was in irreversible decline, but widened, straightened and macadamised highways and reconstructed bridges remained.[2] Nor did the passing of the stage coach and of the turnpike trust presage the disappearance of other horse-drawn transport. Local directories continued to list a host of road carriers, and carts and carriages were to crowd the roads which fed the railways for many years to come.

The same was true of the canals. Much has been written about the internecine strife between canal and railway company – by 1850 nearly 1,000 miles or roughly one-fifth of the navigable waterways of Great Britain had passed into railway ownership.[3] Although the income of some canals, like that of the turnpike trusts, had collapsed, some new canals were dug to act as railway feeders or distributors. The canal mileage was to reach its greatest length in 1858. The mature system was not, however, a national network (Figs. 80 and 106). Administratively, it was an inefficient complex of competing undertakings, and parts of the system – as in the west country – were severed from the central arteries of trunk canals. But its greatest overall weakness was the varying gauges of channels and locks, and the constant changes in the permissible draught. If the narrow boat could go virtually anywhere – except on the tub canals of the west country[4] – the larger barges of the northern and southern canals were confined to their own regions.[5] Waterways, however, continued to compete in the carriage of bulky and heavy industrial commodities, especially those from factories and furnaces in waterside locations. In 1848, for example, the tonnage of goods carried by water between Manchester and Liverpool was twice that carried by rail.[6] The decay of the canals was also suspended by the initial emphasis – especially outside north-eastern England – of the railways on passenger traffic.

But the railway was the truly new feature of the age (Fig. 122). The main trunk lines radiated spoke-like from London: in the south-east, two lines to Norwich were complete, so were lines to the main towns of the south coast – Dover, Brighton, Portsmouth and Southampton; westwards, rail communication had reached Plymouth; in the Midlands, Birmingham

1 Ibid., 222.　　　　　2 C. Singer et al. (eds.), v, 535.
3 C. Hadfield, British canals: an illustrated history (Newton Abbot, 2nd ed., 1966).
4 C. Hadfield, The canals of south-west England (Newton Abbot, 1967), passim.
5 H. J. Dyos and D. H. Aldcroft, 108.
6 W. T. Jackman, The development of transportation in modern England (London, 2nd ed., 1962), 741.

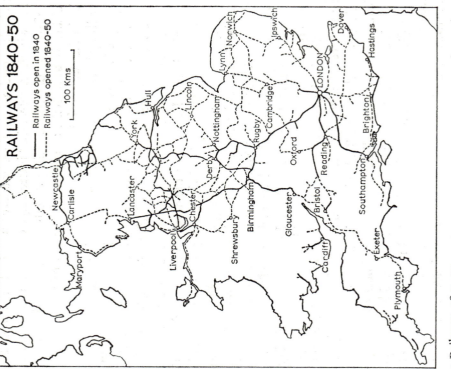

RAILWAYS 1840-50

—— Railways open in 1840
----- Railways opened 1840-50

100 Kms

Maryport
Carlisle
Newcastle
Lancaster
York
Hull
Liverpool
Chester
Lincoln
Nottingham
Derby
Shrewsbury
Birmingham
Rugby
Cambridge
Lynn
Norwich
Ipswich
Cardiff
Gloucester
Oxford
Reading
LONDON
Dover
Hastings
Brighton
Bristol
Southampton
Exeter
Plymouth

Fig. 122 Railways, 1840–50
Based on: (1) H. G. Lewin, *Early British railways* (London, 1952); (2) G.
Bradshaw, *Map of Great Britain showing the railways* (London and Man-
chester, 1850).

was soon to be served by two routes from the capital; and, in the north,
lines from London flanked the Pennines on both east and west. Many
cross-country lines had also been constructed, leaving conspicuous blanks
only in the south-west and in the north Pennines. By 1852, the most
populous English towns not yet served by a railway were Hereford,
Yeovil and Weymouth. On the other hand, the unco-ordinated schemes
of different companies had complicated the basic railway network; sometimes an exceptional gauge, such as the seven-foot of the Great Western,

had resulted; sometimes the lines of two companies both served the same town, but without a convenient interchange of traffic; sometimes a multiplicity of owners controlled a line; and, by 1850, only three companies could claim to have built a continuous stretch of railway more than a hundred miles in length.[1]

The railway builders had contributed a whole range of new features to the landscape. While the shanty towns of navvies' huts, often hastily constructed from turf, timber and tarpaulin, were ephemeral,[2] the great earthworks and engineering structures which carried the iron roads across England were prominent on an unequalled scale. There was much to look at: the dignified tunnel entrances; the viaducts built in brick, in timber, in iron, prominent against the skyline; the great cuttings gashing the countryside, some raw rock, some grass-grown; the high embankments, some newly planted with trees; the well-designed bridges of timber, stone and brick; as well as smaller features such as engine houses and water towers. Even the smallest rural station could add to landscape novelty. As temporary wooden shelters were replaced by permanent buildings, one company architect might perpetuate a traditional building style, another indulge in the exotic: on the Great Western Railway, for example, the stations between London and Reading were classical in design; to the west of Reading, they were Tudor in inspiration, but they were mainly Gothic at the far west of the line.[3] In the spate of new railway structures the aesthetic as well as the utilitarian had received consideration, and some of the English concepts of rural elegance had rubbed off on to railway architecture.

In the towns the railways had been more conspicuous and robust agents of transformation. At Stockport, the great viaduct strode over the town and 'the whole ravine in which it lies'.[4] At ground level, the incision of railways into the central areas of towns often resulted in the creation of

[1] J. Simmons, *The railways of Britain: an historical introduction* (London, 1961), 61; J. H. Appleton, *The geography of communications in Great Britain* (London, 1962), passim; J. A. Patmore, 'The railway network of Merseyside', *Trans. and Papers, Inst. Brit. Geog.,* XXIX (1961), 231–44; J. A. Patmore, 'The railway network of the Manchester conurbation', *Trans. and Papers, Inst. Brit. Geog.,* XXXIV (1964), 159–73; H. J. Dyos and D. H. Aldcroft, 126–45.

[2] T. Coleman, *The railway navvies* (London, 1965), 72–7.

[3] C. Barman, *An introduction to railway architecture* (London, 1950), 30–2, 61–73; C. Barman, *Early British railways* (Harmondsworth, 1950), 32–9.

[4] F. Engels, *The condition of the working class in England in 1844,* translated and edited by W. O. Henderson and W. H. Chaloner (Oxford, 1958), 52.

completely new railway quarters. In central London, road and water had barely felt the challenge of the new railways by 1852,[1] but in the north-west suburbs the building of the Euston terminus (1837) and that of King's Cross (1852) caused a great upheaval. Indeed, King's Cross had carved out a 45-acre site (erasing the old small-pox and fever hospitals and many crowded streets).[2] And around the Great Western terminus, Tyburnia or Paddington New Town had sprung up, so that here, too, the railway had 'altered the whole appearance of the place'.[3] In Liverpool, Manchester and Birmingham, as the detailed researches of J. R. Kellett have shown, the story was much the same. By mid-century, as the railway became a major urban landowner, it had initiated a cycle of demolition (often of slum clearance) and renewal; its inroads into the central business district were matched by a rise in land values, by changing land use, and also by a re-alignment of internal traffic routes. The extent of the railway's land hunger was intensified by the need of competitors each eager to acquire an urban terminus and the manner in which it was satisfied depended significantly on the pre-existing pattern of property ownership.[4] Few towns had resisted the overtures of the railways whose extravagant, often monumental, buildings, befitting the commercial power of their owners, became as nearly a symbol of the Victorian townscape, as castle or cathedral in medieval times.

Thus the importance of the railways in 1850 can hardly be exaggerated: they had not only become 'the iron veins that traverse the frame of our country',[5] but they were also growing into a major industry in terms of the work force and capital employed.[6] Many aspects of life and landscape had felt their transforming touch: towns were assuming new shapes and industry new locations; in the countryside, rural isolation was breaking down, and the farmer, as many contemporaries reported, had acquired new markets. Not least, Englishmen were given new possibilities for the enrichment of life. The railways, as E. J. Hobsbawn has put it, 'transformed the speed of movement...from one measured in single miles per hour to one measured in scores of miles per hour, and introduced the

[1] R. Clayton (ed.), *The geography of Greater London* (London, 1964), 87.

[2] J. Timbs, *Curiosities of London* (London, 1855), 639.

[3] P. Cunningham, *Handbook of London* (London, new ed., 1850), 528.

[4] J. R. Kellett, *The impact of railways on Victorian cities* (London, 1969), 14–15, 125–34, 150–60, 175–88, 244–62.

[5] J. Ruskin, *The seven lamps of architecture* (London, 1849), 182.

[6] H. Pollins, *Britain's railways: an industrial history* (Newton Abbott, 1971), prints statistics for 1850 for the United Kingdom, 49, 56, 65.

notion of a gigantic, nation-wide, complex and exact interlocking routine symbolised by the railway time-table'.[1] One could now 'breakfast in the din of the metropolis' and, by evening, walk on 'the breezy sands of the coast of Devon'.[2] Even the toiling industrial populations were now occasionally set free, by the excursion ticket, for brief, gay, hectic interludes on the occasion of a public holiday. The railways served the nation as lungs as well as arteries.

To transport by road, canal and rail, coastwise shipping must now be added. If this ancient trade had retained only a lesser share of the country's domestic commerce in the face of railway competition, in absolute terms, the volume of coastal cargoes was still increasing in 1850.[3] And, despite the motley fleet which trafficked around England's coasts, with barges, brigs, ketches, schooners and smacks[4] far outnumbering the steamboats sponsored by the railway companies,[5] the quantities handled (including the trade with Ireland) still exceeded the whole of the overseas trade proper. Most commodities were characterised by bulk and a low unit value. As well as the massive trade in coal – some five million tons were shipped annually by the 1850s from the Tyne, Wear and Tees alone[6] – raw materials such as China clay, copper ore and Welsh slate were carried by sea, as were a variety of foodstuffs, including grain and butter. Many of the lesser sea and river ports were also touched by this trade.

Overseas trade

The technological revolution, which was to result in the replacement of wooden sailing ships by iron steamships, was only in its infancy in 1850 but the merchant fleet of the United Kingdom (which totalled some 3,57 million tons at that date) already amounted to one quarter of the world's steam tonnage.[7] The striking progress in long-distance shipping was a direct response to the expansion of England's international trade which, by mid-century, was growing at a faster rate than ever before.[8] The

[1] E. J. Hobsbawn, *Industry and empire* (London, 1969), 110.

[2] M. E. C. Walcott, *A guide to the south coast of England* (London, 1859), xi.

[3] H. J. Dyos and D. H. Aldcroft, 208–10.

[4] B. Greenhill and A. Giffard, *The merchant sailing ship: a photographic history* (Newton Abbot, 1970).

[5] T. R. Gourvish, 'The railways and steamboat competition in early Victorian Britain', *Transport History*, IV (1971), 1–22.

[6] B. F. Duckham, 'The decline of coastal sail: a review article', *Transport History*, II (1969), 75.

[7] H. J. Dyos and D. H. Aldcroft, 232.

[8] B. R. Mitchell and P. Deane, 328.

statistics point not only to the country's manufacturing economy but also to its dependence on foreign trade. The leading imports in 1850, quantities of which were re-exported, were either industrial raw materials (especially cotton, dyewoods and dyestuffs, flax, silk, hemp, wool and timber) or foodstuffs. The nation could no longer feed itself from home agricultural production, and grain, sugar, tea and coffee were among the leading imports. In the export trade cotton goods had by now outpaced woollen manufactures which had for so long been dominant. The cotton export had become nearly three times as valuable as that of woollens (the second commodity) and over four times as valuable as iron and steel which occupied third place; they were followed (in descending order) by hardwares and cutlery, non-ferrous metals and manufactures, clothing and hats, silk goods, and machinery and chemicals.[1]

The robust claim made by Thomas Baines for Liverpool in 1852, that its commercial greatness reached to 'every corner of the globe',[2] could have summed up the geographical pattern of English overseas trade as a whole. In terms of value the strongest commercial links had been forged with North America, but there were also substantial trade flows to and from the underdeveloped parts of the Empire, especially India, and other tropical lands in Asia, in Latin America and Africa. England's trading relationships with the rest of the world were not, however, simply of a bilateral character between an industrial nation and a series of primary producers. There were exceptions. Not least, by 1850, a reciprocal trade in manufactured goods was growing between England and the more advanced industrial countries – especially the U.S.A., and Belgium, France and Germany. As the artificial barriers to Free Trade were dismantled – and its victory was symbolised by the repeal of the Corn Laws[3] – and as the economic structure of the trading nations became more tightly interlocked, England, with its banking, capital, insurance and shipping services, largely controlled the world market.[4]

[1] *Ibid*, 297–303; the statistics, which are indivisible, relate to the United Kingdom as a whole. Much additional commentary on overseas trade appears in J. H. Clapham (1930), 237–50, and (1932), 217–31.

[2] T. Baines, *History of the commerce and town of Liverpool, and of the rise of manufacturing industry in the adjoining counties* (London, 1852), 840; an analysis of Baines' data appears in F. E. Hyde, *Liverpool and the Mersey: an economic history of a port, 1700–1970* (Newton Abbot, 1971), 48–51.

[3] N. McCord, *Free Trade: theory and practice from Adam Smith to Keynes* (Newton Abbot, 1970), 61–97.

[4] H. J. Dyos and D. H. Aldcroft, 233; E. J. Hobsbawn, 134–48.

TOWNS AND CITIES

The 1851 Census revealed how, for the first time, populations classed administratively as 'urban' and 'rural' co-existed in roughly equal numbers. The Census Commissioners found it difficult to demarcate town and country; where, for example, could proper lines be placed in the new suburbs or amidst the congeries of industrial villages? No general definition was stipulated, and in the case of ambiguity it was left to local registrars to nominate and delimit towns. In this empirical fashion the local balance between rural and urban was computed. At one end of the scale, excluding London, only in six registration counties (Lancashire, Warwickshire, Gloucestershire, Staffordshire, the East Riding and Sussex) did the enumerated urban population exceed the rural; at the other end of the scale, in eighteen counties the urban population formed only 30% or less of the total.[1]

In population totals the towns had reached new limits (Fig. 123). We can conveniently define metropolitan London in terms of the Metropolis Management Act of 1855, which covered the area later to be included in the administrative county of 1888. This metropolitan area included about 117 square miles, and contained nearly 2.4 million people (see table on p. 673). It was over six times as populous as Liverpool, the next greatest centre, and there were six other cities, each with over 100,000 people. As in the case of London, it is difficult to define, in geographical as opposed to administrative terms, the limits of these seven provincial cities, but the 1851 Census figures for their municipal boroughs were as follows:

Liverpool	375,955
Manchester–Salford	367,233
Birmingham	232,841
Leeds	172,270
Bristol	137,328
Sheffield	135,310
Bradford	103,310

Around them all, incipient conurbations had sprawled. Furthermore, the 1851 statistics relating to population within the boundaries of municipal boroughs did not, as for Liverpool and Birmingham, include the population of contiguous suburbs. Below these seven provincial cities ranged a hierarchy of smaller places; thirteen other municipal boroughs each had

[1] *Census of 1851: Population Tables, I*, vol. 1, p. 1 (P.P. 1852–3, lxxxv).

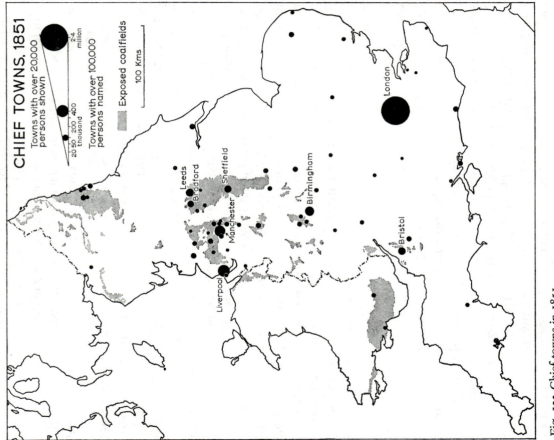

Fig. 123 Chief towns in 1851
Based on *Census of 1851: Population Tables, I*, vol. I, pp. cciv–ccvii (P.P.
1852–3, lxxxv).
The figure for London is that of the Census division which extended into
Middlesex, Surrey and Kent, and which was very similar to the later county
of London.

a population of over 50,000; another thirty-two each had over 20,000 inhabitants. At the other end of the scale, there were small towns each with fewer than 2,000 people (such as the 1,707 in the municipal borough of Chippenham) but such semi-rural places were becoming far less representative of an urban centre in 1850 than were the large industrial concentrations.

The urban areas were not only extensive but infinitely complex, and there was no one 'typical' town which could epitomise the geography of urban England. Administratively, the urban areas were governed by a diverse set of local institutions; the age of buildings ranged from the many surviving enclaves of Tudor and Georgian to the 320,000 or so new houses built in England between 1841 and 1851;[1] the urban plan was often formless, but here and there the town planner had made his mark as at Middlesbrough, Ashton-under-Lyne, Crewe and Birkenhead, or in model 'villages' such as Saltaire.[2] Broadly speaking, however, the towns fell into five categories based upon occupations – country and county towns, manufacturing and mining towns, seaports, coastal and inland resorts, and, in a category by itself, London. But such a grouping can only be very imperfect, for many towns straddled two or even more categories. The larger county towns, in particular, hardly fit the scheme, and some, such as Derby, Leicester and Nottingham, were also important industrial centres.

Country towns and county towns

Rural England, in 1850, was still serviced by a close network of small market towns inherited from the pre-railway age. A number of these places had less than 2,000 inhabitants, but many more had from 2,000 to 5,000 people. The 1851 population of some of the country towns of Norfolk such as Diss (2,419), Downham Market (2,867), East Dereham (3,372), Swaffham (3,858), Thetford (4,075) and Wymondham (2,970) points to the modest scale of their activities. Frequently, their appearance had altered so little since 1800 as to give a quality of timelessness often absent in the Victorian town. The railway, when it came, was often relegated to the edge of a town, where a small station, a gas-lit railway inn and a railway terrace, were relatively unobtrusive. The buildings fronting the cobbled high street and market square were still overwhelmingly of local

[1] *Census of 1851: Population Tables*, I, vol. 1, p. xcv (P.P. 1952–3, lxxxv).
[2] W. Ashworth, *The genesis of modern British town planning* (London, 1954), 118–31.

brick, flint, stone, cob and thatch, although by 1850 the canal and then the railway had sometimes made possible an alien peppering of mass-produced brick and Welsh slate.

Generally speaking, such towns were still in the age of the cart, the carrier, the corn exchange and the retail market, mingling town and country folk once a week in a day of jostling commerce. There were, too, periodic wholesale markets: at Chippenham for Wiltshire cheese, at Chester for cheese sold to the industrial towns of the north-west; at Aylesbury, Evesham and Banbury (to list but a few) for the cattle brought by long-distance drovers whose livelihood the railways would soon capture.[1] In the pages of contemporary trade directories the social and economic role of these towns can be glimpsed: they list not only auctioneers, dealers, innkeepers and tradesfolk, but also varied rural-based industries such as those at Thame, with its fell-mongering and tanning, brewing, basket making, coach building and chair making from beechwood.[2] Yet other craftsmen were noted in the census enumerators' books: Watlington, another small Oxfordshire market town, included braziers, saddlers, wheelwrights, millwrights, coopers and thatchers. The lists of such activities in the small towns of 1850 were very long indeed.[3]

Not surprisingly, however, a number of small towns were quietly reverting to a more completely rural economy. This was the case with six Oxfordshire market towns (Islip, Charlbury, Bampton, Burford, Woodstock and Deddington) which lost their market functions during the nineteenth century.[4] Some towns too had declined as a result of the decrease in stage-coach traffic. By 1845, for example, Staines had lost 'the appearance of bustle and prosperity'. Not half a dozen coaches passed daily instead of the former sixty-eight.[5] On the other hand, not all ancient country towns had shrunk, and many showed clear symptoms of outgrowing their semi-rural status. Market Harborough provided an example. Its inns (like those of Melton Mowbray) were still courting wealthy patrons from amongst the followers of the Quorn and Pytchley hunts; but simultaneously, the town had adopted the gasworks, the railway and some factory industry; off some of its side streets, clusters of back-to-back houses and several

---

[1] C. S. Orwin and E. H. Whetham, 24–8.

[2] W. Guest *et al.* in M. Lobel (ed.), *V.C.H. Oxfordshire*, VII (1962), 183–4.

[3] E. Craster in M. Lobel (ed.), *V.C.H. Oxfordshire*, VIII (1964), 233; *Guides to Official Sources*, No. 2, 28–31.

[4] A. F. Martin and R. W. Steel (eds.), *The Oxford region* (Oxford, 1954), 131.

[5] S. Reynolds in S. Reynolds (ed.), *V.C.H. Middlesex*, III (1962), 25.

narrow and unhealthy courts could be explored; its death rate was more akin to that of an industrial town—so high as to warrant a survey by a Board of Health inspector.[1] This situation was not unusual; where the inn yards, burgage plots and gardens of country towns had been infilled with inferior cottage property, the problems of congestion and sanitation were just as real as in a new cotton town. The places petitioning for the application of the Public Health Act of 1848 included not only Bradford and Dewsbury, but Diss, Eton and Ely: by mid-century, Barnsley and Barnstaple alike were awakening to the insanitary nature of an urban environment.[2]

The most acute problems belonged to those county towns with large industrial populations. Legacies from the past often obstructed modern re-planning. In Nottingham, for example, the imprisonment of the town behind its common meadows, prior to the enclosure of 1845, had resulted in a chronically overcrowded mass of courts, alleys and lanes, but not even the release of new land for building provided a quick solution to the urgent need for decent working-class housing.[3] Coventry (comparable to a large county town) was likewise still hemmed in, in 1849, by about 2,000 acres of Lammas and Michaelmas lands over which freemen and others had rights of pasturage, thus 'depriving the inhabitants from building their residences and places of business in airy situations, generally confining them to the limits of the ancient city'.[4] One result of overcrowding was that old town houses, formerly merchant residences, slid down the social scale, and then suffered swift deterioration as they were split up into workshops and tenements. The evils of dirt, overcrowding, disease and outworn housing, intensified by the urban revolution of the early railway age, were not confined to a few of the larger county towns. By 1850 it would have been hard to find exceptions; Chester, Gloucester, Shrewsbury, Exeter and York were included amongst those places singled out for attention by the Parliamentary Commissioners on the health of towns.[5]

[1] R. A. McKinley in J. M. Lee and R. A. McKinley (eds.), *V.C.H. Leicestershire*, v (1964), 142.
[2] *Returns relating to the Public Health Act of 1848*, 3–13 (P.P. 1852–3, xcvi).
[3] J. D. Chambers, *Modern Nottingham in the making* (Nottingham, 1945), 6–38; S. D. Chapman, 'Working-class housing in Nottingham during the industrial revolution' in S. D. Chapman (ed.), 135–63.
[4] *Second Report of Commissioners for Large Towns*, Appendix, Part II, 262 (P.P. 1845, xviii).
[5] *Ibid.*, Appendix, Parts I and II.

Manufacturing and mining towns

A striking aspect of the urban geography of 1850 was the existence of an inland group of giant provincial towns such as Manchester, Birmingham, Leeds, Bradford and Sheffield, each with over 100,000 inhabitants. Together with Salford, Manchester, 'the masterpiece of the industrial revolution',[1] contained over 400,000 people: the population of the area which is today defined as the Manchester conurbation already included in 1851, 1,063,000 people. Furthermore, the coalescence of nearby towns had begun: the first edition of the Ordnance Survey Map (1848) shows that a continuous line of houses flanked the eight miles of road from Manchester to Oldham; and that Ashton-under-Lyne, Hyde and Dukinfield were also fused by ribbon development.[2] Evidently this pattern of growth had become endemic in the outward sprawl of the large industrial towns. The point was well made by Robert Rawlinson describing Birmingham in 1849. 'Most of our large towns', he wrote, have 'received their increments chiefly from buildings erected along the roads branching out into the country, presenting so many streets radiating from a centre, but leaving the intervening spaces to be irregularly and imperfectly filled up at subsequent periods as change or necessity directed.'[3] In Birmingham, urban extension had, in fact, taken place across the borough boundary along routes leading out of the town by road, railway and canal; Balsall Heath, for example, had grown out from the centre as an elongated limb along Moseley Road and the Birmingham and Gloucester Railway.[4]

Within the expanded urban frontiers important changes had also taken place in the internal geography of some large towns. When Engels visited Manchester he was able to identify 'a fairly large commercial district... almost entirely given over to offices and warehouses'; the lower floors were occupied by shops, there were few permanent residents, and the main streets were congested with an enormous volume of traffic.[5] As streets were improved, as the railways intruded into the towns, and as larger civic and commercial buildings were built to reflect the new municipal status, the competition for central space had grown greater. One

[1] F. Engels, 50.

[2] T. W. Freeman, *The conurbations of Great Britain* (Manchester, 1959), 3, 135.

[3] R. Rawlinson, *Report to the General Board of Health on... Birmingham* (1849); quoted in C. Gill, *History of Birmingham*, I (Birmingham, 1952), 367.

[4] C. R. Elrington and P. M. Tillott in W. B. Stephens (ed.), *V.C.H. Warwickshire*, VII (1964), 16.

[5] F. Engels, 54.

result, apparent in the 1851 Census for some of the central wards of Leeds, Birmingham, Bradford and Manchester, was that the city centres were already dying as residential areas.[1]

Beyond, the more familiar image of the large industrial town was reached in the inner built-up zones where industry and poor-class houses were tightly intermixed. In Manchester, where this belt was, on average, one and a half miles in width, most of the worst features of uncontrolled expansion were to be found. Here was an area crammed with a medley of industrial premises – bone and cotton mills, gas and print works, tanneries and warehouses; it was criss-crossed by road, canal and river, and eaten into by railway sidings and goods yards; the intervening living spaces were characterised by narrow rows of cottages, a chapel here and there, a workhouse, and, in an empty corner, the paupers' cemetery.[2] The back-to-back terraces separated a labyrinth of narrow sunless alleys, cul-de-sacs and courts, where the scavenger seldom took his cart, where the water supply was deficient, and where poverty and filth ruled. Inner Manchester was not, of course, unique in all this. Central Bradford was just as 'dirty and uncomfortable', with its workers' houses at the bottom of the valley packed between high factory buildings.[3] In Leeds, too, near the flood plain of the river Aire and the Timble Beck, was a district of 'narrow, crooked and irregular' streets where 'a great number of dye houses and other manufactories' were interspersed with the dwellings of the working classes.[4] But the town was especially characterised by its disconnected rows of back-to-back houses: in the Ordnance Survey 'Five foot' plan for Leeds, published in 1850, there were no less than 360 streets where these urban cottages, 'one up and one down', often with a ground-floor space of as little as 5 square yards, had been constructed in the years since 1815.[5] And Birmingham, although it was regarded as comparatively healthy, had nonetheless developed extensive slums, built without order, slight and cheap, and only penetrated through a maze of small undrained streets and unpaved courts.[6] In most large towns such jerry-built districts had sprung up piecemeal, often the investment of little speculators,

1 T. W. Freeman, 82, 138, 172.

2 O.S. Plan of Manchester 1/1056, First Edition (1851).

3 F. Engels, 49.

4 *Second Report of Commissioners for Large Towns*, Appendix, Part II, 313 (P.P. 1845, xviii).

5 M. W. Beresford, 'The back-to-back house in Leeds, 1787–1937', in S. D. Chapman (ed.), 95–132.

6 R. Rawlinson, 23.

developing single fields here and there, and lacking even the rudiments of an overall plan.[1]

If the bulk of the working class were still housed near the town centres, the exodus of the well-to-do had started by 1850. Commuting by private carriage and horse omnibus – but not yet in any great numbers by mainline railway[2] – they had taken up residence in the outer suburbs. Thus Engels noted of Manchester, 'The villas of the upper classes are surrounded by gardens and lie in the higher and remoter parts of Chorlton and Ardwick or on the breezy heights of Cheetham Hill, Broughton and Pendleton.'[3] In Sheffield, the larger villas were mostly located on the higher, south-facing slopes;[4] likewise, in Leeds the 'better classes' tenanted the higher parts of the town in 'cheerful open streets'.[5] And yet, although many suburbs contained individual houses laid out in an elegant and spacious manner, overall planning was almost as uncommon as in the industrial neighbourhoods. The Edgbaston estate, where the Birmingham business man carried into 'his retirement a correct taste, not only for the useful, but also for the beautiful and picturesque'[6] was an exceptional development. More characteristically, the built-up area made haphazard contact with older villages, which became the nuclei of new suburban districts, with cores of older buildings forming distinctive inliers in the new landscape. Yet farther afield, beyond the continuously built-up fringe, old agricultural villages were colonised by merchants and factory owners, wealthy and land hungry, so that by mid-century Fallowfield, Withington and Didsbury were already partly suburban to Manchester,[7] and, in effect, Chapel Allerton and Headingley were suburbs of Leeds.[8] Residential segregation by social and economic class had become firmly entrenched in the major towns.

In Bolton, by the late 1840s, there were 61 cotton factories; in Bury, in 1845, there were 12 woollen manufactories and 26 cotton mills; Oldham,

[1] D. Ward, 'The urban plan of Leeds and the factors which have conditioned its growth', unpublished Leeds University M.A. thesis (1960), 78–106.

[2] J. R. Kellett, 356–7.

[3] F. Engels, 55.

[4] A. J. Hunt, 'The morphology and growth of Sheffield' in D. L. Linton (ed.), 237.

[5] J. Smith, 146.

[6] C. Gill, 366, quoting F. White, *General and commercial directory and topography of the borough of Birmingham* (Sheffield, 1855).

[7] T. W. Freeman, 4.

[8] R. Baker, 'On the industrial and sanitary economy of the borough of Leeds', *British Association Report* (1858), 164.

too, had over 100 cotton mills, all worked by steam; and even Wigan, on the western margin of the textile district, had 26 cotton mills by 1846.[1] Appropriately enough, Engels had designated such places 'factory towns'. In east Lancashire and west Yorkshire the Ordnance Survey town plans, published after 1843 at a scale of 1/1056, show parallel rows of workers' cottages, interrupted only by textile mill, chapel and church. The buildings, and in particular the houses, bore the marks of their relatively rapid construction. In Stalybridge many streets appeared 'rather hastily and imperfectly constructed', and St Helens seemed a town 'built in a hurry'.[2] The stone-built towns of the Pennine valleys were more solidly erected and, perhaps, less raw to the eye; but, at their worst, the factory towns were little better than casual accumulations of factories and slums.[2] Their skies were smoke-laden (so much in Lancashire as to have 'blackened the houses of red brick')[3] and there was more filth underfoot. Dr Lyon Playfair's investigations, for example, had many 'melancholy facts': an inadequacy of sewers, polluted streams, a lack of constant water supply, unpaved and unscavenged streets – these were some of the 'nuisances' common to many Lancashire towns.[4] Ashton-under-Lyne, 'the most elegant ordinary town in the country',[5] with its planned streets, its factories segregated from the 'new bright red cottages' gave 'every appearance of comfort',[6] but was almost unique. In the main, improvement had been confined to town centres, where newly erected town halls, mechanics institutes, public libraries and baths, gas-lighting, and a limited 'public' water supply, were now common. Even so, the textile towns remained predominantly huge working-class communities, as a recent study of Macclesfield shows.[7] They were too poorly equipped to be effective service centres; a limited range of retail shops and several score of ale houses alone served the needs of thousands.

[1] S. Lewis, I, 299, 456; II, 377; III, 475–6.
[2] R. Millward, *Lancashire* (London, 1955), 81–91; E. Butterworth, *An historical account of the towns of Ashton-under-Lyne, Stalybridge, and Dukinfield* (Ashton, 1842), 145; *Chambers's Edinburgh Journal*, No. 119, n.s., 11 April 1846, quoted in T. C. Barker and J. R. Harris, *A Merseyside town in the industrial revolution: St Helens 1750–1900* (Liverpool, 1954), 313; G. M. Young (ed.), I, 166.
[3] F. Engels, 49–51.
[4] *Second Report of Commissioners for Large Towns*, Appendix, Part II, *passim* (P.P. 1845, xviii).
[5] E. Butterworth, 54.   [6] F. Engels, 52.
[7] C. S. Davies, *A history of Macclesfield* (Manchester, 1961), 166–8.

The 'mining and hardware' towns were facsimiles, only in part, of other industrial towns. The ubiquitous red rash of inferior dwellings had wrapped around most of them by 1850. But they lacked the disciplined lay-out induced by the factory; a neat framework of streets could not readily be set down on land pitted with subsidence and pocked with furnace and mine. Urban disorder had resulted where houses and industry had vied for possession of the land. Thus, for example, Burslem, Hanley and Longton were reported to be built in an 'irregular and rather dispersed manner'[1] and West Bromwich parish appeared as 'a large straggling town, the buildings being scattered about without much order, but dense enough in some parts to form streets'.[2] Many of the Black Country towns were like this, and it was the same elsewhere. The local improvement acts promoted by municipal corporations – Leeds and Liverpool in 1842, Birkenhead in 1843, Manchester in 1844 and 1845, Nottingham, St Helens and Wallasey in 1845, Newcastle, Burnley and Southport in 1846, and so on[3] – had done little by 1850 to put a brake on the worst evils of urban laissez-faire.

Seaports

The relative importance of the six principal ports of England in 1850 can be seen from table 10.1 below.

Table 10.1 *Vessels, tonnage and exports for the six principal English ports, 1850*

	Inwards		Outwards	
	Ships	Tons	Ships	Tons
London	9,914	1,904,948	6,523	1,384,683
Liverpool	4,531	1,605,315	4,807	1,656,938
Newcastle	2,032	316,297	5,174	849,572
Hull	2,485	466,430	1,764	369,743
Southampton	626	152,117	603	147,519
Bristol	730	137,812	276	79,448

Note: The figures are combined totals for British and foreign ships.
Source: Vessels and Tonnage etc., 1816–1850 (P.P. 1851, lii).

[1] *Second Report of Commissioners for Large Towns*, Appendix, Part I, 9 (P.P. 1845, xviii).
[2] S. Lewis, I, 401.
[3] S. D. Chapman (ed.), 155–6.

If we exclude London – which was more than a great seaport – the pre-eminence of Liverpool, also the largest provincial town of 1850, was un-challenged. And indeed, although the port of London handled more vessels and a slightly larger tonnage, the export trade of Liverpool, as measured by value, was over twice that of London and three times that of Hull. By 1852 Liverpool shipped over 1,000 million tons of coal and 315,000 tons of iron bars, rails, hoops, rods and pig-iron[1] – all commodities reflecting its coalfield in-dustrial hinterland. The most distinctive manifestation of this economy was a complex system of massive docks and warehouses. In 1857, when the Merseyside Docks and Harbour Board assumed control of the port, it acquired no less than 15¼ miles of quayside and 199 acres of enclosed water space. Many docks were newly constructed at mid-century: in 1848 alone five new docks were opened, and, by 1852, a further three;[2] a continuous granite wall impounded the Liverpool shore, to the rear of which stood massive warehouses, some six or seven storeys high, austerely built of cast iron and brick, and already dominating the waterfront.[3] The river now served a city larger than industrial Leeds or Birmingham. It contained some of the most notorious examples of overcrowding in courts and cellars, as well as of undrained and unpaved streets, and of high mortality and sickness rates;[4] at the same time many of its middle-class citizens had already deserted the old town to occupy villas in 'pleasant villages from Bootle to Aigburth', or even as far afield as Hoylake and Southport.[5]

In the other major ports the work of extensive dock improvement had begun somewhat falteringly. At Southampton, the first modern dock had opened only in 1842; and ships under sail still far outnumbered those propelled by steam. At Hull, the construction of the Railway Dock (1846) and the Victoria Dock (1850) more clearly foreshadowed the pattern of its modern dockland; the trade of the port was, moreover, rising steadily and its wharves were among the few in England (including those of Liverpool and the new railway port of Fleetwood) where the tonnage of steam vessels exceeded that of sailing ships. But in some seaports relatively little change was visible in 1850. This was so on the Tyne and the bulk of

[1] W. Smith (ed.), *A scientific survey of Merseyside* (Liverpool, 1953), 157.
[2] J. E. Allison, *The Mersey estuary* (Liverpool, 1949), 29–32.
[3] Q. Hughes, *Seaport: architecture and townscape in Liverpool* (London, 1964), 7–39.
[4] J. H. Treble, 'Liverpool working-class housing, 1801–1851' in S. D. Chapman (ed.), 167–220.
[5] T. Baines, *Liverpool in 1859* (London, 1859), 8.

the coal cargoes still left under sail from Newcastle – as from Sunderland and the other north-east ports. Bristol, notwithstanding its ancient importance, was likewise not keeping abreast of the newer ports serving industrial England.[1]

Apart from the six principal ports, many smaller harbours still shared in the coastal trade. They were the maritime equivalent of the country market town, and the railway ports had not yet captured all their commerce.[2] Indeed, in the statistics for 1850, it was possible to list a further 64 places in England where vessels were registered and where a quantifiable amount of shipping had 'entered and cleared Coastwise'.[3] The smallest harbours, such as Bridport, Chichester and Chepstow, had each cleared less than a tonnage of 30,000 in the year, and, moreover, like a score of other small ports in 1850, including Rye, Penzance and Scarborough, they had never been entered, so the return informs us, by a single steam vessel in the whole of that year.

Coastal and inland resorts

Seaside resorts had expanded more rapidly than any other group of English towns[4] and, in so doing, had overtaken inland spas. Symptomatic of these changes, the population of Brighton came to exceed that of Bath by 1851. A. B. Granville visited and described a number of sea-bathing places around the English coasts in 1839–40;[5] but his itinerary had its gaps, especially along the south coast where many seaside villages were fast becoming 'places of resort'. Seaside places also figured prominently in Spencer Thomson's account of the health resorts of Britain in 1860.[6] Brighton, with a population of 65,569 in 1851, stood apart, far ahead of its rivals in visitors and in built-up area.[7] But Hastings (16,966), Ramsgate (11,838) and Margate (10,099) each exceeded 10,000 inhabitants; Weymouth (9,458) and Torquay (7,903) were not far behind. On the other coasts, only Scarborough (12,915) and Whitby (10,899), which like Dover (22,244), were also small seaports, could match the importance of

---

[1] J. H. Bird, 30.

[2] J. D. Chambers (1961), 59.

[3] *Returns of the Number and Tonnage of Sailing and Steam Vessels*, 4–5 (P.P. 1851, lii).

[4] *Census for 1851: Population Tables*, I, vol. I, pp. xlviii–l (P.P. 1852–3, lxxxv).

[5] A. B. Granville, *The spas of England and principal sea-bathing places*, 2 vols. (London, 1841).

[6] S. Thomson, *Health resorts of Britain; and how to profit by them* (London, 1860).

[7] E. W. Gilbert, *Brighton, old ocean's bauble* (London, 1954), 153–5.

the southern towns. The resorts of Lancashire and Cheshire were still embryonic; Southport (4,765) was the largest; Blackpool had a resident population of only 2,000, and New Brighton, recently planted among sand dunes, comprised little more than a 'few clusters of houses and villas' in a 'perfect desert'.[1] Even less developed were the remote beaches of Cornwall which received only a trickle of visitors. The image of St Ives was that of 'an ugly, narrow, dirty, dull town',[2] and Looe was still supported mainly by its pilchard fisheries. Moreover, the future of the Cornish resorts was regarded as by no means secure, and, although Spencer Thomson considered that 'the lines of rail will by degrees bring into notice other spots suited for invalid residence', he thought 'the extreme distance from the great centres of English life must prove a serious obstacle'.[3]

Each resort did its utmost to imitate Brighton,[4] and this may be one cause of the similarities between the seaside towns of 1850. Piers and promenades were becoming as ubiquitous as bathing machines had been by 1800, and even the boarding houses and hotels were often of a standard type, prompting one critic to admonish the 'interminable terraces, parades, paragons and parabolas of houses...which mere brick-and-mortar speculators have run up'.[5] For leisure, as well as for labour, urban environments had been mass-produced by the early Victorians. Notwithstanding this, almost every resort had some individual character. The varied physique of the coastline had ensured that this was so. In some towns, such as Torquay (where the building levels girdled the semi-circular bay, the lower containing shops, the middle stone-built terraces, the upper detached villas) a restricted site had closely guided the plan of the town.[6] But at others, such as Southport, the flat, sandy shores had enabled the designs of the drawing board to be transferred to the ground more easily.[7] At Bournemouth, Granville concluded that nature had done everything, and that the hand of man had only to fashion 'and suitably and judiciously to convert to its own purpose'.[8] His advice was followed, but by 1850, the future resort was little more than a coastal variant of an upper-class suburb with many detached houses shaded by trees and

[1] A. B. Granville, II, 11.   [2] M. Walcott, 544.   [3] S. Thomson, 164.
[4] E. W. Gilbert (1954), 13.   [5] A. B. Granville, II, 526.
[6] *Illustrated London News*, XVI (1850), 41–2.
[7] F. A. Bailey, 'The origin and growth of Southport', *Town Planning Rev.*, XXI (1950), 297–317; F. A. Bailey, *A history of Southport* (Southport, 1955).
[8] A. B. Granville, II, 532.

beautified by flowering shrubs.[1] Bournemouth was not alone: differences in Victorian social classes were widely reflected at the seaside as well as in the suburbs. Not only did some resorts such as Scarborough (in its new and old towns respectively) have a fashionable and unfashionable end,[2] but whole towns were already characterised by the nature of their clientele. Thus, while on the one hand, the steamer service to Gravesend had made it the resort of the Cockney trippers;[3] on the other, Brighton was more nearly 'part of the "west end" of London maritimized'.[4] Finally, by 1851, a number of seaside resorts had gained a permanent residential population; the invalid and the elderly were amongst the many who had recently migrated to the English coastal towns.

Although A. B. Granville was able to identify 70 spas in England associated with the taking of mineral waters he had to admit, in 1841, that many were 'growing out of fashion' and some of those most in repute at the beginning of the century were 'nearly forgotten'.[5] The map he compiled of these places must also be looked at with a sceptical eye for it included 'spas' such as Shap Wells in Westmorland and Houghton-le-Spring in County Durham which were little more than villages, and towns such as Clitheroe and Thetford, where the springs were a very subsidiary activity.[6] The 1851 Census, in its recognition of four principal watering places – Bath, Leamington, Cheltenham and Tunbridge Wells[7] – gave a more realistic assessment of the contribution of the inland spa to urban life, but it is helpful, in addition, to note the old-established centres of Harrogate and Buxton, which were to acquire a new lease of life with the arrival of the railways. Indeed, the major inland spas, although they were being rapidly overtaken by their coastal competitors, were by no means dead in 1850. Bath and Cheltenham were still the second and third largest of all English resorts and the rate of increase of the major spas was still above the national average for all towns.

The landscape of the spa town was largely a response to medical and social fashion. The focus of its activities were the springs, thermal or chalybeate, and their locations had guided the layouts of their towns. In

[1] C. H. Mate and C. Riddle, *Bournemouth, 1810–1910* (Bournemouth, 1910), 64 *et seq.*
[2] A. B. Granville, I, 176–7.   [3] E. W. Gilbert (1954), 18.
[4] A. B. Granville, II, 565.   [5] A. B. Granville, I, xxxv.
[6] The places recognised by Granville are mapped in J. A. Patmore, 'The spa towns of Britain', in R. P. Beckinsale and J. M. Houston (eds.), *Urbanization and its problems* (Oxford, 1968), 48.
[7] *Census for 1851: Population Tables, I*, vol. 1, p. xlix (P.P. 1852–3, lxxxv).

Harrogate, for example, there were two distinct settlements, High and Low Harrogate, centred on different groups of springs; but at Cheltenham, where the springs were scattered over a considerable area, no less than seven separate spas had left a legacy in the fabric of the town. As well as taking the waters in pump rooms, in saline baths or in 'hydros', visitors were caught up in a social round, manifest in formal parades, in assembly rooms, in large hotels and in public gardens. By 1850, the developments at Leamington provided a coherent expression of these needs. The new town, laid out in 1808, was in its heyday; its spacious streets and dignified squares had matured; and its newly created gardens on the banks of the river Leam were also patronised by a leisured class of visitor. In a different architectural idiom, it had much in common with Bath, the great eighteenth-century spa; both clung to a concept of urban elegance which was fast disappearing from the English scene.[1]

London

Two outstanding characteristics were often emphasised by eyewitnesses of London's geography in 1850. In the first place, there was the simple fact of its size: 'this huge magnitude which drives every other feeling out of mind'[2] was the reaction of one American. Indeed, it was a popular diversion to rise above the great metropolis in a balloon,[3] and look down on the growing city. It filled the Victorian Englishman with pride, albeit occasionally tinged with anxiety, for hardly a day passed without some 'new street takes the place of the green field'.[4] The metropolitan area as defined by the Metropolis Management Act of 1855 (the same, incidentally, as the county of London to be created in 1888) covered some 117 square miles and included a population of nearly 2.4 million. Beyond lay the limits of the Metropolitan Police District up to a radius of 15 miles or so from Charing Cross (Fig. 141); the extent of this 'Greater London' covered some 693 square miles and included a total population of 2.7 million (see table on p. 673). Ont he north bank, the three components of historic London, the City, the West End and the East End, now formed a solid mass of houses for over six miles from east to west; and extensive development on the south bank had given the capital a breadth of four miles from north to south. Beyond, its suburban arms extended, octopus-

[1] J. A. Patmore (1968), 62–5.
[2] W. Ware, *Sketches of European capitals* (Boston, Mass., 1851), 252.
[3] J. H. Banks & Co., *A balloon view of London*: folding map (London, 1851).
[4] J. Weale, 59.

like, creeping into the countryside along main roads and railways (Fig. 140). Places such as Kilburn, Tottenham, West Ham, Lewisham and Balham had now been grafted on to the body of London, as improved public transport – including omnibus, steamboat and train – made a quicker journey to work possible for the growing tide of daily commuters.[1] In the spaces between the ribbon development, detached suburbs, such as Finchley, Norwood and Plaistow (separated by shrinking patches of park and common, woodland and marsh, pasture fields and brick-fields) were only a stone's throw from the new urban boundary.

In the second place, the internal complexity of London was the subject of frequent comment. A bewildered German visitor had decided that 'no other town presents so strong a contrast between its various quarters'.[2] Indeed, the metropolis was a loose aggregate of contrasting settlements. According to one contemporary, by 1850 it had 'swallowed the episcopal city of Westminster, the boroughs of Southwark and Greenwich, the towns of Woolwich, Deptford and Wandsworth, the watering places of Hampstead, Highgate, Islington, Acton, Kilburn,...the fishing town of Barking,' and many 'once secluded and ancient villages'.[3] These older nuclei of the conurbation still administered their own affairs; Greater London, outside the City, was governed by no fewer than 300 bodies, many of parish status.[4] Muddle, mismanagement and local self-interest prevailed; it was 'a province covered with houses'...and so great...its area is so large that each inhabitant is in general acquainted only with his own quarter'.[5]

In so far as London was made up of distinctive communities, each could form the subject of a separate description. But in its diversity London also mirrored the rest of England. In the narrow sense of urban architecture, for example, contemporaries regarded it as a truism that most of the 'distinctive peculiarities' of the provincial towns were present 'in one unnoticed corner or other of the vast metropolis'.[6] But in a broader sense the essence of early Victorian England was distilled in its capital. The

[1] T. C. Barker and M. Robbins, *A history of London Transport: passenger travel and the development of the metropolis* (London, 1964), I, 25–68; J. R. Kellett, 365–71.

[2] M. Schlesinger, *Saunterings in and about London*, English ed. by Otto Wenekstern (London, 1853), 13.

[3] J. Weale, 60.

[4] G. M. Young (ed.), I, 204.

[5] Quoted in A. Briggs, *Victorian cities* (London, 1963), 332–3.

[6] J. Weale, 450.

remodelling of parts of central London from its foundations symbolised the vitality of the mid-century scene; new, grander buildings for government, finance, religion, clubs, City gilds, museums and art galleries were daily altering the skyline. The City, deserted by resident merchants and already beginning to empty of poorer-class population, was the commercial hub of the manufacturing nation – although still set in narrow streets and cramped counting houses. The port of London, a rich 'commercial Aladdin's cave', its docks growing quickly down-river, was England's window to the world, a direct manifestation of expanding imperial and international status. On the other hand, the preservation of the royal parks in London suggested rather a national nostalgia for the older, the more elegant and more rural way of life which Lavergne noted in 1855 – even in the city environment.[1] And in suburbia too, by 1850, the urgent need 'to snatch a clear piece of country from the general fate, and to provide a belt of pure air'[2] around London was recognised. In the final analysis, however, London best epitomised the character and problems of England in this age, by bringing wealth and poverty so sharply together. At one extreme, in the vicinity of the parks, amidst 'a velvety luscious green' stood houses 'like palaces with stone terraces and verandas',[3] but at the other were the squalid rookeries of inner London,[4] the Agar Town so vividly described by Dickens;[5] and the widespread mediocrity of urban environment condemned by Ruskin: 'the pitiful concretions of line and clay...about our capital'.[6] Indeed, the naked facts of London's health problem, marshalled in a string of parliamentary reports,[7] were startling enough: such facts as the 80,000 houses (inhabited by 640,000 people) unsupplied with water, or, the subsoil sodden with decaying sewage, were soon to motivate reform. Meanwhile, the London of 1850, like the nation at large, was testimony to aspiration and indifference, to innovation and conservatism, to solid achievement and abysmal failure – it was a vivid mosaic of black and white. Such were the half-secure foundations from which the geographical changes of the rest of the Victorian age had to spring.

[1] L. de Lavergne, 133.   [2] J. Weale, 465.   [3] M. Schlesinger, 13–14.
[4] G. M. Young (ed.), 1, 175.   [5] *Household Words* (1851).   [6] J. Ruskin, 165.
[7] G. M. Young (ed.), 1, 201–2, lists these; see also A. S. Wohl, 'The housing of the working classes in London, 1815–1914' in S. D. Chapman (ed.), 15–54.

Chapter 11

THE CHANGING FACE OF ENGLAND:
1850–*circa* 1900

J. T. COPPOCK

It was not so much 1900 as 1914 that marked 'the real economic terminus of the nineteenth century'.[1] Over the span of years since 1850 no revolution transformed the face of England as the enclosure movement and the industrial revolution had done during the preceding hundred years; rather there were changes of degree and of emphasis. Nevertheless, they were on a considerable scale: the population doubled, despite large-scale emigration; living standards almost doubled, generating new demands; industrial production quadrupled and became more diversified; and the number of people living in towns trebled. England was becoming increasingly an urban and industrial society.

If an octogenarian alive in 1911 had reflected upon the changes in the landscape of his youth he would have noticed chiefly changes of scale and degree. In the towns he would recall considerable improvements in the quality of the urban environment; streets had been lit, paved and drained; the centres of cities had been extensively rebuilt; and the quality of the new housing had been improved even if in an uninspiring fashion. Schools, churches, public buildings and public parks were all more numerous, although, around the city centres and in stagnating towns, slums remained. Above all, he would have been struck by changes in the scale of urbanisation, by the vast expanses of slate roofs which met his eye. The differences in the countryside would not have been so obvious to him; yet a less prosperous and less well-tended air, a greener landscape, a spatter of derelict fields, occasionally an abandoned farmstead or cottage, some new orchards and plantations would all have indicated quite fundamental changes.

[1] W. Smith, *An economic geography of Great Britain* (2nd ed, London, 1953), 63.

POPULATION

Perhaps the most widespread of all the changes of the age was the re-distribution of population. According to the decennial censuses, the population of England and Wales rose from 17.9 million in 1851 to 36.1 million in 1911, but the number living in rural districts fell from 8.9 million to 7.9 million, and their share of population from about 50% to 22%; at the same time, the urban population, as defined in the census reports, rose by 19.3 million, or 218%, to reach 28.1 million in 1911.[1] These figures underestimate the scale of both rural depopulation and urban growth, for the failure of administrative boundaries to keep pace with changes in the distribution of population resulted in the inclusion of urban populations within so-called rural districts. From a careful examination of the census data, C. M. Law has calculated that the true urban population was underestimated by almost 700,000 in 1851 (and the rural population correspondingly overestimated) and by more than 2,300,000 in 1911.[2]

Rural population. This redistribution was largely due to widespread migration from the countryside, encouraged by improved communications, better opportunities for employment elsewhere, and the changing structure of agriculture.[3] Fig. 124 shows the net changes in population density, although this too minimises the extent of areas where population was decreasing; for a rising population in many enumeration districts merely indicates that gains in the towns within those districts exceeded losses in their rural areas.

This loss of population was neither uniform nor continuous; in some districts numbers were already falling by the 1850s and continued to do so in each succeeding decade. More generally, the downward trend began in the 1860s or 1870s and was sometimes reversed by suburban development or by the construction of a new railway line. In south-east England and in the industrial districts of the Midlands and northern England, numbers increased in each decade, although the population of the remoter parishes often declined.

Where the population in rural districts grew, this was due primarily to the expansion of towns beyond their administrative boundaries, and to the

[1] C. M. Law, 'The growth of urban population in England and Wales', *Trans. and Papers, Inst. Brit. Geog.*, XLI (1967), 126.

[2] *Ibid.*, 126, 130.

[3] R. Lawton, 'Rural depopulation in nineteenth century England' in R. W. Steel and R. Lawton (eds.), *Liverpool essays in geography* (Liverpool, 1967), 247–55.

development of dormitory settlements. In Sussex, increasing numbers of those working in the towns chose to live in rural areas; and in Cheshire cottages were tenanted by workers in neighbouring industries.[1] Mining, too, greatly affected numbers in some rural areas. The production of coal was expanding throughout the period, and new pits were being sunk, often among farmland, and new settlements created, for example, north of the Tyne and in east Durham. Ironstone mining similarly contributed to the growth of population in rural parts of the east Midlands. In contrast, the lead dales of the Pennines, and the tin and copper mining districts of Cornwall, were affected by the falling demands for their products in the face of overseas competition. Lead mining was generally expanding in the 1850s and production reached a peak in the early 1880s; thereafter, both employment and output declined, although the blow was softened by the fact that miners with smallholdings could and did become full-time farmers.[2] Production of copper and tin reached their peaks in the 1850s and 1870s respectively and thereafter their mining communities began to decline.[3]

Migration from rural areas was a near-universal feature of this period, irrespective of whether populations were increasing or declining, although the rate of migration fell at the turn of the century as fertility declined and there were fewer potential migrants.[4] Moreover, it was selective, in that it was predominantly a movement of young people between the ages of 15 and 35, in that more women migrated than men (except from areas where mining was declining), and in that employees were more likely to move than employers.[5] Employment in the countryside was generally less well paid than in the towns; industrial wage rates were some 50% higher than agricultural rates throughout the period, and opportunities for employment in rural areas were not only fewer but were declining.[6] For young

[1] *Report on the Decline in the Agricultural Population of Great Britain, 1881–1906*, 24 (P.P. 1906, xcvi); *R.C. on Labour*, vol. I (England), pt. 4, *Reps., by Mr Roger C. Richards*, 97 (P.P. 1893–4, xxxv).

[2] B. R. Mitchell with P. Deane, *Abstract of British historical statistics* (Cambridge, 1962), 166; J. W. House, *North-eastern England: population movements and the land-scape since the early 19th century* (Newcastle upon Tyne, 1959), 33–4.

[3] B. R. Mitchell with P. Deane, 155, 159; G. R. Lewis in W. Page (ed.), *V.C.H., Cornwall*, I (1906), 563–70.

[4] A. K. Cairncross, *Home and foreign investment 1870–1913* (Cambridge, 1953), ch. 4.

[5] J. Saville, *Rural depopulation in England and Wales 1851–1951* (London, 1957), chs. 2 and 3; Lord Eversley, 'The decline in number of agricultural labourers in Great Britain', *Jour. Roy. Stat. Soc.*, LXX (1907), 275.

[6] J. R. Bellerby, 'Distribution of farm income in the United Kingdom 1867–1938', *Jour. of Proceedings, Agricultural Economics Soc.*, X (1953), 135; J. Saville, 20–30.

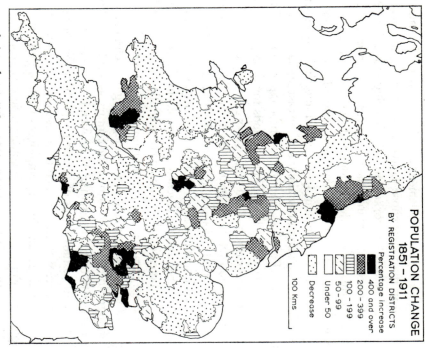

POPULATION CHANGE
1851–1911
BY REGISTRATION DISTRICTS
Percentage increase

400 and over
200–399
100–199
50–99
Under 50
Decrease

100 Kms

Fig. 124 Population change, 1851–1911 Based on R. Lawton in J. W. Watson and J. B. Sissons (eds.), *The British Isles: a systematic geography* (London and Edinburgh, 1964), 232.

women domestic service often provided an escape, and young men were drawn to the mines, the railways, the factories and to the increasing range of other occupations, such as the police force; some emigrated overseas.[1] Change of occupation did not necessarily mean migration, for labourers were attracted to rural industries such as the Peterborough brickworks and to local employment on the railways; but frequently it did. E. G. Ravenstein showed that migrations often took place over short distances and suggested a wave-like sequence of movements; but, while his general thesis is probably correct, there were certainly migrations over long distances.[2]

[1] *Ibid.*, 31–2.
[2] E. G. Ravenstein, 'The laws of migration,' *Jour. Roy. Stat. Soc.*, XLVIII (1885),

Agriculture remained the principal occupation in rural areas, although its share of male employment fell from a quarter in 1851 to a tenth in 1911, and migration was largely confined to agricultural labourers; for despite agricultural depression, numbers of farmers remained remarkably constant.[1] Owing to ambiguities in the census, and uncertainty about the number of farm workers who described themselves as general labourers, it is not possible to say accurately how many labourers moved out of agriculture, although there is no doubt that there was a considerable migration. One careful analysis has suggested a decline of nearly half a million in the number of adult male agricultural labourers in England and Wales between 1861 and 1911, but there was also a reduction in casual labour and a decline in the importance of harvest gangs.[2] To reinforce the attractions of the towns, fewer labourers were required in agriculture; mechanisation of harvesting and haymaking, almost universal by 1906, widespread laying of land to grass, and economy on farm maintenance – all reduced the demand for labour except in areas such as the vale of Evesham where intensive systems of farming were being more widely adopted.[3] There seems to be no clear relationship between agricultural prosperity or type of farming and the scale of rural migration which appears to have been remarkably uniform, affecting both pastoral and corn-growing areas and occurring in times of both agricultural prosperity and depression.[4]

It is not possible to determine the relative importance of the attraction of towns and the falling demands for labour in the countryside. Supply and demand seemed to have kept roughly in balance, although some labourers complained that farmers were deliberately using less labour than the land required.[5] The Assistant Commissioners who reported on agricultural labour to the Royal Commission on Labour in the early 1890s found the supply of labour adequate in 27 of the 38 districts they visited, although there were local difficulties.[6] What is certain is that without migration

198–9; H. C. Darby, 'The movement of population to and from Cambridgeshire between 1851 and 1861', *Geog. Jour.*, CI (1943), 118–25.

[1] 'The agricultural population', *Jour. Board of Agric.* (1904–5), 274; A. L. Bowley, 'Rural population in England and Wales: a study of the changes of density, occupation and ages', *Jour. Roy. Stat. Soc.*, LXXVII (1914), 610; Lord Eversley (1907), 275.

[2] F. D. W. Taylor, 'United Kingdom: numbers in agriculture', *Farm Economics*, VIII, No. 4 (1955), 39; R. Lawton, 249.

[3] *Rep. on Decline in Agric. Pop., 1881–1906*, 14 (P.P. 1906, xcvi).

[4] A. L. Bowley, 616.

[5] *R.C. on Labour*, vol. I (England), pt. 2. *Reps. by Mr Cecil M. Chapman*, 53 (P.P. 1893–4, xxxv).

[6] Lord Eversley (1907), 282.

there would have been widespread unemployment in rural areas. Two other changes contributed to a falling demand for labour in the country-side; the displacement of local manufactures by factory-made products and the decline of the country craftsman. In many parts of the country the incomes of farm labourers had been supplemented by the earnings of their families in industries such as straw plaiting and lace manufacture, as well as by casual employment in agriculture. The extension of factory employ-ment and the importation of cheaper goods led to a falling demand for many of these products, and the introduction of compulsory schooling and the migration of young people reduced the potential labour force. Lace manufacture in Buckinghamshire was almost extinct by 1884 and was in decline in Bedfordshire, and straw plaiting in Essex had ceased to provide employment by 1901.[1] In the Luton area, the employment of outworkers making hats with imported plait in place of straw plaiting survived much longer, but generally such industries were in decline, although some were increasingly factory industries, such as boot and shoe manufacture in Northamptonshire and hosiery manufacture in Leicestershire.[2]

Development of a national market for factory-made goods, and the increasing ease and cheapness of travel to large cities, also affected the demand for the services of local craftsmen; thus, the number of tailors in Rutland fell from 173 in 1851 to 63 in 1911 and that of shoemakers from 236 to 138.[3] Fewer people needed fewer services, and economic and tech-nological changes, such as the disappearance of the stage coach, further diminished the demand for local products and local craftsmen; these losses were not offset by new opportunities for employment in rural areas as policemen, postmen, school teachers and the like.

Population changes also affected rural housing. Modification of the Poor Law in 1863 had encouraged the building of cottages, and by the 1880s the standard of rural housing was said to be much improved.[4] Nor did building cease with the onset of depression, for new cottages were

[1] C. Jamison in W. Page (ed.), *V.C.H. Buckinghamshire*, II (1908), 107; A. Ransom in W. Page (ed.), *V.C.H. Bedfordshire*, II (1908), 123–4; M. Christy in W. Page and J. H. Round (eds.), *V.C.H. Essex*, II (1907), 377.

[2] *V.C.H. Bedfordshire*, II, 107, 121; J. H. Clapham, *An economic history of modern Britain: machines and national rivalries, 1887–1914* (Cambridge, 1938), 179–82.

[3] J. Saville, 74.

[4] J. H. Clapham, *An economic history of modern Britain: free trade and steel, 1850–1886* (Cambridge, 1932), 507–8; R.C. Agric. Depression, Minutes, Evidence of Read, Q. 16,189 (P.P. 1894, xvi, pt. 2).

necessary to retain tenants and labourers. Conditions seemed to have varied with the material used, the severity of depression, and the size of estate. Cottages were said to be best on the estates of large landowners; thus, in the Thakeham Poor Law Union in Sussex there were excellent cottages built in recent years on the estates of the Duke of Norfolk and Lord Leconfield.[1] But in other villages where many of the cottages had been built and occupied by their owners, in areas which were severely hit by depression, and in those where wattle and daub were the usual materials, conditions were often wretched; and on the Cambridge–Sandy road, for example, cottages were rotting away.[2] Depopulation may also have eased the pressure on accommodation, although in some areas surplus cottages were demolished, as on the Bedford estate, where three cottages were said to have been pulled down for each one erected.[3]

Urban population. In 1851 every other Englishman was a townsman, by residence if not by birthright; by 1911 the proportion had increased to four out of every five in a population which had doubled in size.[4] These changes in the urban population were closely associated with the changing urban geography, and interpretation is complicated by the variations in the census returns (four different definitions of urban areas being used between 1851 and 1911), and by the time-lag in adjusting administrative boundaries to keep pace with urban expansion. The rate of urban growth was not uniform throughout this period, for it was only to be expected that it would decline after a century of rapid urban expansion. The urban population increased by more than 20% in each decade between 1801 and 1881, but although the population classified as urban continued to rise, the rate of growth fell to 18.8% in 1881–91, to 17.5% in 1891–1901, and to 12.2% in 1901–11.[5]

Nor was the rate of urban growth uniform throughout the country, although the patterns of distribution did not change greatly during this period (see pp. 655–7 below). In the main, the large towns in 1911 were those that had been large in 1851, but the contrast between large and small

[1] *R.C. on Labour*, vol. 1 (England), pt. 5, *Rep. by Mr Aubrey J. Spencer*, 15 (P.P. 1893–4, xxxv); *Ibid, Rep. by Mr William E. Bear*, 59 (P.P. 1893–4, xxxv).

[2] *R.C. on Labour*, vol. 1 (England), pt. 3, *Rep. by Arthur Wilson Fox*, 16–17 (P.P. 1893–4, xxxv); H. R. Haggard, *Rural England*, 2 vols. (London, 1902), II, 59.

[3] *R.C. on Labour*, vol. 1 (England), pt. 1, *Rep. by Mr William E. Bear*, 21 (P.P. 1893–4, xxxv).

[4] C. M. Law, 126.

[5] *Ibid*, 130.

towns tended to increase during these decades; for while the large towns continued to grow, the population of many small towns in rural areas changed little or even declined, especially where they were not served by a railway or were remote from the main centres of population. C. M. Law's adjustments of the census data to correspond with built-up areas show that the number of towns with populations of 50,000 or more rose from 25 to 79 between 1851 and 1911, and their share of the urban population from 56.5 to 65.6%, while the number of towns with populations between 2,500 and 20,000 rose only from 457 to 764 and their share fell from 30.3 to 21.2% (though this decline was relative and their total population also increased).[1]

THE COUNTRYSIDE

Although the contribution of agriculture and forestry to the national income fell from a fifth in 1851 to barely a twentieth in 1911, changes in the countryside were certainly the most widespread of all the changes in the geography of England in the second half of the nineteenth century.[2] In part they were due to the growth of towns and industry, which both attracted the agricultural labourer and provided new and expanding markets for the farmer; but they were largely a consequence of improvements in international transport and falling freight rates which brought English farmers increasingly into competition with overseas producers farming better land or under better climates. The volume of wheat imported trebled between 1850 and 1875 and doubled again between 1875 and 1914; imports of maize and other feeding stuffs also increased sharply, and those of meat and animals, which had been only a seventh by value of cereal imports in 1854, were nearly as large by 1900.[3] Enforced adjustment to this changed situation was the keynote of rural change.

From an agricultural viewpoint, the period can broadly be divided into three: (1) mid-century until the late 1870s, when the impact of rising imports was delayed by a combination of circumstances; (2) the late 1870s to the mid-1890s, when prices, especially of cereals, were falling and sectors of the agricultural economy were acutely depressed; (3) the years from the 1890s to the First World War, when prices made some recovery,

1 *Ibid*, 130, 135, 141.
2 J. D. Chambers and C. E. Mingay, *The agricultural revolution 1750–1880* (London, 1966), 210.
3 B. R. Mitchell with P. Deane, 298–300.

although many of the trends established in the previous twenty years continued. For this analysis, therefore, the last two periods will be treated together.

Reclamation and improvement, 1851–circa 1875

In many ways the 1850s and 1860s formed an Indian summer in English arable farming, and are sometimes spoken of as the era of high farming. In these two decades the physical resources of English agriculture were raised to their highest level, especially through land drainage and, to a lesser extent, through the re-equipment and re-organisation of farms. Between 1847 and 1870, some twelve million pounds was advanced under various statutory schemes of land improvement, three-quarters of it for drainage; and James Caird estimated that private landowners provided a much larger sum.[1] The Duke of Northumberland alone spent £992,000 on his estates between 1847 and 1878, a sum raised entirely out of estate revenue; and other great landowners invested large sums, although the movement was by no means universal.[2]

Reclamation in the sense of bringing into agricultural use land formerly not used for farming must have played only a limited part. Fen and bog still made their contribution, although on a small scale. Whittlesey Mere was successfully drained in 1851 and crops were growing in 1853 where there had been water only two years before.[3] Very little mossland remained in Lancashire by 1880, and improvements ranged from the large-scale reclamation of Chat Moss to that of the farmer at St Michael's on Wyre, who reclaimed 30 acres of moss on his 160-acre holding in the course of his tenancy.[4] Other reclamation was taking place along the coast; on the Norfolk side of the Wash, for example, various additions to the cultivated area were made between 1858 and 1882.[5] According to the Ordnance Survey, 35,444 acres had been added to the total land area of England and Wales between successive surveys, covering some thirty-five years, and 4,692 acres had been lost by coastal erosion; but there was no record of the

---

[1] J. Caird, *The landed interest and the supply of food* (London, 1878), 83, 87.

[2] F. M. L. Thompson, *English landed society in the 19th century* (London, 1963), 250.

[3] W. Wells, 'The drainage of Whittlesea Mere', *Jour. Roy. Agric. Soc.*, XXI (1860), 141.

[4] *R.C. on Agriculture, Reps. of Assistant Commissioners: Mr Coleman's reports*, 30, 33, 40 (P.P. 1882, xv).

[5] J. E. G. Mosby, *Norfolk*, in L. D. Stamp (ed.), 'The land of Britain', pt. 70 (London, 1938), 237.

acreage brought into agricultural use.[1] Still more agricultural land was created through the clearing of woodland. Some of the most extensive areas reclaimed in this way were a product of disafforestation. The best known example is Hainault Forest, disafforested in 1851, where large areas, including the King's Wood, were grubbed up and converted into farmland to yield the Crown a revenue eight times that from the woodland it replaced.[2] Woods in other counties were similarly cleared to provide new farm land; thus 1,970 acres of Wychwood Forest in Oxfordshire were cleared in 1856–8, and converted into seven new farms.[3]

More usually, reclamation meant the improvement of downland, heath, moor or other rough land already grazed by livestock. It is clear that extensive areas were being improved in this way, although it is debatable whether these gains offset the losses to towns and railways. On Salisbury Plain, a large area of maiden down was said to have been broken up, much of it in the 1860s, to result in the 'bakelands' whose cultivation was to prove so difficult when seasons were bad and prices low.[4] In Sussex, large tracts of the high chalk downs were ploughed up, and in the Breckland much land was also brought into cultivation in the 1850s.[5] In the uplands, too, moorland was being improved, and on Exmoor the years from 1851 to 1866 were a period of active reclamation as rough pasture was converted to permanent grass.[6] Moorland in Yorkshire was being reclaimed at £16 per acre, but reclamation did not necessarily mean ploughing.[7]

Reclamation of rough land often accompanied the enclosure of commons and manorial waste, although the two processes were not synonymous; while some enclosures led to arable cultivation or to improved grazing, others involved little more than a re-organisation of existing uses, and in others land was withdrawn from agriculture to be planted with woods or to provide sites for houses. The great bulk of arable enclosures

[1] R.C. on Coast Erosion, Reclamation of Tidal Lands and Afforestation in the U.K., Final Report, 43 (P.P. 1911, xiv).

[2] Lord Eversley, Commons, forests and footpaths (rev. ed., London, 1910), 83.

[3] C. Belcher, 'On the reclaiming of waste land as instanced in Wichwood Forest', J.R.A.S., XXIV (1863), 274.

[4] R.C. Agric. Depression, Rep. by Mr R. Henry Rew on the Salisbury Plain District of Wiltshire, 13 (P.P. 1895, xvi).

[5] H. C. K. Henderson and W. H. Briault, Sussex (L. of B., pts. 83–4, London, 1942), 502; R.C. Agric. Depression, Rep. by Mr R. Henry Rew on the County of Norfolk, 11 (P.P. 1895, xvii).

[6] C. S. Orwin, The reclamation of Exmoor Forest (Oxford, 1929), 83.

[7] R.C. on Agriculture, Reps. by Assistant Commissioners, 228 (P.P. 1881, xvi).

had been completed by 1850, but there were still probably between two and three million acres of unenclosed common land remaining in England and Wales, of which less than 250,000 acres were common arable.[1] Some 618,000 acres were enclosed between 1852 and 1888 under the General Enclosure Acts, and small additions, totalling 30,751 acres were made between 1878 and 1914 under the Commons Act of 1876.[2] An unknown acreage was also enclosed privately. The acreage of awards approved by the Inclosure Commissioners averaged some 45,000 acres a year between 1852 and 1861, but only some 33,000 acres between 1862 and 1871. From 1868 onwards the Commissioners deliberately discouraged further proposals in expectation of new legislation, and after 1876 it became increasingly difficult to enclose.[3]

Without detailed examination of the enclosure awards it is not possible to say how much of the land enclosed was arable, how much meadow and how much waste, although the location and size of the enclosures and the description in the Commissioners' reports give some indication. According to Gilbert Slater, awards involving the enclosure of common arable in England after 1845 totalled 139,517 acres, or about a fifth of all land so enclosed.[4] The acreages enclosed under individual awards varied widely, ranging from 8 acres in Bromsberrow (Gloucestershire) to 15,200 acres in Dent (West Riding), and in general enclosures were larger in predominantly upland counties; for example, the average in Cumberland was 1,322 acres, compared with 120 acres in Kent.[5] Enclosures were widespread, but there were large concentrations in Devon, in the central Pennines, in Surrey and in Norfolk. Enclosures of common fields were largely confined, as they had been earlier, to the counties extending across the Midlands from the East Riding to Dorset, while enclosures of waste were numerous in the uplands and on some of the poorer soils of the lowlands.

The precise effects of such enclosures must await further investigation. In many instances in the lowlands, as at Totternhoe (Bedfordshire), enclosed in 1892, regular fields bounded by hedges or fences appeared, but in some parishes, particularly those on lights soils or where enclosure came late and ownership had been concentrated into a few hands, there was

[1] *Inclosure Commissioners: Thirtieth Ann. Rep.*, 3 (P.P. 1875, xx).

[2] Calculated from the annual reports of the Inclosure Commissioners.

[3] *Ann. Rep. of proceedings under the Tithe etc. Acts, 1914, 24–5* (P.P. 1914–16, v).

[4] G. Slater, *The English peasantry and the enclosure of common fields* (London, 1907), 191.

[5] *H. of C. Papers*, Nos. 359, 360 and 363 of 1906 (P.P. 1906, xcviii).

little outward evidence of change and the landscape remained open.[1] In the uplands the pattern was more variable. Enclosure might mean new roads, new farmsteads and rectilinear fields, as at High Bishopside (West Riding), or merely the subdivision of moorland into large enclosures separated by straight fences, often of post and wire, as on the moors on either side of Dentdale in the same county.[2]

By far the most important improvement to the land itself during this period was the extension of sub-soil drainage. James Caird, in evidence before the Lords' Committee on Land Improvement in 1873, had said that about two million acres had probably been drained; while another witness estimated the figure at three million acres out of twenty million requiring drainage.[3] Whatever the truth, the area drained was large.

Other improvements were also taking place. New farmsteads and cottages were being built, fields enlarged and field boundaries straightened. Wire fencing was almost unknown in the 1860s, but it had become common by the 1880s; sometimes it was barbed. There was said to be a vigorous campaign against unnecessary hedgerows, and on one farm at Cricklade in Wiltshire five miles of hedges were removed and thirty-six fields reduced to nine.[4] But none of these changes appears to have been on a sufficient scale to transform any extensive area of the countryside.

Although there were annual fluctuations, the 1850s and 1860s were generally years of prosperity, when rents were being increased to pay for improvements, when the benefits of mechanisation were beginning to make themselves felt, and when prices were generally more favourable than they had been since the Napoleonic Wars. The steam plough had been introduced in the 1850s, though only some 200,000 acres were thought to be cultivated by it in 1867; and an improved reaper-binder had appeared in 1850, so that by the 1860s most of the bigger farmers at least had their hay and corn cut by machine.[5] In many localities the railway was beginning to affect the pattern of agricultural activity as the droving of livestock for fattening disappeared, as crops and livestock products began to move to market by rail, and as it became possible for the fertility of

---

[1] Evidence from successive editions of Ordnance Survey 6" maps.

[2] Ibid.

[3] Rep. S.C. of the House of Lords on the Improvement of Land, Report, iii; ibid., Minutes, Evidence of Caird, Q. 4,125–6; Evidence of Denton, Q. 830 (P.P. 1873, xvi).

[4] J. H. Clapham (1932), 267, 503; R. Molland in E. Crittall (ed.), V.C.H. Wiltshire, IV (1959), 86.

[5] J. H. Clapham (1932), 268; see, for example, H. Evershed, 'Agriculture of Hertfordshire', J.R.A.S., xxv (1864), 299.

light soils to be improved by large quantities of manure which were transported 50 miles or more from the great cities, especially from London. Transport was thus coming to play a much more important role in agriculture, both by permitting agricultural produce to move more freely within the country and also, by abolishing the protection against overseas competition which distance had formerly provided.

Agricultural changes, 1875 – circa 1914

Although in mid-century James Caird urged that livestock should play a more important role, and although livestock products probably did increase in importance, many farmers and landowners retained an almost mystical attachment to wheat.[1] Although the proportion of foreign wheat in the nation's bread supply rose steadily from a quarter in 1850 to three-quarters in 1894, and although the price of wheat was falling, it was not until a succession of bad harvests in the late 1870s that the changed situation was made dramatically clear, even though unfavourable seasons distracted attention from the need for fundamental adjustment in the farming systems themselves.[2] A Royal Commission sat during 1879–82 to investigate the situation and, when cereal prices continued to fall and then those of livestock and livestock products, a second Commission was appointed in 1893. The reports and evidence of these Commissions provide a mine of information, but they are often coloured by the assumption that depression was universal.

At first, the burden of lower prices and poor harvests was largely absorbed by landowners who granted remissions of rent and then permanent reductions from the high values of 1860s and 1870s; but, as prices continued to fall, more fundamental adjustments had to be made. The report of the Assistant Commissioners made it clear that the situation was worst on heavy land or on poor light land in the eastern part of the country, where crops were less reliable, alternative systems of farming less easily adopted, and cultivation difficult, at least on heavy land. But in western and northern counties, where wheat occupied a minor place in farming systems, in areas easily accessible by rail from large cities or with local markets, as in the colliery villages of the Durham coalfield, depressed conditions either did not exist or were much less severe, although rents

[1] J. Caird, *English agriculture in 1850–51* (London, 1852), 476, 480–9.

[2] J. H. Clapham (1932), 3; C. S. Orwin and E. H. Whetham, *History of British agriculture 1846–1914* (London, 1964), 259.

were usually reduced, if on a smaller scale. The crisis affected above all the arable farmer for whom wheat was the lynch-pin of his system. Thus in north Devon there had often been no reductions in rent by the 1890s; in Cumberland, rents fell between 15–25% while in Norfolk the fall, even on the best land, was 25–35% reaching 40–60% on land of medium quality.[1] Yet even within the eastern counties, location could make a critical difference; at Much Hadham, in Hertfordshire, rents in 1901 ranged from 30s. an acre on the best land in a good position to between 7s. 6d. and 10s. on land more than three miles from a station.[2]

There can be little doubt that depressed conditions did exist widely in the periods 1879–82 and 1893–5, and that depression was most acute in 'the corn growing counties', although it is often difficult to separate the special and temporary difficulties presented by bad seasons from the long-term trends. Comparison of returns of the gross annual value of land in 1879–80 and in 1894–5 show a marked contrast between eastern counties, where reductions reached 39.9% in Suffolk and 39.7% in Essex, and northern and western counties such as Cheshire and Cornwall, where the reductions were only 6.4 and 6.6% respectively.[3] There were numerous reports of land falling out of cultivation in both periods, and special returns were made in 1884 and 1887 of abandoned farms and abandoned land, although their usefulness is impaired by uncertainties among farmers about their meaning. In the Breckland, fields did tumble down to grass and were soon overrun by heather; and in Essex and Cambridgeshire fields had been invaded by thorn, briar and bramble;[4] but more commonly land 'out of cultivation', was arable land which had become covered with coarse self-sown grass, and Pringle's 'terrible map dotted thick with black patches' of south-east Essex showed such land which had either been put down as temporary pasture or had sown itself (Fig. 125).[5] Some land may have been let go simply to avoid payment of rates and tithes, and such

[1] R.C. on Agric. Depression. Rep. by Mr R. Henry Rew on North Devon, 12 (P.P. 1895, xvi); ibid, Rep. by Mr Wilson Fox on the County of Cumberland, 25 (P.P. 1895, xvii); ibid, Rep. by Mr R. Henry Rew on the County of Norfolk, 19 (P.P. 1895, xvii).

[2] H. R. Haggard, 1, 535.

[3] R.C. Agric. Depression, Statement showing the Decrease and Increase in the Rateable Value of Lands, 1870–1894, 10 (P.P. 1897, XV).

[4] R.C. Agric. Depression, Minutes, Evidence of Read, Q. 16,042 (P.P. 1894, xvi, pt. 2); ibid, Rep. by Mr R. Hunter Pringle on the Ongar ... Districts of Essex, 129 (P.P. 1894, xvi, pt. 1); H. R. Haggard, II, 59.

[5] J. H. Clapham (1938), 79; R.C. Agric. Depression, Rep. by Mr R. Hunter Pringle on the Ongar ... Districts of Essex, 48, 129 (P.P. 1894, xvi, pt. 1).

self-sown pasture was often let at rents of a shilling or two per acre as sheep run, particularly on heavy clay or on very light soil.[1] The Breckland and south-east Essex illustrate the nadir of depression. In the Breckland in the 1890s many farms were unlet and large areas had been let go, with only a few fields around the house still cultivated; in Essex, an enormous area was said to be worthless and covered with coarse herbage.[2] In other areas of light or heavy land in lowland England there were similar difficulties; farms could be had for nothing on the Cotswolds by a tenant willing to pay rates; on Salisbury Plain the tops of the farms had gone out of cultivation, and C. S. Read thought that all the land there that had been reclaimed in his lifetime had gone back.[3] Tenants changed frequently, and many farms were taken in hand either because no tenants could be found or to prevent further damage to the land by those who had lost heart or resources, although it was usually said that tenants could be found for good farms or good land; at one time Lord Wantage had 13,000 acres in hand which he farmed as a single unit.[4] Even where land was not out of cultivation, standards of farming had fallen; for labour costs were the most important item of expenditure and the one which could be most easily reduced. Economy ruled in farm management, and land went out of high cultivation, hedges and ditches were neglected and the appearance of farms was allowed to deteriorate.[5]

As the evidence of rent and land values has shown, the situation was very different in areas which did not depend on arable farming. Few farms were unlet, rent reductions were smaller and, although agriculturalists often complained about depression, there was little evidence of it; Cornwall had not suffered as elsewhere; practically no farms were unlet in

[1] *R.C. Agric. Depression, Rep. by Mr Jabez Turner upon the Frome District of Somerset and the Stratford-on-Avon District of Warwickshire,* 28 (P.P. 1894, xvi, pt. 1); *ibid, Rep. by Mr R. Hunter Pringle on the Counties of Bedford, Huntingdon and Northampton,* 5, 22 (P.P. 1895, xvii).

[2] *R.C. Agric. Depression, Rep. by Mr Wilson Fox on the County of Suffolk,* 31, 51 (P.P. 1895, xvi); *ibid, Minutes, Evidence of Read,* Q. 16,032 (P.P. 1894, xvi, pt. 2); *ibid, Minutes, Evidence of Pringle,* Q. 8,610 (P.P. 1894, xvi, pt. 1).

[3] *R.C. Agric. Depression, Minutes, Evidence of Adams,* Q. 41,950 (P.P. 1894, xvi, pt. 3); *ibid, Rep. by Mr R. Henry Rew on the Salisbury Plain District of Wiltshire,* 15 (P.P. 1895, xvi); *ibid, Minutes, Evidence of Read,* Q. 16,031 (P.P. 1894, xvi, pt. 2).

[4] Lady Wantage, *Lord Wantage, V.C., K.C.B., A memoir* (London, 1907), 377–9.

[5] J. H. Clapham (1938), 83; *R.C. Agric. Depression, Rep. by Mr R. Henry Rew on North Devon,* 14 (P.P. 1895, xvi); *R.C. on Labour,* vol. 1 (England), pt. 2, *Reps. by Mr C. M. Chapman,* 52 (P.P. 1893–4, xxxv).

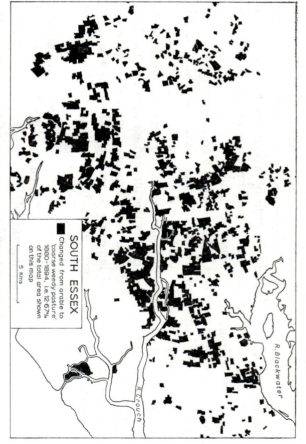

Fig. 125 Abandoned arable land in south Essex, 1894 Based on *R.C. Agric. Depression, Report by Mr R. Hunter Pringle on the Ongar...Districts of Essex*, map (P.P. 1894, xvi, pt. 1).

Devon; there was good competition for farms in Northumberland, and little visible sign of depression in Cumberland.[1]

Neglect, economy, and remission or reduction of rent were spontaneous reactions, but it gradually became clear that fundamental changes in farming systems would have to be made. Such changes had often been anticipated and many were pioneered by migrants from other parts of the country, notably from Scotland and south-west England, who brought with them new systems and new attitudes. To the farmers from Ayrshire, the low rents and ease with which farms could be got made farming in southern England attractive, and agents, anxious for good tenants, advertised vacant farms in Scottish newspapers.[2] The Ayrshire farmer

[1] *R.C. Agric. Depression, Minutes*, Evidence of Collins, Q. 37,162 (P.P. 1894, xvi, pt. 3); *ibid, Rep. by Mr R. Rew on North Devon*, 14 (P.P. 1895, xvi); *ibid, Minutes*, Evidence of Scott, Q. 39,141 (P.P. 1894, xvi, pt. 2); *ibid, Rep. by Mr Wilson Fox on the County of Cumberland*, 22 (P.P. 1895, xvii).
[2] *R.C. Agric. Depression, Rep. by Mr R. Hunter Pringle on the Ongar...Districts of Essex*, 43 (P.P. 1894, xvi, pt. 1); P. Connell, 'Experiences of a Scotsman on the Essex clays', *J.R.A.S.*, 3rd ser, II (1891), 311.

brought a knowledge of dairying (and sometimes even his herd of Ayrshires) and of the cultivation of temporary grass and the potato; one such farmer in Essex had 247 acres of temporary grass and only 71 acres of tillage on a 636-acre farm that fifteen years before had nearly all been cropped.[1] Although he had many lessons to learn, the Scottish farmer often did well, in part at least because he depended largely on family labour and worked hard and lived hard.

Such migrants came in considerable numbers. On one Hertfordshire estate Scots outnumbered English, and so numerous were the newcomers when Rider Haggard made his survey of Hertfordshire that he was led to ask 'But where are the home people?'[2] On the Essex clays in 1894 there were about 120–130 Scots, and one witness claimed that this migration was the chief reason for Essex becoming a dairy county.[3] Other farmers migrated from Devon and Cornwall, bringing with them a knowledge of grassland husbandry and stock farming, and they and the Scots were found throughout southern England.

Conversion to pasture. Perhaps the most widespread agricultural change was the laying down of land to permanent grass, both because such farming was less expensive of labour and because livestock products were less affected by falling prices. An increase in grassland had been advocated in the 1850s and, although land was generally being ploughed up in this period, there is some evidence that arable was being converted to pasture even in the 1860s and 1870s; on the Raby Castle estate in County Durham, for example, 100–150 acres a year had been laid to grass between 1860 and 1880.[4] From the late 1870s, however, arable land was being laid to grass throughout the country. Some of this was land which had tumbled down to grass and become permanent pasture by default; some was sown as a ley and allowed to remain in the hope that conditions would improve. The agricultural returns show an increase in both the acreage and the proportion of arable in temporary grass in the 1880s and 1890s, and a subsequent reduc-

[1] *R.C. Agric. Depression, Rep. by Mr R. Hunter Pringle on the Ongar…Districts o, Essex,* 45 (P.P. 1894, xvi, pt. 1).

[2] *R.C. Agric. Depression, Rep. by Mr Aubrey Spencer on the Vale of Aylesbury and the County of Hertford,* 14 (P.P. 1895, xvi); H. R. Haggard, I, 510.

[3] *R.C. Agric. Depression, Rep. by Mr R. Hunter Pringle on the Ongar…Districts of Essex,* 44 (P.P. 1894, xvi, pt. 1); *ibid, Minutes,* Evidence of Strutt, Q. 13,890 (P.P. 1894, xvi, pt. 1).

[4] F. M. L. Thompson, 255; J. Caird (1852), 480–9; *R.C. on Agriculture, Reps. by Assistant Commissioners,* 226 (P.P. 1881, xvi).

tion in the 1900s as such grass came to be reclassified as the permanent grass it had in fact become. Much of it was therefore of poor quality.

There were both natural and physical obstacles hindering this change. The drier conditions in the eastern parts of the country, where the need to reduce costs was greatest, were thought to make the establishment of good permanent pasture difficult; and it was claimed that up to fourteen years might be necessary for the making of a good sward.[1] Light soils presented particular difficulty and neither on Salisbury Plain nor on Lincoln Heath was it easy to establish good pasture; similar difficulties were also experienced on some heavy land.[2] The high cost of establishing good pasture was also an obstacle to both tenant and owner, although wealthy landowners such as the dukes of Bedford and Northumberland could help by laying down grass for their tenants on a large scale.[3] Another difficulty was the obligation to restore land to arable on the expiry of a tenancy, although this was unlikely to be enforced; for in many areas only farms with a considerable proportion of pasture could be let.

It is impossible to know exactly how much land was converted to permanent pasture. The evidence of the agricultural statistics shows an increase of two million acres in the acreage returned as permanent pasture between 1875 and 1900 and of a further half million by 1914; but the amount of permanent pasture had almost certainly been understated in the statistics by at least a million acres in 1875,[4] and there was a considerable withdrawal of land from agricultural to urban and industrial uses during this period. In view of the confusion between permanent and temporary grass it is probably wisest to examine the decline in the acreage under tillage, which was near-universal, with only Lancashire and Cheshire recording an increase. This anomaly was possibly due to the use of local units of measure in earlier years; for although depression was not serious, land was certainly laid to grass in both counties, and the expansion of towns alone might have been expected to lead to a reduction in the area

[1] J. A. Caird, 'Recent experience in laying down land to grass', *J.R.A.S.*, 2nd ser., xxiv (1888), 148–9.

[2] *R.C. Agric. Depression, Rep. by Mr R. Henry Rew on the Salisbury Plain District of Wiltshire*, 16 (P.P. 1895, xvi); *ibid, Minutes*, Evidence of Epton, Q. 36,138 (P.P. 1894, xvi, pt. 3); *R.C. Agric. Depression, Rep. by Mr R. Hunter Pringle on the Ongar... Districts of Essex*, 62 (P.P. 1894, xvi, pt. 1).

[3] Duke of Bedford, *A great agricultural estate* (London, 1897), 197; F. M. L. Thompson, 313.

[4] R. H. Best and J. T. Coppock, *The changing use of land in Britain* (London, 1962), 73.

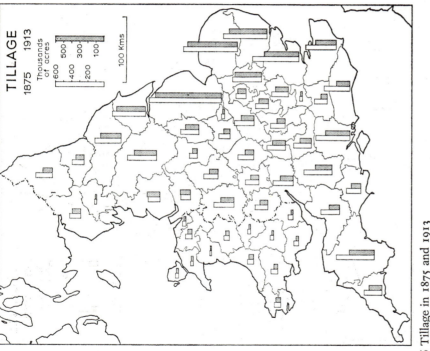

TILLAGE

1875   1913

Thousands
of acres

600 500
400 300
200 100

100 Kms

Fig. 126 Tillage in 1875 and 1913
Based on: (1) *Agricultural Returns, 1875* (P.P. 1875, lxxix); (2) *Agricultural Statistics, 1913* (P.P. 1914, xcviii).

under tillage crops. The relative reduction in the acreage under tillage was greatest in those western counties where arable land had been least important and in those Midland and southern counties where there were large acreages of heavy land (Fig. 126). Within any farm, it was generally the most difficult soils and the least accessible fields which were laid to grass.

On the diminished area of ploughed land there were also changes in cropping. Restrictive covenants governing rotations were usually incorporated in farm leases, but it is difficult to know how far they had been enforced under high farming. Whatever the practice, it seems likely that rotations now became more flexible; for only by allowing considerable

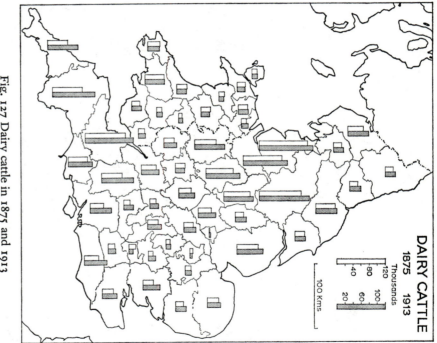

Fig. 127 Dairy cattle in 1875 and 1913
Sources as for Fig. 126.

latitude could landowners retain tenants.[1] The proportion of tillage devoted to wheat fell steadily and its place was largely taken by oats; the change was proportionately greatest in those western and northern counties where little wheat was grown. With the exception of the potato, labour-demanding root crops also became less important, although the mangold gained ground at the expense of the turnip. The growing of potatoes increased on deep friable soils and where communications were good, although competition from both foreign and other domestic producers was felt towards the end of the century. While the replacement of arable by pasture was primarily due to the

[1] J. H. Clapham (1932), 275.

need to reduce costs, it also implied changes in livestock, although numbers of livestock increased relatively more slowly than the acreage of grass, perhaps in consequence of the reduced resources of farmers and the poor quality of much of the new grass. Most of the livestock on arable farms had either been cattle fed in yards to produce manure, or sheep folded on the arable to consolidate and improve the soil; but the former were now unprofitable and the latter expensive of labour. Consequently the rearing and fattening of cattle on grass became more important, and fewer sheep were kept throughout the lowlands, especially in areas of light arable land.

Dairying. The principal change in livestock farming was the expansion of dairying which enjoyed a large measure of natural protection (Fig. 127). In the 1850s the production of milk for sale had largely been confined to stall-fed cows in town dairies and to the immediate periphery of towns; most of the other dairy cattle were kept in the traditional dairying areas of Cheshire, Derbyshire and Somerset, where they provided milk for the making of butter and cheese. By the early 1860s milk was already being sent by rail to London from more distant farms, although the quantity was small and the quality supposedly inferior.[1] The cattle plague of 1865, which reduced the cattle population of the London cow houses from some 24,000 to 14,000 made it immediately necessary to bring supplies from farther afield and the gallonage supplied increased rapidly, much of it by the diversion of milk from butter and cheese production.[2] A growing urban population, an improved railway network, falling costs of transport, and an increasing area of grassland stimulated both the spread of dairying to new areas and the further diversion of milk from cheese and butter making in the traditional dairy districts. In detail, dairying was encouraged by local markets, such as the colliery villages of County Durham, or by good rail communications to large cities; it was hindered by the absence of suitable buildings, by inadequate supplies of drinking water and by the belief of some landowners that dairying harmed the soil.[3]

The most striking changes were those resulting from the need to supply London with milk. Even before the onset of depression the quantity of milk reaching London by rail had risen from 7 million gallons in 1866 to

[1] E. H. Whetham, 'The London milk trade, 1860–1900', *Econ. Hist. Rev*, XVII (1964–5), 370; D. Taylor, 'London's milk supply, 1850–1900: a re-interpretation', *Agric. History*, XLV (1971), 33–8.
[2] F. A. Barnes, 'The evolution of the salient patterns of milk production and distribution in England and Wales', *Trans. and Papers, Inst. Brit. Geog.*, XXV (1958), 179–80.
[3] P. McConnell, 312.

20 million in 1880, and the radius within which supplies were drawn had increased to 100–150 miles.[1] The provision of depots at stations by wholesalers, and the introduction of fast milk trains, facilitated the switch from butter or cheese making to supplying the urban milk market, and permitted farmers in new areas to take up dairying. Farmers in Dorset, north Wiltshire and Berkshire began to deliver milk to the station instead of the factory, and milk churns could be seen at every station between Maidenhead and Faringdon.[2] The catchment area was gradually extended, especially to the west, and by 1892 some 83% of London's milk was thought to come by rail; the maximum regular journey on the Great Western network (which carried more milk than any other line) was 130 miles, but consignments came from much farther afield, especially in cooler weather.[3]

Although London was the biggest single market and probably the greatest single influence on the growth of dairy farming, milk production extended in response to other markets, either by the diversion of milk from butter and cheese making, as with Manchester's supplies from north Cheshire, or by the adoption of dairying as a new enterprise. Dairying developed on the small farms of the industrial Pennines in east Lancashire and around the colliery villages and industrial towns of north-east England.[4] Milk was sent from Leicestershire to Leeds and Newcastle as well as to London; there was even a summer market for milk in the growing holiday resorts, such as Margate and the Lake District.[5]

The Scots and West Country farmers played an important part in encouraging the extension of dairying; so did the railways and, increasingly, the wholesale companies. Although there were many complaints about prices and railway charges, dairying provided many farmers with a welcome

[1] E. H. Whetham, 372.

[2] R.C. Agric. Depression, Rep. by Mr R. Henry Rew on the County of Dorset, 22 (P.P. 1895, xvii); E. H. Whetham, 374–5; P. H. Ditchfield and W. A. Simmons in P. H. Ditchfield and W. Page (eds.), V.C.H. Berkshire, II (1907), 337.

[3] R. H. Rew, 'An inquiry into the statistics of the production and consumption of milk and milk products in Great Britain', Jour. Roy. Stat. Soc., LV (1892), 265; E. A. Pratt, The transition in agriculture (London, 1906), 11–12.

[4] R.C. on Agriculture, Reps. of Assistant Commissioners, Mr Coleman's Reports, 54 (P.P. 1882, xv); T. W. Fletcher, 'The great depression in English agriculture, 1875–1896', Econ. Hist. Rev., XIII (1961), 23; R.C. Agric. Depression, Rep. by Mr R. Hunter Pringle on South Durham etc., 23 (P.P. 1895, xvi).

[5] C. Whitehead, 'A sketch of the agriculture of Kent', J.R.A.S., 3rd ser., x (1899), 451; R.C. Agric. Depression, Rep. by Mr Wilson Fox on the Garstang District of Lancashire etc., 9 (P.P. 1894, xvi, pt. 1); ibid. Minutes, Evidence of Rolleston, Q. 13,315 (P.P. 1894, xvi, pt. 1).

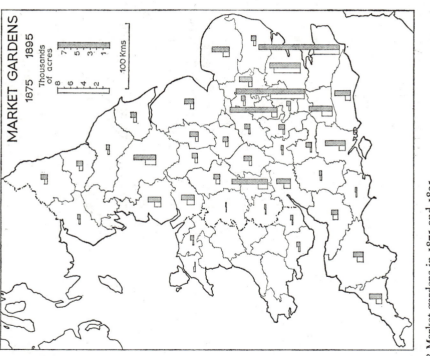

MARKET GARDENS

1875   1895

Thousands
of acres
8
7
6
5
4
3
2
1

100 Kms

Fig. 128 Market gardens in 1875 and 1895

Based on : (1) *Agricultural Returns, 1875* (P.P. 1875, lxxix); (2) *Agricultural Returns, 1895* (P.P. 1896, xcii).

The negligible amounts for Rutland, Westmorland and a number of Welsh counties cannot be shown. It must be remembered that part of the increase was due to differences in definition and interpretation between 1875 and 1895 – see R. H. Best and J. T. Coppock, *The changing use of land in Britain* (London, 1962), 60–1.

relief from falling cereal prices. In Wiltshire, where milk was successfully challenging the traditional sheep and corn on Salisbury Plain, one farmer sent 1,500 gallons daily to London and another near Winchester said that milk was his most profitable enterprise.[1] Although there were fears of over-production, farmers with easy access to a station fared better than

[1] *R.C. Agric. Depression, Rep. by Mr R. Henry Rew on the Salisbury Plain District of Wiltshire*, 18, 28 (P.P. 1895, xvi); *ibid., Minutes*, Q. 6,464 (P.P. 1894, xvi, pt. 1).

many others; for, as one Scots farmer in Hertfordshire put it, to lack communications was 'agricultural death'.[1] Numbers of dairy cattle increased everywhere (although this was partly due to better enumeration), with the highest percentage increases around London (Fig. 127).

The growing of fruit and vegetables. The growing of fruit, vegetables and flowers was also adopted as a means of combating falling prices, although, as with milk-selling, the increase in production began before the onset of depression. Fruit and vegetables had long been grown in and around the large cities, notably around London, where orchards and market gardens were gradually being displaced outwards as the built-up area expanded; but flower growing was largely a new undertaking. The acreage under fruit and vegetables increased partly by planting in and around existing areas and partly by the rise of new centres (Fig. 128). Here, too, good communications were essential, both to carry produce to market and to bring back supplies of town manure to areas of light soil such as mid-Bedfordshire, where a rapid expansion of horticulture followed upon the opening of the Great Northern line in 1850; so critical was access by rail in this area that market gardening gave way to ordinary farming at distances greater than two miles from a station.[2] In the vale of Evesham, where there were only 500 to 600 acres of market gardens in the 1850s, the five railway stations provided a similar stimulus, and the acreage under market-garden crops trebled between 1883 and 1908, when it was estimated at 15,000 acres.[3] In Kent, where market gardening was also expanding rapidly, there were thought to be some 20,000 acres of vegetables and flowers in 1890; transport of produce by road was still important near London, and manure and produce also travelled by water to and from coastal areas.[4]

While market gardening was increasing around the larger towns, as on the mosslands of Lancashire, new centres were also developing farther afield and the urban markets were coming to draw their supplies from a much wider range of both home and foreign producers. Growers in Kent began to send produce to Midland England and Evesham became

---

[1] H. R. Haggard, I, 511.
[2] F. Beavington, 'The change to more extensive methods in market gardening in Bedfordshire', *Trans. and Papers, Inst. Brit. Geog.*, XXXIII (1963), 89; H. Evershed, 'Market gardening', *J.R.A.S.*, 2nd ser, VII (1871), 432.
[3] J. Udale, 'Market gardening', *J.R.A.S.*, LXIX (1908), 95.
[4] C. Whitehead (1899), 431.

a major supplier of Manchester.[1] The railways offered new possibilities in the Fens, where first-class soils had helped farmers to weather the depression, and where flowers, and vegetables such as carrots, were beginning to be grown by both smallholders and large farmers, and in Cornwall, where special trains transported vegetables (especially broccoli) to London, the Midlands and north England.[2] In some of these new centres the stimulus to expansion was provided by migrants from established horticultural areas, for example at Wisbech by a fruit grower from Kent.[3] Personal links were also important in marketing, and the expansion of carrot growing around Chatteris was said to owe much to Chatteris men in the London market.[4]

Two other, though quite different, systems of growing flowers and vegetables were also developing. In the main arable areas farmers began to grow vegetables on a field scale as part of a rotation, as with the production of green peas in Essex; and it became so difficult to distinguish between market gardening and the growing of horticultural crops by arable farmers that the collection of the acreages of market gardens was abandoned in 1897.[5] The second development was a rapid increase of greenhouses, especially around Worthing and in the Lea valley; growth here also owed much to individuals, such as the Rochford family who owned 86 acres of glass in the late 1890s, a quarter of the acreage in the Lea valley.[6] Tomatoes, grapes and flowers were the chief crops, a clear indication of the influence of the growing urban markets.[7]

Parallel to, and often associated with, the spread of market gardening and of the growing of flowers and vegetables was an increase in the production of fruit (Fig. 129). The acreage recorded under orchards rose from 141,000 acres in 1873 to 237,000 in 1914, and that under small fruit from 33,000 to 76,000 acres between 1888 and 1914, although both of

[1] *Ibid.*, 469; *Departmental Cttee on the Fruit Industry of Great Britain, Minutes, Evidence of Wise*, Q. 4,979 (P.P. 1906, xxiv); J. Page, 'The sources of supply of the Manchester fruit and vegetable markets', *J.R.A.S.*, 2nd ser, xvi (1880), 476–7.

[2] *R.C. Agric. Depression, Rep., by Mr Wilson Fox on the County of Cambridge*, 5 (P.P. 1895, xvii); E. A. Pratt, 171.

[3] J. E. G. Mosby, 182.

[4] *R.C. Agric. Depression, Rep. by Mr Wilson Fox on the County of Cambridge*, 5–6 (P.P. 1895, xviii).

[5] E. A. Pratt, 127; R. H. Best and J. T. Coppock, 61.

[6] W. E. Bear, 'Flower and fruit farming in England', pt. 2, *J.R.A.S.*, 3rd ser, ix (1898), 528; *ibid.*, pt. 4, *J.R.A.S.*, 3rd ser, x (1899), 268, 283.

[7] *Ibid.*, pt. 4 (1899), 286.

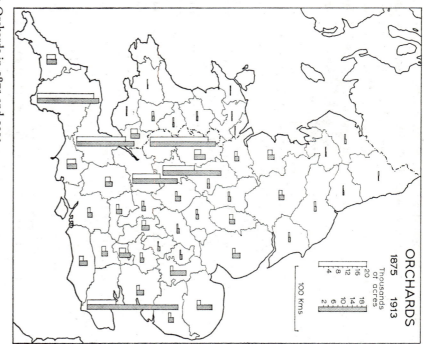

ORCHARDS

1875   1913

Thousands of acres

20
18
16
14
12
10
8
6
4
2

100 Kms

these increases were in part due to better enumeration.[1] Much of the large acreage of cider and perry orchards in the West Country was in poor condition and, although there was some planting of cider orchards in Hereford-shire, western counties were little affected by the expanding market for fruit.[2] New orchards were largely confined to the eastern counties. The negligible amounts for Rutland and a number of Welsh counties cannot be shown. It must be remembered that part of the increase was due to differences in definition and interpretation between 1875 and 1913 – see R. H. Best and J. T. Coppock, *The changing use of land in Britain* (London, 1962), 60–2.

[1] All unattributed acreages are derived from the annual volumes of agricultural statistics.

[2] *Dept'l Cttee Fruit Industry, Minutes*, Evidence of Pickering, Q. 4,045 (P.P. 1906, xxiv); J. H. Clapham (1938), 91.

Fig. 129 Orchards in 1875 and 1913
Sources as for Fig. 126.

biggest gain was in Kent, where the acreage under orchards rose by 200% between 1875 and 1913, and that under small fruit by 92% between 1888 and 1914; and there was also a large increase in the Evesham district.[1] New areas were also developing; there were only a few orchards around Wisbech in 1880, but as many as 6,000 acres by 1899; and some 3,000 acres of fruit were grown within a ten-mile radius of Histon, Cambridgeshire, twenty-one years after the opening of Chivers jam factory in 1875, two-thirds of it attributable to the market offered by the factory.[2] The jam factory at Tiptree in Essex likewise seems to have acted as an incentive to fruit growing in that locality.[3] The growing of small fruit was often associated with orchards, although in some localities, such as the strawberry-growing area of Hampshire, small fruit alone was produced. Small fruit was also sometimes introduced as a field crop by arable farmers.

The growing of hops provides an interesting example of a short-lived and highly localised development.[4] By the 1850s, hops were largely confined to the west Midlands and south-east England. The removal of hop duty in 1861 led to a rapid rise in the acreage under hops, which reached a maximum of 71,000 acres in 1878.[5] This expansion was often achieved by planting on less suitable sites, and, in the face of rising imports and higher yields from the home crop, the trend was almost immediately reversed. In Kent, the leading county, the distribution of hop growing reverted to its previous pattern and many hop fields were converted into orchards; in Worcestershire, by contrast, the gains of the 1860s and 1870s were maintained.[6]

The expansion of fruit, flower and vegetable growing added further variety to the pattern of farming and helped arable farmers to find alternative cash crops in place of wheat. It was not all gain, for in some areas the development of new centres harmed local producers, as in Cheshire, where farmers growing early potatoes faced severe competition from growers in Cornwall and the Channel Isles who could now compete with

[1] W. E. Bear, 'Flower and fruit farming in England', pt. 3, *J.R.A.S.*, 3rd ser., x (1899), 63.

[2] *R.C. Agric. Depression, Rep. by Mr Wilson Fox on the County of Cambridge*, 6 (P.P. 1895, xvii).   [3] W. E. Bear, pt. 3 (1899), 71–2.

[4] D. M. Harvey, 'Locational changes in the Kentish hop industry and the analysis of land-use patterns', *Trans. and Papers, Inst. Brit. Geog*, xxxiii (1963), 123–44.

[5] D. C. D. Pocock, 'England's diminished hop acreage', *Geography*, XLIV (1959), 14.

[6] A. D. Hall and E. J. Russell, *The soils and agriculture of Kent, Surrey and Sussex* (H.M.S.O., 1911), 28; C. Whitehead, 'Fifty years of hop farming', *J.R.A.S.*, 3rd ser. I (1890), 325.

them;[1] but in general the greater acreages under field vegetables and soft fruit were a major benefit to the arable farmer.

Other changes. There were also changes in the structure of farming, although the evidence is conflicting. In areas of light soil it seems likely that large farms were increasing in importance as individuals created farming empires out of farms which could not be let; one of the most striking examples was a holding of 14,000 acres in Wiltshire which was rented from seven landowners.[2] Elsewhere, there is evidence of both amalgamation and subdivision; for while larger farms were often difficult to let, many occupiers of smallholdings lacked resources and suffered severely. In localities which depended on horticultural crops, such as the vale of Evesham, many more smallholdings were created.[3] Smallholdings were the subject of much discussion, but legislation that enabled local authorities to provide such holdings had little effect before the Act of 1908.

Evidence about the land itself is ambiguous. While investments in improvement certainly diminished, they did not cease; new buildings were necessary for new systems of farming, especially for dairying, and new cottages were required to retain agricultural labourers. There is evidence, too, of land reclamation 'in the hilly districts', but its location and extent are unknown, for statistical data are absent or unsatisfactory and map evidence uncertain; comparison of successive editions of the Ordnance Survey 6" maps suggests that, if anything, improved land was reverting. What is certain is that the agricultural area as a whole was diminishing; between 1851 and 1871 it was thought that some 700,000 acres of land went out of agricultural use in England and Wales as a result of the growth of towns and the railway network, and losses must have continued at similar rates in succeeding decades.[4]

Woodland

The fate of the woods and forests during these years is even less certain; there are few contemporary references and statistical data are scanty and unsatisfactory. The principal change in the 1850s and 1860s, at least in the

[1] R.C. on Agriculture, Reps. by Assistant Commissioners, Mr Coleman's Reports, 63 (P.P. 1882, xv).

[2] R.C. Agric. Depression, Rep. by Mr R. Henry Rew on the Salisbury Plain District of Wiltshire, 28 (P.P. 1895, xvi).

[3] R.C. on Labour, vol. I (England), pt. 2, Rep. by Mr Aubrey J. Spencer, 95–6 (P.P. 1893–94, xxxv).

[4] Inclosure Commissioners: Thirtieth Ann. Rep., 3–4 (P.P. 1875, xx).

lowlands, appears to have been the conversion of woodland into agricultural land. Yet in some localities, such as Sherwood Forest, the area under woods was increasing and in the uplands, too, there was afforestation; in Teesdale, for example, a good deal of planting was said to have been done in years before 1881.[1] Some of the proposals for enclosure recommended by the Inclosure Commissioners also referred to the possibility of afforestation, as at Loweswater Common in Cumberland, where it was claimed that the land would be 'much improved' by planting.[2]

In the remaining decades, afforestation seems to have been the general rule, although the statistical evidence is not very reliable. From 1871 the acreage of woodland in each parish was recorded at intervals by the inland revenue officers responsible for the collection of the agricultural returns, the source of information being at first the rate books and subsequently enquiry among landowners. The acreages so recorded rose from 1,314,000 acres in 1871 to 1,683,000 in 1905.[3] These increases are unlikely to be an accurate measure of any change, for it is probable that the estimates became progressively more complete; but the remarks of observers and the evidence of the Ordnance Survey maps suggest that the trend they indicate was real. It was generally recognised that the state of British woodlands was unsatisfactory and that this was due in part to the abolition of duty on imported timber and to the impoverishment of landowners;[4] but some professional foresters often blamed the high esteem in which field sports were held, claiming that this handicapped the proper management of woodland.[5] Consideration of visual amenity also affected the use of private woodlands, and some, such as Burnham Beeches and Epping Forest, were now being acquired for public recreation.[6] The acreage of private parkland was also increasing and probably reached its maximum extent during this period.[7] Yet, despite these criticisms, Lord Ernle could claim that at last British woodlands were coming to be treated commercially.[8] There were many

[1] R.C. *Agriculture, Reps. by Assistant Commissioners*, 229 (P.P. 1881, xvi).

[2] *Special Rep. of Inclosure Commissioners, 1861*, 4 (P.P. 1861, xx).

[3] *Agricultural Returns, 1871*, 55 (P.P. 1871, lxix); *Rep. on census of woodland, 1924* (H.M.S.O., 1928), 6.

[4] C. E. Curtis, 'The management and planting of British woodlands', *J.R.A.S.*, LXIV (1903), 16.

[5] E. P. Stebbing, *Commercial forestry in Britain* (London, 1919), 35–7.

[6] Lord Eversley (1910), 46.

[7] H. C. Prince, *Parks in England* (Shalfleet, I.O.W., 1967), 10.

[8] R. E. Protheroe (later Lord Ernle), *English farming past and present* (1st ed., London, 1912), 383.

reports of new plantings, often on poorer land thrown out of cultivation. At Langham, in Essex, woods of larch and fir had recently been planted when Rider Haggard made his tour in 1901–2.[1] In Surrey there had been extensive afforestation with larch and pine on poor land, and on the Hampton Lodge estate near Farnham, for example, a considerable area had been planted between 1880 and 1905, chiefly with Scots pine, but with some Douglas fir, Corsican pine, spruce and larch.[2] These woods were mainly for ornamental and sporting purposes, but large tracts of Scots pine had been planted on adjacent areas to supply hop poles.[3] Comparison of successive editions of Ordnance Survey maps reveals numerous plantings in other areas, often in small plots on difficult terrain, as on the steep slopes in the Oxfordshire Chilterns, or on poor soils, as on the Greensand outcrop in Bedfordshire, where 666 acres were planted on the Woburn estate between 1897 and 1907.[4] The evidence of the large-scale maps also suggests that some lowland heaths and commons were being colonised by trees, although the different styles of the first and second editions of the six-inch maps make comparison difficult. Large tracts of Hindhead Common, which were bare of trees in 1871, were shown as woodland in 1894–6 and, not far away, on Witley Common scattered trees had become more numerous; but the trend was by no means universal and trees had apparently disappeared from some commons. Afforestation must also have continued in upland areas, where the planting of catchments was being encouraged, while on the Grizedale Hall estate in Lancashire, oak coppice was being converted into high forest by allowing single saplings to grow and cutting back the remainder.[5]

What is clear from the Ordnance Survey maps, from contemporary accounts and from the 1924 census of woodland, is that many of the trees planted were conifers, although interpretation of the census data is complicated by extensive fellings during the First World War. The abnormal age structure of the broadleaf forest shows that it was not being replaced by new plantings, while conifers accounted for a much higher proportion of the woodland planted between 1884 and 1914 than their share of the mature high forest would warrant; coniferous high forests represented

---

[1] H. R. Haggard, I, 445–6.
[2] J. Nisbet in H. E. Malden (ed.), *V.C.H. Surrey*, II (1905), 576, 578.
[3] *Ibid*, 578.
[4] J. C. Cox in W. Page (ed.), *V.C.H. Bedfordshire*, II (1908), 146.
[5] W. Farrer in W. Farrer and J. Brownbill (eds.), *V.C.H. Lancashire*, II (1908), 465–6.

only 26% of high forest in England, but accounted for 45% of new plantings, and for more than 50% in such different counties as Buckinghamshire, Surrey and Westmorland.[1] Whatever the ambiguities of the data, afforestation, much of it with conifers, was certainly taking place; but its scale was small, and the English woodlands remained some of the least extensive and least productive in Europe.

Other rural land uses

Agriculture and forestry were not the only activities in rural areas. In addition to mining, other demands were increasingly made on rural land. Some of these, like field sports, were long-established; but most were new, at least in the form in which they appeared, as with water-gathering in the uplands and with public recreation. Hunting, shooting and fishing had long been features of rural life, but they were generally the preserve of landowners and their friends, and involved little in the way of land management. From the middle of the nineteenth century the shooting of driven grouse became an increasingly important aspect of land management on heather moor, and the renting of shootings became fashionable. Some indication of the changing importance is given by the increase in the size of bag on the duke of Devonshire's 14,000 acres of grouse moor at Bolton Abbey, where the average bag of grouse increased from less than 200 brace a year between 1800 and 1850 to over 3,600 brace by the end of the century.[2] In the lowlands, too, the impoverishment of both landowners and land, and the rise of a wealthy, non-landowning clientele, led to the commercial letting of shootings and often of a landowner's residence.[3] This trend was particularly marked in areas of poor land such as the Breckland, where shooting became the principal activity and where land was kept in cultivation as much for game birds as for its agricultural output; the partridge was said to have been 'the salvation of Norfolk farming' and an estate such as that of Lord Iveagh near Thetford became principally a sporting property.[4] Large stretches were given up to game and rabbits and many owners were no longer resident, but had let their houses to

[1] *Rep. on census of woodland, 1924* (H.M.S.O., 1928), 33.
[2] A. S. Leslie (ed.), *The grouse in health and disease* (London, 1911), 384.
[3] *R.C. Agric. Depression, Rep. by Mr R. Hunter Pringle on the Ongar...Districts of Essex*, 72 (P.P. 1794, xvi, pt. 1); H. R. Haggard, II, 2.
[4] *R.C. Agric. Depression, Rep. by Mr R. Henry Rew on the County of Norfolk*, 58 (P.P. 1895, xvii); H. R. Haggard, II, 2.

shooting tenants.[1] Other poor land which had tumbled down was let for sport as well as for rough grazings.[2]

A middle-class counterpart of these interests in field sports was the growing concern throughout the second half of the nineteenth century with outdoor recreation and with what was coming to be called amenity, the visual quality of the countryside. The General Enclosure Act of 1845 had made provision for the allotment of land for recreation, but by 1876 only 1,742 acres had been allotted out of a total of 618,800 acres enclosed under this Act.[3] Problems were arising over the management of commons around London, and there was a growing feeling, epitomised by the formation in 1865 of the Commons Preservation Society, that the remaining commons ought not to be enclosed but kept for public recreation. Attention was mainly concentrated on commons near towns, but there was also resistance to encroachment on rural commons, dramatised in 1866 by the train load of labourers brought out by night to remove the iron fences erected around Berkhamsted Common in Hertfordshire.[4] The Metropolitan Commons Act of 1866 provided machinery for the regulation of commons within fifteen miles of Charing Cross and the Enclosure Act of 1876 laid much greater stress on the recreational needs of the public at large, although in practice it was not as effective as the Commons Preservation Society would have wished.[5] In addition there were several special acts, such as that for the preservation of the Malvern Hills. Finally the Enclosure Act of 1899 made provision for regulated commons, to which the public could be granted access, and some large rural commons were opened to the public in this way. Attempts to provide rights of access to all moorland areas were unsuccessful, although the growing concern with amenity gained a further outlet with the formation of the National Trust in 1895.

The uplands were also being required to provide water for the growing towns around their flanks. The first considerable reservoirs had been those begun by Manchester Corporation in the Longdendale valley in the 1840s, but gradually all suitable areas in the southern Pennines were taken for gathering grounds and towns had to look farther afield:[6] to the north of the Wharfe, where Leeds and Bradford established reservoirs in the

[1] R.C. Agric. Depression, Rep. by Mr Wilson Fox on the County of Suffolk, 31, 38 (P.P. 1895, xvi).

[2] Rep. on Decline in Agric. Pop., 1881–1906, 35 (P.P. 1906, xcvi).

[3] E. C. K. Gonner, 93.   [4] Lord Eversley (1910), 46.   [5] Ibid., 198.

[6] R. C. S. Walters, The nation's water supply (London, 1936), 99.

Washburn and Nith valleys respectively; to the Lake District, where Manchester acquired Thirlemere; and to Wales, which supplied Birmingham and Liverpool. Other upland areas were required to make their contribution and by 1904 the gathering grounds, mostly moorland, held by local authorities in England amounted to at least 82,050 acres and in Wales to at least 6,350 acres.[1]

Another, though more localised, feature of the countryside was the diversion of land to military use, and in the 1890s the War Office acquired a large tract of land on Salisbury Plain for training grounds.[2] Heathland on the Surrey–Hampshire border was also purchased and extensive barracks and rifle ranges constructed; in Pirbright parish, for example, nearly three-quarters of the parish was military land.[3]

INDUSTRY

The broad features of the industrial geography of England which had emerged by 1850 were still recognisable at the outbreak of the First World War in 1914 and, with the exception of iron and steel and shipbuilding, no major changes in the location of the different branches of industry occurred in the intervening period. The most obvious development was the growth of industry, whether measured by the size of the industrial population, which doubled, or by the volume of output, which increased fourfold between 1851 and 1911.[4] It is true that the character of existing industry altered, that new industries developed and that several new industrial towns appeared on the map of England, but the principal centres of the various branches of manufacturing industry were those established during the industrial revolution; the chief differences were that they were bigger and more complex.

Unfortunately, the data available for the study of changes in the industrial geography of this period are very unsatisfactory, particularly for the new, expanding and technically complex industries such as engineering. Classifications of occupations in the decennial censuses, which are the main (and in many instances the only) source, were frequently changed and, although comparisons can be made between the censuses of 1851, 1861 and 1871 and between those from 1881 to 1911, no reliable view of

---

[1] *Reconstruction Ctee, Forestry Subctee: Final Report*, 93 (P.P. 1917–18, xviii).
[2] C. R. Clinker in E. Crittall (ed.), *V.C.H. Wiltshire*, IV (1959), 290.
[3] D. M. Sprules in H. E. Malden (ed.), *V.C.H. Surrey*, III (1911), 363.
[4] B. R. Mitchell with P. Deane, 60, 271–2.

changes over the period as a whole is possible. Even if data were comparable the use of the occupational tables as a basis for comparison would be unsatisfactory, as changes in productivity varied widely from industry to industry and throughout the period, so that the longer the time-scale the broader and less certain must any conclusion be. Because of their complexity and the lack of data about their distribution, many branches of manufacturing industry will be under-represented in the account which follows.

Comparisons between different industries are particularly difficult, but Hoffman's coefficients of growth for the United Kingdom for the period 1855–1913, which, if acceptable, must broadly reflect conditions in England as a whole, show above-average values in the making of paper, furniture, cotton piece-goods, tobacco and woollen cloth among consumer goods and of aluminium, zinc, rubber products, iron and steel manufactures, timber and copper products among producer goods.[1] Some industries, on the other hand, such as silk manufacture and lead production, declined, and others grew more slowly than industry as a whole. Production by machinery in factories extended its hold, and most remaining kinds of domestic industry were eliminated; by 1901, handwork and outwork, which had been characteristic of hosiery, boot and shoe, nail and cutlery manufacture and many other industries in 1850, were largely confined to the tailoring and clothing trades.[2]

Coalmining

Coal was the principal source of power throughout these decades, although latterly electricity, gas and oil became alternatives, and increasing efficiency in the use of coal was weakening the pull of the coalfields. Nevertheless, coal remained a major factor in industrial location, and coalmining was itself one of the principal growth industries; output of the English fields rose fairly steadily from some 48 million tons in 1854 (the first year for which complete figures are available) to 185 million in 1913.[3] Not all coal was consumed by industry; increasing quantities were taken by railways and households, and the share of exports rose from little more than a twentieth in 1850 to a fifth in 1903 and nearly a third by 1913.[4]

1 W. G. Hoffman, *British industry, 1700–1950* (Oxford, 1955), 69, *passim*.
2 *Rep. S.C. on Home Work*, 238 (P.P. 1907, vi).
3 *Return of the Quantities of Coal etc.*, 2 (P.P. 1856, lv); *Mines and Quarries: General Report, with Statistics, for 1913*, 220 (P.P. 1914–16, lxxx).
4 J. H. Clapham, *An economic history of modern Britain: the early railway age, 1820–1850* (Cambridge, 1926), 430, 484; J. H. Clapham (1938), 63; W. Smith, 275.

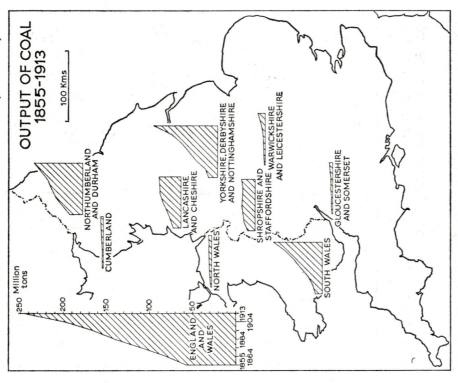

Fig. 130 Output of coal, 1855–1913
Based on: (1) R. Hunt, *Mineral statistics, 1855* (Mem. Geol. Survey, 1856); (2) *Mines and Quarries: General Report with Statistics for 1913* (P.P. 1914–16, lxxx).
The negligible output for Kent in 1913 (59,000 tons) cannot be shown.

Only one new field was discovered, the concealed Kent coalfield, and this was only on the verge of production in 1913, but there were major changes in the share of output from different coalfields and in the distribution of mining within each field (Fig. 130).[1] In 1850, Northumberland and Durham, with their considerable sea-coal trade to London, had been the principal sources of coal, but although their output rose steadily, they were overtaken in the 1890s by the most extensive and most easily worked

[1] J. H. Clapham (1938), 166–7; B. R. Mitchell with P. Deane, 115–16.

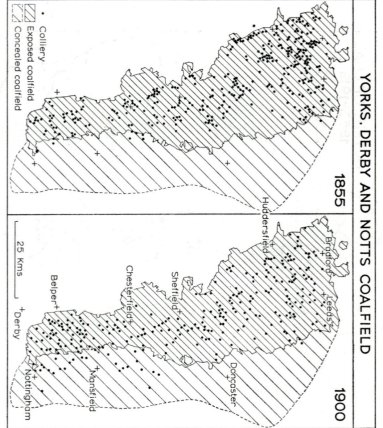

YORKS. DERBY AND NOTTS COALFIELD

1855

1900

Huddersfield
Bradford
Leeds
Sheffield
Chesterfield
Belper
Derby
Nottingham
Mansfield
Doncaster

25 Kms

· Colliery
Exposed coalfield
Concealed coalfield

Fig. 131 Yorks, Derby and Notts coalfield in 1855 and 1900
Based on: (1) R. Hunt, *Mineral statistics, 1855* (Mem. Geol. Survey, 1856);
(2) *List of mines, 1900* (H.M.S.O., 1901).

of the coalfields – that of Yorkshire, Derbyshire and Nottinghamshire. Industrial growth, the increasing concern of the coastal fields with the export trade, the development of the railway network, and competition between the railway companies for the coal trade, were considerably extending the markets of the interior fields. At the same time, the coal resources of some of the old and smaller fields were either approaching exhaustion (with contemporary standards of technology), or becoming more difficult to work; in the south Staffordshire field, the working out of the more accessible seams of the famous 'thick coal' and difficulties of drainage led to an increasing reliance on mines in Cannock Chase, which accounted for a third of joint output of both fields in 1880 and a half in 1900.[1] The relative, and after 1872 the absolute, decline of production on

[1] M. J. Wise, 'The Cannock Chase region', in M. J. Wise (ed.), *Birmingham and its regional setting* (Birmingham, 1950), 279–80.

the south Staffordshire coalfield also stimulated the expansion of mining in east Warwickshire, which had good rail and canal communications with Birmingham and the Black Country.[1]

Changes in the distribution of mining within individual fields were less striking. Many additional shafts were sunk in existing areas and some pits reorganised or deepened, but the most interesting changes geographically were the abandonment of small shallow pits in the outcrop of the Coal Measures, and the increasing number of deeper shafts sunk through over-lying strata to reach concealed fields. This trend may be illustrated by the Yorkshire, Derbyshire and Nottinghamshire field, where many mines along the western margin went out of production between 1855 and 1900, although some of these were very small pits producing only ganister (Fig. 131). Expansion into the concealed field was most marked in Nottinghamshire where the first deep shaft was sunk in 1859 and where sixteen pits were opened between 1870 and 1880.[2] Developments elsewhere were similar but in some fields there were technical difficulties; in south-east Durham, extension of mining was hindered by the difficulty of penetrating water-bearing sands, but a technique of freezing was employed after 1900 to permit the sinking of new shafts into the concealed field.[3] In other areas, new markets or communications extended mining; there was a considerable expansion of mining in the Ashington area of Northumberland, following the development of the port of Blyth, where coal shipments rose from 150,000 tons in 1883 to 4,730,000 tons in 1913.[4]

The iron and steel industries

Iron-ore mining. The distribution of iron-ore mining underwent a much more radical transformation and, by contrast with coal, home production faced increasing competition from imported, high-grade ores (especially from Spain). Imports, which were negligible in 1850, had reached 7·4

[1] J. C. Mitcheson in M. J. Wise (ed.), 296.

[2] G. D. B. Gray, 'The south Yorkshire coalfield', *Geography*, XXXII (1947), 113–31; K. C. Edwards, 'Coal', in K. C. Edwards (ed.), *Nottingham and its region* (Nottingham, 1966), 279–80.

[3] G. Poole and A. Raistrick, 'Extractive industries', in P. C. G. Isaac and R. E. A. Allan (eds.), *Scientific survey of north-eastern England* (Newcastle upon Tyne, 1949), 88–9.

[4] A. E. Smailes, *North England* (London and Edinburgh, 1960), 174; J. H. Clapham (1938), 388.

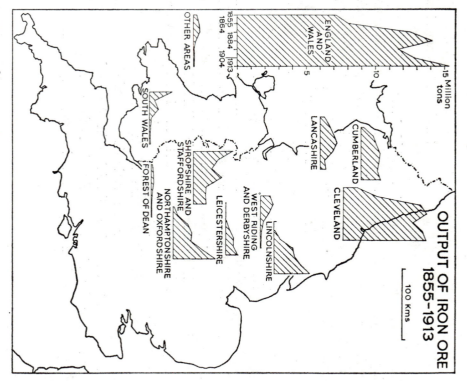

Fig. 132 Output of iron ore, 1855–1913.
Sources as for Fig. 130.
'Other areas' include Cornwall, Devon, Durham, Northumberland, Somerset, Warwick, Wiltshire, Worcester, North Wales and, in some years, Cheshire, Hampshire, Nottingham and Westmorland.

million tons by 1910.[1] The principal internal change was the decline in the production of Coal Measure ores and the exploitation of those not associated with the coalfields (Fig. 132). In 1850, the Coal Measure ores were estimated to provide 95% of output, but by 1913 they contributed only 10% of a much larger total of domestic ore.[2] Their place was taken mainly

[1] H. G. Roepke, *Movements of the British iron and steel industry 1720 to 1951* (Urbana, Illinois, 1956), 61. [2] *Ibid.*, 61, 63; W. Smith, 120, 322.

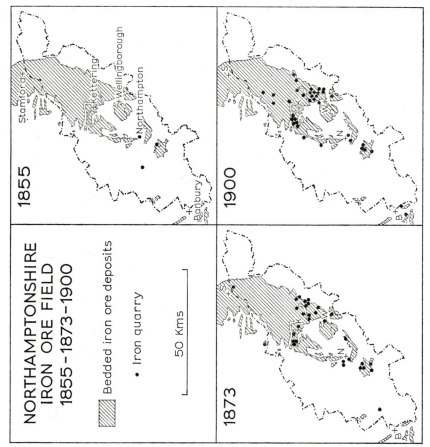

NORTHAMPTONSHIRE IRON ORE FIELD 1855–1873–1900

Bedded iron ore deposits

• Iron quarry

50 Kms

1855

1873

1900

Fig. 133 Northamptonshire ironfield in 1855, 1873 and 1900
Based on: (1) R. Hunt, *Mineral statistics, 1855* (Mem. Geol. Survey, 1856);
(2) R. Hunt, *Mineral statistics, 1873* (Mem. Geol. Survey, 1874); (3) *List of quarries, 1900* (H.M.S.O., 1901).

by the rich haematite of Cumberland and north Lancashire and by the low-grade bedded ores which occur in the Jurassic rocks between Oxfordshire and the North Riding, although other ore bodies played a part, such as the magnetite of Rosedale in the North York Moors.[1]

Mining of haematite was given a strong stimulus by the demand for non-phosphoric ores for Bessemer steel. The output of ore expanded sharply between 1850 and 1870, especially in the Barrow area where a large ore body had been discovered in 1850; but from the 1880s difficulties of mining, competition from imported haematite, and the exploitation of

[1] J. C. Carr and W. Taplin, *History of the British steel industry* (Oxford, 1962), 13.

J. T. COPPOCK

Fig. 134 Blast furnaces in 1855
Based on R. Hunt, *Mineral statistics, 1855* (Mem. Geol. Survey, 1856).
Only furnaces in blast are shown.

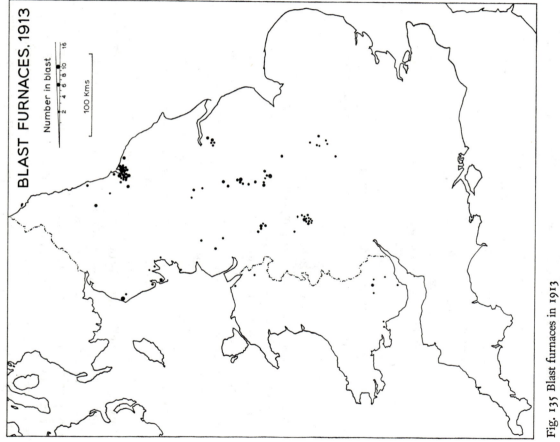

BLAST FURNACES. 1913

Number in blast

2 4 6 8 10   16

100 Kms

Fig. 135 Blast furnaces in 1913
Based on *Mines and Quarries: General Report with Statistics for 1913* (P.P.
1914–16, lxxx).

more easily worked Jurassic ores, led to a decline in production.[1] The existence of ores in the Jurassic rocks had been known much earlier, but they began to be worked extensively only from mid-century. In Cleveland, following the opening of mines at Eston in 1850, there was a rapid expansion of output until the peak year of 1883; thereafter output stagnated, and increasing quantities of ore were imported for steel production on Teeside.[2] Following the rediscovery of ironstone beds in Northamptonshire, a consignment was sent to Staffordshire in 1852, although at first the Midland ironmasters were said to be prejudiced against these ores.[3] By 1864 their prejudices had been overcome, and output grew rapidly in the next decade, reaching nearly 1½ million tons in 1873; the ore was worked in both mines and quarries and, although a steam navvy was introduced in 1895, production was mainly by hand until after the First World War.[4] The Lincolnshire ores, similarly rediscovered, were mined from 1859, and production had reached 2.6 million tons by 1913.[5] Output from east Midland fields accounted for a rising proportion of home production and, by 1913, 43% of the output of domestic ores came from these sources.[6] Changes in the location of mines and quarries within fields was much more narrowly confined than in coalmining; only in Northamptonshire was there any great extension of the area affected by mining (Fig. 133).

Pig-iron production. Between 1855 and 1913 the quantity of pig-iron made in England rose from 1½ to 8 million tons, and there were also changes in the location of pig-iron production. These resulted from developments in mining, and also from the changing demand for its products (Figs. 134 and 135).[7] The major feature was the declining importance of sites on the coalfields and the rise of new centres on the orefields and at the coast. In 1852 three-quarters of all pig-iron was made in the west Midlands, but output was already declining in 1865 when the number of blast furnaces reached its peak.[8] As local ores were exhausted there was an increasing

---

1 *Ibid,* 12; B. R. Mitchell with P. Deane, 129–30.

2 H. G. Roepke, 54, 63; J. H. Clapham (1932), 48; B. R. Mitchell with P. Deane, 129–30.

3 S. H. Beaver, 'The development of the Northamptonshire iron industry 1851–1930', in L. D. Stamp and S. W. Wooldridge (eds.), *London essays in geography* (London, 1951), 34.

4 *Ibid,* 38–9, 49.

5 D. C. D. Pocock, 'Stages in the development of the Frodingham ironstone field', *Trans. and Papers, Inst. Brit. Geog.,* XXXV (1964), 105; H. G. Roepke, 64–5.

6 *Ibid,* 66.

7 *Ibid,* 64–5.

8 B. R. Mitchell with P. Deane, 131; B. L. C. Johnson and M. J. Wise in M. J. Wise (ed.), 237.

reliance on ores from other domestic fields, especially from the east Midlands, but this was only a temporary phase, for economies in the use of coal and the high cost of iron manufacture on coalfield sites were reducing their advantages. It is true that some ironmasters were acquiring leases of orefields in the east Midlands, but production of pig-iron on the coalfields gradually declined.[1] By 1911 there were only two major and eleven smaller furnaces remaining in south Staffordshire, and there was a similar, though somewhat later, decline in the production of pig-iron on other coalfields. In north Staffordshire and west Yorkshire the peak was reached in the 1870s, and by 1913 the number of furnaces in blast on the former field had declined from thirty-six to thirteen.[2]

The most successful of the new areas was Teeside. Its initial advantages of easily worked ore and close proximity to Durham coke were reinforced by the larger size and greater efficiency of the Teeside furnaces compared with older plant elsewhere, and by its estuarine site when Spanish ore began to be imported in 1861, although Cleveland ore still accounted for 84% of the ore used in 1883 and 60% in 1913.[3] The first furnaces were blown in 1852 and by 1869 there were 92; production expanded rapidly, rising to nearly 248,000 tons in 1860, and to 766,000 tons in 1870.[4] For a brief period iron-making spread more widely in the North Riding, but many of the works outside the Tees valley did not survive long, and nearly all pig-iron was produced at estuarine sites. By 1913, some 2,639,000 tons were being made.[5]

Elsewhere on the Jurassic outcrop, the development of pig-iron making came later and the industry grew more slowly; for these fields were much less accessible to sources of coking coal and to the ports, and the ores were not suitable for steel making until the development of the basic process. At first all the Northamptonshire ore was sent to furnaces on the coalfields, but in 1853 pig-iron was made at Wellingborough from local ores; thereafter, the number of blast furnaces increased steadily, although half

[1] S. H. Beaver (1951), 43.

[2] B. L. C. Johnson and M. J. Wise in M. J. Wise (ed.), 243; S. H. Beaver, 'The Potteries', *Trans. and Papers, Inst. Brit. Geog.*, XXXIV (1964), 17; F. J. Fowler, 'West Yorkshire', in J. B. Mitchell (ed.), *Great Britain, geographical essays* (Cambridge, 1962), 364.

[3] J. C. Carr and W. Taplin, 51, 152; J. H. Clapham (1932), 50; W. Smith, 327.

[4] A. A. L. Caesar in J. B. Mitchell (ed.), 445; B. R. Mitchell with P. Deane, 131–2 (the figures are for the North Riding).

[5] B. R. Mitchell with P. Deane, 132. For the north-east coast, the figure was 3,869,000.

the ore continued to be sent to other areas.[1] Pig-iron was first made at Scunthorpe in 1864.[2]

The sequence of events in north-west England was different again, although this area was also handicapped by its remoteness from both markets and supplies of coke, despite the opening of a line from the Durham coalfield in 1861.[3] At first, all ore was sent elsewhere, but a blast furnace was blown at Barrow in 1857, and the increased demand for an iron suitable for Bessemer steel led to a sharp rise in the output of pig-iron between 1865 and 1870; production reached its peak of nearly 1.6 million tons around 1900.[4]

Steel manufacture. In the third quarter of the nineteenth century most pig-iron was still made into wrought or puddled iron, with the Midlands, Sheffield and, increasingly, north-east England as the major centres.[5] Between 1860 and 1875 the number of finished-iron works and puddling furnaces grew but, although puddled iron was able to hold its own for a while, it faced increasing competition from steel.[6] From 1882 both the output of puddled iron and the number of puddling furnaces declined steadily, the industry surviving longest in those areas, such as the west Midlands, where there was a large local market for its products; but even there iron was giving way to steel.[7]

Until the revolutions in steel making associated with the names of Bessemer and Siemens made steel competitive with wrought-iron, steel manufacture was on a small scale and was almost entirely confined to Sheffield, which became, until 1879, the principal area for both Bessemer and open-hearth steel.[8] The Bessemer process was the first to be developed commercially, in 1856, although until the invention of the Thomas–Gilchrist basic process, whereby phosphorus was eliminated from phosphoric ores, the ores in the Jurassic rocks were unsuitable for steel making. The Siemens brothers' 'open-hearth' furnace for steel making, though

[1] J. C. Carr and W. Taplin, 89; S. H. Beaver (1951), 39, 41.
[2] D. C. D. Pocock, 'Iron and steel at Scunthorpe', *East Midland Geographer*, III, pt. 3 (1963), 125; H. G. Roepke, 69, 71.
[3] A. E. Smailes, 184.
[4] J. C. Carr and W. Taplin, 12; H. G. Roepke, 77.
[5] J. H. Clapham (1932), 52, 57.
[6] J. C. Carr and W. Taplin, 38.
[7] J. H. Clapham (1932), 52.
[8] J. C. Carr and W. Taplin, 11; K. Warren, 'The Sheffield rail trade', *Trans. and Papers, Inst. Brit. Geog.*, XXXIV (1964), 131.

patented in 1861, was not in commercial use until 1868, but because of the better control which this method offered and its ability to use scrap, open-hearth steel gradually replaced Bessemer steel; 80% of the steel made in Great Britain in 1880 was Bessemer steel, but by 1900 the proportion was only 36%.[1] With the successful development of the Thomas–Gilchrist process in 1880, an increasing proportion of the steel made by both methods was basic steel.

Accompanying these technical changes were changes in the geography of steel making. The first major market for steel was in the manufacture of rails, where it gradually displaced iron during the 1870s.[2] Rail making was concentrated in Sheffield, which had seventeen of the forty-nine Bessemer converters producing steel in 1869, and in south Lancashire, which had twenty.[3] The suitability of the haematite ores of Cumberland and Lancashire for Bessemer acid steel led to a rapid expansion of steel making in the north-west, where Barrow had, for a while, the largest steelworks in the country.[4] This area accounted for two-fifths of all steel made in England in 1880, but it was handicapped by the high cost of mining the ore and by the lack of both coking coal and a major local market for its steel; although its output had risen to over 700,000 tons by the 1890s, its share of English production declined steadily, to less than a tenth in 1913.[5]

The development of steel making on the north-east coast, especially on Teeside, was delayed by the large investment which had been made in the iron industry since 1850, by the unsuitability of the local ores for acid Bessemer steel, by the large local market for wrought iron, and by un-favourable experience with steel in shipbuilding.[6] Nevertheless, it rapidly became a major centre for Bessemer steel and established an early lead in the production of basic Bessemer; with the widespread acceptance of both the open-hearth process and of sheet steel for ship plates, the north-east became the major centre for steel production in England, accounting for over 2 million tons in 1913, or nearly half the total output.[7]

The west Midlands were slow to adopt Bessemer steel, largely because

[1] *Ibid*, 34; W. A. Sinclair, 'The growth of the British steel industry in the late 19th century', *Scottish Jour. of Polit. Econ.*, VI (1959), 44–5; H. G. Roepke, 85.

[2] J. C. Carr and W. Taplin, 34.

[3] H. G. Roepke, 84.

[4] *Ibid*, 83; J. C. Carr and W. Taplin, 84.

[5] H. G. Roepke, 83. The figure for Great Britain was 5%.

[6] *Ibid*, 28; D. L. Burn, *The economic history of steel-making 1867–1939* (Cambridge, 1940), 173–4.

[7] W. A. Sinclair, 35; H. G. Roepke, 83. The figure for Great Britain was 27%.

of the region's heavy commitment to the wrought-iron trades, and never became a major steel-producing area; for it lacked both good coking coal and ready access to either domestic or imported ore.[1] Production in Sheffield rose almost threefold between 1880 and 1913 to reach 879,000 tons, although its share declined from a third to a fifth.[2] The only new centre to appear was Scunthorpe, where steel production began in 1890; but, unlike continental Europe, there was no major development on the Jurassic orefields, mainly because of the ease with which high-grade ores could be imported and the lack of readily accessible supplies of coking coal.[3]

Engineering industries. Users of iron and steel were often to be found in the same localities as blast furnaces and steelworks, and the steel-rail trade in particular was closely associated with steel making. But there were differences; the old-established districts such as the west Midlands and Sheffield continued as important centres for the manufacture of iron and steel products, but no large iron- and steel-using industries developed inland along the Jurassic outcrop, and the market for iron and steel was much more widely dispersed than their manufacture. Engineering, which had been largely confined at mid-century to the making of textile machinery, stationary steam-engines, locomotives and rolling stock, now became increasingly diversified, especially after 1875;[4] Marine engineering from the 1850s, the manufacture of gas engines from the 1870s – all provide examples of branches of engineering which rose to prominence in this period; and each in turn required the making of machine tools, which were becoming increasingly important for mass production of articles like guns and sewing machines.[5] Not all were widely distributed, although the increasing use of electrical and gas engines enlarged the choice of possible sites; marine engineering was mainly associated with shipbuilding, especially in north-east England, agricultural machinery with towns such as Ipswich and Lincoln in the leading arable districts and the cycle industry with the west Midlands. The manufacture of motor cars, which began in 1896 with

[1] W. K. V. Gale in M. J. Wise (ed.), 208; G. C. Allen, *The industrial development of Birmingham and the Black Country* (London, 1929), 238–43; J. C. Carr and W. Taplin, 79.

[2] H. G. Roepke, 83. For Great Britain, from 28% to 11%.

[3] J. H. Clapham (1938), 147; J. C. Carr and W. Taplin, 156; D. C. D. Pocock, 129.

[4] J. B. Jefferys, *The story of the engineers 1800–1945* (London, n.d.), 52; G. C. Allen, *British industries and their organisation* (London, 4th edn, 1959), 135.

[5] G. C. Allen (1959), 135.

the formation of the Daimler Motor Company, was established in the west Midlands partly by accident, although an area with such varied manufactures, especially in small metal trades, had obvious attractions for an assembly industry of this kind. Employment in car making rose from 151 in 1900 to 34,000 in 1913, partly in newly established firms and partly in firms which had turned to car manufacture from cycle production, and by 1913 the two industries had become the largest single trade in the west Midlands.[1]

The increasing scale of operation, and the mechanisation of many branches of engineering, led to larger factories and to some changes in location; for example, the hand-made nail trade of the Black Country rapidly declined in face of competition from factory nail-making established in Leeds and elsewhere.[2] The major locational change in engineering was in shipbuilding. Before iron was commonly used in the construction of ships, shipbuilding had been widespread, with London and the Thames estuary as the leading area.[3] When iron and, after 1880, steel became the chief material, and increased quantities of coal were needed in shipbuilding, London was at a disadvantage, and the organisation of the industry on Thames-side also helped to raise costs of production.[4] The cheaper classes of ships came to be made in increasing numbers in northern shipyards on the Clyde, the Tyne, the Wear and the Tees, and the Admiralty also began to place orders there. London's share declined both relatively and absolutely, and it became dependent on ship-repairing and the making of expensive or experimental ships.[5] Between 1854 and 1914, employment in shipbuilding on the north-east coast increased tenfold, and by 1914 some 87% of the tonnage launched in England was built in this area.[6]

Textile industries

The distribution of the textile industries underwent no fundamental change, although their relative importance declined as other branches of manufacturing were developed, and their share of total employment almost halved between 1851 and 1911.[7] Employment in cotton manufactures

[1] *Ibid.*, 174; G. C. Allen (1929), 297; S. B. Saul, 'The motor industry in Britain to 1914', *Business History*, v (1962–4), 24.
[2] G. C. Allen (1929), 227, 272–4.
[3] S. Pollard, 'The decline of shipbuilding on the Thames', *Econ. Hist. Rev.*, 2nd ser., III (1950–1), 72.   [4] *Ibid.*, 72; G. C. Allen (1959), 157.
[5] S. Pollard, 76, 78, 85.   [6] A. E. Smailes, 182; W. Smith, 391.
[7] B. R. Mitchell with P. Deane, 60.

DHG

grew, as did the number of spindles, and factories increased in size and productivity; thus, the average cotton weaver managed 1.6 looms in 1850 and 2.2 in 1882.[1] Lancashire's dominant role remained unchallenged, but between 1841 and 1884 the two main branches of the cotton industry became segregated into a southern spinning district, based on Oldham, and a northern weaving district, based on Blackburn, largely as a consequence of the earlier development of spinning as a factory industry in south Lancashire and the co-existence of suitable physical resources and a large reserve of workers with experience of textiles among the handloom weavers of Rossendale and east Lancashire when weaving was being mechanised.[2] By 1884, 62% of the looms were in northern districts and 78% of the spindles in southern districts, and this regionalisation, which was accompanied by specialisation of plant, subsequently became more marked.[3]

Employment in the woollen and worsted industries changed relatively little, although the number of power looms more than doubled between 1856 and 1874 and the consumption of wool trebled between 1850–4 and 1909–13.[4] Wool was not predominantly a factory industry until about 1860 and the concentration of manufacturing into factories was not completed until the last quarter of the nineteenth century.[5] Yorkshire, which already accounted for 70% of employment in wool and worsted in 1851, remained the leading area;[6] indeed, worsted weaving became even more concentrated in the West Riding, until by 1904 more than 93% of looms were to be found there. In the early years of the twentieth century, however, worsted spinning was growing rapidly in the east Midlands to meet the needs of hosiery knitters, although its share of output remained small.[7] Most of the remaining outliers of woollen manufacture were gradually eliminated, with the exception of the Gloucestershire industry based on Stroud, and even this declined. Increasing specialisation by area was also a feature of the Yorkshire textile industry, but this was based, not upon function (as in Lancashire), but on product, with worsted manufacture concentrated in the north-west around Bradford and Halifax, and woollens

[1] J. H. Clapham (1932), 8; J. H. Clapham (1938), 175.
[2] J. Jewkes, 'The localisation of the cotton industry', *Economic History*, II (1930), 95–6; W. Smith, 474–5.
[3] J. Jewkes, 96; S. J. Chapman and T. S. Ashton, 'The sizes of business, mainly in the textile industries', *Jour. Roy. Stat. Soc.*, LXXVII (1914), 491–2.
[4] B. R. Mitchell with P. Deane, 198; G. C. Allen (1959), 254.
[5] *Ibid*, 253.
[6] W. Smith, 135.
[7] *Ibid*, 435, 449, 451.

in the south-east between Huddersfield and Wakefield; there was also a contrast in organisation, with the great majority of firms in the woollen industry combining spinning and weaving, while those in worsted manufacture exhibited a high degree of specialisation.[1]

The silk industry became a less important branch of textile manufacture after the abolition of protective duties in 1861; in the town of Derby, for example, the twenty-one manufacturers of silk in 1864 had been reduced to two in 1912, and the number of employees from 4,760 in 1851 to 300 in 1911.[2] Artificial silk did not take its place until 1885, and the first important factory was opened only in 1900.[3] On the other hand, the hosiery industry grew steadily and was transformed from an almost entirely domestic craft into a factory industry. In 1860, domestic workers, scattered in Leicestershire villages, outnumbered factory workers by more than twelve to one, but between 1850 and 1892 there was a tenfold drop in the number of hand-frames, although some 5,000 remained in 1892 and there were still 25,000 outworkers in 1907.[4] The transformation of the clothing trade from a domestic craft to a factory-based activity was also in progress during this period, especially in the main textile areas; in Leeds, for example, there were some 7 or 8 factories in 1881 and 54 in 1891.[5] Nevertheless, clothing remained the branch of manufacturing industry with the largest number of outworkers; in 1901 these constituted 80% of all those working at home.[6]

Other industries

Boot and shoe manufacture also became a factory industry in this period, although for several decades the industry remained heavily dependent on outwork, and much of what was done in factories was not mechanised. In the 1850s the making of boots and shoes was a widespread rural craft, although it already gave employment to proportionately four times as many people in Northamptonshire as in the country as a whole.[7] Its development in this county was helped by the decline of other domestic

[1] Ibid., 440–7.
[2] D. M. Smith, 'The silk industry of the east Midlands', East Midland Geographer, III, pt. 1 (1962), 29.
[3] J. H. Clapham (1938), 181.
[4] J. H. Clapham (1932), 85–6; J. H. Clapham (1938), 178–9.
[5] J. H. Clapham (1938), 183.
[6] Rep. S.C. on Home Work, 238 (P.P. 207, vi).
[7] P. R. Mounfield, 'The footwear industry of the east Midlands: (iii) Northamptonshire, 1700–1911', East Midland Geographer, III, pt. 8 (1965), 434.

crafts and by the absence of competing industries. Factories were built in Northampton and subsequently in other towns, notably in the Ise valley.[1] Leicestershire also became an important industrial county, especially for lighter footwear, although here the predominance of a single centre, Leicester, was more marked.[2] In 1871, 71% of those working in boot- and shoe-making were still outworkers, but by 1904, most of those employed in this trade were working in factories.[3]

The chemical industry, which roughly doubled its labour force between 1851 and 1911, also rose to prominence, although its great diversity and the lack of satisfactory statistical data make it difficult to measure changes.[4] New branches, such as that of dyestuffs, came into being, and old branches, such as soap-making, were transformed; but the principal regional change was the decline of the established heavy chemical industry in north-east England and the development of alkali manufacture in Lancashire and Cheshire. In 1867 more than half the output of alkali was manufactured on Tyneside by the Leblanc process and production of chemicals here was at its maximum in 1880; by 1895 only four works survived out of the twenty-five which had been in production in the early 1870s.[5] By 1864, there were already twice as many chemical factories in Lancashire and Cheshire as on Tyneside, and the granting of a licence for the Solvay process to the firm of Brunner Mond and Company gave this area an added advantage, for this was replacing the Leblanc process; the growth of Widnes provides a striking example of industrial specialisation.[6] Production of salt from the Cheshire brine fields, on which both areas depended, grew rapidly until the 1880s, but the Tees-side field, although discovered in 1859, was not worked until 1881, too late to save the chemical industry in the north-east.[7]

Many other new and old-established industries grew in this period. Rubber manufacture, which had developed under the protection of the textile industry, was greatly stimulated by the demand for insulating

[1] Ibid., 437.

[2] P. R. Mounfield, 'The footwear industry of the east Midlands: (iv) Leicestershire, 1700–1911', East Midland Geographer, IV (1966), 9, 17.

[3] J. H. Clapham (1932), 119; J. H. Clapham (1938), 181.

[4] J. H. Clapham (1938), 171.

[5] P. C. G. Isaac and R. E. A. Allan (eds.), 162; L. F. Haber, The chemical industry during the 19th century (Oxford, 1958), 152.

[6] J. H. Clapham (1932), 106; S. Gregory et al. in W. Smith (ed.), Scientific survey of Merseyside (Liverpool, 1953), 255.

[7] K. L. Wallwork, 'The mid-Cheshire salt industry', Geography, XLIV (1959), 174; J. H. Clapham (1932), 515; L. F. Haber, 155.

material and for bicycle and car tyres. Its labour force grew fourteen-fold between 1851 and 1901, and it became increasingly localised in Lancashire and Cheshire, which provided less than a fifth of all employment in 1871 and more than a third in 1901.[1] The repeal of paper duties in 1861 gave a fillip to paper manufacture and production increased eightfold between 1865 and 1907, with the Home Counties and Lancashire as the chief centres. Paper box and bag manufacture also grew rapidly.[2]

Many new industries were concerned with the making or processing of foodstuffs, often at ports. One of the most important and interesting geographical changes was in flour milling. In 1851, milling, largely in wind and water mills, had been widely distributed in the grain-growing counties, and half the mills each employed only one or two people.[3] New methods of milling, using metal rollers, made large mills more economical to run, and the increasing volume of imports of hard wheats for bread-making favoured the erection of large mills at the ports, especially London, Liverpool, Hull and Bristol.[4] By 1912, port mills produced nearly two-thirds of all flour used in the country, and, although some inland mills, such as those at Cambridge and York, were saved by mechanisation, many of the older mills were abandoned.[5] In the country as a whole there was only one mill grinding wheat in 1906 for every nine in 1884; and in Colchester, where there were twelve windmills in 1866, only one survived in 1906.[6]

Although the economies of the coalfield industrial regions continued to be dominated by the staple industries in which they had specialised, manufacturing became more widespread and more diversified, especially in Greater London, which retained its position as the largest single centre of manufacturing industry, and in the west Midlands, like those of London, which had long been very varied. The size of plant also tended to grow and large industrial empires, embracing a range of manufactures in different parts of the country, were appearing, especially in engineering.[7] Yet, although employment in manufacturing industry

---

[1] W. Woodruff, *The rise of the British rubber industry during the 19th century* (Liverpool, 1958), 118.

[2] A. D. Spicer, *The paper trade* (London, 1907), 100, 104, 173.

[3] J. H. Clapham (1932), 34, 88; J. H. Clapham (1938), 185.

[4] J. H. Clapham (1932), 89.

[5] W. Smith, 552; J. H. Clapham (1938), 185.

[6] *The Tariff Commission*, vol. 3, *Rep. of the agricultural committee* (London, 1906), para. 323; W. Marriage in W. Page and J. H. Round (eds.), *V.C.H. Essex*, II (1907), 447.   [7] J. H. Clapham (1938), chap. 4.

continued to grow and its share of all employment rose slightly, it ceased to be the only or, in some instances, the major cause of urban expansion; for employment in service industries was rising rapidly.

TRANSPORT AND TRADE

In the growth of the economy, and especially in its geographical manifestations, improved communications played a major part, internally by speeding the flow of goods and services, chiefly through the extension of the railway system, and externally by the development of the steam-ship, which brought both markets and sources of raw materials closer in time. Changes in the geography of communications chiefly concerned the development of the railway net, for the canal and the turnpike had already been eclipsed by 1850 and, while the tram provided effective competition within the larger towns towards the end of the century, the motor vehicle was not a potential threat until after 1900.

Railways

Except along the Welsh Borderland and in the south-west, the main outlines of the railway net had already been established by 1850, and the subsequent activities of railway companies were largely concerned with its elaboration and intensification (Fig. 136). The scale of these later changes can too readily be minimised; the route mileage in England and Wales (separate figures are not available) rose from 5,132 miles in 1850 to 11,789 miles in 1875, an increase of 130%, and reached 16,223 miles in 1912 (216%).[1]

The pattern of growth was extremely complex. Railways were promoted for a variety of reasons by many companies, and the details were often to be explained by quite ephemeral local circumstances;[2] but a number of large companies with much wider perspectives came increasingly to predominate. By the early 1860s, the North Eastern Railway had acquired a territorial monopoly in Durham and the East and North Ridings; and in 1862 amalgamation in East Anglia led to another regional enterprise, the Great Eastern, which also enjoyed a virtual monopoly of the

[1] Rep. of the Commissioners of Railways for the year 1850, VII (P.P. 1851, xxx); General Rep. by Captain Tyler in regard to . . . Railway Companies, 2 (P.P. 1876, lxv); Railway Returns . . . for the year 1912, xxi (P.P. 1913, lviii).

[2] J. H. Appleton, 'The railway network of southern Yorkshire', Trans. and Papers, Inst. Brit. Geog., XXII (1956), 169.

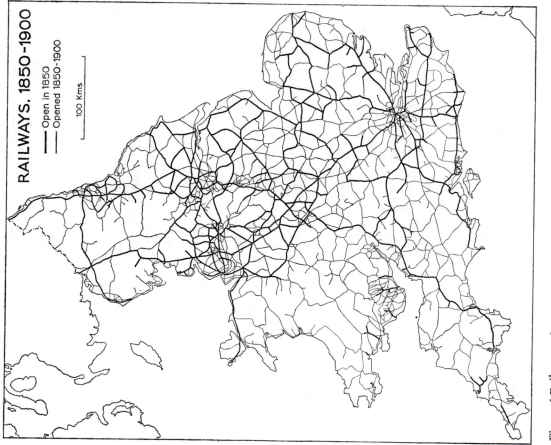

RAILWAYS, 1850-1900

Open in 1850
Opened 1850-1900

100 Kms

Fig. 136 Railways, 1850–1900
Based on: (1) G. Bradshaw, *Map of Great Britain showing the railways*
(London and Manchester, 1850); (2) J. G. Bartholomew, *The survey atlas of
England and Wales* (Edinburgh, 1903), plates 6 and 7.

territory it served.[1] Elsewhere, other groupings emerged – the Great Western, the London & South Western, the London & North Western, the Midland, and the Great Northern, although their predominance was in part achieved indirectly through the leasing of running rights over other lines, which they had often been instrumental in promoting, or through representation on the boards of nominally independent companies. No single company emerged in the south-east where there were three smaller groups, the London, Chatham & Dover, the South-Eastern and the London, Brighton & South Coast, although the last also enjoyed a local monopoly within its more limited area. Competition between them was most acute where they came into contact in areas which still awaited development or which provided lucrative traffic.

The role of competition should not be exaggerated, since there were increasingly informal agreements and understandings, but it undoubtedly played a major part in shaping the railway network by encouraging the construction of alternative routes and stations. Central London's sixteen terminal stations, several of them in close proximity, Manchester's five, Salisbury's three and Tunbridge Wells' two, were largely the product of such competition. Rivalry between the Great Western and the London & South-Western, which was aggravated by a difference in gauge, led to alternative routes from London to Plymouth, both of them completed after 1850; while competition between the South-Eastern and the London, Chatham & Dover similarly provided alternative routes between London and Dover.[2] Duplications were most striking at the local scale, as in the Leen valley (Notts.) which carried three rival lines, or in the Avon valley downstream from Bath, where there were two.[3] Some lines were built primarily to thwart competitors, as with the Hurstbourne–Fullerton line, constructed by the London & South-Western to prevent the Didcot, Newbury & Southampton Railway from reaching Bournemouth.[4] Others were intended to make routes shorter and more competitive, such as the South-Eastern line through Sevenoaks which was built (despite a two-mile tunnel and deep cuttings through the chalk) to reduce the time of the journey to Dover, formerly reached via Redhill.[5] Some lines were

---

[1] C. E. R. Sherrington, *Economics of rail transport in Great Britain* (London, 1928), 123; J. Simmons, *The railways of Britain* (London, 1961), 22.

[2] J. Simmons, 24; H. P. White, *A regional history of the railways of Great Britain,* II (*Southern England*), (London, 1961), ch. 3.

[3] J. H. Appleton, *The geography of communications in Great Britain* (Hull, 1962), 158–9.

[4] H. P. White, 143.

[5] *Ibid*, 44.

promoted because services were considered poor or expensive in the absence of competition, as with the Hull, Barnsley & West Riding line.[1]

The reasons for the construction of lines were legion and often local. Only four acts, all before 1850, had authorised the laying of more than a hundred route-miles by a single company, and many main lines were created piecemeal at different dates by a variety of promoters; the Paddington–Penzance line was built by eight companies in fifteen instalments spread over sixty-eight years.[2] Between 1850 and 1870, many of the main lines in the west Midlands and south-west England were completed, as well as a large number of branch lines. In 1868 a further trunk route from London was opened from St Pancras to Bedford, promoted by the Midland Railway to overcome congestion on the Great Northern lines from Hitchin to King's Cross, and by the mid-1870s no important English town was without a reasonably direct route to London.[3] Subsequent additions were mainly branch lines or cut-offs, designed to speed traffic by avoiding some of the now-unnecessary meanderings of the earlier routes, such as the Sway line, opened in 1888 to shorten the journey from London to Bournemouth, or the Badminton cut-off, linking Paddington more directly with South Wales via the Severn Tunnel, which had been opened in 1886.[4] Other gaps were filled by light railways, of which 1,542 miles were authorised between 1896 and 1914, to carry agricultural produce, minerals and passengers (although this figure also includes some tramways).[5] Two further trunk lines were also constructed, the one promoted by the Midland Railway Company between Settle and Carlisle and opened in 1876 to provide a third route to Scotland, the other, the Great Central, opened in 1899 and following a circuitous route from St Marylebone to Sheffield.[6] By 1914, only the remote moorland areas of the west and north lay more than five miles from a railway.

Few lines were initially built to carry passengers for short distances, and the spacing of stations, on the main lines at least, was often too great for them to fulfil this need; thus, West Drayton, $13\frac{1}{4}$ miles from Paddington, was for some time the first station west of the terminus.[7] As demand grew

[1] J. H. Clapham (1932), 183–4.   [2] J. H. Appleton (1962), 160–1.
[3] C. E. R. Sherrington, 97–8; C. H. Ellis, *British railway history*, 2 vols. (London, 1959), II, 13.
[4] H. P. White, 161; C. H. Ellis, 230
[5] C. Klapper, *The golden age of tramways* (London, 1961), 35.
[6] C. E. R. Sherrington, 99, 134.
[7] H. P. White, *A regional history of the railways of Great Britain*, III (*Greater London*), (London, 1963), 110.

around London, other stations were added and new lines constructed to produce an elaborate network.[1] Electrification was introduced on surface lines in south London in 1906, and 109 miles had been electrified and 127 miles partly electrified by 1912.[2] Other less complex suburban networks developed around the northern cities, notably Liverpool, Manchester and Newcastle, and some of these lines, too, were electrified between 1903 and 1906.[3]

The extension of the railway network was not a continuous process; the rate of activity was determined both by the state of the economy and by the financial circumstances of individual companies. Lines were sometimes left uncompleted for several years, as that between Basingstoke and Salisbury, and gaps, such as that between Horsham and Leatherhead, lay unfilled for long periods.[4] Lines were also closed as better routes were developed, as amalgamation removed competition, as traffic declined, or simply in recognition of the fact that too many railways had been provided; but while some lines were closed in almost every year between 1850 and 1914, the total mileage affected was comparatively small.[5]

The creation of new routes was not the only way in which the railway's influence extended; additional track capacity was also important. Many lines were opened with only a single track, a second being added as demand grew. Physical difficulties sometimes caused delay; doubling of the Great Western line from Exeter to Plymouth was not completed until 1905 for this reason.[6] On the broad-gauge lines, doubling sometimes accompanied the conversion to standard gauge, and the opportunity was also taken to straighten or improve routes; thus several stretches of the original line from Plymouth to Penzance were abandoned.[7] As traffic increased on busy lines, additional tracks were laid, among the earliest being a third track from Bletchley to London, opened in 1859.[8] By 1900, most of the trunk lines out of London had four tracks, at least in part, and on

1 Ibid., 82.
2 P. Hall in J. T. Coppock and H. C. Prince (eds), Greater London (London, 1964), 66; Railway Returns...for the year 1912, 96 (P.P. 1913, lviii).
3 C. E. R. Sherrington, 107, 125.
4 H. P. White (1962), 105, 156.
5 J. A. Patmore, 'Railway closures in England and Wales 1836–1962', Trans. and Papers, Inst. Brit. Geog., XXXVIII (1966), 107.
6 D. St J. Thomas, A regional history of the railways of Great Britain, 1 (The West Country), (London, 1960), 54.
7 Ibid., 106.
8 C. E. R. Sherrington, 90.

very busy sections there were as many as six. On the Great Western route from Paddington, four tracks had been laid to Southall by 1877, to Maidenhead by 1884 and to Didcot by 1889.[1] On the Great Northern route from King's Cross the need for additional tunnels handicapped quadrupling, but much of the line to Peterborough had four tracks by 1900.[2] Where additional tracks were not possible a new line was sometimes constructed, for example south of Croydon.[3] Along a few routes, diminished traffic led to a reduction in the number of tracks, but in general track mileage increased nearly as fast as route mileage.[4] Intermediate stations were also opened to tap additional traffic.

The railway network did not grow haphazardly, but more lines were constructed than were required on economic grounds, and routes were often excessively tortuous because they were conceived in a variety of local circumstances. Its development was not only of economic significance; it represented one of the major landscape changes of the period, whether in imposing structures such as Brunel's bridge at Saltash or Scott's St Pancras station, or unobtrusively, as along many rural lines which, mellowed and blending with the countryside, no longer merited Wordsworth's protest against the 'rash assault'.[5]

Canals and roads

Railways dealt a heavy blow to canal promoters, for the canals' share of traffic fell steadily (although the total tonnage carried changed little). Individual canals often fared very differently. On some, such as the Aire and Calder, trade expanded; on the dense network of the industrial Midlands it held its own; but on most rural canals it declined. On the Oxford Canal, for example, receipts halved between 1848 and 1858, and on the Kennet & Avon the tonnage carried fell to less than one half between 1858 and 1898.[6] Statistics submitted to the Royal Commission on Canals in 1905 showed that there had been 3,162 miles of canal open in 1898 (the latest available returns), and that several hundred miles of canal had been

---

[1] E. T. MacDermot, *History of the Great Western Railway*, II (London, 1931), 322–93.

[2] H. P. White (1963), 160; C. E. R. Sherrington, 129.

[3] H. P. White (1963), 80.

[4] D. St J. Thomas (1960), 138.

[5] A. J. George (ed.), *The poems of Wordsworth* (Boston and New York, 1932), 778.

[6] *R.C. on Canals and Inland Navigations of the United Kingdom*, Appendix I, statement I (P.P. 1906, xxxii); C. Hadfield, *British canals* (London, 1950), 194; C. Hadfield, *The canals of southern England* (London, 1955), 299.

closed; but an official witness admitted that the latter figure included only those canals for which a certificate of abandonment had been issued, and many more miles must have been neglected, or under-used, or illegally abandoned.[1] Some canals were converted for use by the railways, and lines were constructed along their banks; at least one, the Bude Canal, was adapted for water supply.[2] Some canals had been opened after the beginning of the railway age, but only one major canal was constructed, the 35½ mile Manchester Ship Canal which, opened in 1894, transformed Manchester into a major seaport.[3]

The railways quickly acquired the long-distance passenger traffic which the turnpikes had carried, and from 1860 dis-turnpiking was actively pursued; only 184 trusts remained in 1881 and the last disappeared in 1895.[4] The care of the turnpikes then reverted to the parish, which had long been responsible for the much greater mileage of unturnpiked roads. Standards of maintenance were uneven, but the roads were generally neglected and the best turnpikes were the first to deteriorate.[5] In 1888, responsibility for main roads passed to the new county councils and there was some improvement, although not all turnpikes became main roads.[6] Yet there was no national road network; the roads carried only local traffic and, until motorists became numerous, there was no powerful lobby to press for their improvement.[7]

Matters were different in the towns where granite setts and other improved surfaces appeared, where the first experiments in tarred surfaces were made, and where new roads were built, such as Shaftesbury Avenue and Charing Cross Road in central London.[8] In the large towns, trams, at first horse-drawn, began to compete with the long-established horse buses and, later, with short-distance travel by rail.[9] Their use was facilitated by the 1870 Tramways Act and by various technical developments; horses were replaced by steam power and, in the 1890s, by electric traction, and

1 R.C. on Canals, Evidence of Jekyll, Qs. 14,76, 86; ibid., Appendix 1, statements 2 and 5 (P.P. 1906, xxxii).
2 C. Hadfield (1955), 321, 327.
3 C. Hadfield (1950), 226.
4 S. and B. Webb, English local government: the story of the king's highway (London, 1920), 222.
5 R.C. on Transport, Final Report, 1931, 10 (P.P. 1930–31, xvii).
6 C. J. Fuller, 'The development of roads in the Surrey–Sussex Weald and coastlands between 1700 and 1900', Trans. and Papers, Inst. Brit. Geog., XIX (1953), 47.
7 Ibid. 49; R.C. on Transport, Final Report, 11 (P.P. 1930–31, xvii).
8 J. H. Clapham (1938), 375; P. Hall (1964), 57.   9 H. P. White (1963), 79.

the tramway mileage in England and Wales increased from 194 in 1878 to 1,187 in 1902 and had reached 2,213 by 1913–14.[1] From 1903 motorbuses began to replace horse-drawn vehicles. In 1905 the London General Omnibus Company had 70 motorbuses out of its 1,417 buses, and had a further 700–800 on order; elsewhere, changes were inhibited by the fact that local authorities had dual roles as tramway operators and as vehicle-licensing authorities.[2]

Outside the towns there was little mechanised transport, although some trams ran between towns in north England, for example between Leeds and Wakefield.[3] Approximately two thousand miles of new roads had been made under enclosure awards after 1845, but little was done to the existing routes.[4] Improvements in rural road surfaces came only with the development of motor transport, and then only slowly. Although most motor vehicles were restricted to a maximum of four miles per hour before 1896, to twelve between 1896 and 1903 and to twenty thereafter, their numbers grew rapidly, reaching 175,588 in 1911–12.[5] They created so much dust that tarred surfaces became essential; by 1908 there were thought to be 1,269 miles of tarred roads in England, and by 1913 there was a larger mileage of dustless roads than in any other country.[6]

These developments in transport were not the principal changes in the geography of England after 1850; nevertheless, by facilitating and cheapening the movements of both goods and people, they played a major part in the agricultural, industrial and population changes, weakening the barriers of distance and allowing a wider separation of produce and market, mine and factory, residence and workplace.

Maritime trade, coastwise and overseas

Railways were also challenging the immemorial traffic by sea around the coasts of Britain. By 1880, over 60% of the coal imported into London arrived by railway and not by coastal collier; but the challenge was met by faster and larger colliers and by improved unloading facilities along the Thames estuary. 'All in all,' concludes one study, 'coastal shipping

[1] *Return of street and road tramways and light railways, 1914*, 4 (P.P. 1914, lxxvii).
[2] *Rep. R.C. on Motor Cars*, vol. I, p. 6 (P.P. 1906, xlviii); J. H. Clapham (1938), 376–7.
[3] J. H. Clapham (1938), 375; C. Klapper, 125.
[4] *Inclosure Commissioners, Thirty-second Ann. Rep.*, 4 (P.P. 1877, xxvi).
[5] *Rep. R.C. on Motor Cars*, vol. I, pp. 2–3 (P.P. 1906, xlviii); K. G. Fenelon, *The economics of road transport* (London, 1925), 22.
[6] J. W. Gregory, *The story of the road* (London, 1931), 254; S. and B. Webb, 248.

withstood railway competition exceptionally well. It accounted for a falling proportion of the country's internal trade from 1845 onwards but in absolute terms its volume increased, particularly in the forty years after 1865.'[1] Even so, it was the larger ports that benefited. The coming of the railway and the increasing size of ships meant the decline of many smaller harbours serving local hinterlands; ports such as Whitby, Boston, King's Lynn and Rye were falling into reduced circumstances,[2] Moreover, the railways themselves were entering the port and shipping business, and came to be large dock owners in such ports as Grimsby, Hull and Southampton, ports with growing overseas connections.

Vital for the economy of the country as a whole was the growth in foreign trade and the developments in shipping which had made 'transport costs little heavier between continents than they had once been between counties'.[3] In 1833 sailing ships had accounted for 94% of British merchant tonnage, but while the tonnage under sail remained constant between 1850 and 1885, that of steamships rose by 4 million, and their capacity had become twice that of the whole merchant fleet in 1850.[4] By the end of the century virtually all ships were being built of steel and powered by steam; such extra capacity was needed to carry a greatly expanded volume of trade, which rose approximately sevenfold between 1850 and 1913.[5] Throughout the period most imports were of food and raw materials, but these contributed a declining proportion; by 1913 manufactures accounted for a quarter by value, for Great Britain was no longer the workshop of the world.[6] The share of manufactures in exports declined, largely because of the rise in shipments of coal, which accounted for 80% of all exports by weight in 1913, and was also a major contributor to coastwise trade.[7]

London remained the leading port, though its share of trade declined (Figs. 138 and 139). In 1843 it had accounted for over half the customs receipts, but it handled only a third of imports in 1913, and the value of its export trade was exceeded by that of Liverpool, the second port. Hull

[1] H. J. Dyos and D. H. Aldcroft, *British transport: an economic survey from the seventeenth century to the twentieth* (Leicester, 1969), 210.
[2] J. H. Clapham (1932), 522-3.
[3] J. H. Clapham (1938), 73.
[4] J. H. Clapham (1932), 519; J. H. Clapham (1938), 158-9.
[5] A. G. Kenwood, 'Port investment in England and Wales', *Yorks. Bull. Econ. and Soc. Research,* XVII (1965), 156.
[6] W. Smith, 169.
[7] *Ibid.,* 636.

had reached third place, assisted by the railways, the coal trade and the increasing size and number of steam trawlers, but it handled little more than a tenth.[1] Such a large increase required heavy investment at both new and established ports.

TOWNS AND CITIES

Between 1851 and 1911 the area occupied by towns in England and Wales almost certainly increased more rapidly than the population they contained. As living standards improved, new communications were developed and new activities were undertaken in towns; the urban area is estimated to have reached two million acres by 1900.[2] This expansion was achieved mainly by the growth of existing towns; and at both dates most of the larger towns were to be found in the industrial areas established on or near the coalfields on the Pennine flanks, in north-east England, in the west Midlands and around the coast. Throughout the period, London was by far the largest centre, with 2,685,000 inhabitants in 1851 within what was later to be defined as the conurbation, and 7,252,000 in 1911 within the same boundary, an increase of 170%.[3] The most striking feature of Fig. 137, which shows changes in the distribution of towns with 20,000 or more inhabitants in 1911 (as listed in the census reports), is the development of towns with between 20,000 and 50,000 inhabitants in locations away from the coalfields, notably on Tees-side, in the east Midlands, in the Home Counties, and around the coasts; for by 1880 the powerful pull of the north had weakened.[4] Outside these areas were to be found mainly stagnating market towns and decaying ports.

Expansion and redevelopment

The most obvious physical expression of urban expansion was the extension of the built-up areas of existing towns, although its complexities are such as almost to defy generalisation; for each town, and each development in it, was a special case, in which personal initiative and personal preference played a large part. Small towns grew by accretion, by the addition of single houses and small estates around their boundaries, and on land left undeveloped in earlier periods. This process has been studied in detail in the Northumberland town of Alnwick, where 349 houses covering 64

[1] *Ibid*, 165, 170.   [2] R. H. Best and J. T. Coppock, 229.
[3] J. T. Coppock in J. T. Coppock and H. C. Prince (eds.), *Greater London* (London, 1964), 34.
[4] W. Ashworth, *The genesis of modern British town planning* (London, 1954), 88.

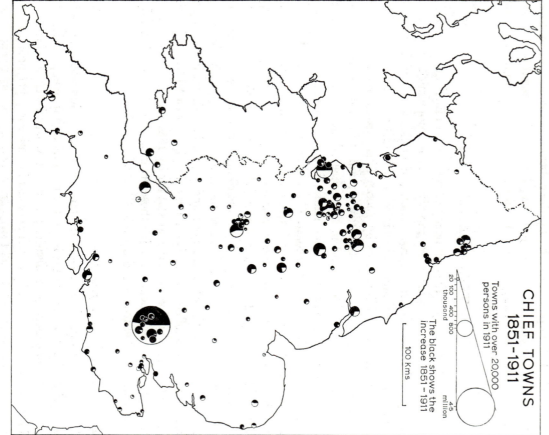

CHIEF TOWNS 1851–1911

Towns with over 20,000 persons in 1911

20
100
400
800
thousand

The black shows the increase 1851–1911

4·5 million

100 Kms

Fig. 137 Chief towns, 1851–1911
Based on: (1) *Census of 1851: Population Tables, I* (P.P. 1852–3, lxxxv, lxxxvi); (2) *Census of 1911*, vol. 1, pp. 10–40 (P.P. 1912–13, cxi). The smaller symbols within that for the county of London represent the Greater London boroughs that lay outside the county. Two provincial towns (Bath and Macclesfield) decreased in population, and are accordingly represented by incomplete circles.

acres were built between 1851 and 1891 in developments ranging in size from a single house to 70 houses, although the population remained at about 7,000 throughout the period.[1] In the larger towns, urban expansion was more complex and was greatly affected by the revolution in transport. Towns which grew rapidly in the early phase of the industrial revolution had remained compact, for modes of transport and standards of living had dictated that most people should live near their places of work. The cab and the horse-drawn omnibus, which became common in the 1830s, were essentially forms of transport for the middle classes, and major changes came only when the railways and tramways (and later the motor-driven omnibus) began to provide cheap travel for a working population whose standards of living were also rising. This was especially so in London and in the great provincial cities of the Midlands and north England, where the tramways permitted long ribbons of housing to develop along the main roads and so link surrounding towns and villages.

Changes in housing standards and housing tastes also affected the outward spread of towns and cities. A substantial villa with half an acre or more of ground increasingly became the aim of the professional or business man, and even the terrace house of the clerk or skilled worker had a garden, however small. But as the tide of building spread outwards, more intensive use was often made of land, and the small parks established earlier were sometimes subdivided to accommodate smaller houses at higher densities. After the Public Health Act of 1875, local authority bye-laws, although not universal and often inadequately enforced, prescribed maximum densities and affected street layouts. From this period date many of the rows of terrace houses laid out in grid pattern which are particularly well shown in rapidly growing industrial towns such as Middlesbrough. Such houses, however unattractive aesthetically, represented a great advance on the back-to-back housing of the early phases of the industrial revolution. Their uniformity was accentuated by the increasing use of similar materials, for the railways enabled builders to use other than local materials, and roofs of Welsh slate were very common in houses built during this period; but the uniformity of buildings should not be exaggerated, for areas were often developed piecemeal in a variety of styles, even if the differences were largely superficial. Building by local authorities, though possible after 1851, was not important, and, although they built more houses in the 1890s than in the four previous decades, the total

[1] M. R. G. Conzen, *Alnwick, Northumberland, Trans. and Papers, Inst. Brit. Geog.,* XXXVII (1960), 49, 75, 82–5.

remained small.[1] New churches, shops, schools and other public buildings also contributed to the enlargement of the built-up area; thus, fifteen new churches and chapels were built in Oldham between 1870 and 1888.[2]

Although outward extension into farmland was the most obvious feature of urban growth, it was complemented by extensive redevelopment in the older parts of towns as houses were replaced by shops, offices, warehouses, theatres, hotels, railway stations and the like, and as some slum clearance was begun. Like suburban development, this was especially a feature of the larger towns. The railways were a major cause of redevelopment for, while a railway station could be placed on the outskirts of a small settlement and new building take place around it, as in St Albans and Watford, the already large size of London, Manchester and other great cities made this impossible and houses had to be cleared; thus, a new goods station at Manchester involved the destruction of some 600 houses.[3] The construction of both roads and railway lines also contributed to urban redevelopment, as did the erection of civic buildings, especially in the northern industrial towns; for example, many halls were built in the 1880s and 1890s. Even in established residential areas land had also to be found for the new schools required by the 1870 Education Act.

Such redevelopment was accompanied by the dispersal of other activities such as private schools, charitable institutions and the like, which demanded space and which were unable to compete with industry and commerce for valuable land in the centre of cities. Redundant churches and ancient burial grounds were similarly converted to other uses: in central London, for example, nine churches were sold between 1860 and 1894.[4] Nevertheless, residential accommodation was the principal loser from the redevelopment of central areas, and the scale of destruction was considerable; thus in Liverpool more than 20,000 houses were demolished between 1873 and 1911.[5] Furthermore, many of the houses which remained were converted to other uses, especially offices and workshops. Slum clearance also contributed to the loss of houses and Birmingham provided one of the earliest and largest examples of municipal slum clearance in the nineteenth century.[6] Yet such clearance was expensive and was never undertaken on

[1] W. Ashworth, 91, 93.
[2] R. Millward, *Lancashire* (London, 1955), 86.
[3] T. W. Freeman, *The conurbations of Great Britain* (Manchester, 1959), 138.
[4] W. Besant, *London in the nineteenth century* (London, 1909), 249.
[5] J. P. Lewis, *Building cycles in Britain's growth* (London, 1965), 134.
[6] W. Ashworth, 97.

a very large scale in this period; large areas of slums remained and in Birmingham, for example, there were still 43,366 back-to-back houses in 1913.[1] It is not, therefore, surprising that the centres of all the larger cities began to lose population at a faster rate than it was replaced by natural increase. In Birmingham, Leeds, Liverpool and Manchester, population in the central areas began to fall during the 1840s, in Newcastle during the 1850s and in Bradford between 1871 and 1881.[2]

The rate of housebuilding varied more widely than that of population growth, although the latter was clearly a major factor. Transport developments (especially in London), the level of industrial activity, rates of emigration, and household formation were other major causes of fluctuations in the rate of building. House construction reached a peak in the late 1870s and again around the turn of the century; in the intervening troughs it proceeded at only half the peak rates.[3] There were, however, regional variations; in Bradford, for example, the earlier peak occurred in the late 1860s and early 1870s, in Liverpool in the late 1870s and early 1880s, and in Newcastle in the late 1880s.[4] The expansion of an urban area was clearly not an even or continuous process; nevertheless, between 1851 and 1911, there was a net addition of 4,118,000 to the stock of houses in England and Wales, 3,332,000 of them in urban registration districts.[5]

All these developments tended to promote social segregation within towns, although few areas could house only a single class; for middle-class residents required shops and other services, and accommodation was needed in working-class areas for doctors and other professional men. Nevertheless, both rebuilding and urban growth led to particular areas being occupied by different social groups. As the wealthy moved out to new suburbs and beyond, the houses they left were often occupied by those with lower incomes, and were subdivided into flats and tenements.

The growing towns

Although most of the major towns were established by the 1850s, urban growth was not uniform and there were several cases of very rapid urbanisation, of which Barrow-in-Furness, Crewe, Middlesbrough and Swindon were among the most striking examples. The development of Barrow and Middlesbrough was chiefly associated with changes in sources of iron ore, the discovery of a body of haematite near Barrow and the rediscovery of the low-grade Jurassic ores in Cleveland. Dependence on

[1] T. W. Freeman, 85.   [2] *Ibid*, 82, 172, 179, 187.   [3] J. P. Lewis, 316–17.
[4] *Ibid*, 323.   [5] *Ibid*, 332.

local supplies was short-lived; but, whereas Middlesbrough continued to expand and to rely on an increasing volume of imported ore, the fate of Barrow, in a cul-de-sac some distance from its source of coking coal, lay for long in the balance. Barrow's expansion began in the 1850s and it was claimed by one observer that not even in America was there another town which grew as fast.[1] The opening of a steelworks in 1858, of a rail link to the Durham coke supplies in 1861, and the building of one of the most modern shipyards, led to a rapid development and its population had reached nearly 18,000 by 1871. Unfortunately its prosperity did not last; ore began to run out, the market for its rails collapsed, and it survived only because of financial support from the duke of Devonshire.[2] In Middlesbrough, the need for imported high-grade ore for Bessemer steel and the increasing importance of shipbuilding led to a concentration of activity around the estuary of the Tees and population rose from 18,992 in 1861 to 91,302 in 1901.[3] On a smaller scale, Scunthorpe provided another example of rapid growth resulting from developments in the iron and steel industry.[4]

The development of both Crewe and Swindon was closely linked with the location of railway workshops. Unlike continental Europe, the railway age in England followed industrialisation, and the railways served to link and strengthen established towns rather than to create new ones; the chief exceptions arose from the building of workshops and the creation of what were, in effect, company towns. By the 1850s, Crewe, where a company village had been established in the 1840s, had become not only a major railway junction but also an important manufacturing centre.[5] A rail rolling mill was opened in 1853, the construction of the London & North Western's locomotives was localised there in 1857, and a Bessemer steel plant erected in 1864.[6] At this time the company had more than 3,000 workers on its pay roll, but as population grew from 4,500 in 1851 to 42,000 in 1901, there was increasing industrial diversification.[7] Swindon was chosen by the Great Western Company as the site for the interchange

[1] S. Pollard, 'Barrow in Furness and the seventh duke of Devonshire', *Econ. Hist. Rev*, 2nd ser., VIII (1955–6), 125.

[2] S. Pollard (1955–6), 216–17.

[3] R. H. Best and J. T. Coppock, 200–2.

[4] D. C. D. Pocock (1963), 124–38.

[5] W. H. Chaloner, *The social and economic development of Crewe 1780–1923* (Manchester, 1950), 44.

[6] *Ibid*, 69–71.

[7] T. W. Freeman, 138.

of engines, and it came to house the principal locomotive workshops of the company.[1] The village which arose around the workshops gradually expanded to link with the old town of Swindon and the two became an administrative whole in 1900, with a population of more than 45,000, and with four-fifths of its working population in the company's employ.[2] Ashford, Darlington, Derby and, on a smaller scale, Wolverton were other towns where rapid growth was stimulated by the opening of railway workshops.

Another type of settlement to experience rapid growth during this period, particularly in south-east England, was the dormitory settlement, which was made possible by the improvements in rail transport, although long-distance commuting predates the railway age, and the heyday of the dormitory town came after 1918. The distinction between a dormitory settlement and a dormitory suburb is somewhat arbitrary, for many suburbs began as separate settlements and were later absorbed into the adjacent city; but the wide separation of workplace and residence certainly became more common in the sixty years after 1850. Long-distance commuting was actively promoted by some of the railway companies, for example, the London & North Western, which offered a free first-class season for a year to all who bought houses of more than £50 rateable value in certain towns served by the company and lying between ten and thirty miles from London (though the railway companies did not generally concern themselves with estate development).[3] Provided land was available for building, and this was not always the case, the opening of a railway station often led to the erection of large detached and semi-detached houses, especially if there was easy access to place of work. Sometimes there was a considerable delay; at Radlett in Hertfordshire, for example, there was little development between 1868 and 1894, after which improvements in rail services and a changed attitude on the part of landowners led to rapid expansion.[4] Such dormitory settlements were often grafted on to existing small towns, such as Dorking and Sevenoaks, but their development is most striking where no previous building had occurred, as at Three Bridges in Sussex or at Amersham on the Hill in Buckinghamshire. Similar settlements developed on a smaller scale around other cities, as at West Kirby in Wirral or at

---

[1] L. V. Grinsell *et al*, *Studies in the history of Swindon* (Swindon, 1950), 99–100.

[2] *Ibid*, 124; T. W. Freeman, 280.

[3] E. Course, *London railways* (London, 1962), 199; H. Pollins, 'Transport lines and social divisions', in R. Glass (ed.), *London: aspects of change* (London, 1964), 42.

[4] J. T. Coppock (1964), 281.

Altrincham in Cheshire. As in the suburbs, the residents of such dormitory settlements sometimes opposed improvements in services or the introduction of cheap fares because they feared the social changes that might result.[1]

Holiday resorts also expanded rapidly during this period.[2] Some had already been established by the middle of the nineteenth century, notably Brighton, with a population of 65,569 in 1851 which reached 131,237 by 1911; but others grew almost from nothing or from small ports and fishing villages.[3] Here, too, the railways played an important part. One of the most striking examples of rapid growth was Bournemouth, which had been planned as a resort, but which had acquired a population of only 691 by 1851; its development was particularly marked after the opening of a direct line to London in 1870, and by 1881 its population had reached 16,859, which grew further by 1911 to 78,674.[4] Blackpool and Southport provided interesting examples of the effect of landownership on development, for the former's rapid growth was facilitated by the subdivision of an estate into numerous lots, while Southport's more orderly expansion was largely under the control of a single estate which planned its development as a garden city.[5]

Other new or greatly expanded settlements were to be found in the mining areas, as larger and deeper pits were sunk in the concealed coalfields, and as mining moved into rural areas as in mid-Northumberland, south Nottinghamshire, Cannock Chase, and in east Durham where four mining settlements came into being between 1850 and 1900.[6] But such settlements often remained little more than villages, with most of their populations dependent on a single industry.

An interesting category of settlement comprised experiments in new styles of urban living. These derived their inspiration from the enlightened industrialists who built model villages for their workpeople, or from the prophets of the garden city movement, personified by Ebenezer Howard, and given expression in the First Garden City Company, which promoted the building of Letchworth on an almost virgin site in 1904.[7] Such

[1] H. Pollins, 43–5.
[2] E. W. Gilbert, *Brighton, old ocean's bauble* (London, 1954), ch. 2.
[3] E. W. Gilbert, 'The growth of inland and seaside health resorts in England', *Scot. Geog. Mag.*, LV (1939), 17–35.
[4] C. H. Mate and C. Riddle, *Bournemouth 1810–1910* (Bournemouth, 1910), 98, 136–7.
[5] R. Millward, 86, 102–3.
[6] A. E. Smailes, 173.
[7] W. Ashworth, 141–3.

developments might either form completely new settlements or take their place among the suburbs surrounding existing towns and settlements. Sir Titus Salt's Saltaire was the first example of an industrial village after 1850; on a site near Bradford he built a model village which had grown to 820 dwellings by 1871 and which included chapel, church, school, dispensary, bath-house, club and institute, as well as a 14-acre park.[1] Other examples include Port Sunlight, begun in 1888, Bournville, begun in 1879 and in effect an essay in town planning rather than a company village, and Hampstead Garden Suburb (although work on this did not begin until 1907).[2] By 1914 there were over fifty schemes for garden suburbs building or in being.[3]

Railway towns, dormitory settlements, seaside resorts and mining villages were among the fastest growing settlements. Between 1851 and 1911 the population of mining settlements increased by 1,194% and that of resorts by 291%, compared with an average urban growth rate of 213%.[4] Established industrial towns also continued to expand, although their rate of growth depended on their earlier industrial history and on their industrial structure, so that averages are misleading. A town might temporarily lose population, as did Coventry in the 1860s after the decline of the silk trade, only to expand again with a change in its industrial structure.[5] Development in the smaller industrial towns was often piecemeal, with new plant being established, and speculative builders or the industrial firms themselves providing housing nearby. In the cotton towns, where the first half of the century had been the main period of expansion, growth was comparatively slow and some of the smaller towns were even losing population at the close of the century; in the chemical district of south Lancashire and north Cheshire, on the other hand, the rapid growth of the industry was paralleled by rapid expansion of towns, epitomised by Widnes, which doubled in population between 1865 and 1875.[6]

Many ports, too, grew rapidly, stimulated by the great expansion of trade and the increasing importance of processing imports, although the establishment of improvement commissions and the activities of the railway companies also played a part (Figs. 138 and 139). Grimsby, for example, owed its rapid rise after 1850 to the activities of the Manchester, Sheffield & Lincolnshire Railway Company, which built new docks and

[1] *Ibid*, 126–8.   [2] *Ibid*, 132–4, 160–3.   [3] *Ibid*, 163.

[4] C. M. Law, 139.   [5] J. H. Clapham (1932), 96.

[6] D. W. F. Hardie, *A history of the chemical industry in Widnes* (Birmingham, 1950), 63.

J. T. COPPOCK

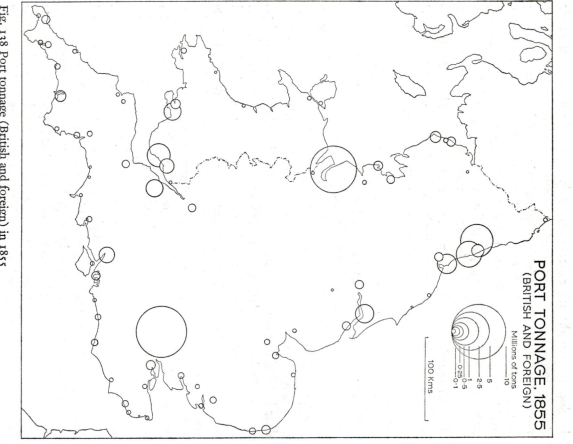

PORT TONNAGE, 1855
(BRITISH AND FOREIGN)

Millions of tons

10
5
2.5
1
0.5
0.25
0.1

100 Kms

Fig. 138 Port tonnage (British and foreign) in 1855
Based on *Trade and Navigation* (*U.K.*), *1855* (P.P. 1856, lvi).
The figures refer to foreign trade only and do not include coastwise traffic.

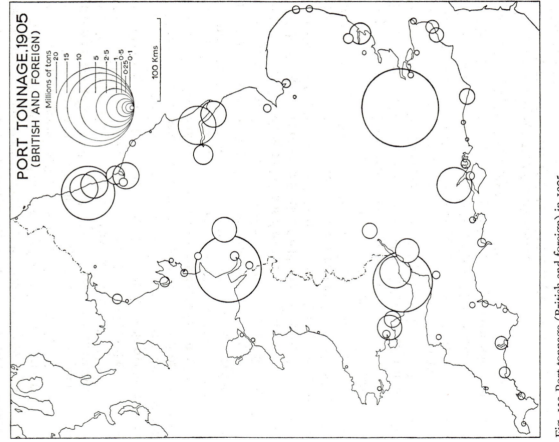

PORT TONNAGE, 1905
(BRITISH AND FOREIGN)

Millions of tons
20
15
10
5
2·5
1
0·5
0·25 0·1

100 Kms

Fig. 139 Port tonnage (British and foreign) in 1905
Based on *Annual Statement of the Navigation and Shipping of the United Kingdom for the year 1905* (P.P. 1906, cxvii).
The figures refer to foreign trade only and do not include coastwise traffic.

even houses.[1] Both the packet ports, which linked railway passenger services with continental Europe, and the coal ports grew rapidly, and new docks and new industries, especially flour milling, contributed to the growth of the larger ports. The ports of north-east England, in particular, rose swiftly to prominence in this period, aided by the coal trade and by the expansion of steel making and shipbuilding.[2] By contrast, Southampton developed as a liner port with the construction of the Test and Itchen Quays and the transfer of the American Line and later the White Star Line from Liverpool; and its population rose from 60,051 in 1881 to 119,012 in 1911.[3] New or greatly expanded cargo ports also appeared. Birkenhead was transformed from little more than a dormitory settlement for wealthy Liverpool merchants into a considerable port; Avonmouth and Portishead were developed in the late 1870s to meet the problem posed by Bristol's shallow tidal access, and Manchester became a major seaport in 1894 with the opening of the Manchester Ship Canal.[4]

Never-the-less, despite all these changes, the principal feature of the urban geography of the period was the continued growth of great cities, and the development of what Patrick Geddes was to recognise in 1915 as conurbations.[5] It is true that their rates of growth were not as rapid as those of the mining settlements, of the resorts or of some of the industrial towns; but they maintained their share of the total population. Sending out tentacles of development along major routeways, absorbing surrounding towns and villages, they came to resemble 'something which has burst an intolerable envelope and splashed'.[6] Their occupational and social composition was also distinctive in the high proportion of employment in service industries and in the large number of immigrants both from elsewhere in the same region and from farther afield (although this latter feature was even more characteristic of the rapidly growing towns such as Birkenhead, 62% of whose population were immigrants in 1861, compared with 49% in Liverpool).[7]

[1] A. G. Kenwood, 156; W. Ashworth, 129–30.
[2] J. Bird, *The major seaports of the United Kingdom* (London, 1963), 345, 384.
[3] *Ibid*, 160–2.
[4] J. H. Clapham (1932), 523–4; J. Bird, 190; A. G. Kenwood, 165.
[5] P. Geddes, *Cities in evolution* (London, 1915), 34.
[6] H. G. Wells, *Anticipations* (London, 1902), 45.
[7] R. Lawton in W. Smith (ed.), 125.

Greater London

These changes in the urban geography of England, especially those resulting from improvements in transport, were most marked in the London area, partly because of its sheer size and its disproportionate share of office and service employment (Fig. 140). The number of local journeys trebled between 1881 and 1901, but the individual railway companies differed greatly in the extent to which they promoted local travel.[1] At first, most companies were concerned only with long-distance traffic, and stations were too widely spaced to serve the needs of those wishing to travel to work; but, gradually, interest in short journeys grew, especially among those companies with a smaller share of long-distance goods and passenger traffic, and intermediate stations were opened, for example, in 1879 at South Hampstead and Queen's Park on the London & North Western line.[2] For this reason, the companies serving south London actively promoted commuter traffic, and numerous branch lines were constructed for this purpose; while those companies whose lines ran north and north-west to the Midlands were less active in this respect; commuter traffic thus grew more rapidly in the south than in the north-west. The railway companies often anticipated urban growth and extended lines 'confident that population would follow', as with the north-western extensions of the Metropolitan Railway in the 1870s;[3] but sometimes the initiative came from intending passengers, and residents in Camberwell were protesting in the 1890s at the inadequacy of rail services.[4] More commonly, speculative builders erected houses near a station, and gradually the land between this new nucleus and the outer fringes of the then built-up area would be built over. The tramways and omnibuses, with their greater flexibility in setting down and embarking passengers, and their commitment to existing roads, made for much more continuous development.

The attitudes of the railway companies and the routes taken by their lines were not the only factors guiding suburban development. Access to place of work, especially to the City, which was receiving a quarter of a million workers daily by 1891, was also important, and the smaller scale of development in north-west London can in part be attributed to the fact

[1] W. Ashworth, 151.   [2] P. Hall (1964), 64.
[3] H. C. Prince in J. T. Coppock and H. C. Prince (eds.), *Greater London* (London, 1964), 132.
[4] H. J. Dyos, *Victorian suburb: a study of the growth of Camberwell* (Leicester, 1961), 79.

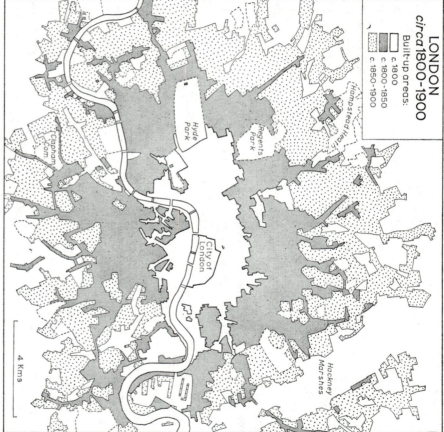

LONDON
circa 1800-1900
Built-up areas:
c. 1800
c. 1800-1850
c. 1850-1900

Hampstead Heath

Regents Park

Hyde Park

Clapham Com.

City of London

Hackney Marshes

4 Kms

Fig. 140 London *circa* 1800–1900
Based on: (1) T. Milne, *Plan of the cities of London and Westminster, circumjacent towns and parishes etc., laid down from a trigonometrical survey taken in the years 1795–1799* (London, 1800); (2) H. G. Collins, *London in 1851* (London, 1851); (3) Ordnance Survey, One-Inch Sheets (1894–1903).

that the City was generally less easily reached from stations along these lines.[1] Availability of land was also a major consideration, for a necessary requirement was a landowner willing to sell or lease and a developer anxious to build; the country to the south-west of Hampstead long remained open because it belonged to an entailed estate.[2] It is probable that physical conditions also had some effect, both by making some areas more

[1] P. Hall (1964), 66.　　[2] H. C. Prince (1964), 130.

attractive than others and by creating difficulties for the builder; the late development of the extensive areas in Middlesex underlain by London Clay was probably due in part to physical obstacles of this kind. Availability of water mains and trunk sewers was also increasingly important, but these, and the availability and physical character of land, affected chiefly the detailed pattern of growth; the broad strategy was determined principally by considerations of accessibility.

Travel to work by rail was at first a middle-class preserve; but the introduction of cheap workmen's fares, obligatory from the Cheap Trains Act of 1883 (though a condition of parliamentary consent to legislation affecting railways in the London area for twenty years before), made it possible for at least the clerk and the better-paid manual worker to live some distance from their places of work; but in some occupations, as in the docks, the nature of their work tied employees to the older parts.[1] By 1882, some 25,000 workmen's tickets were sold daily in the London area.[2] The Great Eastern Railway Company played a major role in this respect, and its cheap fares made possible the great expanses of working-class suburbs which were built in north-east London in the last quarter of the century; a twopenny fare on this company's lines could purchase a journey four times as long as that available for the same fare on the London & North Western Railway, and trains were more frequent.[3] Not all these working-class suburbs grew as a result of cheap fares; house-building was also a response to the development down-river of new docks and the eastward migration of noxious industries, such as the manufacture of chemicals and paint.[4] Nor must the scale of daily movement be exaggerated; less than a quarter of a large sample of trade unionists in south London in 1897 travelled any distance to work.[5] Railway policy together with the attitudes of landowners influenced the social character of new suburbs, and this, once established, tended to be self-perpetuating. Thus, the great spreads of working-class housing in Stamford Hill, Tottenham and Edmonton, which the railways had done much to promote, made these areas unattractive to middle-class families.[6] Conversely, land-

[1] H. Pollins, 42; H. J. Dyos (1961), 77; J. P. Lewis, 134.
[2] H. J. Dyos (1961), 77.
[3] P. Hall (1964), 65–6.
[4] W. Ashworth, 'Types of social and economic development in suburban Essex', in R. Glass (ed.), 63–4.
[5] H. J. Dyos (1961), 62.
[6] *R.C. on the Housing of the Working Class, Minutes*, Evidence of Birt, Q. 10,217 (P.P. 1884–5, xxx).

owners often prescribed restrictive covenants or specified the minimum value of houses to be built, and both landowners and residents could combine to resist developments which were socially unacceptable; thus the labouring classes were effectively prevented from entering St John's Wood and the residents of sedate villas in Acacia Road slept peacefully.[1]

As elsewhere, the development of the suburbs was accompanied by changes within the built-up area. H. T. Dyos has calculated that in the London area some 76,000 people were displaced by the building of termini and other railway developments between 1853 and 1901.[2] At least the houses demolished in this way were often of poor quality, as in the slums of Agar Town, pulled down in the 1860s to provide a site for the new Midland terminus at St Pancras, and it was said that the railway companies often chose routes through working-class areas to minimise costs. Certainly railway construction through middle-class suburbs sometimes involved heavy outlays in compensation and in onerous obligations to minimise damage or loss of amenity, as the directors of the Great Central Railway Company found in bringing their new trunk line to St Marylebone in the 1890s.[3] In central London, this problem was avoided by the construction of underground railways, the first of which, the Metropolitan, was opened between Paddington and Farringdon in 1863.[4] The first deep-level 'tube', the City & South London, followed in 1890, and a complex network of underground railways was soon developed to provide shuttle services within the City.[5]

London also provides some of the best examples of the contribution of road-making to urban redevelopment. The extensive programme of road-work undertaken by the Metropolitan Board of Works in the 1870s and 1880s led, through the construction of Shaftesbury Avenue and Charing Cross Road, to the replacement of the Seven Dials slums and to the erection of numerous shops and offices along the new street frontages.[6] The need for central locations for hotels, theatres, offices and shops similarly led to the destruction of older houses; for example, the Cecil, Savoy, Grand, Metropole, Victoria and Carlton hotels were erected in

[1] H. C. Prince (1964), 134.
[2] H. J. Dyos, 'Railways and housing in Victorian London', *Jour. Transport History*, II (1955–6), 14.
[3] P. Hall (1964), 62–3; H. C. Prince (1964), 127, 133–4.
[4] H. P. White (1963), 82.
[5] *Ibid.*, 98.
[6] P. Hall (1964), 57; T. W. Freeman, 39.

central London in the 1880s and 1890s, and many theatres in the 1880s and early 1900s.[1] Slum clearance also contributed to the loss of houses, although many of these were subsequently replaced. The Metropolitan Board of Works dealt with 22 schemes, covering 59 acres.[2]

Not all the inhabitants of these and other demolished houses were resettled elsewhere. New tenements were constructed in some areas by commercial companies, by charitable institutions and later, to some extent, by local authorities. Such rehousing was at high densities, and on the Peabody estates in London densities of 700 persons per acre were achieved.[3] Nevertheless, such rehousing was on a relatively small scale, and it is not surprising that the population of the City began to fall in the 1840s and that of the remaining central boroughs followed suit in the following decade; the area of declining population grew steadily and by 1911 even the population of the county of London as a whole had begun to fall.[4]

The port of London was also changing. Its centre of gravity shifted downstream with the opening of the Royal Victoria Dock in 1855, the Royal Albert Dock in 1880, and the Tilbury Docks in 1886; the site of this last development was far enough downstream to enable the steamships of the time to get into the docks on one tide. The Port of London Authority was constituted by Act of Parliament in 1908, and, in the following year, the Authority took over control of the entire port, together with the docks and properties of various existing companies.[5]

In the meantime, the size and complexity of this vast and expanding area was recognised by the Metropolis Management Act of 1855. This defined a metropolitan area to be administered for certain purposes by a Metropolitan Board of Works, and in 1888 the area was constituted an administrative county; thus the London County Council came into being. It included the City together with portions of Middlesex, Surrey and Kent and, in 1899, its administration was reorganised by the London Government Act. The new county now comprised the City together with 28 metropolitan boroughs, including that of Westminster which was dignified as a 'city' in 1900 (Fig. 141). The county that thus came into being covered an area of about 117 square miles, and its population in

[1] D. F. Stevens in J. T. Coppock and H. C. Prince (eds.), *Greater London* (London, 1964), 192–3.

[2] W. Ashworth (1954), 95.

[3] J. Parsons, *Housing by voluntary enterprise* (London, 1903), 49.

[4] J. T. Coppock (1964), 33–4.   [5] J. H. Bird, 345–6, 384–8.

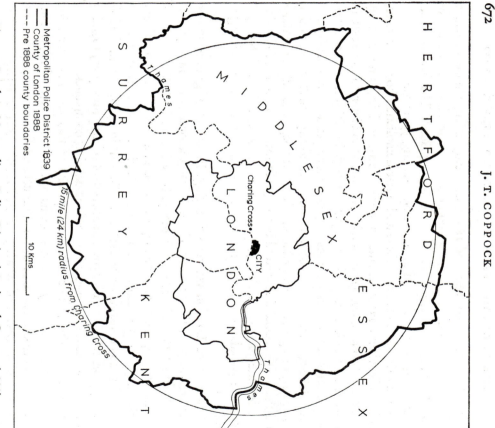

Fig. 141 London: Metropolitan Police District (1839) and County (1888)
Based on *Report R.C. on London Traffic*, plan no. 1 (P.P. 1906, xlv).

— Metropolitan Police District 1839
— County of London 1888
--- Pre 1888 county boundaries

1911 was just over 4.5 million. But the built-up area had extended far beyond the county of London to form an outer ring of suburbs in neighbouring counties. If this 'Greater London' be defined in terms of the Metropolitan Police District of 1839 (i.e. an area extending within a radius of 15 miles or so from Charing Cross), it covered nearly 693 square miles, and included in 1911 a total population of 7.3 million. By this time one Englishman in five was 'a Londoner'. Some idea of the changes over the century as a whole may be obtained from table 11.1 in which the figures have been adjusted to the areas indicated.

Table 11.1   *The growth of Greater London, 1801–1911*

Figures are adjusted to the county and metropolitan police area of 1911.
Figures for the county include those for the City.

	City	County	Outer ring	Greater London
Area in sq miles	1.1	117	576	693
	Population in thousands			
1801	128	959	155	1,114
1851	129	2,363	322	2,685
1881	51	3,830	936	4,767
1901	27	4,536	2,045	6,581
1911	20	4,522	2,730	7,252

Source: Census of 1911, vol. I, pp. xxiii–xxiv, 646 (P.P. 1912–13, cxi); *Census of 1901, County of London*, p. 1 (P.P. 1902, cxxi).

Beyond the limits of Greater London lay a fringe of towns, not only serving their immediate countryside, but also acting as dormitory settle-ments for those who worked in London itself. Little wonder that con-temporaries described it as 'the greatest city in the world'.[1]

[1] A. J. Herbertson and D. J. R. Howarth (eds.), *The British Isles* (Oxford, 1914), 549.

22

DHG

Chapter 12

ENGLAND *circa* 1900

PETER HALL

The Victorian era survived just three weeks of the twentieth century. At 6.30 in the evening of 22 January 1901 the Queen died peacefully at her beloved Osborne, and the Earl Marshal put the country into deep mourning until 6 March.[1] The news was received with gloom in Yeovil, where the staple trade was the manufacture of light tan gloves; 'the death of the late Queen', the factory inspector later reported, 'had the effect of producing unusual depression', though 'with the Coronation next year, manufacturers are looking forward to doing a brisk spring trade'.[2] The war in South Africa ground on, bringing slack times for industries which shipped to the Cape Town market; though in Bridport, on the Dorset coast, the ancient flex and twine trade revived with a flood of orders for balloon netting.[3]

POPULATION

Still in half mourning, Englishmen and English women submitted themselves to the eleventh decennial census of population on the last midnight in March. The 1901 Census had peculiar significance as an epitaph for a century and an age; for the first census had been taken in 1801, while the first reasonably reliable count by the registrars had marked the start of the Victorian era in 1841. The enumerators, 'clergymen, professional men, schoolmasters, and others of exceptional education when available',[4] counted a population that had nearly quadrupled in the century — from 8.9 million to 32.5 in England and Wales together. In absolute terms the last decade of the century had recorded the greatest increase of all: 3.6 million between 1891 and 1901. But, the Registrar-General observed, the proportionate rate of increase was one of the lowest recorded in the century.[5]

[1] *The Times*, 25 January 1901, p. 8, and 29 January, p. 10.
[2] *Ann. Rep. Factories, 1901*, 5 (P.P. 1902, xii).
[3] *Ibid.*   [4] *The Times*, 3 January 1901, p. 5.
[5] *Census of 1901: General Report*, 15 (P.P. 1904, cviii).

[674]

This growth had been diversely spread across England to produce a very uneven distribution of population (Fig. 142). As in the previous decade, between 1891 and 1901 'the counties showing the highest rates of increase mainly include those around London, as Middlesex, Essex, Surrey, Kent and Hertfordshire; counties in which the chief industry is coal-mining, as...Northumberland, Durham..., and to some extent Staffordshire and Derbyshire; or counties which are mainly manufacturing, as Nottinghamshire, Leicestershire, Northamptonshire, the West Riding of Yorkshire and Lancashire'.[1] But ten English counties had lost population; and they were mainly agricultural.[2] Analysing the returns in more detail, the Registrar General found that the rural districts – with which he grouped the smallest urban districts, because they were mostly market towns – were increasing as a group very slowly, and that many individual districts were declining.[3]

The fact was that England had 'now been provided by the railroads with a system of veins and arteries, and by the telegraphy and penny post with a nervous organization', the inevitable result of which was 'a more rapid circulation of labour and an excessive growth of population in some parts of the kingdom at the expense of that of others'.[4] The agricultural labourer now sensed 'the contagion of numbers, the sense of something going on, the theatres and the music halls, the brightly lighted streets and busy crowds: all, in short, that makes the difference between the Mile End fair on a Saturday night, and a dark and muddy country lane, with no glimmer of gas and with nothing to do. Who could wonder that men are drawn into such a vortex, even were the penalty heavier than it is?'[5] So wrote Sir Hubert Llewellyn Smith, in Booth's great survey of the life and labour of the people in London. The statistics of migration proved him right. In his paper on the laws of migration in 1885, Ravenstein had shown that the counties of dispersion – those with enumerated populations less than the number of their natives scattered throughout the kingdom – were gathered in a great belt which ran diagonally across the agricultural heart

[1] *Ibid.*, 21.
[2] *Ibid.*
[3] *Ibid.*, 25. For an extended treatment of the pattern of rural migration losses, see R. Lawton, 'Population changes in England and Wales in the later 19th century: an analysis of trends by registration districts', *Trans. and Papers, Inst. Brit. Geog.*, XLIV (1968), 55–74.
[4] H. L. Smith, 'Influx of population (East London)', in C. Booth (ed.), *Life and labour of the people in London*, III (London, 1892), 76.
[5] *Ibid.*, 75.

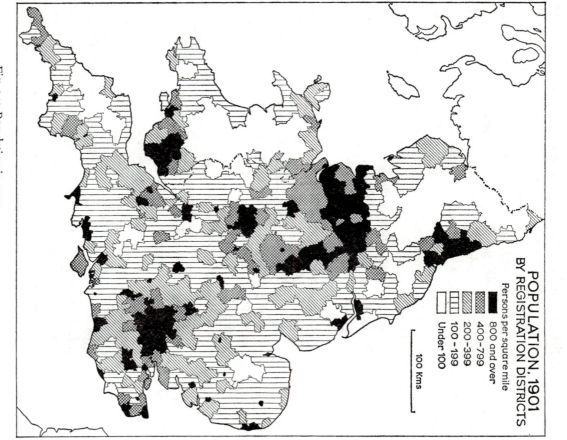

POPULATION, 1901
BY REGISTRATION DISTRICTS

Persons per square mile

- 800 and over
- 400-799
- 200-399
- 100-199
- Under 100

100 Kms

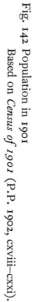

Fig. 142 Population in 1901
Based on *Census of 1901* (P.P. 1902, cxviii-cxxi).

of England from Cornwall to Lincolnshire; while the counties of absorption – those with enumerated populations greater than the numbers of their natives living elsewhere – were gathered in two great groups, one representing the northern and Midland industrial counties from Northumberland to Warwick, the other the metropolitan corner of England from Essex through to Hampshire.[1]

The most important explanation of these differences is found in later pages of the census report, which record changes in the work Englishmen did. The total agricultural population of England and Wales had been falling continuously since 1851 – from over 1.7 million to 1.2 million, or by nearly one third, in half a century.[2] Jobs in coalmining on the other hand had increased prodigiously – by 24.7% in the decade 1891–1901 alone. But this increase was highly concentrated in certain coalfields, above all in Northumberland and Durham; while some of the great coalfields of the first half of the nineteenth century, like the Black Country, were now mere shadows of their former importance.[3] Cotton, the great staple of the early industrial revolution, had shown in the last decade a decline in employment for the first time; woollens declined by no less than 13.5%, and the whole textile group showed an overall reduction.[4] Clothes makers showed only a modest increase; shoemakers hardly any at all.[5] In all these trades, workers were feeling the consequences of mechanisation; in some, foreign competition was beginning to bite. But to compensate, other industries were recording spectacular increases: paper and printing showed 26.9% in a decade, chemicals 40.9%, engineering with machine making 67.4%, general and local government 37.3%, insurance 79.4%.[6]

The trouble was that they did not really compensate because different jobs were done in different places. After two decades of agrarian depression, Englishmen were used to news of migration from the rural districts. They might remember their parents' stories about the fate of the handloom weavers sixty or seventy years before. But they were not yet fully familiar with the notion that long-term, secular decline in a staple industry

[1] E. G. Ravenstein, 'The laws of migration', *Jour. Roy. Stat. Soc.*, XLVIII (1885), 184.

[2] *Census of 1901: General Report*, 101 (P.P. 1904, cviii). But see p. 679 below.

[3] *Ibid.*, 105–6.   [4] *Ibid.*, 118–20.

[5] *Ibid.*, 123–4.

[6] *Ibid.*, 91, 97, 109, 115, 116. For an extended analysis of changes in employment patterns at the national level, see P. Deane and W. A. Cole, *British economic growth 1688–1959* (Cambridge, 2nd. ed., 1967), 142, and W. Ashworth, 'Changes in the industrial structure 1870–1914', *Yorks. Bull. Econ. and Soc. Research*, XVII (1965), 61–74.

could bring a whole manufacturing area into lingering depression and decay. Yet such was beginning to be the case – or at least the prospect. The decline in cotton was already a problem in some Lancashire towns; the problem was not lessened by news that chemicals were booming on the middle Mersey. Shoemakers in Leicestershire and Northamptonshire, seeing their jobs menaced by machinery, drew small comfort from the fact that office jobs were on the increase in London.[1]

These changes naturally affected urban districts of every order of size. Urban England generally was growing; and by 1901, on a rigorous basis of computation, 78% of the population of England and Wales was already urban.[2] Yet when the urban areas were grouped and analysed by the census-takers, it was discovered that the fastest-growing English towns were the medium-sized ones; the greatest cities, in contrast, were growing relatively slowly. The average rate of increase for the urban areas of England and Wales, during the decade 1891–1901, was 15.2%. But towns of between 20,000 and 50,000 people recorded 20.3%, and those of between 50,000 and 100,000 recorded 23.2%;[3] while the four biggest cities all recorded less than 10%: the county of London 7.3%, Liverpool 8.8%, Manchester 7.6%, Birmingham 9.2%.[4] Curiously, the census noted that the English towns with the highest rate of growth – East Ham with 193.5%, Walthamstow with 105.3%, Kings Norton and Northfield with 101.8% – were suburbs of either London or Birmingham;[5] so that 'a falling off in the rate of increase, or even an actual decline, in the population of a great town is not necessarily an indication of a corresponding decline in its prosperity'.[6]

Three main sets of changes, then, dominated the economic and social geography of England as recorded by the census at the outset of the twentieth century. They were:

1. First, the agonised adaptation of English agriculture in the face of foreign competition, coupled with the continued exodus from the country-side.

2. Secondly, the changing fortunes of the great industrial staples and of the areas which depended on them, coupled with the success or failure of these areas to adapt their industrial structures to changing demands.

---

[1] R. Lawton, 62–3, gives detailed evidence.
[2] C. M. Law, 'The growth of urban population in England and Wales, 1801–1911', *Trans. and Papers, Inst. Brit. Geog.*, XLI (1967), 132.
[3] *Census of 1901: General Report*, 26 (P.P. 1904, cviii).
[4] *Ibid.*, 27.
[5] *Ibid.*, 28–9.
[6] *Ibid.*, 29.

3. Thirdly, the suburban explosion which was beginning to liberate large numbers of English people from overcrowded and insanitary homes in the great cities.

THE COUNTRYSIDE

Agricultural depression

The last twenty years of the nineteenth century had witnessed an agricultural depression on a scale never before known in the English countryside. Its precise results are difficult to chart statistically, because both the figures of acreage and the figures of agricultural population are suspect. The contemporary verdict was that in the United Kingdom between 1870 and 1900 land under grain crops had fallen by 40%, and that under green crops by one sixth, while grassland had increased by nearly one third.[1] It is almost certain that the actual declines were greater than this because all the time more land was being recorded in the returns.[2] As to the agricultural population, contemporary experts differed widely. It was generally agreed that the figures for women were not to be taken seriously because before 1871 they had included farmers' wives. It was noticed, too, that the earlier figures had probably included much seasonal labour, but that since 1881 the increase in gardeners had partly helped to offset declines in the agricultural population proper.[3] The most careful estimate for England and Wales, by Lord Eversley in 1907,[4] is that the male agricultural population over 20 years of age had declined by 66,650 between 1881 and 1901, and in all by 245,120 between 1861 and 1901.[5] The general belief – it was officially expressed – was that the decline had been exceptionally severe in the arable districts.

Contemporaries were in no doubt that the great agricultural depression had borne more hardly on the arable east than on the pastoral west of England. The impact of foreign wheat had been felt earlier and more severely than the impact of foreign meat. As late as 1885–7, British farmers had supplied nearly 34% of the wheat for British bread; by 1900–2

---

[1] *The Tariff Commission*, vol. 3, *Report of the Agricultural Committee* (London, 1906), paras. 353–4.

[2] J. T. Coppock, 'The accuracy and comparability of the agricultural returns', in R. H. Best and J. T. Coppock, *The changing use of land in Britain* (London, 1962), 57–9 and *passim*.

[3] W. H. Bear, 'The agricultural population at the Census of 1901', *Jour. Roy. Agric. Soc.*, LXIV (1903), 123, 128, 137; and Lord Eversley, 'The decline in the number of agricultural labourers in Great Britain', *Jour. Roy. Stat. Soc.*, LXX (1907), 270–4.

[4] Lord Eversley, 275.

[5] *Ibid.*

the proportion was down to 22.5%.[1] In meat, the British farmer kept a bigger share, which was estimated at 66.0% for the first half of the 1890s and at 55.3% for the first five years of the new century.[2] Aided by trans-continental railroads, lake steamers and big ocean steamships, the price of North American imported wheat had begun to decline in Britain from about the late 1870s;[3] while the first wholly successful shipment of frozen meat from Australia had been made in 1879, regular shipments had started only in the eighties, and successful chilling of meat had not been applied on a regular reliable basis by 1901.[4]

The results were plain to see on the face of the ground. The Royal Commission of 1897 on Agricultural Depression said: 'Broadly speaking, it may be concluded that the heavier the soil, and the greater the propor-tion of arable land, the more severe has been the depression' (Fig. 143). In the eastern and in some of the southern counties, the situation was 'un-doubtedly a grave one'. But in the pastoral areas of Great Britain, the depression was 'of a milder character'.[5]

If there was one county that had suffered more severely than any in England, in the Commission's view, it was Essex. 'Essex as a whole', an Assistant Commissioner had told them, 'is a corn growing county':[6] the soil was right and the climate was right. But on the stiff, deep, tenacious London Clay soil in the south, which was described as 'stiff, tough, numb, dumb and impervious'[7] there were 'large farms... half or threequarters of which have either been permitted to run wild or have been sown down in despair'.[8] Areas which twenty years before had grown fine crops of wheat and beans and clover were 'lying in wretched pasture, next to useless'.[9]

Rider Haggard, setting out on his tour of rural England at the turn of

[1] R.C. on the Supply of Food and Raw Material in Time of War, Appendix 1, Summary Table B (2), 78 (P.P. 1905, xl). For other figures which are compatible with these Board of Trade estimates see R. F. Crawford, 'An inquiry into wheat prices and wheat supply', *Jour. Roy. Stat. Soc*, LVIII (1895), 81; and for contradictory figures, *Tariff Commission, Report* (1906), para 353.

[2] R. H. Hooker, 'The meat supply of the United Kingdom', *Jour. Roy. Stat. Soc.*, LXXII (1909), 332.

[3] *Tariff Commission, Report* (1906), para 114.

[4] J. T. Critchell and J. Raymond, *A history of the frozen meat trade* (London, 1912), 30-1, 41, 248.

[5] *R.C. Agric. Depression, Final Report*, 21 (P.P. 1897, xv).

[6] *R.C. Agric. Depression, Rep. by Mr R. Hunter Pringle on the Ongar...Districts of Essex*, 38 (P.P. 1894, xvi, pt. 1).

[7] *Ibid*, 37.

[8] *Ibid*, 42.

[9] *Ibid*.

Fig. 143 Lowland England: distribution of heavy soils
Based on miscellaneous sources.

Main areas of heavy soil
Outline of fen and marsh

50 Kms

the century, told the same story. 'Between Billericay and Althorne we saw hundreds, or rather thousands, of acres of strong corn lands which have tumbled down to grass. I can only describe the appearance of this land as wretched: it did not look as if it would support one beast upon ten acres.'[1] Near Maldon, on land which a local resident called 'the finest wheat-growing country in England', capable of yielding 40 bushels to the acre, 'few of the fields seemed to produce a crop of grass high enough to hide a lark; but such as it might be, that was their produce for the year'.[2]

There were two responses to this sorry problem. Around Laindon and Basildon in 1894, there was much cutting up of land into smallholdings of

[1] H. R. Haggard, *Rural England*, 2 vols. (London, 1902), I, 466.   [2] *Ibid.*, I, 470.

between one and ten acres.[1] Temporary and often ramshackle dwellings on the new holdings marked the process.[2] But much land suffered a better fate – argue about it as contemporaries might. As early as 1880, Scotsmen were beginning to move on to the south Essex clays, especially around Brentwood and Chelmsford.[3] Many of them came from the Kilmarnock district of Ayrshire, where the farmers understood dairying but were finding their rents too high.[4] They came south bringing their Ayrshire cattle with them;[5] and 'when we are advising a "brither Scot" who is on the outlook for land in the South, we always advise him to beware of light land, and take the heavy in preference'.[6] In the view of the Scots, the experience of the English farmers was too limited; it was not that the dairying paid better so much as that the Scots understood their trade well.[7]

Northwards in Suffolk, the Commission reported that farmers were also 'in great straits'.[8] Near Bury St Edmunds, a farmer told the Tariff Commission, a few years later, that the majority of farmers were only farming the best part of their land and leaving the other land down for three or four years' ley, and then breaking it up and substituting oats and barley.[9] In Norfolk the story was one of 'struggle to make both ends meet'.[10] In Cambridgeshire, much of the heavy land in the south of the county had 'simply tumbled down' to grass and was 'as good as abandoned';[11] while northwards, between Cambridge and Huntingdon, there was 'a deplorable state of things'.[12] 'The soil', wrote Haggard, 'is for the

[1] R.C. Agric. Depression, Minutes, Evidence of Rutter, Q. 34,288–9 (P.P. 1894, xvi, pt. 3).
[2] B. E. Cracknell, Canvey Island: the history of a marshland community (Leicester, 1959), 42.
[3] R.C. Agric. Depression, Rep. by Mr R. Hunter Pringle on the Ongar...Districts of Essex, 43 (P.P. 1894, xvi, pt. 1).
[4] P. McConnell, 'Experiences of a Scotsman on the Essex clays', J.R.A.S., 3rd ser., ii (1891), 311.
[5] R.C. Agric. Depression, Rep. by Mr R. Hunter Pringle on Ongar...Districts of Essex, 44 (P.P. 1894, xvi, pt. 1).
[6] P. McConnell, 314.
[7] Ibid., 313.
[8] R.C. Agric. Depression, Rep. by Mr Wilson Fox on the County of Suffolk, 82 (P.P. 1895, xvi).
[9] Tariff Commission, Report (1906), para. 883.
[10] R.C. Agric. Depression, Rep. by Mr R. Henry Rew on the County of Norfolk, 23 (P.P. 1895, xvii).
[11] R.C. Agric. Depression, Minutes, Evidence of Fox, Q. 61,329, 61,330 (P.P. 1896, xvii).
[12] Ibid., Q. 61,331.

most part a heavy clay, and much of it has gone down into an apology for pasture, often so thickly studded with wild thorns and briars, that it looks like a game covert which has been recently planted.'[1] Yet in the Fenland, observers agreed, some corn districts had hardly depreciated at all: they were fertile, easy to work, and less likely to suffer from drought.[2]

In adjacent Hertfordshire things were a little better than in south Cambridgeshire: 'Being near London and there being plenty of railway accommodation,' a witness told the Commission in 1895, 'I think it has kept up rather better than other districts of a similar character of soil.'[3] But there was depression in the more remote parts where the land was less good.[4] In 1902 Haggard came to the same conclusion: 'In Hertfordshire prosperity is, in the main, confined to the neighbourhood of a railway line. Where means of communication are lacking, as a Scotch gentleman said to me, there is "agricultural death".'[5] Here, Scots and Cornish immigrants had moved near the railways where they could get cheap London manure for their mixed farming; native Hertfordshire men were suffering depression in the backwoods.[6]

A little to the north in the south Midland counties, the heavy soils of the Kimmeridge and Oxford Boulder Clays were in little better shape than the London Clay soils of south Essex. In the north of Bedfordshire, in much of Huntingdonshire and southern strip of Northamptonshire, for example, conditions on the clay were much the same as in Essex; always on strong clay there was 'the same sort of desperate condition'.[7] But apart from the poorest clays, these lands would carry grass and they were not suffering wholesale abandonment.[8] Nevertheless, the Commissioners concluded of this area, 'the ordeal through which the farmers of strong clay have passed ever since 1879 almost beggars description'.[9]

[1] H. R. Haggard, II, 58.
[2] *R.C. Agric. Depression, Rep. by Mr Wilson Fox on the County of Cambridge*, 25 (P.P. 1895, xvii); H. R. Haggard, II, 2; and for a later view, see A. D. Hall, *A pilgrimage of British farming 1901-12* (London, 1913), 76-7.
[3] *R.C. Agric. Depression, Minutes*, Evidence of Spencer, Q. 46,561 (P.P. 1896, xvii).
[4] *Ibid.*, Q. 46,614.
[5] H. R. Haggard, I, 511.
[6] *Ibid.*, I, 569.
[7] *R.C. Agric. Depression, Minutes*, Evidence of Pringle, Q. 47,508 (P.P. 1896, xvii).
[8] *Ibid.*, Q. 47,494; *R.C. Agric. Depression, Minutes*, Evidence of Brown, Q. 35,716 (P.P. 1894, xvi, pt. 3).
[9] *R.C. Agric. Depression, Rep. by Mr R. Hunter Pringle on the Counties of Bedford, Huntingdon and Northampton*, 37 (P.P. 1895, xvii).

Northwards it was the same story. In Lincolnshire, the clay lands had 'suffered both from the effects of low prices and bad seasons more than any other class of land';[1] the four-course rotation was broken, farmers let the grass seeds lie, and the grass degenerated into sheep-walk of almost no value. In Holderness, they were suffering very much more than on the Wolds; 'in fact, they are suffering more there than in any other district in Yorkshire'.[2] A farmer at Skerne, on the edge of the Wolds and the Holderness clay plain, told Haggard in 1902 that 'half of his farm was wold and half clay, and he wished he were rid of the strong land'.[3]

It was notable though that the light lands had suffered less. In Lincolnshire the Wolds, the Heath and the Cliff were in good condition.[4] At the depth of the depression, 'the Wolds were as highly farmed as ever', while if the Cliff and the Heath had gone back they had not gone back far.[5] Here the witnesses contradicted each other somewhat, both before the Commission and afterwards: in 1906 the Tariff Commission found that 'corn does not pay at all to grow' and that it had not paid 'to farm highly during the last few years'.[6] On the Heath, Sir Daniel Hall was surprised that 'the output was considerable from such poor, thin-looking soil'; the secret was peas plus a two-year ley.[7] Admittedly, light land did not do well everywhere on the drier arable half of England. On the Hampshire Downs, former 'admirable sheep pasture'[8] had been broken up eighty years before during the Peninsular Wars because of high wheat prices, but now 'it would take many years to re-establish it as pasture of the same quality as that which it originally was'.[9] Yet on the Wiltshire Downs, not many miles away, 'throughout the district there was evidence of a general quiet prosperity among the farmers';[10] the holdings were big, twelve or fourteen hundred acres, and the farmers cut their expenses by having few hedges or

[1] *R.C. Agric. Depression, Rep. by Mr Wilson Fox* (P.P. 1895, xvi).

[2] *R.C. Agric. Depression, Minutes*, Evidence of Riley, Q. 36,488 (P.P. 1894, xvi, pt. 3).

[3] H. R. Haggard, II, 364.

[4] *R.C. Agric. Depression, Rep. by Mr Wilson Fox on the County of Lincolnshire*, 6–7 (P.P. 1895, xvi).

[5] *Ibid*, 37.

[6] *Tariff Commission, Report* (1906), para. 863.

[7] A. D. Hall, 96.

[8] *R.C. Agric. Depression, Minutes*, Evidence of Fream, Q. 11,814 (P.P. 1894, xvi, pt. 1).

[9] *Ibid.*, Q. 11,812.

[10] A. D. Hall, 6.

gates. In such a prairie farming system corn and sheep were the staple elements. In Gloucestershire, the Cotswolds had weathered the depression better than the heavy clays of the vale of Berkeley: the lighter lands were 'almost the only lands that can be made to pay anything at all for cultivation', a witness told the Commissioners.[1] These were chalk and limestone districts. On the very lightest lands, on the sands of East Anglia or of the Weald, there was often reversion to waste. Almost all the light land in Norfolk broken up from sheep-walk and rabbit warren in the previous fifty years, a witness told the Commission in 1894, had gone back;[2] in Suffolk 'in the light lands hundreds of acres are going out',[3] though they were not entirely out of cultivation for they were still broken up every two to three years. A skilled cultivator though could cut his losses on the sandy lands. Thus the earl of Leicester, proprietor of Holkham and descendant of the celebrated Thomas Coke, explained to Haggard how he laid down the light lands to sheep-walks for as long as sixteen or twenty years, then broke them up to take four crops without manure. 'My system of temporary pastures is to throw the light lands out of cultivation for as long as it pleases you; if not productive of much gain, it entails no more loss beyond the rent.'[4] But according to a local Norfolk expert giving evidence to Haggard, the rentals of large light land farms had fallen by just the same amount as the rents of the stiff arable land – by no less than half – during the depression.[5]

In a wide zone of Midland claylands, arable eastern England passed almost imperceptibly into pastoral western Britain (Fig. 144). In the south, in Wiltshire and Dorset and Gloucestershire, the experience had been mixed.[6] In the Midland triangle, in Shropshire, Herefordshire, and Worcestershire, the depression had not been so severe except on the strong clays, but there the results of the depression could be dire, for strong clays in the Midlands were strong indeed. Around Stratford upon Avon, strong wheat and bean land had 'laid itself down' into grass:[7] tillage was being

[1] *R.C. Agric. Depression, Minutes*, Evidence of Stratton, Q. 34,914 (P.P. 1894, xvi, pt. 3).

[2] *R.C. Agric. Depression, Minutes*, Evidence of Read, Q. 16,032 (P.P. 1894, xvi, pt. 2).

[3] *R.C. Agric. Depression, Minutes*, Evidence of Simpson, Q. 16,775 (P.P. 1894, xvi, pt. 2).

[4] H. R. Haggard, II, 466.   [5] *Ibid.*, II, 528.

[6] *R.C. Agric. Depression, Final Report*, 16–17 (P.P. 1897, xv).

[7] *R.C. Agric. Depression, Minutes*, Evidence of Turner, Q. 11,728–9, 11,737, 11,743 (P.P. 1894, xvi, pt. 1).

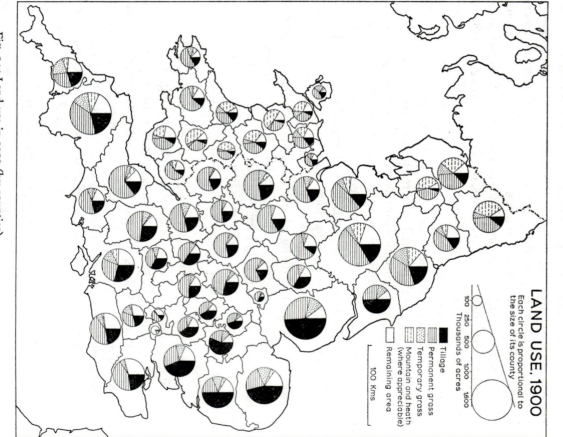

Fig. 144 Land use in 1900 (by counties)
Based on *Agricultural Returns, 1900*, 4–21 (P.P. 1901, lxxxviii).

'supplanted by bad pasture', said an Assistant Commissioner, 'simply because the yields and prices obtainable from arable land do not pay the expense of cultivation'.[1] Good red marl land produced wheat which had sold for 6s. or 7s. a bushel in 1876–7, but only for 3s. by 1893, when the owner laid it down to grass.[2] But in Leicestershire even the heaviest land, 'stiff yellow clay of a very tenacious character, and not to be ploughed except with three horses',[3] was yielding a return under grass, and it was said that many thousands of acres had been converted to grass in the last quarter-century, and that milk prices had fallen, but not as much as those for grain.[4] The farmer benefited too from cheaper manure and feeding stuffs.[5]

Northwards, around Garstang to the north of Preston, 'those who are exclusively engaged in the sale of milk have felt the depression by far the least'; farms which formerly made cheese were now selling milk direct to the big towns, which 'appear to take an unlimited supply'.[6] Northwards again, 'the position of the farmers in Cumberland', an Assistant Commissioner reported in 1895, 'is, I believe, more satisfactory than in any other upon which I have reported'.[7] The Cumberland farmer had felt the drop in cereal prices relatively little, depending as he did upon the sale of cattle, sheep, butter, cheese, poultry and eggs.

The same story was repeated in the south-west. In Devon, arable acreage had declined rapidly, and the agricultural labour force had diminished at about the same rate.[8] But, according to an Exeter auctioneer who talked to Haggard, the agricultural position was nothing like as bad as that in the eastern counties; the fall in grain prices was less crushing, and so great was the dependence on pasture that less labour was needed.[9] In Somerset,

---

[1] *R.C. Agric. Depression, Rep. by Mr J. Jabez Turner upon the Stratford on Avon District of Warwickshire*, 24 (P.P. 1894, xvi, pt. 1).

[2] H. R. Haggard, I, 419–20.

[3] T. Stirton, 'Select farms in the counties of Leicester and Rutland', *J.R.A.S.*, 3rd ser., VII (1896), 524.

[4] *R.C. Agric. Depression, Minutes*, Evidence of Parker, Q. 10,409, 10,609 (P.P. 1894, xvi, pt. 1).

[5] J. Bowen-Jones, 'Typical farms in Cheshire and North Wales', *J.R.A.S.*, 3rd ser., IV (1893), 618–19.

[6] *R.C. Agric. Depression, Rep. by Mr Wilson Fox upon the Garstang District (of Lancashire)*, 7 (P.P. 1894, xvi, pt. 1).

[7] *R.C. Agric. Depression, Rep. by Mr Wilson Fox on the County of Cumberland*, 28 (P.P. 1895, xvii).

[8] *R.C. Agric. Depression, Minutes*, Evidence of Harris, Q. 3,415–16 (P.P. 1894, xvi, pt. 1).

[9] H. R. Haggard, I, 199.

Haggard could conclude that 'on the whole, agriculture still prospers'.[1] Here, especially on the heavier clays, around Frome, very rapid conversion of arable land had taken place during the eighties and early nineties, with very varied results.[2]

Farming for stock did not pay everywhere. In Wharfedale and Airedale, in remote north-west Yorkshire, was an area almost wholly in grass, producing butter, beef and mutton, but 'the condition of the grass land farmers was such that some few are pretty well bankrupt'.[3] Yet, only a few miles away, the Darlington area had not experienced severe depression, and 'the answer is in one word – "meat"'.[4] Here too, though, there was a threat: meat prices had fallen 1s. to 1s. 6d. a stone in four years. As refrigerated meat came in increasing quantities into British ports, it was brought home to farmers that in meat, as in wheat, distance no longer afforded automatic protection.

Dairying, fruit-growing and market gardening

There was, however, one product for which refrigeration could not yet conquer distance. From all over the pastoral half of England, reports were coming of a switch to fresh milk production. In the northern dales of Yorkshire, where for centuries Wensleydale cheese had been the staple product of the small farmers, they were finding that they got a better year-round income from fresh milk, and the Northallerton Pure Milk Society had set up a works on the Danish model.[5] In northern Derbyshire, improved railway facilities allowed farmers to send their milk direct to the large towns, and 'not one-twentieth of the milk that used to be made into cheese and butter is made today'.[6] Near Macclesfield, for instance, the staple cheese-making industry had suffered a severe blow from Canadian competition after 1880; now the milk went to Manchester and other towns.[7] Here, labour shortage was a powerful contributory cause: the young men were leaving farm work and were flocking to the large towns, and 'if at

[1] *Ibid*, 1, 230–1.
[2] R.C. *Agric. Depression, Rep. by Mr Jabez Turner on the Frome District of Somerset*, 7–8 (P.P. 1894, xvi, pt. 1).
[3] R.C. *Agric. Depression, Minutes, Evidence of Broughton*, Q. 15,294 (P.P. 1894, xvi, pt. 2).
[4] J. H. Dugdale, 'Select farms in the Darlington district', *J.R.A.S.*, 3rd ser., vi (1895), 527.
[5] L. Jebb, *The small holdings of England* (London, 1907), 35–47.
[6] *Tariff Commission, Report* (1906), para. 924.
[7] *Ibid*, para. 1230.

the present time milk had to be made up into cheese in the farmhouse as formerly, it is doubtful where the dairy maids would come from'.[1]

In all the great northern cities and towns, rising living standards and improving knowledge of nutrition were helping to increase the demand for milk. But the biggest single market was London, and this had a profound influence on the agriculture of southern England. Since much of the area around London was not natural pasture land, London's milkshed extended principally to the west, where the Great Western Railway was responsible for shipping fully half of all the milk that arrived in the capital.[2] In twelve years from 1892 to 1904, shipments by the Great Western rose by nearly two-thirds, to about $17\frac{1}{2}$ million gallons of milk a year. Most of the milk came from main and branch line stations between Reading and Chippenham, on average about 80 miles from London and at the most 130 miles away. From Swindon, the milk came by express into Paddington, 77 miles distant, in less than two hours. In cooler months the catchment area could be extended: the Great Western brought milk over 320 miles from St Erith in Cornwall to Paddington.[3] The Great Northern was more modest: it tapped the north Staffordshire producers, collecting their milk at Egginton Junction near Burton on Trent, 154 miles from King's Cross. The line which really specialised in long-distance shipments was the London & North Western; in the autumn of 1905 it was bringing milk from Goold's Cross, in County Tipperary, on the Great Southern & Western Railway of Ireland, 430 miles into Euston.[4]

Fresh milk was one special product in which the British farmer was still sure of enjoying a natural monopoly of supply. But there were other products which combined in varying measure the essential attributes of bulk and perishability; and the most dramatic developments in English agriculture at this time were found in those areas which had turned within recent years from grain or stock to the intensive production of fruit or vegetables for the urban markets. Fruit, a departmental committee could say in 1905, was 'the only form of agriculture which has exhibited any sign of progress in recent years',[5] and the reason was 'the extraordinary growth of the taste for fruit on the part of the public'.[6] The orchard acreage of Great Britain had expanded by nearly two-thirds between 1873 and 1904, almost

[1] W. J. Skertchly, 'Agriculture in Derbyshire', *J.R.A.S.*, LXVI (1905), 223.
[2] E. A. Pratt, *The transition in agriculture* (London, 1906), 11.
[3] *Ibid*, 12.   [4] *Ibid*, 13, 16.
[5] *Report of the Departmental Committee upon the Fruit Industry of Great Britain*, 2 (P.P. 1905, xx).   [6] *Ibid*, 3.

all in England.[1] Precisely because of the bulk and perishability of the product, it was natural that most fruit-growing areas should be within easy reach of their markets; and the more perishable the product, the more evident this was. Of the big orchard counties, some were near the towns – Kent, Worcestershire, Gloucestershire, Herefordshire; but some were not – Somerset, Devon (Fig. 129). In small fruit, the leading counties were all reasonably close to the big London or Midlands markets – Middlesex, Worcestershire, Cambridgeshire, Norfolk, Hampshire, Essex.[2]

Some of the outstanding districts were naturally found where the sprawling suburbs of London ended. For fruit, Kent was outstanding. Its nearness to London gave it not only an assured market, but also – for the motor vehicle had not yet banished the horse from London streets – an assured supply of stable manure.[3] Hops for the great London breweries had been Kent's staple, and they had kept up well during most of the depression, but at the turn of the century the experience was much more mixed. At Wye, in Kent, the Tariff Commissioners were told in 1906, 'On an average the last years have been as good as any other consecutive years';[4] at Selling, the acreage had increased, and 'the last 10 years have been as profitable as the 10 previous years'.[5] But in the Tonbridge area 'the acreage under hops... has been reduced by quite one-third since 1886'[6], while around Tenterden hops were 'much reduced in acreage' and profits had 'disappeared'.[7] Some parts of Kent – like that around Swanley – were near enough to London to send their produce daily by road.[8] Here, and extending north to Erith and west to Bexley, the last thirty years had seen a great growth of forced strawberry cultivation in cold houses by small growers.[9]

West of London, conditions were much the same. 'The agriculture around West Drayton', a witness reported in 1898, 'has greatly changed during the last twenty years.'[10] Wheat had given way to fruit trees or market gardening. The west Middlesex fruit and market-garden area started as near to London as Chiswick, extending then through Brentford and Isleworth as far as the Colne Valley. It produced top fruit such as

[1] Ibid, 2.   [2] Ibid.   [3] Ibid, 6.
[4] Tariff Commission Report (1906), para. 997.
[5] Ibid, para. 1003.   [6] Ibid, para. 1021.   [7] Ibid, para. 1028.
[8] Rep. Deptl. Ctee Fruit Industry (1905), 6 (P.P. 1905, xx).
[9] W. E. Bear, 'Flower and fruit farming in England', pt. III, J.R.A.S., 3rd ser., x (1899), 300–1, 305–6.
[10] J. L. Green, The rural industries of England (London, 1895), 96.

apples, pears, plums and damsons; soft fruit such as strawberries, raspberries, gooseberries and currants; roots such as potatoes, beetroot, parsnips and carrots; and every type of vegetable.[1] From here too most of the produce went to Covent Garden by road; for 'whether it costs the same as rail transport or not, there is less handling involved than in the transfer to and from the railway waggons'. 'So it is that residents on the highways and main thoroughfares leading from West Middlesex into London have to sleep as best they can to the constant rumbling of long processions of market-garden carts, which, leaving the farms in the evening, do their twelve, fifteen, or twenty mile journey, and arrive at the London markets with their loads any time between midnight and three o'clock in the morning.'[2] This was an exceedingly labour-intensive form of farming: a typical farm of 50 to 100 acres might employ 60 to 120 hands, compared with a score in the old days of grain.[3]

In the same way, the Lea Valley had been opened up for market gardening since about 1880. Here, along the old road to Waltham Cross, 'a vast and overwhelming extension of London northwards' in the preceding twenty years had 'made of the road almost a continuous street as far as the border of Herts'.[4] The road was the route to the London markets; the villages along it supplied the necessary labour, the canalised river brought in coal to heat the glasshouses; the good gravel and brick-earth soils and the lack of strong winds made the valley ideal for the new form of farming.[5] In Edmonton, a witness said in 1898, there were in 1870 only 10 acres under glass in the whole parish; now there were a hundred, and land was renting at £10 the acre while nearby farm land would fetch only £3.[6] It was quite possible, one writer thought, that the area under glass in Cheshunt alone was equal to that in the whole of the rest of England and Wales outside Middlesex and Kent.[7]

Slightly farther afield, the railway facilities became critically important. Worthing was the capital of the Sussex glass hot-house industry, and within the borough were 3,500,000 square feet of greenhouses.[8] Here, in just over twenty years, the average shipments on a busy day in the season had risen from 100 to 2,500 packages in 1906. The London Brighton &

[1] E. A. Pratt (1906), 99–101.   [2] *Ibid.*, 100–1.   [3] *Ibid.*, 102.
[4] H. S. Vaughan, *The way about Middlesex* (London, 1896), 22.
[5] E. C. Willatts, *Middlesex and the London region* in L. D. Stamp (ed.), 'The land of Britain', pt. 79 (London, 1937), 222.
[6] W. E. Bear, 'Flower and fruit farming in England', pt. II, *J.R.A.S.*, 3rd ser., IX (1898), 525.   [7] *Ibid.*, 306.   [8] E. A. Pratt (1906), 90–1.

South Coast Railway took the fruit, gathered in the morning, by a special train which got to London Bridge by 2.30 in the afternoon, and the fruit was in the markets between 4 and 5 o'clock.[1]

In mid-Bedfordshire, around Sandy and Biggleswade, the Great Northern and the Midland Railways had helped create 'a stretch of country', about fifteen miles long and four or five broad',[2] almost entirely devoted to market gardening. Originally this was poor heath country,[3] but the soils were warm, and the Great Northern line brought stable manure cheaply from London.[4] It was said, in 1906, that the output of market-garden crops – of every sort, but especially of brussels sprouts, carrots and vegetable marrows – had quadrupled in the last 20 years, principally because of increased demand from the towns;[5] and though Biggleswade sold chiefly to London, the produce from Sandy went to the north and even as far as Glasgow.[6] This was smallholders' country; the majority of holdings were 10 to 15 acres, many were only 7 or 8 and some were as little as 2 or 3 acres;[7] the smallholders were shrewd enough to have weathered the storms of competition. For even here, perishability was no necessary protection from growers just across the Channel or North Sea. The family farmers of north Holland, in the Netherlands, had crushed the Biggleswade onion trade,[8] but in its place the production of brussels sprouts was developing into a major industry.[9]

In the vale of Evesham too, fruit and market gardening were developments of the last thirty years of the century; the market-garden acreage had risen from 300–400 acres in the early years of the century to some 2,000 acres in 1870, but by the first years of the twentieth century it had become 15,000 acres.[10] This too was smallholding land: the usual farm was between 1 and 8 acres, and 75% of the gardeners had started life as labourers.[11] 'Land practically derelict 20 years ago was taken by working men at £1 per acre in 2 and 3 acre lots for growing asparagus. Many now hold 6 acres, they own their houses, and some of them have sufficient means to support them for life.'[12] The men sold much of their produce at one or two auction markets in Evesham; most of it went to Birmingham

1 Ibid., 92–5. 2 Ibid., 103–4. 3 A. D. Hall, 424. 4 Ibid.

5 E. A. Pratt (1906), 104; L. Jebb, 84.

6 E. A. Pratt (1906), 104; L. Jebb, 84–5.

7 E. A. Pratt (1906), 104; L. Jebb 85.

8 Tariff Commission Report (1906), paras. 938–9. 9 E. A. Pratt (1906), 106.

10 Ibid., 134; Tariff Commission Report (1906), para. 956.

11 L. Jebb, 57. 12 Tariff Commission Report (1906), para. 956.

or the northern towns, very little to Covent Garden, for, said one, 'if we had to depend on London, we should soon be in the workhouse'.[1] True, there were disadvantages in the smallholding system; distribution was inefficient, the men could not easily make up the two-ton lots the railways demanded, and there was a lack of canning facilities.[2] Cooperation was the answer and by the beginning of the century it was 'becoming a distinctly active force'.[3]

Though both Biggleswade and Evesham were near big urban markets, their produce went far afield. Specialised fruit and vegetable growers necessarily sold to distinct markets. On the fertile warp soils of the Isle of Axholme, wheat had suffered a catastrophic fall in the depression,[4] but here, Haggard wrote, 'is one of the few places I have visited in England which may be called, at any rate in my opinion, truly prosperous in the agricultural sense, the low price of produce notwithstanding, chiefly because of its assiduous cultivation of the potato'.[5] The story was the same on the coastlands between Boston and Spalding. At the small village of Kirton alone, on a peak day they were sending away 95 wagon loads, with 285 tons of potatoes, mainly to London but also to Sheffield.[6] Another great staple of the Lincolnshire coastlands was celery, and in a belt sixteen miles long centred on Boston, half the total area was being given up to celery in the first years of the new century.[7] The Isle of Axholme, too, profited from celery. In the village of Haxey, in 1906, a visitor found celery being grown not only by the stationmaster but by his goods-yard foreman and the foreman's eight-year old boy.[8]

The most striking case of agricultural specialisation for a distant market was, however, to be found at the extreme south-west tip of England. The Isles of Scilly were already exploiting their climatic advantage over the rest of the country by producing potatoes for the London market by 1865. But in that year the first experimental shipment of flowers to Covent Garden had taken place; and since then, despite the heavy freightage – 8s. to 9s. 6d a hundredweight – the trade had boomed, eating into former coastland potato land.[9] In 1885, some 65 tons were dispatched, and

[1] E. A. Pratt (1906), 137.
[2] *Ibid*, 139–40; *Tariff Commission Report* (1906), para. 1071.
[3] E. A. Pratt, *The organization of agriculture* (London, 1904), 302.
[4] *R.C. Agric. Depression, Rep. by Mr R. Hunter Pringle on the Isle of Axholme*, 16 (P.P. 1894, xvi, pt. 1).
[5] H. R. Haggard, II, 186.
[6] E. A. Pratt (1906), 126; L. Jebb, 28–9.
[7] E. A. Pratt (1906), 126.
[8] *Ibid*, 124–5.
[9] W. E. Bear (1898), 298–300.

by 1900 this had become 575 tons.[1] And already, competitors were springing up on the mainland of England – above all in Lincolnshire, where, by the beginning of the new century, a group of villages along the railway near Spalding were virtually living on the sale of flowers.[2]

By then, the early vegetable trade had become centred at the end of the Cornish mainland, in a highly specialised and highly concentrated zone around the little village of Gulval on Mounts Bay. Here, over 300 miles from the London market, soil and aspect were critical. The best land was found about half a mile from the sea on the gently sloping southern slopes, where light soils had developed on the 'killas' or china clay slates. The smallholders, on their 5- to 15-acre plots, got two early crops a year – broccoli between Christmas and February, potatoes in May.[3] The older men here could remember the day when broccoli was sent from Hayle to Bristol by boat, but by the turn of the century the railway dominated operations here, taking the produce as far as the north of England and even up to Dundee. Reckoning the whole of west Cornwall together, over 900 trucks were sent off in a single week.[4]

By the turn of the century, then, the urban Englishman's food came from a bewildering variety of sources. Foreign produce, Haggard pointed out, 'often can be delivered in our market at a lower cost of carriage than must be incurred to despatch it from one part of England to another'.[5] Or in any case for very little more. Apples could be brought from North America to Manchester in 1897 for 35s. upwards a ton, while to bring them from Somerset cost 26s. 8d.; onions took 15s. 10d. to bring from east Yorkshire but only 17s. to bring from Egypt.[6] Here, the cheap freight for vegetables was already telling against local market gardeners. Potatoes came to Manchester at different times of the year from the Canaries, the Channel Islands, Cheshire, Scotland, Bedfordshire, Yorkshire, Worcestershire and Lincolnshire; onions from Spain, France, Holland, Germany, Portugal and Egypt.[7]

The outlook for English agriculture, as the twentieth century began, was far from good. In Haggard's verdict, 'Many circumstances combine to threaten it with ruin, although as yet it is not actually ruined'.[8] Here and there, on small enclaves of favoured land, even small men were making

[1] E. A. Pratt (1906), 73.　　　[2] Ibid., 76.　　　[3] A. D. Hall, 341–2.
[4] E. A. Pratt (1906), 117, 121.　　[5] H. R. Haggard, II, 536.
[6] W. E. Bear, 'The food supply of Manchester', J.R.A.S., 3rd ser, VIII (1897),
226–7.
[7] Ibid., 212–15.　　　[8] H. R. Haggard, II, 536.

good livings from specialised produce; here perhaps, land was actually supporting more people than it had thirty years before. But in general the prospect was one of massive and continuing migration from the land. 'Everywhere the young men and women are leaving the villages where they were born and flocking into the towns.'[1] A Herefordshire schoolmaster said that of a hundred boys who had passed through his hands in six years, not more than a dozen had stayed; those that came back to visit their friends said 'that nothing would induce them to leave what they call "town sweets"'.[2] There was too little to induce them to stay; 'England', as an Essex vicar suggested, being 'hardly merry England for the farm labourer'.[3] Many witnesses said that education was unsettling the farm labourer, and were wont to complain of the yokel who referred to a badly built haystack as 'a most egregious blunder', or to the teamster who talked of 'my colleague the cattleman';[4] but the causes were more complex. General Booth, of the Salvation Army, summed it up for Haggard: 'the smallness of the supplies, the poorness of their food, the struggle there is to make things meet, and nothing before them at the end but the probability of pauperism', coupled with education 'and the newspapers and periodicals and the glory of war. You have all these to contend with, and the railways and the telegraphs and the penny postage which are put down as such great advantages. All these things are against your keeping a man on the land, especially if he can't get enough to eat.'[5]

INDUSTRY AND TOWNS

Not long after 1900 a Royal Commission on coal supplies was discovering that the proved coal resources to a depth of 4,000 feet in the United Kingdom amounted to 100,914,668,167 tons – enough to last 435 years at current rates of consumption; some 60 thousand out of this 101 thousand million tons lay in England itself.[6] True, this coal was being won at an increasing cost; the pioneers of the early industrial revolution had reaped the benefit of the most accessible coal, and by 1900 new deeper shafts and more difficult seams were being mined.[7] The fastest expanding coalfields

[1] *Ibid*, II, 539.   [2] *Ibid*, I, 295.   [3] *Ibid*, I, 493.
[4] P. A. Graham, *The rural exodus* (London, 1892), 34; see H. R. Haggard, I, 251.
[5] H. R. Haggard, I, 495–6.
[6] *R.C. Coal Supplies: Final Report: Part I, General Report (1905)*, 2–3 (P.P. 1905, xvi).
[7] J. E. Williams, *The Derbyshire miners: a study in industrial and social history* (London, 1962), 174.

tended to be the concealed coalfields of the east Midlands where large new pits were being sunk; big increases in output were taking place in central and northern Nottinghamshire and in south Yorkshire, between Mansfield and Doncaster. The Durham–Northumberland and Lancashire coalfields, which had experienced major growth only a few years earlier, were now showing much more modest increases in production, though there was expansion in Northumberland and south-east Durham. The fastest growing of all fields were small Midland fields like those in Warwickshire and Leicestershire, although the Staffordshire field was stagnating.[1]

Coal and steam power provided the basis for England's industrial strength. In the first years of the twentieth century, it was true, there were frequent reports of industries like cotton textiles turning from steam to electricity; but the electricity was produced locally on coalfields, and it would have been costly to transport even if then it had been technically feasible. Most contemporaries would have agreed with the French observer Lozé who said, in 1900, that the greatness of a nation could be measured by its coal production.[2]

Well over one half of the towns with populations of 50,000 and over were situated on or near coalfields. T. A. Welton analysed the results of the 1901 Census to discover what were the primary activities of each of these large towns.[3] By grouping the constituent towns of conurbations such as London and Merseyside he arrived at a total of 65 towns (60 in England) to which he added Oxford and Canterbury. Anticipating the modern distinction between basic and non-basic industries, he distinguished six major types of 'primary occupations' which could support the population of a town, providing it with an economic *raison d'être*.[4] Nineteen of the English towns, from London downwards, were described as basically 'commercial'; mainly ports, they also included four seaside resorts – Bournemouth, Brighton, Hastings and Bath. Eighteen were metal towns; they included major cities such as Birmingham, Sheffield, Newcastle upon Tyne, Leeds and, perhaps surprisingly to some of Welton's readers, Manchester. Fifteen towns, mainly in the north, depended on textiles and clothing. Three towns depended on defence – the two naval

1 *Ibid*, 175.

2 E. Lozé, *Les charbons britanniques et leur épuisement* (Paris, 1900), 17.

3 T. A. Welton, 'Memorandum on primary occupations in 1901', *Jour. Roy. Stat. Soc.*, LXVI (1903), 360–5.

4 Welton first developed the idea in *Statistical papers based on the Census of England and Wales, 1851, and relating to the occupations of the people and increase of population 1841–51* (London, privately printed, 1860).

ports of Portsmouth and Plymouth (with Devonport) and also Canterbury which Welton described as 'a very quiet place, if the soldiers be left out of account'. One town (Wigan) depended primarily on mining. Finally, there was a sixth group of towns with manufactures, 'employing a large section of their inhabitants, which were neither metallic nor textile'. These towns contained 'special industries on a large scale': Hanley (potteries), Burton on Trent (brewing), St Helens (glass, coal and chemicals), Walsall (saddlery and metals), Reading (biscuits), and Oxford (printing). In making this sixfold grouping, Welton recognised that two types of primary occupations were 'universally present, namely the commercial and metal working divisions', and that others were often absent or nearly so, although the class of 'other manufactures' could 'hardly fail to exist in some measure'.

The census also showed how varied were the industries on each coalfield. It would be a mistake to think of a major industrial area as combining coalmining with some staple manufacture alone; each coalfield also sustained a host of varied crafts and trades. Even so, on a broad view two groups of industries were especially associated with particular coalfields – the metal industries, particularly iron and steel, and the textile industries. To these two groups must be added a third outside the coalfields. Industrial London, with its many and varied manufactures, stood in a category of its own.

The iron and steel industries

Whether expanding or not, the coalfields continued to attract the rapidly growing complex of iron, steel and engineering industries. This fact might have surprised a foreign visitor who had noticed the relative shift of iron and steel industries from Ruhr coal to Lorraine ore or Pittsburgh coal towards Duluth ore.[1] Actually, British steelmaking was also shifting away from coal; for the amount of coal needed to make a ton of pig-iron had progressively declined to 2.02 tons by 1899, and steel was a product very economical in its use of coal compared with the old wrought-iron. But there was one remarkable feature which distinguished British steelmaking and its location: it was the failure to exploit the large resources of low-grade Jurassic ores in Lincolnshire and Northamptonshire, which the Thomas–Gilchrist 'basic process' had made available after 1878, and the consequent almost complete failure of the industry to shift towards this

[1] W. Isard, 'Some locational factors in the iron and steel industry since the early nineteenth century', *Journal of Political Economy*, LVI (1948), 205, 207.

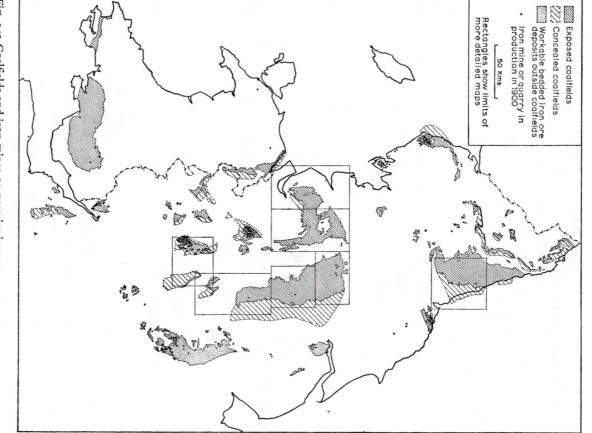

Fig. 145 Coalfields and iron mines or quarries in 1900
Based on: (1) *List of mines, 1900* (H.M.S.O., 1901); (2) *List of quarries, 1900*
(H.M.S.O., 1901).

Exposed coalfields

Concealed coalfields

Workable bedded iron ore
deposits outside coalfields

• Iron mine or quarry in
production in 1900

Rectangles show limits of
more detailed maps

50 Kms

ore.[1] The attitude of British steelmasters, indeed, remained that summed up by Sir Lowthian Bell in his comments on the Bessemer process in 1886: 'It is calculated that ore can be brought from Bilbao and converted into steel rails at Middlesbrough at the same or even a less cost than the same article can be made of iron from the ironstone of the Cleveland ores, lying almost at the gates of the rail-mills.[2] True, as Bell pointed out, 'in Lincolnshire and Northamptonshire the ironstone is obtained cheaply, which enables these districts to furnish, out of their abundance, after supplying the needs of their own furnaces, large quantities to works situated in Staffordshire, Derbyshire, etc.'[3] But these quantities were modest indeed compared with those furnished by the similar minette ores of Lorraine, despite the fact that the Jurassic ores were the cheapest to mine in the country, at a fraction of the cost of imported ore.[4] The Northamptonshire ore increased its price three times on a trip to Middlesbrough; the minette ore only doubled on its trip to the Ruhr.[5] British steelmasters reflected that 'what they gained on the ore, therefore, they might lose on the trains'.[6]

But the reasons were more complex than this, for only one fifth of British steel was of the 'basic' variety, by which advantage was taken of the Thomas–Gilchrist process. Steelmasters in the old iron districts were suspicious of 'basic steel'; they were unfamiliar with it, and they were too often incapable of the scientific methods needed to produce a high-quality product.[7] Their distrust was shared by British structural engineers, and, as a manager of Dorman Long put it in 1905, more rapid development must await the 'technical education of Great George Street'.[8] And the new Jurassic orefields only hesitantly produced a steelmaking tradition: they were making between 6% and 7% of the total make of pig-iron by the turn of the century,[9] but in steel Scunthorpe started only in 1890[10] and Northamptonshire did not produce basic steel until 1927.[11]

[1] D. L. Burn, *The economic history of steelmaking 1867–1939: a study of competition* (Cambridge, 1940), 167; and T. H. Burnham and G. O. Hoskins, *Iron and steel in Britain 1870–1930* (London, 1943), 116.

[2] I. L. Bell, *The iron trade of the United Kingdom* (London, 1886), 17.

[3] *Ibid*, 10. [4] T. H. Burnham and G. O. Hoskins, 110.

[5] *Ibid*, 112. [6] D. L. Burn, 181. [7] *Ibid*, 172–8.

[8] *R.C. on the Supply of Food and Raw Materials, Minutes*, Evidence of Bell, Q. 9,699 (P.P. 1905, xxxix).

[9] H. G. Roepke, *Movements of the British iron and steel industry 1720–1951* (Urbana, 1956), 64–5.

[10] D. C. D. Pocock, 'Iron and steel at Scunthorpe', *East Midland Geographer*, III (1963), 129.

[11] T. H. Burnham and G. O. Hoskins, 116–17. For iron mining in Northamptonshire

Yet it is not true simply to say that iron and steel stayed on the coalfields. Certain important coal and iron districts of the early nineteenth century, like the Black Country, were now of minor importance in iron or steel making.[1] It would have been truer to say that manufacture was being attracted to two sorts of ore supply. One was the high-grade, non-phosphoric haematite of north-west England; the Cumberland coast works made close on 18% of British pig-iron in 1900, and 13% of the steel.[2] But there the ore was said to be 'a vanishing quantity';[3] it was expensive to bring Durham coal across the Pennines,[4] and there was no local market for engineering products (Fig. 145). The other, and much more important, source was imported high-grade ore from Spain or Sweden.

The north-east (Fig. 146). The classic case of a seaboard industry dependent on imported ore was the north-east coast, which was by far the most important single steelmaking region in 1900, with 35% of the country's pig-iron production and 27% of the steel.[5] On Tees-side, seventy years before, 'scarcely a house broke the solitude from Stockton to the sea, and nothing but the smoke from the infrequent farm chimney rose up into the ether'.[6] But a description in 1888 ran: 'Now the sight in the passage up or down the river by day is startling, by night it is spectacular. The clang of the riveters in the shipyards, the roar of the blast furnaces, and the thud of the steam hammers at the rolling mills and the engineering works fill the ear. A long halo of light spreads over a large part of the scene from the Tees Bridge Iron Company's works at Stockton Bridge down to the huge steel works at Eston, and tongues and darts of flame ride up into it.'[7]

All this was the work of iron; for though Middlesbrough had started life as a coal port, prosperity did not come to it until after the 1850s, when the Cleveland ore began to pour into the town's blast furnaces. At the banquet for Queen Victoria's Jubilee in 1887, it was said that 'with coal, and

see S. H. Beaver, 'The development of the Northamptonshire iron industry, 1851–1930', in L. D. Stamp and S. W. Wooldridge (eds.), *London essays in Geography* (London, 1951), 33–58; and for the manufacturing industry, F. Scopes, 'The development of Corby, Part I – before 1918', *Northamptonshire Past and Present*, III (1963), 125–30.

[1] H. G. Roepke, 64–5, for ironmaking. Steel was never important here.
[2] H. G. Roepke, 64–5, 83.
[3] *R.C. Supply of Food etc. Minutes,* Evidence of Milton, Q. 5,757 (P.P. 1905, xxxix).
[4] W. Isard, 212.
[5] H. G. Roepke, 64–5, 83.
[6] Anon, *Industrial rivers of the United Kingdom* (London, 1888), 252.
[7] *Ibid.,* 252–3.

without iron, Middlesbrough would not have achieved its present preeminence'.[1] As it was, 'the growth of the town made the first planned settlement of the "Middlesbrough Owners" seem like a tiny frontier encampment strategically placed alongside the river';[2] a railway line and station now separated it from the new town centre around the town hall and municipal buildings, which had been erected in Jubilee year. No longer did the town owe its *raison d'être* to the ore in the hills behind: throughout the nineties the orefield had been declining, and its miners were moving down into Middlesbrough to find work processing the ore that now came from Bilbao.[3]

Some miles farther north, Newcastle kept its distinctive character, preserved since Tudor times, as a coal-shipping port. Coal shipments, which amounted to over 12 million tons in a typical year in the nineties, made the port of Tyne the third port of England, after London and Liverpool, in respect of tonnage handled.[4] As the trade had increased, exports to foreign countries had steadily become more important relative to the coastwise trade, accounting for seven-twelfths of the tonnage of total exports in the 1890s; and with the growth of steam navigation, the bunkering trade had also increased prodigiously.[5] But on these unrivalled facilities for transport a great and complex manufacturing industry had also been built up. Progress had been remarkable: 'The Tyne in less than half a century,' ran one account, 'has quadrupled its coal shipments, has tripled the quantity of tonnage owned in the port, has practically created an import trade, has covered its banks with iron shipyards and engineering works, has developed the most important coal bunkering trade on the east coast'.[6] And the area had witnessed heavy immigration during the 1890s.[7]

The most distinctive Tyneside industry was representative, for it had evolved directly out of the supply of coal and iron and out of the presence of the waterway. In Palmer's Yard at Jarrow, 'coal and iron ore come in at one end of the works; ships are launched, with steam up, at the other

[1] *Ibid.*, 247.
[2] A. Briggs, *Victorian cities* (London, 1963), 276.
[3] J. W. House, *North-eastern England: population movements and the landscape since the early 19th century* (Newcastle upon Tyne, 1954), 34–5.
[4] R. W. Johnson, *The making of the Tyne* (London and Newcastle upon Tyne, 1895), 43–4.
[5] *Ibid.*, 197, 207.
[6] *Ibid.*, 48.
[7] A. G. Kenwood, 'Residential building activity in north-eastern Eng[land] 1913', *The Manchester School, Econ. and Soc. Studies*, XXXI (1963) 2[1].

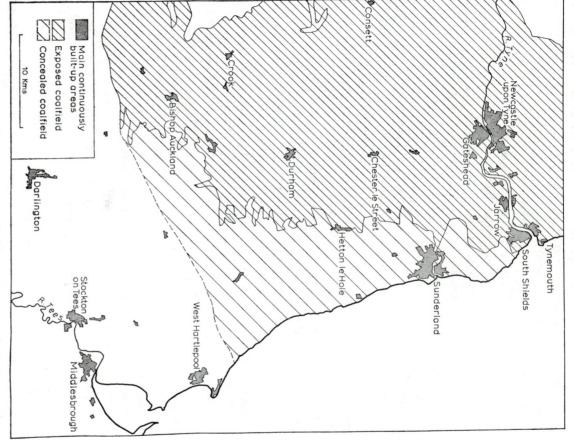

Main continuously
built-up areas

Exposed coalfield

Concealed coalfield

10 Kms

R. Tyne

Newcastle
upon Tyne

Consett

Crook

Bishop Auckland

Durham

Chester-le-Street

Gateshead

Jarrow

Tynemouth

South Shields

Hetton le Hole

Sunderland

Darlington

Stockton
on Tees

R. Tees

West Hartlepool

Middlesbrough

Fig. 146 The north-east: built-up areas *circa* 1900
Based on: (1) J. G. Bartholomew, *The survey atlas of England and Wales*
(Edinburgh, 1903); (2) Ministry of Town and County Planning map of
Coal and Iron (Ordnance Survey, 1945).

without iron, Middlesbrough would not have achieved its present pre-eminence'.[1] As it was, 'the growth of the town made the first planned settlement of the "Middlesbrough Owners" seem like a tiny frontier encampment strategically placed alongside the river';[2] a railway line and station now separated it from the new town centre around the town hall and municipal buildings, which had been erected in Jubilee year. No longer did the town owe its *raison d'être* to the ore in the hills behind: throughout the nineties the orefield had been declining, and its miners were moving down into Middlesbrough to find work processing the ore that now came from Bilbao.[3]

Some miles farther north, Newcastle kept its distinctive character, pre-served since Tudor times, as a coal-shipping port. Coal shipments, which amounted to over 12 million tons in a typical year in the nineties, made the port of Tyne the third port of England, after London and Liverpool, in respect of tonnage handled.[4] As the trade had increased, exports to foreign countries had steadily become more important relative to the coastwise trade, accounting for seven-twelfths of the tonnage of total exports in the 1890s; and with the growth of steam navigation, the bunkering trade had also increased prodigiously.[5] But on these unrivalled facilities for transport a great and complex manufacturing industry had also been built up. Progress had been remarkable: 'The Tyne in less than half a century,' ran one account, 'has quadrupled its coal shipments, has tripled the quantity of tonnage owned in the port, has practically created an import trade, has covered its banks with iron shipyards and engineering works, has de-veloped the most important coal bunkering trade on the east coast.'[6] And the area had witnessed heavy immigration during the 1890s.[7]

The most distinctive Tyneside industry was representative, for it had evolved directly out of the supply of coal and iron and out of the presence of the waterway. In Palmer's Yard at Jarrow, 'coal and iron ore come in at one end of the works; ships are launched, with steam up, at the other

[1] *Ibid.*, 247.

[2] A. Briggs, *Victorian cities* (London, 1963), 276.

[3] J. W. House, *North-eastern England: population movements and the landscape since the early 19th century* (Newcastle upon Tyne, 1954), 34–5.

[4] R. W. Johnson, *The making of the Tyne* (London and Newcastle upon Tyne, 1895), 43–4.

[5] *Ibid.*, 197, 207.

[6] *Ibid.*, 48.

[7] A. G. Kenwood, 'Residential building activity in north-eastern England, 1853–1913', *The Manchester School, Econ. and Soc. Studies*, xxxi (1963)

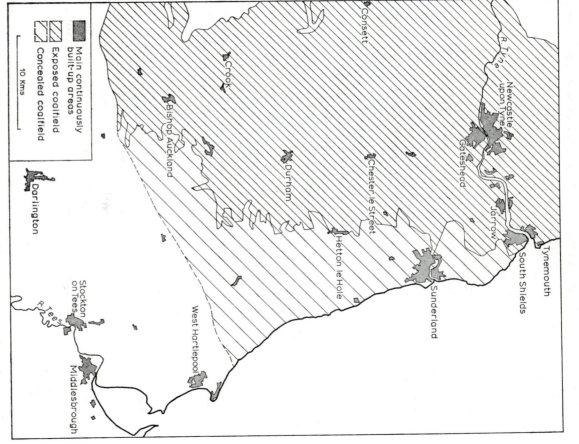

Fig 146 The north-east: built-up areas *circa* 1900
Based on: (1) J. G. Bartholomew, *The survey atlas of England and Wales*
(Edinburgh, 1903); (2) Ministry of Town and County Planning map of
Coal and Iron (Ordnance Survey, 1945).

Main continuously
built-up areas

Exposed coalfield

Concealed coalfield

10 Kms

Consett

Crook

Bishop Auckland

Durham

Chester le Street

Hetton le Hole

Darlington

Stockton
on Tees

R. Tees

Middlesbrough

West Hartlepool

Sunderland

R. Tyne

Newcastle
upon Tyne

Gateshead

Jarrow

South Shields

Tynemouth

end'. Here, 'magnificent blast furnaces extract the iron from the ore, and deliver it in the form of pig iron. Puddlers, working half naked in front of scorching furnaces, convert the cast iron into malleable iron; steam hammers and the rolling mill turn it into plates and bars fit for shipping purposes.' Engines were made here in a large separate factory. Palmer's also made all sorts of ships, 'from the Atlantic mail-steamer down to the tiniest river-craft, from the iron-clad war ship down to the small but dangerous torpedo vessel'.[1] That was a Tyne tradition generally; though its speciality was the manufacture of moderate-sized cargo-carrying boats, which were so greatly in demand to carry the rising tide of world commerce.[2]

South Yorkshire (Fig. 147). Shipbuilding and marine engineering were natural outgrowths of the coal and iron industry on a highly productive coastal coalfield facing the markets of Europe. But in the great interior coal and metal districts of England, districts which had seen rapid development in the first decades of the industrial revolution a hundred or more years before, the process of evolution was more complex and sometimes more difficult. The two most striking examples were Sheffield and the west Midlands.

The traditional Sheffield steel trades, as was fitting before Bessemer made steel a mass product in 1855, were handicraft trades. In the typical cutlery shop could be found 'a few workmen filing, hammering, drilling and fitting parts together, with a deftness and quickness which only long practice, aided by hereditary aptitude, could give. The tools they use are old-fashioned; the shop is meagre of all appliances and aids to labour; their own manner shows surly independence and reserve.'[3] Extreme division of labour was the rule: one shop converted iron to steel by the old cementation process; another sheared it into shape; a third shop might specialise in cast-iron.[4] This handicraft trade was already beginning to give way, in the sixties and seventies, to factory production; by the 1890s table blades, scissors and files were largely made on the factory system.[5] But in Sheffield this was a relative term; in scissor grinding for instance, even in the factories, men must pay for tools, appliances and steam power.[6] Even at the beginning of the twentieth century there were 160 'tenement factories',

[1] Anon, *Great industries of Great Britain* (London, 1886), I, 24.
[2] *Ibid.*, I, 102.   [3] *Ibid.*, III, 120.   [4] *Ibid.*, III, 220-1.
[5] *R.C. on Labour, Minutes, Group A*, Evidence of Wordley, Q. 19,300 (P.P. 1892, xxxvi, pt. 1).
[6] *Ibid.*, Evidence of Holmshaw, Q. 19,394-7.

or small flatted workshops, with about 12,000 workmen, in the city; the number of separate tenements in the same building might vary from three or four to seventy or eighty. Here sometimes only one occupier might be found, but more frequently several. The grinder would rent a proportion of a 'wheel'; there he would employ labour, 'to grind, bolster, buff, glaze or finish' the cutlery; he would provide his own grindstone and tools and would pay for power.[1] And the system applied not merely to cutlery, but to other 'light trades': silver polishing, scale cutting, brass turning. At the turn of the century the progress of these light trades was uneven. Cutlery showed no increase in volume; implements, tools and files were growing more slowly: production of plate grew rapidly.[2]

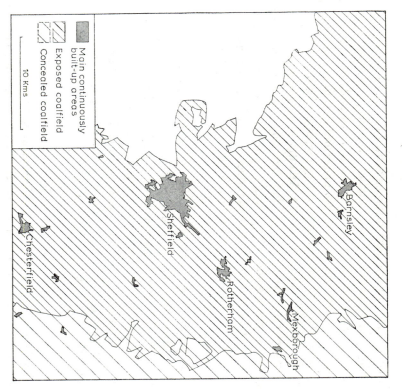

Fig. 147 South Yorkshire: built-up areas *circa* 1900
Sources as for Fig. 146.

[1] *Ann. Rep. Factories, 1902*, 78 (P.P. 1903, xii).
[2] S. Pollard, *A history of labour in Sheffield* (Liverpool, 1959), 202.

In 1900, the old hand trades were still found at the western edge of the town centre of Sheffield, where they had established themselves a hundred years before, and in the residential quarters just outside – in Hillsborough, Walkley, Crookes and Heeley.[1] But on the lower, flatter land in the Don valley to the east, in Brightside and Attercliffe, the Bessemer revolution had created a new Sheffield in the last forty years of the nineteenth century.[2] The division was not complete: 'there are grinding wheels in Brightside, which is not very bright', a witness told the Labour Commission in 1892, but 'I should decidedly say a great majority of the men are not grinders as in another part of the town…It is heavier labour, the great castings, the armour plates, and guns and armaments generally, boiler making and so on.'[3] It was this 'East End' which had witnessed a dramatic increase in industrial employment during the latter half of the century; between 1851 and 1891, employment in the heavy trades had risen over 300%, in the light trades only 30%. The cause was the Bessemer revolution; but because of the inland position of the city, the effect was different from that in other great iron and steel producing areas of England. 'Only by the high value of its steel per ton could Sheffield neutralise its unfavourable geographical position';[4] but it did so triumphantly, retaining a virtual monopoly in the special steels which were in rapidly increasing demand. Around the supply of these steels, too, developed a whole range of heavy engineering products including boilers, railway equipment, and armaments.

The west Midlands (Fig. 148). In the other great inland metal district – the west Midlands – the advent of the Bessemer process had been accompanied by rapid decline. The south Staffordshire coalfield had reached its peak as an iron-making area about 1840, when local iron and local coal united to give Staffordshire no less than 29% of total British production.[5] But the switch to imported iron dealt the area, isolated as it was from tidewater on a remote inland plateau, a particularly grievous blow. The Great Depression of the 1830s, coupled with waterlogging of the coalmines, marked the end; by the 1890s the whole of Staffordshire accounted for only about 7% of British pig-iron production.[6] In 1889 the historian of the Black Country, F. W. Hackwood, was already looking back with nostalgia on

[1] *Ibid.*, 89–90; A. J. Hunt in D. L. Linton (ed.), *Sheffield and its region: a scientific and historical survey* (Sheffield, 1956), 234.

[2] S. Pollard, 89–90, 159; A. J. Hunt, 234.

[3] *R.C. on Labour, Minutes*, Evidence of Holmshaw, Q. 19,556–9 (P.P. 1892, xxxvi, pt. 1).　　[4] S. Pollard, 159.　　[5] H. G. Roepke, 28.　　[6] *Ibid.*, 64.

DHG

23

the days, only thirty years earlier, when there were 155 blast furnaces in work in the area: 'To stand on Church Hill [in Wednesbury] at night and gaze around on the busy scene of myriad fires belching forth like so many volcanoes was a sight not easily forgotten. But all this has passed away now...'[1] For Hackwood the causes were not far to seek: 'The need for canals to the sea-board, and the burden of heavy royalties on nearly exhausted and water-logged mines, are cankerous problems.' But 'another form of decay' lay in the 'antiquated ideas' of the local ironmasters, who failed 'to adopt all these newer methods which characterised the strength of their new rivals and competitors who sprang up on the northern sea-boards and other favoured places'.[2] Wrought-iron had been the Stafford-shire staple and it could not withstand the competition of Bessemer steel. In fact one of the first basic Bessemer converters in Britain, using the phosphoric 'Puddlers tap' which was in such liberal supply here, opened in Wednesbury in 1882,[3] but this was an isolated exception to a dismal general rule. By the early years of the twentieth century whole plants were removing to coastal sites;[4] between 1895 and 1903 eight iron mills had 'been dismantled and three others ceased working with little prospect of restarting'.[5]

Contemporaries like Hackwood realised that the real strength of the Black Country lay in the lighter, more highly manufactured iron goods for which the area's distance from ore supplies put it at little disadvantage.[6] But even there, some of the old staples — staples on which the industrial revolution in the area had been based — were now in serious plight. The wretched hand nailers of Dudley were suffering from the competition of machine-made nails from Birmingham and (more importantly) from the north of England.[7] The same story came from Halesowen and Bromsgrove to the south, where the women were a majority in the flimsy sheds which formed the home workshops.[8] They got their materials from, and gave

[1] F. W. Hackwood, Wednesbury workshops (Wednesbury, 1889), 118.   [2] Ibid., ii.
[3] G. C. Allen, The industrial development of Birmingham and the Black Country, 1860–1927 (London, 1929), 239.
[4] Ann. Rep. Factories, 1902, 38 (P.P. 1903, xii).
[5] Ann. Rep. Factories, 1903, 43 (P.P. 1904, x).
[6] F. W. Hackwood (1889), passim.
[7] G. P. Seven, A handbook to the industries of the British Isles and the United States (London, 1882), 17; see also F. W. Hackwood, The story of the Black Country (Wolverhampton, 1902), 16.
[8] R.C. on Labour, Minutes, Group A, Evidence of Price, Q. 17,596; Evidence of Powell, Q. 18,409 (P.P. 1892, xxxvi, pt. 1); J. L. Green, 123–4.

their nails back, to the 'foggers' or middlemen,[1] and in the heavier branches of the trade, such as the manufacture of 'railway dogs', their work 'is more like blacksmith's labour than labour for females'.[2] As late as 1892 it was estimated that there were still 6,000–7,000 hand nailers in Halesowen, about 100 in Sedgley and 1,500 in Bromsgrove.[3] The extinction of the trade, Hackwood judged, 'will not be matter for much regret', for it had 'never brought either profit or renown upon any locality in which it has been seated'.[4] In the neighbouring Cradley Heath area, the staple chain-making trade still employed a large number of equally wretched female home workers as late as 1902.[5] Ten years earlier it was estimated that perhaps one half the work in the area was done at home; the female workers, perhaps a quarter of the whole labour force, all worked at home.[6] The merchants or foggers 'have no factories, they have no machinery, and they do not lay out anything whatever. They only simply (*sic*) buy the iron and deliver it out to the workers, and the workers have to find their own tools and their own firing, to manufacture the iron to chain...I do not know that you could call them anything bad enough, they are sweaters and foggers.'[7] Up in Gateshead they also made chains, and could claim a higher price, because 'the carriage on the Tyne' allowed them to deliver the work 15s. a ton cheaper.[8]

In Halesowen the women also worked in home workshops to forge small bolts, fetching the iron from the sweater's workhouse, half a hundredweight at a time. Here was to be seen 'a poor woman carrying a bundle of iron on her shoulder, or rather in her arms, a bag of work on her head, and a child about four years of age dragging at her dress; probably she would have to carry that iron and work a distance of half a mile'.[9] The pay was so low in the nineties that the home workshops could still under-cut the factory machinery.[10]

It was the same story up to the north of the Black Country, where about nine in every ten of Britain's lock and key makers were found.[11] Of about 1,800 workers in Wolverhampton and Willenhall, about a quarter were

[1] *Ibid.*, Evidence of Price, Q. 17,628–34.   [2] *Ibid.*, Q. 17,617.

[3] *Ibid.*, Q. 17,586–8, and Evidence of Powell, Q. 18,384.

[4] F. W. Hackwood (1902), 16.

[5] *Ann. Rep. Factories, 1902*, 39 (P.P. 1903, xii).

[6] *R.C. on Labour, Minutes, Group A*, Evidence of Homer, Q. 16,918–19 and 16,927 (P.P. 1892, xxxvi, pt. 1).

[7] *Ibid.*, Q. 16,920, 17,186.   [8] *Ibid.*, Q. 17,047.

[9] *Ibid.*, Evidence of Juggins, Q. 18,067.

[10] *Ibid.*, Q. 17,817–18.   [11] *Ibid.*, Evidence of Day, Q. 18,104–10.

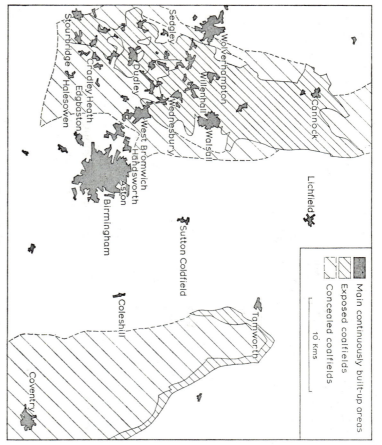

Fig. 148 West Midlands: built-up areas *circa* 1900
Sources as for Fig. 146.

outworkers, 'men who work in bedrooms occasionally, the bed at the back and the vice in the window, and in the washhouses and all about'.[1] The real homework, of this order, was associated especially with Wolverhampton; but 'instead of being ones or twos, as in those little cribs, in Willenhall it is on a greater scale';[2] the small masters provided workshops. Here, 'locks of every variety of principle and quality' were produced;[3] the rim lock, the cabinet lock, the stock or fine plate lock, the drawback lock, the dead lock, the mortice lock, the loose padlock and others. If a worker let fall the lock he was making, he never stooped to pick it up; he could make another in less time. In the 1890s, cash-box locks had been made for 1d. each and padlocks for 1½d. a dozen,[4] but employment and output were both

¹ *Ibid.*, Q. 18,128–31.   ² *Ibid.*, Evidence of Martin, Q. 18,330.
³ F. W. Hackwood, *The annals of Willenhall* (Wolverhampton, 1908), 161.
⁴ *R.C. on Labour, Minutes, Group A*, Evidence of Day, Q. 18,195 (P.P. 1892, xxxvi, pt. 1).

declining in the face of German competition. By 1900, factory competition was making its mark: output was up, but the work force stayed about the same.[1]

In all these trades then the sweated home worker was engaged in desperate competition with the factory product, and most contemporary observers realised that he must eventually lose the battle. But meanwhile the old traditions of the small workshop offered a unique advantage in the development of new industrial traditions. Nowhere was this more evident than in the capital of the whole district. 'Birmingham, in the nineties, was poised between the town of a thousand workshops and the home of the great industries of the twentieth century. The hitherto busy activity of craftsmanship in innumerable trades was to take second place, but it provided a large body of skilled workers and a great variety of manufacturers.'[2] Contemporary observers noticed the adaptability of Birmingham industry in the face of rapid change: 'it has many different trades, and if some are depressed and slack others may be active and prosperous', wrote a chronicler of the city in 1900; 'Birmingham...is pretty smart at taking up new ideas, and does not let new manufacturing industries go begging for a home'.[3] There were clear advantages too, for the worker, who could easily find alternative employment if a factory shut its doors.[4]

Admittedly, in 1900, Birmingham's two most important staples were still doing well. Jewellery was in increasing demand from an increasingly affluent middle class, and there was a shift towards the better type of product; the brass trade, if it experienced a falling demand from the gas industry, found compensation in the new electrical and motor industries.[5] Other trades experienced varying fortunes. The manufacture of high-quality buttons had received a blow from the advent of the long tie, but the women's trade was increasing;[6] in the holloware trade, the tinplate section suffered from the multiplication of modern bathrooms, and the japanned ware section was hard hit by the increasing use of china teapots, cash registers and leather bags.[7] The demand for guns was limited by the long years of world peace and by restrictions made by foreign governments;[8] but the steel-pen trade, aided by the spread of literacy, went from

[1] G. C. Allen, 256-7.
[2] P. W. Kingsford, 'The Lanchester Engine Company Limited 1899-1904', *Business History*, III (1961), 107.
[3] T. Anderton, *A tale of one city: the new Birmingham* (Birmingham, 1900), 141.
[4] E. Deiss, *A travers l'Angleterre industrielle et commercielle* (Paris, 1896), 15.
[5] G. C. Allen, 251-2. [6] *Ibid*, 264. [7] *Ibid*, 261. [8] *Ibid*, 265-6.

strength to strength,[1] while the plating trade benefited from the rising demands of the hotel industry.[2] In general though the traditional industries were declining, or were stagnant, or at best were increasing only slowly.[3]

These traditional trades were still tightly concentrated in the inner parts of the city, and certain of them were extremely localised in industrial quarters: armaments, jewellery, copper, pens, beds, buttons,[4] Most notable was the jewellery quarter, which had been established in the Vyse Street area north of the city centre since 1865; at the turn of the century it employed some 300 master jewellers with several thousand workers.[5] But new industrial traditions were now being rapidly superimposed on the old, and from the first they tended to be factory traditions, carried on in new areas where space was available for expansion at the edge of the growing city.

The new traditions were centred around engineering. It is a myth that the industrial revolution in Birmingham was built on engineering products: the happy combination of Watt and Boulton and Murdock at the Soho works was an isolated phenomenon, and up to 1880 machine-building expertise was concentrated especially in Lancashire and Yorkshire, where the demand existed for textile machinery; in certain of the coastal coalfields, which developed a shipbuilding tradition; and in London, which had its old skills in the manufacture of precision machinery.[6] In the Black Country on the other hand there was a very limited local demand. Yet during the years between 1880 and 1914 the West Midlands became one of the most important engineering areas of Britain; and a disproportionate part of the new effort was concentrated in Birmingham. The sources of the development were technological innovation and new demands. The modern bicycle was really invented between 1885, when the first workable safety bicycle was developed, and 1888, when the pneumatic tyre appeared; Coventry at first attracted the trade, both because of personal factors and because a depression in the sewing-machine trade had produced available factory buildings. The forged parts came from Birmingham, and here the industry migrated progressively during the cycling boom of the 1890s.[7]

[1] *Ibid.*, 256.
[2] C. J. Woodward, 'Manufacturing industries', in *British Association handbook of Birmingham* (Birmingham, 1886), 203.
[3] *Ibid.*, 291.
[4] E. Lozé, 708.
[5] E. Deiss, 40; M. J. Wise, 'On the evolution of the jewellery and gun quarters of Birmingham', *Trans. and Papers, Inst. Brit. Geog.*, XV (1949), 70.
[6] G. C. Allen, 292.
[7] *Ibid.*, 294–6.

Visiting the area about 1896, Deiss noted that workshops in Coventry fell from 50 to 43 between 1892 and 1896, while in Birmingham they rose from 114 to 153.[1] By 1897 the trade in the Midlands was estimated to employ 66,000.[2] Bicycle making led naturally to motor-car manufacture, especially when combined with the tradition of gas-engine manufacture in Birmingham and Wolverhampton.[3]

An advertising brochure of 1897 could prophesy with satisfaction: 'In commenting on the merits of MOTOR CARS, we are of necessity compelled to regard them as being at present only in an experimental stage. That they are destined to become the means of conveyance in the future...is admitted without doubt by all far-seeing men...It is very satisfactory to reflect upon this fact that greater progress has been made here in Birmingham in this direction than in any other part of the world.'[4] By 1905, the big firms were 'at high pressure' and employment was 'rapidly increasing',[5] and by the outbreak of the First World War Birmingham and Coventry were together established as the first centres of the new industry in Britain.[6] In the verdict of a later historian: 'It is questionable whether apart from the desire to be near the centre of the grapevine, there were serious economic factors involved';[7] William Morris flourished outside the area though he got his components from it,[8] and there were successful firms in Scotland. The Midlands provided a suitable environment though; for many of the firms were new firms, engaged in evolving a new industrial tradition, while the light metal tradition provided necessary materials.

The manufacture of motor cars soon became a large-scale industry which needed space, and so it rapidly found itself on vacant land on the edge of Birmingham: Austins were already at Longbridge, on the Bristol Road south-west of the city, by 1906.[9] The same thing happened in

[1] E. Deiss, 10.　[2] *Ann. Rep. Factories, 1896*, 36 (P.P. 1897, xvii).
[3] G. C. Allen, 297. See also G. Maxcy and A. Silberston, *The motor industry* (Cambridge, 1959), 11.
[4] Anon, *Birmingham 1837–97: a souvenir of the Diamond Jubilee Year* (Birmingham, 1897), no page number.
[5] *Ann. Rep. Factories, 1905*, 64 (P.P. 1906, xv).
[6] Evidence in S. B. Saul, 'The motor industry in Britain to 1914', *Business History*, v (1962), 30.　[7] *Ibid.*
[8] *Ibid.*; see P. W. S. Andrews and E. Brunner, *The life of Lord Nuffield* (Oxford, 1955), 62, 66.
[9] M. J. Wise and P. O'N. Thorpe, 'The growth of Birmingham 1800–1950', in M. J. Wise (ed.), *Birmingham and its regional setting* (Birmingham, 1950), 226–7.

electrical engineering, where the General Electrical Company established a big plant at Witton, in the Tame valley north-east of the city, in 1901.[1] It was notable that the new industries reversed a Birmingham tradition: the older industries, like Watt's factory and foundry, had established themselves north-west of the city, adjacent to the coalfield,[2] as at Witton, at Longbridge, developments were away from the city, but now the or at Cadbury's model village of Bournville, which had been established as early as 1882 on vacant land south of the city.[3]

This was the direct effect of the new industry. But both cars and electrical goods also created a vast demand for materials and components of every kind, which revolutionised the whole economy of the city and of the area around, and brought a new lease of life to many small and medium-sized firms. Bicycles demanded great quantities of weldless steel tubes; Wednesbury, which had specialised in the welded variety, failed to respond to the challenge and so remained industrially moribund,[4] but other towns were to the fore. Cars demanded not merely tubes but also sheets, wires, screws, locks, windscreens and lighting equipment.[5] Even the saddlery industry of Walsall received a new lease of life, first from bicycle saddles, then from car upholstery.[6] The tremendous demand for rubber led Dunlops to concentrate their activities in the city.[7] The electrical industry stimulated instrument making, brass founding, wire and tube making, and general engineering.[8] All these developments created a great demand for machine tools, which soon became a separate industry; power presses in particular came to be used in a great variety of industries, including the old ones which became mechanised, like locks and hollow ware and brass. Such interdependence allowed the Black Country and the city to retain their traditional small-scale form of industry, so that the motor-car industry, for instance, became a classic case of 'a great number of independent medium sized or small firms, each specializing on some single process of service and depending on a general background of metal processes and services'.[9]

1 *Ibid*, 224.   2 G. C. Allen, 324.

3 *R.C. on Labour, Report by Miss Clara E. Collet on the Conditions of Work in Birmingham, Walsall, Dudley and the Staffordshire Potteries*, 54 (P.P. 1893–4, xxxvii, pt 1); W. J. Ashworth, *The genesis of modern British town planning* (London, 1954), 132–3.

4 J. F. Ede, *A history of Wednesbury* (Wednesbury, 1962), 271; S. J. Langley, 'The Wednesbury tube trade', *Univ. of Birmingham Hist. Jour.*, II (1949–50), 173.

5 G. C. Allen, 298.   6 *Ibid*, 299.   7 *Ibid*, 301.   8 *Ibid*, 305–6.

9 P. S. Florence, *Investment, location, and size of plant* (Cambridge, 1948),

The close continuity of these plants within the city and the neighbouring area was especially economical where production was in small lots.[1]

In the Black Country engineering traditions developed on a coalfield that was well on the way to extinction. They might as easily have developed off the coalfields altogether. An observer looking for industry up the agrarian eastern side of Britain would find many examples of towns like Boston, where 'industry at the turn of the century was small scale and was related to local needs rather than national markets'.[2] He would find everywhere small market towns and village where the old rural artisans were suffering badly from the agrarian depression.[3] He would find a few examples of localised village crafts in decay, such as the straw plaiting of south Bedfordshire.[4] He would happen on small-town industries catering for a national market, which continued precariously despite their inefficiency, like the Luton straw-hat industry, 'one street in Luton being known as Rotten Row due to its number of failures'.[5] But in the bigger market towns of the eastern counties, and above all in Lincolnshire, he would find a flourishing engineering industry which in the course of a century had grown from serving purely local needs to meeting the demands of a national and even a world market.[6] The biggest centre of all was Lincoln, where six big firms specialised in traction engines: 'From the engine which drives the merry-go-round or the wild-beast show to that which hauls heavy warlike and other stores in the country fairs, to that which hauls heavy warlike and other stores in the Sudan or South Africa, every type of road locomotive is made in Lincoln.'[7]

Alike on active coalfields, on worked-out coalfields and on no coalfield at all, therefore, the new engineering industries were rapidly evolving and were already contributing to British industry its distinctive twentieth-century character. What was important, of course, was not coal or the lack

---

[1] *Ibid.*, 59–60.
[2] F. H. Molyneux, 'Industrial development in Boston, Lincolnshire', *East Midland Geographer*, III (1964), 269.
[3] L. M. Springwall, *Labouring life in Norfolk villages 1834–1914* (London, 1936), 97.
[4] J. L. Green, 57; R. Trow-Smith, *The history of Stevenage* (Stevenage, 1958), 74.
[5] J. G. Dony, *A history of the straw hat industry* (Luton, 1942), 128. On the domination of the Luton economy at this time by the hat industry, see J. Dyer *et al.*, *The story of Luton* (Luton, 1966), 170–1.
[6] R. H. Clark, *Steam-engine builders of Lincolnshire* (Norwich, 1955), *passim*; Lord Aberconway, *The basic industries of Great Britain* (London, 1927), 80–9.
[7] Lord Aberconway, 83–4.

of it, but the existence of an industrial tradition that could be transmuted: most commonly the coalfields had thrown up such a tradition, but it could readily evolve in a different environment. Lincoln showed that; Birmingham, which had never directly depended on its neighbouring coalfield, showed it too; but the best example was the capital city and chief port which still, as it had a hundred or fifty years before, dominated the industrial map of England.

The textile industries

South Lancashire (Figs. 149 and 150). By 1900 the major industrial supports of the British urban economy were experiencing very different fortunes. Reversals in fortune were just beginning to occur, and their precise significance escaped contemporary observers, whose verdicts frequently contradicted each other. Perhaps the most striking case was provided by cotton textiles. Cotton still employed one half of all the textile workers in England and Wales, or indeed one in every 25 of the recorded occupied population of the country. And, as the Registrar-General recorded in his General Report on the 1901 Census, 'The decline here shown of 3.1 per cent in the number employed in an industry of such magnitude is unquestionably a matter of serious concern.'[1] There was one part of England above all where that concern should have been manifest. Throughout the nineteenth century the cotton industry had become progressively more concentrated in Lancashire, until by the turn of the century that county counted over 80% of all the cotton workers in England and Wales.[2] Yet there was a contradiction here, which was reflected in the reports of the Lancashire factory inspectors themselves. On the one hand there was unprecedented prosperity, record dividends and boundless confidence in the future: 'The year 1900 has so far as cotton spinning is concerned in this district been one of unprecedented profit making,' was the report from Oldham.[3] In 1905, in Lancashire as a whole, 57 new mills with five million spindles were opened or in course of erection.[4] Yet in the same period, on the other hand, it could be reported that 'many women weavers have earned less than 7s. per week', and 'many cotton workers have keenly felt the pinch of poverty'.[5]

1 *Census of 1901: General Report*, 119 (P.P. 1904, cviii).
2 S. J. Chapman, *The Lancashire cotton industry* (Manchester, 1904), 148–9.
3 *Ann. Rep. Factories*, 1900, 278 (P.P. 1901, x).
4 *Ann. Rep. Factories*, 1905, 148 (P.P. 1906, xv).
5 *Ann. Rep. Factories*, 1903, 101 (P.P. 1904, x).

There was one good explanation for this: increasing mechanisation and increasing productivity. Schulze-Gaevernitz, a German visitor, had been astonished in 1895 by the high productivity of the British industry in comparison with its European competitors: Oldham had 2.4 operators to a thousand spindles, but Mulhouse, in Alsace, used 5.8, and the factories in the Vosges 8.9; in weaving there was one operative to 4.6 looms, in Alsace one to 1.5.[1] Nevertheless, by the beginning of the new century there was already a cloud on the Lancashire industrialist's horizon. 'The question of the growth of cotton in the British dominions is one which is occupying the attention of the trade considerably, and is of great importance, looking at the large number of mills which are being erected in America, Russia, etc., every year,' remarked the superintending inspector for the north-western district in 1902.[2] And, taking the representative period 1881–1911, in east Lancashire and north-east Cheshire, 'population was stagnating in many areas, this reflecting in some measure a slackening in the rate of growth of the economy of many of the textile towns'.[3] Premonitions of future depression were thus already in the air.

Some parts of south-east Lancashire were already less concerned with cotton than in years gone by. 'Manchester is constantly becoming more and more simply the seat of the export trade,'[4] commented Schulze-Gaevernitz in 1895. An observer looking out from the new town hall tower saw no longer factories, but 'immediately around him, and receding to a considerable distance, are the business streets, made up of warehouses, banks, insurance and other offices, shops, clubs and institutes of various kinds'.[5] On a closer look, indeed, the subdivisions of the 'Manchester trade' had localised themselves in particular commercial quarters within this great commercial complex; the dealers in yarn and grey cloth were mainly round the Exchange, the calico printers were more centralised than the shippers, the home trade houses were round the infirmary.[6] Essentially, central Manchester was the great emporium for the surrounding cotton districts; its lifelines were the new Ship Canal docks or the great goods depot at London Road railway station, where 2,000 tons of goods trundled

[1] G. von Schulze-Gaevernitz, *The cotton trade in England and on the Continent* (London and Manchester, 1895), 96, 98, 208.

[2] *Ann. Rep. Factories, 1902*, 96 (P.P. 1903, xii).

[3] R. Lawton, 'Population changes in Lancashire and Cheshire from 1801', *Trans. Hist. Soc. Lancs. and Cheshire*, CXIV (1962), 197, 201.

[4] G. von Schulze-Gaevernitz, 74.

[5] J. Mortimer, *Mercantile Manchester past and present* (Manchester, 1896), 76.

[6] *Ibid.*, 95–6.

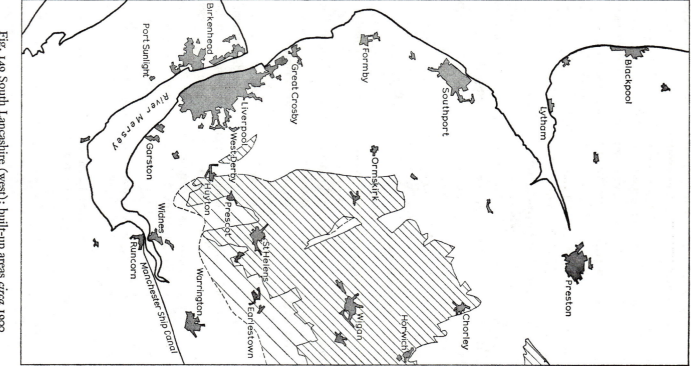

Fig. 149 South Lancashire (west): built-up areas *circa* 1900
Sources as for Fig. 146.

ENGLAND *circa* 1900

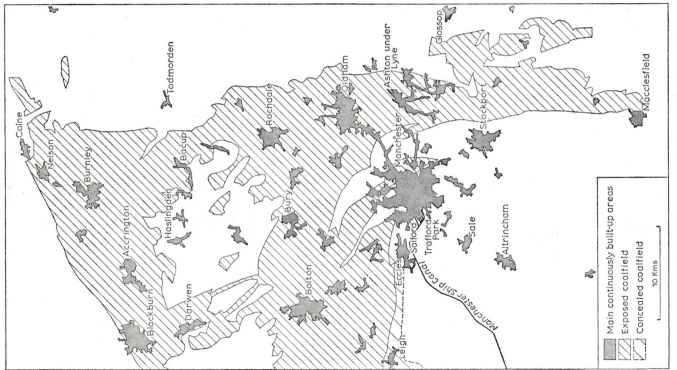

Fig. 150 South Lancashire (east): built-up areas *circa* 1900
Sources as for Fig. 146.

in or out, carried by the horse-drawn 'lurries' over the cobbles, every working day.[1]

The Ship Canal had been opened in 1894, and according to many Manchester observers that was not a moment too soon. In the newspapers of the late eighties there had been constant complaints of, 'an exodus of merchant houses such as was never too seen before'.[2] 'That Manchester is rapidly losing her supremacy as a distributing centre,' wrote a representative of the traders in 1890, 'will hardly be denied.'[3] 'There is no doubt that we have suffered severely from the effects of the high rates of transit, both for import and export.'[4] The Ship Canal brought a dramatic change. By 1900 the port of Manchester already had a trade of over 3 million tons, and was handling sea-going vessels of 4,000 tons upwards.[5] The railways were forced to lower their rates in competition, and one alderman estimated that the community was saving half a million pounds or more a year in reduced charges;[6] rateable values and post office business were booming.[7] The benefit extended into Salford and other surrounding towns.[8]

The Mancunian promoters of the canal were no doubt hoping that it would chiefly save their cotton trade. But one of the most immediate effects was that 'in the wake of the Canal, all kinds of industries have sprung up'.[9] Adjacent to the docks, the 1,200-acre park of Sir Humphrey de Trafford, 'until quite recently, owing to its privacy, a home for wild birds and fowl of all descriptions',[10] had been bought by a development company and was being developed by an industrial estate. The Westinghouse Company, later Metropolitan Vickers, was already building its factory at Trafford Park, thus helping powerfully to give it a distinctive character as a centre of the new, fast-growing electrical engineering industry.[11] Here indeed Manchester was following an old tradition; for the

---

[1] Ibid, 108, 117–18.

[2] 'Home trader', The home trade of Manchester (London and Manchester, 1890), 45.

[3] Ibid, 46.

[4] Ibid, 59.

[5] K. Baedeker, Great Britain (London and Leipzig, 1901), 356.

[6] A. W. Fletcher, 'The economic results of the Ship Canal on Manchester and the surrounding district, 1', Trans. Manchester Stat. Soc. (Session 1896–7), 92.

[7] Ibid, II (1898–9), 158–9.

[8] J. S. McConchy, 'The economic value of the Ship Canal to Manchester and district', Trans. Manchester Stat. Soc. (Session 1912–13), 10–12.

[9] Ibid, 17.

[10] Ann. Rep. Factories, 1900, 279 (P.P. 1901, x).

[11] T. H. S. Stevens, Manchester of yesterday (Altrincham, 1958), 148; W. H. Chaloner, 'The birth of modern Manchester', in C. F. Carter (ed.), Manchester and its region (Manchester, 1962), 145.

city had developed a natural monopoly in textile machinery manufacture, and an important position in the making of steam engines to drive the spindles and looms, by the last two decades of the eighteenth century; by 1850 it had also become a major centre for heavy machine tools, locomotives and boilers.[1] As new demands evolved, Manchester engineers met them; the old-established textile machinery firm of Mather and Platt went into dynamos, and many small firms were just experimenting with motorcar manufacturing so that by 1914 indeed Manchester looked like becoming a major centre of the British motor industry.[2]

In 1900 Manchester, therefore, was then a textile city only in a very special sense. Its great rival, Liverpool, had never had more than a very indirect relationship with Lancashire's staple industry. Essentially Liverpool was the great commodity port for the whole of northern and much of Midland England, and because of its command of an industrial hinterland it took from London pride of place as first export port of the kingdom. Its total tonnage had risen from 450,000 tons in 1800 to well over 11 million tons at the end of the nineteenth century.[3] 'The most unimaginative individual could hardly fail to be impressed with the scene that meets his gaze when coming for the first time into the port of Liverpool by sea', wrote a Liverpudlian in 1896, welcoming delegates to the British Association meeting. 'Stretching for a distance of six or seven miles he beholds a line of the finest docks in the world, behind which rises the city, which, with its recent additions and its suburbs, contains nearly three-quarters of a million inhabitants.'[4] From the Herculaneum Dock in the south to the Hornby Dock in the north stretched 60 dock basins with 26 miles of quays.[5] At its centre, the great Landing Stage, largest structure of its kind in the world, had been extended to a length of half a mile to cater for 'the recent great development of the Atlantic passenger trade'.[6]

In the general mood of euphoria there were more cautious voices to be heard. Both in absolute and general terms, Liverpool's trade had been falling for much of the 1890s; and so, with the sole exception of Manchester,

[1] W. H. Chaloner, 141–2; Lord Aberconway, 129–33.
[2] W. H. Chaloner, 145.
[3] E. Deiss, 261.
[4] H. S. H. Shaw and H. P. Boulnois in W. A. Herdman (ed.), *Handbook to Liverpool and the neighbourhood* (Liverpool, 1896), 102.
[5] K. Baedeker, 346.
[6] H. S. H. Shaw in W. Herdman (ed.), 113.

had that of virtually all the Mersey ports. The ports feeding London in contrast had gained trade.[1] Nor was Liverpool distinguished as an industrial city. In chemicals, though many important processes had first been developed here, the city itself had 'long ceased to be the principal seat of these manufactures';[2] they had migrated outwards to St Helens, Widnes and Runcorn. The old eighteenth-century pottery had long since been closed.[3] And 'the first thing which the inquirer discovers about Liverpool is that there is no one important industry, and that there are few large factories...Apart from dressmaking, tailoring, and the clothing trade generally, by far the most important industry in Liverpool and the neighbourhood is the manufacture of tobacco – an industry which has greatly increased in recent years.'[4] Port industries, depending on imported raw materials and the fact of break in bulk, were in fact Liverpool's staples. The heavier, early stages of processing – soap making, flour milling – were concentrated along the river, next to the docks. The later stages, such as jam and confectionery making, tended to concentrate in industrial areas on the outer edge of the city, often adjacent to the railways that provided the means of distribution to regional and national markets.[5] Unlike the textile factories of interior Lancashire, these factories employed relatively low proportions of women, for 'it is unusual for the more respectable working-class women to go to the factories after marriage'.[6] And although the trade of Liverpool had given rise to an imposing commercial centre – the Exchange with its associated financial houses and brokers' offices, and the shipping lines[7] – the contribution to the city's employment was not impressive. The Booth Line, with its assets of half a million pounds and profits averaging £100,000 a year, in 1900, had a total office staff of only seventeen.[8]

Between the two giants of Lancashire, a new industrial region was in process of developing around the tidal flats at the head of the Mersey

[1] E. D. Jordan and M. J. B. Baddeley (eds), *Black's guide to Liverpool* (Liverpool, 1900), xv.

[2] C. A. Kohn in W. Herdman (ed.), 137.

[3] E. Deiss, 321.

[4] A. Harrison, *Women's industries in Liverpool* (Liverpool, 1904), 13.

[5] W. Smith, 'The location of industry', in W. Smith (ed.), *A scientific survey of Merseyside* (Liverpool, 1953), 178–9.

[6] A. Harrison, 14.

[7] G. Chandler, *Liverpool's shipping: a short history* (London, 1960), 39.

[8] A. H. John, *A Liverpool merchant house: being the history of Alfred Booth and Co.,* 1863–1958 (London, 1959), 104–6.

estuary. The beginnings of the chemical revolution had been felt in Liverpool eighty years before, and soda manufacture had been a staple of St Helens and Widnes since that time. Now, however, new technologies were causing rapid changes in fortune. The new processes for making caustic products – the ammonia soda and the electrolytic process – had helped to industrialise the Cheshire saltfield and to bring prosperity to Widnes and Runcorn, but it killed the industry of St Helens, based on the old Leblanc process, which had migrated out from Liverpool about 1820.[1] Glass saved St Helens: the development of new processes for window glass and bottle manufacture allowed Pilkingtons to pull ahead of their rivals, and, increasingly, St Helens and Pilkingtons were becoming synonymous terms.[2] On the south side of the Mersey, in 1887, W. H. Lever had bought the stretch of desolate waste ground which became Port Sunlight. By the first decade of the new century here were 90-acre works, railways, sidings, docks, wharves, and a 140-acre model village. By 1900, the factory was turning out not only the original Sunlight soap but also Lifebuoy, Monkey Brand soap and Sunlight Flakes – which in this year, by a happy inspiration, Lever rechristened Lux.[3]

West Yorkshire (Fig. 151). The other great textile district of northern England had suffered even poorer fortunes than the cotton towns in the last years of the century. Though total employment in the woollen and worsted industries had fallen by nearly 14% in a decade, still in 1901 over 150,000 people in the West Riding depended on it, and in the city of Bradford, where 46,000 of them were concentrated,[4] a contemporary observer could say: 'The foundation of Bradford is wool. It has grown out of wool as Manchester grew out of cotton and Middlesbrough has grown out of iron.'[5] Probably, in the estimation of this witness, at least five-sixths of all the wool manufactured or partly manufactured in the country was at some stage the subject of a bargain in the Bradford wool exchange, or in some Bradford merchant's warehouse.[6] But in the last

---

[1] L. F. Haber, *The chemical industry during the nineteenth century* (Oxford, 1958), 153–9; T. C. Barker and J. R. Harris, *A Merseyside town in the industrial revolution: St Helens, 1750–1900* (London, 1959), 444–6.

[2] T. C. Barker and J. R. Harris, 447–52.

[3] C. Wilson, *The history of Unilever: a study in economic growth and social change*, I (London, 1954), 34–6, 56–7.

[4] F. Hooper, *Statistics relating to the City of Bradford and the woollen and worsted trades of the United Kingdom* (Bradford, 1904), 60.

[5] A. R. Byles, 'Industries', in British Association, *Handbook to Bradford and the neighbourhood* (Bradford, 1900), 46.   [6] *Ibid.*

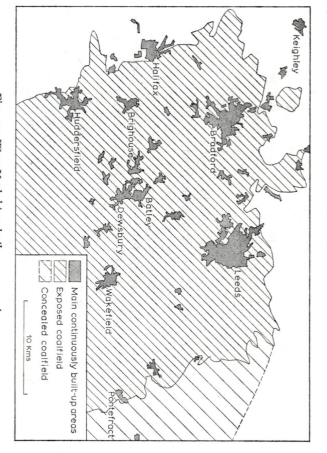

Fig. 151 West Yorkshire: built-up areas *circa* 1900
Sources as for Fig. 146.

decade of the century, wool had brought Bradford only a 14,000 increase in population. Leeds, in contrast, had evolved quite as fundamentally away from wool as Manchester had from cotton. 'At the beginning of the century a clothier in Leeds was a manufacturer of cloth,' Clara Collet could write in 1891, 'at the end of the century a clothier is a manufacturer of clothes.'[1] And the city's population was booming; between 1891 and 1901 alone it had increased by 61,000 or 14%. The secret was constant adaptation of its industrial structure to changing needs. Leeds had 'a very diversified trade', an observer could tell the Royal Commission on Labour in 1892;[2] the clothing trade had grown out of the old staple woollen and worsted trade; though the old flax and linen trades had nearly died out, Leeds was the biggest leather centre of the kingdom; above all, the engineering trades were flourishing and now provided employment for 29,000 people.[3]

[1] C. E. Collet, 'Women's work in Leeds', *Econ. Jour.*, 1 (1891), 467.
[2] *R.C. on Labour, Minutes, Group, C*, Evidence of Beckworth, Q. 15,513 (P.P. 1892, xxxvi, pt. 2).
[3] *Ibid.*, Q. 15,503–12.

The distinctive Leeds industry, by 1900, was the manufacture of ready-made clothing in factories. The sewing machine had been introduced into the city's workshops about 1855; the band-saw, which could cut many thicknesses of cloth simultaneously, about ten years later. 'By the cutting machine some two dozen double thicknesses were given out at a time for trousers, the work was fixed and given out to women to do at home.'[1] The Leeds system put a much greater proportion of the total job into factories than in London, and Clara Collet hoped that 'the factory system has such immense advantage over the domestic system that there is good ground for hoping that East London will either lose its clothing entirely, or save it by adopting the much more economical factory system'.[2] In Leeds, juvenile suits, trousers and waistcoats were made chiefly by female labour, either in factories or at home. Coats needed more skill, and, after being cut up in the factories, they went to Jewish workshops where the work was minutely subdivided, some parts – the finishing and the machining for instance – being performed by women, other parts – the buttonholing and the felling – by Jewish workmen.[3]

The Jews had started to come into Leeds in large numbers in the 1880s, probably because the city was on the way from Hull, where they landed from eastern Europe, to Liverpool, where they hoped to embark for America. Including a small number of an older generation, the Jewish population of the city was estimated at 15,000–20,000 at the start of the century,[4] and one district – the Leylands – had been taken over entirely by the immigrants.[5] Even by the year 1888 there were at least 64 Jewish clothing workshops, employing more than 2,000 people.[6] In some of these complete vertical integration of the processes had already been achieved: 'highly paid skilled designers prepare work for the costly "cutting out" guillotines, and hundreds of women guide the pieces through self-acting

1 C. E. Collet, 467.

2 *Ibid.*, 469.

3 *Board of Trade, Report on the Volume and Effects of Recent Immigration from Eastern Europe into the U.K.*, 117 (P.P. 1894, lxviii); *R.C. on Alien Immigration, Minutes of Evidence*, Evidence of Marston, Q. 14,269–94 (P.P. 1903, ix); J. Thomas, *A history of the Leeds clothing industry* (Yorks. Bull. Econ. and Soc. Research, Occasional Paper No. 1, 1955), 20–3.

4 *R.C. Alien Immigration, Minutes*, Evidence of Marston, Q. 14,327–30 (P.P. 1903, ix).

5 *Ibid.*, Evidence of Connellan, Q. 15,018. For a description, see E. Krauz, *Leeds Jewry: its history and social structure* (Cambridge, 1964), 21–2.

6 *B.O.T. Rep. on Immigration*, 116 (P.P. 1894, lxviii).

sewing and button-sewing machines, to be finally pressed into the "smart new suit" of the city clerk.'[1]

With an increasing number of clerks looking for a 'smart new suit', the Leeds clothing industry had a bright future. But other parts of the city's economy were already past their peak. The great complex of leather industries had reached its apogee in the 1890s, when it 'consisted of a score of tanneries, four score boot manufacturers and over four hundred boot and shoe makers'.[2] Even so, it was already on the decline due to its reluctance to adopt new techniques like tanning with hemlock bark and chrometanning for the new lighter shoes.[3] One part of the shoemaking trade, as in London, had passed into the hands of the Jewish immigrants; they were said to make slippers in workshops, which were also living rooms by night, at half the price of English makes.[4]

Most of Leeds' staple industries at the start of the new century, then, were outgrowths of the relatively simple technical innovations of the early industrial revolution. But less clearly evident to most contemporary visitors, another great complex of industries was rapidly growing. Leeds had developed an engineering tradition, as had Manchester, due to the local demands of the textile industry in the early nineteenth century. Already by the mid-century general machinery shops had been starting up, and in the intervening decades they had evolved into large factories with a wide range of products.[5] The firm of Greenwood and Batley was typical: founded in 1856, by the early twentieth century it already had hydraulic, steam turbine, electrical, textile, ordnance and machine tools divisions.[6] The same process of diversification into engineering was observable in this period in nearby Huddersfield.[7]

The east Midlands (Fig. 152). There was one other area of England where the textile and dress industries were a staple trade, and where these industries were suffering changes of fortune by 1900. In a belt of the east

[1] B. Webb, 'Women and the factory acts', in S. and B. Webb, *Problems of modern industry* (London, 1898), 98.

[2] W. G. Rimmer, 'Leeds leather industry in the nineteenth century', *Publications of the Thoresby Society*, XLVI (1961), 151.

[3] *Ibid.*, 155–64.

[4] R.C. Labour, *Minutes*, Group C, Evidence of Ingle, Q. 13,937 (P.P. 1892, XXXVI, pt. 2).

[5] J. Buckman, 'Later phases of industrialisation, to 1918', in M. W. Beresford and G. R. J. Jones (eds.), *Leeds and its region* (Leeds, 1967), 161–2.

[6] Lord Aberconway, 100–4.

[7] R. Brook, *The story of Huddersfield* (London, 1968), 213–17.

Midlands stretching from the mid-Nottinghamshire coalfield in the north to Northampton in the south, up to 1880 the basic industry had been the manufacture of hosiery, or footwear, or – as in Leicester and the surrounding villages – of both. In the northern part of the belt from Nottingham to Leicester, hand-knit hosiery, contracted out from the cities through middlemen, had been a basis of the economy of the village for centuries. But the new automatic knitting machinery needed mechanical power and could not be economically installed in the cottages.[1] By 1894, a Leicester witness could claim that 99% of the product was made by machinery;[2] merely a few hand looms existed for glove making.[3] Four years later a Nottinghamshire observer noted: 'The main indoor industry in and all around Arnold is frame-work knitting, which has seriously declined as an outdoor occupation. One used to hear "Clikketty, clikketty, churrrr" (the noise of the machinery) from almost every other house. Now such a sound is rarely heard, the work having passed into factories.'[4] And there the tendency, wherever possible, was to bring in cheap female labour to operate the new machines.[5]

The villages benefited from new developments. In the mid-Nottinghamshire area around Mansfield, Sutton in Ashfield and Kirkby, in the 1890s, a great deal of seaming and finishing of underclothing was done in the homes.[6] And here was an expanding trade, for plain and fancy hosiery makers were moving extensively into making up ladies' blouses, skirts and all types of ladies' and children's clothing.[7] Even before this, the hosiery trade in Leicester had been overhauled by the boot and shoe trade, which (a witness said in 1894) had been established, only some twenty-five years before but was now the biggest industry of the place.[8] Not long before this it had been a village trade, concentrated in the villages of Northamptonshire. The material had been cut up in London and then sent to North-

[1] *Ann. Rep., Factories, 1903*, 42 (P.P. 1904, x).

[2] *R.C. on Labour, Minutes, Group C*, Evidence of Oscroft, Q. 13,466 (P.P. 1892, xxxv, pt. 2).

[3] *Ibid*, Evidence of Holmes, Q. 12,672–3.

[4] J. L. Green, 103.

[5] *Ann. Rep. Factories, 1902*, 39 (P.P. 1903, xii).

[6] *R.C. on Labour. The Employment of Women: Report by Miss May E. Abraham on the Conditions of Work in the Confectionery, Hosiery and Lace Trades*, 159 (P.P. 1893–4, xxxvii, pt. 1).

[7] *Ann. Rep. Factories, 1902*, 39 (P.P. 1903, xii); *ibid, 1903*, 42 (P.P. 1904, x).

[8] *R.C. on Labour, Minutes, Group C*, Evidence of Wates, Q. 12,526–33 (P.P. 1892, xxxvi, pt. 2).

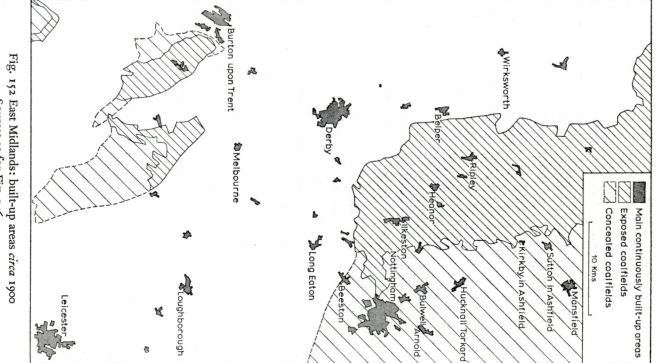

Fig. 152 East Midlands: built-up areas *circa* 1900
Sources as for Fig. 146.

ampton and the surrounding villages, where small masters with old-established skills performed the process of 'closing' before the shoes were sent back to a London factory to be made up.[1] But by the turn of the century, so it was said, 'the continued introduction of labour-saving machinery, and the consequent sub-division of labour, is slowly but surely directing the trade into the hands of large employers or limited companies, and with them the little man has but very slight chances of competing either in the home or foreign markets'.[2] The change was associated with the fact that, as a union official said in 1892, 'a comparatively new class of people entered the trade. Among the most successful manufacturers in Leicester were hat manufacturers and so on.'[3] With increasing demand, the change did not mean the decay of the industry in the villages. In contrast to the Leicestershire hosiery industry, the Northamptonshire villages acquired their own shoe factories. 'Several villages, the homes of innumerable outworkers some ten or fifteen years ago, are now the centres of large factories, quite up-to-date in every respect, and equipped with all the latest developments of labour-saving machinery.'[4]

To the north in Nottingham, shoemaking did not become a staple industry; but here, if anything, the course of industrial evolution was even more representative of the contemporary changes in the national economy. At the turn of the century, hosiery and lace, which in 1851 had employed the great majority of the Nottingham work force, were still flourishing; lace in particular was a major industry with some 20,000 work people in the early years of the new century.[5] But new industrial traditions were developing, exemplified by three great enterprises which had become concentrated in the city for very varied reasons. Boots had become a manufacturing drug company in 1888, and by the end of the century had 60 shops in 28 towns; Players had come into existence in 1877 when John Player bought the tobacco factory established fifty years earlier; the small Raleigh Street cycle works had been taken over by Frank Bowden in 1887,

[1] *S.C. House of Lords on the Sweating System, Minutes*, Evidence of Pocock, Q. 11,521–3 (P.P. 1888, xxi).

[2] *Ann. Rep. Factories, 1902*, 39 (P.P. 1903, xii).

[3] A. Fox, *A history of the National Union of Boot and Shoe Operatives, 1874–1957* (Oxford, 1958), 26.

[4] *Ann. Rep. Factories, 1904*, 5–6 (P.P. 1905, x). See P. R. Mounfield, 'The footwear industry of the east Midlands (iv), Leicestershire to 1911', *East Midland Geographer*, iv (1966–9), 13–14.

[5] F. A. Wells, 'Nottingham industries: a hundred years of progress', in J. D. Chambers *et al*, *A century of Nottingham history 1851–1951* (Nottingham, 1951), 33.

just as the great boom in cycling was about to start.[1] There were good reasons why these and other new industries should have found it exceptionally easy to grow in Nottingham. The city was centrally sited, being within fifty miles of Birmingham, Coventry, Sheffield or the Potteries. The manufacturers of lace-making machinery turned readily to cycles; indeed, during the two last decades of the century more than one third of the cycle makers had earlier been occupied in making lace machinery. The tradition, long established in the lace trade, of renting a 'standing' made it relatively easy for a small man to start up in a new line.[2] The new industries helped swell the city's population by no less than 25% in the last two decades of the old century.[3] Soon, the more successful plants were finding space too constricted within the city of Nottingham itself. Beeston, a frame-knitting village to which lace making had been attracted in the late 1870s, found itself first the home of the Humber cycle works and then, just after the turn of the century, of British Ericsson telephones.[4] Despite the growth of engineering, other developments like ready-made clothing helped to preserve Nottingham's reputation as a city of women workers.[5]

Industrial London (Figs. 153 and 154). London's industrial domination would be easily missed by the casual observer. A French observer noted how London's industries, scattered in the sea of houses, did not show their importance very visibly.[6] It was not merely a matter of scatter, though; it was the lack of a clear industrial base, which confused visitors accustomed to simpler industrial traditions. 'London has no single staple industry. We find in it no dominant trade or group of trades...Ship-building may leave the Thames; silk-weaving decline in Spitalfields; chair-making desert Bethnal Green; books be printed in Edinburgh or Aberdeen; and sugar-refining be killed by foreign fiscal policy; but the industrial activity of London shows no abatement.'[7] This was the judgement of Ernest Aves in Booth's survey of *London life and labour* just before the turn of the

---

[1] *Ibid.*, 32; J. M. Hunter, 'Sources of capital in the industrial development of Nottingham', *East Midland Geographer*, II, No. 16 (1961), 39.

[2] J. M. Hunter, 'Factors affecting the location and growth of industry in Greater Nottingham', *East Midland Geographer*, III (1964), 338–42.

[3] R. A. Church, *Economic and social change in a Midland town: Victorian Nottingham, 1815–1900* (London, 1966), 236.

[4] D. M. Smith, 'Beeston: an industrial satellite of Nottingham', *East Midland Geographer*, II, No. 14 (1960), 48–9.

[5] H. R. Haggard, II, 281.

[6] E. Deiss, 8.

[7] E. Aves, 'London as a centre of trade and industry', in C. Booth (ed.), v (London, 1895), 48.

century. Writing in a nineteenth-century context, Aves could not help being surprised by London's success. True, it was the greatest market of the kingdom and of the Empire; it was the place where materials, machines, tools and labour were most readily found; it had unrivalled communications and the greatest port in the world.[1] But 'the disadvantages are grave, and if London had to start again, would prove insuperable':[2] London had neither coal nor iron, ample running water nor fresh air, nor sufficient light and space, nor good workers at low wages. Therefore, Aves concluded, the economic hold of London would always be weak where fuel, iron or steel entered largely into production costs; where materials were bulky or weighty; or where the processes demanded much space.[3]

The conglomeration of London trades, then, was largely a series of residual categories. Apart from purely local trades like baking, brewing and newspaper production, there were the trades where prompt execution was necessary or the direct and constant supervision of the buyer or his agent was desirable, as in jewellery or precious instruments. There were the later stages of production, especially where manufacture was associated with sales, as in the dress trades; there were the repairing trades. There was a large group depending on a pool of cheap unskilled labour – cheap furniture, ready-made clothing, boots and shoes, ropes and sacks, paper and cardboard boxes, envelope making. There was a small group dependent on local raw materials – soap, glue, size, leather. And there was a group of trades simply found in London due to inertia; it included the making of clocks, pianos, baskets, saddlery and harness, portmanteaus and leather bags, and carriages.[4]

In truth most London industries were located where they were because of a complex of reasons. Aves observed how many London industries, like those of Birmingham, were gathered in small industrial quarters, where there was economic advantage 'in grouping around the main processes of an industry those allied and subsidiary trades and processes, which, combined with adequate means of distribution, go to secure the maximum of aggregative efficiency'.[5] These quarters were to be found mainly throughout a district which ran round the northern and eastern and southern edges of the central area from the West End to Bow, east of St Paul's, and from there south of the river to Southwark and Lambeth and Battersea. Here were found the packing-case makers of the City, the carriage builders of the West End, the heavy-van makers of east and south

---

[1] *Ibid*, v, 87–9. [2] *Ibid*, v, 90. [3] *Ibid*, v, 91. [4] *Ibid*, v, 93–5.
[5] *Ibid*, v, 97.

Fig. 153 Industrial London (west), 1898
Based on *Report R.C. on London Traffic*, plate F (P.P. 1906, xlii).
This map and Fig. 154 give a very imperfect view of London's industries in
that they omit the large number of small workshops, e.g. for clothing in the
East End. For the names of the boroughs see Fig. 156.

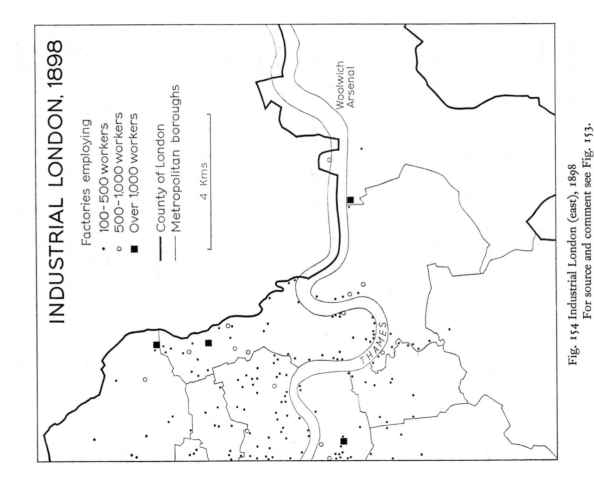

INDUSTRIAL LONDON, 1898

Factories employing
• 100–500 workers
○ 500–1,000 workers
■ Over 1,000 workers

—— County of London
—— Metropolitan boroughs

4 Kms

Woolwich
Arsenal

THAMES

Fig. 154 Industrial London (east), 1898
For source and comment see Fig. 153.

London, the envelope makers of Southwark, the watchmakers of Clerkenwell, the furniture makers of Shoreditch, and the boot makers of Hackney.

Within this complex of trades one stood out: the clothing industry, which (including the manufacture of footwear and fur) accounted for nearly one third of all London's 793,000 manufacturing workers in 1901.[1] Part of this trade – especially in dressmaking – was concentrated in the West End, where it satisfied the whims of wealthy customers.[2] But ready-made tailoring and footwear manufacture was the virtual monopoly of the East End, where there had grown up 'a new province of production, inhabited by a peculiar people, working under a new system, with new instruments, and yet separated by a narrow and constantly shifting boundary from the sphere of employment of an old-established native industry'.[3] This was small-scale industry, and in the public view it was peculiarly associated with the sweating or subcontract system. But, as Beatrice Webb pointed out, there was no more subcontracting here than in any trade dependent on wholesale warehouses for orders.[4] And that was the *raison d'être* of the system; the great warehouses on the city's eastern edge, outgrowth of a commercial tradition, had pioneered the transition from second-hand clothes dealing to new, ready-made clothes making between 1840 and 1860, aided by technological innovations like the band-saw and sewing machine. But the East End trade had reached its peak only in the years immediately before 1900, when thousands of Jews poured into east London from persecution in Russia and Poland. They settled in the areas close to where they landed, which happened to be the traditional homes of London's Jewish community since the seventeenth century; and they took up clothes manufacture because little skill or capital was needed.[5] By the 1901 Census, of the 95,245 Russian- and Polish-born immigrants in the United Kingdom, 53,537 were in London county, and 42,032 of these (or 45% of the total for the whole country) were in the borough of Stepney.[6] Here 'the portion bounded on the City side by the Minories, Houndsditch and Bishopsgate, north by the Great Eastern Railway and Buxton Street, and south by Cable Street, forms the central Jewish area.'[7] They were migrating outwards along 'the path of least resistance', along the

[1] Figures in G. Pasquet, *Londres et les ouvriers de Londres* (Paris, 1914), 211.

[2] *Ibid.*

[3] B. Potter (i.e. B. Webb), 'The tailoring trade', in C. Booth (ed.), IV (London, 1893), 37.

[4] *Ibid.* IV, 56–7.

[5] *Ibid.* IV, 60.

[6] *R.C. Alien Immigration, Report*, 14 (P.P. 1903, ix).

[7] C. Russell and H. S. Lewis, *The Jew in London* (London, 1900), xxxviii.

great radial arteries of the Whitechapel and Commercial Road.[1] Travelling incognito as a machinist in search of work, Beatrice Webb had penetrated this area in 1888. 'It is mid-day...For a brief interval the "whirr" of the sewing machine and the muffled sound of the presser's iron have ceased. Machinists and pressers, well-clothed and decorated with heavy watch-chains; Jewish girls with flashy hats, full figures and large bustles; furtive-eyed Polish immigrants with their pallid faces and crouching forms; and here and there poverty-stricken Christian women – all alike hurry to and from the mid-day meal.'[2]

The advantages of small-scale production in certain of the clothing trades were manifest. There was the need for rapid execution of orders; there was the limited amount of capital needed, for 'with £1 in his pocket any man may rise to the dignity of a sweater',[3] there was the possibility of introducing the division of labour principle, which the English journey-man had neglected. 'The English tailor would take one hour to put in one pocket, and the Jew tailor, with the sub-division, puts in four in 20 minutes, or four in a quarter of an hour.'[4] Of course division of labour was no monopoly of the small system, and 'the change in the character of the industry which is generally associated with Jewish labour is, in fact, due to industrial conditions which (in London) the Jew has been the first to see and take advantage of'.[5] But the proprietors of the big clothing fac-tories of Leeds had seen and taken advantage too. Admittedly, Beatrice Webb was right in saying that the London system depended on 'a class of workers...*with an indefinitely low standard of life*', and that without it the trade would cease to exist.[6] But even though the immigrants might become richer and more knowledgeable and more skilled, the Leeds factories provided the guarantee that the system of divided labour would remain. This message was already plain in the east London of the early twentieth century, from the example of other trades where the factory system was already triumphing: in footwear, where the small master system of the East End was giving way before the factories of Leicester and North-ampton;[7] and in furniture, where the more successful of the small shops

[1] *Ibid.*, xl.

[2] B. Potter in C. Booth (ed.), IV, 1.

[3] *Ibid.*, 60.

[4] *R.C. Alien Immigration, Minutes*, Evidence of Wright, Q. 19,722 (P.P. 1903, ix).

[5] C. Russell and H. S. Lewis, 71–2.

[6] B. Potter in C. Booth (ed.), IV, 66.

[7] P. G. Hall, 'The east London footwear industry; an industrial quarter in decline', *East London Papers*, V (1962), 3–21.

of Shoreditch and the Hackney Road were already outgrowing their cramped sites and were considering an outward move.[1]

Some of the most important trades of London's overcrowded inner industrial ring, therefore, were facing major economic challenges at the turn of the century. Particularly this was true of the engineering trades which had settled in this ring with the development of engineering itself, eighty or more years before. South of the river a pronounced concentration in Lambeth dated from that time; north of it, the watchmakers and precision-instrument makers of Clerkenwell originated in an even earlier age. But though most sections of engineering were recording rapid increases in employment, the Booth survey noted that 'for some years past the tendency throughout the engineering and metal trades has been for London to become more and more exclusively a repairing shop . . . Circumscribed for space, with heavier rent, rates and taxes, greater cost of labour, and in most cases more to pay for the carriage of raw material, the London manufacturer finds himself severely handicapped in competing for work with his provincial rivals of the North and Midlands.'[2] Even so, 'electrical engineering had made enormous strides; automatic machinery, from a gasmeter to a money-box, has become a craze; cycles are a requisite of youth, and no household is complete without its sewing machine'.[3] Similarly, though watchmaking had declined in Clerkenwell, due to the reluctance of the makers to adopt mass production methods, the men could readily find new work in expanding trades like electrical machinery.[4]

One of the most dramatic developments was the rise of motor engineering. In 1903, a factory inspector reported: 'The chief development in West London is I think the motor-car industry. One new factory started at the beginning of the year, and has since had to be enlarged. Another one is in course of erection and there are other factories that turn out new cars. Also almost very month we receive several notices of small new repair factories for motor cars, scattered all over the London portion of the district.'[5] In this particular year an important move occurred. Napiers, the printing machine and minting machine engineers, had been established in Lambeth since the 1830s. But in 1903, with the rapid expansion of the

---

[1] P. G. Hall, *The industries of London since 1861* (London, 1962), 90.

[2] J. Argyle, 'Engineering, iron-ship building, and boiler-making', in C. Booth (ed.), v (London, 1895), 294.

[3] *Ibid*, 296.

[4] G. H. Duckworth, 'Watches and clocks' and 'Surgical, scientific and electrical instruments', in C. Booth (ed.), VI (London, 1895), 27 and 41–2.

[5] *Ann. Rep. Factories, 1903*, 8 (P.P. 1904, x).

motor-car side of their business, they moved to a new works at Acton, and by 1904 they were employing more than 500 men there.[1]

Thus the new industries could evolve in the small workshops of inner London; but the move represented by Napiers was becoming more typical, for the new type of engineering soon demanded more space. In 1901 the west London inspector reported: 'the chief things to be noted are the great increase in the use of electrical motors, and the extension of factories towards the west outside London'.[2] 'There appears to be a general tendency', it was observed, 'for the larger factories to move out of London into the country westward to Hayes, Southall, and even so far as Reading.'[3] To the north of London too, 'owing to the prohibitive rents of buildings in the City, and the increasing facilities for obtaining labour outside the centres', there was also 'a tendency to remove factories further into suburban districts'.[4]

Technological developments helped to make this possible. The increasing use of electricity in west London factories had been noticed as early as 1897,[5] and by 1902 the inspector was speculating on its effects: 'It seems probable that electricity will have nearly as important an influence in the twentieth century on the industries of this Kingdom as steam had in the nineteenth…It is to be hoped that in the new age our manufacturing towns will be able to transplant many of their industries into the country …Surely London is big enough already without wanting to grow any more.'[6] Only four years later the north London inspector was commenting on another development: 'A few of the country factories now rely almost entirely on motor waggons for the carriage of their manufactures to their warehouses in London, etc., and this has proved so successful that it will give in the future much wider freedom of choice when fixing a site for a new factory.'[7] These years therefore saw the beginnings of the new suburban industrial areas of London, which many later observers mistakenly interpreted as a phenomenon of the interwar years of 1918–39. Between 1900 and 1914, Park Royal – a show-ground which proved unsuccessful and was redeveloped for industry – attracted nine factories, Hayes and Southall eleven, the Chiswick–Isleworth belt eleven; while the

[1] C. H. Wilson and W. Reader, *Men and machines: a history of D. Napier and Son, Engineers, Ltd, 1808–1958* (London, 1958), 80.

[2] *Ann. Rep. Factories, 1901*, 4 (P.P. 1902, xii).   [3] *Ibid.*, 5.   [4] *Ibid.*, 4.

[5] *Ann. Rep. Factories, 1897*, 118 (P.P. 1898, xiv).

[6] *Ann. Rep. Factories, 1901*, 3–4 (P.P. 1902, xii).

[7] *Ann. Rep. Factories, 1905*, 5 (P.P. 1906, xv).

Lea Valley, earlier established as an area of working-class housing, attracted no less than thirty-seven factories, nineteen of them migrants from inner London.[1]

THE SPREAD OF THE SUBURBS

But it was not industry alone which was seeking space in the suburbs. In 1899 Adna Ferrin Weber had published his definitive analysis of the growth of cities in the nineteenth century. Between 1881 and 1891 (and, for that matter, between 1891 and 1901), he said, the largest growth was 'in the cities of from 20,000 to 100,000 inhabitants, with the classes 100,000 to 250,000 and 10,000–20,000 in close company'. The biggest cities had increased less than the general population.[2] During the years 1880–90 of the century Liverpool appeared to have lost 6.2% of its population. 'Yet nobody believes that Liverpool is decaying; the explanation of the matter is simple enough: the growing business of the city requires the transformation of dwellings into stores and the dispossessed persons move away from the centre of business. As there is little more room within the municipal limits most of these people live in the environs, but are no longer counted in Liverpool.'[3] That was partly remedied in the decade that Weber wrote: the city took in Walton, Wavertree and some of Toxteth Park in 1895; it later took in Garston in 1902 and Fazakerley in 1904.[4] By 1914, Liverpool had jurisdiction over all the effective urban area on the Lancashire side of the Mersey, save Bootle; and she had lost Bootle only after a desperate parliamentary battle.[5]

Both in Liverpool and in Manchester, suburban growth was causing loss of population in the central cities while outer areas like West Derby or Birkenhead, and Chorlton, continued to grow. By 1900 in Manchester, 'the "cottontots" were prosperous inhabitants living in a wide range of suburbs from Victoria Park to Alderley Edge and Wilmslow, the less prosperous but solid middle classes occupied smaller houses in suburbs such as Moss Side, Withington and Fallowfield on the south or

[1] Evidence in D. H. Smith, *The industries of Greater London* (London, 1933), 41, 106–9. See also P. G. Hall (1962), 127 and 130.

[2] A. F. Weber, *The growth of cities in the nineteenth century* (New York, 1899), 50. See also J. S. Nicholson, *The relations of rents, wages and profits in agriculture and their bearing on rural depression* (London, 1906), 138. See p. 678 above.

[3] A. F. Weber, 51.

[4] B. D. White, *A history of the Corporation of Liverpool, 1835–1914* (Liverpool, 1951), 172.

[5] *Ibid*, 175.

Cheetham Hill on the north, and were also colonising the growing suburban areas of Cheshire'.[1] There was progressive displacement out from the centre; the innermost ring of house property was liable to become a zone of blight as the commercial core extended outwards, but eventually, it was said, 'as the trade continues to grow, this property rises again in value, being occupied for manufacturing purposes or for salerooms, and so recoups in large measure, if not fully, all past losses'.[2] Thus the 'conglomeration of people spreads over an unusual space, the working class here preferring small self-contained houses to barrack-like tenements, in which respect Manchester resembles Philadelphia, that huge 'city of homes'; and the houses 'extend long tentacles along the roads leading to the neighbouring towns, which often seemed to be joined to the central mass'.[3] The physical growth of the conurbation, here as elsewhere, outran the administrative map, which however laboured to keep up: Manchester swallowed Harpurhey, Bradford and Rusholme in 1885, Blackley, Moston, Crumpsall, Newton Heath, Clayton, Openshaw, Kirkmanshulme and West Gorton in 1890, Withington and Moss Side in 1904.[4]

Thus in Merseyside and south-east Lancashire, the suburban spread of a central city had already produced a proto-conurbation in 1900, fully ten years before Patrick Geddes in Edinburgh coined the name to express the reality. In the very different conurbation of west Yorkshire the same process was already taking place, and the visitor, in 1900, found 'one vast manufacturing hive, in which city verges on city, and one village merges into another, so that a person travelling by night from Kildwick on the north to Holmfirth on the south would never be out of sight of the gas-lamps, with a population increased more than ten-fold in numbers and a hundred-fold in real wealth and comfort of life' since 1800.[5] Here, as elsewhere in the north, improved urban transport had brought about a great change. Until late in the nineteenth century the privately run horse trams in Leeds had catered badly for the needs of workers and had failed to colonise new suburban areas – especially the working-class ones. But in the early years of the new century the new municipal electric trams had

---

[1] T. W. Freeman, 'The Manchester conurbation' in C. F. Carter (ed.), 56–7.

[2] 'Home trader' (1890), 32.

[3] A. R. Hope Moncrieff (ed.), *Black's Guide to Manchester* (London, 1900), 3.

[4] A. Redford, *The history of local government in Manchester* (London, 1939–40), II, 318, 322–3; III, 18.

[5] British Association, *Handbook to Bradford and the neighbourhood* (Bradford, 1900), 49–50.

24

remedied this and were connecting the city centre with distant suburbs for twenty hours in the twenty-four.[1]

In the west Midlands it was the same story: 'A very large scheme is gradually being developed for covering practically the whole of the midlands with a network of tram lines, electrically equipped',[2] and already 'it might be said that the continuous roads and houses from Aston on the east to Wolverhampton on the west, covering as they do, various municipalities and urban districts, are quite as much entitled to a single name as is Greater London'.[3] Birmingham in particular was 'swallowing up its immediate suburbs',[4] and whole villages like Moseley and Handsworth were in course of being submerged in the urban flood. 'These now old villages often present some curious anachronisms. A grey old church, partly buried by a hoary fat churchyard, is surrounded by the most modern of shops and stores; and a primitive little bow-windowed cottage, with a few flower pots in the window, has, perchance, a glaring ginshop next door.'[5] Erdington, for instance, grew from a village of 2,599 people in 1881 to a major suburb with a population of 16,368 in 1901.[6] The trams were the agents of suburban growth, though Edgbaston, 'the Birmingham Belgravia', was still successfully resisting trams along the Harborne road in the early years of the new century.[7] Farther out, in the neighbourhood of Acocks Green and Northfield, the railway was the agent of colonisation.[8] In Birmingham administrative reality ran steadily behind physical reality; large areas like Aston, Erdington, Handsworth and many of the southern suburbs were not incorporated until 1911.[9] And a year after that, when the Handsworth trams were still not completely integrated physically with those of Birmingham, a historian could write that 'Perry Barr is still rural of aspect, and some of the old-world garden lore lingers there yet'.[10]

[1] G. C. Dickinson, 'The development of suburban road passenger transport in Leeds, 1840–95', *Jour. Transport History*, IV (1959–60), 219–22.

[2] *Ann. Rep. Factories, 1900*, 245 (P.P. 1901, x).

[3] *Cornish's stranger's guide through Birmingham* (Birmingham, 1902), 17.

[4] T. Anderton, 119.   [5] *Ibid.* 116.

[6] A. Briggs, *History of Birmingham, borough and city 1865–1938*, II (Oxford, 1952), 139.

[7] T. Anderton, 89; A. Briggs, II, 140.

[8] M. J. Wise and P. O'N. Thorpe in M. J. Wise (ed.), 224.

[9] A. Briggs, II, 155.

[10] F. W. Hackwood, *Handsworth: old and new* (Privately printed, Handsworth, 1908), 78.

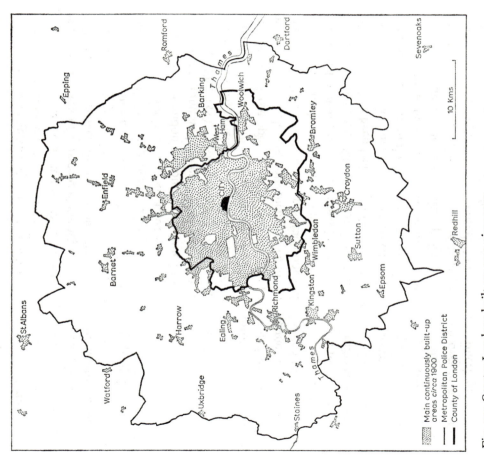

Main continuously built-up
areas *circa* 1900

— — — Metropolitan Police District

——— County of London

Fig. 155 Greater London: built-up area *circa* 1900
Based on: (1) Ordnance Survey One-Inch Sheets (1894–1903); (2) *Report
R.C. on London Traffic*, plan no. 1 (P.P. 1906, xlv).

In a smaller city, Bristol, expansion of the old port industries had
caused prodigious growth, and the citizens were bemused. 'LOST, STOLEN
OR STRAYED', wrote one of them in 1902, 'an ancient city known as
Bristol. Not heard of for some time. When last seen was wearing well and
running uneventfully over a number of hills.'[1] Bristol had no Edgbastons;
when the street tramways were promoted in the seventies, 'a few persons
avowedly opposed the lines from a dread of the influx into the fashionable

[1] L. Cowen, *Greater Bristol* (Bristol, 1902), 11.

suburbs of working men and their families on holidays. The tramways were, however, sanctioned.'[1] And following this, the workmen began to penetrate the new areas on a more permanent basis. 'Several firms, such as manufacturers of tobacco, clothing, corsets, rope, etc., have recently built fine, modern factories on the ourskirts of the town,' wrote the factory inspector in 1899, 'and round these factories have sprung up rows of neat cottages, in some cases built and owned by the factory proprietors, who rent them to their workpeople.'[2] The city extended itself in 1894 and again in 1897, bringing its area from 4,461 to 11,500 acres.[3]

Above all, London's suburbs continued to grow,' 'stretching out into the country long and generally unlovely arms', wrote Henrietta Barnett in 1905 (Fig. 155). 'Most people have seen the homes inhabited by the middle class with their small villas side by side, their few yards of garden carefully cherished, the monotony of mediocrity unbroken by fine public buildings or large open spaces.'[4] Less familiar were 'the districts occupied by the industrial classes, with their rows and rows of mean houses, every one alike, with limited back-yards, each one only divided from the other by a wall'.[5] Foreign observers commented on the low density and the low height of London's residential areas in contradistinction to those of their own cities.[6] Between 1881 and 1901 the whole area within six or so miles of the centre had been losing by migration.[7] This area, which became the county of London in 1888, experienced a slowing down in its rate of population growth in the last thirty years of the century.[8] But that meant no diminution in the attractive power of the metropolis; outside the county, but inside the metropolitan police area, the so-called 'outer ring' had been growing at the rate of more than 50% each decade since 1861 (see table on p. 673).[9] Here the population, wrote an American observer in 1895, 'now exceeding a million and a half, may in an approximate fashion be regarded as Londoners who are dependent upon the suburban trains of the great railways, in so far as their daily work takes them from circumference to center'.[10] 'Suburbia,' as one suburban child remembered it from that time, 'was a railway state...a state of existence within a few

[1] J. Latimer, *The annals of Bristol in the nineteenth century* (Bristol, 1887), 463.
[2] *Ann. Rep. Factories, 1899*, 145 (P.P. 1900, xi).
[3] J. Latimer, *The annals of Bristol concluded, 1887–1900* (Bristol, 1902), 38, 50.
[4] H. O. Barnett, 'A garden suburb at Hampstead', *Contemporary Review*, LXXXVII (1905), 231.
[5] *Ibid.*
[6] G. Pasquet, 40–1.
[7] *Ibid*, 23–4.
[8] A. Shaw, *Municipal government in Great Britain* (London, 1895), 295.
[9] *Ibid.*
[10] *Ibid.*

minutes walk of the railway station, a few minutes walk of the shops, and a few minutes walk of the fields.'[1] Of course it did not remain near the fields for long. By 1899, 'to all intents and purposes Croydon and London are now continuous';[2] and two years later, an observer in Middlesex noted of Hendon that 'it is hardly too much to say that it is a suburb of London'.[3] 'One may go east or north or south or west from Charing Cross,' wrote the American visitor of 1895, 'and almost despair of ever reaching the rim of the metropolis.'[4]

The new suburbs of the 'industrial classes', which Henrietta Barnett had noted, were a more recent and a more limited phenomenon. Still in 1901, some 16% of the population of the L.C.C. area was recorded as 'overcrowded', in the sense of living more than two to a room; and the worst boroughs, gathered just north and east of the centre, all recorded more than 25%[5] – Holborn, Finsbury, Shoreditch, Bethnal Green and Stepney (Fig. 156). Such a slum district 'represents the presence of a market for local casual labour. So the slums of Drury Lane are the creation of one market for casual labour – the theatres; and another market for labour which is riveted to the spot – the early work of Covent Garden.'[6] Similarly 'the fringe of misery radiating on the east and north-east of the City'[7] represented the demands of the low tailoring trade, the fur workers, the City outworkers, the small furniture trade, the match-box makers, the sack makers, and the Docks. 'If we could abolish all such casual labour,' this writer continued, 'then the poor labourer might do what the aristocracy of labour does now – migrate to the suburbs, or to outside areas like Tottenham, where there are cottages to be had with gardens.'[8] But before this could happen, redevelopment was eating into the slum areas. Land near the Bank of England was said to be running up towards £1 million an acre, and 'even a mile or two away the commercial value is so great that the residential population is steadily and rapidly vanishing'.[9] The process of displacement, said a witness in 1903, was taking place in all the boroughs adjoining the city – in Southwark, in Holborn, in Finsbury, in Shoreditch; but it was particularly serious in Stepney.[10] The effect was that

[1] J. Kenward, *The suburban child* (Cambridge, 1955), 74.
[2] E. A. Martin, *Croydon: old and new* (London, 1900), 6.
[3] J. P. Emslie, 'A walk on the banks of the Brent', *Home Counties Magazine*, III (1901), 122.
[4] A. Shaw, 231–2.   [5] G. Pasquet, 128–9.
[6] B. F. C. Costelloe, *The housing problem* (Manchester, 1899), 48.
[7] *Ibid.*, 49.   [8] *Ibid.*   [9] *Ibid.*, 53.
[10] *R.C. Alien Immigration, Minutes*, Evidence of Gordon, Q. 17,681 (P.P. 1903, ix).

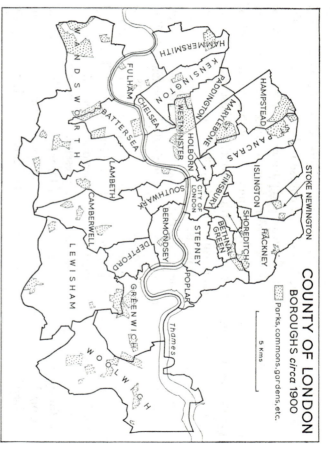

Fig. 156 The county of London and boroughs *circa* 1900 Based on J. G. Bartholomew, *The survey atlas of England and Wales* (Edinburgh, 1903), plate 80.

workmen's rents might rise 70% or 80% in a short time;[1] and analysis by the L.C.C. showed that the average weekly rent of new working-class houses was 3s. 3½d. in the central area but only 2s. in the areas outside the county altogether.[2] While the price of land in the central area made it quite uneconomic to rehouse the working classes at rents they could afford without a loss, in the outer areas it was still possible.[3]

Charles Booth in 1901 logically argued that large numbers of the working class were not casually employed and were free to move; he urged 'a large and really complete scheme of railways underground and overhead, as well as a net-work of tram lines on the surface', so that 'London will spread in all directions'.[4] To some extent this was happening.

[1] B. F. C. Costelloe, 53.
[2] *R.C. on London Traffic, Report,* 9 (P.P. 1905, xxx).
[3] *Ibid.*
[4] C. Booth, *Improved means of locomotion as a first step towards the cure of the housing difficulties of London* (London, 1901), 15–16, 17.

'Between 1877 and 1904 new building advanced in a great crescent beyond the Lea, covering the Flood Plain and Taplow Gravels from Highim (*sic*) Hill to Ilford and extending southwestwards onto the alluvium to link with Canning Town.[1] Between 1861 and 1901 the population of Walthamstow rose from 7,137 to 95,131; on the opposite side of the Lea in Enfield it rose from 10,930 to 61,892.[2] Of course these were one-class suburbs; developed on the basis of the cheap workmen's fares of the Great Eastern Railway, they would not attract the middle class.[3] The real trouble was that there were not enough railways like the Great Eastern; the really cheap workmen's fares only served very limited areas of London. Thus in south London, while workmen's fares had become more general, they had not achieved their specific aim of decentralisation; but when a really cheap fare offered the opportunity, as along the South London where a 2*d.* fare was available, there the working class had concentrated.[4] Shortly after the turn of the century the L.C.C. statistical officer compared rents and workmen's fares in central with suburban districts; he found that by using the cheap L.C.C. trams the suburban workman gained a clear advantage, but with railways the comparison was far more evenly balanced, and indeed in some cases the central location had the advantage.[5] Small wonder then that in 1897, less than one-quarter of the trade union members of south London used public transport to get to work.[6] In London, as elsewhere, commuting in 1900 was still a phenomenon affecting only a minority.[7]

By the beginning of the twentieth century, the impact of improved transport upon the physiognomy of London was far from complete. True, the central area had been losing residential population for some time, while it gained daytime population. The decline in the night population of the City itself had begun forty years before: from 112,063 in 1861 it was down to 26,923 in 1901. But by this time the estimated daytime population was 359,940.[8] The West End was still in a state of transition: in

---

[1] H. Rees, 'A growth map for northeast London during the railway age', *Geog. Rev.*, xxxv (1945), 462.

[2] *R.C. London Traffic, Report*, 14 (P.P. 1905, xxx).   [3] *Ibid*, 15.

[4] H. J. Dyos, 'Workmen's fares in south London 1860–1914', *Jour. Transport History*, I (1953–4), 15.

[5] *Rep. London Traffic Branch of the Board of Trade*, 14 (P.P. 1910, xxxvi).

[6] H. J. Dyos, *Victorian suburb: a study of the growth of Camberwell* (Leicester, 1961), 62.

[7] J. R. Kellett, *The impact of railways on Victorian cities* (London, 1969), 95.

[8] *R.C. London Traffic, Report* 5–6 (P.P. 1905, xxx).

contrast to the City it was still a residential area, yet, here too, younger workers were beginning to commute in to work.[1] Altogether it was estimated that in inner London – embracing the City, Westminster, St Marylebone, Holborn, Finsbury, Shoreditch, Bethnal Green, Stepney, Bermondsey and Southwark as well as parts of St Pancras and Lambeth – the 'day population' was 2,427,000 and the 'night population' 1,387,960.[2] These figures are suspect because the recorded movement by public transport into the area during the morning rush period between 7 a.m. and 10 a.m. was only 419,000, of which the great majority, 326,000, came by train.[3] Including those who walked or used bicycles, perhaps between one-half and three-quarters of a million people commuted into inner London every morning.

Important as the railways were in this process, they had not yet fully realised their potential function. Though by 1901 over two million people lived in the outer ring of Greater London, between six and fifteen miles from the centre, and the entire population of that area had reached 6,600,000,[4] and though the rate of growth in the inner area (i.e. the county of London) was slowing down,[5] nevertheless the last decade of the old century still saw about two-fifths of the total population growth of Greater London taking place within the six- or seven-mile ring which represented the effective radius of horse tram or bus.[6] The railway system was still a very imperfect system for shutting the commuter to and fro; it was particularly deficient in the West End, where the Inner Circle Line made a very wide circuit and the commuter had to rely on the horse bus, which was cheap but not very comfortable.[7] The first tube line under the West End, the Central London Line, had remedied this to some extent for east–west travellers by 1900; but it connected with only two surface termini. Even by 1905 the Royal Commission on London Traffic could estimate that the average commuter spent two hours per day in transit.[8] An improvement

[1] G. Kennman, *Der Verkehr Londons mit besonderer Berücksichtigung der Eisenbahnen* (Berlin, 1892), 17. On the growth of retailing in the West End, see the useful chronological table in A. Adburgham, *Shops and shopping 1800–1914* (London, 1964), 283–7.

[2] *R.C. London Traffic, Report of Advisory Board of Engineers*, 12, 13 (P.P. 1906, xlv).

[3] *Ibid*, 22.

[4] G. Cadoux, *La vie des grandes capitales* (Paris and Nancy, 1908), 110.

[5] See p. 740 above.

[6] T. C. Baker and M. Robbins, *A history of London transport*, 1 (London, 1963), 221–2, 272.

[7] G. Kennman, 30.

[8] *R.C. London Traffic, Report of Engineers*, 27 (P.P. 1906, xlv).

had to await a multiplication of the tubes in central London and an electrification of the suburban surface lines; though the first was achieved rapidly between 1906 and 1910, the second had to wait until after the First World War.

Slowly, therefore, Englishmen were becoming more mobile. This was reflected not only in their pattern of work but also in their pattern of play. The turn of the century represented perhaps the golden age of the English seaside resort. Like the commuting habit, the holiday habit had been pioneered by the middle class, and until late in the nineteenth century the new resorts remained largely their preserve; but by 1900 the excursion train habit was ushering in a momentous change. 'We have occasionally heard that Scarborough is vulgar, in consequence of the numerous day excursions that run into it', complained a brochure of 1891; but there could be 'no more delightful spectacle than to witness the exuberant spirits, the ludicrous efforts at enjoyment, and the utter disregard of the proprieties and conventionalities of society. But they do not frequent the Spa Grounds, or interfere with the pleasures of genteel society.'[1] By this time, certain resorts had begun to mushroom without the patronage of genteel society. Blackpool had been built on the earnings of a new class, the skilled workmen who had benefited from the rising productivity and rising living standards of the last decades of the nineteenth century. 'Ninety per cent of these people', a Lancashire observer told Schulze-Gaevernitz on the promenade at Blackpool in 1895, 'are mill-hands . . . the remainder mostly machine workers . . . That slender, caring-for-himself man, who with his better-half came on the trip, is a mule-spinner who can earn £2 and more weekly. That stronger-looking man next to him is probably a representative of the machine-making everywhere planted alongside of the cotton industry, and then certainly a member of the well-known Amalgamated Society. That full-grown girl, who evidently places great value on herself and her outward appearance, may be one of those four-loom weavers so numerous in our town, with weekly wages of 24s. or more, and fair savings.'[2]

For many Lancastrians, such prosperity was recent and still not quite familiar. At a sumptuous boarding house high tea, the German observer was able to talk to an old man who could well remember a youth of bitter poverty: '"But if you want to see the true sign of this change . . ., it lies before you on the table, the strength of Lancashire", as he raised with

---

[1] Anon, *The Yorkshire health resorts illustrated* (Scarborough, 1891), 9.
[2] G. von Schulze-Gaevernitz, 198–9.

triumphant bearing a piece of wheaten bread. Cobden was a sacred being to the old man. "We fought the battle, and we have won." With these words he closed his story."[1] Whether quite fair or not, the point was well made. Blackpool's visitors, and its residents, were the beneficiaries of half-a-century of free trade, of the policy of ruthless geographical division of labour, which had essentially been the creation of Cobden and his fellow Anti-Corn-Law campaigners. Ruining many British farmers in the process, this policy had yet made Britain the acknowledged workshop of the world. Even as the old man looked back, Germany and the U.S.A. were closely threatening this unique role. Yet as Lancashire and England played through the late Victorian and Edwardian summer, outside competition must have seemed a remote cloud in an otherwise untroubled sky.

[1] *Ibid*, 201.

INDEX

DA 600.D35
ledl,circ
A new historical geography of England,

3 1862 002 748 966   C.2

University of Windsor Libraries